ALBERT A. NEILSON

ELEMENTARY
STEAM POWER ENGINEERING

TRENTON CHANNEL STEAM POWER PLANT

Power economically developed in steam power plants, and applied to industry and commerce, has played a large part in developing our present civilization.

ELEMENTARY
STEAM POWER ENGINEERING

BY

EDGAR MacNAUGHTON, M.E.

*Professor of Mechanical Engineering, Tufts College Engineering School;
Member American Society of Mechanical Engineers and
The Society for the Promotion of Engineering Education*

Edgar MacNaughton

SECOND EDITION
Tenth Printing

NEW YORK
JOHN WILEY & SONS, Inc.
LONDON: CHAPMAN & HALL, LIMITED

Copyright, 1923, 1933, by
EDGAR MacNAUGHTON

All Rights Reserved

This book or any part thereof must not be reproduced in any form without the written permission of the publisher.

SECOND EDITION
Tenth Printing, November, 1947

Printed in U. S. A.

DEDICATED
TO MY WIFE
MARY ROSS MacNAUGHTON

PREFACE TO SECOND EDITION

The cordial reception given the first edition of this work has encouraged the author to thoroughly revise the text, in order to increase its usefulness, and to cover the advances made in steam power engineering during recent years.

The same presentation has been followed as was used in the previous work. Some chapters have been enlarged and relocated, and valuable suggestions received, from users of the first edition, have been incorporated in the revision. The book has been entirely reset, obsolete material removed, and new material, illustrations and problems added.

The major additions are: use of Keenan's steam tables; a discussion of intrinsic energy, work done under the expansions used in steam power apparatus, theoretical steam power cycles, boiler performance, furnace refractories, automatic combustion control, handling of ashes from pulverized coal furnaces, heat balancing, trends in boiler rating, fan performance, pump performance, modern power plant trends, power plant location and costs, power plant buildings, and the mercury-vapor steam cycle; and descriptions of desuperheaters, purifiers, reheaters, pulverized coal preparation and burning systems, oil and gas burning systems, motor controlled valves, welded piping, feedwater heaters, purifiers, evaporators, high-pressure and multi-cylinder turbines, high-pressure pumps, recent types of condensers and cooling towers, air pumps, industrial and hotel power plants, the steam accumulator, and steam generating units.

The author acknowledges his indebtedness to users of the first edition for their suggestions, and especially to Professor Grant K. Palsgrove, of Rensselaer Polytechnic Institute for material used in the revision, and to the various manufacturing companies for illustrations without which the text would be much less valuable.

EDGAR MACNAUGHTON.

TUFTS COLLEGE, MASS.,
May 1, 1933.

PREFACE TO FIRST EDITION

The purpose of this book is to present in a clear and concise manner a discussion of the fundamental principles underlying the construction and operation of steam power plant equipment.

The book is the result of the author's experiences while teaching courses in steam engineering at several engineering colleges. The material presented has been used successfully in classroom work, and although intended primarily for men of college grade, it is hoped that it will prove satisfactory for vocational schools and for the general reader seeking information about steam power plants. With this end in view, the mathematical discussion has been kept at a minimum.

The arrangement of the material differs from that of other books in the same field in that an effort has been made to introduce the practical phases of the subject previous to the theoretical. This has been done by having a description of the apparatus precede the theory. This plan has been followed throughout the book with the exception of the chapter on stokers, where it seemed preferable to place the discussion of combustion first. The use of the more important equations is illustrated by examples, and in addition a number of review questions and problems are placed at the end of each chapter in order to give the reader a ready means of testing his knowledge of the material contained in the chapter.

The first chapter discusses the relationship existing between the various equipment used in common types of steam power plants. The remaining chapters naturally divide themselves into two parts. The first part deals with representative types of steam boilers and their auxiliaries. Included in this part is a discussion of physical units; properties of steam; fuels; combustion; methods of burning coal, oil and pulverized fuel; smoke prevention; feedwater purification, and boiler testing. The second part takes up the various types of steam engines, steam turbines, pumps, and condensers. Included in this part is a description of the method used to set the valves of engines; a discussion on the testing of engines and turbines, and a description of the methods used to cool the circulating water for condensers. The last chapter contains a brief description of the equipment used in several modern steam power plants.

The author takes this opportunity to thank and acknowledge his indebtedness to the various manufacturers who so willingly supplied bulletins and illustrations, in many cases at considerable expense to themselves; and regrets that because of the large number they cannot be listed here.

The author also thanks the following for assistance rendered during the preparation of the text: Dean G. C. Anthony; Professors A. C. Willard; A. P. Kratz; Wm. H. Severns; C. H. Chase; F. E. Seavey; J. A. Polson and R. U. Fittz.

Special credit is due the following for permission to use certain material from publications of which they are the authors: Professors J. A. Moyer; C. F. Hirshfeld; C. F. Gebhardt; C. H. Fessenden; E. M. Shealy; Mr. J. F. Cosgrove; F. R. Low, and for permission to use portions of the Power Test Code, The American Society of Mechanical Engineers.

Suggestions or criticisms that will make the book more useful, or for corrections that will make the book more accurate, will be appreciated by the author.

EDGAR MacNAUGHTON.

TUFTS COLLEGE, MASS.,
   July, 1923.

CONTENTS

Chapter		Pages
I.	Elementary Steam Power Plants	1–16
II.	Steam Boilers and Settings	17–62
III.	Physical Units and their Measurement	63–80
IV.	Properties of Air, Gases, Water, and Saturated Steam	81–115
V.	Superheated Steam, Superheaters, Desuperheaters, Reheaters, Purifiers and Steam Cycles	116–131
VI.	Fuels	132–157
VII.	Combustion, Flue Gas Analysis, Boiler Losses	158–180
VIII.	Smoke Prevention, Methods of Burning Fuels, and Refractories	181–219
IX.	Rating, Efficiency and Testing of Steam Boilers	220–237
X.	Pipe Systems, Pipe, Valves, and Pipe Accessories	238–262
XI.	Heat Saving Equipment, Heaters, Economizers, and Station Heat Balancing	263–279
XII.	Boiler Feedwater Conditioning, Evaporators and Deaërators	280–296
XIII.	Draft and Methods of Producing Draft	297–320
XIV.	Coal and Ash Handling Equipment	321–336
XV.	Reciprocating Steam Engine Parts — Simple Engine	337–356
XVI.	Slide Valve Engines, Valve Diagrams, and Slide Valve Setting	357–397
XVII.	Multi-valve and Unaflow Engines	398–416
XVIII.	Steam Engine Indicator, Engine Efficiencies and Losses	417–442
XIX.	Compound and Multi-expansion Engines	443–451
XX.	Methods of Lubrication and Engine Accessories	452–464
XXI.	Steam Engine Testing	465–476
XXII.	Steam Turbines	477–535
XXIII.	Steam- and Power-driven Pumps	536–557
XXIV.	Condensers and Condenser Auxiliaries	558–583
XXV.	Modern Power Plants, and Factors Related to their Design and Operation	584–625
	Miscellaneous Tables	625–628

NAME AND LOCATION OF TABLES

Table		Page
1.	Whyte's System of Locomotive Classification	15
2.	Densities of Common Substances	64
3.	Volume and Weight of Air at Various Temperatures	81
4.	Thermal and Physical Properties of Common Gases	84
5.	Volume and Weight of Water at Various Temperatures	90
6.	Specific Heat of Water	90
7.	Properties of Dry and Saturated Steam	94
8.	Properties of Superheated Steam	118
9.	Classification of Coal by Rank	138
10.	Classification of Coals by Carbon Content and Carbon-hydrogen Ratio	139
11.	Analyses of Typical Coals	140
12.	Sizes of Anthracite, or "Hard," Coal	140
13.	Sizes of Bituminous, or "Soft," Coal	141
14.	Composition and Heat Value of Typical Fuel Oils	153
15.	Volume and Calorific Value of Various Gases	154
16.	Temperatures at which Various Combustibles Ignite	159
17.	Atomic and Molecular Weights of Substances Entering into Combustion	161
18.	Weight and Volume of Oxygen and Air Required for Combustion	164
19.	Stoker Data	205
20.	Lap-welded, Charcoal-iron Boiler Tubes	223
21.	Feedwater Log	231
22.	Data and Results of a Boiler Test	234
23.	Dimensions of Standard and Extra-heavy Wrought-iron and Steel Pipe	248
24.	Analyses of a Boiler Feedwater	286
25.	Loss of Draft in Boilers	298
26.	Draft Between Furnace and Ashpit to Burn Coal	298
27.	Diagram Factors for Simple Engines	433
28.	Rankine-cycle Efficiencies and Theoretical Water Rates	437
29.	Values of Rankine-cycle Ratio for Typical Steam Engines	438
30.	Cylinder Ratios for Compound and Multi-expansion Engines	448
31.	Heat Balance for an 8-in. by 18-in. Corliss Engine	473
32.	Data and Results of Reciprocating Steam Engine Test	474
33.	Losses in a 200-kw. De Laval Turbine Generator	532
34.	Performance of Modern Steam Turbines at Rated Capacity	533
35.	Comparison of Steam Rates for 500-kw. Turbine Operating Condensing and Non-condensing at 3600 R.P.M.	533
36.	Comparison of Size and Capacity of High Pressure Boilers — Edgar Station	586

Table	Page
37. Equipment of Edgar Station, Boston	596
38. Equipment of Trenton Channel Plant, Detroit	604
39. Equipment of Long Beach Station No. 3	610
40. Areas of Circles	625
41. Common Logarithms	626
42. Decimal Equivalents of Fractions of an Inch	628

ELEMENTARY STEAM POWER ENGINEERING

CHAPTER I

ELEMENTARY STEAM POWER PLANTS

1. Foreword. — The present is an industrial age built upon the extensive use of power, which is supplied by power plants using energy in the form of gas, oil, water and steam. The type of plant to be used is generally determined by the following factors:

1. Kind of service.
2. Location with regard to fuel and water.
3. Space available.
4. Reliability in operation.
5. Cost to produce a commercial unit of power, when all factors entering into the cost are considered.

The **water power,** or **hydro-electric, plant** may produce a unit of power during a part of the year at a lower cost than a steam power plant of the same capacity, but the water supply may vary sufficiently to lower its capacity during the remainder of the year and thus make necessary an auxiliary source of power. Furthermore, on account of the higher initial cost of the water power installation and the long distributing lines, the cost of power, which would otherwise be low, is increased. The **gas or oil power plant** has a high **thermal efficiency,** or *ability to convert the heat supplied into work*, but depreciation, repairs and high cost of fuel generally offset this advantage. The **steam power plant,** because of its dependability and all-round efficiency, can produce power in large quantities at the lowest cost and *is used to supply* **nearly three-quarters** *of the total power now used* in manufacturing, heating, lighting and railway service in the United States.

2. Classification. — Modern steam power plants may be classified:

1. According to location with reference to distribution of output as { Isolated, Central Station

2. According to the method used to dispose of the steam exhausted from the *main power* units as
 { Condensing
   Non-condensing

3. According to kind of service as
 { Stationary
   Locomotive
   Marine

An *isolated station is a power plant that supplies light, heat, and power to a single building or group of buildings situated near each other.* Manufacturing, office building, hotel and traction plants are usually of this type.

The *central station is a power plant that supplies light and power to* **public** *consumers more or less widely scattered and often far distant from the station.* This type of station uses highly efficient apparatus in order to obtain high heat economy.

When the central station forms a part of an interconnected or **super-power system,** it is spoken of as a **total load,** a **base load, peak load** or **standby station,** depending upon whether the plant supplies the entire load, the main load, the peak load or the emergency load.

By interconnecting and thus pooling their power facilities, the various power plant systems may exchange current so that districts having a high demand may draw on those where the demand is low. This effects large savings on duplicate plant equipment, assures constant service, and assists in reducing costs.

In the non-condensing plant, exhaust steam from the engine or turbine is discharged at or near the pressure of the atmosphere. The exhaust steam may be wasted to the air, or it may be used for commercial purposes. This type of plant is commonly used in hotels, office buildings and factories where the exhaust steam can be used for heating and various manufacturing processes. *In the condensing plant, the exhaust steam may be discharged into a* **surface condenser,** a vessel closed to the atmosphere and containing a large number of small tubes through which cooling water is forced to circulate, condensing the steam which surrounds the tubes, and lowering the pressure in the vessel; *or the exhaust steam may be discharged into a* **jet condenser,** a closed vessel in which the steam is condensed by coming into direct contact with jets of cooling water which are sprayed into the chamber. The condensed steam is known as **condensate.**

Condensing equipment is always used in central station power plants, where the amount of exhaust steam is in excess of that required for heating and where the maximum amount of power obtainable is desired.

3. Types of Power Plants to be Considered. — The following types of steam power plants will be briefly described, and the arrangement of the apparatus will be shown in outline form to make clear the relation existing between parts of the equipment:

NON-CONDENSING ISOLATED STEAM POWER PLANT 3

1. Non-condensing isolated plant.
2. Condensing isolated plant.
3. Central station having complete equipment for economical operation.
4. Locomotive power plant.
5. Marine power plant.

4. Essential Steam Power Plant Equipment. — Every steam power plant must have a **furnace** in which fuel is burned to generate heat; a **boiler,** or steel vessel containing water, to utilize the heat generated in the furnace and convert the water into steam; an **engine** or **turbine,** called the **prime mover** or **main unit,** to use the heat energy stored in the steam and perform work; and suitable **piping** to convey the steam. Besides the above equipment, the plant requires numerous **auxiliaries** and **accessories,** the type of which depend upon the location of the plant, the fuel and water available, the service for which the plant is intended, and the economy desired.

5. Non-condensing Isolated Steam Power Plant. — The arrangement of the apparatus in a typical plant of this type is shown in Fig. 1. The boiler is a **fire-tube** boiler of the **return tubular type**; that is, the hot gases from the furnace pass along the boiler shell and return through **tubes** which are surrounded by water. It is located in a **boiler setting** made of brick. The setting serves to confine the heat to the boiler, forms a passage through which the gases pass and often forms a support for the boiler, but in the best modern practice boilers are suspended from steel I-beams supported by steel columns.

Coal is delivered to the front of the boiler by a car pushed along a track by hand, and is shoveled upon a **stationary grate** under the front half of the boiler. The ashes resulting from burning the coal fall to the **ashpit** below the grate, and the smoke and hot gases from the furnace pass to the rear of the setting, then through tubes to the **smoke connection** at the front of the boiler. While passing through the tubes the gases give up some of their heat to evaporate the water, which surrounds the tubes. From the smoke connection at the front of the boiler, the gases pass, through a circular or rectangular **sheet-metal breeching,** to the **chimney** or **stack,**[1] which produces sufficient **draft,** or difference of pressure above and below the fire, to supply the air necessary to burn the fuel. The draft is regulated to meet the demand for power by a **damper** placed in the smoke connection to the breeching.

The steam formed in the boiler passes through a **steam nozzle** and **steam pipe** to a **steam header,** from which the **steam lead** to the engine is taken. Part of the steam flowing through the pipe is condensed and carried along by the rapidly moving steam. To prevent this surplus water from entering

[1] The word **chimney** is generally used to denote brick and concrete construction, and **stack** to denote steel construction.

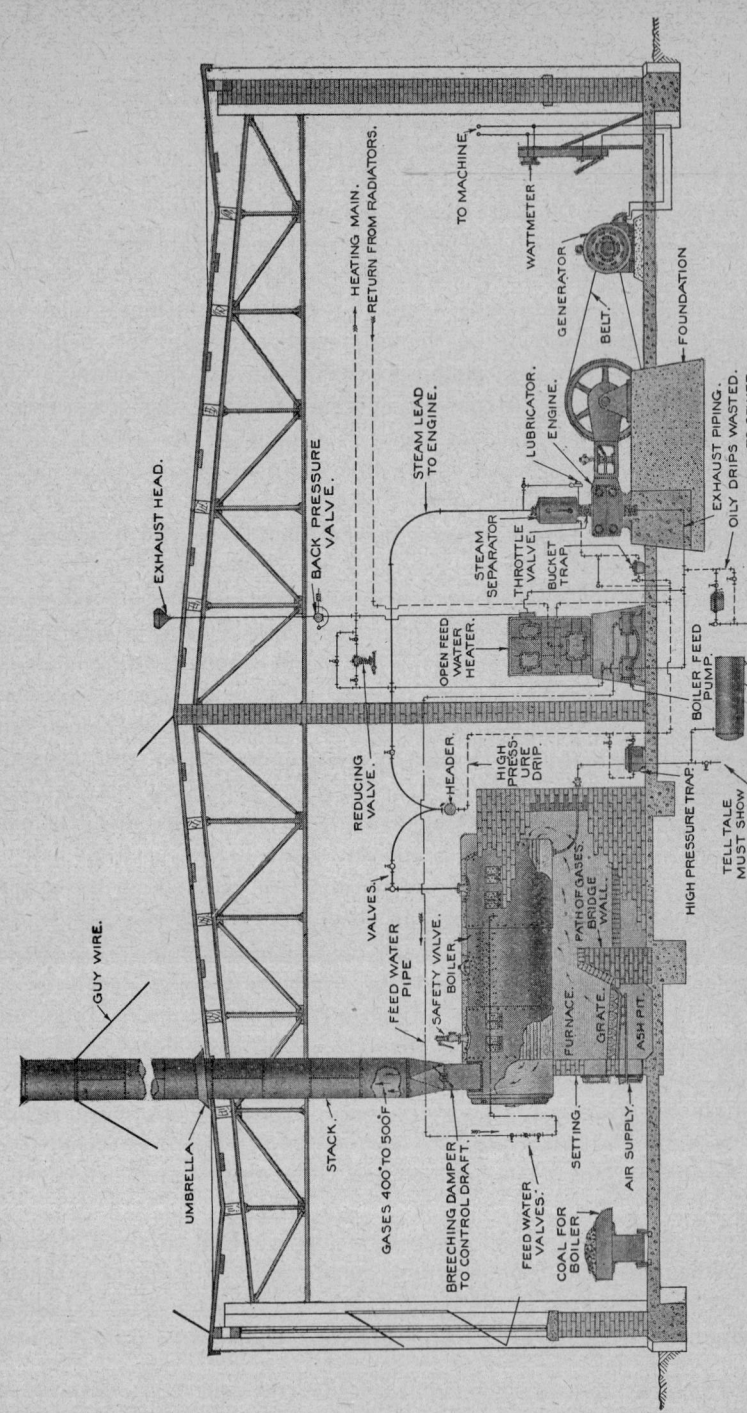

Fig. 1. — Non-condensing Isolated Steam Power Plant.

NON-CONDENSING ISOLATED STEAM POWER PLANT

the cylinder of the engine and causing damage, it is removed by a **separator** placed in the steam lead near the engine cylinder. The water which collects in the separator is removed automatically by a **steam trap** with a minimum loss of steam.

The steam is used in a **reciprocating engine,** which drives an **electric generator,** by means of a belt, and furnishes current for lighting and other useful purposes. The steam exhausted from the engine cylinder passes through the **exhaust piping** either to an **open feedwater heater** where it comes into direct contact with the feedwater and raises its temperature, or directly to the outside air through a **back-pressure valve** which opens to relieve excessive pressure in the exhaust pipe system.

An **exhaust head** is located at the top of the exhaust pipe and above the roof of the building to remove oil from the escaping steam and prevent the oil from being thrown over the roof and the surrounding buildings.

Any oil carried from the engine cylinder by the **exhaust steam** is removed before entering the feedwater heater by an **oil separator,** which is often a component part of the heater.

A **direct-acting boiler feed-pump** driven by **live steam** from the main steam line takes feedwater from the feedwater heater and forces it into the boiler. The heater is located above the pump and supplies water to the pump under a small pressure or **head,** because a pump will not lift hot water through any appreciable distance. Exhaust steam from the pump is discharged into the exhaust pipe running to the feedwater heater.

During the winter months, exhaust steam from the engine and pump is discharged into the heating system. The back-pressure valve is then set to open at a pressure above that required for heating. As the oil which is picked up by the steam in the engine cylinders is detrimental to the heating system, it is removed before entering the piping to the radiators. In case the amount of exhaust steam from the engines is insufficient to supply the heating requirements, live steam from the main steam line can be admitted to the heating system through a **reducing valve,** which lowers the pressure of the steam from that of the boiler to that required for heating.

Steam passes to the radiators from the exhaust piping near the feedwater heater, and, after being condensed in the radiators or heating coils, is discharged through **thermostatic valves** into a pipe line which discharges into the feedwater heater or the **returns tank.**

To operate the equipment shown in Fig. 1, certain **accessories, trimmings,** or **fittings** are necessary. The *boiler* requires the following fittings: a **safety valve** to limit the pressure which the boiler may carry; a **water column** and a **water gage** to show the water level in the boiler; a **steam gage** to indicate the pressure of the steam in the boiler; a **blow-off pipe** and **valves** to discharge the sediment which collects in the boiler shell and to empty the boiler; and feedwater piping containing a **check valve,**

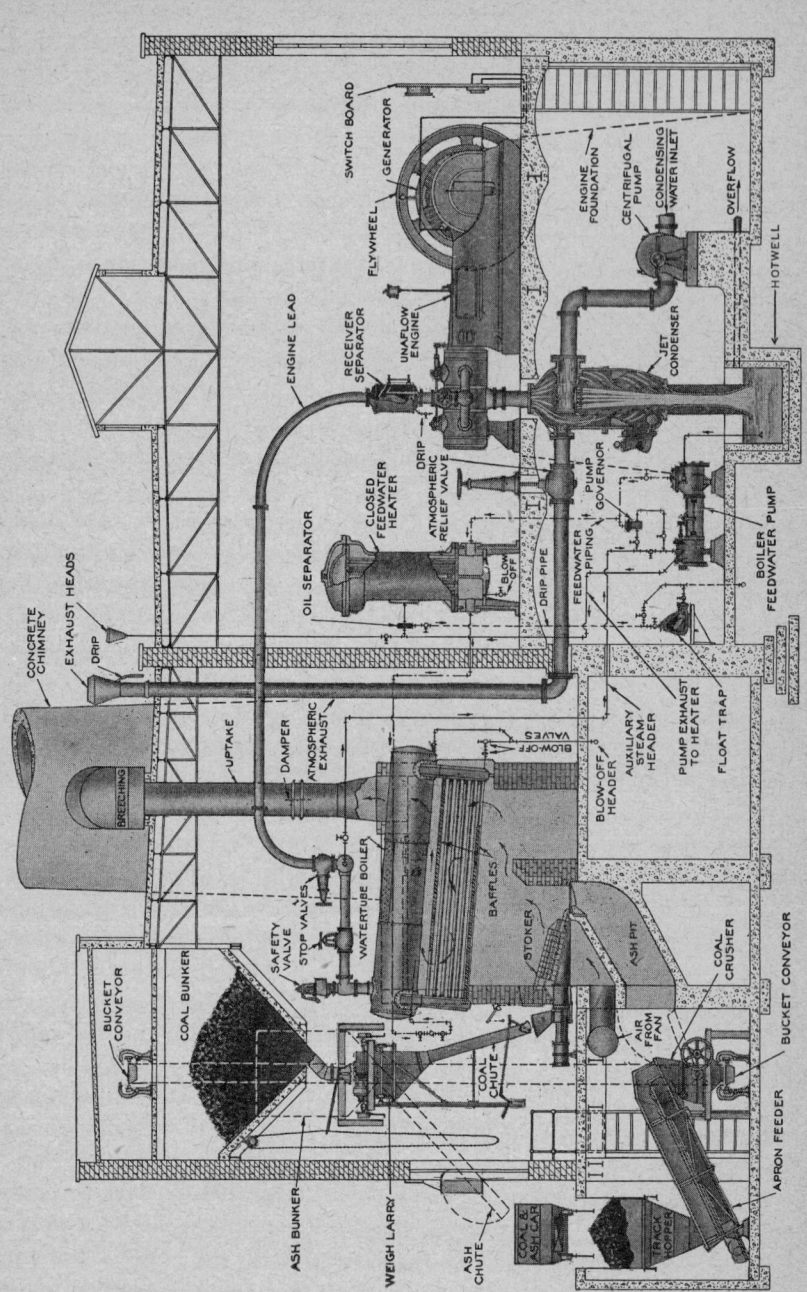

Fig. 2. — Isolated Condensing Steam Power Plant.

and the proper control valves. The check valve prevents the return of water from the boiler in case the boiler-feed pump is not operating.

The *steam unit* requires a **throttle valve** to control the flow of steam, a **governor** to regulate the speed, **oilers** to supply lubrication to the sliding and rotating parts, and **piping and valves** to drain the condensed steam from the cylinders. As a part of the engine, a **flywheel** is required to prevent excessive fluctuation in speed.

6. Condensing Isolated Steam Power Plant. — The apparatus required for the operation of this type of power plant is shown in Fig. 2. The boiler shown is a **water-tube boiler;** that is, it has a series of tubes containing water, around which gases from the furnace pass on their way to the stack. **Baffle walls** force the gases to take the path shown by the arrows.

Coal is delivered to a **track hopper** directly from a coal car, and is then discharged upon a **flight conveyor** for delivery to a **bucket conveyor,** which elevates the coal and discharges it into an **overhead bunker.** From the overhead bunker, the coal passes to **automatic scales** and then through a **coal chute** into the hopper of a **stoker** having a steam-operated plunger to feed the coal into the furnace. The ashes collect on a **dump plate** at the rear of the furnace and after being dumped into the ashpit under the stoker are delivered by a bucket conveyor to an overhead **ash storage bin.**

Steam is used in a reciprocating **unaflow engine** which is directly connected to an electric generator. The exhaust steam from the engine passes directly into a multi-jet condenser from the lower part of which the mixture of condensed steam, air, non-condensible gases and condensing water is discharged through a combining tube to the **hot well.** Water jets create the necessary vacuum by condensing the steam, and maintain it by removing the air and non-condensible gases, thus eliminating the necessity for a separate **air** or **vacuum pump.** In case it becomes necessary to stop the injection water pump, provision is made to automatically discharge the exhaust steam, through an **atmospheric relief valve,** to the atmosphere, the engine then operating non-condensing.

The exhaust steam from the steam-driven boiler feed pump is discharged into a **closed feedwater heater,** in which the steam does not come into direct contact with the feedwater. This heater is generally so constructed that the steam surrounds tubes in which the feedwater is circulating. The remaining equipment is similar to that described under the non-condensing plant.

7. Comparison of Non-condensing and Condensing Plants. — The thermal efficiency of a non-condensing plant, like that shown in Fig. 1, is low. Ordinarily it does not convert more than 4 per cent of the heat energy stored in the coal into work. This may be increased to 6 or 8 per cent during the winter when the exhaust steam is used for heating purposes. A condensing plant like that shown in Fig. 2 is capable of

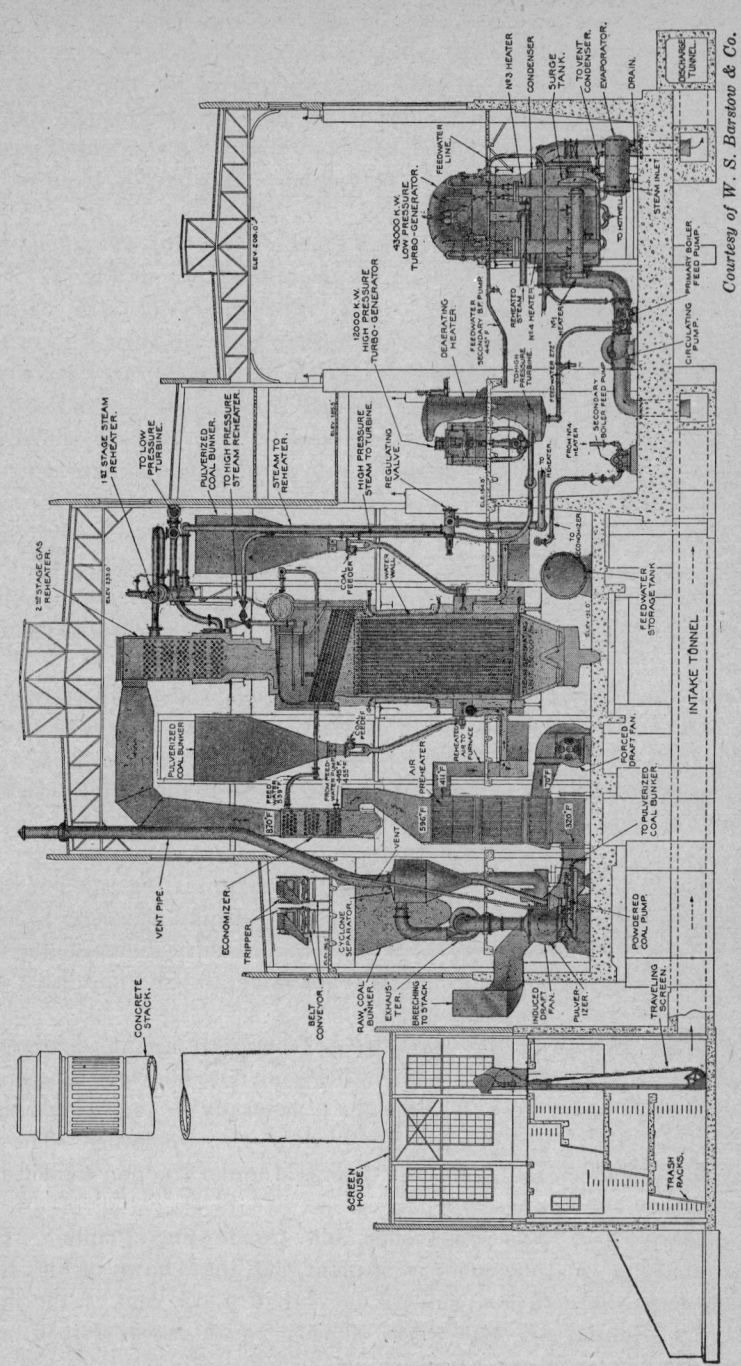

Fig. 3. — Central Station Steam Power Plant — Holland Station.

converting 10 per cent of the heat value of the coal into work, and in exceptional cases may run as high as 12 per cent.

In the non-condensing and condensing plants shown in Figs. 1 and 2, no effort was made to bring about all the saving possible by use of special heat-saving equipment. There are several kinds of heat-saving equipment, the installation of which would not be advisable in small power plants, but which might bring about large savings when installed in large central stations.

8. Central Station Steam Power Plant. — The general layout of a modern central station fully equipped with heat saving apparatus is shown in Fig. 3. Coal from the raw coal bunker is fed by gravity to a **pulverizing mill**, having rolls which reduce the coal to a fine powder, since coal can be burned in this form at its maximum efficiency. An **exhauster** discharges the pulverized coal into a **cyclone separator** in which the air is separated from the coal and discharged to the open air through a **vent pipe**. From the separator the pulverized coal drops by gravity to a motor-driven **coal pump**, and is pumped to the pulverized coal storage bunkers. Coal from the bunkers is fed to two **pulverized coal burners** located on opposite sides of the furnace, which has **water-cooled** side walls and a **dry ashpit** protected by a special **water screen**. The two walls equipped with burners consist of bare tubes set in the brickwork. Air for combustion is supplied to the furnace by a **forced draft fan** which takes air at 70 deg. fahr. and forces it over a plate type **air-preheater** and into the furnace at a temperature of 411 deg. fahr. The furnace gases are removed by an **induced draft fan** and after passing successively through the **second stage reheater, economizer** and **air-preheater** pass to the chimney at 320 deg. fahr.

The boiler is a **cross-drum** water-tube boiler, having a capacity of 250,000 lb. of steam per hour, and generates steam at 1400 lb. per sq. in. pressure. Steam, leaving the boiler, passes either to the **first stage reheater** where it is used to raise the temperature of the exhaust steam from the **high-pressure turbine**, or to a **superheater** which consists of a series of bent tubes connected at each end by a header and located in the path of the hot furnace gases. The gases, having a temperature higher than that of the steam, transmit some of their heat to the steam and raise the temperature above that existing in the boiler. When in this condition steam is called **superheated steam**.

Leaving the superheater the steam flows through suitable piping to the **high-pressure steam turbine**, where a part of its energy is converted into work. The exhaust steam then passes in succession through two reheaters, the first stage reheater using steam at boiler pressure and having a **thermostatic control** to maintain constant reheat, while the second stage reheater uses the flue gases as they leave the boiler. These two reheaters are so proportioned that at full load the steam reheater is nearly cut out, the flue gas reheater supplying all the heat required. The steam leaves the second-stage reheater at the pressure and temperature desired in the **low-pressure turbine.** After being used in the low-pressure turbine, the steam is exhausted to a **single-pass** surface condenser. The cooling or circulating water for the condenser is taken from the **inlet tunnel** by a motor-driven **centrifugal pump,** and after passing through the condenser is discharged into the **discharge tunnel.** The centrifugal pump has a rotating member, or **impeller,** and receiving water near the shaft discharges it at the periphery of the impeller. A **screen house,** having revolving screens, removes débris from the cooling water and is located at the entrance of the inlet tunnel.

The changes which take place in the working substance (water and steam) as it changes from water to steam and then back to water is called the

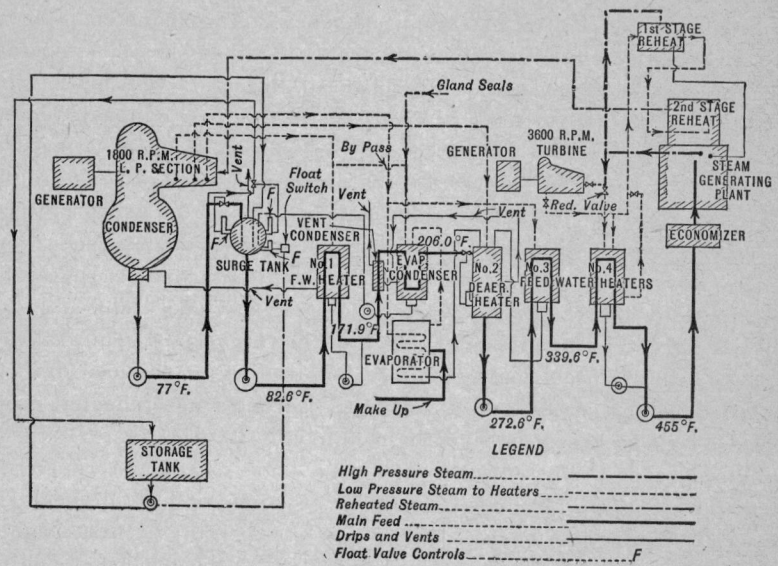

Fig. 4. — Heat-cycle Diagram — Holland Station.

heat cycle. This cycle and the arrangement of the auxiliary equipment is indicated diagrammatically in Fig. 4. The condensate collects in the **condenser hot well** and is pumped to a **surge tank** which also receives the make-up water from the **evaporator condenser.** A float valve located in the surge tank diverts any excess water to the **storage tank.** Three centrifugal pumps acting in series deliver the feedwater from the surge tank to the boiler.

The *first feedwater pump,* operating against 70 lb. pressure, delivers the water to a **de-aërating heater** at 273 deg. fahr., in which oxygen is removed from the water by violent boiling.

The *second feed pump* operates against 600 lb. maximum pressure, and forces the water through (1) the third-stage heater, taking steam from the sixth stage[1] of the turbine in parallel with the make-up feedwater evaporator, and (2) the fourth-stage heater, which is supplied from the crossover header, receiving the exhaust from the high-pressure turbine.

The *third feed pump*, working against a maximum head of 1600 lb. takes the water, which reaches a temperature of 455 deg. fahr. at full load, and discharges it to the boiler through the economizer at about 539 deg. fahr.

The **deaërator heater** is supplied with steam from the tenth stage of the turbine and is equipped with float control for regulation of the water supply at that point. By the addition of this heat-saving equipment the temperature of the feedwater is raised from 77 deg. fahr. to 539 deg. fahr., thus saving a large quantity of heat which would otherwise be wasted, and in addition there is an appreciable saving by using preheated air in the furnace. The thermal efficiency may thus be increased under the best conditions to about 28 per cent.

9. Locomotive Steam Power Plant. — An external view of the locomotive steam power plant is shown in Fig. 5, and a vertical section through the plant in Fig. 6. The plant is of the non-condensing type, and the equipment is not essentially different from that installed in the same type of stationary plant. The space in which

Fig. 5. — Locomotive Steam Power Plant — External View.

[1] See page 479, Chapter XXII, for explanation of the term stage as used in a turbine.

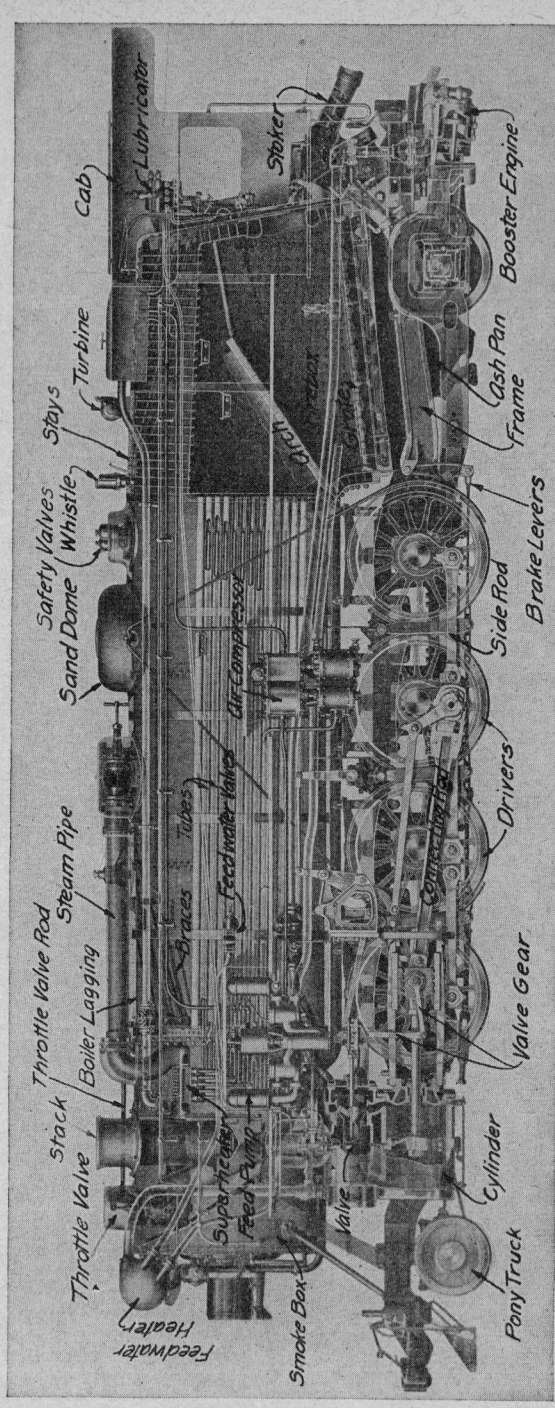

Fig. 6. — Locomotive Steam Power Plant — Vertical Sectional View.

the equipment is installed is restricted, however, and the equipment must be accommodated to the space. The thermal efficiency of the best modern locomotive plants varies from 5 to 8 per cent at normal operating pressures of 250 lb. per sq. in. Attempts are being made to increase the efficiency by using higher steam pressures, stream lining, and, in some European countries, by using turbine-driven locomotives in which the efficiency is claimed to be 15 per cent.

The boiler is a fire-tube boiler of the **internally-fired** type; that is, the furnace is within the boiler shell. An intensely hot fire is obtained in the furnace, by a strong draft produced by exhausting the steam from the engine cylinders through a nozzle, which discharges the steam into a short stack attached to the smoke box. This reduces the pressure in the smoke box and causes an increased flow of air through the coal on the grate. Coal is stored in the U-shaped portion of the tender formed by the inner walls of the water storage tank, and is fired into the furnace by a screw type mechanical stoker. In smaller locomotives coal is generally fired by a hand shovel. The gases rising from the burning coal pass through the tubes to the smoke box.

Ashes from the **shaking grate** fall into an ash pan located under the grate. The amount of air admitted under the grate is controlled by dampers attached to each end of the ash pan and operated by rods and levers from the engineer's cab.

Steam generated in the boiler is fed, through an insulated pipe that runs along the boiler, to the superheater which consists of a system of piping in the smoke box and flues. From the superheater the steam passes to the **throttle valve** which controls the flow of steam to the engine cylinders, and is operated by a lever in the engineer's cab. The steam pipe branches at the throttle valve, one branch going to each cylinder of the engine. In order to obtain greater starting power, a **booster engine** operates upon the rear truck-wheels during the starting period.

Feedwater, from the water-storage tank in the tender, is forced into the boiler through a feedwater heater by a steam-operated boiler-feed pump. An **injector;** that is, a pump operated by a jet of steam, is often installed for emergency use.

A **vertical steam-driven air-compressor** is supported at the side of the locomotive boiler. This air compressor supplies air for operation of the brakes and sometimes for the valve mechanism on heavy locomotives. A small steam-turbine-driven generator, attached to the top of the locomotive boiler or to the side of the frame, furnishes electricity for the headlight.

Strictly speaking, the boiler does not have a setting. Its front end is attached to a saddle formed by the cylinder castings. At the rear end, brackets riveted to the side of the fire-box rest upon the side frame of the locomotive. The outside of the locomotive is covered with blocks of

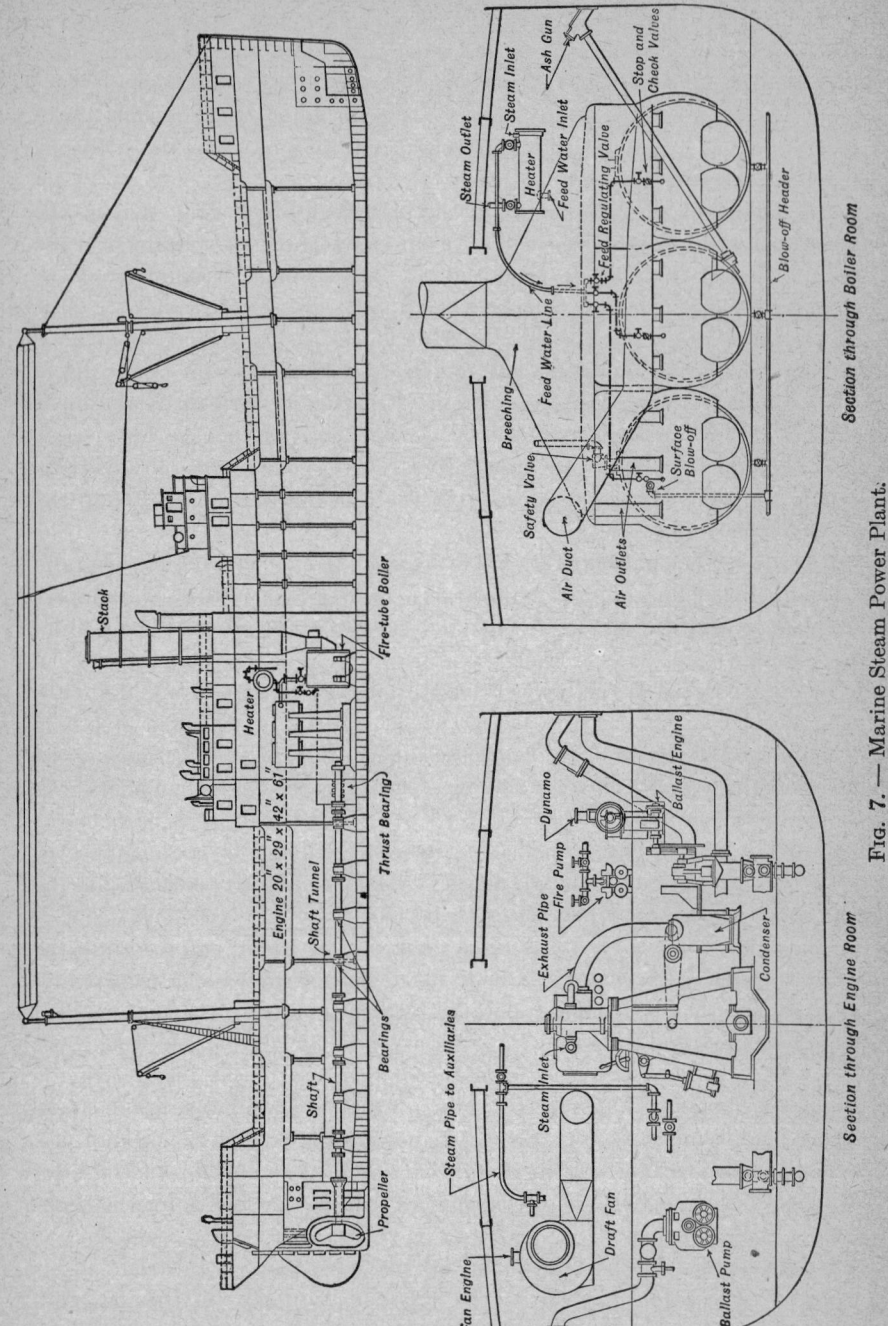

Fig. 7. — Marine Steam Power Plant.

85 per cent magnesia, an insulating material, to reduce the loss of heat, and a sheet-metal covering is placed over the magnesia blocks.

Locomotives are designated by two systems, which have reference to the arrangement of the wheels. One method gives an arbitrary name to the arrangement of the wheels, such as **Pacific, Prairie, Mogul, Mallet, Santa Fe.** The more common method is the **Whyte system,** which classifies locomotives by giving the number of wheels on the forward, or **pony truck,** the number of **driving wheels,** and finally the number of wheels on the rear, or **trailing truck.** Thus, a 4-4-0 locomotive has 4 front wheels, 4 drivers, and 0 rear wheels. Table 1 gives a few typical wheel arrangements, with the Whyte classification and the corresponding arbitrary name.

TABLE 1. — WHYTE'S SYSTEM OF LOCOMOTIVE CLASSIFICATION

Symbol	Wheel Arrangement	Type	Service
0-4-0		*4-wheel Switcher*	*Yard*
0-8-0		*8-wheel Switcher*	*Yard*
2-6-0		*Mogul*	*Freight*
2-6-2		*Prairie*	*Passenger*
4-6-4		*Hudson*	*Passenger*
2-8-0		*Consolidation*	*Freight*
2-8-2		*Mikado*	*Freight*
2-10-2		*Santa Fe*	*Freight*
4-6-2		*Pacific*	*Passenger*
4-8-2		*Mountain*	*Freight*
0-8-8-0		*Mallet articulated*	*Freight*
2-8-8-8-2		*Mallet articulated (Triplex)*	*Freight*

10. Marine Steam Power Plant. — Like the locomotive power plant, the marine power plant must occupy a small space. The steam marine plant is run " condensing "; it uses a surface condenser on ocean steamships in order to have pure water for use in the boiler, since sea water is not suitable for boiler feed, because of the salt it contains. Jet condensers are used on lake and river steamers. Figure 7 shows the location of the equipment on a typical lake steamship, together with transverse elevations through the engine and boiler rooms. The compact arrangement of the apparatus is apparent.

The boilers are of the marine type and have fire-tubes or water-tubes. Those shown are of the fire-tube type. Coal is generally fired by hand, although stokers of the underfeed type are sometimes used. Oil is burned as a fuel under the boilers, in many ships.

The connection between the stack and the furnace is short. The stack is made of sheet steel; since it is relatively short as compared with a land installation, a fan, located as shown, is necessary to produce sufficient draft to burn the fuel. The fan may force air, under pressure, into the boiler room or ashpit of the boilers, or it may withdraw the gases by suction and deliver them to the stack.

The engines are **vertical marine engines** and they drive the propeller shaft direct; that is, the engine shaft is directly connected to the propeller shaft, with a **thrust bearing,** placed at the engine end of the propeller shaft, to transfer the thrust of the propeller to the frame of the ship. When a turbine is used, the speed is reduced to that required by the propeller shaft; it is then necessary to use either **mechanical reduction gearing** or a **motor** connected to the propeller shaft and driven by current from the main turbo-generator.

REFERENCES

Steam Power Plant Engineering, GEBHARDT.
Power Plant Engineering, FERNALD and ORROK.
Steam Power, HIRSHFELD and ULBRICHT.
Shipbuilding and Locomotive Cyclopedia, SIMMONS-BOARDMAN PUB. CO.
Development of Turbo-Locomotives — VOL. 51, MECHANICAL ENGINEERING.
High Steam Pressure in Locomotives — VOL. 48, MECHANICAL ENGINEERING.
Holland Station — VOL. 32, GENERAL ELECTRIC REVIEW.
Gilbert Station — VOL. 35, NATIONAL ENGINEER.

REVIEW QUESTIONS

1. Name the essential equipment in (a) a non-condensing power plant, (b) a condensing plant.

2. Distinguish between an isolated power plant and a central station plant.

3. Explain what is meant by a superpower system.

4. Make an outline sketch showing the relative location of the essential equipment in a steam power plant operating (a) non-condensing, (b) condensing.

5. About what per cent of the heat in the coal is actually converted into power in (a) a condensing plant, (b) a non-condensing plant.

6. Name the heat-saving equipment which is often used in a steam-power plant.

7. Trace the course of the feedwater from the hot well to the boiler, Figs. 2 and 3.

8. Trace the course of the steam from the boiler to the hot well, Fig. 4.

9. In what respect does the steam power plant of a locomotive differ from a stationary steam power plant?

10. What type of steam power plant is generally used on ocean going ships? Why?

CHAPTER II

STEAM BOILERS AND SETTINGS — BOILER AND SETTING FITTINGS

11. Foreword. — The form and arrangement of the parts on a modern boiler are substantially the same as they were a decade ago, except for structural alterations necessary to adapt it to the requirements of the high pressures and temperatures now being used. This is, however, not true of the so-called steam generator and high-pressure types discussed on page 617, Chapter XXV.

The present tendency in **power boiler** operation is toward higher steam pressures. A few years ago 100 pounds per square inch was the prevailing gage pressure. Today a pressure of, at least, 400 pounds per square inch is being used in the majority of Central Stations, and in a few instances 1400 pounds is being used with apparently no more operating troubles than are encountered with the lower pressures. One industrial plant is using 1800 pounds.

Boilers carrying 25 pounds per square inch gage pressure, or less, are usually classed as **heating boilers**.

12. Material. — The material used in boiler construction should conform to requirements stated in the A. S. M. E. (AMERICAN SOCIETY OF MECHANICAL ENGINEERS) BOILER CODE.[1] Steel plate used in the boiler shell must be *flange or firebox steel made by the open-hearth process* and must conform to certain chemical and physical specifications. Its tensile strength must be from 55,000 to 65,000 pounds per square inch, and its crushing strength 95,000 pounds per square inch. When the part is under pressure and exposed to the fire or hot gases only firebox steel can be used.

Forged steel, when used to form boiler drums, must be made by either the open-hearth or electric process and must have a tensile strength between 60,000 and 75,000 pounds per square inch.

The above materials when bent cold around a pin having a specified diameter must not show cracks. For material 1 inch or under the pin diameter equals the plate thickness.

Cast iron cannot be used for flanges or nozzles attached directly to the boiler but can be used for boiler connections such as fittings and valves when the steam temperature does not exceed 450 deg. fahr., or the pressure 250 pounds per square inch.

---

[1] The A. S. M. E. Boiler Code may be obtained by addressing the Secretary of the American Society of Mechanical Engineers, 29 West 39th St., New York City.

13. Classification. — A classification of steam boilers is difficult, because of the large number of variations in a few fundamental types. The following classification is suggestive:

1. According to relative positions of water and hot gases
   { Water-tube
   { Fire-tube

2. According to location of furnace
   { Externally fired
   { Internally fired

3. According to use
   { Stationary
   { Portable
   { Locomotive
   { Marine

4. According to direction of principal axis
   { Horizontal
   { Inclined
   { Vertical

14. Typical Steam Power Boilers. — In order to give a clear idea of boiler details, the following types and makes, which include the most common types of boilers generally used for power purposes, will be described:

1. Stationary boilers, as exemplified by
   - Fire-tube
     { Horizontal return tubular
     { Vertical tubular
   - Water-tube
     { Babcock and Wilcox
     { Heine
     { Stirling
     { Wickes

2. Locomotive boiler

3. Marine boilers, as exemplified by
   { Fire-tube — Scotch Marine
   { Water-tube — Babcock and Wilcox

15. Boiler Nomenclature. — The following terms are common to the various types of boilers, and their application should be thoroughly understood. The application of each term is illustrated in Fig. 8.

Shell. — The boiler shell consists of several steel plates bent into cylindrical form and riveted together. Each separate cylindrical ring is called a **course**. A shell made up of two rings is called a two-course shell, and one made of three rings, a three-course shell. The ends of the shell are closed by means of boiler heads, which are made of flat, convex or concave boiler plate having flanged edges to permit riveting to the shell. The shell, together with the heads, forms the **drum**. When used for high-pressures the drum is often made from a steel forging.

Setting. — The boiler setting is usually made of brick. It provides a means of support for some types of boilers and, at the same time, forms the walls of the furnace and combustion chamber.

Grate. — The grate consists of cast-iron bars or plates upon which the fuel is burned and is so made that air for combustion can pass through the grate to the fuel. The area of the surface upon which the fire rests is the **grate surface,** and is usually expressed in square feet. A grate 5 feet wide and 6 feet long would have 30 square feet of grate surface.

Furnace. — The furnace, or firebox, is the space above the grate and below the boiler shell, in which the fuel is burned. It is enclosed by the side and front walls of the setting. The **combustion space** is that part of the furnace in which the volatile matter and combustible gases are burned.

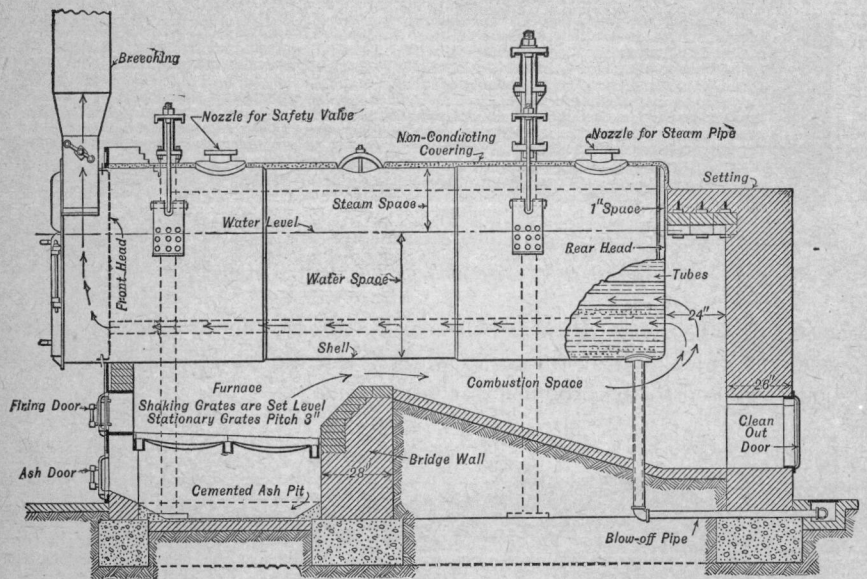

Fig. 8. — Boiler and Setting.

Water and Steam Space. — The water space is the volume of the shell that is occupied by the water. The space occupied by water and tubes in a return-tubular boiler is about two-thirds the volume of the entire shell. The steam space is the volume of the entire shell not occupied by water and tubes and usually varies from 0.65 to 1.00 cubic feet per boiler horsepower. (For definition of boiler horsepower, see Art. 204, page 220.)

The level at which water stands in the boiler shell is known as the **water level.** It is subject to considerable fluctuation, but for best operation the variation in level should be small. The height at which water stands in the boiler shell is indicated by a **water-gage glass** located where it may be easily seen, or by **gage cocks** attached to the **water columns** or directly to the boiler head or shell.

The area of the water surface from which steam is separated is called the **disengaging surface.** This area should be sufficient to prevent water from being carried from the boiler along with the steam. Steam carrying a large amount of water is said to be **primed**.

Heating Surface. — The **water heating surface** is that surface of the boiler which is exposed to hot gases on one side and water on the other. **Superheating surface** is that surface having steam on one side and hot

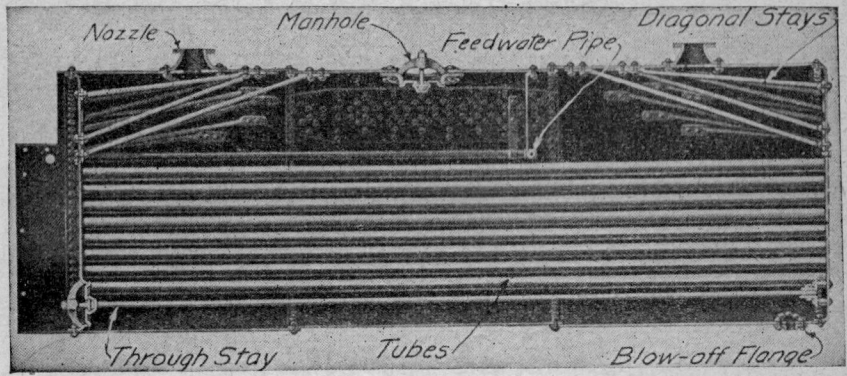

Fig. 9. — Longitudinal section H. R. T. Boiler.

gases on the other. Heating surface is expressed in square feet and is generally calculated at the mean water level and for that surface of the boiler exposed to the products of combustion.

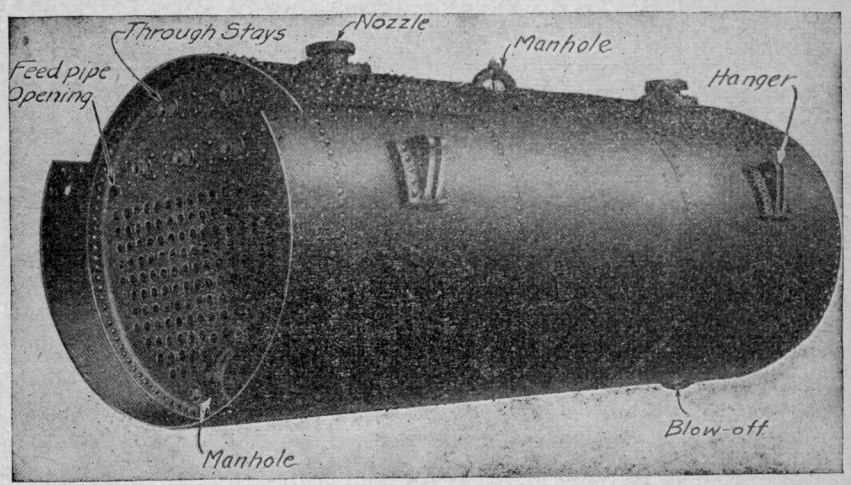

Fig. 10. — Outside View of H. R. T. Boiler.

16. Horizontal Return-tubular Boiler. — This type of boiler, Fig. 9, is commonly used in small power plants in capacities up to 300 horsepower, because of its low first cost, high evaporative capacity, and compactness. As its name implies, the gases from the furnace pass to the rear along the under side of the boiler shell and return to the smoke connection at the front by passing through the tubes.

The shell is ordinarily made of three courses as shown in the longitudinal section, Fig. 10. The separate courses are fastened together by **lap joints,** No. 1, Fig. 11. The seam or row of rivets thus made around the circumference of the boiler is known as a **ring** or **girth seam.** The free ends of

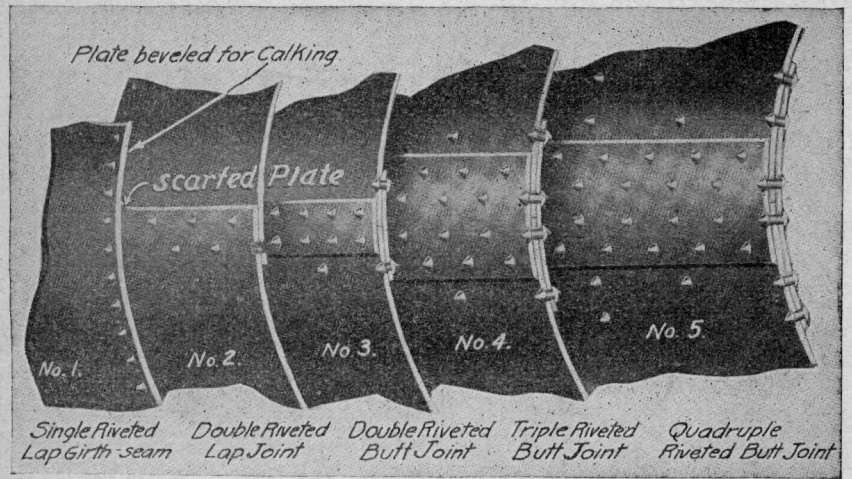

Fig. 11. — Types of Boiler Joints.

the plate forming each course are brought together to form a **butt joint,** as shown in No. 3, Fig. 11. A **covering strip,** or **welt,** is placed outside, and a wider covering strip inside. These covering strips hold the ends of the plate together by means of rivets. The joints uniting the ends of each course are called **longitudinal joints** and must be arranged as described. A butt joint may be double, triple or quadruple riveted.

The joint made by fastening No. 1 and No. 2, Fig. 11, together illustrates how the outside strap is **scarfed,** or **drawn out,** and carried under the ring seam to produce a tight joint where the longitudinal and ring seams meet. The ends of all outside plates are beveled at an angle of 15 degrees before they are rolled into shape. After being riveted, the edges are **calked,** that is, hammered against the plate to which they are riveted, thus making a joint that will not leak. For calking, a tool, Fig. 12, having a rounded edge is used in order not to injure the plate.

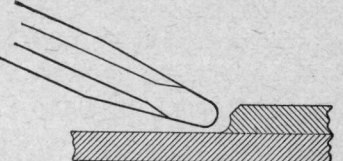

Fig. 12. — Round Nose Calking Tool.

A smoke connection is attached to that part of the front course which remains after the top front portion is cut away to make an opening for the passage of the gases to the breeching. The ends of the shell are closed by two heads, containing openings into which the **tubes** are arranged as shown in Fig. 10. The holes in the tube sheets are made $\frac{1}{8}$-inch larger than the diameter of the tubes, and the tubes are expanded to a tight fit in the tube sheets by some form of expander. Figure 13 shows the Dudgeon and Fig. 14 the Prosser expander, together with the appearance of the tubes after being expanded by each type of expander. After expansion, the

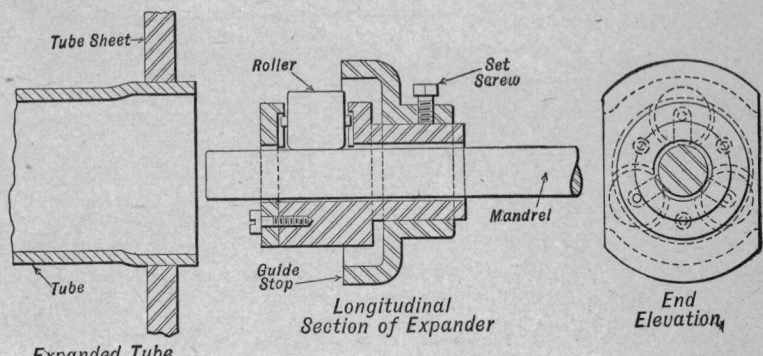

Fig. 13. — Dudgeon Expander.

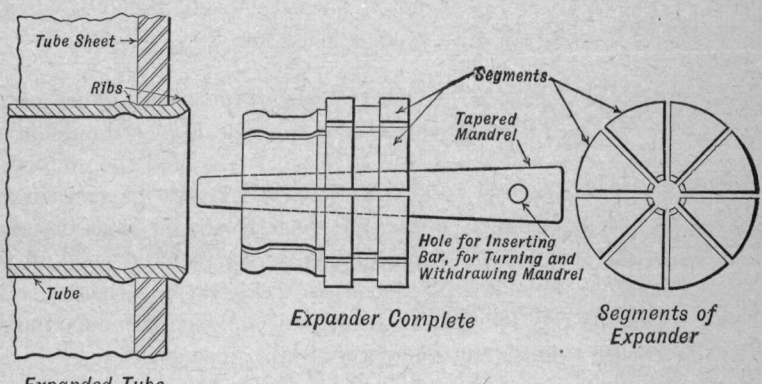

Fig. 14. — Prosser Expander.

projecting ends of the tubes are rounded over against the plate by a beading tool, Fig. 15.

The flange of the front head is turned outward, to permit riveting by machine and make construction easier. On the so-called New York boiler, this flange is turned in. The rivets forming the joint are thus protected by water from possible overheating, but riveting is made more difficult during construction. A **manhole** or **handhole** is located in the front head, below the tubes, to give access to the lower part of the shell for inspection or cleaning, and the tubes are arranged in the head as shown in Fig. 10, to give as free a passage for water circulation as possible.

The area of the heads above the tubes may be stayed by (1) **diagonal stays,** Fig. 9, or (2) heavy angles or channels and through braces, Fig. 44, page 51. The angles or channels support the boiler head with rivets, and are in turn supported by through braces from head to head. These braces, or longitudinal stays, pass through the boiler heads and are threaded for nuts inside and outside of the heads. That part of the head area occupied by the tubes does not require bracing, because the tubes act as braces. The

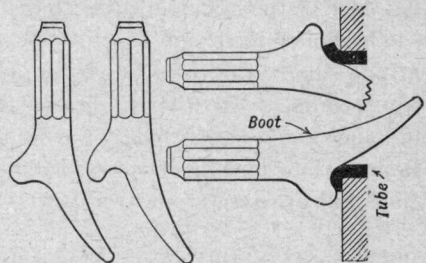

FIG. 15. — Beading Tool and Method of Use.

area of the heads below the tubes is braced by a special form of through brace, as shown in Fig. 9. It is attached to the rear head by means of a pin connection through two angles, which are riveted to the head, but are separated from it by nipples in order to prevent overheating of the plates.

The use of the diagonal brace below the tubes is not permitted, because the pads or enlarged flattened shoulder of the brace, when riveted to a shell, which is surrounded by hot gases, would not conduct heat away with sufficient rapidity to prevent the shell from being overheated.

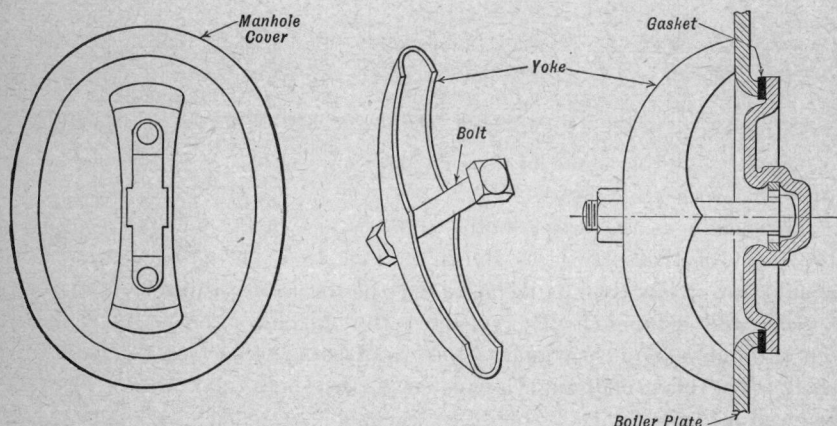

FIG. 16. — Manhole and Cover with Yoke.

Steam nozzles, made of cast iron or pressed steel, are riveted to openings in the top of the front and rear courses. A safety valve is attached to one nozzle, and the boiler stop valve and main steam line to the other. An 11 by 15-inch elliptical manhole, which gives access to that part of the boiler above the tubes, is located in the middle course at the top of the shell. A **manhole ring** is riveted to the shell, or the boiler plate is flanged over, as shown in Fig. 16, and machined to form a seat for the manhole cover, which is held in position by bolts and yokes. A **gasket**, made of lead or rubber, is generally placed between the manhole ring and the flat edge of the manhole cover to make a tight joint.

The **boiler feed pipe** is screwed into a **brass** or **steel bushing**, Fig. 17, at one side of the front head, or into a **flanged connection** on top of the front course. An internal brass feed-pipe is screwed into the bushing or flange and is supported by a strap from the boiler shell or from a through brace. This pipe passes toward the rear, turns across to the middle of the shell and discharges downward with an open end between the central row

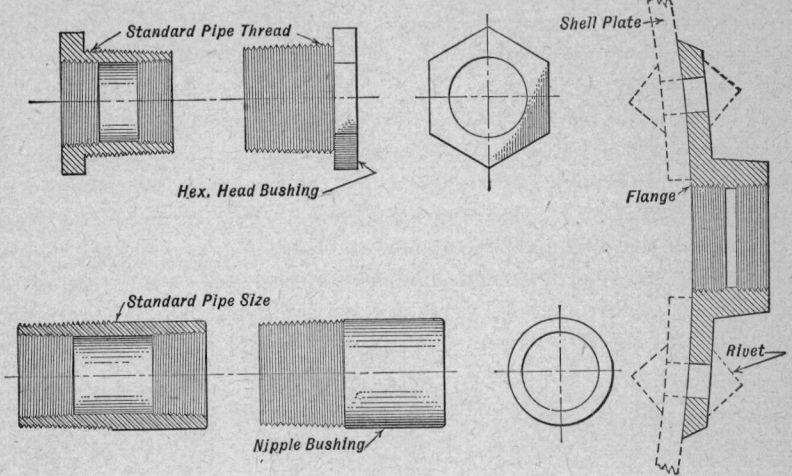

FIG. 17. — Typical Feedwater Pipe Bushings.

of tubes below the water level, and about three-fifths of the length of the tubes from the front head.

The height of the water in the boiler is shown by a water column at the front of the setting, as shown in Fig. 18. The lowest water level should not be less than $3\frac{1}{2}$ inches above the top row of tubes.

Scale and sediment, which collect in the bottom of the shell, are blown out through a blow-off connection at the bottom of the rear course of the shell. The **blow-off flange** is riveted to the shell and the blow-off pipe screwed into the flange.

The boiler shown in Fig. 18 is supported by brackets riveted to the upper half of the boiler shell. The brackets must be arranged in pairs on each side of the boiler, on each end course, near the ring seams of the middle course. The front brackets rest on a plate in the setting. The front end is fixed, and when the boiler expands, the rear end moves on rolls placed between the bracket and a plate resting on the setting.

Steel columns and I-beams built into the setting, Fig. 19, are also used

Fig. 18. — H. R. T. Boiler supported by Brackets — Flush-front Setting.

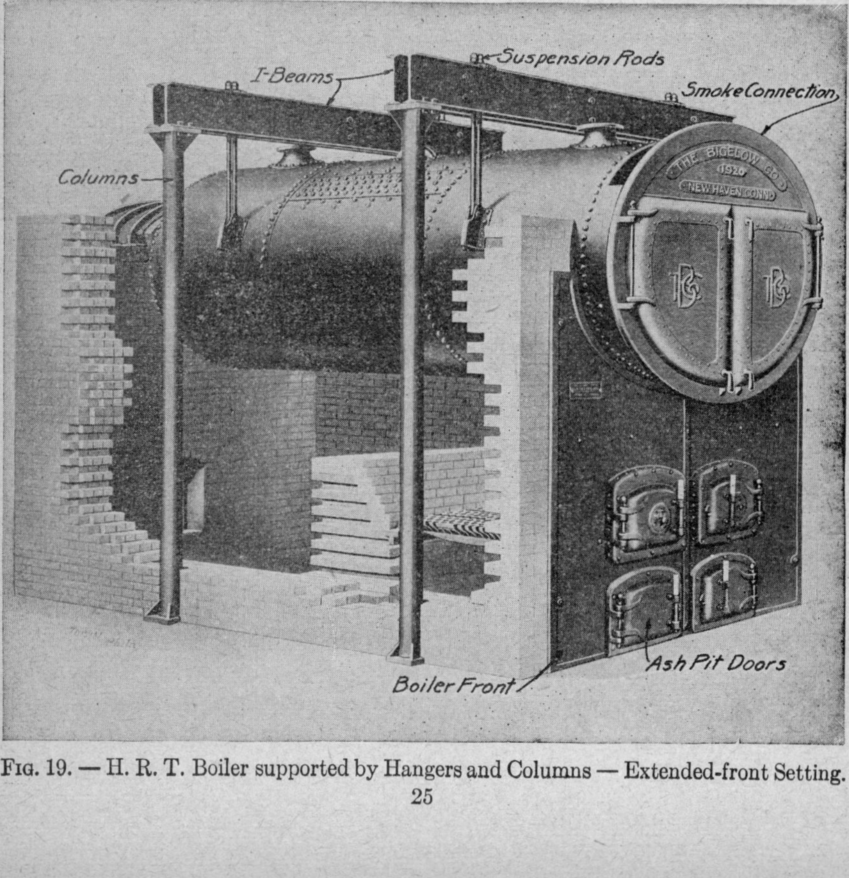

Fig. 19. — H. R. T. Boiler supported by Hangers and Columns — Extended-front Setting.

to support this type of boiler. The boiler is hung from I-beams by rods which hook into **hangers** riveted to the upper part of the boiler shell. The second method is preferable to the first, as the boiler is entirely free from the setting and will not injure it when contracting or expanding. A third method of support often used is known as the **three-point suspension.** This method, proposed by Mr. Woolson, is most satisfactory, as the boiler is evenly supported. The front of the boiler is suspended by hangers and hanger rods on each side of the shell. The suspension rods at each side of the rear course of the boiler are swung from the ends of an equalizing lever, which is supported at its center by the I-beam overhead construction, Fig. 8, page 19.

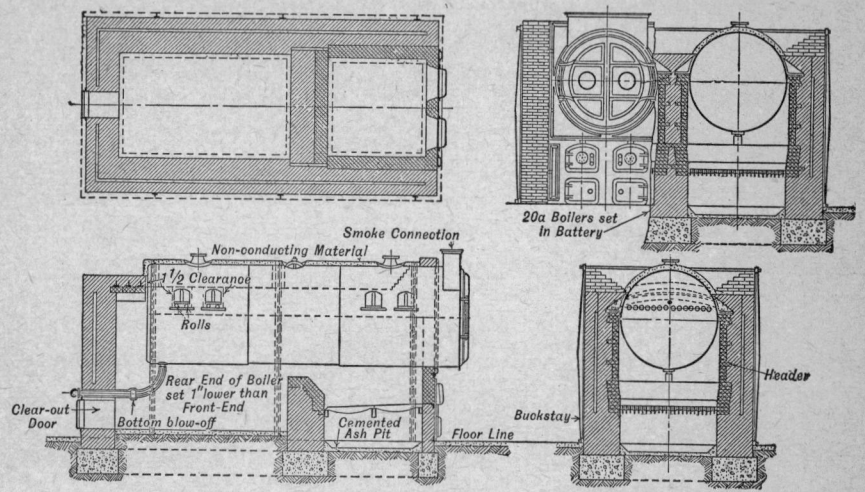

FIG. 20.—Setting for Horizontal Return-tubular Boiler.

17. Setting for Horizontal Return-tubular Boiler. — Boiler settings are ordinarily made of brick. Such settings are comparatively porous and often crack, thus permitting air to leak into the furnace. Settings are also made of **silocel,** an insulating brick of low heat conductivity. Silocel is made from a rock which contains silica and, as constructed, has a high percentage of small air spaces. The brick may also be covered with a steel casing lined with magnesia block riveted to the plate, or by a covering of wire and asbestos. The steel covering is expensive, but allows small air leakage, while settings made of silocel are liable to crack in much the same way as brick settings.

Top, front, and side views of a brick setting are shown in Fig. 20. The setting rests upon a concrete foundation which should have sufficient depth to prevent settling. The side and rear walls shown are made with an inner and an outer wall having an air space between. The inner and warmer wall, being separate from the outer wall, can expand independently. *The space between the walls, when used, should be filled with a heat insulating*

SETTING FOR HORIZONTAL RETURN-TUBULAR BOILER

material, such as ashes or soot, to prevent air circulation within the space and the consequent loss of heat. The outer wall is made of hard-burned strong brick and the inner wall, which comes in contact with the hot gases, is made of **firebrick**, a material especially prepared to resist high temperature. Firebrick walls are often laid with every sixth row of brick as a header, to permit repair of the wall between any two headers without disturbing the remainder. Experience indicates that the best results are obtained when the brick are laid up as closely as possible, after being dipped in a slurry and then rubbed and tapped into place — this slurry being made of about 60 per cent calcined clay and about 40 per cent raw clay, both of the same composition as the material used in making the brick. The walls may be made solid and their total thickness varies from 16 to 24 inches, depending upon the size of the boiler. The side walls are carried above the center line of the boiler as shown in Fig. 20, and the space between the boiler and inner wall is filled with asbestos material to prevent leakage from or into the furnace. The top of the boiler is covered with non-heat-conducting material. The return-tubular boiler is ordinarily set 1 inch lower at the rear than at the front, with *sufficient clear space left in front of the setting to permit removing and replacing defective tubes.* When two boilers are placed in a setting having a common wall separating the furnaces, or in battery, the construction is as shown in Fig. 20a.

The furnace shown in Fig. 20 is externally fired; that is, the fire is external to the boiler shell. The grates are supported by **bearer bars** fastened to the side walls and are set from 3 to 5 inches lower in the rear than at the front. The distance between the boiler shell and the grate should be at least 30 inches for anthracite coal and 36 inches, or preferably more, for coal which gives off large quantities of gas.

At the rear of the grate a **bridge wall** extends between the side walls, and projects a distance above the level of the grates. This wall often supports the rear end of the grate, retains the fuel on the grate, and assists in mixing the gases and air passing over it. The connection between the bridge and side walls should be so made that there will be room for expansion of the bridge wall; otherwise, cracked side walls will result. The top of the bridge wall should be about 12 inches from the shell for anthracite coal, and the distance should be increased for coals that evolve large amounts of gas.

The distance between the rear wall and the rear of the shell varies from 18 to 24 inches. The rear wall is carried above the top row of tubes, and the space between the rear wall and shell is bridged over, to prevent escape of the furnace gases, by a horizontal or curved firebrick wall, or arch, strengthened and supported by T-bars. As the expansion of the boiler may be as much as $\frac{1}{2}$-inch, a movable connection is made between the shell and the wall. Two methods of making this connection are shown in Fig. 21, the first of which is known as the **Woolson arch.**

The front wall consists of two **fire-door arches** made of hard-burned fireclay or of tile with a cast-iron fire-door arch. The arch rests upon a dead plate made to hold firebrick, or upon a bearer plate supported by the side walls. Above the fire-door arches is a hard brick wall having a

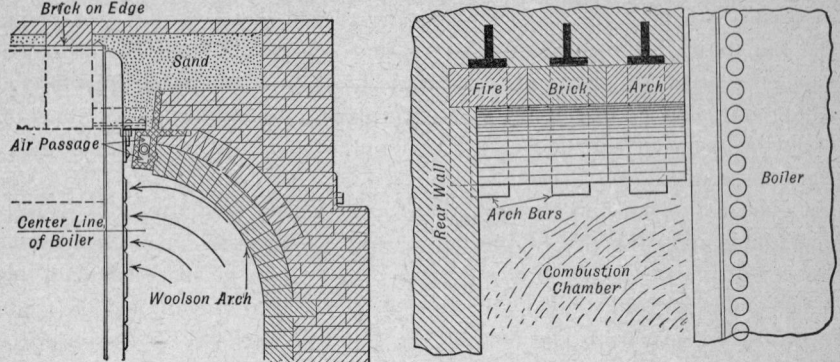

Fig. 21. — Two Methods of making Back Connection to Boiler.

firebrick lining. Every other layer of firebrick, between the fire-door arch and the shell, is arranged as a header.

A **clean-out door** in the rear wall gives access to the inside of the setting, for cleaning. The blow-off pipe is carried through a **thimble** in

Fig. 22. — Setting Fixtures.

the rear wall, and the space in the thimble not occupied by the blow-off pipe is filled with asbestos wool, to prevent leakage from or into the combustion space.

18. Setting Fixtures. — Certain castings and forgings, Fig. 22, are needed to complete the setting. A **cast-iron** or **steel casing** covers the front of the boiler and setting. This casing may be a one-half, three-quarter, or full-front casing. In an **overhung front setting,** the smoke box extends out from the boiler front over the fire doors, as in Fig. 19, page 25. In this respect the overhung front setting differs from the **flush front setting.**

The **rear-wall brace** and **buckbars** are used to prevent spreading of the rear and side walls. The **front** and **back bearing bars** support the grate bars, with the front bearing bar also serving as a support for the fire-door arch. The **return plate** and **rib** are used in making the connection between the boiler and the rear setting wall. The **fire-door liner** protects the fire door from the direct heat of the fire and thus prevents it from overheating and warping. The **damper** and **damper frame** are located in the breeching connection, and the damper turns upon trunnions carried by the breeching or the damper frame.

19. Vertical Tubular Boiler. — This type of boiler requires small floor space per boiler horsepower and is well adapted to carry the steam pressures now used. It furnishes superheated steam, that is, steam at a temperature above that corresponding to the steam pressure. It is built in sizes from 50 to 500 boiler horsepower.

A vertical tubular boiler of the Manning type is shown in Fig. 23. The shell is made of three courses, the lower course serving as the outside sheet of the firebox. A smoke box is attached to the top course, and has an opening to which the breeching connection is attached. Between the lower outside course and the middle course, a reduction in size is made by means of a **reverse flange,** or **Ogee ring.** This permits a smaller diameter of shell with no decrease in the area of the firebox. *The inside sheet of the firebox is fastened to the outside sheet by stay bolts, which are the only stays used in the Manning boiler.* The reverse flange is a source of weakness and, to do away with it, the shell is often made straight from top to bottom, or a tapered course is placed between the outer sheet of the firebox and the next course. These types of construction are known as **straight-shell** and **tapered-shell vertical tubular boilers.** The lower part of the water leg[1] is closed by a wrought-iron mud ring, while handholes are provided just above the mud ring for cleaning purposes.

The lower head forms the crown sheet, and is the support for the lower ends of the tubes. It is riveted to the top of the firebox sheet. The upper head is riveted to the upper course of the shell, with the flange turned outward. The tubes vary from 2 to $2\frac{1}{2}$ inches in diameter. They are expanded into each head and are well beaded to give maximum support to the crown sheet.

Handholes are provided on a level with the crown sheet, opposite each row of tubes, for removing dirt and sediment which may have collected on the crown sheet. The fusible plug is screwed into an extra thick tube

[1] See Art. 31, page 46.

on an outside row, about one-third the length of a tube above the crown sheet; and the opening in the shell for removing and replacing the plug is directly opposite. The outer sheet is re-inforced by a steel band where the handholes are located. Tapered and straight-shell types generally have manholes in the middle course. A fire-door opening is flanged in the outer firebox sheet, and the inner sheet is riveted to it in such a manner that a leaky connection at this point is avoided.

The water column is attached to the front of the boiler, and the steam nozzle to the upper part of the top course of the shell. The external feedwater pipe is screwed into a flange riveted to the upper part of the middle course, and an internal distributing pipe, running between a row of tubes, is screwed to the flange from the inside. A blow-off connection is attached to the lower part of the waterleg.

This type of boiler does not require the usual setting. A cast-iron base, resting on a concrete foundation, forms the ash-pit arch and serves as a support for the boiler and also for the circular grate. A brick base is often used instead of a cast-iron base. In this case the brick base is made about 20 inches high, and a plate, which supports the grate and upon which the boiler rests, is placed upon the brick.

The furnace of the vertical tubular boiler is **internally fired,** the hot gases coming into contact with the tubes and the inside plates of the fire-

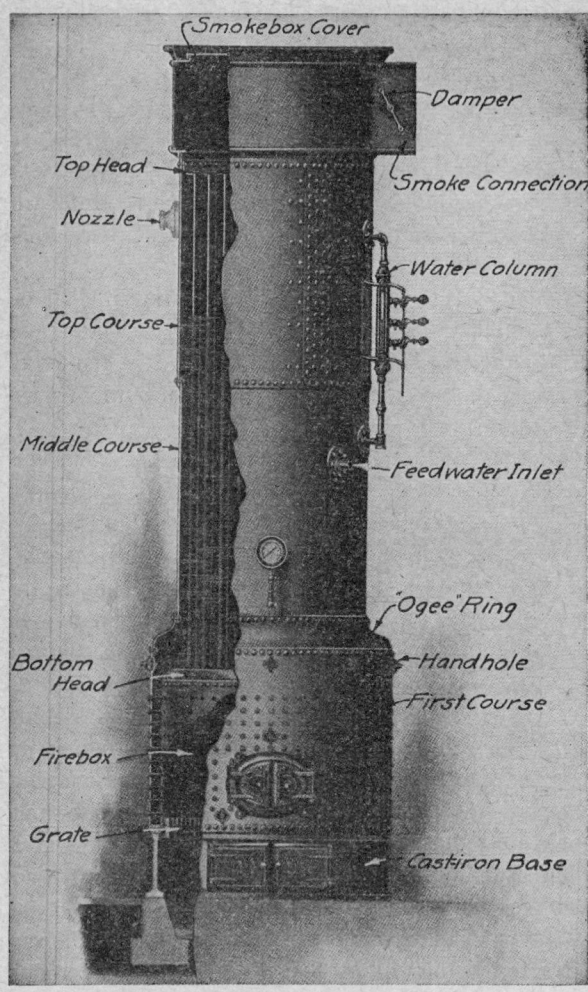

Fig. 23. — Manning Vertical Fire-tube Boiler.

box. The outside plates may then be made sufficiently thick to withstand high pressures, and the inside firebox plates, which are made thinner to transmit heat readily, are easily stayed from the outer shell.

The gases pass straight up through the tubes to the smoke box. The water-heating surface consists of the inside area of the firebox and as much of the tube surface as is surrounded by water. The tube surface above the water level is superheating surface.

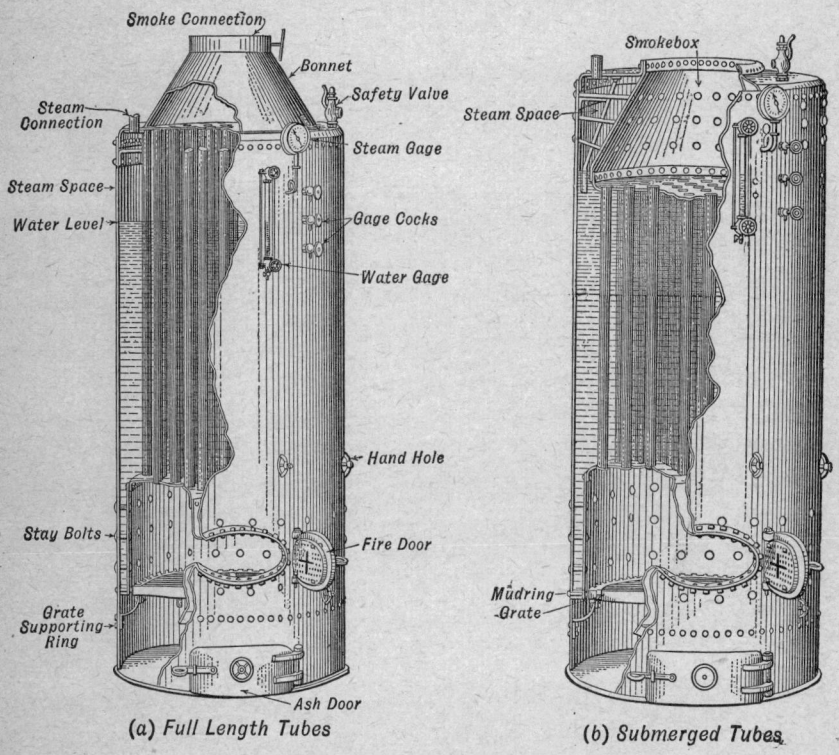

FIG. 24. — Small Vertical Fire-tube Boilers.

20. Small Vertical Tubular Boiler. — This type of boiler is made in sizes from 3 to 100 horsepower and is used on construction jobs where a portable boiler is necessary. The steam space is small because the water level comes within a short distance of the top head.

Two types of small, vertical fire-tube boilers made with a **one-course** shell are shown in Fig. 24. The firebox in each type is formed as in the larger types of vertical fire-tube boilers. The regular upright type frequently gives trouble from tubes leaking at their point of connection to the top head. This difficulty is overcome by using the submerged type, in which the tubes are surrounded by water throughout their length. This design gives a restricted area at the top, for separation of steam from the water; and the cone forming the smoke chamber above the tube sheet, being subjected to pressure, sometimes becomes leaky. A water column is ordinarily not used on this type of boiler, the water level being determined by gage cocks and a water glass attached directly to the shell.

21. Water-tube Boilers. — Water-tube boilers have water inside the tubes and hot gases surrounding the tubes, and are used extensively be-

Fig. 25. — Babcock and Wilcox Longitudinal-drum Water-tube Boiler.

cause they can be built in large capacities, are safe, quick steaming, and flexible in operation. They may be classified as follows:

1. According to service $\begin{cases} \text{Stationary} \\ \text{Marine} \end{cases}$

2. According to position of drum $\begin{cases} \text{Vertical} \\ \text{Cross} \\ \text{Longitudinal} \end{cases}$

3. According to form of header — Sectional or Box
4. According to the type of tube used — $\begin{cases} \text{Straight tube} \\ \text{Bent tube} \end{cases}$

22. Babcock and Wilcox Longitudinal-drum Water-tube Boiler. — This type of boiler, Figs. 25 and 26, is made with one or more horizontal drums

Fig. 26. — Front View of B. and W. Boiler showing Headers.

having a cross box, Fig. 27, riveted to the end courses of each drum and connected to a series of headers by short tubes, or nipples, expanded into openings in the headers and cross boxes. Each front and rear header is

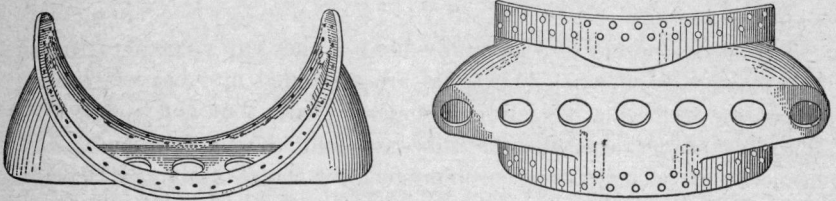

Fig. 27. — Forged-steel Cross Box.

connected with a single row of straight tubes, the unit being called a **section**. These sections may be arranged single deck, as in Fig. 25, or double or triple deck, as in Fig. 132. The capacity of the boiler depends upon the number of sections and tubes used. Each drum is made of a single course consisting of either one or two plates joined by a longitudinal butt-strap as required by the diameter of the boiler. The heads are forged by hydraulic press and are dished to a radius equal to the diameter of the shell. When so dished the heads are seldom stayed, although it is considered better practice, by some engineers, to use stays. In recent large-capacity units, the drums extend crosswise with respect to the tubes, and the boiler is called a **cross-drum water-tube boiler.**

A manhole is flanged in at the center of each head, and the flanged edges are machined to a flat surface. The manhole plates are machined to fit

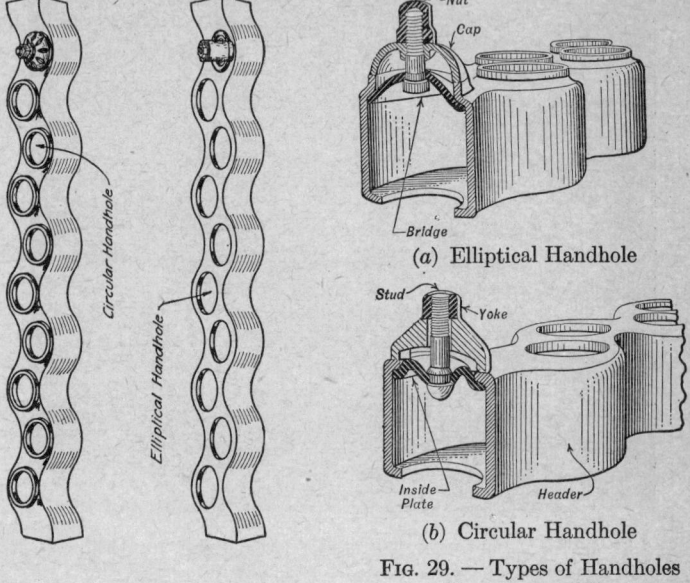

Fig. 28. — Vertical Steel Headers.

(a) Elliptical Handhole

(b) Circular Handhole

Fig. 29. — Types of Handholes for Wrought Steel Header.

the manhole opening and are held in place by forged-steel yokes and bolts. Flat surfaces are provided in the front head for water column and feed-water connections.

Two steam nozzles are generally used. They are riveted to the shell and located as shown. The tubes are expanded into headers having a serpentine form, Fig. 28, and the ends are flared but not beaded. This shape of header arranges the tubes so that they are staggered. The headers may be made of cast-iron or steel, the choice of material depending upon the pressure which the boiler is to carry, and may be vertical, as in Fig. 25, or inclined. Opposite the end of each tube an elliptical or circular handhole is located. It must be of sufficient size to permit cleaning or removal of a tube, and, if the handhole covers are to be removed through the circular handholes, one handhole must be made larger than the others.

LONGITUDINAL-DRUM WATER-TUBE BOILER

The handhole covers are forged plates, fitted inside, and shouldered to center in the opening. A nut and yoke, or a special clamp and nut, Fig. 29, hold the covers in place. The joint between the hole and cover is made tight either by having the surfaces milled true and the joint made metal to metal or by placing a thin gasket between the surfaces.

A forged steel **mud drum**, $7\frac{1}{4}$ inches square, is attached to the bottom of the rear headers by means of wrought-iron nipples expanded into the mud drum and the headers. The mud drum is tapped for blow-off connections, and handholes are provided for cleaning.

The boiler is suspended from a steel-girder frame by steel rods called **straps** passing around each end of the drum and is thus entirely independent of the brickwork of the setting. This method of support permits free expansion and contraction without injury to the boiler or brickwork, and also makes it possible to repair the brickwork without disturbing the boiler or its pipe connections.

The feedwater pipe enters the front head as shown in Fig. 30 and opens into the boiler several feet from the front head. The water circulation is from front to rear of drum, downward through connecting tubes to the rear headers, then forward through the tubes to the front headers, and up into the drum again. A **deflection plate,** or **baffle,** is placed above the header connections at the front, to throw back the water carried by the steam, since the steam formed in the passage through the tubes is liberated when it reaches the front of the drum.

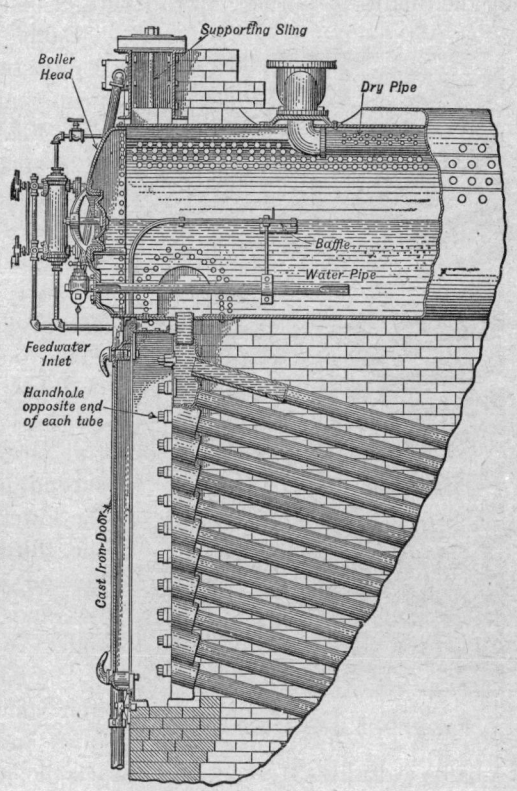

Fig. 30. — Section showing Construction of Front End of B. and W. Boiler.

The steam formed in the boiler passes to the steam pipe line through a **dry-pipe,** which is a perforated pipe attached to the steam nozzle inside the boiler. Its purpose is to prevent the water carried by the steam from passing into the pipe line.

23. Setting for Babcock and Wilcox Boiler. — The setting for this type of boiler resembles the setting described in Art. 17, page 26. It is higher, because the horizontal drum has to be hung at a height suitable for the grade of fuel to be burned. Ordinarily, the distance from the grate to the lowest row of tubes should be at least 6 feet. The side and front walls of the setting, together with the **arch** or bridge wall, enclose the furnace space. The bridge wall can be located to give any depth of furnace demanded by the fuel to be burned, up to length of boiler.

FIG. 31. — Turner Baffle Wall.

Baffle plates or **walls** separate the combustion chamber into compartments called **passes.** The front baffle wall, Fig. 25, is supported by the tubes at the rear wall of the furnace, and the rear baffle wall is attached to a **curtain wall,** which is supported by a special **cross girder.**

The front baffle forces the gases to pass around the forward portion of the tubes to a chamber beneath the drum or drums. The gases then turn downward over the front baffle into the central portion of the tubes, called the **second pass,** until they pass around the lower end of the second baffle into the **third pass** and across the rear portion of the tubes to the damper box and flue connection in the rear wall.

The baffle walls are formed of cast-iron baffle plates lined with special firebrick and held in position by clamps. A baffle wall, Fig. 31, made of firebrick tile, separated by a positive expansion joint, may be used.

The space between the bridge wall and the rear boiler wall, and beneath the tubes, forms a pocket into which much of the soot from the gases in their passage through the second pass is deposited.

Leakage of air into the furnace through the space between the header sections is prevented by filling this space with asbestos cement or asbestos rope.

Cleaning doors, Fig. 25, page 32, are provided, to give access to the tubes for cleaning purposes. Small **dusting doors** are located in the side of the setting, to permit the cleaning of all parts of the heating surface, and sufficient space is allowed between settings for this purpose. The front of the boiler is enclosed by an ornamental cast-iron front, Fig. 26.

This figure also shows the top supporting girder and the method of making the water connection when two boilers are set in battery.

24. The Heine H-type Water-tube Boiler. — This boiler, Figs. 32 and 34, is typical of a large number of makes. It consists essentially of *one or more drums connected to waterlegs or box headers with tubes connecting the headers.* It is built in capacities up to 1200 horsepower and pressures as high as 500 pounds per square inch.

The drum is constructed of three courses, with the longitudinal seams of double strap construction. The heads are dished to a radius equal to the diameter of the shell, with a manhole provided in the rear head. The shells

FIG. 32. — Heine Two-drum, Two-pass, Water-tube Boiler from Front Waterleg.

are made in diameters from 30 to 48 inches and in length from 17 to $21\frac{1}{2}$ feet. The main steam nozzle is attached to the front course of the shell, and a throat opening for the waterleg connection is cut in the bottom of the shell near each end. **Throat stays** are riveted across this opening to compensate for the metal thus cut away. The feedwater pipe passes through the top of the middle course and empties into a **sheet-steel mud drum,** running parallel to the shell and near its bottom. This mud drum is closed, with the exception of a small opening at the top, near the front end, and impurities which may be in the water are deposited in the mud drum and are removed through a blow-off pipe passing from the mud drum through the rear head.

Over the throat opening at the front of the shell is a deflection plate closely fitted to the head and the sides of the drum, and having the dry pipe directly above it.

The waterlegs, Fig. 33, are made of two flat plates, called respectively the tube sheet and the handhole sheet, which are flanged and joined together, except at the top, by a "**butt strap**."[1] *The flat surfaces of the waterlegs are stayed by hollow steel staybolts,* screwed into each sheet and riveted over against the sheet. The tubes are fastened to the tube sheet by being expanded and then slightly flared to increase their holding power. The bottom of the rear waterleg is connected to the blow-off piping. A recent type of **handhole cap,** used on this boiler to replace the ordinary type of handhole cover, is also shown in Fig. 33. The claims for this type of cap are that it does not require a gasket, is self-tightening and non-leaking, and saves time and labor when the boiler is cleaned. The water column is attached at the top to the upper part of the front head and at the bottom to the top of the front waterleg.

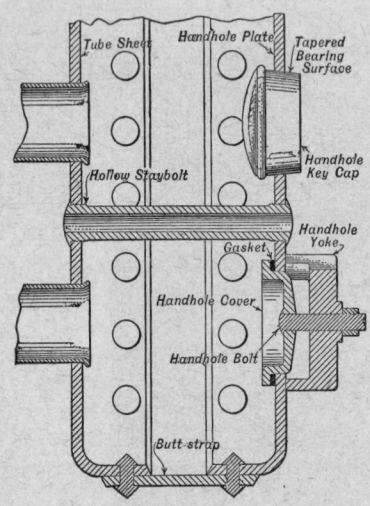

Fig. 33. — Waterleg of Murray Boiler.

25. Setting for Heine Boiler. — As ordinarily set, the tubes and drum incline downward from the front to the rear with a slope of 1 in 12, Fig. 34. The boiler in the smaller sizes for hand firing is usually supported in front on cast-iron tee-section columns which are a part of the firing front and form a support for the fire and ash door frames. The rear waterleg of the boiler rests on rollers supported by a structural steel backstand. The larger size boilers and also those having stoker settings are usually supported, both front and rear, by a U-loop, hung from steel overhead supports and passing under that part of the drum which projects beyond the waterleg.

The setting may be made as high as the fuel and furnace conditions require. When erected ready for service the boilers are inclosed in whatever kind of setting may be preferred, this usually being the ordinary refractory-lined brickwork which extends in height to above the middle of the drums. Cast-iron cover plates support a brick deck extending from the side walls to the drum and also between the drums in multi-drum boilers. At the rear of the setting, the walls and a cross wall over the drums are run up to form the burned gas uptake.

The gases leaving the furnace are ordinarily forced to take the path shown by the arrows using horizontal baffles, which may be arranged in a

[1] Also called **wrapper strap.**

great variety of ways in order to proportion the areas of the gas passages to the kind of fuel to be used, the probable capacity at which the boiler will run, and the draft available. For low capacity, hand firing and with natural draft, in the smaller sizes, the single-pass arrangement is ordinarily used, while for larger capacities, a two-pass arrangement, Fig. 34, is used. This consists of one baffle on the lower, second, or third row of tubes as may be desired, with an opening at the back, another baffle on the

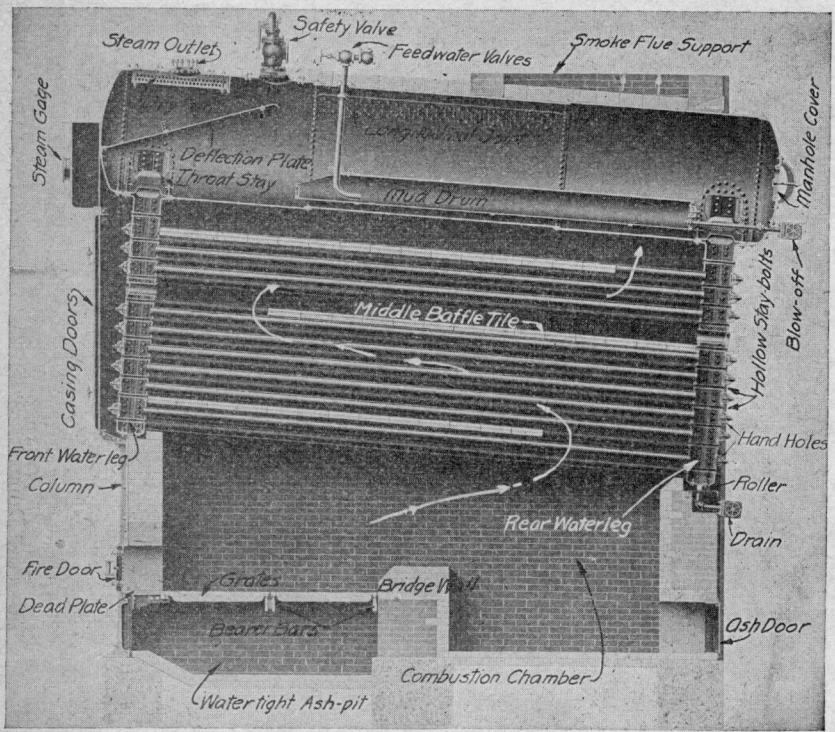

Fig. 34. — Section through Heine Boiler and Setting.

top row of tubes with the opening at the back, and a middle baffle with an opening at the front. This method of arrangement exposes the entire length of the lower rows of tubes so that the maximum absorption of radiant heat may be obtained while at the same time some absorption of heat results from the actual contact of hot gases. A bridge wall suited to the type of furnace or grate should always be provided.

The direction of circulation of the water is the same as in the Babcock and Wilcox boiler.

Access to the setting, for cleaning, is obtained through a rear cleaning door and two side cleaning doors at the top of the setting. Soot and dust are blown from the tubes by a **soot blower** consisting of a number of small

Side Elevation showing Arrangement of Tubes and Drums.

Partial Front View and Section through Front Drum.

FIG. 35. — Stirling Water-tube Boiler.

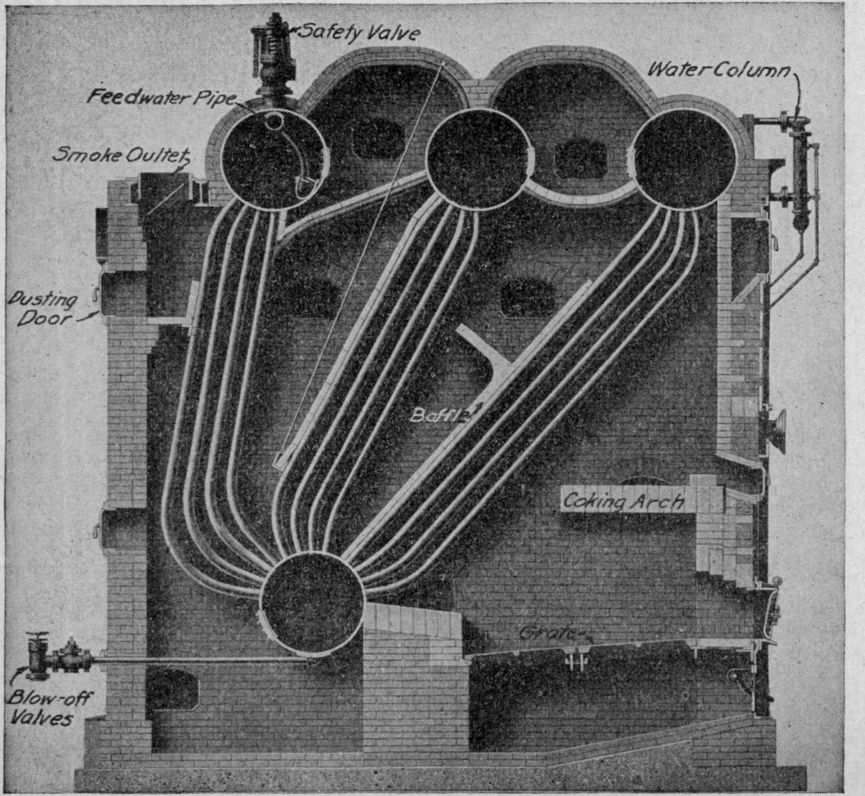

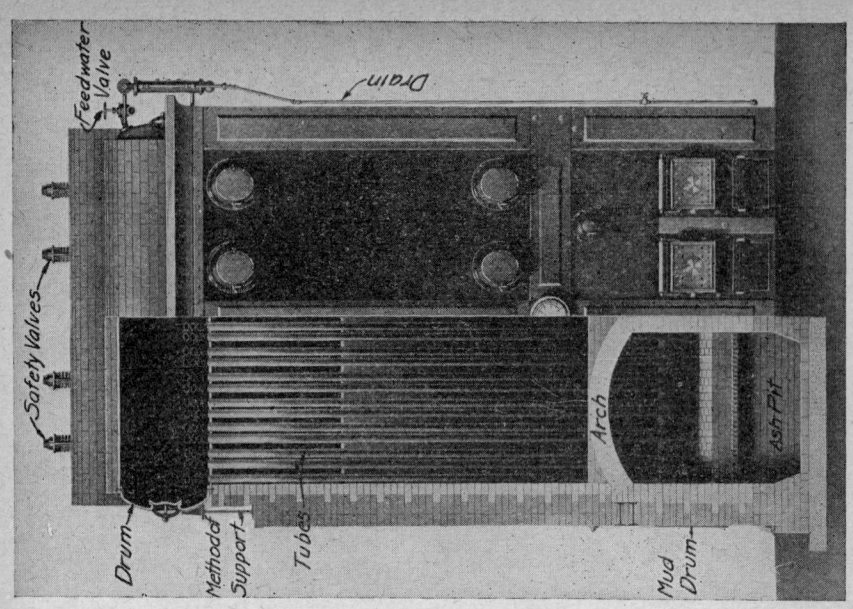

steam or air nozzles inserted into the hollow stay bolts. The stay boltholes not occupied by the soot-blower nozzles are closed by wooden or cast-iron plugs.

26. Bent-tube Water-tube Boiler. — This type of boiler has one or more parallel steam drums above connected by bent sloping or vertical tubes to one or more parallel water drums below. The various types of bent-tube boilers result from an effort to obtain a steady water level and dry steam. Some boilers have all of the upper drums partially above the water line, such as the *Stirling, Wickes, Heine, and Erie City;* or one or more of the upper drums completely submerged, such as the *Connelly, Kidwell, and Badenhausen.* The bent-tube boiler has a large amount of heating surface, obtained by using a large number of tubes, and consequently is suitable for high capacities. The drums are generally made single course with the ends closed by dished heads containing the manholes. The tubes are bent to enter the drums radially and are expanded to fit the tube holes, but are not beaded.

A section, taken through the four drums and the walls of the setting of a **Stirling boiler,** is shown in Fig. 35. The three upper drums contain water and steam and are set at the same level. They are connected to the mud drum by tubes, so curved as to enter the tube sheets radially. The center drum is equidistant from the front and rear drums and its steam space is connected to the steam spaces of the front and rear drums by a row of curved **steam-circulating tubes.** The water spaces of the front and center drums are connected by rows of **water-circulating tubes.** The water space of the center drum is connected with the mud drum by one-half the tubes of the front row of the rear bank, which support a baffle protecting the rear steam drum.

The main steam outlet is placed on top of the rear drum, which also carries the safety valves. The feedwater pipe enters the rear of this drum and discharges into a **perforated trough,** which distributes the water over a relatively large portion of the area of the drum.

The water column is attached to one end of the center drum and the blow-off pipe to the bottom of the mud drum.

Each drum is made of a tube sheet, riveted by butt-and-strap longitudinal seams to a drum sheet. The drum heads are of forged steel, one head in each drum being provided with a manhole. The upper drums are supported at both ends by lugs resting on a rectangular structure of rolled steel and entirely independent of the brickwork; the lower drum is suspended from all of the steam-and-water drums by the water tubes, swinging entirely free of the setting. The leakage of air around the ends of this drum is prevented by soft asbestos packing placed between it and the brickwork.

27. Setting for Stirling Boiler. — The entire boiler is surrounded by four walls having clean-out doors at the side of each bank of tubes, with the front of the setting covered by an ornamental cast-iron and steel front, as shown by the partial front elevation in Fig. 35. The brick walls of the combustion chamber are faced with firebrick, and a **coking arch** is sprung over the front of the grate, with the grate directly under the arch. The furnace gases are directed, by firebrick baffle tiles, to pass from the grate along the first bank of tubes, then down the middle bank and up the rear bank, passing out through the smoke connection at the rear. In some recent installations of this type of boiler the baffle tiles are arranged to run crosswise to the tubes, Fig. 36. This arrangement gives a better balanced area for the passage of the gases. The distribution of heating service, in each pass of a typical installation, is as follows: 45.85 per cent in the *first pass*, 29.4 per cent in the *second pass* and 25.75 per cent in the *third pass*. The front portion of the front baffle is usually located between the second and third tubes in the first bank. This gives a better heat transfer and is preferable because of slag formation and dust deposits with certain fuels.

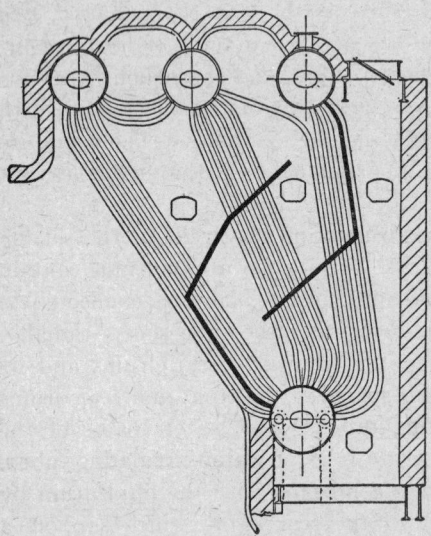

Fig. 36. — Cross-baffle Bent-tube Boiler.

The water, which is fed into the rear steam-and-water drum, passes downward through the rear bank of tubes to the mud drum, thence upward through the front bank of tubes to the front steam-and-water drum. The steam formed during the passage upward through the front bank of tubes becomes separated from the water in the front drum and passes through the steam circulating tubes into the middle drum and then, with the steam generated in the middle bank of tubes, into the rear drum, from which it passes through the dry pipe into the steam main. The water from the front drum passes through the water circulating tubes into the middle drum and thence downward through the middle bank of tubes to the mud drum, from which it again passes up the front bank to retrace its course.

28. Wickes Vertical Water-tube Boiler. — A Wickes boiler is shown in Fig. 37. It consists of *an upper and a lower drum joined by straight tubes about* 22 *feet in length*. The tubes are arranged in parallel rows, are expanded into the drum heads and are flared to increase their holding power. The top drum is the steam drum. It contains a manhole in the top drum head, a steam nozzle, and feed pipe and water column connections. The bottom drum is the mud drum. The boiler is supported by brackets riveted to the mud drum and resting upon plates fastened in the walls of the setting. The blow-off connection and a manhole are located in the lower head of the mud drum. This boiler does not require stays.

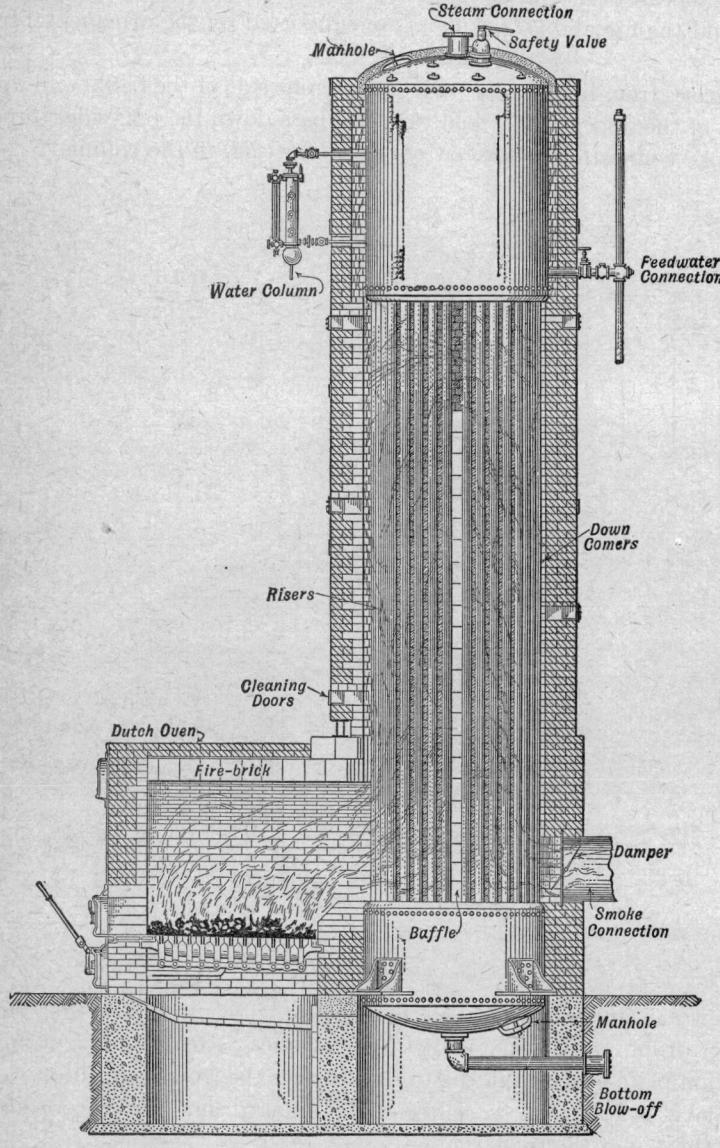

Fig. 37. — Wickes Vertical Water-tube Boiler.

29. Setting for Wickes Boiler. — The entire boiler is enclosed by a circular brick setting having a firebrick lining. A firebrick baffle wall extending nearly to the upper drum divides the setting vertically into a front and rear compartment. The tubes in front of this wall are called "**risers**" and those in the rear "**downcomers**." The wall of the setting is prevented from spreading by bands of steel. The brickwork is often covered by sheet iron to make the setting air-tight. Doors are provided for cleaning purposes and the fire and ashpit doors are supported by the ornamental front casing.

The gases from the furnace pass up the front side of the baffle wall to the bottom of the upper drum, then turn and pass down the rear side, through the smoke connection at the bottom of the setting, to the chimney.

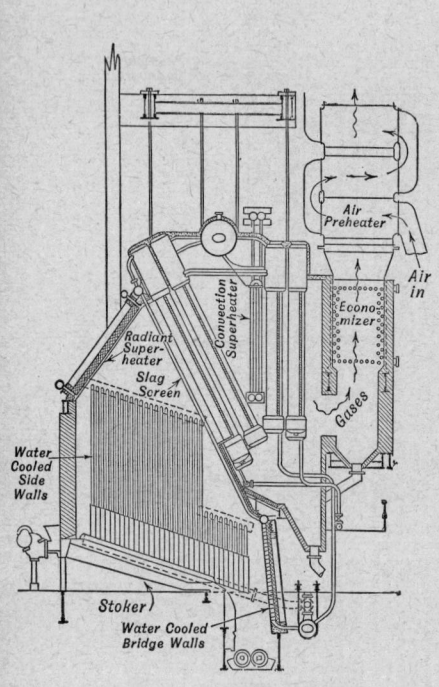
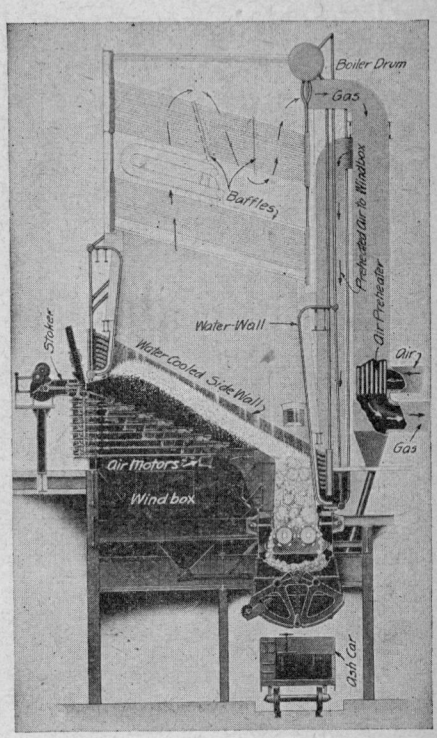

FIG. 38. — Bigelow-Hornsby.   FIG. 39. — Taylor.
Large Capacity Boiler Units.

The water circulation is from the bottom drum up the "risers," across the top drum, then through the "downcomers" to the mud drum. A deflection plate is placed in the top drum over the front set of tubes. The water level is about at the center of the upper drum and hence is always above the top of the tubes.

30. Large Capacity Boiler Units. — The interconnection of power stations has made it possible to produce steam more economically by using larger boiler units operating at high steam pressure. These large units decrease the station cost per unit of capacity and offer opportunities for

LOCOMOTIVE BOILER 45

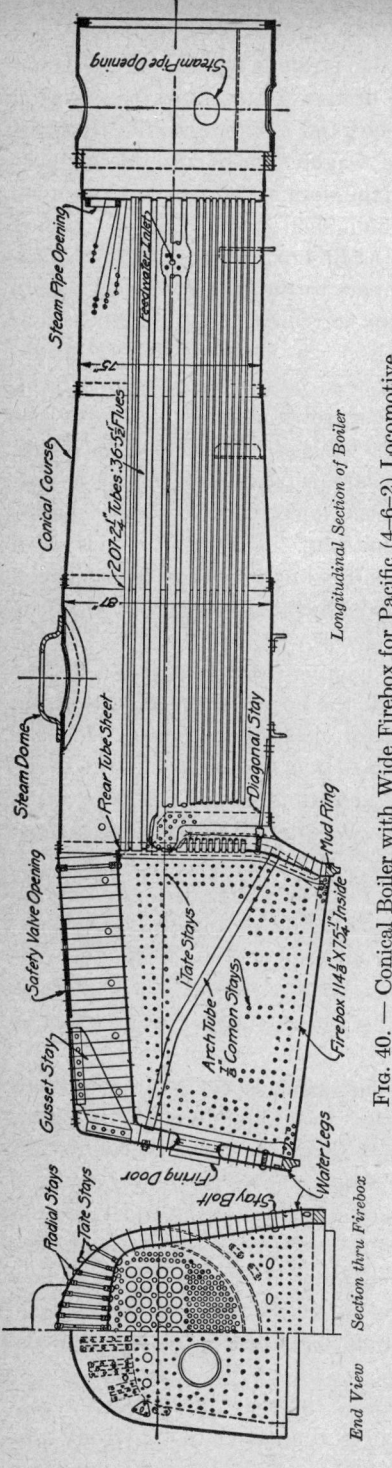

FIG. 40. — Conical Boiler with Wide Firebox for Pacific (4-6-2) Locomotive.

higher fuel economies, lower maintenance costs, greater ease of operation, and improved methods of control by using refinements that could not be considered in small units. Two large capacity stoker-fired units are described here and additional pulverized fuel units are described in Chapter XXV, page 617. The Bigelow-Hornsby boiler unit with the parts making up the unit clearly labeled is shown in Fig. 38. It is a straight tube multi-drum boiler designed to produce 250,000 pounds of steam per hour. An interesting feature of the unit is the special slag screen in front of the boiler tubes.

The Taylor unit, Fig. 39, is capable of producing 500,000 pounds of steam per hour and is arranged to so control fuel and air that large fuel-burning capacity is obtained. It consists of a cross-drum boiler, air preheater, water-cooled side walls and a Taylor extended surface stoker, using an electro-hydraulic system for feeding the coal, the ashes being discharged into a deep ashpit to which preheated air is admitted to burn out the carbon in the ashes.

31. Locomotive Boiler. — A longitudinal and a transverse section of a modern locomotive boiler of three courses are shown in Fig. 40. The locomotive boiler is a *straight fire-tube boiler having an internal firebox*. It requires a large amount of heating surface and a grate upon which coal can be burned at a rapid rate. This is obtained by using a large number of staggered tubes and a strong draft induced by the steam exhausted from the engine cylinders. Locomotive boilers are commonly classified by the shape of the shell; as, **straight top,** having the cylindrical shell of uniform diameter from the firebox to the smoke box; **wagon top,** having the steam dome over the firebox and a sloping course from the firebox to the cylindrical shell; **extended wagon top,** having one or more cylindrical courses between the firebox and the sloping course which tapers on the top and sides to the diameter of the main shell; **conical,** Fig. 40, having one or more cylindrical courses between the firebox and the sloping course and a conical connecting course which tapers to the diameter of the main shell, to the rear course of which the firebox is riveted. It forms the chamber within which the fuel is burned, and also forms a support for the grates, which are at the bottom of the firebox. *The firebox consists of inside, front, rear and side sheets riveted together and to the* **crown sheet,** which forms the top of the firebox. The crown sheet and side sheets are made of one piece, whenever possible. The outside sheets are separated from the inside sheets by a space, forming the **waterleg,** the lower end of which is closed by a steel ring called the **mud** or **foundation ring.** The front outside sheet is riveted to the rear course of the shell and is known as the **throat sheet.** The outside top or **roof sheet** and the side sheets are generally made in one piece. The outside rear sheet is flanged inward and is riveted to the roof and side sheet. There is a flanged opening in this sheet for the fire door. **Handhole openings** on a level with the crown sheet and **clean-out plugs** above the mud rings are provided for cleaning purposes. *The best heating surface of the boiler is that which surrounds the firebox.*

Locomotive boilers are often classified according to the construction of the firebox, as **wide firebox, narrow firebox, Wootten, Belpaire,** and **Jacobs-Shupert.**

The wide firebox illustrated in Fig. 40 rests on the frames and extends out beyond the driving wheels at the sides, while the narrow firebox extends down between the frames and the driving wheels. The Wootten firebox is very wide and shallow and has a curved crown sheet of large radius. It is used on locomotives, burning anthracite coal, to give a large grate area.

The Belpaire firebox, Fig. 41, has a flat crown sheet joining the side sheets by curves of short radii. The roof sheet and upper part of the outer side sheets are flat and parallel to those of the inner firebox.

The Jacobs-Shupert firebox is one in which the usual arrangement of flat sheets is replaced by inner and outer sets of channel-shaped sections riveted together, with flanges away from the fire. Stay bolts are replaced by **stay sheets,** one at each joint of the channels and secured by the same rivets that hold the channels. The stay sheets are partially cut away in the waterleg to allow horizontal circulation of the water. All seams and joints are submerged.

The front inside sheet of the firebox forms the **rear tube sheet;** it contains openings for the tubes by which it is connected to the **front tube** sheet which is flanged outward and riveted to the front course. The latter also contains an opening for the steam pipe passing to the cylinders. The *tubes* are made of the best grade of charcoal iron and ordinarily have an outside diameter of 2 to $3\frac{1}{2}$ inches, although tubes as large as $5\frac{1}{2}$ inches are sometimes used, and are then known as **flues.** *Locomotive boilers have from* 300 *to* 500 *tubes and flues.*

To the front course of the shell, an additional ring, known as the **smoke box,** is riveted, the front of which is closed by a casting called the **smoke-box front.** The smoke pipe or stack is attached to the top of the smoke box, and the openings for steam pipes are at the bottom. There are screens in the smoke box which prevent the escape of large cinders. On top of the shell and in front of the firebox is an opening over which is a dome-shaped chamber known as the **steam dome.** It is made of sections of curved steel plates riveted to the shell. Steam is taken from this elevated part of the boiler in order that it may contain as small an amount of water as possible. A **sand dome,** having no opening into the boiler

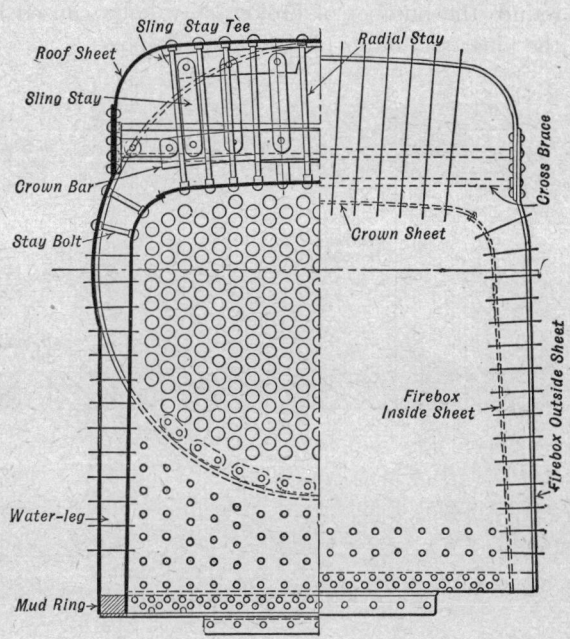

Fig. 41. — Belpaire Firebox.

and holding sand for use under the driving wheels, is attached to the shell in front of the steam dome.

Radial stays or braces, Fig. 40, are used to brace the crown sheet and thus prevent failure of the crown sheet when under pressure. These stays are round steel bars threaded at each end, set radial to the curvature of the crown and roof sheets, and screwed into the surfaces to be stayed. Nuts may be placed on the projecting ends, or the ends may be riveted over.

The front end of the crown sheet is often supported by a **sling stay,** Fig. 41, which permits slight relative movement of the supported and supporting surfaces and prevents breaking of stays. It consists of straps of steel fastened at each end to an **angle** or **tee bracket,** by a pin joint. A bracket is riveted to each surface to be stayed, and short pieces of pipe are placed around the rivets between the lower bracket, or **crown bar,** and the crown sheet to allow water to circulate between the bracket and plate.

Stay bolts, Fig. 42, are used to brace the flat surfaces forming the waterlegs. The stay bolt ordinarily used on locomotive boilers is like the radial stay except that it is shorter. A hole $\tfrac{3}{16}$-inch in diameter is often drilled in the outer end of each stay bolt to a depth slightly beyond the inside of the plate. A broken stay bolt may then be detected by the escape of steam and water through the opening. Flexible stay bolts may be used to allow relative movements of the two plates which they secure, and thus reduce the number of broken stay bolts caused by unequal expansion of the inner and outer plates of the firebox.

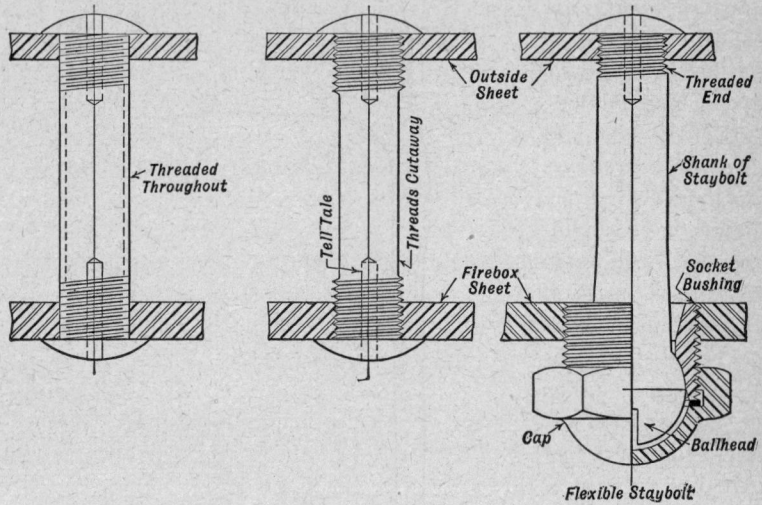

FIG. 42: — Common and Flexible Stay Bolts.

Diagonal stays brace that part of the front head above and below the tubes, and the outside rear firebox sheet above the crown sheet. These stays are made of steel and are riveted to the shell and head. **Gusset stays,** Fig. 40, are often used in place of diagonal stays. The gusset stay is a steel plate riveted to angle irons, which are riveted to the stayed surfaces.

A safety valve, which prevents the pressure in the boiler from becoming dangerously high, is placed in the top of the steam dome, or is fastened to a nozzle riveted to the top of the outside firebox sheet. **A fusible plug,** another safety device, is screwed into the crown sheet and protects the boiler against low water level. A blow-off pipe is attached to the outside of the waterleg at its lowest point.

32. Marine Boilers. — Boilers used on shipboard are of either the fire-tube or water-tube type. The fire-tube boiler is used where extreme lightness or high speed are not essential, as in the merchant marine. The water-tube type is used where rapid steaming and high pressure qualities are essential, as in naval vessels and rapid passenger service.

The most common type of marine fire-tube boiler is the Scotch marine boiler. It is self-contained, requires low head-room and is manufactured in units up to 2000 boiler horsepower.

There are many classes of marine water-tube boilers. The majority have one or more top cylindrical drums and one or more drums below. The upper and lower drums are connected by straight or curved tubes. The Almy, Yarrow, Thornycroft, and Dyson express boiler are typical makes. The Babcock and Wilcox marine boiler has the lower drum replaced by headers with connecting tubes. This make is known as the free-circulation type and is much used. Units as large as 4500 boiler horsepower are manufactured.

33. Scotch Marine Fire-tube Boiler. — Longitudinal and vertical half sections are shown in Fig. 43. The boiler consists of a two-course shell ordinarily varying in diameter from 7 to 16 feet and in length from 8 to 11 feet. The heads are made of two plates, an upper and a lower, riveted together and flanged for riveting to the shell. Heads up to $15\frac{1}{2}$ feet in diameter are now being made from single plates.

The shell ordinarily contains from one to four cylindrical, **corrugated-steel furnaces**, 42 to 48 inches in diameter, attached at the front end to a flanged opening in the front head and at the rear end to the front wall of the combustion chamber, into which the furnace opens. These furnaces are internally fired, are entirely surrounded by water, and, because of the corrugations, require no staying. The grates and bridge wall occupy the front part of the furnace, and the space in each furnace beneath the grates forms the ashpit. At the rear end of the furnace is the combustion chamber; it is entirely surrounded by water and does not require a firebrick lining. The rear wall of the combustion chamber is flat and is supported from the rear head by stay bolts. The side walls of the combustion chamber conform to the curvature of the shell and are attached to it by stay bolts. The front wall forms the rear tube sheet. A crown sheet is riveted to the upper ends of the walls and forms the top of the combustion space. **Girder stays** or **crown bars** support the crown sheet. The walls of the combustion space are the best heating surface of the boiler.

The tubes which are arranged as shown in the half-sectional view, are expanded into the tube sheet and beaded. **Stay tubes** which are heavier than ordinary tubes, are often used in this type of boiler. They may be threaded at the ends and are sometimes protected at the firebox end by ferrules. The area of the front and rear heads above the tubes is braced by channels and through braces. The area of the front head below the furnace is braced by a special form of through stay fastened at the rear end to the front wall of the combustion space by angle irons and a pin. Such bracing permits some flexibility.

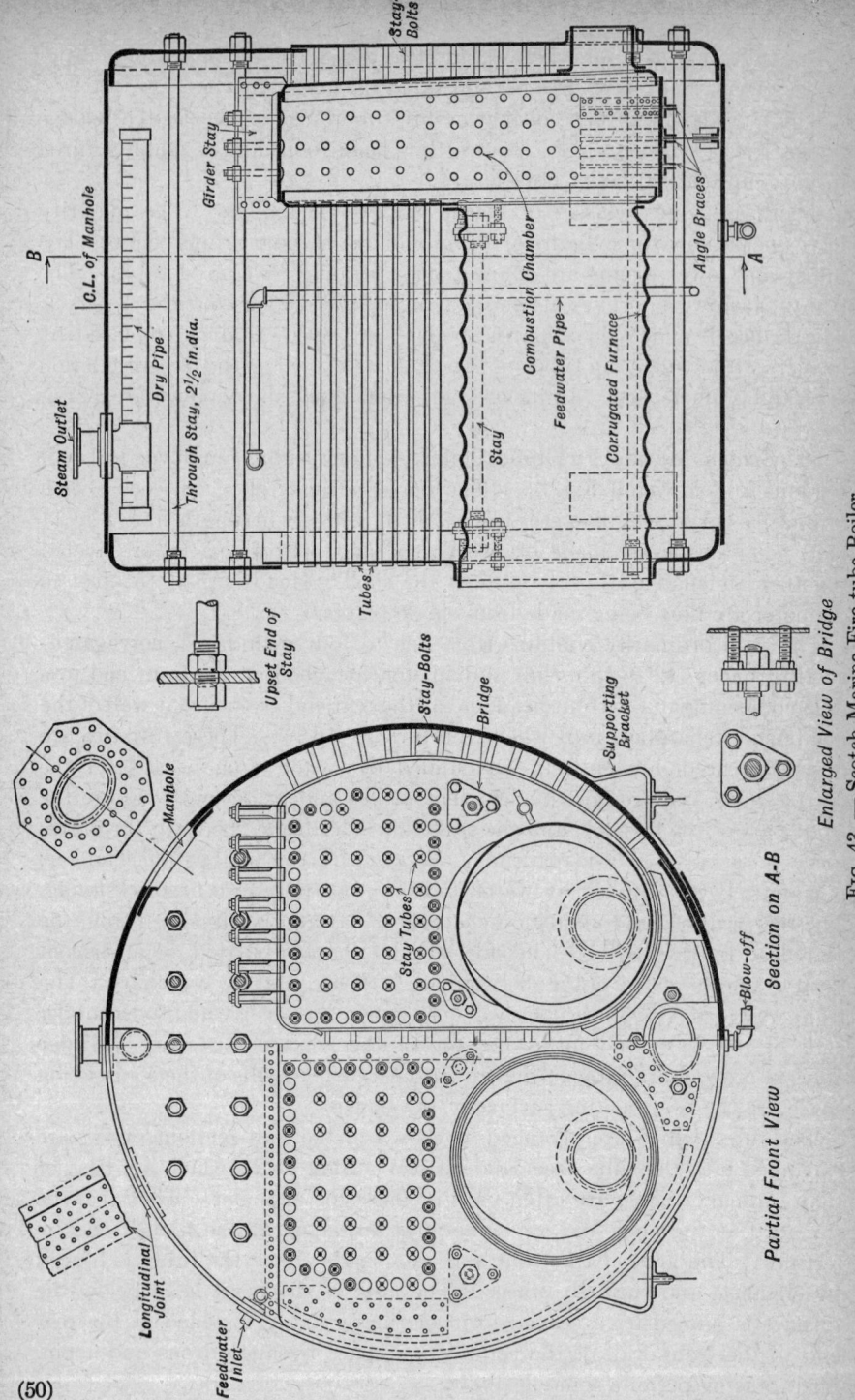

FIG. 43. — Scotch Marine Fire-tube Boiler.

SCOTCH MARINE FIRE-TUBE BOILER

The openings in the upper half of the boiler shell are for a manhole and a steam nozzle. The feed pipe enters the side of the shell and passes down around the side of the shell, discharging about midway the length of the boiler. The water column is generally placed in an inclined position, though the connections are the same as for any fire-tube boiler. Several manholes are provided in the front head, and there is an opening into the combustion chamber in the rear head.

This boiler *does not require a setting*. It is supported by a cradle, or **saddle,** securely fastened to the frame of the ship. Adjustable turn-buckle stays hold the boiler in place in the saddles.

The gases pass from the furnace over the bridge wall and into the combustion space, returning through the tubes to the smoke connection at the front.

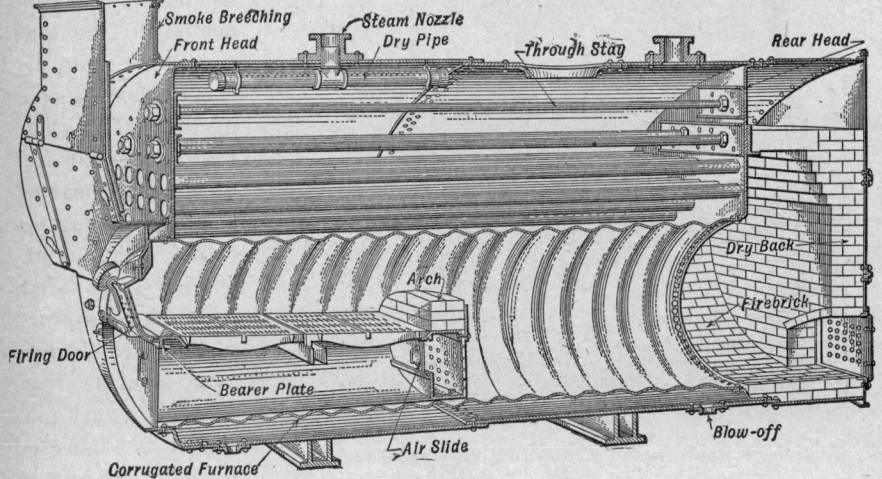

Fig. 44. — Dry-back Scotch Boiler.

The water level is about 8 inches above the top row of tubes. The water circulation is down the sides and up the middle when two furnaces are in the shell. It is often necessary to force the circulation by a steam jet and a series of nozzles placed near the bottom of the boiler.

When more than one furnace is used, each may have a separate combustion space, or all may be connected to one combustion space. The boilers may be either single ended or double ended. A **double-end boiler** consists essentially of two single-end boilers placed back to back, with the back heads removed, the shells joined, and the rear sheets of the combustion chambers stayed together.

A modification of the Scotch marine boiler much used for office and hotel buildings, because of the low headroom required, is shown in Fig. 44. The combustion chamber is made of firebrick carried by an extension of the shell. This type is known as a **dry-back marine boiler.**

34. Babcock and Wilcox Marine Water-tube Boiler. — This boiler, Fig. 45, is constructed along the same general lines as the Babcock and Wilcox longitudinal water-tube boiler. The main difference is in its shape and in the arrangement of the heating surfaces. The drum is located at the front of the boiler above the headers and crosswise of the tubes. It is made of a shell having a single sheet with the heads dished and containing the manhole openings. The steam nozzle is located at the top of the drum with the main and auxiliary feedwater connections and gage glasses attached to the front side of the drum. The tubes are extra heavy and are arranged in sections, each section having a front and rear header. The front headers are connected to the bottom of the drum by short tubes 4 inches in diameter. The rear headers are set higher than the front headers so that the tubes, which are expanded and flared, have an upward slope of 15 degrees from front to rear. The top of each rear header is connected to the upper part of

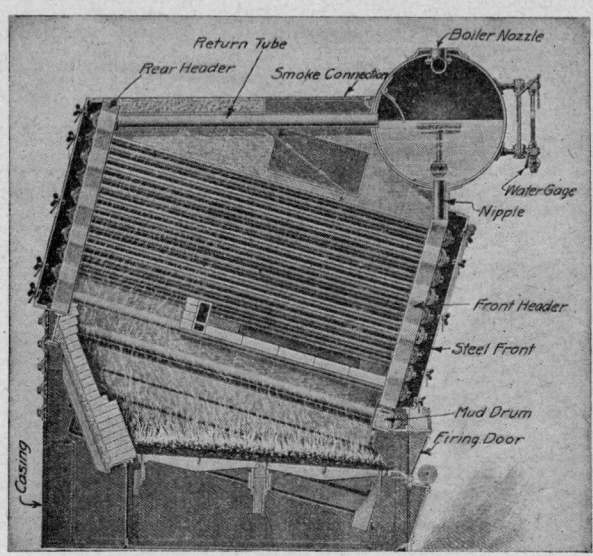

Fig. 45. — Babcock and Wilcox Marine Water-tube Boiler.

the drum by a 4-inch return tube, and the drum is reinforced at the points where the tubes enter by a strip of steel riveted to the inside of the shell. The side headers are carried down to a level with the grate, and the lower tubes are replaced by forged steel boxes 6 inches square, which form the side walls of the furnace and maintain a cool side casing. At the bottom, the headers are attached to a 6-inch square mud drum to which the blow-off connection is attached.

This boiler is supported by a structural iron framework, to which the outer casing is attached, and a brick setting is not required. The spaces between the outer row of tubes are filled with asbestos fiber and the whole covered with a steel casing to make an air-tight covering.

A firebrick wall extending from a level with the grate to the bottom of the rear headers makes the rear wall of the furnace. The gases from the furnace are forced to pass to the rear by a horizontal baffle of firebrick, along the lower row of tubes, then upward to the return tubes. Here a

WATER COLUMN

baffle plate deflects the gases downward between a set of baffle walls to the lower row of tubes. They then pass upward around the lower end of the front baffle wall to the drum and the breeching connection at the top.

The water circulation is down the front headers, or **downtakes,** through the tubes to the rear headers, or **uptakes,** returning to the drum through the

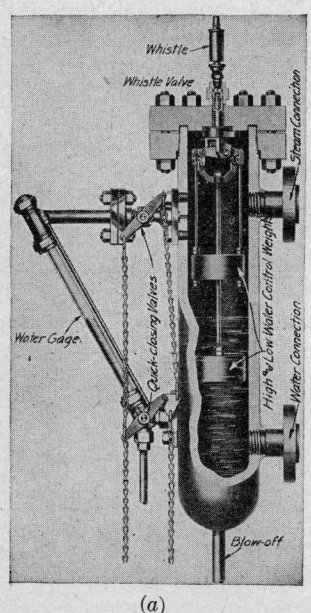 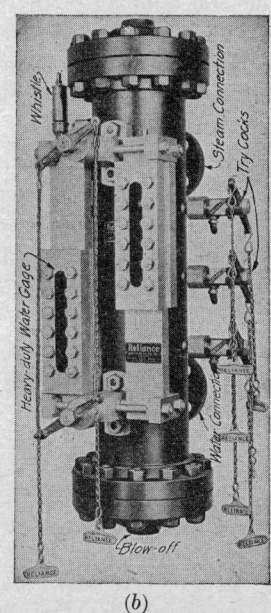

(a) (b)

Fig. 46. — Water Columns.

return tubes. A baffle plate at the open ends of the return tubes deflects the water and liberates the steam, which passes around the ends of the baffle plate and through the dry pipe to the steam nozzle. The water level stands at about the center of the drum.

Hinged to the framing at the front and rear of the boiler are large doors giving access to the handhole plates. Dusting doors, with each opening covered by a shutter sliding vertically, are provided in the side walls, for use of a **steam lance.** (See page 278.)

35. Boiler Fittings. — The fittings commonly supplied with a new boiler are a **water column, fusible plug, safety valve, steam gage, and blow-off and feedwater connections.**

36. Water Column. — An external and sectional view of a water column is shown in Fig. 46. It is essentially a hollow cast-iron or forged steel vessel having two connections to the boiler. The top connection enters the steam space of the boiler, either through the upper part of the head or through the top of the shell. The bottom connection enters the water space at least 6 inches below the lowest permissible water level. The connecting pipes should be at least 1-inch in diameter, with the water connection made of brass. Valves are not ordinarily allowed in the water-column connections. Mud and sediment are removed through a ¾-inch blow-off pipe connection attached to the lower end of the water column.

A **water gage** makes the water level visible from the boiler room floor. It is attached to the front of the water column; the lowest visible part of which must be 2 inches or more above the lowest permissible water level. The water gage consists of a strong glass tube which is connected to two needle valves by stuffing boxes and is protected by guards from being accidentally broken. The lower needle valve has a pet-cock connection used to clean the gage glass by blowing steam through it. When used on boilers where the boiler drum is located at a considerable distance above the floor, the water gage is often inclined, as in Fig. 46a, in order to make the water line visible from any position, and in addition, hydraulic and electric operated water-level indicators, located at the operating floor level, are being used, to make the water level more easily observed. In a few installations a periscope is used to observe the level of the water by indirect means. For high pressures, the water gage is made with flat prismatic glass as shown in Fig. 46b.

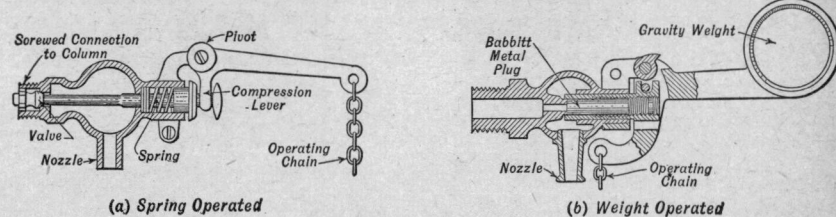

(a) Spring Operated   (b) Weight Operated

Fig. 47. — Gage Cocks.

Because the pipes, which connect the water column to the boiler, may become clogged, or the stuffing boxes become leaky and the water gage thus indicate incorrectly, three valves, Fig. 47, called **gage** or **try cocks** are attached to the water column within the visible range of the water glass. The middle gage cock should be at the mean water level of the boiler, with the other two gage cocks located at equal distances above and below the middle cock, the distance varying from 3 to 5 inches according to the size of the boiler. The lower cock should be above the lowest safe water level. Vertical and cross-drum boilers often have the water column omitted and the water gage and gage cocks attached directly to the shell as shown in Fig. 24, page 31.

A boiler may be provided with two water gages located not less than 3 feet apart on the same level, in which case the gage cocks may be omitted.

The water column is sometimes provided with a **high and low water alarm;** seamless copper floats, or solid weights inside the column, are so arranged that they will admit steam and blow a whistle, when the water level becomes too high or too low. Many engineers prefer not to use a high and low water alarm but to rely upon constant attention of the water tender.

37. Fusible Plug. — The fusible plug, Fig. 48, is used to protect the boiler against low water level. It consists of a bronze casing threaded on one end and having a conical hole from end to end, the taper of the hole being not less than $\frac{5}{8}$-inch to the foot. The hole must be reamed and tinned, before being filled, with an alloy having 99 per cent tin. The melting point of the alloy must be above the temperature corresponding to the steam pressure and below the temperature of the hot furnace gases. *The location of the fusible plug in any boiler must be such that the highest surface of the boiler exposed to heat of the gases will be protected from danger of overheating.* Water ordinarily covers the plug, and its temperature is not above that of the water. In case the water level should become low enough to uncover the plug, the alloy will melt and permit steam to escape, thus attracting the attention of the fireman.

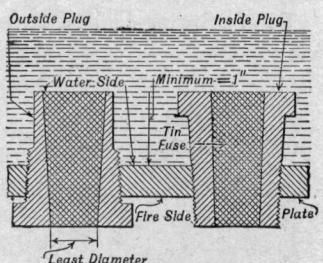

FIG. 48. — Outside and Inside Types of Fusible Plugs.

The location of the fusible plug in a few typical boilers is as follows:

Horizontal return-tubular boiler: — In the rear head, not less than 2 inches above the upper row of tubes, and projecting through the sheet not less than 1 inch.

Locomotive type of boiler: — In the highest part of the crown sheet.

Vertical fire-tube boiler: — In an outside tube not less than one-third the length of the tube above the lower tube sheet.

Water-tube boiler, Heine type: — In the front course of the drum, not less than 6 inches above the bottom of the drum.

For location in other types of boilers, consult the A. S. M. E. BOILER CODE.

The fusible plug is not entirely reliable. The water end may become coated with scale from the water, or the gas end may become coated with incrustations from the gases. Both scale and incrustation are poor conductors of heat; the scale would cause the alloy to melt before it should, and the incrustation would prevent the alloy from melting when it should. In any case, the end exposed to the gases should be kept clean. *It must be replaced, at least once a year.*

38. Safety Valves. — A safety valve is used to protect boilers against excessive pressure, by automatically discharging steam when the pressure rises above a definite point, at which the valve is set to open.

The safety valve should be bolted directly to the steam nozzle, without pipe, bends or valves. *It should be large enough to discharge the maximum amount of steam that the boiler is capable of generating, without building* up the discharge pressure more than 6 per cent above the maximum allowable working pressure. If a discharge pipe from the safety valve is used, it should be properly dripped and should have an open end. Each valve should have its own discharge pipe.

There are three principal types of safety valves, **the direct loaded, the lever and weight,** and **the direct spring loaded.** The latter type is the only one permitted by the A. S. M. E. BOILER CODE.

Direct Spring-loaded Nozzle Type "Pop" Safety Valve. — A sectional view of the Crosby nozzle-type safety valve, suitable for high pressures and temperatures, is shown in Fig. 49 with the various parts named. The spring is made of spring steel, and is located outside the body and separated from it in order to prevent overheating. The **body** is fitted with a forged alloy-steel nozzle tube, the upper end of which is beveled at an angle of 45 degrees to form the valve seat. This end also supports the **nozzle ring,** which is locked in position by a set screw. Valve seats are sometimes made flat. The lower end of the spindle is provided with a hardened steel ball which rests in a hardened steel cup held in the alloy-steel disc. The **disc** is fastened to the spindle by a spring lock clip that engages a shoulder on the spindle, and is guided by a nicoloy guide. The **adjusting ring** screws on to the lower end of the guide, and is locked in position by a set screw. The shape of the nozzle and disc is such that the flow of the steam is parallel to the seat, thus minimizing the tendency of the steam to cut the seat and disc. The full nozzle capacity is developed with a lift equal to approximately one-quarter of the nozzle diameter. A **lifting lever** is provided so that the valve may be operated by hand and should be capable of lifting the valve from its seat at least $\frac{1}{16}$ of an inch. *The valve should be lifted from its seat once a day in order to be sure that it will operate properly.* This may be done by hand, but it is better practice to raise the steam pressure until the valve opens, and at the same time note the steam pressure at which it opens. This valve opens with a sharp, clear pop at the pressure for which it is set, with practically no preliminary warning, or simmering. If the pressure accumulates above the set pressure, the lift will increase rapidly until the maximum lift and valve capacity are obtained at an accumulation of not more than 3 per cent above the set pressure. When the pressure drops below the set pressure, the lift will gradually diminish until the valve closes sharply, with a **blow down,** i.e., the *difference between*

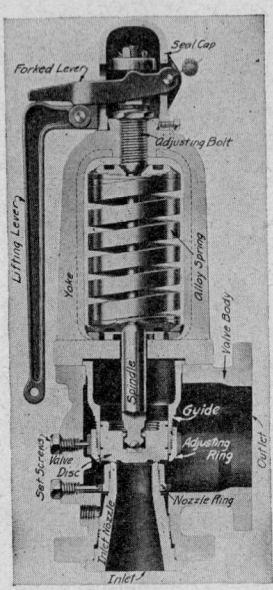

FIG. 49. — Crosby Safety Valve.

SAFETY VALVES

popping and closing pressures, of not more than 4 per cent of the set pressure and with no simmering or leaking after closing. The **valve operation** is caused by two actions working together to produce a continuous pop. The **principal action**, causing the high lift characteristic of the valve, *is the downward deflection of steam by the adjusting ring,* as shown in Fig. 50a. To assist in starting the principal action, the small jet of steam escaping, at lifts up to about $\frac{1}{16}$-inch, is deflected by a small lip at the outer edge of the valve disc, as shown in Fig. 50b. *The position of the adjusting ring determines the amount of blow down.* Raising the ring makes the blow down less; lowering it makes the blow down more. A secondary adjustment is provided by the *nozzle ring,* the lowering of which decreases the power of the lift action and vice versa. This ring is also used to correct simmering at opening and closing without sensibly affecting the blow down. The method of adjustment for blow down differs in the various makes of spring-loaded safety valves. The principle, however, is the same in all.

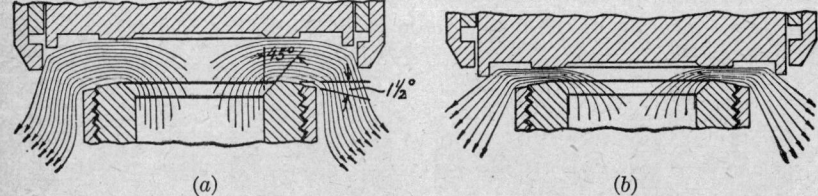

Fig. 50. — Valve Disc and Seat for Crosby Safety Valve.

The adjustment for the popping pressure is made by tightening or loosening the pressure of the spring upon the valve disc by means of the adjusting bolt which is provided at the top of the spindle for this purpose. This adjustment also affects the blow down slightly. Raising the popping pressure lengthens the blow down and lowering the popping pressure shortens it.

Twin valves, that is, two spring-loaded valves having a common Y base, are sometimes used in places where more than one valve is required, and a single connection to the boiler is desired.

The size or capacity of a spring-loaded safety valve can be determined by the formulae proposed in the A. S. M. E. BOILER CODE. The calculations may be based on the heat units in the fuel or on the amount of steam generated. In any case it should be capable of delivering all the steam the boiler can generate without having an excessive pressure built up within the boiler.

The discharge capacity in pounds of steam per hour may be found as follows:

$W = 110 \times P \times D \times L$ for bevel seats at 45 degrees . . . . . (1)

$W = 155 \times P \times D \times L$ for flat seats . . . . (2)

$W = 50 \times P \times A$ for seats at any angle . . . . (3)

in which W = weight of steam that a safety valve will handle, lb. per hr.
P = absolute boiler pressure, lb. per sq. in.
D = inside diameter of valve seat, in.
L = vertical lift of valve disk, measured with 3 per cent excess pressure, in.
A = relieving area in sq. in. = 3.1416 × D × L sine of seat angle.

Example 1. — Find the size of a "pop" safety valve required to discharge 10,150 lb. of steam per hour, if the valve rises from its seat a distance of 0.11 in. Seat is beveled at an angle of 45° and the absolute boiler pressure is 239.7 pounds per square inch:

Solution. — Rearranging Equation (1) and substituting the following values. W = 10,150, P = 239.7 lb. per square inch, D = required diameter, L = 0.11 in.

$$D = \frac{W}{110 \times P \times L} = \frac{10{,}150}{110 \times 239.7 \times 0.11} = 3\tfrac{1}{2} \text{ in. diam.}$$

39. Feedwater Connections and Valves. — The feedwater pipe may enter the boiler through the head or through the top of the front course, with the connection to the boiler made as shown in Fig. 9, page 20. The

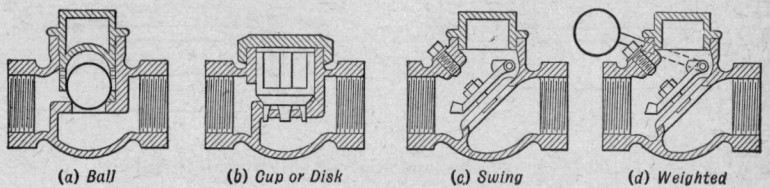

(a) Ball (b) Cup or Disk (c) Swing (d) Weighted

Fig. 51. — Types of Check Valves.

feed pipe line requires the following valves: a **check valve**, Fig. 51, which permits flow in one direction only and automatically prevents the back flow of water from the boiler when neither the feed pump nor the injector is working; a **valve** or **cock**, which is placed between the check valve and the boiler, and which ordinarily remains wide open except when it is desired to inspect or repair the check valve; and sometimes a **globe valve** (Art. 250, page 257) placed between the check valve and the source of water supply to regulate the amount of entering feedwater.

40. Bottom Blow-off Connection and Valves. — The blow-off connection to the boiler is made to the lowest water space practicable. The connection to the boiler should be by a screwed flange riveted to the shell, as shown in Fig. 10, page 20. *The blow-off pipe and fittings should be extra strong and should be protected from the hot gases of the furnace, by either a firebrick wall, a substantial cast-iron removable sleeve, or a covering of non-conducting material.* In spite of this protection the blow-off may burn out. To prevent this, the water in the blow-off pipe is often made to circulate, by means of a pipe which connects the water space of the boiler to the blow-off pipe outside the setting. The joint where the blow-off pipe passes through the setting should permit free expansion and contraction.

If the pipe discharge is hidden from view, a tell-tale should be provided as a guard against leaks. *Outside the setting, a blow-off valve and cock, Fig. 52, are required in the blow-off piping when the pressure is above 125*

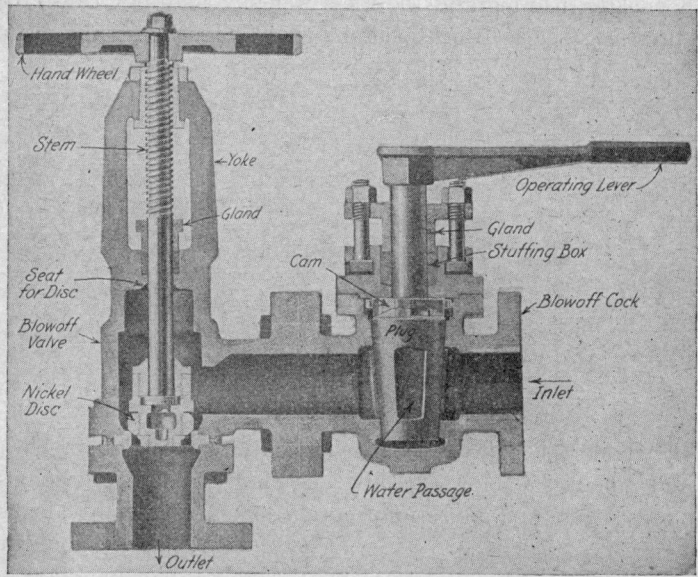

Fig. 52. — Blow-off Valve and Cock.

pounds per square inch. When the pressure is below 125 pounds per square inch, either a blow-off valve or a cock may be used.

The blow-off cock must have the plug held in place by a guard or gland, and the upper end of the plug marked in line with the opening in the plug. A blow-off valve larger than $2\frac{1}{2}$ inches or less than 1-inch cannot be used.

41. Surface Blow-off. — To remove scum and other floating impurities, a surface blow-off, Fig. 53, is sometimes placed at the water level. The inner and outer pipe must be screwed into a brass bushing in the boiler head in such a manner that it makes a smooth passage. The size of pipe used must not be larger than $2\frac{1}{2}$ inches.

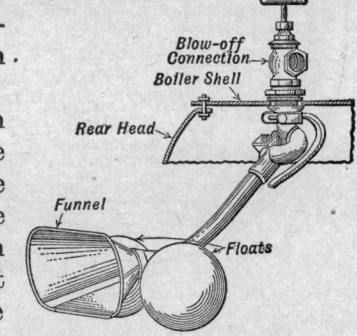

Fig. 53. — Surface Blow-off.

42. Blow-off Tank. — When the location of boilers is such that they cannot be blown down directly into the open, the discharge is made into a tank made of steel plate. The tank has a manhole, an open vent pipe, and inlet and outlet pipes connecting with the blow-off pipe and the sewer.

Sufficient water may be blown off at one time to fill the tank. The water is then allowed to cool and, when cool, is discharged into the sewer.

The discharge of hot water into the sewer is not permitted in most cities, because hot water disintegrates the tile sewer pipe and may cause trouble if under pressure. Measuring boiler blow-off helps materially in main-

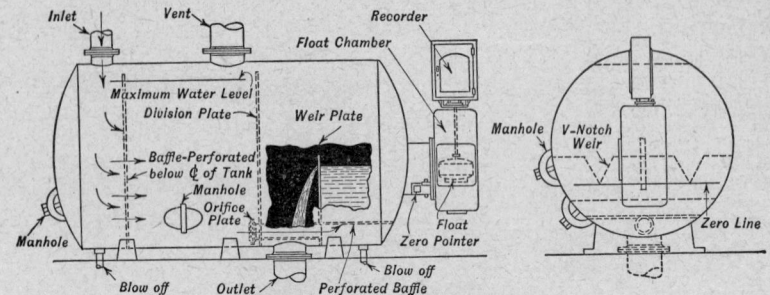

Fig. 54. — Cochrane Blow-off Meter.

taining high boiler economy. It gives a continuous record of both frequency and amount of blow-off valve discharge, including leakage and regular blow downs, and gives information indispensable in calculations of heat balance and efficiencies. A blow-off tank combined with a V-notch meter is shown in Fig. 54.

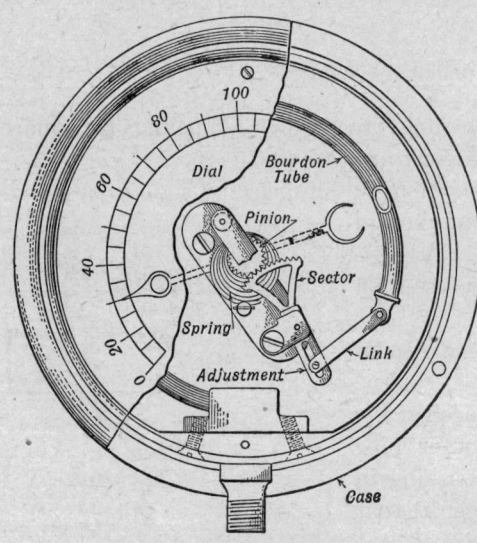

Fig. 55. — Bourdon-tube Steam Gage.

43. Steam-pressure Gage. — Each boiler must have a **steam-pressure gage** connected to the steam space, or to the steam connection of the water column. The type of gage ordinarily used for indicating steam pressure is known as the **Bourdon-tube steam gage,** which is shown in Fig. 55, with the dial removed. Inside the case of the gage is a tempered copper or steel tube, or spring, of oval cross section, bent into an arc of a circle. One end of the tube passes out through the case to which it is attached, and is provided with a $\frac{1}{4}$-inch pipe thread for making the pressure connection. The other end is closed and is free to move under change in pressure. This free end of the tube is connected by a small link to an adjustable arm of a **toothed sector,** which moves about a pivot. The sector engages with a small pinion mounted on the shaft to which the **pointer**, or **hand**, is attached. The hand moves over **a dial graduated in pounds per square inch.**

Pressure applied to the inside of the tube causes the section of the tube to become more nearly a true circle. This changes the radius of the arc to which the tube is bent, and moves the free end outward, thus rotating the shaft to which the pointer is fastened. A **hair spring** attached to the pinion shaft keeps the teeth of the sector and pinion in contact and compensates for lost motion.

Bourdon tube gages which are subjected to jar, such as locomotive gages, are made with the tube supported in the center and the free ends up. They are then called **double spring** gages. When the gage is thus supported the vibration of the needle caused by jarring is reduced to a minimum.

A **water siphon,** Fig. 56, is used to prevent steam from coming into contact with the inside of the gage tube and thus destroying its accuracy. It consists of a chamber holding sufficient water to completely fill the tube. It is placed between the gage and the boiler and should be of such a form that the water will not be easily drained from it. The pipe connecting the gage to the boiler should be of either brass or steel, depending on the pressure, and should have a T- or L-handled gage cock which will line up with the pipe connection when open. The

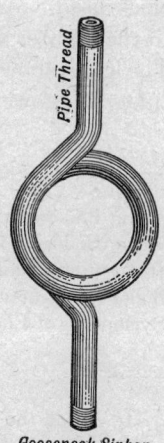

Gooseneck Siphon

FIG. 56. — Gage Siphon.

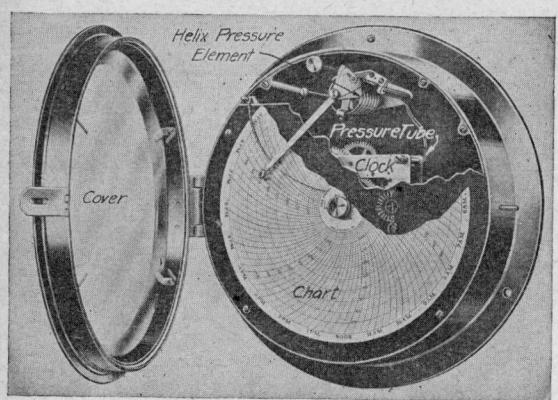

FIG. 57. — Pressure Recording Gage.

cock should preferably be below the siphon to prevent the water escaping from the siphon if the cock should leak.

Diaphragm Gage. — This type of gage is not as common as the Bourdon tube gage, from which it differs in that a corrugated diaphragm, held between two flanges, replaces the Bourdon tube. The pressure acting on one side of the diaphragm deflects it, and the deflection is communicated to the needle through a suitable system of levers. The deflection of the diaphragm is proportional to the pressure acting upon it.

Recording Gages. — To obtain a graphic record of the variation of pressure in the boiler, a recording gage, Fig. 57, is used. Its construction is similar to that of an ordinary steam gage, with the addition of a **clock mechanism attached to the dial.** The dial may make a complete revolution in 24 hours or some other period of time, depending on the type of clock used. A chart is attached to the movable dial, and a pen attached to the pointer of the gage records the variation in pressure upon the chart. The gage shown has a pressure element in the form of a **helix**; Bourdon tubes are however often used.

REFERENCES

Steam Boilers, PEABODY and MILLER.
Steam Boilers, CROFT.
Steam Power Plant Engineering, GEBHARDT.
Boiler Construction Code of the A. S. M. E.
Cyclopedia of Engineering, SIMMONS-BOARDMAN Co.

REVIEW QUESTIONS

1. Name the fundamental types of steam boilers considering the relative position of water and hot gases.
2. Define (*a*) boiler shell, (*b*) setting, (*c*) furnace, (*d*) water level, (*e*) heating surface.
3. State whether the following are water-tube or fire-tube boilers: (*a*) Locomotive, (*b*) Babcock and Wilcox, (*c*) Heine, (*d*) Wickes, (*e*) Stirling.
4. Name three types of fireboxes used on locomotive boilers.
5. Name five kinds of bracing, or staying, used in boiler construction and state a possible application of each.
6. Give the names of eight setting fixtures and state the purpose of each.
7. Name the classes of water-tube boilers according to header construction.
8. At what point does the feedwater enter (*a*) a return tubular boiler, (*b*) horizontal water-tube boiler?
9. Describe the construction of the Scotch Marine boiler.
10. Name six boiler fittings and state the function of each.
11. Explain the operation of the pop safety valve.
12. Name the requirements for steel used in making boiler plate.
13. Name the advantages gained by using large capacity boiler units.
14. What is the purpose of using cross-baffles on a Stirling boiler?
15. Describe the Taylor Unit mentioned in Article 30.

CHAPTER III

PHYSICAL UNITS AND THEIR MEASUREMENT

44. Foreword. — Engineering has become a science to such an extent that those practicing power engineering, or who desire to understand its application must have a thorough knowledge of those fundamental principles which underlie physics, thermodynamics, chemistry, hydraulics and mechanics.

45. Matter. — Matter may exist in the solid, the liquid, or the gaseous state. In each of these states it is conceived to be made up of minute particles called **molecules,** which are further subdivided into smaller parts called **atoms.** According to modern theory, atoms consist of a number of extremely small particles, partly positive and partly negative in character. The positive particles are called *protons*, and the negative *electrons*. All matter consists ultimately of these protons and electrons, which are held together by mutual attraction.

The molecules existing in matter are considered as being in continuous motion and as exerting an attraction between themselves. This attraction is strongest in the solid, because, even though the molecules are in constant motion, the form of the solid remains unchanged. In the liquid state the attraction is less, and the molecules conform to the shape of the containing vessel. In the gaseous state the molecules spend much of their time outside the range of each others attraction; and the gas, or mixture of gases, conforms to the shape and size of the containing vessel.

Matter has **mass** and **inertia.** *A force is required to put it in motion; to change its direction of motion or its speed; or to bring it to rest when in motion.* This property of matter is called **inertia.**

Matter may be changed from one form to another, but the total quantity remains unchanged; upon this fact is based the law known as the **Law of the Conservation of Matter.** It may be stated thus: *The total quantity of matter in the universe remains constant.*

46. Units of the F. P. S. System. — The system of units generally employed in engineering work, for measurement of matter and energy, is the **Foot Pound Second System,** often called for the sake of brevity, the **F. P. S.** system.

The **unit of time,** T, according to this system, is the second, or the $\frac{1}{86,400}$ part of a mean solar day. Time is often expressed in minutes and hours.

64 PHYSICAL UNITS AND THEIR MEASUREMENT

The **unit of length**, L, is the foot. ($=0.3048$ meter.)
The **unit of weight**, W, is the pound. ($=0.4536$ kilogram.)
The **unit of area**, A, is the square foot or the square inch, as preferred.
The **unit of volume**, V, is the cubic foot. Volume equals the product of the cross-sectional area and the length. In calculations involving the quantity of air, Q is often used for the number of cubic feet.

The **unit of force** is weight of the standard pound; centrifugal force is the force which a body exerts by reason of its rotation, and equals $\frac{Wv^2}{gr}$, in which W = weight in lb., v = velocity ft. per sec., $g = 32.2$ and r = radius in feet.

Pressure is the force acting on a body per unit of area.

The **density**, D, of a substance is the weight of a unit volume. The density of a few common substances is given in Table 2.

TABLE 2. — DENSITIES OF COMMON SUBSTANCES

	Substance	Specific Gravity	Weight per Cu. Ft., Lb.	Temp. ° F.	Weight per Cu. In., Lb.
Liquids	Mercury.............	13.6	848.7	60	0.4906
	Water, max. density...	1.00	62.43	39	0.036
	Water...............	0.958	59.83	212	
	Water, sea...........	1.02	64.0		
	Petroleum...........	0.87	54.0		
	Kerosene............	0.78 to 0.82	50		0.0289
	Gasoline.............	0.70 to 0.75	46		
Solids	*Coal, anthracite.....	0.75 to 0.93	47 to 58		
	Coal, bituminous.....	0.64 to 0.87	40 to 54		
	Coal, coke...........	0.37 to 0.51	23 to 32		
	Coal, ashes..........		40 to 45		
	†Coal, anthracite.....	1.4 to 1.8	97		
	Coal, bituminous.....	1.2 to 1.5	84		
	Ice..................	0.88 to 0.92	56		
	Concrete.............	1.5 to 2.4	100 to 144		
	Sand or gravel.......		60		

* Coal as piled in bin. † Coal in solid form.

Speed is the rate of motion of a body, measured by the space passed over in a unit of time, and is usually expressed in feet per second. When equal distances are passed over in equal times the motion is **uniform** and when the distances are unequal the motion is **non-uniform**.

Velocity, v, differs from speed in that it involves the direction as well as the rate of motion. The velocity of a rotating particle may be expressed as: (1) **tangential**, or **linear velocity**, that is, the velocity in feet per second at which a point at a given radius r is traveling; or (2) **angular velocity**, that is, the number of unit angles per second through which the radius turns. The **unit angle**, or **radian**, is the angle subtended by an arc equal to the radius. Thus, if N = number of revolutions per minute, and r = radius in feet:

$$\text{Tangential velocity} = \frac{2\pi rN}{60} \quad \ldots\ldots\ldots\ldots \quad (4)$$

$$\text{Angular velocity} \quad = \frac{2\pi rN}{60\,r} = \frac{\pi N}{30} \quad \ldots\ldots\ldots \quad (5)$$

Acceleration, a, is the rate of change of velocity and is expressed in feet per second per second, generally written "ft. per sec.^{2}." Acceleration may be either positive or negative. *It is positive if the velocity of the body is increasing, and negative if it is decreasing.* The acceleration of gravity, g, at sea level and at latitude 45 degrees is 32.174 ft. per sec.^{2}. It is generally taken as 32.2 ft. per sec.^{2}.

Mass, so called, is the weight of a body divided by the acceleration of gravity, or $\frac{W}{g}$. The **unit of mass** is a derived unit which equals the quantity of matter to which a unit force (1 lb.) will give an acceleration of 1 ft. per sec.^{2}. A pound, to the physicist, means one pound mass, whereas to an engineer it means a pound weight, or in other words, a force, since the weight of a body equals the force with which its mass is drawn toward the earth. The size of a unit mass is therefore the mass of a standard pound multiplied by 32.2.

47. Relation between Velocity, Acceleration, Time, Force and Mass. — For bodies starting from rest, the relation existing between velocity, acceleration, and time is written $V = at$. The force acting on a body is proportional to the acceleration produced; hence, the relation between force, mass and acceleration is written $F = Ma$, or substituting $\frac{W}{g}$ for M,

$$F = \frac{W}{g} a \quad \ldots\ldots\ldots\ldots\ldots \quad (6)$$

in which F equals the force in pounds and the other units are as defined in Art. 46. Since acceleration equals the velocity divided by the time, Equation (6) may be written $F = \frac{Wv}{gt}$.

48. Energy. — Energy may be defined as the **ability to overcome resistance.** It exists in a great variety of forms, such as light, heat, sound and electricity. A body which has the ability to perform work is said to possess **mechanical energy,** which is measured in foot-pounds and may exist in either of the following forms,

1. Potential energy, or energy possessed by reason of position or deformation.
2. Kinetic energy, or energy possessed by reason of motion.

Potential and kinetic energy are interchangeable and when one form is converted into the other the amount of energy of the second form exactly equals that of the first form. This fact is well illustrated by the pendulum of a clock; the pendulum at the top of its swing possesses potential energy; as it swings downward, its potential energy is given up to produce velocity; at the lowest point in the travel of the pendulum the potential energy has been converted into kinetic energy, and as the pendulum swings upward again the kinetic energy is converted into potential energy.

Kinetic energy is measured in foot-pounds and is calculated by the equation $K = \dfrac{Wv^2}{2g}$, in which K = ft-lb. of kinetic energy, W = weight of body in lb., v = velocity in ft. per sec., and g = acceleration of gravity = 32.2 ft. per sec.^{2}.

Electrical energy is usually measured in **joules**. A **joule** is the work done when one coulomb of electricity is conveyed between two points which differ in potential by one volt. It is a mechanical unit adapted to electrical purposes.

49. Work. — Work may be defined as the **overcoming of a resistance through space.** The unit of work is the quantity of energy expended by a force of one pound when acting through a distance of one foot in the line of action of the force. Unless there is motion, work is not performed in a mechanical sense. Work is expressed in foot-pounds and is calculated by the equation

Work in foot-pounds = Force in pounds × a distance in feet.

50. Work Diagram. — Work may be represented by a diagram, that is, an area bounded by lines representing force and the distance through which it is exerted, or by lines representing pressure and change in volume. *The area enclosed represents work and is generally expressed in foot-pounds.*

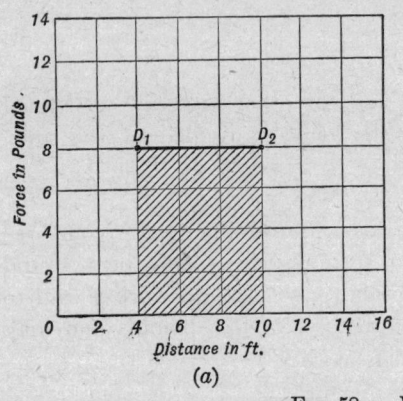
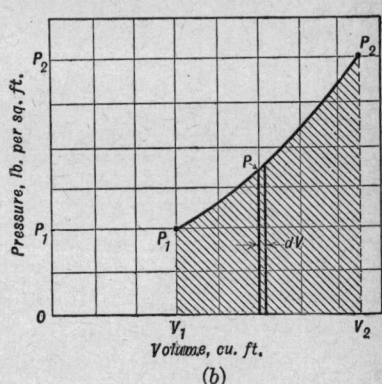

Fig. 58. — Work Diagrams.

Force Constant. — The path traveled by a body, when under the action of a constant force of 8 pounds, may be represented graphically as in Fig. 58a. The work diagram, in this case, is a rectangle, and the work required to move the body from D_1 to D_2 equals the area 4 $D_1 D_2$ 10

Work = $(D_2 - D_1) \times 8 = 6 \times 8 = 48$ ft-lb.

Force Changing. — If the force or pressure varies, as represented by the line P_1P_2, Fig. 58b, while the work is being done, the work area $V_1P_1P_2V_2$ may be considered as made up of a series of very narrow sections of width (dV) for each of which there is a pressure P in pounds per square foot. The area $P\,dV$ represents the work (dW) done during the small change in volume dV. The summation of these elementary areas, dW, from V_1 to V_2 while the pressure is changing from P_1 to P_2 gives the work area $V_1P_1P_2V_2$, and is usually expressed in a mathematical form, thus:[1]

$$\text{Work in ft-lb.} = \int_{V_1}^{V_2} P\,dV \quad \ldots \ldots \ldots \quad (7)$$

51. Power. — Power is the **rate of performing work,** and hence involves time as a factor. It is equal to the **amount of work performed** divided by the **time.** The unit of mechanical power is the **horsepower** (hp.) which equals 550 foot-pounds of work per second, or 33,000 foot-pounds per minute. The horsepower relation may be expressed:

$$\text{Horsepower} = \frac{F \times v}{550} \quad \ldots \ldots \ldots \quad (8)$$

in which F = the force in pounds and v = velocity in feet per second. A mechanical unit of power exerted continuously for one hour is known as a **horsepower-hour** and is a unit of work or of energy.

The unit of electrical power is the **watt,** which equals the product of the **volt** *times the* **ampere.** The **volt** is the unit of electrical pressure, or difference in potential, and the **ampere** the unit of electrical current. A watt is equivalent to one joule of work per second. The **kilowatt,** a larger unit of electrical power, is equal to 1000 watts. A kilowatt of power delivered continuously for one hour is called a **kilowatt-hour.**

52. Heat. — *Heat is a form of energy.* The heat of a body is the combined energy of the moving molecules of which every substance is composed. Heat can only be observed and recorded by its effect upon matter, producing change in shape, volume, and internal stress; change of state, as ice to water; change of temperature; and electrical and chemical effects. Conversely, heat may be obtained from mechanical and electrical energy, from chemical changes, and from changes of physical state.

Heat which changes the temperature of a body is called **sensible heat.** Its intensity may be measured by a thermometer. The heat that is used in changing the state of the body, as in changing ice into water and water into steam, is called respectively the **latent heat of fusion** and the **latent heat of vaporization,** and while it is being added, no temperature change is indicated.

Since it would be impossible to represent all the forms of energy mentioned above in a simple equation, it is customary to eliminate the energy

---

[1] Consult page 86, for further discussion of this subject.

terms which are constant during the process considered, and to consider only the following **effects** as being **produced by the application of heat to a substance:** (1) *a change in the temperature*, (2) *a change in the physical state*, and (3) *the performance of external work by or upon the substance*.

The *first two effects* represent the change in **internal,** or **intrinsic energy,** of the body and the *third effect* represents the heat that was absorbed in order to perform the external work by increasing the volume against external forces. This may be expressed as an equation, by denoting the heat added, sometimes called the **total heat,** by the symbol dH, the intrinsic energy by dU, the work performed by dW, and the reciprocal of 778[1] by A. Thus

$$dH = dU + A\, dW$$

$$H = \int_{U_1}^{U_2} dU + A \int_{V_1}^{V_2} P\, dV \quad \ldots \ldots \ldots \quad (9)$$

The intrinsic energy of a body depends only upon the physical state of the body and not on the method of bringing it to that state, hence it may be said to be independent of the **path,** or *series of intermediate states,* through which the body passes in reaching its final state. It is, therefore, an exact differential[2] and can be directly integrated,[2] that is, $\int_{U_1}^{U_2} dU = U_2 - U_1$.

This is not true of the external work term $A \int_{V_1}^{V_2} P\, dV$ which is dependent upon the path as can be seen from a work diagram, for as the path changes the area under the path or curve changes and this area represents work, Art. 50, page 66. It is, therefore, evident from Equation (9) that the total heat H added to a body is dependent upon the path. *The most common path in power engineering work being of the form* $PV^n = constant$ *and is called a* **polytropic path.**

It should be noted that the *internal energy* of a perfect gas is a function of its sensible heat, and depends only upon the change of temperature. For any weight (w) of a gas, it may be expressed as follows:

$$\text{Change of internal energy in B.t.u.} = w \int C_V\, dT$$

in which T refers to absolute temperature, and C_V is the specific heat at constant volume. The latter term is always used in dealing with internal energy changes, since under constant volume conditions external work is not performed and consequently the energy change is entirely internal. For an increase of internal energy the above equation may be written as follows:

$$U_2 - U_1 = wC_V(T_2 - T_1). \quad \ldots \ldots \ldots \quad (10)[3]$$

[1] See page 74.

[2] The reader is referred to any of the many books on calculus for explanation of these terms.

[3] See also Equation 29, page 87.

53. Temperature. — Temperature is a manifestation of the intensity of heat in a body, and is an indication of the rate of molecular activity. If a body is capable of transmitting heat to another body unaided, the first body is said to be at a higher temperature than the second body. When two bodies are at the same temperature, neither has any tendency to transmit heat to the other.

Temperature is ordinarily measured by instruments called thermometers and pyrometers, of which there are many forms. The **mercury thermometer,** which depends upon the uniform expansion and contraction of mercury to indicate temperature change, is most commonly used to measure temperature. It consists of a glass tube, or stem, with a small uniform bore, having its lower end enlarged to form a bulb. All air is removed from the stem, and the bulb and part of the stem are filled with mercury. The tube is marked in equal divisions, called degrees, which are numbered according to the scale of temperature to be employed. Since mercury expands when heated and contracts when cooled, the temperature of a body is obtained by placing the thermometer in contact with it and noting the height at which the mercury stands.

The unit of temperature measurement is the **degree.** It is capable of exact determination, provided two points can be obtained at which the intensity of heat is always constant. The melting-point of ice and the boiling-point of water, at atmospheric pressure (14.7 pounds per square inch), are the points usually selected. The thermometer is first placed in melting ice, and the height at which the liquid stands in the stem is marked. Then the thermometer is immersed in steam, from water boiling at atmospheric pressure, and the point reached by the top of the liquid is marked. The distance between these two points is divided into 180, 100 or 80 divisions, or degrees, according to whether the scale of temperature to be used is the Fahrenheit, Centigrade, or Réaumur. The last-named scale is seldom used at the present time.

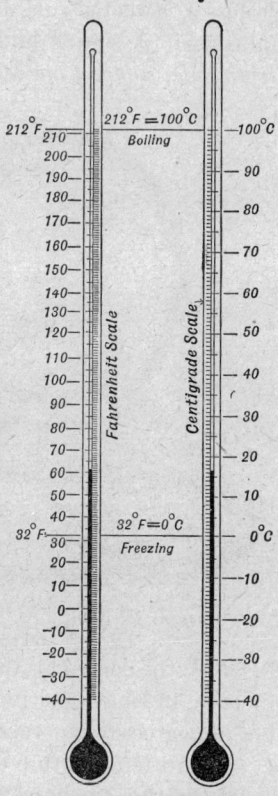

Fig. 59. — Thermometer Scales.

The **Fahrenheit scale,** Fig. 59, largely used by engineers in America, takes the temperature of melting ice as 32 deg. fahr., and the temperature of boiling water, at atmospheric pressure, as 212 deg. fahr. The **Centigrade scale,** commonly used by scientists throughout the world, makes the melting-point of ice 0 deg. cent., and the boiling-point of water, at atmospheric pressure, 100 deg. cent. Since 100 deg. cent. equals 180 deg. fahr., one degree Fahrenheit equals five-ninths of a degree Centigrade, and one degree Centigrade equals nine-fifths of a degree Fahrenheit. The following equations may be used to convert temperatures from one scale to the other:

$$\text{Fahrenheit degrees} = \tfrac{9}{5}\text{ Centigrade degrees} + 32 \quad \ldots \quad (11)$$

$$\text{Centigrade degrees} = \tfrac{5}{9}\text{ (Fahrenheit degrees} - 32) \quad \ldots \quad (12)$$

Professor Sweet's rule for converting degrees Centigrade into degrees Fahrenheit is simple and easily applied. *"Double the number of degrees Centigrade, subtract one-tenth of this value, and add 32."*

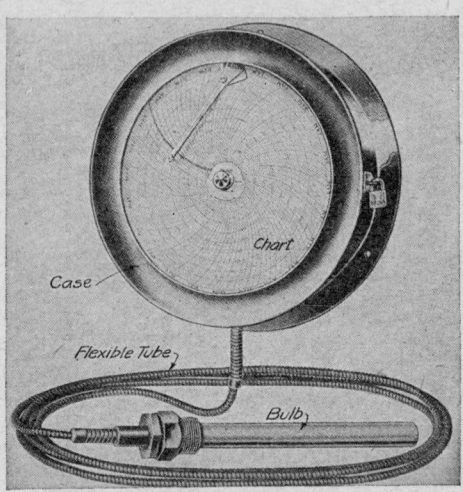

FIG. 60. — Recording Thermometer.

The glass from which thermometers are made may undergo small changes from time to time, making it necessary, in order to obtain accurate results, to **calibrate the thermometer,** by comparing it with a standard thermometer, and noting its variations. Besides calibrating the thermometer, it is necessary to make a correction for stem exposure when extreme accuracy is desired. This correction is given by the equation:

$$\text{Stem correction in degrees} = 0.000085\, N\,(t - t_s) \quad \ldots \quad (13)$$

in which the decimal is the difference between the coefficient of expansion, or increase in length per inch per degree, of mercury and of glass, N = number of degrees of emergent mercury column, t = observed temperature and t_s = mean temperature of emergent stem.

In addition to the common form of thermometer, **recording** and **resistance thermometers** are extensively used. The *Brown recording thermometer,* Fig. 60, consists of a copper bulb connected to the instrument by a capillary copper tube, 20-thousandths (0.020) of an inch in diameter, which is protected by a heavy flexible bronze re-inforcing tube. The capillary tube connects with either a helix, or bourdon, spring in coil form, located within the instrument and to which the pen is connected. Nitrogen gas is used in the bulb for temperatures up to 1000 deg. fahr., and as the temperature changes the volume of gas increases or decreases, thus moving the pen across the chart exactly in proportion to the temperature change. Since the coefficient of expansion for nitrogen is uniform, the chart is evenly graduated.

The *resistance thermometer* consists of a nickel or platinum resistance unit enclosed in a protecting tube. The operation depends upon the principle that the change in the resistance of the wire forming the bulb is proportional to the change in the temperature of the bulb. The change in the resistance of the wire is ordinarily measured by a **galvanometer,** an instrument for indicating small electrical currents. It is connected into the circuit of a wheatstone bridge, and is calibrated to read in temperature units.

54. Absolute Temperature and Absolute Zero. — Besides the above temperature scales, there is another called the absolute scale of temperature. It is used in all calculations in which the temperature of gas volumes is involved. This scale is based on the so-called "absolute zero of temperature," or the point at which a perfect gas is considered to have zero volume. Scientists have found that a perfect gas, of which air is taken as a type, expands or contracts $\frac{1}{491.6}$ of its volume at 32 deg. fahr. for each deg. fahr. change in temperature. The absolute zero, therefore, may be taken as 491.6 deg. fahr. below the melting-point of ice. This equals 459.6 degrees below zero on the Fahrenheit scale, or 273 degrees on the Centigrade scale. The absolute temperature (deg. fahr.) of a substance is found by adding 459.6 to the observed temperature. If the observed temperature were 60 deg. fahr. the absolute temperature would be 459.6 + 60, or 519.6 deg. fahr. The value 460 is used in ordinary engineering calculations, instead of 459.6.

55. Pyrometers. — Temperatures above 500 deg. fahr. are measured by **pyrometers.** High-grade thermometers, having nitrogen under pressure enclosed in the tube above the mercury, can be used for temperatures as high as 1000 deg. fahr. Such a thermometer is termed a mercurial pyrometer. The most common types of pyrometers are **expansion, thermoelectric, radiation, optical,** and **Seger cone.**

Metallic Pyrometer. — The metallic pyrometer consists essentially of two metal rods having widely different rates of expansion, such as iron and brass, and so connected as to move a pointer over a graduated scale during a change in temperature. Such pyrometers should not be used for temperatures above 1000 deg. fahr.

Thermo-electric Pyrometer. — This type of pyrometer, Fig. 61, comprises a thermocouple, an indicating or recording device, and suitable connecting wires. The thermocouple, Fig. 62, consists of two wires of dissimilar metals, and of different electrical conductivity, welded together at one end. When the weld or **" hot junction,"** is heated and the other ends joined to form a **" cold junction,"** an electric current, which is proportional to the difference in temperature between the hot and cold junctions, will flow. A **galvanometer,** or **milli-voltmeter,** is generally placed in the electric circuit, and the current is read from it. It is calibrated to read in degrees, by comparison with a standard thermometer. In actual operation, the "cold junction" is sometimes immersed in an ice bath or buried in the ground, in order to maintain it at a constant temperature and the "hot junction," or furnace end, is placed in an iron or porcelain tube to protect it from breakage and deterioration. Recent instruments are so constructed that they automatically compensate for changes in temperature of the cold junction.

Base metal thermocouples of low resistance are made of $\frac{1}{8}$-inch wires of nickel steel and copper, nickel steel and chromium or No. 8 gage iron and constantin, and are satisfactory for temperatures up to 1800 deg. fahr. For temperatures below 3000 deg. fahr., **high resistance thermocouples,**

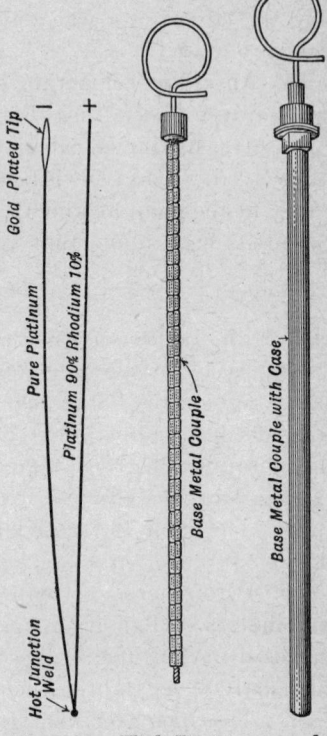

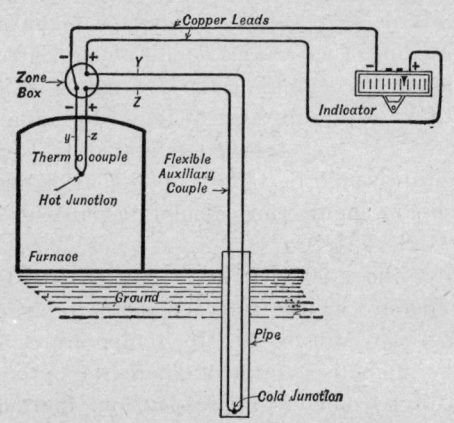

Fig. 61. — Thermo-electric Pyrometer.

Fig. 62. — High Resistance and Base Metal Thermocouples.

made with one wire of pure platinum and the other 90 per cent platinum and 10 per cent rhodium, are used.

Radiation Pyrometer. — Temperatures above 2500 deg. fahr. are measured by a radiation pyrometer, which consists of a cylindrical tube containing a concave mirror and a lens which is focused on the hot object. The mirror concentrates the rays upon a small thermo-electric couple connected to a galvanometer. The temperature reading is obtained in the same way as with a thermo-electric pyrometer.

Optical Pyrometer. — The measurement of temperature with this instrument is based on the fact that light varies in a definite manner with changes of temperature. Red light is separated from all the other light emitted by the incandescent body, and its intensity is compared with the intensity of light of the same color from a standard source of light, such as a special lamp having a tungsten filament. The eye is sensitive to differences in brightness between superimposed surfaces. The optical pyrometer, Fig. 63, varies the intensity of the standard light by changing the current supplied, until the filament appears of the same brightness as the hot object when viewed through the eye piece. When a balance has been obtained, the reading of the milli-ammeter is made and the corresponding temperature is read from a curve supplied with the instrument.

Seger Cone Pyrometer. — Furnace temperatures may be obtained by the use of metals having different melting-points. The oxides of the metals are made into cones which are graded to melt at temperatures differing by 100 to 200 deg. fahr. Several cones are placed in the furnace, and the temperature is nearest that corresponding to the melting temperature of the cone, the top of which has just bent sufficiently to touch the plate upon which the cones rest. Seger cones can be used for temperatures from 500 to 1900 deg. fahr.

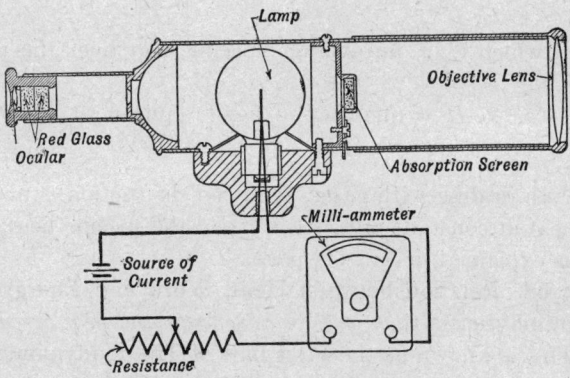

FIG. 63. — Optical Pyrometer.

Accuracy of Pyrometers. — The mercurial pyrometer is the most accurate for low-temperature measurements. The electrical pyrometer is the best for high-temperature measurements. Expansion pyrometers are subject to wide variations and should be used only after careful calibration. On high-temperature measurements the deviation from accuracy of a pyrometer may be as high as 40 deg. fahr. The location of the measuring element is important since the radiation error may be large.

56. Quantity of Heat. — Heat may be expressed by the usual energy units, such as foot-pounds or joules. It is the custom in engineering work to express the quantity of heat by a separate unit known as the **British thermal unit, B.t.u.,**[1] which is **the amount of heat required to raise the temperature of one pound of pure water one degree Fahrenheit,**[2] often taken from 62 deg. fahr. to 63 deg. fahr. The **mean,** or **average,** B.t.u. is most commonly used in engineering calculations. It is the average amount of heat per degree required to raise the temperature of one pound of water from 32 deg. fahr. to 212 deg. fahr.

[1] 1 B.t.u. = 252.2 calories = 1054 watt seconds.
[2] The International Steam Table Conference has accepted the **kilocalorie** as a heat unit which is defined as *the heat equivalent of* $\frac{1}{860}$ *kilowatt hour.*

57. Specific Heat. — *The specific heat of a substance is the quantity of heat necessary to raise the temperature of one pound of the material one degree.* It varies with the physical properties of the substances and with the temperature of the substance. Its numerical value is obtained by comparison with water as a standard. Two specific heats are recognized:

1. The **"true"** **specific heat** measured at the temperature stated.
2. The **"mean"** **specific heat** which is an average over the temperature range considered.

The mean specific heat may be expressed by an equation as follows:

$$C = \frac{H}{W(t_2 - t_1)} \qquad (14)$$

in which C = an average specific heat over the temperature range from t_1 to t_2.

H = quantity of heat required to raise W pounds of substance from t_1 to t_2.

In dealing with gases, a further distinction is made between the specific heat at constant pressure, C_p, and the specific heat at constant volume, C_v, as explained in Art. 66, page 82.

58. Relation between Heat, Work and Energy. — It has been shown by physicists that *mechanical energy and heat are mutually interchangeable.* This is known as the **First Law of Thermodynamics**. The relation which exists between heat and mechanical energy was first determined by Joule, and is known as **Joule's equivalent**. The most recent value states that 1 B.t.u. = 778.57 foot-pounds. The value 778 is generally used for engineering calculations. Expressed mathematically:

$$1 \text{ B.t.u.} = 778 \text{ foot-pounds} \qquad (15)$$

$$1 \text{ foot-pound} = \frac{1}{778} \text{ B.t.u.} \qquad (16)$$

59. Relation between Heat, and Electrical and Mechanical Power. — The unit of mechanical power is the horsepower, which is equal to 2545 B.t.u. per hour, or 33,000 × 60 = 1,980,000 foot-pounds per hour.

The unit of electrical power is the kilowatt and is equal to 3413 B.t.u. per hour. The following relations therefore exist and should be memorized, as they are of much use in engineering calculations:

$$\begin{aligned}
1 \text{ kilowatt-hour} &= 3413 \text{ B.t.u.} \\
1 \text{ horsepower-hour} &= 2545 \text{ B.t.u.} \\
1 \text{ horsepower-hour} &= 0.746 \text{ kilowatt-hour} \\
1 \text{ kilowatt} &= 1.34 \text{ horsepower}
\end{aligned}$$

60. Heat Transmission. — There are three methods of transmitting heat; namely, conduction, convection and radiation.

Conduction is a molecular transmission of heat through the substance, from molecule to molecule. Such transmission will take place between any two parts of a substance which are at different temperatures.

All substances conduct heat, though the rate at which heat is conducted varies with the material. Substances which transmit heat readily are called good conductors, and those which transmit heat slowly, poor conductors. The heat transmitted by conduction varies with the thickness of the material, the area in contact and the temperature difference.

Convection is the transmission of heat by circulation of a fluid or gas over the surface of the hotter or colder body. This circulation may be due to natural causes or it may be produced mechanically. The quantity of heat transferred by convection does not depend upon the nature of the material or its absolute temperature. It does depend on the velocity of the moving fluid or gas, the form and dimensions of the body, and the temperature difference between the moving substance and the contact surface of the body.

Radiation is the transmission of heat through an agency commonly known as ether, which is assumed to occupy all intermolecular space. It takes place in straight lines and obeys the same laws as light. The rate at which radiant heat is emitted or absorbed depends upon the character of the surface of the hot or cold body, the temperature difference between the surface and surroundings, and the absolute temperature. It does not depend upon the form of body, however, unless there are re-entrant surfaces to intercept the rays.

61. Measurement of Pressure. — The unit of pressure is usually the *pound per square inch* and is written "lb. per sq. in." Low pressures are often expressed in inches of mercury or inches of water. Pressure is ordinarily indicated by a gage, Fig. 64. Gages indicating pressures above that of the atmosphere are called **pressure gages** and those showing pressures below atmospheric pressure are called **vacuum gages**.

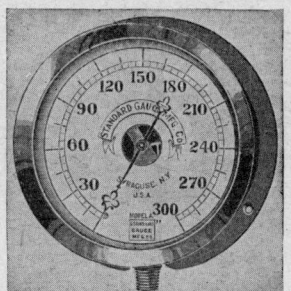

FIG. 64. — External View of Pressure Gage.

Gages are subject to variations in use and therefore require frequent **calibration**, i.e., *comparison with a standard*. For this purpose a dead-weight tester or a mercury column is used. When using the dead-weight tester the gage is subjected to pressures produced by standard weights and the readings on the dial, as indicated by the hand of the gage, are compared with the pressures in pounds per square inch exerted by the standard weights. For many purposes, standard test gages, known to be correct by calibration with the dead-weight tester or mercury column, are used for comparison. The piping connections below a boiler gage are required to provide for the connection of a test gage.

The simplest type of pressure-measuring device is the manometer which is a glass tube bent into the form of a U and partially filled with water or mercury, Fig. 65a. One side, A, of the tube is attached to the vessel in which the pressure is to be measured, the other side, B, is left open to the atmosphere. With equal pressures on each leg of the U, the surfaces of the liquid remain stationary. Upon the application of pressure to leg A, the

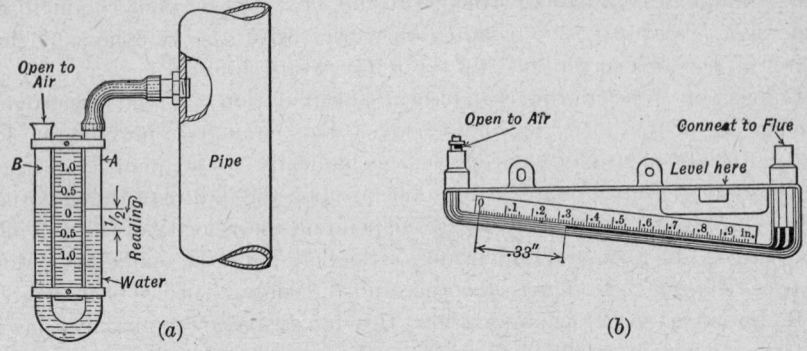

Fig. 65. — Manometer and Draft Gage.

surface of the liquid will fall in leg A, and rise in leg B. The distance between the level of the liquid in A and B, multiplied by the weight in pounds per cubic inch of the liquid, gives the difference in pressure between A and B, expressed in pounds per square inch.

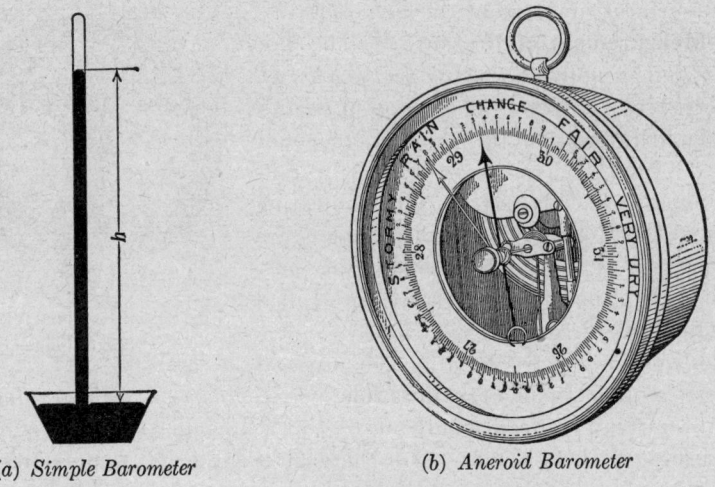

(a) *Simple Barometer*   (b) *Aneroid Barometer*

Fig. 66. — Barometers.

Manometers using water should be read at the bottom of the meniscus and those using mercury should be read at the top of the meniscus, to make the error from capillarity as small as possible. *Manometers are only used for low pressures.*

Manometers used for measuring the difference in pressure between the inside and outside of a chimney or furnace are called **draft gages,** a common form being shown in Fig. 65b. If one leg of the U tube is arranged on an incline, the distance moved by the liquid is increased for a given pressure change. The inclined leg ordinarily rises 1 inch in 10. A light oil is used in the gage to reduce the effect of capillarity and to give a greater deflection than can be obtained by the use of water. The gage is calibrated to read equivalent inches of water, by comparison with a gage using water, and can easily be read to 0.001 inch.

62. Barometers. — The pressure of the atmosphere is measured by means of barometers, of which there are two principal types, the mercurial barometer, Fig. 66a, and the aneroid barometer, Fig. 66b. The latter is similar to a delicate pressure gage and is calibrated by comparison with the mercurial barometer.

The Mercurial Barometer. — One type of this barometer, shown in Fig. 66a, consists of a glass tube about 3 feet long sealed at one end. After all air has been excluded, the tube is filled with mercury and is inverted in a bath of mercury. The mercury in the tube then drops to a certain point, where it remains. This height, h, represents the pressure of the atmosphere in inches of mercury and varies about 0.01 inch for each 10 feet of variation in altitude above sea level.

A commercial, or standard, type of this instrument, shown in Fig. 67, consists of a glass tube closed at the top and enclosed in a brass tube having a movable scale at the top. The lower end of the glass tube dips into a glass cup having a leather bottom resting on a movable disk. The level of the mercury in the cup can be raised and lowered by means of an adjusting screw until its surface touches a fixed ivory point which is located at the zero of the measuring scale.

Standard atmospheric pressure is defined as the pressure of a column of pure mercury, 29.92 inches high, at a temperature of 32 deg. fahr. Expressed in pounds per square inch, it is 14.7. The decrease in pressure for each 1000 feet in altitude is about $\frac{1}{2}$ pound per square inch. To correct observed barom-

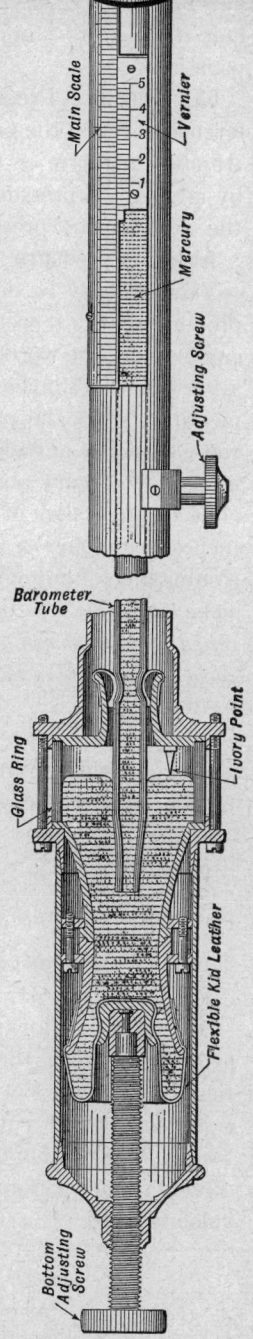

Fig. 67. — United States Weather Bureau Barometer.

eter readings to standard conditions, see method of correction, Art. 520, page 569.

63. Absolute Pressure and Vacuum. — Pressure gages show the difference between the pressure in a vessel and the pressure of the atmosphere. *Absolute pressure is the pressure above the zero of pressure.* It is found by adding the pressure shown by the barometer to the pressure shown by the gage, both expressed in the same units.

Absolute pressure = gage pressure + atmospheric pressure.

Vacuum may be defined as the absence of pressure, a vessel having no pressure within it would have an absolute vacuum. As ordinarily used in engineering, the word vacuum means only a partial vacuum. For instance, a vacuum of 18 inches means that the pressure corresponds to 18 inches of mercury below the pressure of the atmosphere; if the barometer reading were 30 inches of mercury, the absolute pressure would be (30 − 18), or 12 inches of mercury absolute.

64. Conversion of Pressure. — It is frequently necessary to convert inches of mercury or water to pounds per square inch. Since the weight of a cubic inch of mercury at 70 deg. fahr. is 0.4906 pound, and of water at the *same temperature is* 0.0360 *pound, pressure in inches of mercury can be converted to pounds per square inch by multiplying by* 0.491, *and pressures in inches of water can be converted to pounds per square inch by multiplying by* 0.036.

The pressure expressed in feet of water existing on a system is ordinarily known as **head.** When so expressed, the pressure in feet may be converted to pounds per square inch by multiplying by 0.434. The factor 0.434 is obtained as follows: one cubic foot of water weighs 62.4 pounds at 60 deg. fahr., hence, a column of water one foot high, at 60 deg. fahr., and one square inch in cross section, will exert a pressure of $\frac{62.4}{144}$ or 0.434 pounds.

To convert pounds per square inch to feet of water, multiply by the reciprocal of 0.434, or 2.31. *This value varies with the temperature of the water.*

Water or air flowing in a pipe has a velocity produced by the pressure existing between the two points between which flow takes place. The relation existing between the velocity and the pressure, in this case, is expressed by the equation $v = \sqrt{2gh}$, in which v = velocity in feet per second, h = head in feet of fluid flowing and g = 32.2. This equation has a direct application in pump problems. The factor, h, is known as the **velocity head.**

REFERENCES

Power Plant Testing, MOYER.
Mechanical Equipment of Buildings, Vols. I and II, HARDING and WILLARD.
Bulletins of Society for Promotion of Engineering Education, Oct. 1920, Dec. 1920.
American Math. Monthly, Vol. 15, Jan. 1918.

Mechanics, FRANKLIN and McNUTT.
Heat for Advanced Students, EDSER.
Magnetism and Electricity, HADLEY.
Theory and Practice of Mechanics, SLOCUM.
Mechanical Engineering, Vol. 52, pages 132, 139–141.

PROBLEMS AND REVIEW QUESTIONS

1. The side of a square coal bin is 14 ft. and its height is 16 ft. Express its volume in cubic meters.

2. One cubic foot of solid anthracite coal weighs 97 lb. Express this weight in kilograms.

3. How many more cubic feet of space will be required to store 1000 tons of bituminous coal than to store the same amount of anthracite? Their respective weights, as placed in a bin, are 52.8 lb. and 53.4 lb. per cubic foot.

4. A locomotive is traveling West at 20 miles per hour. Find its velocity in feet per second.

5. A locomotive exerts a pull of 45 tons on a train, when traveling at 5 miles per hour. What is the horsepower delivered by the locomotive?

6. An electric car starting from rest attains a speed of 30 miles per hour in five minutes. Find the acceleration.

7. A pound of steam, at 20 lb. per square inch absolute pressure, has a volume of 20.10 cubic feet. What is its density?

8. A body weighs 190 lb. when weighed on a platform scale. Find its mass (so-called).

9. A revolving body weighs 200 lb. and has a linear velocity of 2000 ft. per min. Find the kinetic energy of the body.

10. Define work and express as an equation. A compound duplex steam pump lifts 432,000 gallons of water, at 70 deg. fahr. through 300 feet, in 24 hr. A gallon of water contains 231 cubic inches. Find the work done per minute. (For density of water consult Table 5, page 90.)

11. Draw a diagram representing the work done by a constant force F acting through a distance d. Explain how the area of the diagram represents the work done by the force.

12. A loaded box car weighs 60,000 lb. Find the force required to give it an acceleration of 2 ft. per second, per second.

13. The temperature of water entering a feedwater heater is 70 deg. fahr. Express as degrees Centigrade.

14. The temperature of water leaving a condenser is 50 deg. cent. Express as degrees Fahrenheit, using Professor Sweet's rule.

15. Express as absolute temperature, using the Fahrenheit scale, 30 deg. cent., 90 deg. fahr., — 10 deg. fahr.

16. Name five types of pyrometers and explain the method of measuring temperature, using one of the types named.

17. The water in a tank weighs 200 lb. The temperature of the water is changed from 60 deg. fahr. to 212 deg. fahr. Express the quantity of heat required in B.t.u. Mean specific heat of water taken as 1.00.

18. The mean specific heat of iron between 80 deg. and 2000 deg. fahr., is 0.126. Find the quantity of heat in B.t.u. given up, if 4 pounds of iron at 2000 deg. fahr. are placed in sufficient water to lower the temperature of the iron to 80 deg. fahr. Express this quantity of heat as ft-lb.

19. A mechanical horsepower is equal to 2545 B.t.u. per hour. Show how this value is obtained.

80 PHYSICAL UNITS AND THEIR MEASUREMENT

20. The turbo-generator at the State Line Power Station of the Commonwealth Edison Company is rated at 208,000-kw. What is the horsepower rating of this machine?

21. A Corliss engine delivers 24 horsepower continuously for six hours. How many horsepower-hours is it delivering?

22. The steam gage attached to a boiler reads 210 lb. per sq. in. The barometer reading is 29.92 in. of mercury. What is the absolute gage pressure in lb. per sq. in.?

23. The pressure of water in a water pipe is 500 lb. per sq. in. What "head" of water would be required to produce this pressure if the water temperature is 65 deg. fahr.?

24. A vessel is under a pressure of 3 ft. of water at 120 deg. fahr. The density of water at 120 deg. fahr. is 61.71 lb. per cu. ft. Find the pressure in lb. per sq. in. caused by the water.

25. The reading of a draft gage at the base of a stack in a power house is 1 inch of water. What pressure does this represent in lb. per sq. in., if the temperature of the air surrounding the gage is 70 deg. fahr.?

26. The velocity of water flowing in a water pipe is 1800 ft. per minute. Find the pressure in lb. per sq. in. required to produce this velocity.

27. The factor for converting pressure in lb. per sq. in. into feet of water is 2.31 at 70 deg. fahr. Explain how this factor is obtained.

28. The permissible rise in temperature of the windings of an electric generator is given as 40 deg. cent. What is the corresponding temperature rise in deg. fahr.?

29. Determine the gage reading in lb. per sq. in., at an altitude of 14,000 ft., which corresponds to standard conditions at sea level.

30. In determining the temperature in a furnace, by using a fire-clay egg and a pail of water, the following data were taken: weight of water, 15.5 lb.; weight of pail, 3.3 lb.; specific heat of pail, 0.11 B.t.u.; weight of fire-clay egg, 0.6 lb.; specific heat of egg, 0.20 B.t.u.; initial temperature of water in pail, 45 deg. fahr.; final temperature of water, 59 deg. fahr. Find the temperature of the furnace.

CHAPTER IV

PROPERTIES OF AIR, GASES, WATER, AND DRY AND SATURATED STEAM

65. Foreword. — A knowledge of the properties of air, gases, water and steam is essential to an understanding of the operation of power plant equipment. Only sufficient material is here given as will make clear the following chapters of the book.

66. Air. — Pure air is a mechanical mixture of oxygen and nitrogen in the following proportions:

	Per cent by Volume	*Per cent by Weight*
Oxygen	20.91	23.15
Nitrogen	79.09	76.85

As ordinarily found, air is not pure, but contains impurities, such as carbon dioxide, ozone, water vapor, and dust.

The **specific density,** or weight per cubic foot, of air decreases with an increase in temperature, and the specific volume therefore increases with the temperature. The density and volume of air at various temperatures are given in Table 3.

TABLE 3. — VOLUME AND WEIGHT OF AIR AT VARIOUS TEMPERATURES.

Barometer 29.92 Inches Mercury.

Temperature °F.	Volume Cu. Ft. per Lb.	Weight Lb. per Cu. Ft.	Temperature °F.	Volume Cu. Ft. per Lb.	Weight Lb. per Cu. Ft.	Temperature °F.	Volume Cu. Ft. per Lb.	Weight Lb. per Cu. Ft.
32	12.39	.0807	160	15.62	.0640	340	20.15	.0496
50	12.84	.0779	170	15.87	.0630	360	20.66	.0484
55	12.97	.0771	180	16.12	.0620	380	21.16	.0473
60	13.10	.0763	190	16.37	.0611	400	21.66	.0462
65	13.22	.0756	200	16.62	.0602	425	22.29	.0449
70	13.35	.0749	210	16.88	.0593	450	22.92	.0436
75	13.47	.0742	212	16.93	.0591	475	23.55	.0424
80	13.59	.0735	220	17.13	.0584	500	24.18	.0414
85	13.72	.0729	230	17.38	.0575	525	24.81	.0403
90	13.85	.0722	240	17.63	.0567	550	25.44	.0393
95	13.98	.0715	250	17.88	.0559	575	26.07	.0384
100	14.10	.0709	260	18.14	.0551	600	26.70	.0374
110	14.36	.0697	270	18.39	.0543	650	27.96	.0358
120	14.61	.0685	280	18.64	.0537	700	29.22	.0342
130	14.86	.0673	290	18.89	.0529	750	30.48	.0328
140	15.11	.0662	300	19.14	.0522	800	31.74	.0315
150	15.36	.0651	320	19.65	.0509	850	33.00	.0303

Air has a specific heat at constant pressure, C_p, and a specific heat at constant volume, C_v. The former is the quantity of heat required to raise the temperature of a unit weight of the gas one degree Fahrenheit at constant pressure. It varies from 0.2375 to 0.2430, but for most engineering calculations 0.24 is sufficiently accurate. *A gas expanding at constant pressure performs a certain amount of external work, and the heat equivalent of this work is included in the value of the specific heat at constant pressure.* The **specific heat at constant volume** is the quantity of heat required to raise the temperature of a unit weight of a gas one degree Fahrenheit, the volume remaining constant. As the gas does not change in volume, external work is not performed, and hence, the *specific heat at constant volume is less than the specific heat at constant pressure by the amount of work done in the first case.* This fact is sometimes written, $C_p - C_v = R$, in which R is the amount of external work done, all quantities being expressed in the same units. The ratio of C_p to C_v is usually represented by either the letter k or the Greek letter γ (gamma). The former will be used in this book.

The specific heat at constant pressure for a mixture of gases is obtained by multiplying the specific heat of each constituent gas by the percentage weight of the gas in the mixture, adding the results and dividing the sum by 100.

67. Laws of Perfect Gases and Their Application to Air. — **Boyle's,** or **Mariotte's Law** refers to the relation existing between the pressure and volume of a gas when the temperature remains constant. The law is stated as follows: *when the temperature is constant, the volume of a given weight of a gas varies inversely as the absolute pressure.* Stated as an equation:

$$\frac{V_1}{V_2} = \frac{P_2}{P_1}, \quad \text{or} \quad P_1 V_1 = P_2 V_2 = \text{a constant} \quad \ldots \ldots \quad (17)$$

in which P_1 = initial absolute pressure, lb. per sq. ft.
P_2 = final absolute pressure, lb. per sq. ft.
V_1 = initial volume of a given weight of gas, cu. ft.
V_2 = final volume of a given weight of gas, cu. ft.

When a change occurs in a gas, such that the product of pressure and volume is constant with the temperature also remaining constant, the change is called **isothermal.**

Example 2. — Two hundred cubic feet of air at atmospheric pressure is compressed at constant temperature until the final pressure is 80 lb. per sq. in., gage. What volume will it occupy?

Solution. — Using Equation (17) and making proper substitutions:

$$P_1 V_1 = P_2 V_2, \quad \text{or} \quad V_2 = V_1 \frac{P_1}{P_2} = \frac{200 \times 14.7 \times 144}{(80 + 14.7) \times 144} = 31 \text{ cu. ft.}$$

$P_1 = 14.7 \times 144;\quad P_2 = (80 + 14.7) \times 144;\quad V_1 = 200 \text{ cu. ft.};\quad V_2 = \text{required}.$

Charles', or **Gay-Lussac's Law** refers to the relation between pressure, volume and temperature of a gas when either the pressure or the volume remains constant during the change in absolute temperature. It may be stated thus: *upon the addition of heat to a gas, the volume of a given weight varies directly as its absolute temperature when the pressure remains constant; or the pressure of a given weight of a gas varies directly as the absolute temperature when the volume remains constant.* Stated as an equation:

$$\frac{V_1}{V_2} = \frac{T_1}{T_2} \quad \text{or} \quad \frac{P_1}{P_2} = \frac{T_1}{T_2} \quad \ldots \ldots \ldots (18)$$

in which T_1 = initial absolute temperature, deg. fahr.
T_2 = final absolute temperature, deg. fahr., and other symbols as given in Equation (17).

Example 3. — One pound of air at sea level and atmospheric pressure has a volume of 12.387 cubic feet when the temperature is 32 deg. fahr. Find its volume at 62 deg. fahr. and atmospheric pressure.

Solution. — Using Equation (18) and making proper substitutions:

$$\frac{V_1}{T_2} = \frac{V_2}{T_2}, \quad \text{or} \quad V_2 = V_1 \frac{T_2}{T_1} = \frac{522 \times 12.387}{492} = 13.14 \text{ cu. ft.}$$

V_2 = required; V_1 = 12.387 cu. ft.
$T_1 = (32 + 460) = 492$ deg. fahr. abs.; $T_2 = (62 + 460) = 522$ deg. fahr. abs.

68. The Combined Law of Gases. — When the pressure, volume and temperature, P_1, V_1, and T_1, of a quantity of air or gas are all changed, the resulting values P_2, V_2, and T_2 may be found by the combined laws of Boyle and Charles', giving the equation:

$$\frac{P_1 V_1}{T_1} = \frac{P_2 V_2}{T_2} = \text{a constant} \ldots \ldots \ldots (19)$$

Considering, for convenience, that this action takes place in two steps:

1. With the absolute temperature T_1 constant, let the pressure be changed to P_2 according to Boyle's Law. The resulting volume V_n can then be obtained as follows:

$$\frac{P_1}{P_2} = \frac{V_n}{V_1}, \quad \text{and} \quad V_n = V_1 \frac{P_1}{P_2} \quad \ldots \ldots (A)$$

2. From the condition P_2, V_n, T_1, let the change be made to a final temperature T_2, according to Charles' Law. Then

$$\frac{V_n}{V_2} = \frac{T_1}{T_2}, \quad \text{and} \quad V_n = V_2 \frac{T_1}{T_2} \quad \ldots \ldots (B)$$

Equating the values of V_n from Equations (A) and (B):

$$\frac{P_1 V_1}{P_2} = \frac{T_1 V_2}{T_2}, \quad \text{or} \quad \frac{P_1 V_1}{T_1} = \frac{P_n V_n}{T_n} = \text{a constant}$$

If the constant per pound of air or gas is called R, the resulting equation for one pound of gas may be expressed $PV = RT$, and for any weight W of a gas

$$PV = WRT \quad \ldots \ldots \ldots \ldots (20)$$

in which P = absolute pressure, lb. per sq. ft.
  V = volume, cu. ft.
  T = absolute temperature, deg. fahr.
  R = constant for any given gas. Values of R are given in Table 4.

TABLE 4. — THERMAL AND PHYSICAL PROPERTIES OF COMMON GASES

Gas	Molecular Chemical Symbol	Specific Gravity Air = 1	Weight per Cu. Ft. Lb.*	Specific Heat		$k = \dfrac{C_p}{C_v}$	R
				Constant Pressure C_p	Constant Volume C_v		
Air...............		1.000	0.0807	.2375	.1689	1.406	53.37
Oxygen...........	O_2	1.053	.0892	.2175	.1551	1.402	48.55
Nitrogen..........	N_2	0.967	.0783	.2438	.1727	1.412	55.32
Hydrogen.........	H_2	0.069	.0056	3.409	2.412	1.413	775.66
Carbon Dioxide...	CO_2	1.529	.1227	.2169	.167	1.299	38.82
Carbon Monoxide	CO	0.967	.0781	.2450	.174	1.408	55.24
Methane..........	CH_4	0.558	.0447	.5930	.450	1.320	96.31
Ethylene..........	C_2H_4	0.967	.0780	.4040			
Ethane............	C_2H_6	1.075	.0838				
Acetylene.........	C_2H_2	0.920	.0725	.350	.270	1.280	59.37
Sulphur Dioxide .	SO_2	2.264	.1786	.154	.123	1.250	24.10
Ammonia.........	NH_3	0.682	.2156	.5202	.4011	1.297	89.36

* At 32 deg. fahr. and atmospheric pressure of 14.7 lb. per sq. in.

The numerical value of R is the number of foot-pounds of external work done when one pound of gas is raised in temperature one degree at constant pressure, and it may be obtained by calculation using: (1) the universal gas constant; (2) the constant volume, and constant pressure specific heats of the gas, as $R = 778 \, (C_p - C_v)$; or (3) the observed volumes, under standard conditions, in the equation $PV = RT$, in which V would be the specific volume of the gas.

For further study along this line the reader is referred to any of the many books on thermodynamics. (See page 113.)

Example 4. — Using the first method above, find the value of R for air, which has a molecular weight of 28.93.

Solution. — Applying Equation (20) to a unit weight of gas and multiplying both sides of the equation by the molecular weight, m, of the gas, it may be written

$$mVP = mRT$$

in which mV is the volume of one **mol** (see page 162) of the gas, and is constant for all gases when they are at the same pressure and temperature. This is according to **Avogadro's Law** *which states that equal volumes of all gases at the same pressure and temperature contain the same number of molecules.*

Since mV is constant, mR must also be constant and is known as the **universal gas constant** with a value of 1544. This value may be found by solving the above equation using oxygen, for which $mV = 358.7$ at 32 deg. fahr. and 14.7 lb. per sq. in. abs., $m = 32$, and the density = 0.0892 lb. per cu. ft. Under these conditions, then

$$mR = \frac{mVP}{T} = \frac{358.7 \times 144 \times 14.7}{492} = 1544$$

Hence the value of R for a perfect gas can be found by dividing 1544 by the molecular weight (m) of the gas. For air, therefore, $R = 1544 \div 28.93 = 53.37$.

69. Gas Expansions, and the Pressure Volume or P-V Diagram. — The expansions which are commonly dealt with in steam engineering are: (1) **Constant pressure expansion,** often called isobaric, in which sufficient heat from an outside source is added to do the work of the expansion and to raise the temperature of the gas; (2) **Isothermal expansion,** in which the temperature is maintained constant and the work of expansion is done by heat added from an external source; (3) **Adiabatic expansion,** in which heat is not added from an external source, the necessary heat to perform the work of the expansion coming from the internal energy of the gas, causing a drop in both pressure and temperature. This is a special form of the more general **polytropic expansion,** in which heat may be added during the expansion.

Graphical representations of the above expansions which show simultaneous values of pressure and volume, are called **P-V curves** and if a series of these curves are plotted in such a way that they form a diagram thus showing the changing states of the working substance, the diagram is called a **P-V diagram** or more generally an **indicator diagram,** page 361. Since the area bounded by such a diagram represents work it is useful in investigating the performance of machinery using a gas or a vapor as the working medium. The amount of work performed[1] under the various expansion curves, and the energy changes which take place are discussed in Articles 70, 71, and 72.

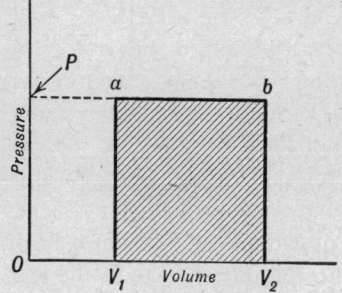

Fig. 68. — Work at Constant Pressure.

70. Work at Constant Pressure. — The work in foot-pounds done when a gas expands at constant pressure is equal to the pressure in pounds per square foot, multiplied by the change in volume in cubic feet. As an example, consider the initial volume of a gas as represented by the point a in Fig. 68, and the final volume, after the addition of heat, by point b. The work performed is represented by the shaded area and equals:

$$\text{Work in ft-lb.} = P(V_2 - V_1) \quad \ldots \ldots \quad (21)$$

in which P = pressure on the gas, lb. per sq. ft.

V_2 and V_1 = larger and smaller volume of gas respectively, cu. ft.

The change in the *internal energy* for this expansion can be obtained by using Equation (10), page 68.

---

[1] The work equations developed in Arts. 70–73 inclusive may also be used to determine the work done during the compression of a gas, but when so used the expressions will have a negative value; that is, work must be done on the gas to decrease its volume.

71. Work done during Isothermal Expansion.

— The temperature remains constant during an isothermal expansion of a perfect gas, and the expansion takes place according to Equation (17) or

$$PV = P_1V_1 = C = \text{a constant}$$

Considering P_1 and V_1 as known quantities, the work done may be expressed by Equation (23), and is represented graphically by the shaded area in Fig. 69.

$$\text{Work in ft-lb.} = W = \int_{V_1}^{V_2} P \, dV = P_1 V_1 \int_{V_1}^{V_2} \frac{dV}{V} \quad \ldots \ldots \ldots \quad (22)$$

$$= P_1 V_1 \log_e \frac{V_2}{V_1} = 2.3 \, P_1 V_1 \log_{10} \frac{V_2}{V_1} \quad \ldots \ldots \quad (23)$$

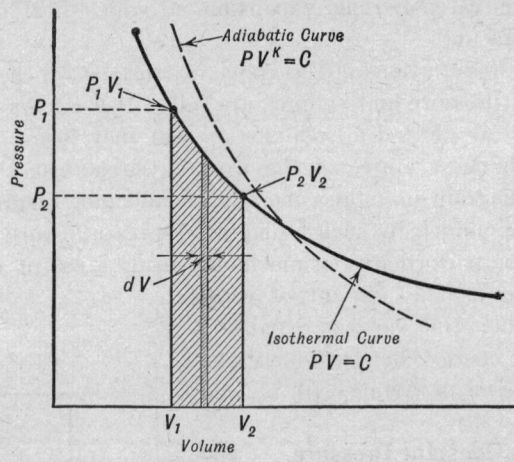

Fig. 69. — Isothermal and Adiabatic Curves, showing Work under Isothermal Curve.

in which $\log_e$ = natural logarithm = 2.3 log base 10, and P_1 and V_1 refer to the initial state of the gas, while P_2 and V_2 refer to the final state.

Since $PV = WRT$, Equation (23) may also be written

$$W = WRT \log_e \frac{V_2}{V_1} \text{ (ft-lb.)} \quad \ldots \ldots \ldots \quad (24)$$

When a gas expands isothermally, the heat added equals the work done, and there is no change in the *intrinsic energy* of the gas.

Example 5. — A cubic foot of air at a pressure of 150 lb. per sq. in. abs. expands isothermally until its pressure is 50 lb. per sq. in. abs. Calculate the work done during the expansion.

Solution. — Using Equation (23) and making proper substitutions

$$\text{Work} = P_1 V_1 \log_e \frac{V_2}{V_1} = 150 \times 144 \times 1 \times \log_e \frac{3}{1}$$
$$= 150 \times 144 \times 1 \times 1.0986$$
$$= 23{,}729 \text{ ft-lb.}$$

To find V_2 use the relationship $P_1V_1 = P_2V_2$, then

$$V_2 = \frac{P_1V_2}{P_2} = \frac{150 \times 144 \times 1}{50 \times 144} = 3 \text{ cu. ft.}$$

72. Work done during Adiabatic Expansion. — As mentioned in Art. 69, the work done by a gas expanding adiabatically is at the expense of its intrinsic energy. The relation between pressure and volume for any adiabatic expansion is written:

$$PV^k = P_1V_1^k = P_2V_2^k = C = \text{a constant} \quad \ldots \ldots \quad (25)$$

in which $k = C_p \div C_v = 1.4$ for air or other perfect gas. P and V refer to the pressure and volume for any point on the expansion curve, P_1 and V_1 refer to the initial condition, and P_2 and V_2 to the final condition of the gas.

The work performed may be expressed as follows:

$$Work\ in\ ft\text{-}lb. = W = \int_{V_1}^{V_2} P\,dV = P_1V_1^k \int_{V_1}^{V_2} \frac{1}{V^k}\,dV \quad \ldots \quad (26)$$

Integrating and simplifying

$$W = \frac{P_1V_1 - P_2V_2}{k-1},\ (\text{ft-lb.}) \ldots \ldots \ldots \quad (27)$$

Since $PV = wRT$, Equation (27) may also be written:

$$W = \frac{wR(T_1 - T_2)}{k-1},\ (\text{ft-lb.}) \ldots \ldots \ldots \quad (28)$$

It was shown in Chapter III, page 68 that the change in internal energy for a perfect gas is independent of the path, consequently a *general expression for the change in internal energy* can be obtained by considering the adiabatic.

Using Equation (9), page 68 for the energy change, with $H = 0$, there results

$$U_2 - U_1 = -W = -\int_{V_1}^{V_2} P\,dV \ldots \ldots \ldots \quad (C)$$

Substituting the value of W from Equation (27) in Equation (C), the change in the internal energy of a gas, in passing from a state P_1, V_1, T_1 to a final state P_2, V_2, T_2 is:

$$U_2 - U_1 = \frac{P_2V_2 - P_1V_1}{k-1},\ (\text{ft-lb.}) \ldots \ldots \quad (29)$$

Example 6. — Three cubic feet of air at 100 lb. per sq. in. abs. expands adiabatically to 30 lb. per sq. in. abs.

Find: (a) the final volume of the air; and (b) the amount of external work done, and the change in the internal energy of the gas, during the expansion.

Solution. — (a) Using Equation (25)

$$P_1V_1^k = P_2V_2^k$$

$$V_2^{1.4} = \frac{100 \times 144 \times 3^{1.4}}{30 \times 144} = 3.333 \times 3^{1.4}$$

$$1.4 \log_{10} V_2 = \log_{10} 3.333 + 1.4 \log_{10} 3$$

$$\log_{10} V_2 = 0.5227 + 0.6679 = \frac{1.1906}{1.4} = 0.8504$$

$$V_2 = 7.087 \text{ cu. ft.}$$

$P_1 = 100 \times 144$, $P_2 = 30 \times 144$, $V_1 = 3$, $V_2 =$ to be found, $k = 1.4$

(b) Using Equation (27)

$$W = \frac{P_1V_1 - P_2V_2}{k - 1}$$

$$= \frac{(100 \times 144 \times 3) - (30 \times 144 \times 7.087)}{1.4 - 1} = \frac{12,580}{0.4}$$

$$= 31,450 \text{ ft-lb.}$$

Values of P_1, P_2, V_1, V_2, and k taken from (a) above.
The change in internal energy equals the work done = 31,450 ft-lb.

Relation between Volume, Pressure and Temperature for Adiabatic Expansion. The values of P, V, and T vary during an adiabatic expansion and the relation existing between them can be found by using the following equations:

$$P_1V_1^k = P_2V_2^k \quad \cdots\cdots\cdots\cdots \text{(D)}$$

$$\frac{P_1V_1}{T_1} = \frac{P_2V_2}{T_2} \quad \cdots\cdots\cdots\cdots \text{(E)}$$

By using Equations (D) and (E), the final condition of pressure, volume and temperature can always be obtained when two initial conditions and one final condition are given.

For example, to find P_1, having given V_1, P_2, and V_2.
Using Equation (D) thus:

$$\frac{P_1}{P_2} = \left(\frac{V_2}{V_1}\right)^k \quad \cdots\cdots\cdots\cdots \text{(30)}$$

or suppose it is desired to find T_1, having given V_2, V_1, and T_2, or P_1, P_2, and T_2, combining the Equations (D) and (E) given above

$$\frac{T_1}{T_2} = \left(\frac{V_2}{V_1}\right)^{k-1} = \left(\frac{P_1}{P_2}\right)^{\frac{k-1}{k}} \quad \cdots\cdots\cdots\cdots \text{(31)}$$

The equations developed in Art. 72 can be used for a *polytropic expansion*, which is represented by the equation, $PV^n =$ a constant, provided k in the various equations is replaced by n, which ordinarily has a value between unity and k.

Example 7. — Three cubic feet of air in a tank, at a pressure of 120 lb. per sq. in. abs., and a temperature of 70 deg. fahr., are expanded adiabatically to 50 lb. per sq. in. abs. Calculate the temperature at end of expansion.

Solution. — Using Equation (31)

$$\frac{T_1}{T_2} = \left(\frac{P_1}{P_2}\right)^{\frac{k-1}{k}}$$

Rearranging and solving

$$T_2 = \frac{T_1}{\left(\frac{P_1}{P_2}\right)^{\frac{k-1}{k}}} = \frac{530}{\left(\frac{120 \times 144}{50 \times 144}\right)^{\frac{1.4-1}{1.4}}} = \frac{530}{2.4^{0.286}}$$

$\log_{10} T_2 = \log_{10} 530 - 0.286 \log_{10} 2.4 = 2.6165$

$T_2 = 412.6$ deg. absolute, or -47.4 deg. fahr.

$T_1 = 460 + 70,$ $\quad P_1 = 120 \times 144,$ $\quad k = 1.4$

$T_2 =$ to be found, $\quad P_2 = 50 \times 144$

73. Water. — Pure water is a chemical combination of two elements, hydrogen and oxygen. These elements when combining chemically always do so in the proportion of two parts, by volume, of hydrogen to one part, by volume, of oxygen. If the two gases are mixed while cold in the above proportions, the mixture is purely a mechanical one until, by the influence of heat, they combine chemically. If the union of the two gases is brought about in a vessel so arranged that the resulting water is maintained at a high temperature, it will retain its gaseous condition and will form two volumes of steam. Conversely, two volumes of steam may be dissociated by the application of heat into its constituent elements, two volumes of hydrogen and one volume of oxygen.

Water has been universally adopted as the standard by which the relative weights of other liquids and solids are determined, this relation being expressed by the term **specific gravity.** The specific gravity of any body indicates its weight as compared with the weight of an equal volume of pure water. The volume and weight, per cubic foot, of water changes with the temperature, as shown in Table 5. Its density is 62.427 pounds per cubic foot at 39.2 deg. fahr., the point of maximum density.

A United States standard gallon has a volume of 231 cubic inches, and a weight of $8\frac{1}{3}$ pounds at 60 deg. fahr.

Water is but slightly compressible, and for all practical purposes may be considered non-compressible. Its boiling-point varies with the pressure; the higher the pressure, the higher the boiling-point. See Table 7, page 94.

The specific heat of water varies with the temperature, as shown in Table 6.

Because of this variation in specific heat, the heat required to raise one pound of water through a given temperature range, known as the **heat of liquid,** will depend upon the mean value of the specific heat over this range. For ordinary calculations the specific heat of water may be taken as unity.

Water as found in nature is never pure, being always more or less contaminated by impurities, which often have serious effects when the water is fed into boilers. The impurities most commonly found in water are earthy

PROPERTIES OF AIR, GASES, WATER AND STEAM

TABLE 5.—VOLUME AND WEIGHT OF WATER AT VARIOUS TEMPERATURES

(GOODENOUGH)

Temperature ° F.	Volume Cu. Ft. per Lb.	Weight Lb. per Cu. Ft.	Temperature ° F.	Volume Cu. Ft. per Lb.	Weight Lb. per Cu. Ft.	Temperature ° F.	Volume Cu. Ft. per Lb.	Weight Lb. per Cu. Ft.
32	0.01602	62.42	210	0.01670	59.88	390	0.0186	53.84
39.2	.01602	62.43	212	.01672	59.83	400	.0187	53.42
40	.01602	62.43	220	.01677	59.63	410	.0189	52.99
50	.01602	62.42	230	.01684	59.37	420	.0190	52.55
60	0.01603	62.37	240	0.01692	59.11	430	0.0192	52.11
70	.01605	62.30	250	.01700	58.83	440	.0194	51.66
80	.01607	62.22	260	.01708	58.55	450	.0195	51.20
90	.01610	62.11	270	.01716	58.26	460	.0197	50.70
100	.01613	62.00	280	.01725	57.96	470	.0199	50.20
110	0.01616	61.87	290	0.01735	57.65	480	0.0201	49.70
120	.01620	61.71	300	.01745	57.32	490	.0203	49.20
130	.01625	61.55	310	.01755	56.98	500	.0205	48.70
140	.01629	61.38	320	.01766	56.62	510	.0208	48.20
150	.01634	61.20	330	.01778	56.24	520	.0210	47.60
160	0.01639	61.00	340	0.01790	55.85	530	0.0212	47.10
170	.01645	60.80	350	.01803	55.46	540	.0215	46.50
180	.01651	60.58	360	.01816	55.06	550	.0218	45.90
190	.01657	60.36	370	.01829	54.66	560	.0221	45.20
200	.01663	60.12	380	.01843	54.24	600	.0235	42.60

matter, bi-carbonates of lime and magnesia, iron, sulphate of lime, chlorides and sulphates of magnesia, carbonate of soda in large amounts, acids, dissolved carbonic acid and oxygen, grease and organic matter. The bi-carbonates usually cause "**temporary**" hardness which can be removed by

TABLE 6. — SPECIFIC HEAT OF WATER

Temperature ° F.	Specific Heat
30	1.0098
55	1.0000
100	0.9967
160	1.0002
210	1.0050

boiling, and the sulphates produce "**permanent**" hardness which can only be removed by chemical treatment. *The degree of hardness is usually stated in grains of solids per U. S. gallon, or in parts per million.* Water containing 1 to 10 grains would be **soft**; 10 to 20 grains, **moderately hard**; and above 25 grains, **very hard**. Hardness, when stated in parts per million, may be converted to grains per U. S. gallon by dividing by 11.7. For a discussion of the effects of these impurities and methods of removing them, consult Chapter XI, page 263.

FORMATION OF STEAM 91

74. Steam. — All substances exist in either a solid, a liquid or a gaseous form, depending upon the pressure and temperature to which they are subjected. For instance, water under an atmospheric pressure of 14.7 pounds per square inch absolute, is a solid for temperatures below 32 deg. fahr., a liquid for temperatures between 32 and 212 deg. fahr., and a vapor at 212 deg. fahr.; upon further rise in temperature it approaches a gaseous state. *Steam is water vapor*[1] *and as such does not obey the laws of perfect gases.* It exists as a vapor because sufficient heat has been added to the water from which it is formed to supply the latent heat of evaporation and change the liquid to a vapor. *This change takes place at a definite and constant temperature which depends solely upon the pressure under which the change takes place.*

75. Formation of Steam. — The effect of heat, when applied to water in a boiler is best explained by considering 1 pound of water at 32 deg. fahr. as enclosed in a vertical cylinder having a cross-sectional area of 1 square inch. The pressure of the atmosphere at sea level may, for convenience, be considered as replaced by a frictionless piston weighing 14.7 pounds.

If heat is now applied to the cylinder, the rate of vibration of the molecules of the water will be increased, and the resulting increase in kinetic energy will be shown by a rise in temperature of 1 deg. fahr. for each B.t.u. added to the water, considering the specific heat of water as unity. This increase in temperature will continue until a point is reached at which the attraction between the molecules as a liquid is broken down and boiling, or evaporation, takes place. The point at which this occurs is called the **boiling-point,** and the amount of heat added between 32 deg. fahr. and the boiling-point is known as the **heat of liquid,** or sensible heat.

When the boiling-point is reached, bubbles of steam form at the point of application of heat, and, being lighter than water, rise to the surface and discharge the steam which they contain. The temperature at which boiling takes place depends entirely upon the pressure. In the case being considered, boiling occurs at 212 deg. fahr.; while if the pressure were increased to 200 pounds per square inch absolute, boiling would occur at 381.8 deg. fahr., and if the pressure were lowered to, say, one pound per square inch absolute, the water would boil at 101.8 deg. fahr. *The temperature of the steam formed is in each case the same as that of the water from which it is formed.* Steam at the temperature of evaporation is said to be **saturated.**

After the boiling-point is reached, the temperature remains constant until the entire pound of water is evaporated. During this time, the heat that is added changes the physical state of the water, and the quantity of heat required to bring about the change is called the **latent heat of evaporation.** While evaporation is taking place, the volume of the cylinder occupied by water and steam is increased from the volume occupied by the water alone

---

[1] When substances are changed from a liquid or solid to the gaseous state, the gas first formed does not obey the laws of perfect gases and when in this condition is called a *vapor.*

to that occupied by a pound of saturated steam, thus lifting the piston and performing work. After the entire pound of water has been converted into steam, the addition of more heat to the steam will increase the temperature above the temperature of evaporation and the steam will become **superheated.** At the same time, the volume is increased above that of saturated steam.

The changes taking place during the formation of steam may be illustrated by a diagram, as in Fig. 70, in which the line AB represents the change in the heat of liquid with the change in temperature; BC represents by a horizontal line the latent heat of evaporation and CD the change in the heat required to superheat the steam to some temperature t_s above that of evaporation, t_v.

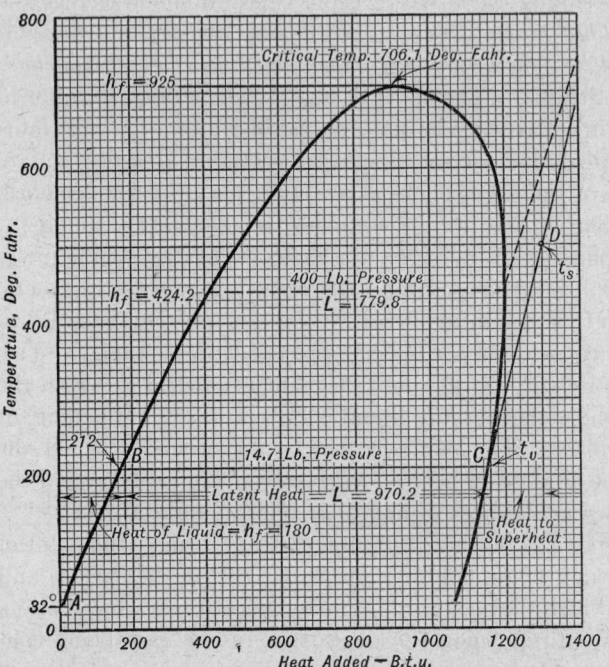

Fig. 70. — Illustrating Changes taking place during Formation of a Pound of Steam.

Several pressures are shown in order to indicate the changes that occur in the heat of liquid and the latent heat of vaporization as the pressure increases. When the pressure is raised sufficiently high a point is reached at about 3226 pounds per square inch absolute above which no pressure, however great, is sufficient to cause the vapor to liquefy. The temperature corresponding to this point is 706.1 deg. fahr. and is known as the **critical temperature of steam.** The **critical pressure** may be defined as the vapor pressure of water at the critical temperature. *At this pressure the density of the water is equal to the density of the steam, and there is no difference between the two states, water and steam.* Thus, if water be heated to the critical temperature while at the critical pressure, it changes suddenly into steam without further application of the so-called latent heat.

76. Dry and Saturated Steam. — In the previous article, *saturated steam was defined as steam at the temperature of evaporation corresponding to the* *pressure.* Saturated steam may be wet or dry, depending on the conditions accompanying evaporation. *A pound of steam existing at the temperature and pressure of evaporation is dry and saturated if all the water from which it is formed has been evaporated;* when in this condition steam is invisible. If the entire pound of water from which the steam is formed is not evaporated, the steam will be **wet;** that is, will contain small particles of water in suspension which have not been evaporated.

77. Steam Tables. — To solve problems pertaining to equipment using steam, it is necessary to know the exact amount of heat supplied per pound of steam used; in addition, the volume of the steam is often desired. The various quantities mentioned in Art. 75 are called the **properties of steam,** and are tabulated in **Steam Tables,** arranged to give values for either saturated or superheated steam. The steam tables are based upon the following considerations: (1) all quantities of heat are given per pound of dry saturated steam, or per pound of superheated steam, as the case may be; (2) all heat quantities are stated in British thermal units above 32 deg. fahr.; and (3) all pressures are expressed as absolute pressure in pounds per square inch when above atmospheric pressure, and in pounds per square inch or inches of mercury when below atmospheric pressure.

78. Properties of Steam. — The numerical values of the properties of steam are the results of experimental data obtained by scientists, and compiled by themselves or by others who have studied their data. A thorough understanding of the use of the steam tables is necessary in order to make the calculations involved in the heat interchanges taking place in all apparatus using steam. The following properties of steam are those given in the **saturated steam table,** on the following pages:

p = steam pressure in lb. per sq. in. absolute.

t_v = temperature in deg. fahr. corresponding to the absolute pressure.

v_g = specific volume of dry steam, cu. ft. per lb.

$\dfrac{1}{v_g}$ = density in lb. per cu. ft.

h_f = heat of liquid, B.t.u. above 32 deg. fahr.

L or h_{fg} = latent heat of evaporation, B.t.u. per lb. of dry steam.

h_g = total heat, heat content, or enthalpy, B.t.u. per lb. of dry steam.

ρ = internal latent heat of evap.,[1] B.t.u. per lb. = $L - 144\,Apu$.

E = internal energy of steam,[1] B.t.u. per lb. = $\rho + h$.

s_f = entropy of water.

$\dfrac{L}{T}$ or s_{fg} = entropy of evaporation.

s_g = entropy of steam.

v_f = volume of one pound of water, cu. ft.

[1] The numerical values of ρ and E are not given in the steam table on page 94.

TABLE 7. — PROPERTIES OF DRY AND SATURATED STEAM
(Prepared by Prof. J. H. Keenan)

Pressure		Temperature	Specific Volume, Cu. Ft. per Lb.		Heat of Liquid, B.t.u.	Latent Heat of Evaporation, B.t.u.	Total Heat of Steam, B.t.u.	Entropy		
Inch of Mercury	Lb. per Sq. In.		Steam	Liquid				of Liquid	of Evaporation	of Steam
*	p	t_v	v_g	v_f	h_f	L or h_{fg}	$h_g = L + h_f$	s_f	s_{fg} or L/T	s_g
	0.0886	32	32.94	0.01602	0	1073.4	1073.4	0	2.1832	2.1832
½	0.295	58.83	1256.9	0.01603	26.88	1058.8	1085.7	0.0533	2.0422	2.0955
1	0.491	79.06	652.7	0.01607	47.06	1047.8	1094.9	0.0914	1.9451	2.0365
	0.5	79.68	643.2	0.01607	47.72	1046.9	1094.6	0.0926	1.9415	2.0339
2	0.982	101.17	339.5	0.01613	69.10	1035.7	1104.8	0.1316	1.8468	1.9784
	1	101.76	334.0	0.01614	69.69	1035.3	1105.0	0.1326	1.8442	1.9769
3	1.473	115.08	231.8	0.01618	82.96	1027.9	1110.8	0.1561	1.7885	1.9446
	2	126.10	174.0	0.01623	93.97	1021.6	1115.6	0.1750	1.7442	1.9192
	3	141.49	118.9	0.01630	109.33	1012.7	1122.0	0.2009	1.6847	1.8856
	4	152.99	90.79	0.01636	120.83	1005.9	1126.8	0.2198	1.6420	1.8618
	5	162.25	73.63	0.01641	130.10	1000.4	1130.6	0.2348	1.6088	1.8435
	6	170.07	62.09	0.01645	137.92	995.8	1133.7	0.2473	1.5814	1.8287
	7	176.85	53.74	0.01649	144.71	991.7	1136.4	0.2580	1.5582	1.8162
	8	182.87	47.42	0.01652	150.75	988.1	1138.9	0.2674	1.5379	1.8053
	9	188.28	42.46	0.01656	156.19	984.8	1141.0	0.2758	1.5200	1.7958
	10	193.21	38.46	0.01658	161.13	981.8	1143.0	0.2834	1.5040	1.7874
	11	197.75	35.18	0.01661	165.68	979.1	1144.8	0.2903	1.4894	1.7797
	12	201.96	32.43	0.01664	169.91	976.5	1146.4	0.2968	1.4760	1.7727
	13	205.88	30.09	0.01666	173.85	974.1	1147.9	0.3027	1.4636	1.7663
	14	209.56	28.08	0.01669	177.55	971.8	1149.3	0.3082	1.4521	1.7604
	14.7	212.00	26.83	0.01670	180.00	970.2	1150.2	0.3119	1.4446	1.7564
	15	213.03	26.32	0.01671	181.04	969.6	1150.6	0.3134	1.4414	1.7548
	16	216.32	24.78	0.01673	184.35	967.4	1151.8	0.3184	1.4312	1.7496
	17	219.43	23.41	0.01676	187.48	965.4	1152.9	0.3230	1.4218	1.7448
	18	222.40	22.18	0.01678	190.48	963.5	1154.0	0.3274	1.4127	1.7402
	19	225.23	21.09	0.01680	193.34	961.7	1155.0	0.3316	1.4042	1.7358
	20	227.96	20.10	0.01682	196.09	959.9	1156.0	0.3356	1.3960	1.7317
	21	230.56	19.20	0.01684	198.72	958.2	1156.9	0.3395	1.3883	1.7278
	22	233.07	18.38	0.01685	201.25	956.6	1157.8	0.3431	1.3809	1.7240
	23	235.49	17.64	0.01687	203.70	955.0	1158.6	0.3466	1.3738	1.7204
	24	237.82	16.94	0.01689	206.05	953.4	1159.5	0.3500	1.3670	1.7170
	25	240.07	16.31	0.01690	208.33	951.9	1160.2	0.3533	1.3604	1.7137
	26	242.25	15.72	0.01692	210.54	950.4	1161.0	0.3564	1.3542	1.7106
	27	244.36	15.17	0.01694	212.67	949.0	1161.7	0.3594	1.3481	1.7075
	28	246.41	14.66	0.01695	214.75	947.7	1162.4	0.3624	1.3422	1.7046
	29	248.40	14.19	0.01697	216.77	946.3	1163.1	0.3652	1.3365	1.7018
	30	250.34	13.75	0.01698	218.73	945.0	1163.7	0.3680	1.3310	1.6990
	31	252.22	13.34	0.01700	220.64	943.7	1164.4	0.3707	1.3257	1.6964
	32	254.05	12.94	0.01701	222.54	942.5	1165.0	0.3732	1.3206	1.6938
	33	255.84	12.58	0.01703	224.32	941.2	1165.6	0.3758	1.3156	1.6914
	34	257.58	12.23	0.01704	226.09	940.0	1166.1	0.3783	1.3107	1.6890
	35	259.28	11.90	0.01706	227.82	938.9	1166.7	0.3807	1.3060	1.6866
	36	260.94	11.59	0.01707	229.51	937.7	1167.2	0.3830	1.3014	1.6844
	37	262.57	11.30	0.01708	231.17	936.6	1167.8	0.3853	1.2968	1.6822
	38	264.16	11.02	0.01710	232.79	935.5	1168.3	0.3876	1.2925	1.6800
	39	265.72	10.76	0.01711	234.38	934.4	1168.8	0.3898	1.2882	1.6780
	40	267.24	10.50	0.01712	235.93	933.3	1169.2	0.3919	1.2840	1.6759
	41	268.74	10.26	0.01713	237.45	932.3	1169.7	0.3940	1.2799	1.6739
	42	270.21	10.03	0.01715	238.96	931.2	1170.2	0.3961	1.2759	1.6720
	43	271.65	9.814	0.01716	240.42	930.2	1170.6	0.3981	1.2720	1.6701
	44	273.06	9.605	0.01717	241.86	929.2	1171.1	0.4000	1.2682	1.6683
	45	274.45	9.404	0.01718	243.28	928.2	1171.5	0.4020	1.2645	1.6665
	46	275.81	9.212	0.01719	244.67	927.2	1171.9	0.4039	1.2608	1.6647
	47	277.14	9.028	0.01720	246.03	926.3	1172.3	0.4057	1.2572	1.6630
	48	278.45	8.851	0.01722	247.37	925.4	1172.7	0.4076	1.2537	1.6613
	49	279.74	8.681	0.01723	248.68	924.4	1173.1	0.4093	1.2503	1.6596

* Symbols recommended by American Standards Committee.

TABLE 7. — Continued. PROPERTIES OF DRY AND SATURATED STEAM
(Prepared by Prof. J. H. Keenan)

Pressure		Temperature	Specific Volume, Cu. Ft. per Lb.		Heat of Liquid, B.t.u.	Latent Heat of Evaporation, B.t.u.	Total Heat of Steam, B.t.u.	Entropy		
Inch of Mercury	Lb. per Sq. In.		Steam	Liquid				of Liquid	of Evaporation	of Steam
*	p	t_v	v_g	v_f	h_f	L or h_{fg}	$h_g = L + h_f$	s_f	s_{fg} or L/T	s_g
	50	281.01	8.517	0.01724	249.98	923.5	1173.5	0.4111	1.2469	1.6580
	51	282.26	8.360	0.01725	251.26	922.6	1173.9	0.4128	1.2436	1.6564
	52	283.49	8.209	0.01726	252.52	921.7	1174.3	0.4145	1.2404	1.6549
	53	284.70	8.063	0.01727	253.76	920.9	1174.6	0.4162	1.2372	1.6533
	54	285.90	7.922	0.01728	254.99	920.0	1175.0	0.4178	1.2340	1.6518
	55	287.07	7.786	0.01729	256.19	919.1	1175.3	0.4194	1.2309	1.6504
	56	288.23	7.655	0.01730	257.38	918.3	1175.7	0.4210	1.2279	1.6489
	57	289.37	7.528	0.01732	258.55	917.5	1176.0	0.4226	1.2249	1.6475
	58	290.50	7.406	0.01733	259.71	916.6	1176.4	0.4241	1.2220	1.6461
	59	291.62	7.287	0.01734	260.86	915.8	1176.7	0.4256	1.2191	1.6447
	60	292.71	7.172	0.01735	261.98	915.0	1177.0	0.4271	1.2162	1.6434
	61	293.79	7.061	0.01736	263.09	914.2	1177.3	0.4286	1.2134	1.6420
	62	294.85	6.954	0.01737	264.18	913.4	1177.6	0.4300	1.2107	1.6407
	63	295.91	6.850	0.01738	265.27	912.7	1177.9	0.4315	1.2080	1.6394
	64	296.94	6.750	0.01739	266.33	911.9	1178.2	0.4329	1.2053	1.6382
	65	297.97	6.652	0.01740	267.39	911.1	1178.5	0.4343	1.2026	1.6369
	66	298.98	6.556	0.01741	268.43	910.4	1178.8	0.4356	1.2001	1.6357
	67	299.99	6.464	0.01742	269.47	909.6	1179.1	0.4370	1.1975	1.6345
	68	300.98	6.375	0.01743	270.49	908.9	1179.4	0.4384	1.1950	1.6333
	69	301.96	6.288	0.01744	271.50	908.2	1179.6	0.4397	1.1924	1.6321
	70	302.92	6.203	0.01745	272.49	907.4	1179.9	0.4410	1.1900	1.6310
	71	303.88	6.121	0.01746	273.48	906.7	1180.2	0.4423	1.1876	1.6298
	72	304.82	6.041	0.01746	274.45	906.0	1180.5	0.4435	1.1852	1.6287
	73	305.76	5.963	0.01747	275.42	905.3	1180.7	0.4448	1.1828	1.6276
	74	306.68	5.887	0.01748	276.37	904.6	1181.0	0.4460	1.1805	1.6265
	75	307.60	5.813	0.01749	277.32	903.9	1181.2	0.4473	1.1782	1.6254
	76	308.50	5.741	0.01750	278.25	903.2	1181.5	0.4485	1.1759	1.6244
	77	309.39	5.671	0.01751	279.17	902.6	1181.7	0.4497	1.1736	1.6233
	78	310.28	5.602	0.01752	280.09	901.9	1182.0	0.4509	1.1714	1.6223
	79	311.16	5.535	0.01753	281.00	901.2	1182.2	0.4520	1.1692	1.6212
	80	312.03	5.470	0.01754	281.90	900.5	1182.4	0.4532	1.1670	1.6202
	81	312.88	5.406	0.01755	282.79	899.9	1182.7	0.4544	1.1649	1.6192
	82	313.74	5.343	0.01756	283.67	899.2	1182.9	0.4555	1.1627	1.6182
	83	314.58	5.282	0.01756	284.55	898.6	1183.1	0.4566	1.1606	1.6173
	84	315.42	5.222	0.01757	285.42	897.9	1183.4	0.4578	1.1586	1.6163
	85	316.25	5.164	0.01758	286.28	897.3	1183.6	0.4589	1.1565	1.6154
	86	317.06	5.107	0.01759	287.13	896.7	1183.8	0.4599	1.1545	1.6144
	87	317.88	5.051	0.01760	287.97	896.0	1184.0	0.4610	1.1524	1.6135
	88	318.68	4.997	0.01761	288.80	895.4	1184.2	0.4621	1.1505	1.6126
	89	319.48	4.944	0.01762	289.63	894.8	1184.4	0.4632	1.1485	1.6116
	90	320.27	4.892	0.01763	290.45	894.2	1184.6	0.4642	1.1465	1.6107
	91	321.05	4.841	0.01764	291.26	893.6	1184.8	0.4652	1.1446	1.6099
	92	321.83	4.791	0.01764	292.07	893.0	1185.0	0.4663	1.1427	1.6090
	93	322.60	4.742	0.01765	292.87	892.4	1185.2	0.4673	1.1408	1.6081
	94	323.37	4.694	0.01766	293.67	891.8	1185.4	0.4683	1.1389	1.6072
	95	324.13	4.647	0.01767	294.47	891.2	1185.6	0.4693	1.1370	1.6064
	96	324.88	4.602	0.01768	295.25	890.6	1185.8	0.4703	1.1352	1.6055
	97	325.62	4.557	0.01768	296.02	890.0	1186.0	0.4713	1.1334	1.6047
	98	326.37	4.512	0.01769	296.80	889.4	1186.2	0.4723	1.1316	1.6038
	99	327.10	4.468	0.01770	297.57	888.8	1186.4	0.4732	1.1298	1.6030
	100	327.83	4.426	0.01771	298.33	888.2	1186.6	0.4742	1.1280	1.6022
	102	329.27	4.344	0.01773	299.83	887.1	1186.9	0.4761	1.1245	1.6006
	104	330.68	4.265	0.01774	301.30	886.0	1187.3	0.4779	1.1211	1.5990
	106	332.08	4.189	0.01776	302.76	884.9	1187.6	0.4798	1.1177	1.5974
	108	333.44	4.115	0.01777	304.19	883.8	1188.0	0.4816	1.1144	1.5959
	110	334.79	4.044	0.01779	305.61	882.7	1188.3	0.4834	1.1111	1.5944
	112	336.12	3.976	0.01780	307.00	881.6	1188.6	0.4851	1.1079	1.5930
	114	337.43	3.910	0.01782	308.36	880.6	1188.9	0.4868	1.1048	1.5915
	116	338.72	3.846	0.01783	309.71	879.5	1189.2	0.4885	1.1017	1.5901
	118	340.01	3.784	0.01785	311.05	378.5	1189.5	0.4901	1.0986	1.5887

96   PROPERTIES OF AIR, GASES, WATER AND STEAM

TABLE 7. — *Continued.*   PROPERTIES OF DRY AND SATURATED STEAM
(Prepared by Prof. J. H. Keenan)

Pressure		Tempera-ture	Specific Volume, Cu. Ft. per Lb.		Heat of Liquid, B.t.u.	Latent Heat of Evapo-ration, B.t.u.	Total Heat of Steam, B.t.u.	Entropy		
Inch of Mercury	Lb. per Sq. In.		Steam	Liquid				of Liquid	of Evapo-ration	of Steam
*	p	t_v	v_g	v_f	h_f	L or h_{fg}	$h_g = L + h_f$	s_f	s_{fg} or L/T	s_g
	120	341.26	3.725	0.01786	312.37	877.4	1189.8	0.4918	1.0956	1.5874
	122	342.50	3.670	0.01788	313.67	876.4	1190.1	0.4934	1.0926	1.5860
	124	343.73	3.615	0.01789	314.96	875.4	1190.4	0.4950	1.0897	1.5847
	126	344.94	3.560	0.01791	316.23	874.4	1190.6	0.4965	1.0868	1.5834
	128	346.14	3.505	0.01792	317.49	873.4	1190.9	0.4981	1.0840	1.5821
	130	347.31	3.451	0.01794	318.73	872.4	1191.2	0.4996	1.0812	1.5808
	132	348.48	3.401	0.01795	319.95	871.5	1191.4	0.5011	1.0784	1.5796
	134	349.64	3.353	0.01796	321.16	870.5	1191.7	0.5026	1.0757	1.5783
	136	350.78	3.306	0.01798	322.37	869.6	1191.9	0.5041	1.0730	1.5771
	138	351.91	3.260	0.01799	323.56	868.6	1192.2	0.5056	1.0703	1.5759
	140	353.03	3.216	0.01801	324.74	867.7	1192.4	0.5070	1.0677	1.5747
	142	354.14	3.173	0.01802	325.91	866.7	1192.6	0.5084	1.0651	1.5735
	144	355.22	3.130	0.01804	327.06	865.8	1192.9	0.5098	1.0625	1.5724
	146	356.31	3.089	0.01805	328.20	864.9	1193.1	0.5112	1.0600	1.5712
	148	357.37	3.049	0.01806	329.32	864.0	1193.3	0.5126	1.0575	1.5701
	150	358.43	3.010	0.01808	330.44	863.1	1193.5	0.5140	1.0550	1.5690
	152	359.47	2.972	0.01809	331.54	862.2	1193.7	0.5153	1.0526	1.5679
	154	360.51	2.935	0.01810	332.64	861.3	1193.9	0.5166	1.0502	1.5668
	156	361.53	2.900	0.01812	333.72	860.4	1194.1	0.5180	1.0478	1.5658
	158	362.54	2.864	0.01813	334.80	859.5	1194.3	0.5193	1.0454	1.5647
	160	363.55	2.830	0.01814	335.86	858.7	1194.5	0.5205	1.0431	1.5636
	162	364.54	2.797	0.01816	336.91	857.8	1194.7	0.5218	1.0408	1.5626
	164	365.52	2.764	0.01817	337.95	857.0	1194.9	0.5230	1.0385	1.5616
	166	366.50	2.733	0.01818	338.99	856.1	1195.1	0.5243	1.0363	1.5606
	168	367.46	2.701	0.01819	340.01	855.2	1195.3	0.5255	1.0340	1.5597
	170	368.42	2.671	0.01821	341.03	854.4	1195.4	0.5268	1.0318	1.5586
	172	369.37	2.641	0.01822	342.04	853.6	1195.6	0.5280	1.0296	1.5576
	174	370.31	2.612	0.01823	343.04	852.7	1195.8	0.5292	1.0275	1.5566
	176	371.24	2.584	0.01825	344.03	851.9	1196.0	0.5304	1.0253	1.5557
	178	372.16	2.556	0.01826	345.01	851.1	1196.1	0.5315	1.0232	1.5548
	180	373.08	2.529	0.01827	345.99	850.3	1196.3	0.5327	1.0211	1.5538
	182	374.00	2.502	0.01828	346.97	849.5	1196.4	0.5339	1.0190	1.5529
	184	374.90	2.476	0.01829	347.94	848.6	1196.6	0.5350	1.0169	1.5520
	186	375.78	2.451	0.01831	348.89	847.9	1196.8	0.5362	1.0149	1.5511
	188	376.67	2.425	0.01832	349.83	847.1	1196.9	0.5373	1.0129	1.5501
	190	377.55	2.401	0.01833	350.77	846.3	1197.0	0.5384	1.0109	1.5493
	192	378.42	2.377	0.01834	351.70	845.5	1197.2	0.5395	1.0089	1.5484
	194	379.27	2.353	0.01835	352.61	844.7	1197.3	0.5406	1.0070	1.5475
	196	380.13	2.330	0.01837	353.53	844.0	1197.5	0.5417	1.0050	1.5467
	198	380.97	2.307	0.01838	354.43	843.2	1197.6	0.5427	1.0031	1.5458
	200	381.81	2.285	0.01839	355.33	842.4	1197.7	0.5438	1.0012	1.5450
	205	383.89	2.231	0.01842	357.56	840.5	1198.1	0.5465	0.9964	1.5329
	210	385.93	2.180	0.01844	359.76	838.6	1198.4	0.5491	0.9918	1.5409
	215	387.93	2.131	0.01847	361.91	836.8	1198.7	0.5516	0.9873	1.5389
	220	389.89	2.084	0.01850	364.02	834.0	1199.0	0.5541	0.9829	1.5369
	225	391.81	2.039	0.01853	366.10	833.2	1199.3	0.5565	0.9786	1.5350
	230	393.70	1.997	0.01856	368.14	831.4	1199.6	0.5589	0.9743	1.5332
	235	395.56	1.955	0.01859	370.15	829.7	1199.8	0.5612	0.9702	1.5313
	240	397.40	1.916	0.01861	372.13	827.9	1200.1	0.5635	0.9661	1.5295
	245	399.20	1.878	0.01864	374.09	826.2	1200.3	0.5658	0.9620	1.5278
	250	400.97	1.841	0.01867	376.02	824.5	1200.5	0.5680	0.9581	1.5261
	260	404.43	1.772	0.01872	379.78	821.2	1201.0	0.5723	0.9504	1.5227
	270	407.70	1.708	0.01877	383.44	818.0	1201.4	0.5765	0.9430	1.5194
	280	411.06	1.649	0.01882	387.02	814.7	1201.8	0.5805	0.9357	1.5163
	290	414.24	1.593	0.01887	390.50	811.6	1202.1	0.5845	0.9287	1.5132
	300	417.33	1.541	0.01892	393.90	808.5	1202.4	0.5883	0.9220	1.5102
	350	431.71	1.325	0.01914	409.81	793.7	1203.6	0.6061	0.8905	1.4966
	400	444.58	1.160	0.01936	424.22	779.8	1204.1	0.6218	0.8625	1.4843
	450	456.27	1.030	0.01956	437.4	766.7	1204.1	0.6361	0.8371	1.4732
	500	466.99	0.9261	0.01976	449.7	754.0	1203.7	0.6493	0.8137	1.4630

TABLE 7. — Concluded. PROPERTIES OF DRY AND SATURATED STEAM
(Prepared by Prof. J. H. Keenan)

Pressure		Temperature	Specific Volume, Cu. Ft. per Lb.		Heat of Liquid, B.t.u.	Latent Heat of Evaporation, B.t.u.	Total Heat of Steam, B.t.u.	Entropy		
Inch of Mercury	Lb. per Sq. In.		Steam	Liquid				of Liquid	of Evaporation	of Steam
*	p	t_v	v_g	v_f	h_f	L or h_{fg}	$h_g = L + h_f$	s_f	s_{fg} or L/T	s_g
	550	476.90	0.8400	0.01996	461.3	741.7	1203.0	0.6616	0.7920	1.4535
	600	486.17	0.7677	0.02016	472.3	729.8	1202.1	0.6731	0.7716	1.4447
	650	494.84	0.7061	0.02036	482.8	718.2	1201.0	0.6840	0.7524	1.4364
	700	503.04	0.6527	0.02056	492.9	706.8	1199.7	0.6943	0.7342	1.4285
	750	510.79	0.6064	0.02075	502.5	695.7	1198.3	0.7041	0.7170	1.4210
	800	518.18	0.5653	0.02094	511.8	684.9	1196.7	0.7135	0.7004	1.4139
	850	525.20	0.5292	0.02114	520.8	674.2	1195.0	0.7225	0.6846	1.4071
	900	531.95	0.4969	0.02134	529.5	663.8	1193.3	0.7311	0.6694	1.4005
	950	538.40	0.4680	0.02154	537.9	653.6	1191.5	0.7394	0.6549	1.3942
	1000	544.58	0.4419	0.02174	546.0	643.5	1189.6	0.7473	0.6408	1.3881
	1100	556.28	0.3960	0.02215	561.7	623.9	1185.6	0.7624	0.6141	1.3765
	1200	567.14	0.3582	0.02258	576.5	604.9	1181.4	0.7764	0.5891	1.3656
	1300	577.32	0.3259	0.02300	590.6	586.3	1177.0	0.7897	0.5654	1.3552
	1400	586.96	0.2983	0.02346	604.3	568.1	1172.4	0.8024	0.5428	1.3452
	1500	596.08	0.2741	0.02392	617.5	550.2	1167.6	0.8146	0.5212	1.3357
	1600	604.74	0.2528	0.02440	630.2	532.6	1162.7	0.8262	0.5003	1.3265
	1700	612.98	0.2338	0.02489	642.5	515.0	1157.5	0.8373	0.4801	1.3174
	1800	620.86	0.2167	0.02541	654.7	497.2	1151.8	0.8482	0.4601	1.3083
	1900	628.39	0.2014	0.02595	666.8	478.9	1145.7	0.8589	0.4402	1.2990
	2000	635.61	0.1875	0.02652	679.0	460.0	1139.0	0.8696	0.4200	1.2896
	2200	649.25	0.1623	0.02775	703.7	420.0	1123.8	0.8912	0.3788	1.2700
	2400	661.94	0.1404	0.02921	729.4	376.4	1105.8	0.9133	0.3356	1.2488
	2600	673.82	0.1205	0.0310	756.7	327.8	1084.5	0.9364	0.2892	1.2257
	2800	684.91	0.1021	0.0333	786.7	272.3	1058.9	0.9618	0.2379	1.1996
	3000	695.25	0.0844	0.0367	823.1	202.5	1025.6	0.9922	0.1754	1.1676
	3200	704.91	0.0601	0.0459	887.0	75.9	962.9	1.0461	0.0651	1.1112
	3226	706.1	0.0522	0.0522	925.0	0	925.0	1.0785	0	1.0785

79. Pressure and Temperature. — The absolute pressure is the quantity upon which all other properties depend; consequently, it is the first item in the steam table. The temperature of evaporation, t_v, is the second item in the table, since it is so closely related to the pressure. Under low pressure the temperature increases rapidly for small increase in pressure, while under higher pressure the increase is less rapid — a fact which is of great importance in the operation of certain apparatus using steam.

80. Heat of Liquid. — *The heat of liquid, h_f, is the amount of heat added to raise the temperature of a pound of water from 32 deg. fahr. to the boiling-point,* and as given in the steam table is calculated from the equation.

$$h_f = C_p (t_v - 32) \quad \ldots \ldots \ldots \ldots (32)$$

in which C_p is the mean specific heat of water or $\frac{1}{180}$th. of the heat required to raise one pound of water from 32 to 212 deg. fahr., and t_v is the temperature of evaporation. For some calculations it is sufficiently accurate to consider C_p equal to 1; then the heat of liquid = $h_f = t_v - 32$. However, for accurate work, the values as given in the steam table should be used. It must be remembered that the heat added is measured from

98   PROPERTIES OF AIR, GASES, WATER AND STEAM

32 deg. fahr., and in case the temperature is other than 32 deg. fahr. the steam table value must be corrected as explained in Art. 88, page 101.

81. Latent Heat of Evaporation. — *The heat added to convert one pound of water into steam under a constant pressure is known as the latent heat of evaporation and is designated by L.* This heat is potential, since it is stored in the steam. Consequently upon condensation of the steam, the same amount of heat will be given up as was originally added during evaporation. At a pressure of 14.7 pounds per square inch the latent heat of evaporation is 970.2 B.t.u. per pound of steam.

82. Internal and External Latent Heat. — The latent heat of evaporation is considered as composed of the internal latent heat,[1] ρ, plus the heat equivalent of the external work 144 Apu, in which $A = \frac{1}{778}$, p = the absolute pressure in pounds per square inch, and u equals the change in volume between one pound of water and one pound of steam at the given pressure, as illustrated in Fig. 71. The internal heat is considered as used in changing the rate and amplitude of vibration of the molecules. The external latent heat is the heat equivalent of the work done in changing the volume from the volume of one pound of water, v_f, to the volume of one pound of dry steam, v_g, the pressure remaining constant during the change.

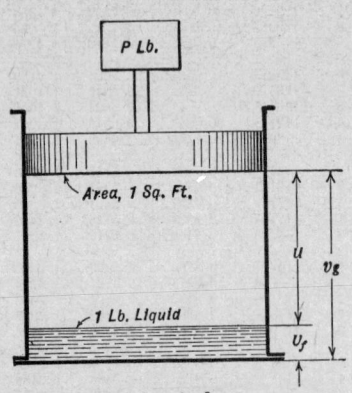

Note:- External work of evaporation $= P(v-v_f) = Pu$ ft.-lb.
or, $\frac{P(v-v_f)}{778} = A\,Pu$ in which $A = \frac{1}{778}$

FIG. 71. — Diagram illustrating APu, or 144 Apu.

83. Total Heat, or Heat Content,[2] per Pound of Dry Saturated Steam. — The total heat per pound of dry saturated steam is represented by the symbol h_g. *It is the total heat required to convert a pound of water at 32 deg. fahr. into dry steam,* and equals the sum of the heat of liquid, the internal latent heat and the external latent heat. In Fig. 70, page 92, it is shown by the horizontal distance from A to C and when written as an equation, is as follows:

$$h_g = h_f + \rho + 144\,Apu, \quad \text{or} \quad h_g = h_f + L \quad \ldots \ldots \quad (33)$$

If the pressure under which evaporation takes place is increased, the temperature of evaporation is increased, the heat of liquid is increased and the latent heat of evaporation decreased. The net result, however, is to increase the total heat per pound of dry steam. (See Fig. 70, page 92.)

Example 8. — Steam is formed in a boiler in which the gage pressure is 115.3 lb. per sq. in. with the barometer reading 29.92 in. mercury. Find the volume, heat of liquid, internal latent heat and the total heat in each pound of dry steam formed.

[1] Can be found by subtracting 144 Apu from the latent heat (L).
[2] Also known as **enthalpy**.

SPECIFIC VOLUME OF DRY SATURATED STEAM

Solution. — Referring to the Steam Table, page 95, the desired numerical values of the properties are found under the proper column and on a horizontal line corresponding to the absolute pressure in the boiler.

Atmospheric pressure = 29.92 × 0.491 = 14.7 lb. per sq. in.
Absolute pressure = 115.3 + 14.7 = 130 lb. per sq. in.
v = 3.46 cu. ft.; h_f = 318.73 B.t.u.; ρ = 789.9 (calculated); h_g = 1191.2 B.t.u.

84. Interpolation from Tables. — The finding of values which lie between those given in the steam tables is known as interpolation. In interpolating, the direction in which the quantities vary must be kept in mind, in order to make the correction in the proper manner.

Example 9. — Find the latent heat per pound of dry and saturated steam when the absolute steam pressure is 91.7 lb. per sq. in.
Solution. — From Steam Table, page 95, it is seen that

L for a pressure of 91 = 893.6 B.t.u.
L for a pressure of 92 = 893.0 B.t.u.
Difference = 0.6

Of this difference, 0.7 × 0.6 or 0.42, must be subtracted from the heat given for L at 91. The latent heat, therefore, equals 893.6 − 0.42 = 893.18 B.t.u.

85. Specific Volume of Dry Saturated Steam. — The specific volume, v_g, of dry saturated steam is the volume occupied by one pound of dry saturated steam. It varies with the pressure, as shown by the curve in Fig. 72. At low pressure the change in specific volume is very rapid. This curve is often called the **saturation curve**, and is useful in determining the condition of a given amount of steam. The volume occupied by a pound of dry saturated steam at any pressure will be twice the volume occupied by half a pound. Neglecting the small volume occupied by water in wet steam,

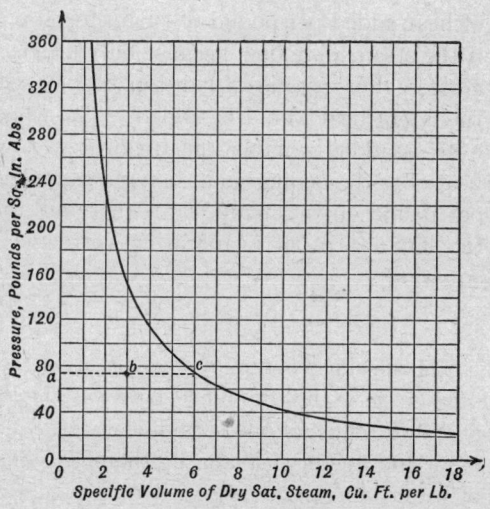

FIG. 72. — Curve showing Variation of Specific Volume with change in Pressure.

the quality of steam at point b in Fig. 72 is $\dfrac{ab}{ca}$, or 50 per cent, since only this fraction of the total has been converted into steam. The reciprocal of the specific volume, or $\dfrac{1}{v_g}$ equals the weight per cubic foot of dry saturated steam and is called its **specific density**.

86. Wet Steam, Moisture, and Quality. — When the evaporation of steam takes place rapidly, as is generally the case in a boiler subjected to the intense heat of the furnace, the bubbles of steam formed are large and, upon arriving at the surface of the liquid, burst violently, thus throwing the film of water which surrounds them into the steam space. These particles of water are light and remain suspended in the steam as a mist, or fog. When in this condition steam is called **wet steam,** and a pound of it is composed partly of water and partly of vapor. The term **moisture,** or **priming,** is applied to that portion of the water that remains unevaporated, and the term **quality,** or **dryness factor,** is applied to that portion of each pound of water actually evaporated into dry steam. Quality and moisture are expressed in per cent of the total weight, and the former is represented by the symbol x. Thus, steam having a quality of 98 per cent is a mixture of ninety-eight parts by weight of steam and two parts by weight of water. The quality of steam can be obtained by subtracting the percentage of moisture from 100 per cent, or vice versa.

87. Total Heat per Pound of Wet Steam. — Equation (33) only applies to dry saturated steam. When the quality is other than unity, the amount of heat added per pound of steam for any given pressure will be less than if the steam were dry, because only a part of the latent heat, L, has been added. For instance if half a pound of water were evaporated one-half of the latent heat would be added. The whole of the heat of liquid, h_f, however, is added whether the steam is wet or dry, since the water must be heated to the boiling-point before evaporation takes place. The total heat per pound of wet steam, h_w, is therefore the sum of the heat of liquid and the latent heat of evaporation corrected for quality, or, written as an equation,

$$h_w = xL + h_f \qquad (34)$$

The specific volume of wet steam, v_w, equals $xu + v_f$ in which $x =$ the quality, $u =$ the difference between the specific volume of dry steam v_g and the volume of one pound of water, v_f. The total volume of wet steam v_w for any weight of steam, W, therefore equals $W(xu + v_f)$, or if the volume of water be neglected $v_w = Wxv$.

88. Nature of General Problem. — In general, the problem involved in calculations using steam tables is not one of determining the amount of heat above 32 deg. fahr. in one pound of steam, but in determining how much heat is necessary to form W pounds of steam from water at some temperature other than 32 deg. fahr., or to find how much heat is released by steam when it is condensed.

If the change occurs at constant pressure, and this is the common case, the heat added to the liquid equals the heat of liquid at the absolute pressure under which the liquid is heated, minus the heat of liquid correspond-

TOTAL HEAT PER POUND OF SUPERHEATED STEAM 101

ing to the temperature of the water at the start of the addition of heat. This is so because sufficient heat has previously been added to the water to raise it from 32 deg. fahr. to the temperature being considered. In other words, the heat added to the liquid equals $h_f - h_{f_1}$, in which h_f = heat of liquid at the absolute pressure in boiler, and h_{f_1} = heat of liquid at temperature of water entering boiler.

Example 10. — Find the quantity of heat required to convert 8 lb. of water at a temperature of 182 deg. fahr. and a gage pressure of 150.5 lb. per sq. in. into steam having a quality of 0.98, barometer reading 29.5 in. of mercury.
Solution. — Pressure of the atmosphere = 29.5 × 0.491 = 14.5 lb. per sq. in. abs.
Absolute steam pressure = 14.5 + 150.5 = 165 lb. per sq. in.
Total heat per lb. of steam, $h_w = (xL + h_f) - h_{f_1}$
From the Steam Tables, page 96, at an absolute pressure of 165 lb. per sq. in.
L = 856.5 B.t.u., h_f = 338.47 B.t.u.
h_{f_1} for a temperature of 182 deg. fahr. = 149.87 B.t.u., x = 0.98.
Total heat in 8 lb. of steam = 8 [0.98 × 856.5 + 338.47 − 149.87] = 8223.7 B.t.u.

Note. — h_{f_1} can be taken as (182 − 32) without serious error.

89. Superheated Steam.[1] — *Whenever steam exists at a temperature higher than the temperature of saturated steam for that pressure, it is superheated.* The difference in temperature between the temperature of saturated steam, t_v, and the actual temperature of the superheated steam, t_s, is the amount, or **degree of superheat.** For instance, if the absolute pressure of the steam is 200 pounds per square inch and a thermometer placed in it shows a temperature of 510 deg. fahr., the degree of superheat is 510 − 381.8 = 128.2 deg. fahr., since the temperature of saturated steam at 200 pounds per square inch absolute pressure is 381.8 deg. fahr.

The volume of superheated steam may be obtained with fair accuracy by using the following equation: $pv^{1.063} = 484.2$, in which p = absolute pressure in pounds per square inch and v = specific volume.

90. Total Heat per Pound of Superheated Steam. — The total heat per pound of superheated steam, h_s, equals the total heat per pound of dry saturated steam plus the heat required to change the temperature from t_v to t_s. The latter amount of heat equals the specific heat of superheated steam, C_p, times $(t_s - t_v)$.

The total heat per pound of superheated steam can be found from Table 8, page 118, or can be computed from the following equation:

$$h_s = h_g + C_p (t_s - t_v) \qquad (35)$$

The value of the specific heat, C_p, is not constant, but varies with the pressure and temperature. For the range of temperatures encountered in present practice, a mean value may be used as shown by the curves in Fig. 73. When the pressure is below 100 pounds per square inch, the variation in the mean specific heat is small.

[1] See also Chapter V.

Example 11. — Find the quantity of heat required to produce 10 lb. of steam, superheated 110 deg. fahr., when the pressure is 160 lb. per sq. in. abs. and the temperature of the water from which it is formed is 60 deg. fahr.

Solution. — Using Equation (35) and the Steam Tables, page 96,
$h_s = h + C_p (t_s - t_v) = 1166.5 + 0.578 (110) = 1230.1$ B.t.u.
h at 160 lb. per sq. in. abs. above 60 deg. fahr. $= h_g - h_{60} = 1194.5 - 28.05 = 1166.5$ B.t.u.; C_p from curve, Fig. 73 $= 0.578$; $t_s - t_v = 110$ deg. fahr.
Heat to produce 10 lb. steam $= 1230.1 \times 10 = 12,301$ B.t.u.

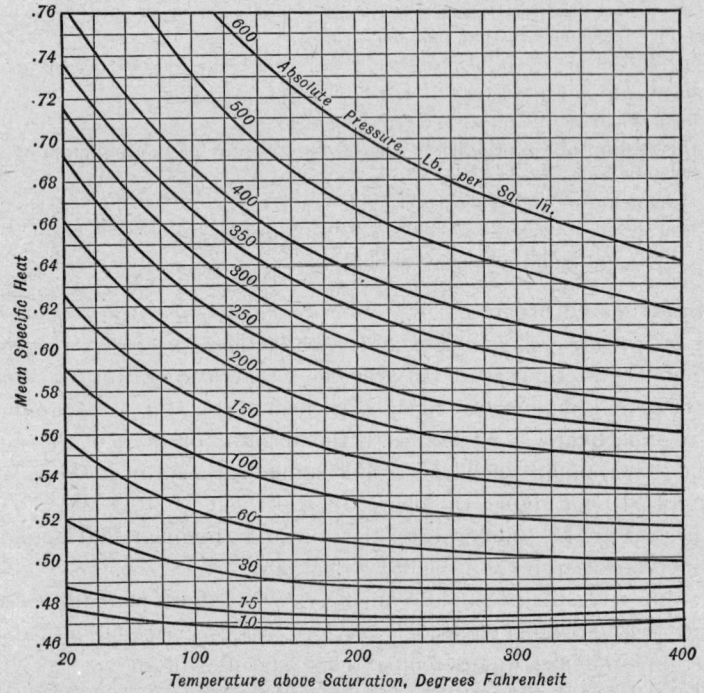

FIG. 73. — Variation of Mean Specific Heat, Superheated Steam.

91. Determination of Quality. — The amount of moisture in a pound of steam is ordinarily determined by a steam calorimeter. The most common types of which, arranged according to their accuracy, are: throttling, combined throttling and separating, separating, electric, and barrel. The throttling and separating calorimeters will be described, since they are the types commonly used in test work.

92. Throttling Calorimeter. — The Peabody type of throttling calorimeter is shown in Fig. 74. It consists of a chamber so shaped that it provides a thermometer well in the center with an orifice at one side through which the steam enters the calorimeter chamber and a discharge opening at the bottom. The pressure inside the calorimeter is measured by a mercury manometer, with a gage cock so placed that the manometer may be cut out of service when necessary. The calorimeter is attached to the steam line by a perforated pipe known as the **sampling nozzle** with a globe valve placed between the calorimeter and the steam line to shut off the calorimeter when desired. The pressure of the steam entering the calorimeter is obtained by a steam gage attached to the main steam line by a water siphon.

Operation. — Steam from the main steam pipe, after passing through the sampling nozzle, is expanded from the pressure in the main steam pipe to that in the calorimeter chamber by passing through the orifice. The temperature of the steam, after passing the orifice, is indicated by a

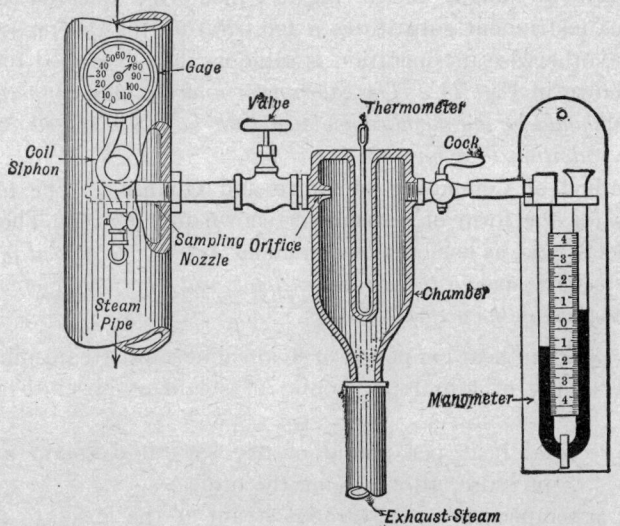

Fig. 74. — Peabody Type of Throttling Calorimeter.

thermometer placed in the thermometer well. The instrument should be thoroughly warmed and the temperature as indicated by the thermometer should become stationary before any moisture determinations are made. The steam is discharged from the calorimeter chamber through suitable piping directly into the atmosphere.

Principle of Operation. — This instrument works upon the principle that steam, in passing through an orifice, such as that of the calorimeter, does so without change of total heat. The quantity of heat in a pound of steam, under the conditions existing in the calorimeter, being the same as that in a pound of wet steam at the absolute pressure in the steam main, provided there is no radiation loss. Expressing the above differently, *a pound of wet steam at a high pressure contains more heat than a pound of wet steam at a low pressure, and as external work is not performed by the steam, when passing through the orifice, the surplus heat evaporates the moisture and superheats the steam.* For instance, take the higher pressure as 150 pounds per square inch absolute and the lower pressure as

14.7 pounds per square inch absolute. The total heat per pound of dry steam at the former pressure is 1193.5 B.t.u., and at the latter, 1150.2 B.t.u., a difference of 43.3 B.t.u. The number of degrees of superheat would therefore be 43.3 B.t.u. divided by the specific heat of superheated steam (0.47) which equals 92.2 deg. fahr. Moisture in the steam at the higher pressure would reduce the amount of the superheat shown above.

The type of calorimeter recommended in the American Society of Mechanical Engineers' testing code, is made of pipe fittings and consists essentially of two tees in which thermometer wells are placed, with a disc held between two flanges, and having an $\frac{1}{8}$-inch orifice, located between the tees. This instrument substitutes a temperature for the pressure before the orifice, otherwise its operation is same as that described for the calorimeter shown in Fig. 74. *The calorimeter and the pipe connections to the steam line should be thoroughly insulated with hair felt, to reduce the losses caused by radiation to a minimum.*

93. Method of Computing Moisture and Quality. — The method explained is for the form of calorimeter shown in Fig. 74. The following relation holds true, as has been shown under Art. 92: *total heat per pound of wet steam before passing the orifice = total heat per pound of superheated steam after passing the orifice.*

Let L = latent heat per pound of steam at absolute steam-pipe pressure.

h_f = heat of liquid per pound of steam at absolute steam-pipe pressure.

h_g = total heat per pound of dry saturated steam at absolute pressure, after passing the orifice.

t_v = temperature of saturated steam at the absolute pressure in the calorimeter.

t_s = temperature of superheated steam in the calorimeter.

C_p = 0.47 = specific heat of superheated steam at atmospheric pressure.

x = proportion by weight of dry steam per pound of steam.

m = proportion by weight of moisture = $1 - x$.

The total heat per pound of wet steam before passing the orifice = $xL + h_f$, and the total heat per pound of superheated steam after passing the orifice = $h_g + C_p (t_s - t_v)$; consequently

$$xL + h_f = h_g + C_p (t_s - t_v), \quad \text{and} \quad x = \frac{h_g + C_p (t_s - t_v) - h_f}{L} \quad . \quad (36)$$

Quality may also be found, without using the above equations, from the total-heat-entropy diagram, as explained under Art. 107, page 113.

Example 12.— The pressure in a steam line to which a calorimeter is attached is 150 lb. per sq. in. gage; pressure in calorimeter atmospheric; temperature in calorimeter, 262 deg. fahr.; barometer, 30.02 in. mercury. Find the quality of the steam.

Solution. — Pressure of the atmosphere = $30.02 \times 0.491 = 14.74$ lb. per sq. in. Absolute steam pressure in line = $150 + 14.74 = 164.74$ lb. per sq. in.

SEPARATING CALORIMETER

Temperature corresponding to calorimeter pressure (t_v) = 212.15 deg. fahr.
Temperature in calorimeter (t_s) = 262.0 deg. fahr.
Specific heat of superheated steam (C_p) at atmospheric pressure = 0.47.
Latent heat (L) corresponding to absolute steam-pipe pressure = 856.7 B.t.u.
Heat of liquid (h_f) corresponding to absolute steam-pipe pressure = 338.3 B.t.u.
Total heat per pound of dry steam (h_g) at absolute calorimeter pressure = 1150.3 B.t.u.
Using Equation (34) with proper substitutions

$$x = \frac{h_g + C_p(t_s - t_v) - h_f}{L} = \frac{1150.3 + 0.47(262 - 212.2) - 338.3}{856.7} = 0.975$$

Moisture = $m = (1 - x)100 = 2.50$ per cent.

94. Sources of Error and Limits, Throttling Calorimeter. — There is a slight error due to the value taken for the specific heat of superheated steam at atmospheric pressure; this error, however, is negligible. The thermometer stem is ordinarily not heated throughout its full length. Moreover, the thermometer may have an initial error, and there may be radiation losses from it.

The **limits** of moisture within which the throttling calorimeter will work are, at sea level, from 2.88 per cent moisture at 50 pounds gage pressure, to 7.3 per cent moisture at 400 pounds pressure.

95. Separating Calorimeter. — The separating calorimeter, Fig. 75, consists of a cast-iron body so constructed that there is an inner and and an outer vessel with a space between. This space forms a steam jacket for the inner cylinder and serves to prevent radiation from the inner vessel, except that which takes place from the gage-glass connections.

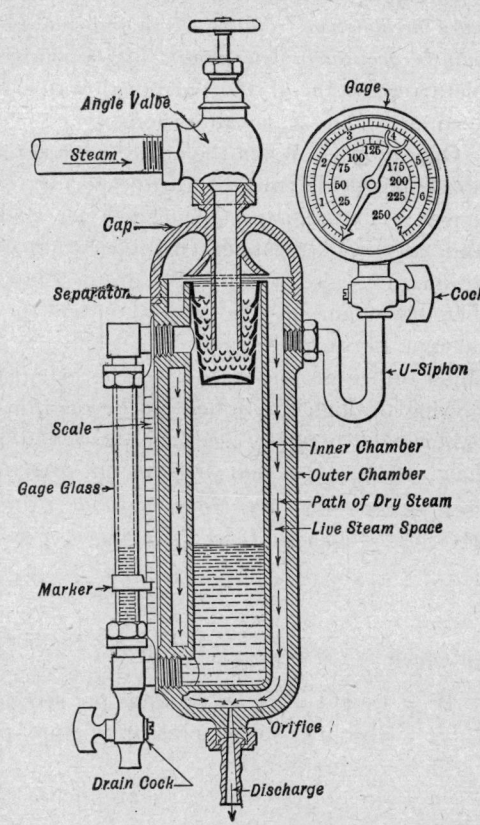

Fig. 75. — Separating Calorimeter.

The inner chamber has no direct connection with the outer jacket space, but is connected at one side to a water glass having an attached scale graduated to read in hundredths of a pound of water. This scale is corrected for expansion, to give correct readings with steam at 100 pounds per square inch gage pressure.

At the top of the inner chamber is a small cup having projecting fins which assist in separating the moisture from the steam. The bottom of the cup is closed, but there is an opening into the inner chamber under each fin. The outer chamber is connected to a steam gage which indicates the pressure in the outer vessel. This gage is graduated, by trial, to read the weight of steam passing through the instrument in ten minutes. At the bottom of the outer chamber is a small orifice of known size, through which the dry steam passes from the calorimeter.

A cap, having a deflecting plate and a pipe which extends well down into the cup, or separator, is attached to the top of the body by a threaded connection. An angle valve connects this cap with the sampling nozzle.

Principle of Operation. — *The separating calorimeter mechanically separates the moisture from the steam and collects it in a chamber where its amount may be accurately determined.* It depends for its accuracy upon the complete separation of the entrained water from the steam, and upon the accuracy of the gage calibrations.

Operation. — When the angle valve is opened, steam from the sampling nozzle passes downward into the perforated cup, where its direction is reversed. The moisture, being heavier than the steam, is left in the cup and falls into the inner chamber where its amount may be read on the graduated scale, or it may be drained into a suitable vessel and weighed. The dry steam, passing upward, enters the outer chamber through a small opening between the inner vessel and the cap. It then flows through the small orifice at the bottom to the exhaust pipe. The amount of steam flowing through the orifice can be read on the gage dial, which should be calibrated whenever used; or its weight may be computed by **Napier's Law,** which states that: *the flow of steam from a higher to a lower pressure is proportional to the higher absolute steam pressure, as long as the lower pressure is less than 0.58 of the higher pressure.* Expressed as an equation

$$W = \frac{Pa}{70} \qquad \qquad (37)$$

in which

W = weight of steam, pounds per second.
P = absolute steam pressure before passing the orifice, pounds per square inch.
a = area of the orifice, square inches.

This equation will give accurate results, provided the orifice is not clogged with sediment and its area has been accurately obtained. The amount of steam passing through the instrument may be determined more accurately by attaching a hose to the steam outlet, the hose extending into a vessel containing water and resting on a scale. The steam flowing will then be condensed by the water and its amount determined by weight.

The weight of moisture in pounds can be read on the graduated scale, or weighed as desired.

The proportion of moisture can be calculated from the following relation:

$$\text{Moisture} = \frac{\text{weight of moisture}}{\text{weight of dry steam plus weight of moisture}} = \frac{w}{W+w} \quad (38)$$

Example 13. — During the test of a 9 × 16 × 24 × 24 in. steam engine, the separating calorimeter attached to the exhaust pipe of the high-pressure cylinder gave the following data: weight of moisture in 10 min., 0.13 lb.; weight of dry steam in 10 min., 1.775 lb.; pressure in calorimeter, 29.5 lb. per sq. in. gage; barometer, 29.45 in. mercury. Find: (a) quality of steam and (b) weight of steam passing orifice in 10 min., using Napier's equation.

Solution. — (a) Using Equation (38) with proper substitutions:

The proportion of moisture $= \dfrac{w}{W+w} = \dfrac{0.130}{1.775 + 0.13} = \dfrac{0.130}{1.905} = 0.068$, or 6.8 per cent.

Per cent quality $= (1 - 0.068) \, 100 = 93.8$

(b) To find weight of steam passing the orifice use Equation (37)

$W = \dfrac{Pa}{70} = \dfrac{44 \times 0.0047}{70} = 0.00295$ lb. per second.

$P = 29.5 + 14.5 = 44$ lb. per sq. in. abs.

$a = 0.0047$ sq. in. (by measurement).

Weight of steam in 10 minutes $= 0.00295 \times 60 \times 10 = 1.770$ lb.

Limits. — Theoretically, this instrument is not limited to any specific range. If it is desired to have a check upon its accuracy, a throttling calorimeter may be attached to the steam outlet; the total moisture will then equal the sum of that given by the two instruments.

96. Sampling Nozzle. — The principal source of error in steam calorimeter determinations is the failure to obtain an average sample of the steam delivered by the boiler, and it is doubtful whether such a sample is ever obtained. The two governing factors in obtaining such a sample are the type of sampling nozzle used and its location.

The Power Test Code of the A. S. M. E. recommends a sampling nozzle made of $\tfrac{1}{4}$- or $\tfrac{3}{8}$-inch pipe, closed at the inner end and having the portion of the nozzle inside the steam main drilled with $\tfrac{1}{8}$-inch holes spaced on $\tfrac{1}{2}$-inch centers for steam mains up to 6 inches in diameter. This nozzle should be located in the pipe in such a position that the holes will directly face the steam flow. The calorimeter should be located as near as possible to the point from which the steam is taken, and the *sampling nozzle* should be placed in a section of the main pipe near the boiler, turbine, or engine under test, and at a point where there is no chance of moisture pocketing in the pipe. The *ideal location* is in a pipe in which the steam flow is vertically downward, and as far removed from a valve, or elbow, in the direction of flow, as it is possible to have it.

97. Entropy. — The general heading "Entropy" is given to the last three columns in the saturated steam table. *Entropy may be defined as the* *change in the quantity of heat per degree of absolute temperature.* During the conversion of water into steam, the heat of liquid and the latent heat of evaporation are given to the pound of water, their sum making the total heat per pound of dry steam. The entropy values to correspond are called entropy of the liquid, s_f, entropy of evaporation, L/T_1, and entropy of steam, s_g.

The significance of entropy may be shown by reference to the heat diagram, Fig. 76a, the area of which represents heat units. It is bounded by lines representing absolute temperature and entropy.

Entropy Change while Temperature is Constant. — If the temperature is constant at T_1 degrees absolute, as during the conversion of water into steam, and L heat units are added, the change in entropy, or heat per

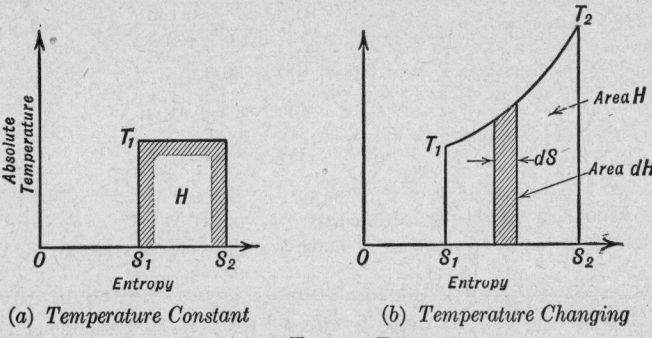

(a) *Temperature Constant*   (b) *Temperature Changing*

FIG. 76. — Entropy Diagrams.

degree absolute, is $s_2 - s_1 = L \div T_1$. In this case $L = H =$ heat added at the temperature of evaporation.

Example 14. — Calculate the change in entropy during the evaporation of 1 lb. of water at 212 deg. fahr.

Solution. — Since the temperature remains constant during the period of evaporation, the change in entropy equals

$$L \div T = 970.2 \div 671.6 = 1.4446$$

Entropy Change while Temperature Changes. — If the temperature changes while heat is being supplied, as during the addition of the heat of the liquid, the heat area, H, Fig. 76b, may be considered as made up of a series of very narrow sections of area dH, for each of which there is a temperature T and a very small width ds. Then the area $dH = T\,ds$, or $ds = dH \div T$; that is, the change in entropy is equal to the small quantity of heat added, divided by the absolute temperature during the addition of the heat. The summation of the elementary heat areas gives the area H. The summation of the ds values gives the total entropy change, $s_2 - s_1$. It can be shown that if the specific heat, C_p, is constant during the addition of the heat, the change in entropy may be written:

$$s_2 - s_1 = C_p \log_e \frac{T_2}{T_1} \quad \ldots \ldots \ldots \ldots \quad (39)$$

in which T_1 is the initial absolute temperature in degrees Fahrenheit, and T_2 is the final absolute temperature in degrees Fahrenheit.

Example 15. — Calculate the change in entropy of water for the short temperature interval from 65 deg. fahr. to 75 deg. fahr., over which the specific heat is nearly constant and equal to 1. Compare this result with the Steam Table value for the same temperature range and find the entropy of steam at 212 deg. fahr.

Solution. — Using Equation (39)

$$s_{75} - s_{65} = C_p \log_e \frac{T_2}{T_1} = 1 \log_e \frac{75 + 459.6}{65 + 459.6} = \log_e 1.019 = 0.1899 \text{ or } 0.019$$

The entropy of water at these temperatures, as given in the Steam Table, is 0.0839 and 0.0651 respectively, and the difference is 0.019, as calculated above.

The *entropy of steam* at 212 deg. fahr. equals the entropy of water plus the entropy of evaporation $= s_f + \frac{L}{T} = 0.312 + 1.4446 = 1.7566$.

98. Temperature-entropy or T-s Diagram. — The temperature-entropy diagram is useful for the graphical solution of problems, involving the prop-

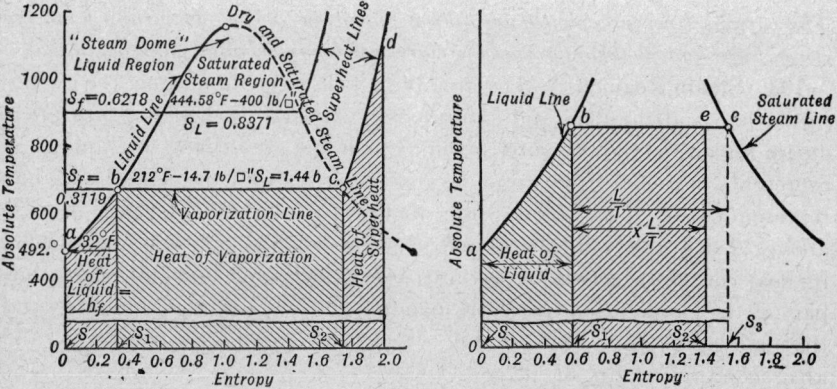

FIG. 77. — T-s Diagram.   FIG. 78. — Quality from T-s Diagram.

erties of steam, that are tedious and difficult by analytical methods. The line drawn upon it represent changes in temperature and the corresponding changes in entropy for one pound of water or steam. As previously explained, the area of a closed figure on this chart represents heat units.

In Fig. 77, *vertical distances represent temperature and horizontal distances represent units of entropy*. The temperature scale may be marked t degrees by the thermometer or T degrees absolute. The path *abcd* shows the changes in temperature and entropy for one pound of water as it is heated in a boiler from 32 deg. fahr. to the temperature corresponding to the pressure maintained, then evaporated at constant pressure into saturated steam at that pressure, and finally superheated by the further addition of heat.

99. Liquid Line. — The line *ab* is known as the **liquid line** and is made up of points representing the entropy and temperature of water at beginning of evaporation. While the water is heated, the entropy increases from s to s_1 and the temperature from 32 deg. fahr. to the boiling-point. The area $s\ ab\ s_1$ represents the **heat of the liquid**.

100. Evaporation, or Latent-heat, Line. — The evaporation line, bc, shows the change in entropy while heat is added at constant temperature until evaporation is complete at the point c. The area $s_1\ bc\ s_2 = T_v\ (s_2 - s_1)$ below the line bc, then represents the latent heat L, which corresponds to the sum of the internal and the external work done during the formation of steam.

101. Saturation Line. — At the point c, the pound of water has been completely evaporated and has taken up all the heat that it can take up at that temperature, and the successive points representing temperature and entropy for complete evaporation make up the saturation or dry-steam line.

102. Superheated-steam Lines. — If more heat is added to the pound of dry steam as it passes away from the boiler, its temperature and entropy increase, and this change is shown by the line cd sloping up and to the right. *The various lines are plotted by taking the proper value of entropy from the steam table and plotting against the corresponding absolute temperature.*

103. Steam Regions. — The area lying below the liquid line, the evaporation line and the dry- and saturated-steam line is called the **saturated-steam region** for the pressures given. Since the area below the liquid line represents the heat of the liquid, h_f, and the area below the evaporation line represents latent heat, L, the sum of these areas represents total heat h_g above 32 deg. fahr. If the pound of steam is only partially evaporated, its heat content is shown by the area below the liquid line and below that part of the evaporation line corresponding to the proportion of the pound that is evaporated. Thus, in Fig. 78, $be \div bc$ equals the quality x, and area $oabes_2$ represents $xL + h_f$.

That part of the temperature-entropy chart lying to the right and above the saturation line is called the **superheated-steam region**. The area below the constant pressure line, cd, Fig. 77, corresponds to the heat added to superheat while the temperature and entropy increase.

104. Adiabatic Changes. — During an adiabatic change, there is no change in entropy, since heat is not added or taken away. Lines at right angles to the axis of entropy represent such changes. If a pound of steam having a quality x_1, pressure p, and temperature t expands adiabatically to pressure p_2 and temperature t_2, its quality x_2 at the end of the expansion may be found by equating the values of the entropies for the two conditions and solving for x_2 in the adiabatic equation.

$$s_{f_1} + x_1 \frac{L_1}{T_1} = s_{f_2} + x_2 \frac{L_2}{T_2} \quad \ldots \ldots \ldots (40)$$

The value of x_1 being known, the other quantities, except x_2 are taken directly from the Steam Tables.

105. Constant-quality Lines. — Constant-quality lines are made up of points denoting the same quality of steam through the range in temperature and entropy shown on the diagram.

Thus, if the point 1, Fig. 79, is at 90 per cent of the length of the evaporation line bc, and similarly points 2, 3, etc., are at 90 per cent of the lengths of their respective evaporation lines, the constant-quality line xx, at 0.90, passes through these points.

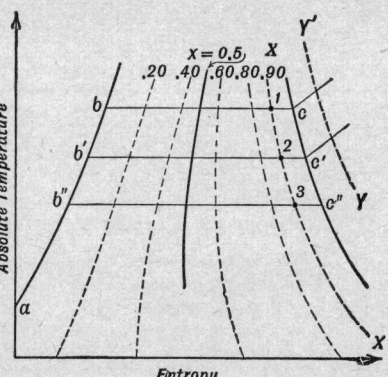

Fig. 79. — Constant-quality Lines.

Similarly, there may be drawn in the superheated-steam region constant-quality lines yy', or lines joining points at which the number of degrees of superheat is the same through the range of temperature and entropy shown on the chart. *The saturation line and the liquid line are special examples of constant-quality lines.*

106. Constant Total-heat Lines. — Constant total-heat lines are drawn in such a way as to represent the changes occurring in temperature and entropy of a pound of steam while the total heat, or heat content above 32 deg. fahr., remains the same. As previously explained, heat units are represented by areas, hence, on the T-s diagram, Fig. 80, the area below the broken line abt represents the heat content of one pound of wet steam H_w. If, for dry steam, the area below aum represents the same value of H and for superheated steam the area below $avwh$ shows the same total heat, all these areas being the same, the line tmh is a line of constant total heat. From the above it is evident that the heat contents which are only sufficient to partially evaporate the steam at the higher temperature and pressure are sufficient to evaporate the steam and produce some superheat at the lower temperature and pressure. Other lines of constant total heat may be drawn in a similar manner. This leads to consideration of the total-heat-entropy diagram, one use for which will be explained.

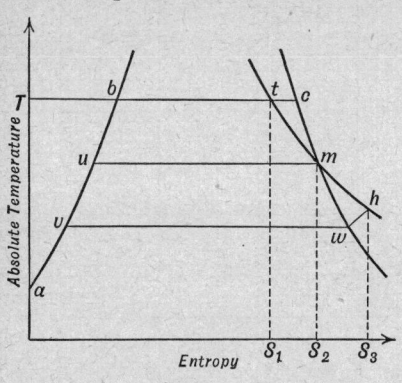

Fig. 80. — Constant Total-heat Lines.

107. Total-heat-entropy Diagram. — *By using values of total heat as vertical distances, and values of entropy as horizontal distances, a graphical diagram can be prepared on which constant-pressure lines and constant-quality lines are drawn, for use in the graphical solution of problems.*

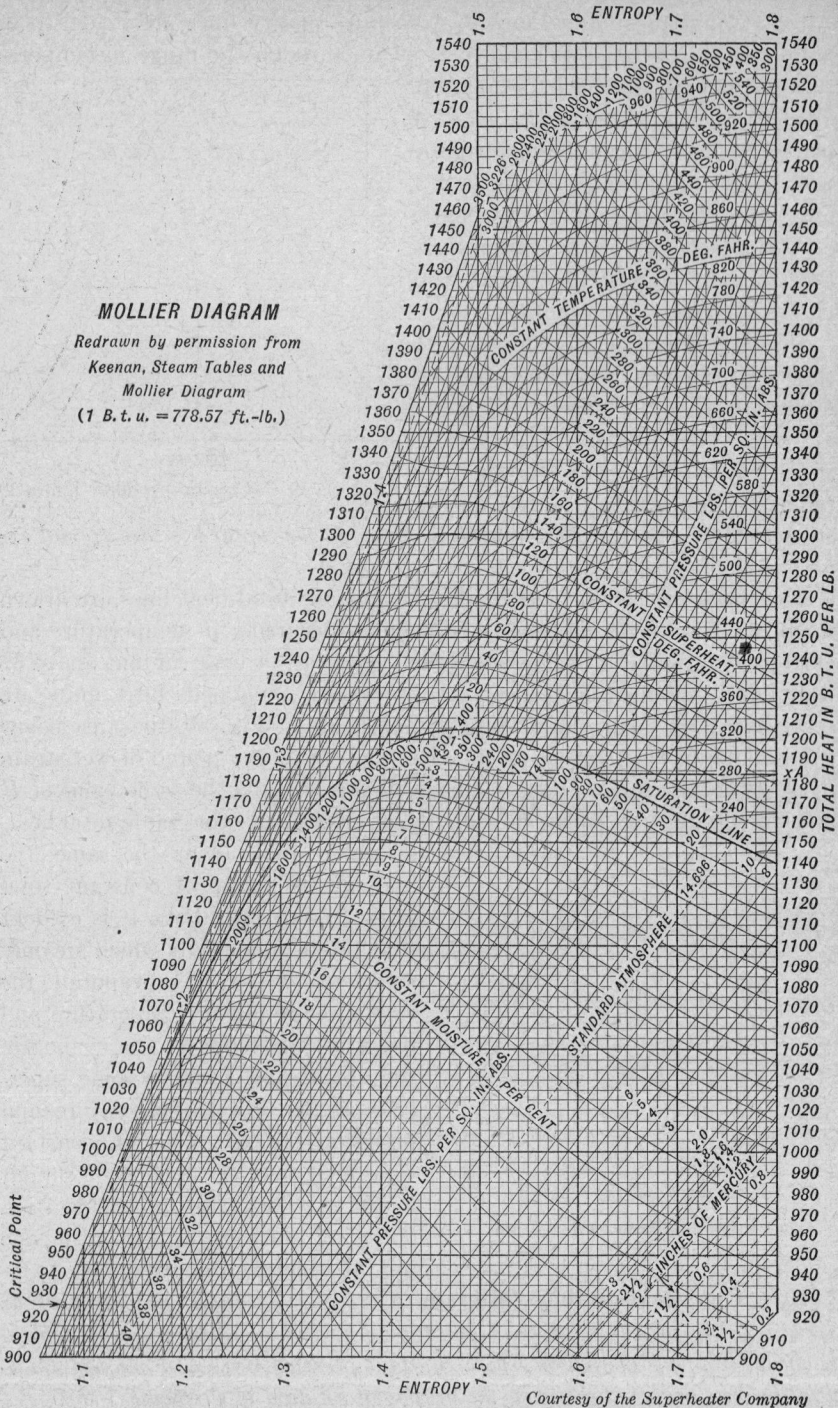

Fig. 81.—Total-heat-entropy, or Mollier Diagram.

A small section of such a diagram is shown in Fig. 81. Horizontal lines are lines of constant total heat, and vertical lines are lines of constant entropy. The saturation line is shown as a heavy line separating the region of saturated steam from that of superheated steam. Lines which slope downward and to the right are constant-moisture, or constant-superheat lines, lines which slope upward and to the right are lines of constant pressure, and in the superheat region lines of constant temperature deg. fahr. are added.

Example 16. — If steam at an absolute pressure of 200 lb. per sq. in. abs. is expanded in a calorimeter without loss of heat to an absolute atmospheric pressure of 14.696 lb. per sq. in., and the temperature in the calorimeter is, 280 deg. fahr., the condition of the steam is shown at xA, Fig. 81. To find its moisture before expansion, follow the constant total-heat line toward the left until it intersects the pressure line at 200 lb. per sq. in. abs. The moisture corresponding is thus found to be 1.8 per cent and the quality, 98.2 per cent.

REFERENCES

Steam Tables, KEENAN.
Properties of Steam and Ammonia, G. A. GOODENOUGH.
Steam Formulas, R. C. HECK, Trans. Am. Soc. M. E., Vol. 42, page 711.
Thermodynamics, EMSWILER.
Elements of Thermodynamics, MOYER, CALDERWOOD, POTTER.
Power Plant Testing, MOYER.
Steam, BABCOCK & WILCOX Co.
Power Test Code, A. S. M. E.

PROBLEMS AND REVIEW QUESTIONS

1. For the complete burning of 1 lb. of carbon, 2.66 lb. of oxygen are required. What weight of air is necessary to supply this amount of oxygen?
2. Find the quantity of heat in B.t.u. required to heat 10 lb. of air from 60 deg. fahr. to 550 deg. fahr. the pressure being constant.
3. Find the weight of 100,000 cu. ft. of air at a temperature of 125 deg. fahr. and at 14.7 lb. per sq. in. abs.
4. What is the change in volume when 100 lb. of air is cooled from 180 deg. fahr. to 60 deg. fahr. at atmospheric pressure?
5. Explain the meaning of the term "isothermal," as applied to heat changes.
6. A perfect gas occupies a volume of 20 cu. ft. at a pressure of 100 lb. per sq. in. abs. and a temperature T. Find the volume if the pressure is increased to 1500 lb. per sq. in. abs. and the temperature remains constant.
7. A cylinder holds a certain volume of a gas at 500 lb. per sq. in. abs. What pressure will be required to reduce the volume of the air one-quarter, if the temperature remains constant?
8. An automobile tire contains air at 60 deg. fahr. and 60 lb. per sq. in. gage pressure. If the tire is allowed to stand in the sun and the temperature is thus raised to 85 deg. fahr., what will be the resulting pressure if the tire does not stretch during the change?
9. Explain what each symbol in the equation $PV = WRT$ represents, and state the units in which each is expressed.
10. Calculate the density of 1 lb. of air at normal atmospheric pressure and temperatures of 100 deg. fahr. and 70 deg. fahr.

11. A gas expands at constant pressure from a volume of 2 cu. ft. to a volume of 10 cu. ft. against a total pressure of 2000 lb. per sq. in. abs. Find the work done in foot-pounds.

12. Using the molecular weight of oxygen as 32, and ammonia as 17, calculate the value of R for these gases.

13. Find the volume of 380 cu. ft. of air measured at 60 deg. fahr. and 30 in. mercury, if the gas is heated to 1038 deg. fahr. and at a pressure of 90 in. mercury.

14. An air compressor compresses 300 cu. ft. of air per min. from a pressure of 14.35 lb. per sq. in. abs. to a pressure of 100 lb. per sq. in. gage. Find: (a) the final volume of the air; and (b) the work done during the adiabatic compression of the air.

15. Calculate the value of R for air by the second and third methods mentioned on page 84. Use values of C_p and C_v as given in Table 4. Volume of 1 lb. of air at a pressure of 14.69 lb. per sq. in. abs. is 12.39 cu. ft. when the temperature is 32 deg. fahr.

16. One pound of air at 60 deg. fahr. is expanded to 120 deg. fahr., the pressure remaining constant. Find: (a) the external work done, (b) the change in internal energy.

17. The capacity of a pump is given as 2000 gallons per hour. Express this as volume in cubic feet per minute and as weight in pounds, if the temperature of water is 70 deg. fahr.

18. Mention some of the impurities found in boiler feedwater.

19. Give the names of the properties of steam as given in the saturated-steam table. Write the symbol used to represent each property.

20. Find the heat required, above 32 deg. fahr., to raise the temperature of 1 lb. of water to the boiling-point when under an absolute pressure of 100, 152, 210, and 300 lb. per sq. in. Would these values be changed if the quality of the steam were other than unity?

21. Same as Problem 20, except that the water temperature before the application of heat was 60, 70, 90 and 82 deg. fahr. respectively.

22. Find the latent heat of evaporation for 1 lb. of dry steam at the pressures in Problem 20. If the quality were 90 per cent, what would be the amount of heat at the above pressures?

23. Find, by calculation, the amount of external work done in converting 10 lb. of water at 60 deg. fahr., into dry steam at 150 lb. per sq. in. gage, with a barometer reading 29.92 in. of mercury.

24. Find the total heat above 32 deg. fahr. in a pound of dry saturated steam at the following absolute pressures: 70, 80, 95, 115, 200, and 300 lb. per sq. in.

25. Find the total heat, above 32 deg. fahr., in a pound of wet steam having a quality of 98 per cent at the pressures in Problem 24.

26. In Problems 24 and 25, find the total heat in each case, when the lower temperature, instead of being 32 deg. fahr., is, respectively, 60, 120, 150, 110, 200, and 230 deg. fahr.

27. The pressure in a steam main is 150 lb. per sq. in. gage. Barometer reads 29.92 in. of mercury. Steam is allowed to expand through an orifice to a pressure of 25 lb. per sq. in. gage. Temperature of steam after expanding is 293 deg. fahr. What was the quality of steam in the main pipe? Check, using Mollier diagram.

28. A steam gage attached to a steam pipe reads 110 lb. per sq. in. What is the maximum amount of moisture in the steam which a throttling calorimeter can detect, if 10 degrees superheat is required and the discharge is directly into the atmosphere, where the barometer reads 29.2 in. of mercury?

29. The temperature in a throttling calorimeter, attached to a steam main, is 275 deg. fahr. The pressure in the steam main is 100 lb. per sq. in. gage. Barometer, 29.5 in. of mercury. Pressure in the calorimeter is atmospheric. Find the quality of steam in the steam main.

30. Derive the equation used in calculating the quality of steam when using a throttling calorimeter. Explain the meaning of each symbol.

PROBLEMS AND REVIEW QUESTIONS

31. State the principle of operation of (a) the throttling calorimeter, (b) the separating calorimeter.

32. The following data apply to a separating calorimeter attached to a steam main: diameter of orifice, 0.04 in.; pressure in calorimeter, 116 lb. per sq. in. abs.; moisture collected in 10 min., 0.2 lb.; barometer, 29.48 in. mercury. Find the quality of the steam.

33. A separating calorimeter is attached to a steam main in which the pressure is 120 lb. per sq. in. abs. The moisture collected in 15 min. is 1.17 lb. Find the quality if the area of the orifice is 0.011 sq. in. and the barometer, 29.9 in. mercury.

34. Explain the meaning of the term *superheated*.

35. Find the heat in a pound of superheated steam, using Equation (35), page 101, at absolute pressures of 75, 100, 150, 95, 200, and 250 lb. per sq. in. The temperature of the superheated steam is respectively 400, 450, 380, 500, 430, and 550 deg. fahr.

36. A steam boiler evaporated 10 lb. of water for each pound of coal used. The steam gage read 150 lb. per sq. in. with the barometer at 30.01 in. of mercury, and the feedwater temperature was 120 deg. fahr. What amount of heat did the steam receive per pound of coal burned?

37. A steam boiler, on test, evaporated 907,500 lb. of water in twenty-four hours. Gage reading, 244 lb. per sq. in.; atmospheric pressure 14.5 lb. per sq. in. abs.; feedwater temperature 238 deg. fahr.; superheat 165 deg. fahr. How much heat was given to the water per hour?

38. Find the entropy of water, evaporation, and total entropy, from the Steam Tables, for pressures of 110, 75, 95, and 200 lb. per sq. in. abs.

39. Steam under the pressures given in Problem 38 has a quality of 98 per cent. Find the entropy of evaporation.

40. Explain the method of constructing a temperature-entropy diagram. For what purpose is it used?

41. Obtain, by using the total-heat-entropy diagram, the quality of steam after expansion from a pressure of 150 lb. per sq. in. abs. and dry steam to a pressure of 50 lb. per sq. in. abs. Assume that the entropy remains constant during the change.

CHAPTER V

SUPERHEATERS, SUPERHEATED STEAM, DESUPERHEATERS, REHEATERS, PURIFIERS AND STEAM CYCLES

108. Foreword. — The use of superheaters, as a means of saving fuel, has become almost universal in the modern steam power station. The application of a high-degree of superheated steam often produces a greater increase in the efficiency of steam power plants than any other single factor in steam engineering today.

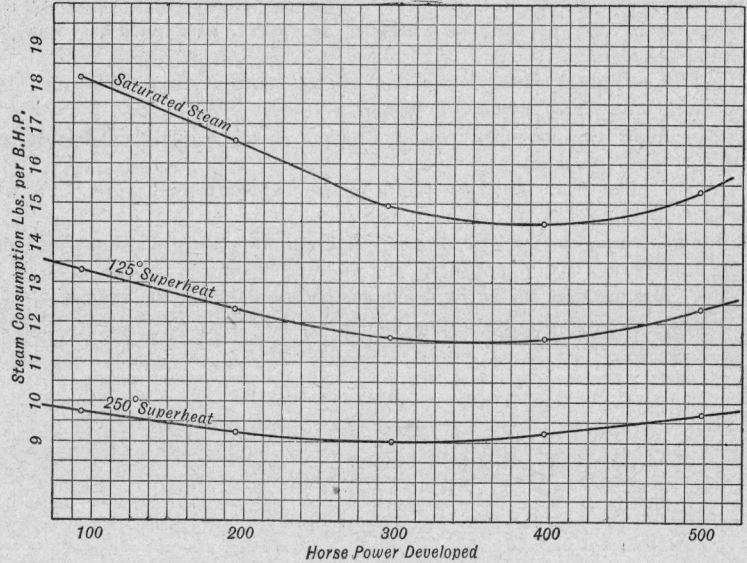

FIG. 82. — Steam Consumption of a 250-Hp. Compound Condensing Engine.

Superheated steam, when used in steam engines, effects a saving in fuel by:
1. Reducing condensation in the steam-pipe lines, and engine cylinder.
2. Reducing the friction of the steam in the steam ports while entering and leaving the cylinder.
3. Producing more work per pound of steam used.

The saving in the steam required to produce a brake horsepower for different degrees of superheat for a 250-horsepower compound engine, operating condensing is shown in Fig. 82. The poorer the original economy of the engine the greater will be the saving.

A high degree of superheat may be used in steam turbines, with a resulting increase in the economy of the turbine of about 1 per cent for each 10 degrees of superheat, the saving dropping off slightly as the amount of superheat is increased. The elimination of entrained moisture decreases the wear on the turbine blading and also makes it easier to maintain the vacuum since a smaller quantity of steam is used.

The overall thermal efficiency[1] of the combined boiler and superheater will be greater than the boiler without the superheater when evaporating the same amount of water, because by the addition of the superheating surface more heat is absorbed and the final uptake gas temperature is lowered. The increase in efficiency varies with the rate of evaporation, and generally will average from 2 to 5 per cent. This result can be accomplished, however, only when the superheater is correctly designed and the flow of the gases through the boiler is not obstructed sufficiently to lower the overload capacity of the boiler.

109. Properties of Superheated Steam. — The properties generally given in the superheated steam table, Table 8, are:

p = pressure in lb. per sq. in. absolute.
h_s = total heat, B.t.u. per lb.
v = specific volume, cu. ft. per lb.
s = entropy per lb.

All quantities are calculated above 32 deg. fahr. as explained under Art. 77, page 93, and for the temperatures as given. The *saturation temperature corresponding to each pressure is given in the parenthesis directly below the pressure*, and various total temperatures in deg. fahr. at the top of each column.

110. Methods of Superheating Steam. — Steam is superheated by adding heat to it after the steam is removed from contact with water. This may occur in: (1) the shell of the boiler, as in the steam space surrounding the top of the fire tubes in large vertical boilers; (2) special forms of superheaters located in the path of the furnace gases and known as **attached superheaters** of the **convection** and **radiant-heat type;** (3) **reheaters** or steam re-superheaters while passing from the high- to the low-pressure stages of a large turbine, with the heating agent either furnace gases or live steam; and (4) superheaters with separate furnaces, known as **separately fired** or portable superheaters.

The attached or integral superheater forms a part of the boiler itself and when of the convection type, for moderate superheats, it may be located in the gas path above the first and second passes of a boiler as shown in Fig. 25, page 32, for a horizontal drum B. and W. boiler. The exact location, however, varies with the type of the boiler. In many of the large boilers the convection superheater is located near the furnace above a

---

[1] See Chapter IX for discussion of boiler efficiency and rate of evaporation.

TABLE 8. — PROPERTIES OF SUPERHEATED STEAM
(Prepared by Prof. J. H. Keenan and reproduced by permission)

Pressure lb. per sq. in. abs. (Saturation temp., deg. fahr.)		Temperature of steam, deg. fahr.								
		400	450	500	550	600	650	700	750	800
20 (227.96)	h_S v s	1238.3 25.42 1.8386	1261.9 26.94 1.8652	1285.5 28.45 1.8906	1309.3 29.95 1.9148	1333.4 31.46 1.9380	1357.6 32.96 1.9604	1382.1 34.46 1.9820	1406.9 35.95 2.0029	1432.0 37.44 2.0232
40 (267.24)	h_S v s	1235.6 12.62 1.7599	1259.7 13.39 1.7871	1283.7 14.16 1.8128	1307.8 14.92 1.8373	1332.0 15.68 1.8607	1356.5 16.44 1.8832	1381.2 17.19 1.9050	1406.1 17.94 1.9261	1431.3 18.69 1.9465
60 (292.71)	h_S v s	1232.8 8.352 1.7128	1257.5 8.879 1.7406	1281.9 9.398 1.7667	1306.2 9.911 1.7915	1330.7 10.42 1.8151	1355.3 10.93 1.8378	1380.2 11.44 1.8597	1405.3 11.94 1.8810	1430.6 12.44 1.9014
80 (312.03)	h_S v s	1229.9 6.216 1.6785	1255.2 6.618 1.7070	1280.0 7.015 1.7336	1304.6 7.406 1.7586	1329.3 7.793 1.7825	1354.2 8.177 1.8054	1379.2 8.558 1.8274	1404.5 8.936 1.8488	1429.9 9.313 1.8694
100 (327.83)	h_S v s	1226.9 4.934 1.6512	1252.8 5.263 1.6805	1278.0 5.585 1.7075	1303.0 5.902 1.7328	1327.9 6.215 1.7569	1353.0 6.525 1.7800	1378.2 6.831 1.8023	1403.7 7.136 1.8238	1429.2 7.439 1.8445
120 (341.26)	h_S v s	1223.8 4.077 1.6283	1250.4 4.359 1.6584	1276.1 4.632 1.6859	1301.4 4.899 1.7116	1326.5 5.162 1.7358	1351.8 5.422 1.7591	1377.2 5.680 1.7816	1402.8 5.935 1.8032	1428.5 6.189 1.8240
140 (353.03)	h_S v s	1220.5 3.465 1.6084	1247.9 3.713 1.6393	1274.1 3.951 1.6674	1299.7 4.182 1.6933	1325.1 4.410 1.7179	1350.6 4.634 1.7413	1376.2 4.857 1.7640	1402.0 5.078 1.7857	1427.8 5.297 1.8066
160 (363.55)	h_S v s	1217.1 3.004 1.5905	1245.3 3.227 1.6224	1272.1 3.440 1.6510	1298.0 3.645 1.6774	1323.7 3.846 1.7022	1349.4 4.043 1.7258	1375.2 4.240 1.7486	1401.1 4.434 1.7704	1427.1 4.627 1.7915
180 (373.08)	h_S v s	1213.5 2.646 1.5742	1242.7 2.849 1.6072	1270.0 3.041 1.6364	1296.3 3.226 1.6631	1322.2 3.407 1.6882	1348.1 3.584 1.7121	1374.2 3.760 1.7351	1400.2 3.934 1.7570	1426.3 4.105 1.7782
200 (381.82)	h_S v s	1209.8 2.358 1.5592	1240.0 2.547 1.5934	1267.9 2.722 1.6231	1294.6 2.892 1.6502	1320.8 3.056 1.6756	1346.9 3.217 1.6997	1373.1 3.376 1.7228	1399.4 3.533 1.7449	1425.6 3.688 1.7662
220 (389.89)	h_S v s	1205.9 2.122 1.5451	1237.2 2.299 1.5804	1265.7 2.462 1.6109	1292.8 2.617 1.6384	1319.3 2.769 1.6641	1345.7 2.916 1.6883	1372.1 3.062 1.7117	1398.5 3.206 1.7339	1424.8 3.347 1.7553
240 (397.40)	h_S v s	1201.9 1.925 1.5317	1234.4 2.092 1.5684	1263.5 2.244 1.5996	1291.1 2.389 1.6276	1317.8 2.529 1.6534	1344.4 2.666 1.6779	1371.0 2.800 1.7014	1397.6 2.933 1.7238	1424.1 3.063 1.7452
260 (404.43)	h_S v s	....	1231.5 1.916 1.5571	1261.3 2.060 1.5890	1289.3 2.195 1.6174	1316.3 2.327 1.6435	1343.2 2.454 1.6682	1370.0 2.579 1.6919	1396.7 2.702 1.7144	1423.3 2.823 1.7360
280 (411.06)	h_S v s	....	1228.4 1.765 1.5462	1259.0 1.902 1.5790	1287.4 2.030 1.6078	1314.8 2.153 1.6343	1341.9 2.272 1.6592	1368.9 2.389 1.6830	1395.8 2.504 1.7057	1422.5 2.617 1.7274
300 (417.33)	h_S v s	....	1225.3 1.634 1.5359	1256.7 1.765 1.5695	1285.6 1.886 1.5988	1313.3 2.002 1.6256	1340.6 2.115 1.6508	1367.8 2.224 1.6748	1394.9 2.332 1.6976	1421.7 2.438 1.7194
350 (431.71)	h_S v s	....	1217.1 1.371 1.5117	1250.7 1.490 1.5476	1280.9 1.598 1.5783	1309.4 1.700 1.6059	1337.4 1.799 1.6316	1365.1 1.894 1.6561	1392.6 1.987 1.6792	1419.8 2.080 1.7013
400 (444.58)	h_S v s	....	1208.3 1.172 1.4891	1244.3 1.283 1.5277	1276.0 1.382 1.5599	1305.5 1.474 1.5884	1334.1 1.562 1.6147	1362.3 1.647 1.6396	1390.2 1.730 1.6631	1417.7 1.812 1.6854

h_S = total heat above 32 deg. fahr., B.t.u. v = specific volume, cu. ft. per lb.
s = total entropy above 32 deg. fahr.

TABLE 8. — *Concluded*. PROPERTIES OF SUPERHEATED STEAM
(Prepared by Prof. J. H. Keenan and reproduced by permission)

Pressure lb. per sq. in. abs. (Saturation temp., deg. fahr.)		Temperature of steam, deg. fahr.								
		400	450	500	550	600	650	700	750	800
450 (456.27)	h_S v s	…… …… ……	…… …… ……	1237.6 1.121 1.5091	1271.0 1.213 1.5430	1301.5 1.298 1.5725	1330.7 1.377 1.5994	1359.4 1.455 1.6248	1387.8 1.530 1.6487	1415.7 1.603 1.6713
500 (466.99)	h_S v s	…… …… ……	…… …… ……	1230.5 0.9907 1.4915	1265.8 1.078 1.5274	1297.3 1.156 1.5579	1327.3 1.230 1.5855	1356.6 1.301 1.6114	1385.3 1.370 1.6356	1413.6 1.436 1.6586
550 (476.90)	h_S v s	…… …… ……	…… …… ……	1222.8 .8829 1.4747	1260.3 .9663 1.5126	1293.1 1.040 1.5443	1323.8 1.109 1.5726	1353.6 1.175 1.5989	1382.8 1.238 1.6236	1411.5 1.300 1.6469
600 (486.17)	h_S v s	…… …… ……	…… …… ……	1214.7 .7922 1.4580	1254.6 .8728 1.4986	1288.7 .9431 1.5316	1320.2 1.008 1.5607	1350.6 1.069 1.5875	1380.3 1.129 1.6125	1409.3 1.186 1.6361
700 (503.04)	h_S v s	…… …… ……	…… …… ……	…… …… ……	1242.4 .7251 1.4719	1279.7 .7906 1.5080	1312.9 .8489 1.5387	1344.5 .9037 1.5666	1375.1 .9562 1.5924	1404.9 1.006 1.6165
800 (518.18)	h_S v s	…… …… ……	…… …… ……	…… …… ……	1228.8 0.6126 1.4463	1270.1 0.6750 1.4862	1305.4 0.7293 1.5188	1338.2 0.7791 1.5477	1369.7 0.8266 1.5743	1400.4 0.8721 1.5992
900 (531.95)	h_S v s	…… …… ……	…… …… ……	…… …… ……	1213.6 0.5233 1.4208	1259.8 0.5841 1.4655	1297.4 0.6358 1.5003	1331.7 0.6820 1.5305	1364.2 0.7258 1.5579	1395.8 0.7674 1.5835
1000 (544.58)	h_S v s	…… …… ……	…… …… ……	…… …… ……	1196.5 0.4539 1.3950	1248.7 0.5107 1.4455	1289.2 0.5602 1.4829	1324.9 0.6039 1.5144	1358.5 0.6450 1.5427	1391.0 0.6837 1.5691
1100 (556.28)	h_S v s	…… …… ……	…… …… ……	…… …… ……	…… …… ……	1236.6 0.4500 1.4257	1280.5 0.4982 1.4663	1317.9 0.5401 1.4992	1352.6 0.5786 1.5285	1386.1 0.6152 1.5557
1200 (567.14)	h_S v s	…… …… ……	…… …… ……	…… …… ……	…… …… ……	1223.4 0.3986 1.4059	1271.3 0.4462 1.4501	1310.6 0.4866 1.4848	1346.5 0.5234 1.5151	1381.1 0.5579 1.5431
1300	h_S v s	…… …… ……	…… …… ……	…… …… ……	…… …… ……	1208.7 0.3537 1.3855	1261.6 0.4018 1.4343	1303.0 0.4412 1.4709	1310.7 0.4775 1.4775	1376.0 0.5095 1.5312
1400 (586.96)	h_S v s	…… …… ……	…… …… ……	…… …… ……	…… …… ……	1192.4 0.3141 1.3643	1251.2 0.3632 1.4186	1295.2 0.4022 1.4574	1333.8 0.4364 1.4900	1370.8 0.4679 1.5199
1500	h_S v s	…… …… ……	…… …… ……	…… …… ……	…… …… ……	1174.3 0.2789 1.3420	1240.2 0.3298 1.4029	1287.1 0.3682 1.4442	1327.2 0.4014 1.4780	1305.4 0.9318 1.5091
1600 (604.74)	h_S v s	…… …… ……	…… …… ……	…… …… ……	…… …… ……	…… …… ……	1228.3 0.2997 1.3870	1278.6 0.3385 1.4313	1320.3 0.3708 1.4666	1360.0 0.4001 1.4986
1800 (620.86)	h_S v s	…… …… ……	…… …… ……	…… …… ……	…… …… ……	…… …… ……	1201.5 0.2482 1.3537	1260.6 0.2880 1.4058	1306.2 0.3195 1.4444	1348.7 0.3472 1.4787
2000	h_S v s	…… …… ……	…… …… ……	…… …… ……	…… …… ……	…… …… ……	1168.9 0.2044 1.3168	1240.7 0.2468 1.3802	1291.4 0.2781 1.4230	1337.0 0.3047 1.4599
3000	h_S v s	…… …… ……	…… …… ……	…… …… ……	…… …… ……	…… …… ……	…… …… ……	1066.3 0.0983 1.2028	1199.3 0.1476 1.3155	1271.1 0.1742 1.3737
3226	h_S v s	…… …… ……	…… …… ……	…… …… ……	…… …… ……	…… …… ……	…… …… ……	…… …… ……	1171.2 0.1271 1.2874	1253.8 0.1572 1.3545

shallow bank of boiler tubes, Fig. 132, page 200. It is then known as an **interdeck superheater** and receives most of its heat by convection but also some by radiation. This arrangement permits a high degree of superheat with a reasonable amount of superheating surface, and also insures a fairly constant temperature at different ratings.

Radiant type superheaters, as the name implies, receive their heat by direct radiation from the furnace and are located in one or more walls of the furnace, Fig. 83. Besides superheating the steam, they perform the additional function of steam-cooling the furnace wall and thus replace considerable perishable brickwork in high-duty furnaces. This type of superheater is generally used where a high degree of superheat is desired, although it has been applied for superheat as low as 40 degrees. Another method of securing high superheat temperatures is to use a convection superheater in series with a radiant heat superheater. By this arrangement a fairly constant temperature is maintained over a wide range. This is so because the tendency of the convection type is to cause slightly increased temperatures with increase in load, while the radiant-heat type produces a reduced temperature with increase in rating.

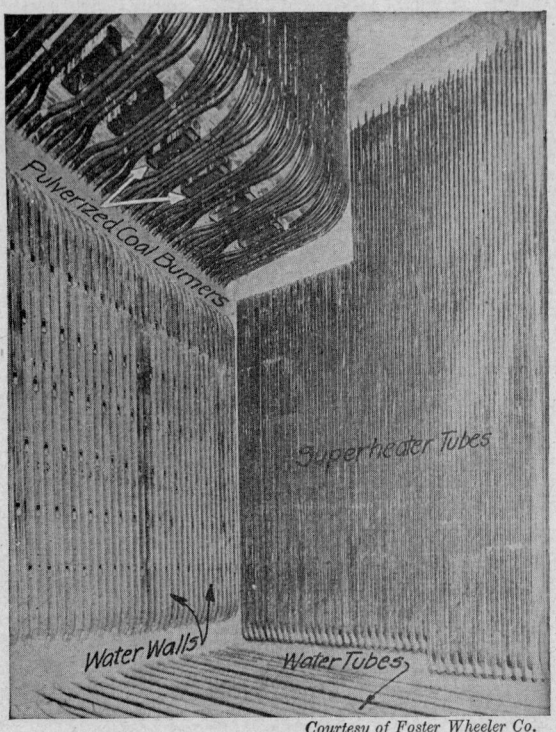

Courtesy of Foster Wheeler Co.

FIG. 83. — Radiant Type Superheater.

The attached superheater has four advantages: besides having (1) lower first cost, and (2) higher operating efficiency, it requires (3) less attention, and (4) less space. As ordinarily installed, it is subject to the fluctuating temperatures of the furnace, which produce a varying degree of superheat, depending upon the manner in which the furnace is being operated. Superheaters are sometimes constructed to maintain a constant degree of superheat automatically, but in most cases the superheat increases slightly with the load. The attached, or integral, type of superheater is most common in standard central station practice. It is used with both fire- and water-tube boilers, and with special construction is adapted to marine and locomotive boilers.

The separately-fired superheater is similar to the attached type, but has a special furnace and setting, not connected with the boiler setting. By this method the steam from a number of boilers may be superheated by piping them all to the superheater. It has the following advantages:

1. Degree of superheat may be varied independently of the load on the boiler.
2. May be placed at any convenient point.
3. Repairs can be easily made without shutting down the boiler.
4. Gives a higher superheat than most superheaters of the integral type.

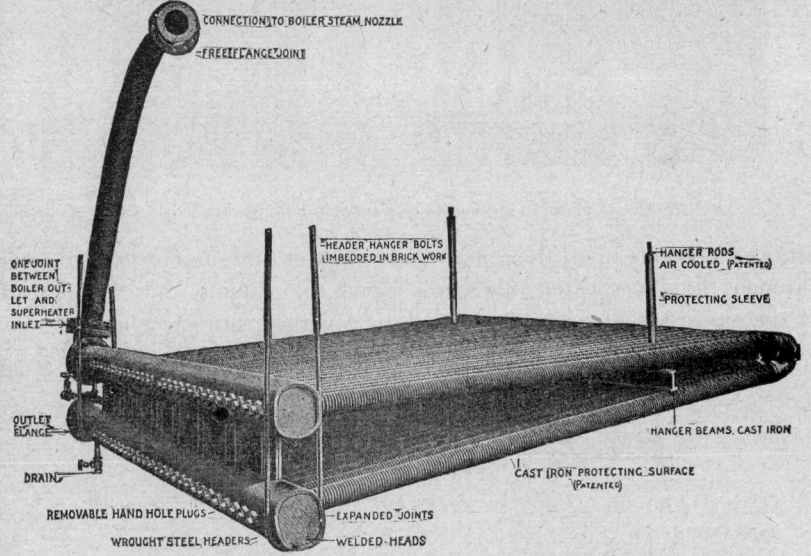

FIG. 84. — Foster Convection Superheater.

Its disadvantages are that it requires: (1) separate attention, and (2) extra space and piping.

111. Foster Convection Superheater. — This superheater, Fig. 84, has an inlet and outlet header, into which are expanded a series of steel pipes through which the steam passes to be superheated. In order to present as large an area as possible to the heat, a closed steel tube is placed inside the straight portion of each pipe, and an annular space is thus formed for the passage of the steam. A series of cast-iron rings are shrunk over the outside of the steel pipes, to protect them from overheating when putting the boiler in service or when there is not a rapid movement of steam through the superheater.

122 SUPERHEATERS

Steam leaves the boiler through the boiler nozzle and enters the inlet pipe of the superheater; as it passes through the superheater, its temperature is raised above that existing in the boiler. The final temperature

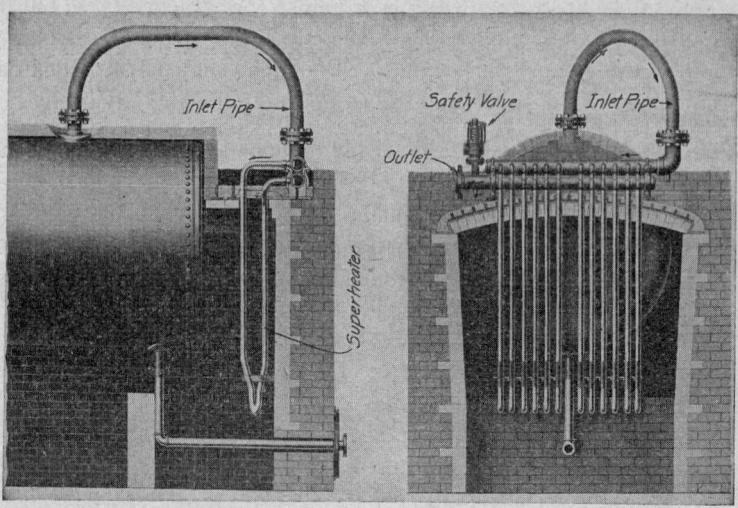

Fig. 85. — Elesco Convection Superheater on H. R. T. Boiler.

attained depends upon the volume and temperature of the steam passing through the superheater tubes, and upon the volume and temperature of the gases flowing over the tubes. In American practice the final tem-

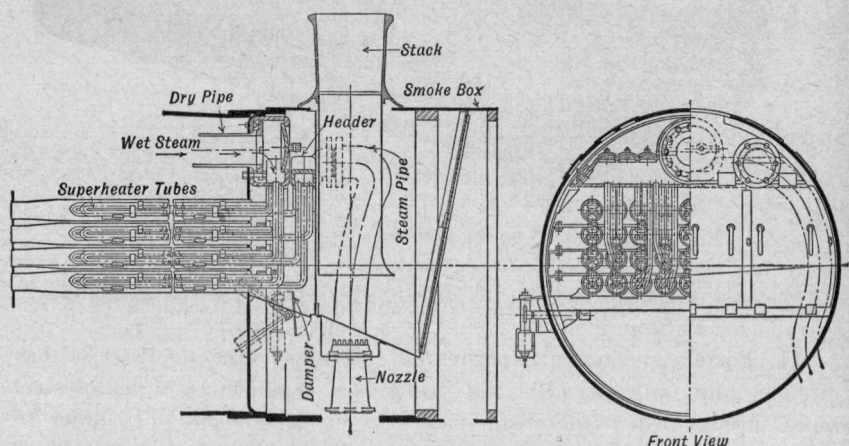

Fig. 86. — Locomotive Superheater.

perature is seldom above 750 deg. fahr., while in European practice the final temperature is sometimes as high as 1000 deg. fahr.

An **Elesco** convection superheater located in the rear pass of a return-tubular boiler setting is shown in Fig. 85.

112. Locomotive Attached Superheater. — A typical locomotive superheater is shown in Fig. 86, located in the flues of a locomotive boiler. It consists of two headers and a number of small tubes connecting the headers.

113. Radiant Superheater. — The Elesco superheater, Fig. 87, consists of units made of steel tubing extending vertically between two headers and arranged along one or more sides of a furnace behind water-cooled furnace walls which afford sufficient protection to insure prolonged life to the superheater units while not shutting off to any appreciable extent the radiation from the furnace. The units are held in position away from the boiler brickwork by high heat-resisting supports and are free to expand and contract without in any way affecting the boiler setting. The headers are outside the setting, which affords ready access to the joints. High steam velocities are used to keep the inside of the units clean and prevent accumulation of any foreign matter which may come over in the steam and which becomes more objectionable with higher steam pressures and temperatures. The steam velocities are ordinarily limited to a minimum of 1500 feet per minute in order to prevent overheating.

114. Steam Purifier. — In order that the saturated steam entering the superheater may be as dry as possible, and the life of the superheater thereby increased, a steam purifier is often installed in high capacity units at the outlet of the boiler, either inside or outside the boiler drum. It may be of either the **centrifugal** or **baffle** type. A typical baffle-type purifier, located between the steam drum and superheater, is shown in Fig. 88 and consists of a horizontal cylinder, at one end of which the steam enters an expansion chamber, where the greater part of the entrained moisture is arrested. The steam then passes through an approach nozzle, which directs it against a baffle specially designed to remove the last finely divided particles or droplets of water. The separated water is drained away by three pipes leading to a **discharger,** which is a large capacity steam trap (page 246) specially designed for high pressures. It is claimed that this purifier will deliver steam containing less than 0.5 per cent of moisture.

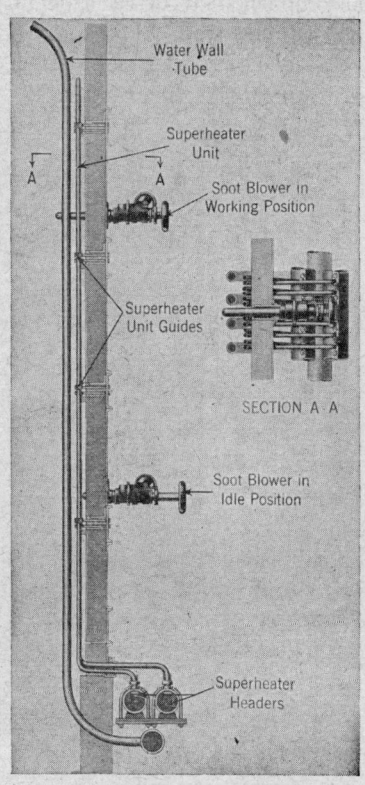

FIG. 87. — Radiant Superheater.

115. Desuperheaters. — In many plants, operated at comparatively low pressure and temperature, new extensions are designed for high pressure and high superheat, and in order to make the operation more flexible and use the new boilers in connection with the existing equipment, desuperheaters are applied. The desuperheater, with water as a cooling medium and in connection with a reducing valve, lowers the temperature and pressure of the steam to that required by the low-pressure equipment. This is accomplished by using either a **spray-type desuperheater** or a **surface or non-contact type**. The spray-type desuperheater is illustrated in Fig. 89.

Fig. 88. — Cochrane Baffle-type Purifier.

It consists of a chamber in which are located single or multiple nozzles which deliver the water in the form of a fine spray into the steam after the pressure has been reduced to the discharge pressure. The superheat is removed by the evaporation of a portion of the water sprayed. Baffles are arranged to catch the moisture carried in suspension by the desuperheated steam and thus have the steam leave the desuperheater in a dry and saturated condition.

When using a desuperheater it is desirable that a constant temperature be maintained on the discharge side of the desuperheater. This is accomplished by means of an automatic temperature control, Fig. 89. When the temperature in the saturated steam header rises, as the result of an increased flow of superheated steam, the temperature regulator reduces the air pressure acting on the diaphragm in the water control valve, and permits this valve to open and admit additional water to the desuperheater nozzles. This immediately reduces the temperature of the steam to normal. When the temperature in the saturated steam header falls below normal, because of decreased steam flow, the flow of water is reduced, thereby restoring the temperature of the saturated steam to normal. Compressed air at not less than 30 pounds is used for the temperature regulator, and a by-pass arrangement for hand control is provided in the water line.

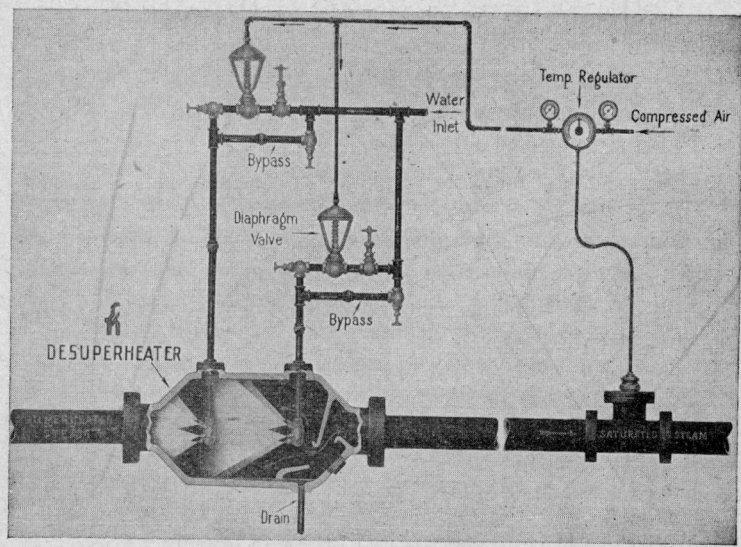

Fig. 89. — Schutte-Koerting Spray-type Desuperheater.

116. Reheaters. — Economical turbine operation limits the maximum moisture content in the exhaust steam to about 10 per cent. With present temperatures limited to below 800 deg. fahr., this point is reached at about 500 pounds pressure for normal turbine operation; consequently for higher pressures it is advisable to resort to re-superheating or *reheating;* that is, taking the entire flow of steam after it has expanded, to about the saturation point in a turbine, and reheating it to its original temperature before passing it through the remaining stages of the turbine. This can be done in both single cylinder and compound units. The latter is the more common since the break between high- and low-pressure units is a natural reheating point.

The use of drier steam throughout the turbine also results in less erosion of the low-pressure buckets, and increased turbine efficiency. The decrease in steam consumption may be as much as 6 or 7 per cent. The gain produced by reheating is mainly the result of the increased turbine efficiency, rather than through any improvement in the steam cycle.

Fig. 90 (*a*) and (*b*) show diagrammatically the condition of steam, in the various stages of a 19-stage turbine, without reheating and with reheating for an initial pressure of 600 pounds and with 250 degrees superheat. The

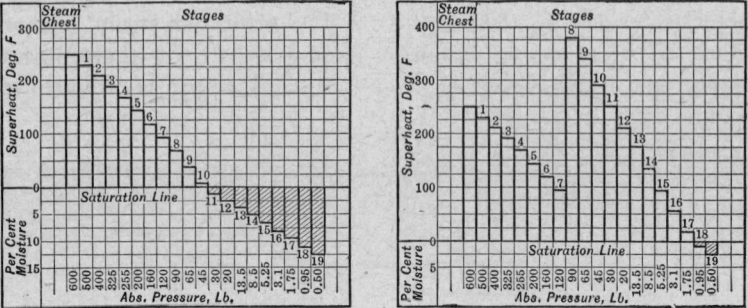

Fig. 90. — Effect of Reheating on Moisture Content of Steam in Turbine.

first figure indicates the large number of stages in which moisture is found, and the considerable percentage of moisture in the last stage when the steam is not reheated. The *second figure* shows the condition of the steam

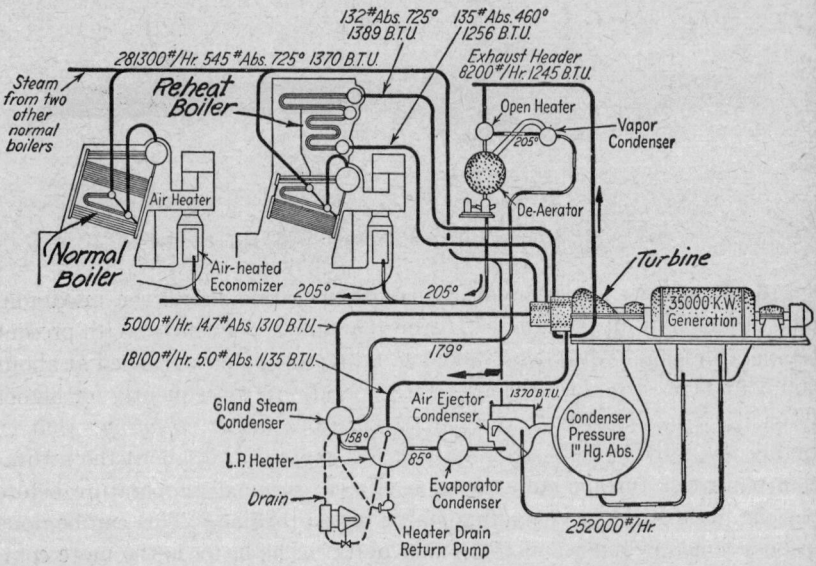

Fig. 91. — Heat Balance Diagram, Normal and Reheat Boilers — Stanton Station.

in the various stages of the turbine, when steam is withdrawn at the 7th stage, and reheated to a total temperature of 700 deg. fahr. before being returned to the 8th stage. The diagrams clearly show that, when using steam at 600 pounds pressure, one reheating makes it possible to have dry steam in nearly all stages of the turbine.

Reheating may be accomplished by using any of the following types of superheaters: (1) separately fired, (2) radiant, (3) convection, (4) steam, or (5) a combination. The radiant, convection and steam reheaters are

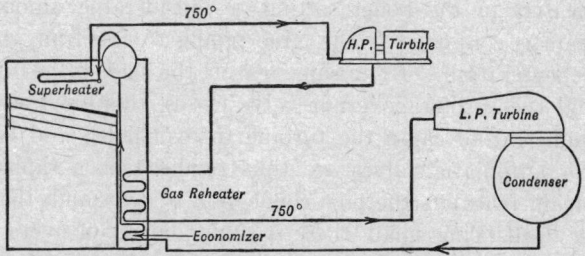

FIG. 92. — Reheating in High-pressure Boiler Setting.

being used to the greatest extent. When using the former, reheating may take place within the boiler setting and by the furnace gases, using a special boiler called a reheat boiler, Fig. 91, which is similar to a normal boiler except that it has a different arrangement of the boiler heating surface. For example, the reheat boilers installed in the *Columbia Power Station* are proportioned as follows: *boiler heating surface*, 6066 square feet at 650 pounds per square inch pressure, *reheater surface*, 17,612 square feet

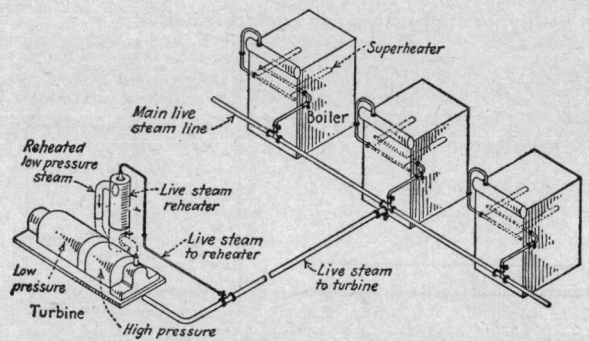

FIG. 93. — Steam-heated Reheater.

at 120 pounds per square inch. Reheating in a high-pressure boiler installation, Fig. 92, may also take place in the same setting with the main boiler, using either a convection or radiant superheater.

A somewhat more simple method of reheating consists of passing the steam through a live-steam reheater, consisting of a series of coils surrounded by the steam to be reheated. When using this method, the heater is placed near the turbine, Fig. 93, and long runs of large piping and many fittings are thus eliminated. A relatively small amount of high-pressure steam is required for reheating by this method. In some cases not more than 5 per cent of the steam used in the turbine will be required for reheating the low-pressure steam 100 deg. fahr. The thermodynamic gain in using steam is, however, less than when the steam is reheated in a boiler to live-steam temperature.

With reheaters in the boiler setting, a considerable amount of low-pressure steam is contained in the large piping to and from the reheater and in the reheater itself. If, for some reason, the load on the turbine drops suddenly and the turbine governor valve closes, this low-pressure steam would be sufficient to cause the turbine to overspeed, and low-pressure regulators or automatic valves are thus required as a safety measure. With live-steam reheaters the pipe connection is short and, the volume of the reheater itself being small, there is slight danger of overspeeding the turbine and no special regulators are required.

117. Theoretical Steam Power Cycles. — Although an actual steam power plant cannot be operated under ideal conditions, the study of 'ideal cycles is of much importance to engineers because of the efforts being made to increase the efficiency of the steam plants by using high-steam pressures, together with various schemes for recovering as much as possible of the heat ordinarily wasted in the engine or turbine exhaust. The exhaust loss in a real plant is larger than all the others combined, so that the best way of utilizing some of this exhaust heat, other than by industrial heating, which is sometimes feasible, becomes a problem of increasing importance as higher fuel prices are encountered.

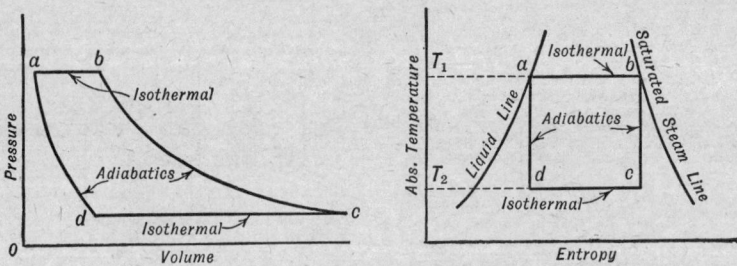

Fig. 94. — Carnot Cycle for Dry and Saturated Steam.

The cycles which are of most use in studying the heat changes occurring in steam power plants, are: (1) the **Carnot**, (2) **Rankine**, and (3) modifications of the Rankine called, (a) the **reheating cycle** and, (b) the **regenerative cycle**.

The Carnot cycle, as applied to a steam engine or turbine, using dry and saturated steam, is shown in Fig. 94, using both pressure-volume and temperature-entropy as coördinates. All the heat is added to the water, which is considered as though it were actually contained within the cylinder of the engine, between a and b at constant pressure, and temperature (T_1).

As soon as the water is all evaporated expansion takes place adiabatically to point c which is at condenser temperature, (T_2), a portion of the steam being condensed during the expansion. The mixture of the steam and water is then compressed to point d at constant pressure and temperature (T_2). Between d and a the moisture is compressed adiabatically to the starting point, all the steam having been condensed at the temperature of evaporation corresponding to a. *The cycle, therefore, consists essentially of two isothermals and two adiabatics, the isothermals in this case being lines of constant pressure.* This cycle is of particular interest because it is a reversible cycle and can be passed through in either direction, and it can be shown that it gives the maximum obtainable efficiency, which can be

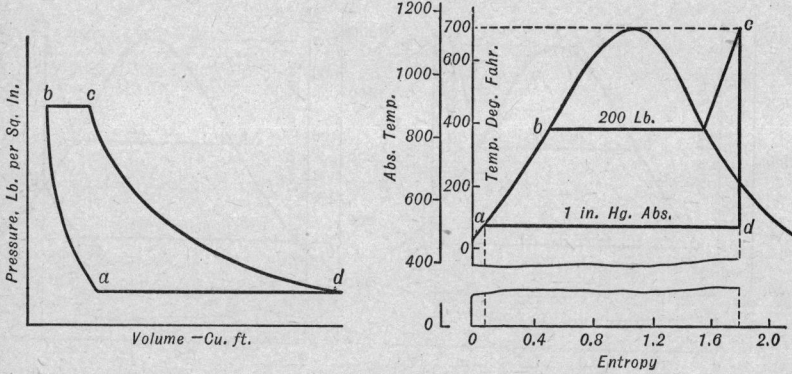

FIG. 95. — Modified Rankine Cycle.

shown to be $(T_1 - T_2) \div T_1$, in which T_1 corresponds to the steam temperature and T_2 to the condenser temperature, both expressed in the absolute scale of temperatures.

The modified Rankine cycle[1] is made up of the processes indicated in Fig. 95. *In this cycle the steam is generated in a boiler at constant pressure and then passes without any loss by throttling or radiation to the engine or turbine*, in which adiabatic expansion, cd, takes place until exhaust pressure is reached as indicated by the state d. Condensation of the entire amount of exhaust steam then takes place at constant pressure as shown by the line da, and is finally pumped back cold to the boiler. *Such a cycle is not reversible and consequently is less efficient than the Carnot.* This cycle is in common use and has much to recommend it from the standpoint of simplicity, but it does not have as great possibilities from the standpoint of thermal economy as some of the following modifications.

The *reheating cycle*, Fig. 96a, is similar to the Rankine cycle except that after the adiabatic expansion from c to d the steam is reheated in a suitable reheater at constant pressure from d to e, after which it is returned

[1] The ideal Watt diagram for the Rankine cycle is discussed on page 436.

to the turbine where a second adiabatic expansion takes place to the pressure of the exhaust as shown from *e* to *f*.

The *regenerative cycle*, Fig. 96b, is so called because steam is bled from the turbine at a number of stages to feedwater heaters, and the heat which is thus recovered is returned to the boiler. In the ideal cycle an infinitely large number of bleeder heaters are assumed to be used so that, if superheated steam is not bled, the line or curve *de* is drawn parallel to the liquid line *af*, using the entire weight of steam considered in the cycle, even though successive portions of the steam are withdrawn from the turbine and condensed thereby heating the feedwater, while the remainder continues to expand adiabatically through the turbine. The efficiency of this cycle can

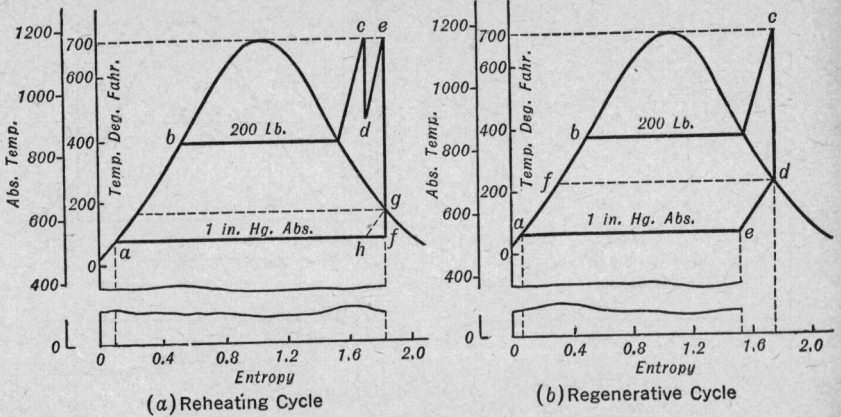

Fig. 96. — Reheating and Regenerative Cycles.

be raised to that of the *Carnot cycle* by heating the feedwater to the boiler temperature using bled steam.

A combination of the last two cycles known as the **reheating-regenerative cycle,** *first* uses steam reheated as in the reheating cycle, and *second* the regenerative principle is applied as indicated by the dotted line *gh* in Fig. 96a.

The practical application of the last three cycles results in complicated piping and station layout, but because of their advantages they are being used in a number of modern stations.

In a paper entitled "High pressure, Reheating and Regenerating Steam Power Plants," presented before the American Society of Mechanical Engineers, C. F. Hirshfield and F. O. Ellenwood state: "From the consideration of the *ideal cycles* alone it appears that the Rankine cycle, which is the one requiring the least apparatus, is the least efficient throughout the entire range of pressures studied, and yields the least energy per unit volume of exhaust steam. At the top of the list, throughout the upper portion of the pressure range considered, and very close to the top in the lower portion, stands the regenerative cycle both as to efficiency and available energy per unit volume of exhaust steam. The extra apparatus which must be used in order to follow the regenerative cycle instead of the Rankine would seem to be simple enough to justify the extra expense in many cases."

REFERENCES

Superheat Engineering Data, THE SUPERHEATER COMPANY.
Steam, BABCOCK and WILCOX Co.
Steam Power Plant Engineering, GEBHARDT.
Power, Vols. 47, 49, 50, and 51, January numbers.
Fuels and Steam Power, 1928 to date.

REVIEW QUESTIONS AND PROBLEMS

1. State the advantages brought about by superheated steam.
2. Describe the construction of an attached type of superheater.
3. Find out to what extent locomotives operating in your neighborhood use superheated steam.
4. Referring to Fig. 82, find the saving in the amount of steam used by raising its temperature from that of saturated steam until it is superheated 150 deg. fahr. Is the percentage saving the same when the superheat is raised from 150 to 250 deg. fahr. as in the former case? What does this show regarding the point at which the greatest saving is made?
5. What is meant by the term direct-fired superheater?
6. Describe the Carnot, Rankine and reheating cycles.
7. Name two types of desuperheaters, and describe the operation of one type.
8. What is the main reason for using a reheater?
9. The Walsh and Weidner cross-drum boiler at the Grand Avenue Station of the Kansas City Power and Light Company, operates at a pressure of 650 lb. per sq. in. abs. and has a convection type superheater which at lowest normal operation adds 250 deg. fahr. superheat to 150,000 lb. of steam per hr.

Find: (*a*) the per cent saved as a result of superheating, feedwater temperature, 353 deg. fahr., (*b*) total B.t.u. saved per hour.

10. Using data from the steam tables replot Fig. 96 (*a*) so as to show the effect of increasing the temperature from 700 deg. fahr. to 800 deg. fahr., then add a 1200 lb. curve and show the effect upon this latter plot of increasing the temperature to 800 deg. fahr.

CHAPTER VI

FUELS

118. Foreword. — A fuel is any material, natural or artificial, that can be burned, and that may be obtained in quantities at a reasonable price. The fuels most commonly used for power purposes exist in the following forms: (1) *solid*, such as coal, wood and vegetable wastes; (2) *liquid*, such as crude petroleum and its distillates; or (3) *gaseous*, such as natural or artificial gas. Coal is the most common fuel for power purposes, with crude oil being used in regions adjacent to oil fields, and on shipboard. Natural gas is used where available, and artificial gases, such as blast-furnace, coke-oven, producer and illuminating, are used where special conditions warrant.

The proper utilization of any fuel calls for a knowledge of its chemical and physical characteristics. This is especially true in the case of coal; consequently when choosing a coal for power plant use, information should be at hand regarding the ash and moisture content, size and storage requirements, and the working qualities of the coal under the conditions where it is to be burned. Regardless of its first cost, the coal which will result in lowest cost per unit of steam when labor, maintenance and ash disposal are included, is the best.

Coal in the pulverized form is used extensively because it can be burned with high economy and efficiency. The burning of this fuel has called for more complete information regarding what takes place during the burning process, and the investigations for this purpose have resulted in increasing the power output per ton of coal.

Recently efforts have been made to produce a smokeless fuel suitable for power and domestic purposes by carbonizing bituminous coal in a retort at low temperatures and thus obtaining semi-coke, and in addition a considerable yield of gas, ammonia, light oil and tar. Two large, low-temperature carbonization plants are in experimental operation in the United States, and several in Europe. This process would have a beneficial effect upon the smoke nuisance. It has not, however, been developed to an extent sufficient to indicate its ultimate effect upon power plant practice.

119. Formation of Coal. — Coal is of vegetable origin, and is the remains of prehistoric forests. Destructive distillation, together with great pressure, *has resolved the organic matter into its ultimate constituents, carbon, hydrogen, oxygen and other substances*, in varying proportions. The factors that have produced the different grades of coal are, (1) time, (2) depth of bed, (3) disturbance of bed, and (4) intrusion of mineral matter.

120. Composition of Coal. — Coal consists of carbon, hydrogen, moisture, ash, oxygen, nitrogen, and sometimes sulphur. The carbon, hydrogen and sulphur are the "combustible" in the coal, a term applied in boiler practice to that part of the fuel which is left after taking away the moisture and ash, thus including whatever nitrogen and oxygen may be present. The remaining substances are impurities which lower the value of a coal as a fuel.

121. Coal Analysis.[1] — Two methods of analysis are used to determine the composition of coal. They are known as the **proximate analysis**, which determines the composition of coal by mechanical processes, and the **ultimate analysis,** which separates coal into its chemical constituents by chemical processes. The analysis to be used depends in a large measure upon the use which is to be made of the data obtained.

122. Proximate Analysis. — In the proximate analysis of coal, as received, five determinations are usually made; namely, *moisture, volatile matter, ash, fixed carbon, and sulphur.* These constituents are expressed in per cent by weight; the sum of the first four items is taken as 100 per cent, with the sulphur separately determined.

Moisture. — This determination does not differentiate between the moisture that comes from external sources and that which is inherent in the coal. The amount of moisture in such coals as anthracite and semibituminous may be obtained by placing a 15 or 20 pound sample in a shallow pan and drying, at about 90 deg. fahr., for twelve hours. A more accurate method is to heat a small quantity of finely ground coal, in a double-walled air bath, to a temperature between 240 and 280 deg. fahr. until the sample reaches a minimum weight. For lignites, a lower drying temperature is used. *The loss in weight is called moisture.*

Volatile matter. — *The volatile matter consists of the carbon combined with the hydrogen, together with the other gas-forming constituents which are driven off by heat.*

One gram of finely powdered coal from which the moisture has been removed is placed in a platinum crucible with a tight cover. The crucible and contents are heated to about 1750 deg. fahr., for seven minutes. The crucible is then cooled and weighed. *The loss in weight is the volatile matter.*

Ash. — *Ash is the incombustible residue from the complete combustion of the coal.* The sample used in the volatile matter determination is heated in an open pan, by a blast lamp, until completely burned. A stream of oxygen may be used to hasten combustion. *The weight of the residue is ash.*

Fixed carbon. — *Fixed carbon is the carbon which is not in combination with any other element.* It does not represent all the carbon in the coal, as a considerable amount is driven off with the volatile matter in combination with hydrogen and oxygen. It is not pure carbon, as it contains some

---

[1] Consult A. S. M. E. Test Code for Solid Fuels.

ash-forming constituents, and approximately one-half the sulphur. *The weight of fixed carbon is obtained by subtracting the weight of moisture, ash, and volatile matter from the weight of the original sample.*

Sulphur. — This element may be separately determined by burning a portion of the sample with a suitable chemical mixture that will combine with the sulphur in such a form that the sulphur can be separated into a sulphur compound and its amount determined by weighing.

123. Ultimate Analysis. — This analysis expresses the composition of a coal in percentage by weight of carbon, hydrogen, nitrogen, sulphur, oxygen, and ash. It requires the careful use of chemical apparatus, and reliable results can be obtained only by those entirely familiar with all details of the work. *The sum of these six constituents, with moisture included as a separate item, is taken as equal to* 100 *per cent.*

124. Method of Reporting Analyses. — Both analyses may be expressed in terms of:

(1) Coal "as received" or "**as fired.**"
(2) Coal "moisture free" or "**dry.**"
(3) Coal "moisture-and-ash-free" or "**combustible.**"

The first term under each method is that used by the BUREAU OF MINES, and the second that used in the A. S. M. E. TEST CODE. The latter will be used in this text.

The A. S. M. E. Test Code for fuels calls for the proximate analysis to be expressed on the "as fired" basis, which represents the condition of the coal as fed to the furnace; and the ultimate analysis to be reported on both the "as fired" and the "dry" basis. The information obtained from these analyses serves to indicate the adaptability of a coal for a given use; furnishes data from which the heat value may be calculated and the various coals classified, as explained in Art. 131, page 138. The fusing temperature of the ash should be included with the analysis in order to give an indication of the working qualities of the coal under furnace conditions.

Coal "as fired" is in the same condition as when it comes from the bunkers. The analysis on a dry-coal basis is found from the analysis "as fired," by dividing each constituent by $(1 -$ proportional weight of moisture$)$. The analysis on a combustible basis is found by dividing the analysis "as fired" by $[1 - $ (proportional weight of ash $+$ moisture)$]$.

Example 17. — The per cent by weight of the constituents of a Kentucky coal "as fired" is given by the following ultimate analysis: ash 4.39, sulphur 1.22, hydrogen 5.43, carbon 77.37, nitrogen 1.83, oxygen 9.76. Moisture from proximate analysis 3.10. Convert this analysis to the dry-coal and the combustible basis.

Solution. — The ultimate analysis of a coal "as fired" when reported as above, contains the free moisture as a part of the hydrogen and oxygen. To obtain the moisture as a separate item, one-ninth of the moisture should be subtracted from the hydrogen and eight-ninths from the oxygen. The reason for this is that water is composed of one

DESCRIPTION OF OXYGEN BOMB FUEL CALORIMETER 135

part hydrogen and eight parts oxygen. The first column in the solution has been corrected by this method.

Dividing the analysis "as fired" by (1 − the proportional weight of moisture) to obtain the analysis on the dry basis, and by [1 − (the proportional weight of ash + moisture)] for the combustible basis, there results:

	Coal as fired	Coal dry	Combustible
Carbon	77.37	79.84	83.63
Hydrogen	5.07	5.24	5.48
Nitrogen	1.83	1.89	1.98
Oxygen	7.02	7.24	7.59
Sulphur	1.22	1.26	1.32
Ash	4.39	4.53	
Moisture	3.10		
	100.00	100.00	100.00

125. Heat Value of Solid Fuels. — *The heat value of a fuel is the amount of heat a pound of the fuel will generate when completely burned.* It may be determined (1) by **calorimeter**, (2) by **computation based on ultimate analysis,** and (3) from **curves based on the proximate analysis.**

126. Description of Oxygen Bomb Fuel Calorimeter. — A standard type of fuel calorimeter, known as the Mahler Bomb Calorimeter, is shown in

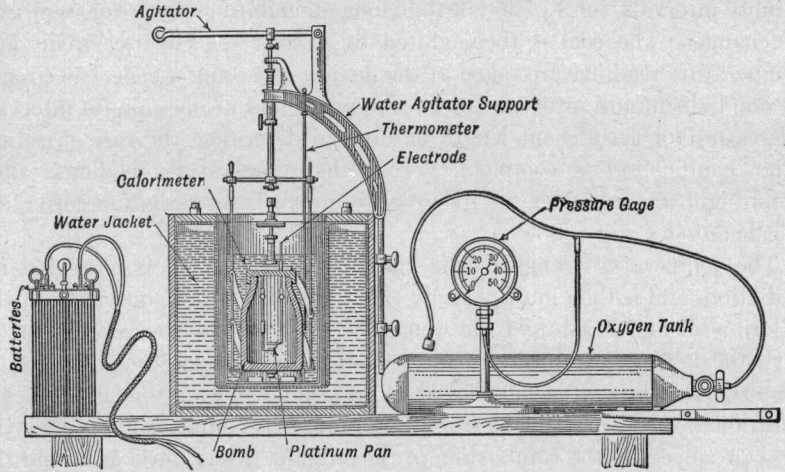

Fig. 97. — Mahler Bomb Fuel Calorimeter.

Fig. 97. It may be used for solid or liquid fuels. The calorimeter has a porcelain- or gold-lined steel bomb having a tight-fitting cover screwed into place. Within the bomb is a small platinum pan for holding the fuel. Two electrodes, connected at their lower end by a fuse wire, extend through the cover of the bomb, and connect the fuse wire to an electric circuit containing a suitable switch and a 12-volt storage battery. The bomb sets into a calorimeter which contains a definite weighed amount of water. A stirring device for agitating the water within the calorimeter is attached to a support connected to the outside of the apparatus. The temperature of the water in the calorimeter is shown by an accurate thermometer graduated to read to one-thousandth of a degree. The calorimeter and bomb are surrounded by an outer vessel covered with asbestos and filled with water at a temperature slightly higher than the surrounding atmosphere. This outer vessel maintains constant conditions and makes the determination of the radiation loss more accurate. An oxygen tank supplies the oxygen required to burn the fuel, and a pressure gage attached to the oxygen tank piping is used to determine the pressure of the oxygen entering the bomb.

Operation. — From an average sample of pulverized coal, approximately one gram is placed in the fuel pan, and a piece of fine iron fuse-wire, which dips into the coal, is connected between the electrodes. The cover of the bomb is then screwed tightly into place and the bomb is placed in the calorimeter, which has been partially filled with a definite weight of water — generally an amount of water equal to the **water equivalent** of the apparatus, as determined by calculation or experiment. Oxygen from the tank is then slowly admitted to the bomb until the pressure is about 25 atmospheres.

Temperature readings of the water in the calorimeter are taken at one-minute intervals, for a period of time long enough to insure a constant rate of change. The coal is then ignited by closing the electric circuit and temperature readings are taken at the instant of closing the electric circuit, at one-half-minute intervals for five minutes, and at one-minute intervals thereafter for a sufficient length of time to determine the rate of change after combustion is complete. From the temperature readings taken before ignition and after the five-minute interval, the radiation correction is calculated.

The temperature range during the period of burning is corrected for radiation, and is then multiplied by the sum of the water equivalent of the calorimeter and bomb and the weight of the water in the calorimeter, to give the heat of combustion expressed in gram-calories, weight being in grams and temperature in degrees Centigrade. From this product are subtracted the heat resulting from the formation of aqueous nitric acid, the heat produced by the combustion of the sulphur to sulphuric acid, and the heat formed by the combustion of the fuse wire.[1] The remainder, divided by the weight of the sample in grams and multiplied by 1.8, gives the heat in B.t.u. per pound.

127. High and Low Heat Value of Fuel. — With fuels which contain hydrogen, the heat value as found by the calorimeter is higher than that realized under working conditions in boiler practice, by an amount equal to the latent heat of evaporation of the water which is formed. This heat

---

[1] These corrections are small and are often neglected.

reappears when the vapor is condensed, but in ordinary conditions of combustion, the vapor passes away uncondensed.

The **"higher" heat value** is that determined by the calorimeter, and should be used in all boiler test calculations. The **"lower" heat value** is the higher heat value less a correction for the heat lost in the escaping steam.

128. Heat Value from the Ultimate Analysis. — The equation most used for calculation of heat values from the composition of coal is that proposed by Dulong, and gives values accurate to within 1 or 2 per cent. It is written:

$$\text{Heat per pound of fuel, B.t.u.} = 14{,}600\,C + 62{,}000\left(H - \frac{O}{8}\right) + 4000\,S \quad (41)^1$$

in which C, H, O, S, = proportion by weight of the carbon, hydrogen, oxygen and sulphur, as shown by the ultimate analysis of a pound of fuel, and the coefficients = the heat evolved when one pound of carbon, hydrogen, and sulphur are completely burned.

Part of the hydrogen in the coal is assumed to be in combination with the oxygen in the coal. The hydrogen available for producing heat is therefore the actual hydrogen minus that already combined with the oxygen to form water. This latter equals one-eighth of the oxygen present, or $\frac{O}{8}$.

The **available, or free hydrogen,** as it is called, is written $\left(H - \frac{O}{8}\right)$.

Example 18. — The ultimate analysis of a coal, expressed in per cent by weight, gave: carbon, 84.45; hydrogen, 4.25; oxygen, 3.04; nitrogen, 1.28; sulphur, 0.91; and ash, 6.07. Find the heat per pound of coal by Dulong's equation.

Solution. — Using Equation (41)

$$\text{Heat per pound of coal} = 14{,}600\,C + 62{,}000\left(H - \frac{O}{8}\right) + 4000\,S.$$

Substituting the value of carbon, hydrogen, oxygen and sulphur, there results

$$\text{Heat per pound of coal} = 14{,}600 \times 0.845 + 62{,}000\left(0.0425 - \frac{0.0304}{8}\right) + 4000 \times 0.0091$$

$$= 14{,}772 \text{ B.t.u.}$$

129. Heat Value from Proximate Analysis. — Calculations of the heat value of a coal based on a proximate analysis are not satisfactory. A curve showing the relation between the heat value per pound of combustible of a coal and the percentage of fixed carbon is given in Fig. 98. It is based on data taken from tests conducted by the United States Geological Survey.

130. Methods of Expressing Heat Value of Coal. — The heat value of a coal may be expressed in terms of (1) *coal as fired*, (2) *dry coal*, or (3) *combustible*.

[1] Bulletins of the Bureau of Mines give Dulong's equation as:

$$\text{B.t.u. per lb.} = 14{,}544\,C + 62{,}028\left(H - \frac{O}{8}\right) + 4050\,S.$$

The heat value per pound of dry coal is obtained from the heat value per pound of coal as fired, by dividing by (1 − *proportional weight of moisture*). It represents the heat that would be generated by burning one pound of coal containing no moisture. The heat value per pound of combustible is found by dividing the heat value per pound of coal as fired, by [1 − (*proportional weight of ash + moisture*)]. It represents the heat that would be generated by burning one pound of coal containing neither ash nor moisture.

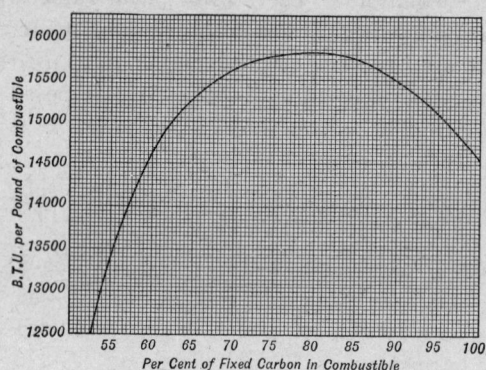

FIG. 98. — Heat Value of Coal using Proximate Analysis.

Example 19. — The heat value per pound of a North Dakota coal *as fired* is 7069 B.t.u., with moisture 35.96% and ash 7.75%. Find the heat value per pound of (*a*) dry coal, (*b*) combustible.

Solution. — Dividing the heat value as fired, by (1 − *proportional weight of moisture*) for the heat value of dry coal, and by [1 − (*proportional weight of ash + moisture*)] for heat value of combustible.

Heat per pound of dry coal $= \dfrac{7069}{1 - 0.3596} = 11{,}031$ B.t.u.

Heat per pound of combustible $= \dfrac{7069}{1 - (0.3596 + 0.0775)} = 12{,}560$ B.t.u.

131. Coal Classifications. — There are two basic methods used to classify coal; the *first* based on the intrinsic chemical and physical properties of the coal itself; and the *second* based on the purposes for which the coal can be used to good advantage.

TABLE 9. — CLASSIFICATION OF COAL BY RANK

Rank, Ash-free		Approximate Moisture Content	Fuel Ratios $\dfrac{FC}{VM}$
Peat		80–90	
Lignite		30–45	
Sub-bituminous		12–30	
Bituminous	most common for power purposes	3–15	3 or less
Semi-bituminous		3–6	3–6
Semi-anthracite		3–6	4–10
Anthracite		2–3	over 10
Superanthracite		2–13	over 10

The *first method* includes: (1) that used by the Bureau of Mines and the U. S. Geological Survey as tabulated in Table 9, and is based on the

PHYSICAL AND BURNING CHARACTERISTICS OF COAL 139

proximate analysis and the chemical and physical properties of the coal. The various boundaries of the ranks for the better grades of coal being set by the "fuel ratio," i.e., the ratio of fixed carbon to volatile matter; (2) that proposed by Seyler, based on the ultimate analysis which gives the composition of the coal substance, namely carbon, hydrogen and oxygen, calculated free from ash, sulphur and nitrogen; and (3) the method recommended by Prof. Parr based on volatile matter and calorific value, both values being given in terms of "unit" coal, that is, coal calculated free from moisture and ash, and corrected for sulphur and water of hydration. Other methods of classification based on the proximate and ultimate analyses are shown in Table 10.

TABLE 10. — CLASSIFICATION OF COALS BY CARBON CONTENT AND CARBON–HYDROGEN RATIO*

Name of Coal	Percentages of Combustible		Carbon — Hydrogen Ratio
	Fixed Carbon	Volatile Matter	
Anthracite............	97 to 92.5	3 to 7.5	30 to 26
Semi-anthracite........	92.5 to 87.5	7.5 to 12.5	26 to 23
Semi-bituminous.......	87.5 to 75	12.5 to 25	23 to 20
Bituminous, East......	75 to 60	25 to 40	20 to 14.4
Bituminous, West......	65 to 50	35 to 50	14.4 to 11.2
Lignite................	50 and under	50 and over	11.2 to 9.3
Peat..................			9.3 to.....

* Or ratio of the total carbon to the hydrogen.

The *second method* includes: classification by *name of coal bed*, as Pittsburgh or Pocahontas coal; *geographically*, as New River or Hocking Valley; by *use*, as gas coal, steam coal or by-product coal; and by *burning properties*, as free burning, coking and smokeless.

The A. S. M. E. code says that "these systems of classification give small information concerning the behavior characteristics of a coal, and can be used only in a limited way to determine where and how it shall be burned and what results to expect. It is well known that coals of the same class have quite different performance characteristics and the physical and working characteristics often are of prime importance in indicating the use."

132. Physical and Burning Characteristics of Coals. — Typical analyses of coals, together with the fusing point for the lower grades of coal, are given in Table 11.

Anthracite coal has a deep black color, and high luster. It is a hard coal, composed almost entirely of fixed carbon, with a specific gravity of 1.3 to 1.9. It is nearly smokeless and burns with a short, bluish flame, retains its original size, and does not swell when burning. The commercial sizes of

TABLE 11. — ANALYSES OF TYPICAL COALS

	Name of Coal						
	Anthracite	Semi-Anthracite	Semi-Bituminous	Eastern Bituminous	Western Bituminous	Sub-Bituminous	Lignite
Proximate Analysis							
Moisture........	3.45	1.57	4.07	8.98	14.43	22.63	36.78
Volatile matter..	2.75	9.40	16.34	34.49	29.48	35.68	28.16
Fixed carbon....	87.90	83.69	68.47	50.30	42.81	37.19	29.97
Ash.............	5.90	5.34	11.12	6.33	13.28	4.50	5.09
Ultimate Analysis:							
Carbon..........	88.86	85.46	76.51	70.50	54.59	54.91	41.87
Hydrogen.......	2.04	3.72	4.27	4.76	5.49	6.39	6.93
Nitrogen........	0.90	1.12	1.00	1.36	1.11	1.02	0.69
Oxygen.........	1.95	3.45	6.59	15.66	21.52	32.59	44.94
Sulphur.........	0.35	0.91	0.51	1.39	4.01	0.59	0.48
Ash.............	5.90	5.34	11.12	6.33	13.28	4.50	5.09
Heat Value:							
Calorimeter.....	13,950	....	13,509	12,417	10,064	9,734	7,002
Dulong's equation	14,103	14,552	13,329	12,084	9,866	9,478	6,944
Classification:							
Carbon-hydrogen ratio..........	42.50	23.0	19.60	14.80	12.30	9.40	9.60
Fusing point of ash*...........	....	....	2600°F.	2300°F.	2100°F.	2200°F.	....

* Approximate values.

anthracite are given in Table 12, and are graded by the size of wire-mesh screen through which the sample will pass or will not pass. A $\frac{1}{4}$-inch mesh screen has openings $\frac{1}{16}$ of a square inch in area.

TABLE 12. — SIZES OF ANTHRACITE, OR "HARD," COAL

Name of Size	Size of Screen through which Coal	
	Will Pass	Will Not Pass
Culm...................	$\frac{3}{32}$-in.	......
Birdseye................	$\frac{5}{16}$-in.	$\frac{1}{4}$-in.
Buckwheat No. 1........	$\frac{1}{2}$-in.	$\frac{1}{4}$-in.
Buckwheat No. 2, or Rice...	$\frac{1}{4}$-in.	$\frac{3}{8}$-in.
Pea....................	$\frac{3}{4}$-in.	$\frac{1}{2}$-in.
Chestnut...............	$1\frac{3}{8}$-in.	$\frac{3}{4}$-in.
Stove or Range.........	2-in.	$1\frac{3}{8}$-in.
Egg (in the East).......	$2\frac{3}{8}$-in.	2-in.
Broken.................	4-in.	$2\frac{1}{2}$-in.

Semi-anthracite coal has an iron-black color, is not as hard as anthracite and has less luster. It is a free-burning coal, and burns with a longer and more luminous flame than anthracite. While not very plentiful, it is the best steam producing coal found in the United States. Its specific gravity is about 1.4.

PHYSICAL AND BURNING CHARACTERISTICS OF COAL

Semi-bituminous coal is a softer and more friable coal than anthracite, with a lower specific gravity and more volatile matter. It kindles readily and gives off a small amount of smoke when burning. This coal is extensively used in steam power plants.

Bituminous coal is soft, having a color ranging from pitch black to dark brown, with a silky or resinous luster. It has a large amount of volatile matter, is brittle, breaks into small pieces when stored, and burns with a long yellow and smoky flame. Its specific gravity is about 1.3.

Bituminous coals may be classified into **coking** or **caking**, and **non-coking** coals. The latter burn freely, do not fuse and are known as free-burning coals. Coking coals swell up, become pasty and fuse together when burning. They are rich in hydrocarbons and hence are valuable for the manufacture of gas. These coals absorb moisture readily. They are sized as in Table 13.

TABLE 13. — SIZES OF BITUMINOUS, OR "SOFT," COAL

Name of Size	Size of Screen through which Coal	
	Will Pass	Will Not Pass
Run of the Mine	Mixture of lumps and fine coal, or slack.	
Sized Coal,* such as:		
Lump	6-in.	3-in.
Egg	3-in.	$1\frac{1}{2}$-in.
Nut	$\frac{1}{2}$-in.	$\frac{3}{8}$-in.
Slack	$\frac{3}{8}$-in.	
Screenings	Smallest sizes	

* The sizes given differ in different parts of the United States.

Cannel coal is homogeneous, has a grayish-black color with a resinous luster, is high in hydrocarbons, kindles readily, and burns with a dense, smoky flame. It is used for the manufacture of gas and in fireplaces. Its specific gravity is about 1.24, and its heating value is low.

Sub-bituminous coal is a grade between true bituminous coal and lignite. It resembles bituminous coal in appearance, is not woody like lignite, and slacks and absorbs moisture readily. It does not coke, burns freely with a long, yellow flame and is likely to produce smoke.

Lignites have a brown color with a pitchy luster and woody structure. Lignite is high in ash, moisture and oxygen, with a specific gravity between 1.2 and 1.23, and cannot be transported far because it breaks so easily. It burns freely with a bright, slightly smoky, yellow flame, and is used principally for the manufacture of gas.

Peat is an accumulation of partly decomposed water plants, mosses, and other vegetable matter. It has a color ranging from yellow to reddish brown, a fibrous texture and a high moisture content, and must be dried before burning. It is seldom used in the United States as a fuel.

133. Location of Coal Fields. — Anthracite coals are nearly all found in northern Pennsylvania, with some distribution in small areas in some of the Western States. Semi-anthracite is only found in small quantities in southern Pennsylvania, West Virginia, and Virginia. The centers of production of semi-bituminous coals are the Pocahontas and New River fields in Virginia and West Virginia, Georges Creek field in Maryland, Windber field in Pennsylvania, and small areas in Arkansas, Washington, and Colorado. Bituminous coal fields are found in the Appalachian mountains and scattered through the Middle Western and Western states. Sub-bituminous fields are located mostly in North Dakota. Lignites are found in North Dakota and most of the Western states, and in Alaska.

134. Purchase of Coal. — The purchaser should be familiar with boiler and furnace room conditions in order to intelligently purchase coal that will be suited to the draft, load and furnace conditions of the plant. Coal costs from 50 to 85 per cent of the total cost of making steam and consequently that coal should be chosen which will give the lowest cost per unit of steam. Coal is ordinarily purchased by weight; but when weight is the only consideration the results are likely to be unsatisfactory, because the heat value of the coal and its working qualities are of prime importance in the operation of the plant. A more satisfactory method of purchase is by specification, the contract to cover coal of a definite analysis and heat content, with suitable allowance in price for variation therefrom.

135. Storage and Weathering of Coal. — The storage of coal has become a necessity, because of market conditions, danger of labor difficulties, and the crowding of transportation facilities. Anthracite is almost an ideal coal for storing; it is not subject to spontaneous ignition and may be stored in large piles. Bituminous coal is likely to ignite if placed in deep piles and will also suffer from disintegration. It is sometimes stored under water to prevent **spontaneous ignition,** that is, ignition resulting from the heat produced by the absorption of oxygen by the hydrocarbons in the coal and by the slow combustion, or oxidation, of carbon, sulphur, and available hydrogen. Spontaneous ignition can only take place when the air supply is sufficient to support oxidation but will not entirely remove the heat produced. Another method used to prevent spontaneous ignition is to pack the coal until it weighs approximately 65 pounds per cubic foot. Causes which are contributory to fires in stored coal are the size of the coal, the freshness of the coal, and temperature at time of storage.

Coal loses heat value when exposed to air during storage. When coal is stored in piles, its temperature should be closely watched; if it is becoming too high the pile should be moved.

136. High- and Low-ash Coals. — Coals having **over ten per cent of ash** are classed as high-ash coals, and those having **less than ten per cent** as low-ash coals. Coals having a high-ash content require more labor for the removal of ash from the furnace.

EFFECT OF ASH, MOISTURE, SULPHUR, OXYGEN IN COALS

137. Clinkering and Non-clinkering Coals. — Clinker is formed by the mechanical combination of the constituents which form the ash, or by the fusing of the ash; it may be hard or soft. Hard clinker, which is produced by the melting of the ash, forms in hard lumps; soft clinker, which is formed by the silica of the ash combining with the fusible constituents, iron and lime, in the ash, remains molten and continues to grow. Ash, having a melting temperature of 2700 deg. fahr., will not form clinker at ordinary furnace temperatures. *The clinkering qualities of a coal may be of more importance than the heat value, for if the fusion temperature of the ash is low for the working conditions in the furnace, the furnace linings, grates, and tubes will be damaged and the up-keep costs increased.*

Clinker causes furnace linings and grate bars to wear out rapidly. It reduces furnace efficiency and capacity, besides entailing considerable work in its removal. *Hard clinker may be formed by poor firing methods.*

Clinker may be prevented by (1) carrying a thin fire, (2) avoiding the improper use of slice bar and rake in mixing ash into the burning fuel, (3) firing in small amounts on thin spots of fire, (4) having ash-pit clean and doors open, (5) using water in the ashpit or steam jets under grates to prevent clinker sticking to grates, (6) avoiding shaking coal that will burn in ashpit, and (7) preventing too high a furnace temperature.

138. Effect of Ash, Moisture, Sulphur, and Oxygen in Coals. — Ash is composed of silica, alumina oxide, iron oxide and other impurities in small amounts. It reduces the heat value of a coal by replacing combustible matter, and adds to the cost of handling and storage. Additional expenses resulting from the presence of ash are the cost of cleaning fires and ashpit and the expense of disposal of the ash. Ash obstructs the flow of air through the grates and may form troublesome clinker under certain conditions. The higher the ash content, the higher will be the percentage of unburned carbon in the refuse. The elimination or removal of the ash which would be discharged into the atmosphere, when burning pulverized fuel, is a problem of great importance.

Moisture lowers the heat value of a coal, because heat is required for its evaporation; it reduces the actual weight of coal purchased by the weight of moisture, and adds to the cost of transportation. Small sizes of coal hold much more moisture than large sizes.

Sulphur occurs in coal as iron pyrite, marcasite, sulphate of lime, iron and alumina, as organic sulphur, and sometimes as free sulphur uncombined with other elements. *As free sulphur it has its full heat value. When it occurs as a sulphate, it has no heat value,* and in any case the available hydrogen must be high for sulphur to have any appreciable heat value. In small amounts it may assist in the prevention of clinker; but if the ratio of the weight of sulphur in the ash to the total ash is high, it may assist in forming a troublesome clinker. It is also a frequent cause of corrosion in equipment coming in contact with the gases, which are cooled below the dew-point temperature of the vapor.

Oxygen has about the same effect upon the heat value of a coal as an equivalent amount of ash. It is an original impurity of which the better grades of coal have small amounts.

139. Briquetted Coal. — This fuel consists of finely ground coal, mixed with a suitable binder, such as pitch, and pressed into briquettes. Such treatment decreases the fuel loss through the grates, increases the heat value of a fuel and permits the use of lower-grade fuels. Although the cost and difficulty of manufacture have prevented the extensive use of this fuel as a source of power, briquetted lignites are used extensively in some regions for domestic purposes. Briquetting permits the weathering, handling and burning of lignites without disintegration.

140. Pulverized Coal. — Pulverized coal is coal that has been reduced to a powder of such fineness that it will float in air, the degree of fineness depending on the kind of coal and its moisture content. Formerly it was thought necessary to have 85 per cent of the coal pass through a 200-mesh screen; it has been found, however, that this is unnecessary, and it is now customary to pulverize so that between 65 and 75 per cent will pass a 200-mesh screen, about 90 per cent through 100-mesh and not less than 99 per cent through 40-mesh screen. There is a real question as to what is the proper fineness for best combustion and greatest operating economy. It is certain, however, that the finer the coal, the more simple will be the problem of delivering it to the burners, as well as that of burning, and handling it in the furnace. With large particles of coal there is a tendency for the carbon loss to become excessive as the result of the coal particles dropping into the ashpit and being cooled before they are burned out.

The burning of pulverized coal for power purposes results in an increase in the overall efficiency of the boiler, because of the complete combustion of the fuel, and the accurate control of the air and coal needed for combustion. It also forms an extremely flexible method of varying the steam rate by varying the rate of firing the fuel. Other advantages to be gained from burning coal in the pulverized form are: the wider range of fuels that can be burned, the prevention of loss from the banking of fires, and the removal of ash in most cases is easier than in other methods of firing.

141. Pulverized Coal Systems. — Either the **bin, or central, system** and the **unit system** are used to fire pulverized coal. In large central power stations there is no distinct trend toward either the central or unit systems, but in the industrial field the unit system is largely predominant.

142. Bin or Central System. — The bin system, Fig. 99, shows this method of firing as used in the Calumet Station of the Commonwealth Edison Company. Coal, after being crushed, is delivered to the raw-coal bunker from which it is fed by gravity to a waste-heat drier through which there is a continuous flow of coal to the ball-pulverizing mill. Preheated air is used for drying, and is drawn through the drier by an exhauster fan and delivered to the cyclone separator or dust collector. The pulverized coal discharged from the pulverizing mill is delivered to a pulverized-coal pump, and is pumped to the pulverized-coal bunker, using compressed air as the conveying medium. From this bin the coal passes through the pulverized-coal feeder to the Calumet type burners located in the furnace walls. Air for combustion is supplied by the forced-draft fan as shown. All the pulverizing equipment is driven by electric motors.

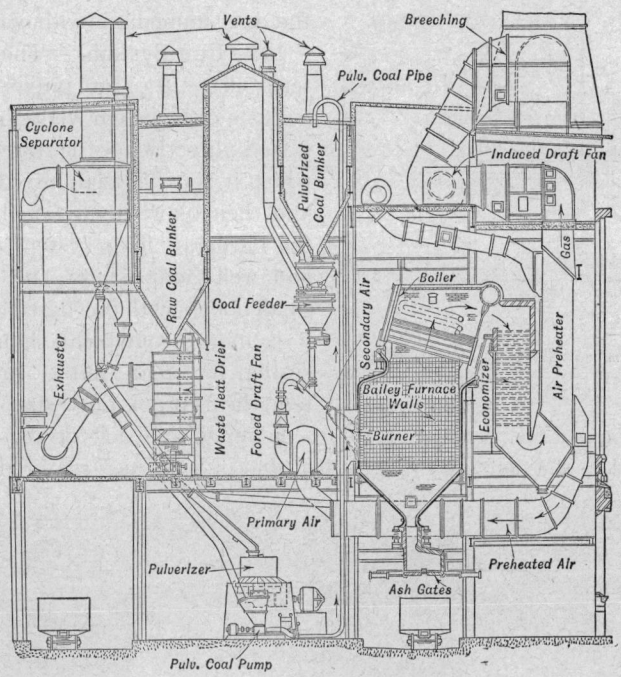

FIG. 99. — Bin System for Pulverized Coal.

Coal is often dried in separately fired rotating horizontal driers, similar to the carbonization retort shown on page 152. The tendency, however, appears to be to use vertical driers which may use waste-flue gases, preheated air, or bleeder steam as the drying agent. Both the horizontal and vertical driers are being used to dry the coal sufficiently to prevent caking during storage. This is especially true in metallurgical and cement furnaces where moisture is prohibitive, but in steam power work drying is now being accomplished by using either flue gases or preheated air directly in the pulverizers.

The **Randolph vertical waste-heat drier** is shown in Fig. 100. The undried coal enters at the top and travels gradually down toward the bottom in a zigzag course, being deflected by the gas ducts and thence into the path of the gases. Drying is accomplished by heat transfer through the gas-duct walls and by direct contact of the gas with the coal as the gases pass from the inlet to the outlet openings. The drier body is built in unit sections mounted one above the other. Each unit has a series of gas-distributing ducts. The gas enters at one end of the lower duct in each section and rising through the coal is discharged from the opposite end of the duct immediately above.

143. Unit System. — The unit system, Fig. 101, for pulverized coal consists of a system wherein raw coal is fed directly to the pulverizer in which it is pulverized, mixed with air, and then blown by a fan directly to the burners. This system is simple, and well adapted to the firing of small boilers with powdered coal, since it is more economical than the bin system for small units. In this system the type of pulverizer is of the utmost importance because it has to pulverize and deliver the coal at a rate equal to the demand of the boiler, and of a fineness required for good

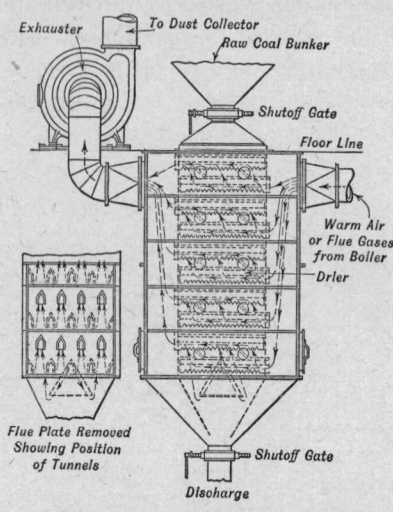

FIG. 100. — Vertical Waste-heat Drier.

FIG. 101. — Unit System for Pulverized Coal.

combustion. The accurate supply of the coal to the furnace, therefore, involves a careful design by means of regulating the flow of coal to the mill.

COAL PULVERIZING MILLS

The central system pulverizer can be operated at constant load, and the fluctuation in demand taken care of by the variation in the rate of flow from the powdered coal bin. With the unit system the velocity of the mixture of coal and air in the conduits to the burner must be maintained in order to prevent the coal and air from separating, since if this occurred, the coal would be delivered in slugs, with a consequent loss in combustion. The *advantages* of the unit system consist mainly in dispensing with (1) the driers, (2) the transport system from the pulverizing house to the boiler-room and (3) the feeders and bins for powdered coal. Its *disadvantages* are (1) the difficulties of distributing the coal evenly to the

FIG. 102. — C-E Hammer Pulverizing Mill — cover removed.

boilers, (2) the higher power consumption for pulverizers operating with a varying load, and (3) a lack of a reserve supply of powdered coal.

144. Coal Pulverizing Mills. — Pulverizing mills may be classified according to the nature of the crushing action into three main types, namely: the *hammer mill, the ball or roller mill, and the tube mill.*

The **hammer mill** operates at high speed and depends upon the hammer or impact action between the rotating paddles and the stationary casing to cause breaking. This action affects primarily the larger particles, but of itself it does not positively eliminate all coarse particles. The Combustion Engineering unit system pulverizing mill is typical of this type, and consists of the parts shown in Fig. 102. The coal enters through a hopper mounted above an automatic **ratchet feeding device** which discharges coal into the grinding chamber formed by the base and top of the mill and at a rate determined by the boiler load. The alloy-steel hammers are attached to a forged steel disc and are rotated by the shaft. They break the coal down to a powder by the force of the impact and by knocking it against the walls of the grinding chamber. Mounted on the same shaft with the hammers is a **fineness regulator,** and an exhaust fan. The fan continuously moves a stream of air through the mill and sweeps from the grinding chamber such particles of fuel as will pass the fineness regulator, and discharges them into the delivery pipe to the burner, the coarser particles remaining in the chamber for further reduction.

The fineness regulator is mounted on the shaft between the hammers and the fan and revolves within a conical housing. The outer edges of the regulator blades are formed at an angle identical to that of the housing or

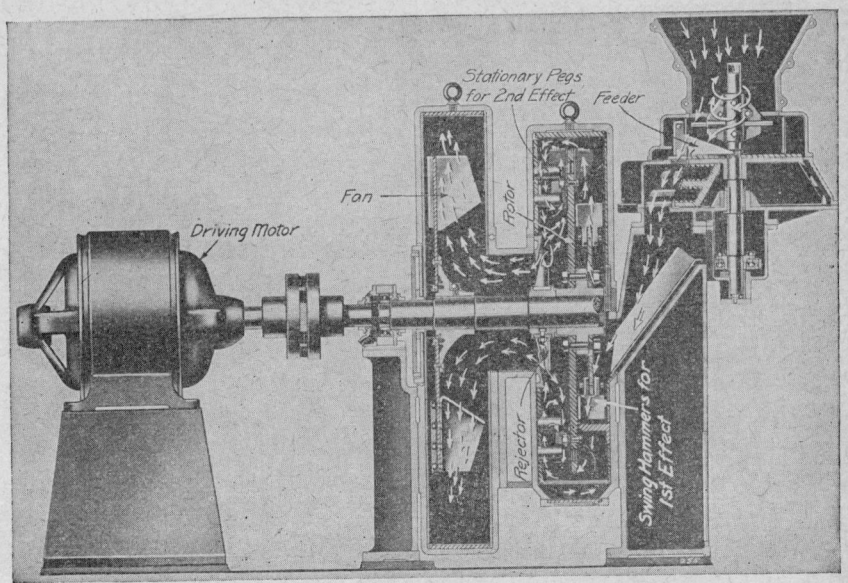

Fig. 103. — Riley Pulverizing Mill — section.

throat, and the desired degree of fineness is obtained by moving the regulator blade along the shaft, toward or away from the housing.

A tramp iron pocket is provided in the bottom of the mill to catch any pieces that are small enough to pass through the feeding arrangement. A recent development in this type of mill is shown in Fig. 103 with all parts clearly named.

The ball or roller mill is characterized by large balls or cylinders running in a die ring or on a revolving plate. Crushing is obtained by pressure on the coal between the rolling ball or wheel and the ring or plate. The mill action is more positive than the hammer type in the elimination of coarse particles. Mills of this type are usually air-swept, thus introducing a second factor in reducing the amount of coarse coal in the final product.

COAL PULVERIZING MILLS

A typical ball mill of the air separator type is illustrated in Fig. 104. It consists of a top or separator section, an intermediate or pulverizing section, and a base or drive section.

The top section is constructed of heavy steel-plate and cast iron. It contains the classifier, coal spout and mechanism for adjusting the grinding pressure. The intermediate section is of heavy-section cast iron, with carefully machined flanges to which the top and base sections are secured, and contains the grinding elements, coal basket and driving yoke

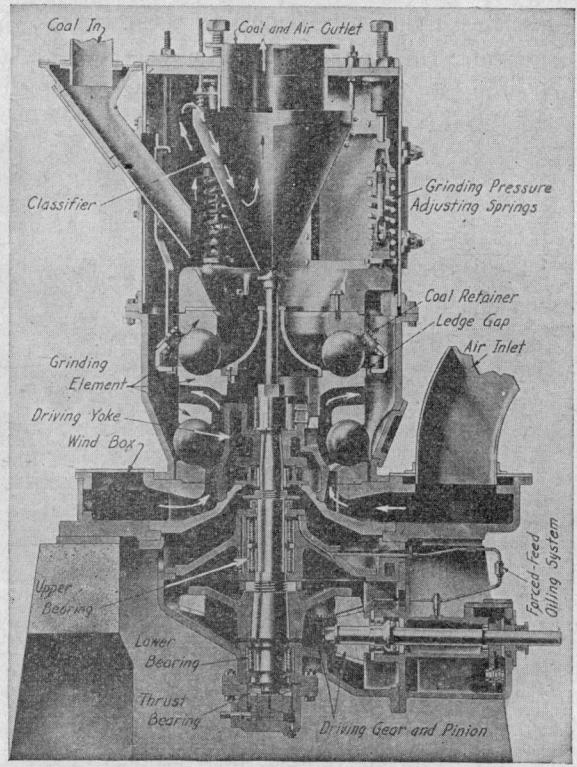

Fig. 104. — Lehigh Ball Type Pulverizing Mill.

The base or drive section provides a rigid support for the mill, and encloses the air-distributing chamber, main and thrust bearings, drive shafts and gears, and the force-feed oiling system.

The grinding elements consist of two rows of large diameter balls, two stationary and one rotating grinding ring. The rows of balls — one row mounted above the other — are separated and propelled by the rotating ring which is driven by and floats on the main driving shaft. Grinding pressure between the balls and rings is applied and kept uniform by externally controlled steel springs mounted on the top section.

The main driving shaft is supported by a self-aligning, heavy-duty Rollway thrust bearing, and all bearings and gears are lubricated by an automatic, force-feed oiling system.

Raw coal, crushed to pass through a $\frac{3}{4}$-inch ring, is fed to the mill from an overhead bunker by means of a drag feeder. The coal drops through the feed spout onto a cone, which rotates inside the upper row of grinding balls, whence it is thrown outward into the path of the grinding balls.

The coal after being ground in the upper row of balls passes through an annular opening between the rotating grinding ring and the coal basket. Air currents, discharged through openings in the rotating grinding ring, entrain the fines and carry them to the top of the mill, while the coarse particles of coal drop into the lower row of grinding balls for further pul-

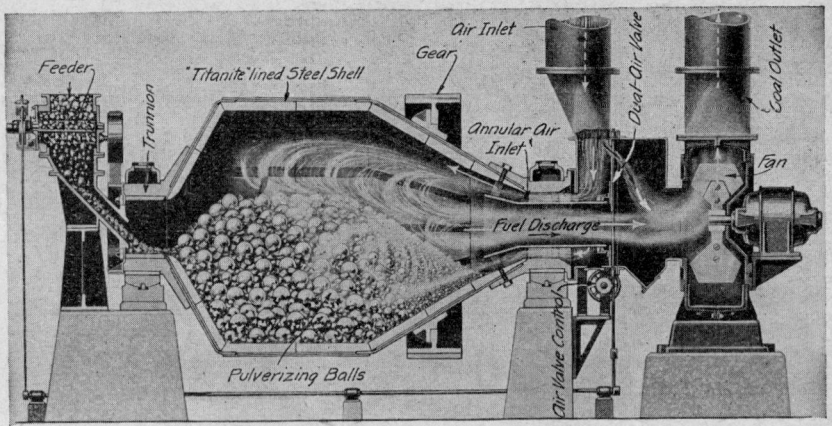

FIG. 105. — Ball-tube Type Pulverizing Mill.

verizing. Any coarse particles, carried by the air stream to the top of the mill, are rejected by gravity or the action of the classifier. These coarse particles return to both rows of grinding balls for further pulverizing.

The **tube mill** includes all slow-speed mills in which the tumbling of balls or slugs within a rotating compartment produces pulverization. In the usual path of flow through a mill the large particles are subjected to crushing action, so that though there is no positive elimination the coarse particles are effectively reduced. Mills of this type are, as a rule, swept with air.

Since grinding results from the cascading of the balls and from the pressure and movements of the balls within the charge, large and small particles alike are crushed by this action; and a large amount of superfine material should be produced for a given sieve fineness.

The *Hardinge Conical Ball-tube Mill* consists of the parts indicated in Fig. 105. The drum consists of two truncated horizontal cones rigidly joined together by a short cylinder. The drum rotates on hollow trunnions placed at the ends. A feed mechanism at one end supplies coal through one trunnion and an exhauster fan at the other end discharges the pulverized coal through the other trunnion. The Hardinge mill may be driven by either motor or turbine, the power being transmitted to a herringbone gear mounted on the longer cone. The grinding is accomplished by the cascading of steel balls and coal within the drum.

In the operation of the mill the quantity of coal entering the mill is automatically adjusted to the amount of coal discharged from the drum. The coal, being pulverized, passes gradually toward the discharge end of the mill where it is picked up by air which is drawn into the mill through a reversed air classifier by a fan located at the discharge end. The classifier is so designed that the coarser particles of coal are forced back toward the feed end, while the finer particles are carried out of the mill. The amount of air entering the mill is controlled by a hand-operated valve, so arranged that all of the air may flow directly from the air inlet to the fan without entering the mill, or all of the air may flow through the mill.

145. Low-temperature Carbonization. — *This process is one in which bituminous coal is distilled in a retort to obtain semi-coke or charcoal, and in addition the by-products of tar, gas and light oil.* A ton of high-volatile American coal will yield about 25 gallons of tar, 3 gallons of light oil, 1400 pounds of 12 per cent volatile semi-coke and 3500 cubic feet of 800 B.t.u. gas. As yet no process has proved itself from a commercial standpoint in connection with power plant practice. The future success of this process will no doubt depend upon the economic disposal of the by-products.

The great obstacles that have hindered progress in low-temperature coking are (a) *mechanical difficulties* and (b) *thermal difficulties*.[1] The mechanical difficulties have arisen through the necessity of handling hot, sticky, plastic, materials and the evil of hard, graphitic deposits. The thermal difficulties arise through the reduced temperature gradients and resultant troubles of heat transfer. Neither of these difficulties has been surmounted, but efficient design is progressing toward the solution of the one, and the recognition of certain principles such as that of using thin layers of coal to facilitate heat transfer is pointing to the solution of the other.

Low-temperature processes may be classified as: (1) externally heated retorts, (2) internally heated retorts and (3) complete gasification processes. The temperatures used vary from 660 to 1300 deg. fahr. The individual processes differ mainly in the distribution of the heat units between coke, gas, tar and radiation.

The type of retorts used are: static or dynamic; internally or externally fired, or both; and vertical, horizontal, or inclined. In some cases the coal is briquetted before coking, in some the coal is not briquetted, and in others is used in a pulverized form.

[1] Economics of Coal Carbonization in the United States by George A. Orrok.

152   FUELS

The **K. S. G.** or **The Kohlenscheidungs Gesellschaft** process, Fig. 106, belongs to that group of carbonizing methods in which coal is distilled as it moves through a rotating externally heated cylindrical retort consisting of two concentric drums, inclined slightly from the horizontal, supported on bull wheels located at each end, and rotated at three-quarters of a revolution per minute. Each retort is surrounded by a steel casing lined with insulating and fire brick. The outer drum is 72 feet long with a diameter of 10 feet, while the inner drum is 85 feet long with a diameter of $5\frac{1}{2}$ feet.

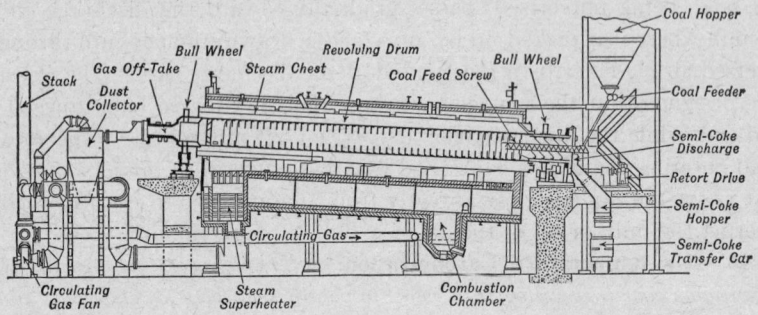

Fig. 106. — K. S. G. Low-temperature Carbonization Retort.

The raw coal is fed continuously from the storage bin into the lower end of the retort through a screw conveyor. It is carried by the helical flanges of the inner drum, which has a 5 per cent pitch, to the upper end, where it spills through open ports into the outer drum, and gravitates to the lower end of the retort. It is here picked up as semi-coke by a series of scoops, dropped on to a receiving plate and carried by reverse helical flanges to the discharge gate, or delivered into hot-coke cars from which the hot semi-coke is raised by means of a skip hoist into a dry-coke quencher, which uses water-cooled inert gases to quench the coke. This results in a better coke structure with a lower percentage of unsalable fines than is obtained with the wet process of quenching. The coal gas escapes through the drum head and the off-take pipe at the upper end of the retort. The time required for this operation is about two hours.

146. Liquid Fuel. — Petroleum is the only liquid fuel sufficiently abundant and cheap to be used for the generation of steam. It is generally classified according to the base it yields upon distillation; that is, (1) **paraffin**, (2) **asphalt**, and (3) **olefine**.

To the first group belong the fuels of the Appalachian Range and the Middle West of the United States. This group yields so many valuable light oils that the price is prohibitive for use as a fuel.

The asphalt group is found in Texas and California. They have a color varying from reddish brown to jet black, and are extensively used for fuel. The olefine group is found in Russia and is used as a fuel oil.

Crude oils consist of carbon and hydrogen, with varying amounts of moisture, sulphur, nitrogen, arsenic, phosphorus and silt. The moisture content varies from 1 to 30 per cent and affects the heat value of the fuel. Analyses of typical fuel oils are given in Table 14.

TABLE 14. — COMPOSITION AND HEAT VALUE OF TYPICAL FUEL OILS

Location of Oil Well	Per Cent Carbon	Per Cent Hydrogen	Per Cent Sulphur	Per Cent Oxygen	Specific Gravity	Flash Point °F.	B.t.u. per Lb.
California	81.52	11.51	0.55	6.92*	....	230	18,667
Texas, Beaumont	84.60	10.90	1.63	2.87	0.924	180	19,060
Pennsylvania	84.90	13.70	....	1.40	0.886	...	19,210
Russia, Caucasus	86.60	12.30	....	1.10	0.938	...	20,138

* Includes nitrogen.

147. Comparison of Coal with Oil. — The **advantages of oil** over coal, when used as a fuel under boilers, are: (1) reduction in cost of handling, (2) saving of labor throughout the plant, (3) reduction in storage space required, (4) higher efficiencies and capacities, (5) easy regulation of load, (6) no loss in heat value when properly stored, (7) cleanliness and freedom from dust.

Disadvantages of oil are: (1) danger of explosions, (2) possibility of larger upkeep, when furnaces are not properly designed and operated.

148. Gaseous Fuels. — These fuels are used in certain localities for steam-generating purposes and in manufacturing enterprises. As previously mentioned, the principal gaseous fuels are: (1) natural gas, and (2) artificial gas, such as blast-furnace, coke-oven, and producer gas.

Natural gas is a product of nature and, in the coal district, consists chiefly of methane and ethylene with some hydrogen and nitrogen. A typical analysis of gas from the Pittsburgh region, in per cent by volume, gave the following results: hydrogen 20.02, methane 72.18, carbon monoxide 1.00, carbon dioxide 0.80, oxygen 1.10, heavy hydrocarbon 4.30. In the South Western oil fields, large quantities are available from gas wells. This "casing-head" gas, as it is called, from which gasoline is sometimes made, consists of about 79 per cent methane and 18 per cent ethylene. *The higher heating value of natural gas varies from 950 to 1000 B.t.u. per cubic foot.*

Blast-furnace gas is the waste product from furnaces used to smelt iron ores. For each ton of iron produced, about 10,000 pounds of gas is obtained. A typical analysis of a blast-furnace gas, expressed in per cent by volume, is as follows: carbon dioxide 9.85, nitrogen 53.92, oxygen 0.36, carbon monoxide 32.73, hydrogen 3.14. *Its heat value is due primarily to carbon monoxide and varies from 85 to 100 B.t.u. per cubic foot* of gas.

By-product coke-oven gas is a product of destructive distillation of coal in a distilling coke oven. The gases from the oven, instead of being burned at the point of origin, as in the bee-hive coke oven, are removed from the oven through an uptake pipe, and cooled, yielding tar, ammonia, illuminating gas and fuel gas, as by-products. This gas resembles natural gas more closely than does blast-furnace gas. A typical analysis, expressed in per cent by volume, is as follows: carbon dioxide 3.20, oxygen 0.4, carbon monoxide 6.3, methane 29.60, hydrogen 41.6, nitrogen 16.1. *The heat value varies from* 400 *to* 500 *B.t.u. per cubic foot.*

Illuminating gas is generally a carburetted water gas, which is made by the decomposition of steam into hydrogen and carbon monoxide by contact with incandescent carbon. Hydrocarbon gases from oil or naphtha are then mixed with it to give the illuminants — particles of carbon which become incandescent.

Producer gas is a mechanical mixture of carbon monoxide, hydrogen, nitrogen, and small amounts of methane, ethylene, oxygen and carbon dioxide. It is made from the incomplete combustion of anthracite coal, bituminous coal, coke or peat, and also from briquettes of anthracite slack and lignite.

149. Heat Value of Gaseous Fuels. — The heat value of gaseous fuels is generally expressed in B.t.u. per cubic foot, and may be obtained by:

1. Calculation, based upon the chemical analysis of the gas.
2. Experiment, made with a gas calorimeter.

Heat Value Based on Analysis. — The most satisfactory method of determining the heat value of a gas is to use the percentages of the constituent gases and the heat value of each constituent gas, as determined by experiment, and as given in Table 15.

TABLE 15. — VOLUME AND CALORIFIC VALUE OF VARIOUS GASES

Name of Gas	Symbol	Volume Cu. Ft. per Lb.	B.t.u. per Cu. Ft.	Volume of Air per Cu. Ft. of Gas
Hydrogen	H	177.90	349	2.41
Carbon Monoxide	CO	12.81	347	2.39
Methane	CH_4	22.37	1053	9.57
Ethylene	C_2H_4	12.80	1675	14.33
Ethane	C_2H_6	11.94	1862	16.74
Acetylene	C_2H_2	13.79	1556	11.93

Example 20. — The analysis of a natural gas taken from an oil well in Anderson, Indiana, gave percentages by volume as follows. Find the heat value of the gas.

Hydrogen, H	1.86	Nitrogen, N	3.02
Methane, CH_4	93.07	Oxygen, O	0.42
Carbon monoxide, CO	0.73	Heavy hydrocarbons	0.47
Carbon dioxide, CO_2	0.26	Hydrogen sulphide, H_2S	0.15

Solution. — Using the heat values of the various combustible constituents as given in Table 15, and multiplying by the weight of each constituent in a cubic foot of the gas, the following results are obtained:

From hydrogen	$0.0186 \times 349 =$	6.50
From methane	$0.9307 \times 1053 =$	980.02
From heavy hydrocarbons	$0.0047 \times 1364 =$	6.41
From carbon monoxide	$0.0073 \times 347 =$	2.53
B.t.u. per cu. ft.		995.46

150. Junker Gas Calorimeter. — This instrument is shown in Fig. 107, and consists of a vertical cylindrical water chamber containing vertical tubes, heated by the gas burned in a Bunsen burner placed underneath the water chamber. The products of combustion pass upward through the combustion chamber and downward through the tubes, while water passes

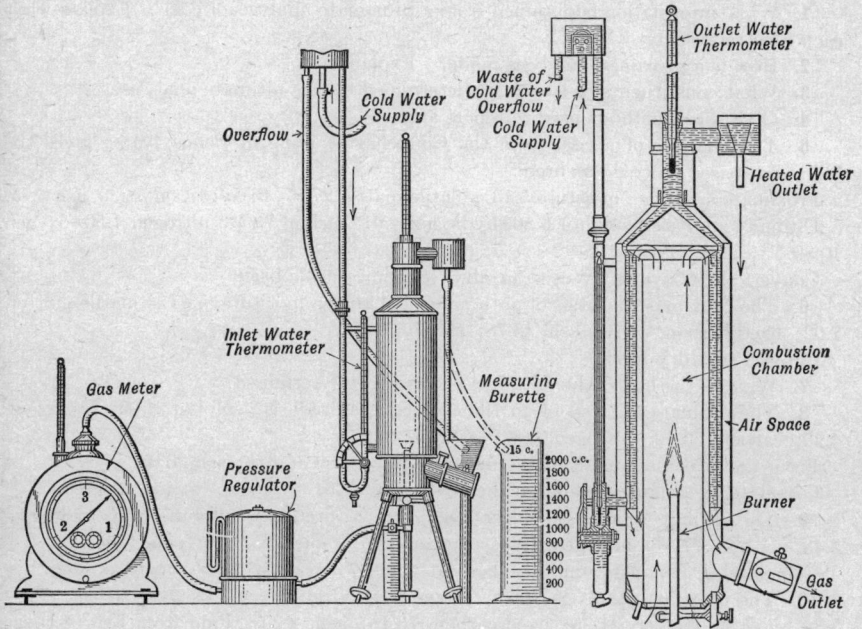

Fig. 107. — Junker Gas Calorimeter, Assembly and Sectional View.

continuously in at the bottom and out at the top. The quantity of gas is measured by a gas meter, and the water that passes is collected and weighed. The temperature of the entering and leaving water is measured by thermometers. The heat of combustion per cubic foot of gas is determined by multiplying the weight of water in pounds by the rise of temperature in degrees Fahrenheit and dividing the product by the volume of the gas used in cubic feet. This result is corrected for the moisture in the gas and reduced to equivalent heat value at 60 deg. fahr. and 30 inches barometer, and is the "higher" heat value. The "higher" heat value is used in all test work, unless otherwise mentioned.

REFERENCES

Coal and its Combustion, COSGROVE.
Steam, BABCOCK and WILCOX Co.
Finding and Stopping Waste in Modern Boiler Rooms, COCHRANE CORPORATION.
Technical Papers, Nos. 80 and 63; } BUREAU OF MINES.
Bulletins Nos. 40, 123, 129, 135, 235
Bulletin No. 116, Engineering Experiment Station, UNIVERSITY OF ILLINOIS.
Economics of Coal Carbonization in the United States, GEORGE A. ORROK.
Fuels and Steam Power, Vols. 50 and 51, AMERICAN SOCIETY OF MECHANICAL ENGINEERS.

PROBLEMS AND REVIEW QUESTIONS

1. What information is obtained from a proximate analysis of coal? Explain what each item represents.
2. How is a proximate analysis made? Explain fully.
3. What constituents of a coal are determined by the ultimate analysis?
4. Give three methods used to report an analysis.
5. The analysis of a coal from the Connellsville mine in Pennsylvania gives the following, based on coal "as fired":
Proximate analysis: moisture 5.13, volatile matter 27.87, fixed carbon 58.29, ash 8.71;
Ultimate analysis: sulphur 0.86, hydrogen 4.91, carbon 73.13, nitrogen 1.50, oxygen 10.89.
Convert the above analyses to (a) dry, (b) combustible basis.
6. The proximate analysis of anthracite coal gives: moisture, 5.41; volatile matter, 7.02; fixed carbon, 71.79; ash, 15.78; sulphur, 0.74.
Convert to a dry basis.
7. What is the heat value of a fuel? How is it determined?
8. The ultimate analysis of an Illinois coal "as fired" is: sulphur, 4.06; hydrogen, 4.95; carbon, 53.40; nitrogen, 0.89; oxygen, 16.61.
From the proximate analysis the moisture was 10.69 and the ash 20.09.
Calculate by Dulong's equation the heat value of (a) dry coal, (b) coal as fired.
9. The ultimate analysis of a Kentucky coal "as fired" is: sulphur, 1.22; hydrogen, 5.43; carbon, 77.37; nitrogen, 1.83; oxygen, 9.76; moisture, 3.10; ash, 4.39.
Calculate by Dulong's equation the heat value of (a) dry coal, (b) coal "as fired."
10. The proximate analysis of New River West Virginia coal "as fired" is: moisture, 3.34; volatile matter, 21.25; fixed carbon, 73.18; ash, 2.23. Find from Fig. 98, page 138, the heat value of the coal per pound of combustible.
11. Coals from various parts of the United States have heat values "as fired" and moisture contents as shown below. Calculate the heat values on : (a) dry, (b) combustible basis.

Locality	Moisture	Ash	Heat value "as fired" B.t.u.
Colorado..........	18.68	5.99	10,143
Illinois..., ...	8.31	10.49	11,727
Maryland.........	3.42	7.09	14,162
Missouri..........	8.33	19.36	10,586
Montana..........	4.13	30.86	9,095

12. Give the physical characteristics of four common coals.
13. How is anthracite coal sized? Bituminous coal?

14. What is powdered coal and what advantages has it?
15. Explain the cause of clinker.
16. What is the effect of the following in a coal: (a) ash, (b) moisture, (c) sulphur (d) oxygen.
17. Describe the K. S. G. process for carbonizing coal.
18. Distinguish between a ball, hammer and tube mill for pulverizing coal.

CHAPTER VII

COMBUSTION, FLUE GAS ANALYSIS, BOILER LOSSES

151. Foreword. — Combustion, or burning, is a chemical process in which a substance unites with oxygen, with the evolution of heat and often of light. At the instant of burning, the fuel must be in the gaseous form, to unite with the oxygen, which is generally supplied by the air. A rapid combustion, such as that which takes place in the cylinder of a gas engine, is called an explosion. An illustration of imperfect combustion is that of a smoldering fire in a peat bog.

The burning, or oxidation, of a gaseous fuel is accomplished by mixing air or oxygen with it in the proper proportions, and at the right temperature. To burn a liquid fuel, suitable means must be provided to vaporize or atomize it. The process of gasifying or atomizing a solid fuel and mixing it with air at the proper temperature, is attended with more difficulty, involving also a method of disposal of ash and waste. When in the powdered form, a solid fuel must be burned while suspended in the necessary air for combustion.

In recent years progress in the burning of fuels, especially coal, has been made by: (1) increasing the size and arrangement of the furnace in order to give a greater transfer of heat to the boiler by radiation and less by convection; (2) using smaller quantities of air which is preheated to temperatures around 400 deg. fahr., thereby increasing furnace temperatures; (3) using equipment better adapted to the fuels being burned; (4) improved design of stoker and pulverized coal equipment; (5) increased use of instruments for indicating furnace conditions, such as the CO_2 recorders, fuel-gas temperature and air-pressure recorders, steam-flow meters, instruments to show the relation between steam produced and air used in burning the fuel to produce the steam; and (6) the application of automatic combustion apparatus to control many of the operations connected with the burning of fuel.

152. Requirements for Perfect Combustion. — There are three conditions necessary for perfect combustion;

1. An ample supply of air.
2. A thorough mixing of the air and gases.
3. A sufficiently high temperature to maintain combustion.

The successful solution of the problem of combustion is dependent upon the fulfilment of these requirements.

153. Stages of Combustion of Solid Fuel. — In the combustion of a solid fuel, three stages are usually recognized, namely:

1. Heating the fuel to the temperature required for ignition, or kindling point, Table 16.
2. Expelling the volatile gases, which are then burned.
3. Burning the solid remainder to ash.

These may be briefly stated as **absorption, distillation** and **oxidation.**

TABLE 16. — TEMPERATURES AT WHICH VARIOUS COMBUSTIBLES IGNITE*

Name of Combustible	Temp. °F.	Name of Combustible	Temp. °F.
Phosphorus	150	Fixed carbon, bituminous coal	766
Sulphur	470		
Carbon monoxide, CO	1210	Fixed carbon, semi-bituminous	870
Hydrogen	1130		
Methane, marsh gas, CH_4	1202	Fixed carbon, anthracite	925
Ethylene, olefiant gas, C_2H_4	1022	Cannel coal	688
Ethane, C_2H_6	1000	Dried peat	435
Acetylene, C_2H_2	900	Lignite dust	300

* From STROMEYER's Marine Boiler Construction, and COSGROVE's Coal.

The combustion of fuel on a grate may be explained by reference to Fig. 108. Consider the fuel to be coal. When fresh coal is thrown on an incan-

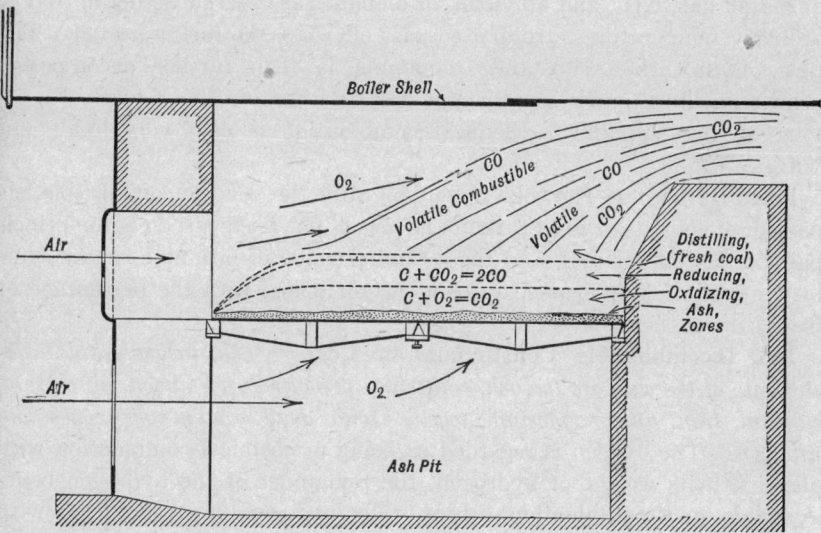

FIG. 108. — Illustrating Changes occurring during Combustion of a Solid Fuel.

descent fuel bed, the moisture in the coal is driven off as steam. The heat then distils off the hydrocarbon gases, which unite with oxygen from the air admitted above the grate, and, under favorable conditions, are completely burned. The fuel which remains consists of fixed carbon and ash. The carbon in the lower layers, where sufficient air is present below the grate, is burned to **carbon dioxide,** CO_2. As the carbon dioxide passes through the middle layer, where there is excess of carbon, it loses oxygen and becomes **carbon monoxide,** CO. *This is called reduction, or loss of oxygen.* Then the carbon monoxide, passing away over the fuel bed, should receive oxygen again from air admitted over the fire and be completely burned to carbon dioxide. *The zones in the fuel bed are the distillation, reduction, oxidation and ash zones.*

When coal is burned in the pulverized form, the coal particles are surrounded with air, and the distillation of the volatile matter takes place, at least partly, in the presence of oxygen. Some of the distilled tar vapors are burned directly, while others are first broken into soot and permanent gases and then burned. The proportion of the tar vapors burned directly depends upon the amount of air supplied with the coal and the initial intensity of mixing. The rate of combustion of the fixed carbon depends upon the rate at which oxygen diffuses through the gas film, which surrounds each carbon particle, and is increased by proper mixing and fineness of the coal.

154. Combustible Constituents of Coal. — The combustible constituents of coal may be grouped as *volatile combustible* and *fixed carbon.*

Volatile Combustible. — The light hydrocarbon gases, called **methane,** or **marsh gas,** CH_4, and **ethylene,** or **olefiant gas,** C_2H_4, existing in coal at ordinary temperatures, are first to pass off when the fuel is heated. The tarry hydrocarbon substance remaining is then further decomposed, principally into marsh gas, olefiant gas, and carbon in the form of soot. Part of the sulphur in chemical combination is also volatilized, and burns.

Fixed Carbon. — The coke remaining after the volatile combustible has passed off consists of fixed carbon and ash. The fixed carbon is the principal combustible of coal, and the steaming value of bituminous and semi-bituminous coals increases, up to a certain point, with the percentage of fixed carbon they contain.

155. Incombustible Constituents of Coal. — *The incombustible constituents of the coal are the ash, consisting principally of silicon, aluminum, calcium, iron, and magnesium, together with small amounts of oxygen and nitrogen.* The oxygen is regarded as being in chemical combination with one-eighth its weight of hydrogen, the remainder of the hydrogen being available for the production of heat. The nitrogen, in passing away with the gases, instead of adding heat value, takes away the heat required to raise its temperature to that of the waste gases. Part of the sulphur remains in combination in the ash.

156. Fundamentals of the Chemistry of Combustion. — A knowledge of the chemical changes resulting from the union of oxygen with carbon,

FUNDAMENTALS OF THE CHEMISTRY OF COMBUSTION 161

hydrogen and sulphur is essential to an adequate presentation of the subject of combustion.

All substances are composed of atoms, which are minute particles of chemical elements, arranged in groups called molecules. The elements are represented, for convenience, by symbols; thus, the symbols, C, H, N, O, S, represent one atom each of carbon, hydrogen, nitrogen, oxygen, and sulphur. These elements, together with iron (Fe), silicon (Si), calcium (Ca), aluminum (Al), and magnesium (Mg), are the principal constituents of coal.

Molecules of the simple gases, such as hydrogen, oxygen, and nitrogen, have 2 atoms each and *are represented by the symbols*, H_2, O_2, *and* N_2, respectively. The atoms of carbon, sulphur, and iron are found in combination with other elements and have no separate molecular symbol given to them.

When substances unite chemically to form a new combination, they unite in definite proportions by weight, new molecules being formed by a redistribution of the atoms.

The molecular weights of the substances formed are the sum of the atomic weights. The atomic weights are purely relative and were arbitrarily established. Hydrogen, as the lightest known substance, was given the atomic weight of 1 and the weights of the other elements were fixed in relation to it. Carbon, which is 12 times as heavy as hydrogen, was given an atomic weight of 12, and oxygen, being 16 times as heavy, was given the atomic weight of 16. Later determinations have shown that oxygen is less than 16 times as heavy as hydrogen, and as a result, hydrogen has been given an exact atomic weight of 1.008. Oxygen, with an atomic weight of exactly 16, is now used as the basis of all atomic weights.

The atomic and molecular weights of substances entering into combustion are given in Table 17.

TABLE 17.— ATOMIC AND MOLECULAR WEIGHTS OF SUBSTANCES ENTERING INTO COMBUSTION

Substance	Atomic Symbol	Atomic Weight	Molecular Symbol	Molecular Weight
Hydrogen................	H	1	H_2	2
Carbon..................	C	12	—	—
Oxygen..................	O	16	O_2	$2 \times 16 = 32$
Nitrogen................	N	14	N_2	$2 \times 14 = 28$
Sulphur.................	S	32	—	—
Iron.....................	Fe	56	—	—
Carbon monoxide.........	—	—	CO	$12 + 16 = 28$
Carbon dioxide...........	—	—	CO_2	$12 + 32 = 44$
Methane, or marsh gas....	—	—	CH_4	$12 + 4 = 16$
Ethylene, or olefiant gas...	—	—	C_2H_4	$24 + 4 = 28$
Gaseous water, or steam...	—	—	H_2O	$2 + 16 = 18$

157. Chemical Reactions. — The chemical union of substances may be expressed by an equation, or reaction, as it is called. Thus, if two molecules of hydrogen unite with one molecule of oxygen, forming water vapor, the equation may be written in the **atomic form** as:

$$2H + O = H_2O \qquad (42)$$

or using the **molecular form,** for reasons to be explained, and writing the molecular weights below each molecule,

$$2H_2 + O_2 = 2H_2O \qquad (43)$$
$$4 + 32 = 36, \text{ or } 1 + 8 = 9$$

From Equation (43) the following information regarding the weight and volume of the substances entering into the reaction may be obtained:

1. **Weight.** — It is evident from these figures that when the coefficients, indicating the relative number of molecules entering a reaction, are multiplied by the molecular weight of the molecule that the proportional weights, involved in a reaction, are obtained. This product is called the **pound-molecular weight,** the **pound-mol,** or just **mol,** *and may be defined as the weight in pounds which equals the molecular weight.* When thus written Equation (43) reads: 2 *mols* of hydrogen combine with 1 *mol* of oxygen to form 2 *mols* of water.

2. **Volume.** — By **Avogadro's Law,** the number of molecules present in a given volume, at the same pressure and temperature, is approximately the same for all gases.

Since in Equation (43) there are twice as many molecules of hydrogen and water vapor as there are of oxygen, the volume required to contain either the hydrogen or the water vapor is twice as large as the volume containing the oxygen. In terms of volume this Equation may therefore be read: 2 pound-molecular volumes of hydrogen combine with 1 pound-molecular volume of oxygen to form 2 pound-molecular volumes of water vapor. The **volume of the mol** for all gases may be taken as 359 cubic feet at 32 deg. fahr. and 14.7 pounds per square inch.

Thus, equations written in the molecular form indicate directly, by the coefficients, the relative volumes entering into a reaction.

158. Weight of Air Theoretically Required for Combustion per Pound of Coal. — The proportion of carbon, hydrogen, and sulphur present in the coal is obtained from the ultimate analysis, and the weight of air required to burn each is obtained as follows:

1. *Weight of air per pound of carbon for complete combustion:* The reaction equation is

1 atom carbon		1 molecule oxygen		1 molecule carbon dioxide	
C	+	O_2	=	CO_2	(44)
12	+	32	=	44	
1	+	$2\frac{2}{3}$	=	$3\frac{2}{3}$	

On a basis of weight, the above equation shows that 1 pound of carbon combines with $2\frac{2}{3}$ pounds of oxygen to form $3\frac{2}{3}$ pounds of carbon dioxide, and on a basis of volume, that carbon burned with one volume of oxygen produces one volume of carbon dioxide. Therefore, the volume of gas, before and after the reaction, is the same, provided the volume is measured at the same pressure and temperature in each case.

The equation also shows that 1 pound of carbon requires $2\frac{2}{3}$ pounds of oxygen for complete combustion; and therefore, since air is 23 per cent oxygen by weight, the weight of air required per pound of carbon burned equals $2.667 \div 0.23 = 11.57$ pounds.

The heat liberated is 14,600 *B.t.u. per pound of carbon burned to carbon dioxide.*

1a. *Weight of air per pound of carbon for incomplete combustion:* The oxygen being insufficient, there is an excess of carbon, and the reaction that occurs is

2 atoms carbon		1 molecule oxygen		2 molecules carbon monoxide	
2 C	+	O_2	=	2 CO	.... (45)
24	+	32	=	56	
1	+	$1\frac{1}{3}$	=	$2\frac{1}{3}$	

Equation (45) shows that $1\frac{1}{3}$ pounds of oxygen are required to burn 1 pound of carbon to carbon monoxide; therefore, the weight of air required per pound of carbon $= 1\frac{1}{3} \div 0.23 = 5.76$ pounds.

The heat liberated is 4380 *B.t.u. per pound of carbon burned to carbon monoxide;* consequently, the loss resulting from incomplete combustion $= 14,600 - 4380 = 10,220$ B.t.u. per pound of carbon burned to carbon monoxide.

2. *Weight of air per pound of hydrogen:* Art. 157 shows that 1 pound of hydrogen unites with 8 pounds of oxygen to form water, and therefore requires $8 \div 0.23 = 34.8$ pounds of air per pound of hydrogen.

The heat liberated per pound of hydrogen, when burned to water, equals 62,000 *B.t.u.*

3. *Weight of air per pound of sulphur:* The combustion of sulphur occurs according to the following equation

1 atom sulphur		1 molecule oxygen		1 molecule sulphur dioxide	
S	+	O_2	=	SO_2	.... (46)
32	+	32	=	64	
1	+	1	=	2	

This equation shows that the weight of air required, per pound of sulphur burned to sulphur dioxide $= 1 \div 0.23 = 4.35$ pounds.

The heat evolved is 4000 B.tu.. per pound of sulphur burned to sulphur dioxide.

The sum of items 1, 2, and 3 gives the following equation for the weight of air per pound of coal:

$$\text{Weight of air per pound of coal, lb.} = 11.6\,C + 34.8\left(H - \frac{O}{8}\right) + 4.35\,S \quad (47)$$

in which C, H, O, and S are the weights of carbon, hydrogen, oxygen, and sulphur per pound of coal, as shown by the ultimate analysis. As noted elsewhere, the "available" hydrogen, $H - \frac{O}{8}$, is less than the amount of hydrogen as shown by the analysis, on account of the appropriation by the oxygen of one-eighth its weight of hydrogen.

If combustion is incomplete, the air required for the proportion of the carbon burned to carbon monoxide must be multiplied by the weight of air required for incomplete combustion of carbon, as given by Equation (45), and the result added to Equation (47).

Example 21. — The ultimate analysis of the coal "as fired," from the test given in Example 22 and expressed in per cent, is: carbon, 83.68; hydrogen, 4.70; oxygen, 4.25; nitrogen, 1.61; sulphur, 0.73. Find the theoretical weight of air per pound of coal "as fired."

Solution. — Substituting the values given above in Equation (47).

$$\text{Weight of air per pound of coal} = 11.6\,C + 34.8\left(H - \frac{O}{8}\right) + 4.35\,S$$

$$= 11.6 \times .8368 + 34.8 \left(.047 - \frac{.0425}{8}\right) + 4.35 \times .0073$$

$$= 11.19 \text{ lb.}$$

The weights and volumes of air and oxygen required to burn the common combustibles, together with the weights and volumes of the resulting products, are tabulated in Table 18.

TABLE 18. — WEIGHT AND VOLUME OF OXYGEN AND AIR REQUIRED FOR COMBUSTION.

At 32° and 29.92 in. mercury

Combustible or Oxidizable Substance Column 1	Chemical Symbol	Reaction Equation	Product of Combustion	Weight required per Lb. of Column 1		Weight of resultants per Lb. of Column 1		Heat Value per Lb. of Column 1 B.t.u.
				Oxygen Lb.	Air = $4.32 \times O$ Lb.	Nitrogen = $3.32 \times O$ Lb.	Gaseous Product Lb.	
Carbon	C	$C + 2O = CO_2$	Carbon dioxide	2.667	11.52	8.85	12.52	14,600
Carbon	C	$C + O = CO$	Carbon monoxide	1.333	5.76	4.43	6.76	4,380
Carbon-monoxide	CO	$CO + O = CO_2$	Carbon dioxide	0.571	2.47	1.90	3.47	10,150
Hydrogen	H	$2H + O = H_2O$	Water	8	34.56	26.56	35.56	62,000
Methane	CH_4	$CH_4 + 4O =$ $CO_2 + 2H_2O$	Carbon dioxide and water	4	17.28	13.28	18.28	23,550
Ethylene	C_2H_4	$C_2H_4 + 3O_2 =$ $2CO_2 + 2H_2O$	Carbon dioxide and water	3.43	14.81	11.38	15.81	1,591
Sulphur	S	$S + 2O = SO_2$	Sulphur dioxide	1	3.32	3.32	5.32	4,050

ACTUAL WEIGHT OF AIR REQUIRED PER POUND OF COAL

TABLE 18. (Continued)

Combustible or Oxidizable Substance Column 1	Volumes of Column 1 entering combination	Volumes of Oxygen Combining with previous column	Volumes of Product Formed	Volume required per Lb. of Column 1		Volume of Resultants per Lb. of Column 1		
				Oxygen Cu. Ft.	Air Cu. Ft.	Products of Combustion Cu. Ft.	Nitrogen = 3.782 × O Cu. Ft.	Gas = sum of two previous Columns Cu. Ft.
Carbon	1 C	2	2 CO_2	29.89	143.10	29.89	112.98	142.87
Carbon	1 C	1	2 CO	14.95	71.55	29.89	56.49	86.38
Carbon-monoxide	2 CO	1	2 CO_2	6.40	30.62	12.80	24.20	37.00
Hydrogen	2 H	1	2 H_2O	89.66	429.07	179.32	339.09	518.41
Methane	1 CH_4	4	1 CO_2, 2 H_2O	44.83	204.97	67.34	169.55	236.89
Ethylene	1 C_2H_4	3	2 CO_2, 2 H_2O	38.80	184.31	42.84	145.81	187.65
Sulphur	1	2	2 SO_2	11.21	53.71	11.21	42.39	53.60

159. Actual Weight of Air Required per Pound of Coal. — The weight of air which should be actually supplied to the fuel depends upon the method of supplying the air, the method of firing the fuel, and the furnace conditions. If the fuel is fired continuously, the air introduced at suitable places, and the opportunity afforded for mixing with the gases at a sufficiently high temperature before cooling against boiler surfaces occurs, the per cent of air in excess of that theoretically required is ordinarily from 30 to 40 per cent for stokers, 20 to 30 per cent for pulverized coal, 12 to 15 per cent for oil, and 8 to 10 per cent for gas. *In general, 50 to 100 per cent excess is required, with the figures running to 200 per cent excess under poor furnace conditions.* Too great an excess of air results in: (1) lower furnace temperatures, (2) increased heat loss in the chimney gases, (3) increased carbon and smoke, (4) more combustible in the ashpit, and (5) more power being required to handle the air and the flue gases. Too small an excess of air results in incomplete combustion, and increased furnace temperatures, which has a bad effect on the refractories, stoker parts and burners. In general the air supply should be reduced to the points corresponding to minimum total operating and maintenance expense per unit of steam generated.

When burning pulverized fuel a portion of the air is supplied along with the fuel and is known as **primary air**. The remainder of the necessary air is supplied after the fuel has entered the furnace and is known as **secondary air**. The primary air may be unheated or preheated.

The volume of air corresponding to its weight is dependent upon its temperature and pressure, as shown by Table 3, page 81, which covers the range of temperatures at normal atmospheric pressure. For other temperatures the volumes may be calculated by the method explained under Art. 68, page 83.

160. Flue Gases. — The products of combustion that pass away, through the flue or uptake connections, to the stack, are called flue gases.

The composition of the flue gas depends on the material burned and the completeness of the combustion. *Flue gas from burning coal ordinarily contains carbon dioxide, carbon monoxide, nitrogen, oxygen, unburned hydrocarbons, carbon in suspension as smoke and ash dust, a small percentage of superheated steam, and some sulphur dioxide.* When combustion takes place under the most favorable conditions, there should be no carbon monoxide, unburned carbon, or hydrocarbons, in the flue gas Sulphur dioxide has a corrosive effect on the steel walls of the smoke flues, air preheaters, and economizers in the presence of moisture resulting from condensation of steam.

161. Flue-gas Analysis. — *This analysis generally determines the proportion by volume of the carbon dioxide, carbon monoxide, oxygen, and nitrogen present in the products of combustion.* The method of making the analysis is to pass a small sample volume of the gas, usually 100 cubic centimeters (cc.), through a series of receptacles in each of which a chemical reagent absorbs one of the principal constituents of the chimney gas. A typical analysis, taken from the gases passing the smoke box in a return-tubular boiler, gave 12.6 per cent CO_2, 5.7 per cent O_2, 0.0 per cent CO and, by difference, 81.7 per cent N_2. An additional test is sometimes made for unburned hydrocarbons, by passing that part of the sample remaining after removal of the CO_2, O_2, and CO into a tube into which oxygen is introduced. The reduction in volume resulting from ignition of the mixture shows the proportion of hydrogen or hydrocarbons present.

The analysis indicates the conditions under which combustion is taking place, and makes it possible to calculate the amount and distribution of the losses. Analyses are made for the following purposes:

1. To determine the best method of firing and handling the fire for the coal in use, with regard to coking, spreading, and leveling to prevent holes in the fuel bed.
2. To establish the best draft, considering thickness of fire and rate of combustion.
3. To discover air leaks in the setting.
4. To find amount of additional air required for complete combustion.
5. To obtain information bearing upon the proper design and construction of the furnace.

162. Sampling of Flue Gas. — Since the sample of gas drawn out for analysis should represent all the gas passing the point at which the sampling device is located, it is difficult to obtain a proper sample. A sampling tube, consisting of a pipe having a closed end and a series of perforations arranged in a lengthwise row, is sometimes used, in the passes of boilers and at various points along the gas stream, to collect the sample. The objections to this form of tube are that the holes may be filled by soot, and that those nearest the suction end will furnish most gas. Besides, the velocity of the gas

GAS ANALYSIS APPARATUS 167

stream varies across its section, making it more difficult to obtain a true sample. These objections may be overcome to some extent by graduating the size of the holes.

The Power Test Committee of the A. S. M. E. recommends that: "the sample should be drawn from the region near the center of the main body of the escaping gases, using a sampling pipe not larger than $\frac{1}{4}$-inch gas pipe. The pipe should contain perforations extending the whole length of the part immersed and pointing toward the current of gas; the collective area of the perforations being less than the area of the pipe." The **Bureau of Mines** recommends a water-cooled tube or quartz tube, as preferable to a metal tube, to overcome any effect of the hot tube on the gas. An open-end tube located in the center of the gas stream gives results accurate within 0.5 per cent of the average composition, if the setting is fairly tight.

163. Location of Sampling Tube. — The location of the tube depends upon the information sought. If it is desired to find the conditions in the combustion space above the fuel bed, as for example in a B. and W. type boiler, the sampling pipe may be introduced from the side clean-out doors, between the tubes of the first pass. Care must be taken to prevent an excess of air entering around the sampling tube. *The usual location of the tube is near the boiler damper on the furnace side, where the effect of leakage of air into the boiler setting will be shown.*

164. Collection of the Sample. — The flue gas is drawn and collected over water, by means of a water aspirator or some form of displacement apparatus. The water should be saturated with gas, to prevent absorption of CO_2 during the analysis. A brine solution absorbs CO_2 less readily than water, and may be used when it is necessary for the sample to stand for several hours before analysis. The gas should be analyzed immediately after being drawn, for best results. Conditions may change rapidly during combustion, and for this reason snap samples drawn directly into the gas analysis apparatus should be taken at short intervals. A continuous sample is necessary, in order to obtain an average that is of value.

165. Gas Analysis Apparatus. — There are two principal types of apparatus used in power plants for analyzing gases, the hand, or portable type, and the automatic-recording, stationary type. The **Orsat** is the most widely used hand apparatus. A description of this instrument, as given by the Bureau of Mines, follows:

"The essential parts of the apparatus, Fig. 109 are: a 100-cc. measuring burette graduated in 100 units with each unit subdivided into fifths, four absorption pipettes, a leveling bottle, and a connection header — all of glass. The measuring burette is usually enclosed in a water jacket, which prevents sudden temperature changes while the analysis is being made. The pipettes are U-shaped glass vessels, and contain solutions for absorbing the gas constituents. *The first contains caustic potash solution for absorbing the carbon dioxide. The second contains an alkaline solution of pyrogallic acid for absorbing oxygen, while the third and fourth contain an ammoniacal solution of cuprous chloride for absorbing the carbon monoxide.* The side of the pipette in which the absorption is to take place is filled with small glass tubes, which increases the surface of liquid in contact with the gas. The pyrogallic acid solution and the cuprous chloride solution absorb oxygen readily, and *air must be kept from coming in contact with them.* A rubber bag is placed on the rear side of both these pipettes or a water-seal, consisting of a fourth pipette, is used. The gas is drawn into the burette and forced out of it by means of a **leveling bottle** attached to the lower end of the burette by a 3-foot rubber tube. The leveling bottle also permits regulation of the pressure on the gas. The header connecting the burette and the pipette is made of $\frac{1}{4}$-inch glass tubing. Each outlet is sometimes fitted with a ground-glass cock, but such cocks require constant attention to prevent them from sticking and leaking. **Pinch-cocks** on the rubber tubing are equally satisfactory."

The preparation of the solutions should be undertaken only by one who has the required knowledge. The solution for determination of CO_2 can be used for about 150 determinations.

Manipulation. — The manipulation of the apparatus may be briefly described as follows:

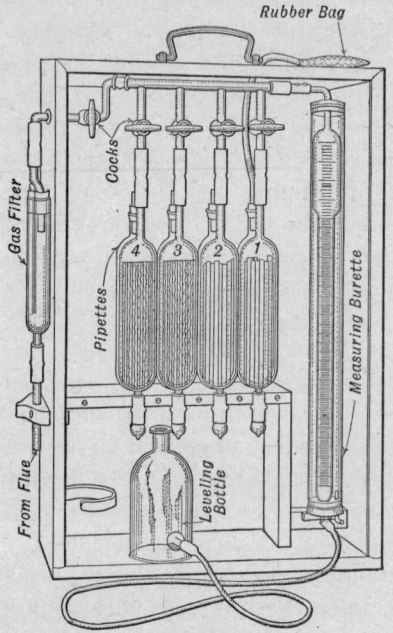

FIG. 109. — Orsat Apparatus.

1. The apparatus is attached to the bottle containing the sample, and 102 cc. of sample gas is drawn into the measuring burette by lowering the leveling bottle. The gas should be drawn through a gas filter placed in the tube furnishing the gas sample, in order to free the gas of mechanical impurities, such as carbon dust. The filter may consist of a glass tube, larger at the center than at the ends and loosely filled with mineral wool or, in some cases, with clean cotton waste.

2. The level of the water in the bottle is raised to a position opposite the 100 cc. mark on the burette, and the three-way cock opened to the air to relieve pressure on the gas. *All readings of the burette are taken with the levels of water in bottle and burette at same height.*

3. The cock to the first pipette where CO_2 is absorbed is opened and the water bottle raised, transferring the gas to the pipette. The gas is allowed to remain in the pipette several minutes and the solution is worked up and down, using the leveling bottle to assist the action. The gas is then trans-

ferred, by lowering the water bottle back to the burette for measurement, after which the gas is again transferred to the pipette and the process repeated until a constant reading is obtained for the CO_2. The same method is then used with the second and third pipettes. The difference in the readings of the burette at the start and finish, for each gas, gives the volume of gas absorbed. The process must be carried out in the order given; otherwise, the results will be incorrect.

The water and the solutions should not mix; to prevent this the movement of the leveling bottle should be stopped when either the water or the solution reaches the mark on the tube attached to the pipette.

Care must be taken, when transferring the sample to the apparatus, to fill the ends of the tubes connecting the Orsat and the sample-container with water, to prevent entrance of air.

166. Automatic Flue-gas Recorders. — These recorders determine only CO_2, and may be classified as: (1) the **mechanical Orsat** uses the principle of the hand-operated Orsat and records an actual chemical analysis, (2) **electric thermo-conductivity recorder,** which depends upon the difference in thermal conductivity of different gases to vary the temperature, and this in turn operates the resistance of one leg of the Wheatstone Bridge, thus comparing the conductivity of the gas with that of air — the difference is calibrated in the per cent of CO_2, (3) **absorption pressure recorder,** which is based on the pressure reduction in the stream of gas from which CO_2 has been absorbed — the difference between this reduced pressure and some other constant pressure of the same gas stream is calibrated in the per cent of CO_2, (4) **gas density recorder** compares the density of the gases with the density of air, the difference being calibrated in per cent of CO_2.

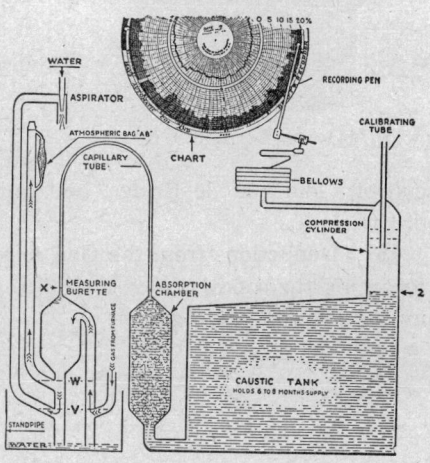

FIG. 110. — Hays CO_2 Recorder.

The absorption pressure and electric type instruments are used to the greatest extent. The **Hays CO_2 Recorder** is of the absorption pressure type and consists of the parts diagrammatically shown in Fig. 110. A continuous stream of flue gas is drawn rapidly from the last pass of the boiler and at regular intervals, the recorder takes a measured sample, causes the CO_2 to be absorbed and the pen to record the amount of the absorption or in other words the CO_2 content.

Water passing through an aspirator draws the flue gases directly through the **measuring burette** and is discharged into the **standpipe,** and as the water level rises to the point V a sample of the gas is trapped. The water continuing to rise sends the surplus gas through a pipe to the **atmospheric bag** so that as soon as the water reaches the point W an accurate sample of gas at atmospheric pressure is contained in the burette. The water continuing to rise pushes the gas from the burette into the absorption chamber where it is brought into intimate contact with the caustic potash solution on the moist surfaces of steel wool. Movement of the gas pushes some of the caustic potash from the absorption chamber and seals the compression chamber of the caustic tank at the same time operating the bellows to move the pen across the chart. When this cycle is completed the water in the standpipe has reached the top of the machine and starts a syphon which empties the machine of water and allows the gas to be again drawn into the pipe head.

FIG. 111. — Electric CO_2 Meter.

The **Brown Electric CO_2 meter** shown in Fig. 111, is essentially a galvanometer which is connected into the circuit of a Wheatstone Bridge, and calibrated to read directly in per cent of CO_2.

167. Deductions from the Gas Analysis. — For an intelligent judgment regarding the conditions of combustion, the readings of CO_2, O_2, and CO must be considered together.

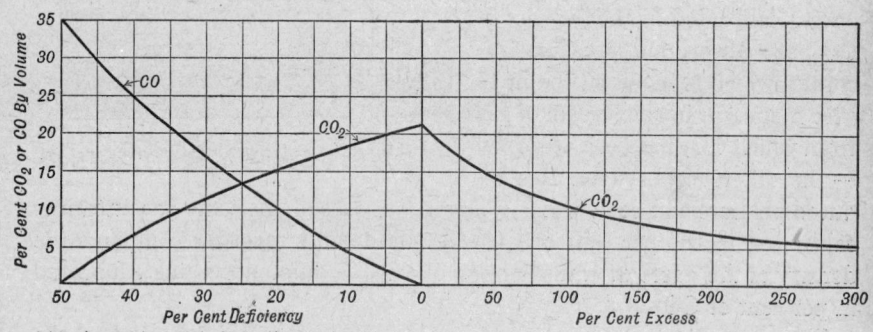

FIG. 112. — Effect of Air Supply on CO and CO_2 in the Flue Gas.

Carbon dioxide, CO_2. — If the fuel were all carbon, perfect combustion would result in 20.91 per cent CO_2, since the volume of CO_2 formed is the same as the volume of oxygen used, as shown in Art. 158. This per cent is reduced, however, on account of the hydrogen in the fuel, because the nitrogen contained in the air for burning the hydrogen remains in the flue gas and dilutes it. A further reduction occurs if hydrocarbon gas

DEDUCTIONS FROM THE GAS ANALYSIS

escapes without being burned, since it is accounted for as nitrogen by difference, if not separately determined. *The net result is that the sum of* CO_2, *O, and CO often equals* 18 *or* 19 *per cent.*

It is generally considered that 4 to 6 per cent CO_2 is very low, 10 to 12 per cent good, and above 12 per cent excellent. Low CO_2 is usually caused by a large excess of air, as shown by the curve, Fig. 112. It may be caused by an insufficient air supply, in which case the CO content is high, and the flue gas approaches a fuel gas in quality. Even with more than sufficient oxygen present, some CO may escape, if its mixture with the gases does not occur at a suitable temperature for combustion before the gases are cooled by the heating surfaces. High CO_2, 12 to 15 per cent, is often accompanied by some loss due to the formation of CO. An analysis showing 0.1 to 0.2 per cent CO, with 12 to 13 per cent CO_2, indicates an excellent efficiency.

The maximum to which CO_2 can attain without formation of CO is dependent upon furnace conditions. Tests by the Bureau of Mines show the presence of CO with 8 to 10 per cent CO_2. "J. W. Hays[1] does not consider that there is danger from CO until about 15 per cent CO_2 is reached." Tests have shown, however, 0.40 to 1.0 per cent CO for 13.8 to 15.6 per cent CO_2, under unusually good conditions. If the draft is proportional to the thickness of the fuel bed, high values of CO_2 can be produced, whether the rate of combustion, pounds per square foot of grate surface per hour, is high or low.

For perfect combustion, the reaction Equation (44), page 162, shows that the volume of carbon dioxide resulting from the combustion of carbon is the same as the volume of oxygen used. The CO_2 per pound of carbon burned should then be in the same proportion to the resulting flue gas as the oxygen is to the air supplied; that is, 20.91 per cent.

If 50 per cent excess air is supplied, then the 1.5 volumes of gas resulting contain 0.209 volumes CO_2, from which $\frac{0.209}{1.5} = 0.139$ or 13.9 per cent is CO_2. If 100 per cent excess air is supplied, the CO_2 in the flue gas is $\frac{0.209}{1.0 + 1.0} = 0.1045$ or 10.45 per cent.

The curve, Fig. 112, shows the values of CO_2 for corresponding values of excess air by volume, with carbon as fuel and perfect combustion.

The excess air required to ensure combustion under good conditions, as given in Art. 159, *varies from* 8 *to* 40 *per cent. Forty per cent excess air corresponds theoretically to a value of* 14.9 *per cent* CO_2.

Carbon monoxide, CO. — In general, the presence of carbon monoxide in flue gases indicates imperfect conditions for combustion. As A. Bement points out, this is a danger signal, not so much because of the loss it represents (never over 1 per cent of the heat value of the fuel) as because it

[1] Finding and Stopping Waste in Modern Boiler Rooms, Cochrane Corporation.

is usually accompanied by the loss of unconsumed hydrogen and hydrocarbons.

The formation of carbon monoxide is due to:

1. Improper methods of firing.
2. Deficient air supply.
3. Improper mixing of the combustible gases with the air.
4. Low temperature of furnace, allowing gases to be cooled below ignition temperature.
5. Furnace not adapted to fuel to be burned, or responsible, on account of poor design, for (3) and (4).

Oxygen, O_2. — A high percentage of O_2 shows directly that a considerable gain in efficiency can be made, when using hand or stoker firing, by, (1) covering holes in the fire bed, (2) stopping the entrance of air through the setting, or (3) adjusting the draft to the thickness of the fuel bed.

168. Combustion Indicator. — Since there is an intimate connection between draft and efficiency, combustion indicators, consisting of a special combination of draft gages, are used to show, by variation from a reading that has been established for standard conditions, the necessity for attention to the fires. The Hays pointer gage, Fig. 113, is typical. It consists of two pointers, one connected to show furnace draft, and the other to show the draft in the last pass of the furnace.

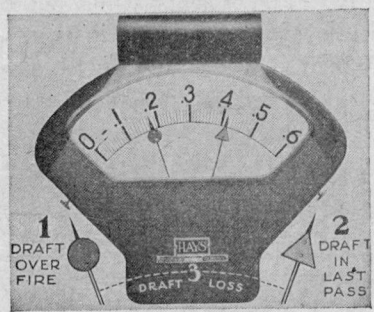

FIG. 113. — Combustion Indicator.

The movements of the pointers, to right or left from the position which has been established, as standard, give the firemen the information necessary for controlling the depth of fuel bed, air supply, and rate of combustion.

169. Weight of Air Supplied per Pound of Coal. — In most cases, the air supplied to burn the coal cannot be measured directly; but the amount of air can be calculated quite accurately, by using an analysis of a representative sample of the flue gas, together with the proportion of carbon burned per pound of coal, as obtained from the ultimate analyses of the coal and ash.

The calculation by this method is based upon the weight of nitrogen, as given by the flue-gas analysis. This nitrogen comes mostly from the air, while a small amount is supplied by the coal. *The nitrogen per pound of coal is determined from the flue-gas analysis, and the nitrogen from the coal is subtracted to give the nitrogen supplied by the air per pound of coal.* This result, divided by the proportion of nitrogen by weight in the air, will give the weight of air per pound of coal. The method of making the calculation follows:

The *percentage by weight* of carbon, hydrogen, oxygen, and nitrogen, as shown by the ultimate analysis of the coal, are represented by their respective chemical symbols C, H, O, and N.

The *percentage by volume* of the constituents of the flue gas, as shown by the flue-gas analysis, are represented by their respective chemical symbols CO_2, CO, O_2, and N_2.

1. *Weight of carbon burned*

$$\text{per pound of coal fired, lb.} = \frac{\text{Total carbon burned, lb.}}{\text{Total coal fired, lb.}}$$

Using symbols this may be written

$$W_c = \frac{WC - W_a C_a}{W} \quad \ldots \ldots \ldots \quad (48)$$

in which

W_c = weight of carbon burned per pound of coal fired, lb.
W = weight of coal fired, lb.
W_a = weight of ash, lb.
C = proportional weight of carbon per pound of coal by ultimate analysis.
C_a = proportional weight of carbon per pound of ash by ultimate analysis.

2. *Weight of nitrogen*

$$\text{per pound of carbon, lb.} = \frac{28\, N_2}{12\,(CO_2 + CO)} \quad \ldots \ldots \quad (49)$$

in which $28\, N_2$ = relative weight of nitrogen in the flue gas.
$12\,(CO_2 + CO)$ = relative weight of carbon in the flue gas.

3. *Total weight of nitrogen*, W_n, *per pound of coal fired* is found by combining Equations (48) and (49), thus giving

$$W_n = \frac{WC - W_a C_a}{W} \left[\frac{28\, N_2}{12\,(CO_2 + CO)} \right] \quad \ldots \ldots \quad (50)$$

The *net weight of nitrogen* supplied by the air per pound of coal burned equals W_n minus the weight of nitrogen in the coal, which is small and is usually neglected.

4. *Weight of dry air*, A_f, *supplied per pound of coal fired* is found by dividing Equation (50) by the proportional weight of nitrogen in the air; that is, $\frac{77}{100}$.

$$A_f = \frac{WC - W_a C_a}{0.77\, W} \left[\frac{28\, N_2}{12\,(CO_2 + CO)} \right] \quad \ldots \ldots \quad (51)$$

If it is desired to find the weight of air supplied per pound of coal fired, uncorrected for loss through the grates, the factor $W_a C_a$ drops from Equation (51), which can then be simplified to give the following equation:

$$A_f = 3.03\, C \left[\frac{N_2}{CO_2 + CO}\right] \quad \dots \dots \quad (52)$$

in which A_f = weight of dry air supplied per pound of coal fired, lb.
C = weight of carbon per pound of coal, lb.

Example 22. — During the test of a return-tubular boiler located at a pumping station, the following data were taken: weight of coal fired, 10,970 lb.; weight of ash and refuse, 670 lb.; air temperature, 72 deg. fahr.; flue-gas temperature, 434 deg. fahr.; relative humidity, 80 per cent; heat value of coal, 14,546 B.t.u. per lb., efficiency 75.5.

Ultimate analysis of coal, as fired, per cent by weight: carbon, 83.68; hydrogen, 4.70; oxygen, 4.25; nitrogen, 1.61; sulphur (volatile), 0.73; ash, 5.03.

Proximate analysis of coal, as fired: moisture, 2.21; volatile matter, 20.31; fixed carbon, 72.35; ash, 5.13.

Analysis of ash: volatile matter, 3.36; fixed carbon, 16.82; ash, 79.82.

Flue-gas analysis, per cent by volume: carbon dioxide, 14.3; oxygen, 3.0; carbon monoxide, 0.80; nitrogen (by difference), 81.9.

Find: (a) the weight of air supplied per pound of coal fired, and (b) the weight of air per pound of coal fired uncorrected for loss through the grates.

Solution. — Using Equation (51) and substituting values as given.

(a) Weight of air per pound of *coal fired*

$$= \frac{WC - W_a C_a}{0.77\, W} \left[\frac{28\, N_2}{12\, (CO_2 + CO)}\right]$$

$$= \frac{10{,}970 \times 0.8368 - 670 \times 0.168}{0.77 \times 10{,}970} \left[\frac{28 \times 81.9}{12\, (14.3 + 0.80)}\right]$$

$$= 13.60 \text{ lb.}$$

(b) Weight of air per pound of *coal fired, uncorrected*

$$= \frac{C}{0.77} \left[\frac{28\, N_2}{12\, (CO_2 + CO)}\right]$$

$$= \frac{0.8368}{0.77} \left[\frac{28 \times 81.9}{12\, (14.3 + 0.80)}\right]$$

$$= 13.75 \text{ lb.}$$

170. Weight of Dry Flue Gas per Pound of Coal. — To determine the heat carried away to the stack, the weight of dry chimney gas per pound of coal is calculated, by using the weight of dry gas per pound of carbon and the proportion of the carbon burned, as determined from the carbon in the coal and ash.

$$\text{Weight of dry gas per pound of carbon} = \frac{\text{Relative weight of dry gases}}{\text{Relative weight of carbon}}$$

$$= \frac{44\, CO_2 + 28\, (CO + N_2) + 32\, O_2}{12\, (CO_2 + CO)}$$

$$= \frac{11\, CO_2 + 7\, (CO + N_2) + 8\, O_2}{3\, (CO_2 + CO)} \quad (53)$$

in which CO_2, CO, N_2, and O_2 = percentages by volume of the constituents of the flue gas.

44 CO_2, 28 CO, 28 N_2, and 32 O_2 = relative weights of carbon dioxide, carbon monoxide, nitrogen, and oxygen.

12 (CO_2 + CO) = relative weight of carbon present.

Since the nitrogen in the flue-gas analysis is found by difference, N_2 = 100 − CO_2 − CO − O_2. Substituting this value of N_2 in Equation (53) and reducing, there results

$$\text{Weight of dry chimney gas per pound carbon, lb.} = \frac{4\,CO_2 + O_2 + 700}{3\,(CO_2 + CO)} \quad . \quad . \quad (54)$$

The weight of dry chimney gas per pound of coal fired is obtained by multiplying Equations (48) and (54) together, thus giving

$$W_d = \frac{WC - W_a C_a}{W}\left[\frac{4\,CO_2 + O_2 + 700}{3\,(CO_2 + CO)}\right] \quad \ldots \ldots \quad (55)$$

in which W_d = the weight of dry chimney gas per pound of coal fired.

171. Total Weight of Gas per Pound of Coal. — This includes the weight of steam formed by combustion of hydrogen (9 H ÷ 100), the weight of moisture shown directly by the analysis, and the weight of dry gas per pound of coal.

Example 23. — Using the data from Example 22, page 174, find: (a) The weight of dry chimney gases per pound of coal burned, (b) the total weight of gas per pound of coal.

Solution. — (a) The weight of dry chimney gases per pound of coal fired

$$= \frac{WC - W_a C_a}{W}\left[\frac{4\,CO_2 + O_2 + 700}{3\,(CO_2 + CO)}\right]$$

$$= \frac{10{,}970 \times 0.8368 - 670 \times 0.168}{10{,}970}\left[\frac{4 \times 14.3 + 3.0 + 700}{3\,(14.3 + 0.80)}\right]$$

$$= 13.86 \text{ lb.}$$

(b) The total weight of gas per pound of coal

= Weight of steam + dry gas per pound of coal + moisture

= 9 × 0.044 + 13.86 + 0.022 = 14.3 lb.

172. Excess Air. — The weight of excess air equals the difference between the weight of air shown by calculation from data of the ultimate analysis, Equation (47), page 164, and the weight found from the flue-gas analysis, Equation (52), page 174.

The excess air supplied is sometimes calculated by the use of the following formula:

$$\text{Per cent excess air} = \left[\frac{N_2}{N_2 - 3.782\,O_2} - 1\right] \times 100 \quad \ldots \quad (56)$$

The values of N_2 and O_2 are taken from the gas analysis. This formula assumes that the oxygen present is in excess of that theoretically required, a condition that may not exist in reality.

Example 24. — Using the results of Examples 21 and 22 find the amount of excess air.
Solution. — From Example 21, the theoretical weight of air is 11.19 lb.
Actual weight of air, as calculated in Example 22, is 13.75 lb.
The weight of excess air = (13.75 − 11.19) = 2.56 lb.

Per cent excess air $\frac{2.56}{11.19} \times 100 = 22.9$.

173. Air Infiltration. — Special attention should be directed to the losses incurred by infiltration of air, in excess of that required for combustion. This air, passing through holes in the fuel bed and through holes or cracks in the furnace or boiler setting, carries away with it a large amount of heat, and reduces the overall efficiency of the boiler plant. This condition is revealed by the gas analysis, by the increase in the volume of oxygen, and the decrease in the volume of CO_2. The BUREAU OF MINES gives the following example:

		CO_2	O_2	CO
Rear of combustion chamber	Av.	14.5	3.3	0
Base of stack	Av.	10.7	8.1	0

174. Boiler and Furnace Heat Balance. — The distribution of the heat in the fuel, between the heat absorbed by the boiler and the losses which occur in the operation of boiler plants are discussed under the following headings: Items 2, 5, and 6 are the chief preventable losses.

1. Heat absorbed by the steam in boiler.
2. Heat carried away in the dry chimney gases.
3. Evaporation of moisture formed by burning the hydrogen.
4. Evaporation of moisture in the fuel.
5. Incomplete combustion of the fuel.
6. Unconsumed carbon in the ash.
7. Heating of moisture in the air.
8. Radiation and all other losses.

1. *The heat absorbed by the boiler equals the heat per pound of fuel times the boiler efficiency.* — It may also be calculated from the actual rate of evaporation times the total heat content of the steam less the heat content of the feedwater, thus:

$$h_1 = w(h - h_f) \quad \ldots \ldots \ldots \ldots (57)$$

in which h_1 = heat absorbed by the boiler per pound of fuel, B.t.u.
w = actual rate of evaporation.
h = total heat per pound of steam, B.t.u.
h_f = B.t.u. per pound of feedwater.

2. *Heat carried away in the dry chimney gases.* — The loss of heat in the chimney gases is usually the largest loss, amounting to 10 or 12 per cent of

the heat generated by the boiler under the most favorable conditions of standard practice, and to 20 or 40 per cent where the supply of air is greatly in excess and there is leakage of air through the boiler setting. The greatest gain in efficiency can usually be made by reducing the air supply to the proper amount for the fuel used, by controlling the draft, by preventing holes in the fuel bed, and by preventing the entrance of air through breaks in the setting.

The heat carried away by the dry chimney gases is calculated by means of the following equation:

$$h_2 = W_g C_p (t_s - t) \quad \ldots \ldots \ldots \quad (58)$$

in which h_2 = B.t.u. lost per pound of fuel.
W_g = weight of dry chimney gas per pound of fuel, lb.
t_s = temperature of escaping gases, deg. fahr.
t = temperature of air entering furnace, deg. fahr.
C_p = mean specific heat of the dry gases = 0.24.

3. *Evaporation of moisture formed by burning hydrogen.* — This loss is unpreventable if there is hydrogen in the fuel. For anthracite coal the loss per pound of combustible is about 2.5 per cent of its heat value, and for bituminous coal it varies from 3 to 4.6 per cent. The water that is formed by combustion passes away as steam superheated to the temperature of the flue gases. The heat loss is

$$h_3 = 9H [212 - t + 970.2 + C_p (T - 212)] \quad \ldots \quad (59)$$

in which h_3 = B.t.u. loss per pound of fuel.
H = weight of hydrogen per pound of fuel, lb.
212 = temperature of evaporation, deg. fahr.
t = temperature of coal as fired, deg. fahr.
970.2 = latent heat at atmospheric pressure.
C_p = mean specific heat of superheated steam = 0.47.

Other symbols as in Equation (58).

4. *Evaporation of the moisture in the fuel.* — Moisture in the fuel reduces boiler efficiency, because of the loss of heat required to evaporate it. In hand-firing some fuels, less coal appears to be lost through the grates if the fuel is wet before firing. *Tests have shown that there is generally no gain in efficiency as a result of wetting the coal.* The possibility of gain in a particular case where considerable hydrocarbon escapes unburned, lies in the avidity with which the nascent oxygen dissociated from the steam unites with the escaping hydrocarbons. The hydrogen from the steam is burned and liberates the same amount of heat as that required to dissociate it.

The loss in heat is

$$h_4 = W_m [212 - t + 970.2 + C_p (T - 212)] \quad \ldots \ldots \quad (60)$$

in which h_4 = B.t.u. per pound of fuel.

W_m = weight of moisture per pound of fuel fired, lb.

Other symbols as in Equation (59).

5. *Incomplete combustion.* — This term usually refers to the loss by incomplete combustion of carbon. The loss due to the escape of unburned hydrocarbons is included in the losses which are unaccounted for. The loss resulting from the formation of carbon-monoxide gas is generally small, and usually does not exceed 2 per cent of the heat value per pound of coal. The heat loss is

$$h_5 = \left[\frac{CO}{CO_2 + CO}\right](14{,}600 - 4380)\,C \quad \ldots \ldots \quad (61)$$

in which C = proportion of carbon in the fuel, and CO and CO_2 are percentages by volume, as from the gas analysis.

6. *Unconsumed carbon in the ash.* — This is due to the method of working the fire or to a grate that is not suited to the fuel. Under good conditions it should not amount to more than 4 per cent of the heat value of the coal.

The loss may be computed thus:

$$h_6 = h_a \frac{W_a}{W_f} C_a \quad \ldots \ldots \ldots \ldots \quad (62)$$

in which h_6 = loss in B.t.u. per pound of fuel.

h_a = heat value per pound of combustible in the dry refuse, B.t.u. (usually taken as the heat value of carbon 14,600.)

C_a = Weight of combustible per pound of dry ash and refuse.

W_a = weight of ash and refuse, pound per hour.

W_f = weight of fuel fired, pound per hour.

7. *Heating of moisture in the air.* — This loss is small and, although frequently neglected, is calculated thus:

$$h_7 = \text{B.t.u. loss per pound of fuel} = MC_p\,(T - t) \quad \ldots \quad (63)$$

in which M = weight of moisture in air used per pound of fuel, lb.

= weight of moisture per pound of air × weight of air per pound of fuel.

= [(relative humidity × weight in pounds per cubic foot of water vapor at temperature, t, of air entering the furnace × volume of one pound of dry air at t deg. fahr.) × (weight of dry air per pound of fuel)].

Other symbols as in Equation (58).

8. *Radiation and all other losses.* — These losses are taken as the difference between the heat value per pound of fuel and the sum of items

1 to 7 inclusive. For well-designed boiler furnaces, these losses vary from 2 to 6 per cent.

$$h_8 = B.t.u. \text{ per pound of fuel} - (h_1 + h_2 + h_3 + h_4 + h_5 + h_6 + h_7) \quad (64)$$

Example 25. — Using the data, Example 22, page 174, compute the boiler losses and tabulate in form of a heat balance, efficiency of boiler 75.52 per cent.
Solution. — Distribution of heat per pound of coal "as fired."

Equation	Name and Calculation of Loss	B.t.u.	Per Cent
	Calorific value of coal "as fired" (by calorimeter)	14,546	
	1. Heat absorbed by the boiler		
Item 1	$0.755 \times 14,546$	10,985	75.52
	2. Heat carried away in dry chimney gases		
58	$13.86 \times 0.24 \,(434 - 72)$	1,204	8.28
	3. Evaporation of moisture from burning hydrogen		
59	$9 \times 0.044\,[212 - 72 + 970.2 + 0.47\,(434 - 212)]$	482	3.31
	4. Evaporation of moisture in the fuel fired		
60	$0.0221\,[212 - 72 + 970.2 + 0.47\,(434 - 212)]$	27	0.19
	5. Incomplete combustion		
61	$0.83 \times (14{,}600 - 4380) \times \dfrac{0.8}{14.3 + 0.8}$	447	3.07
	6. Unconsumed carbon in the ash		
62	$14{,}600 \times \dfrac{670 \times 0.20}{10{,}970}$	178	1.22
	7. Superheating water vapor in the air		
63	$0.80 \times 0.0012 \times 13.14 \times 13.75 \times 0.47\,(434 - 72)$	29	0.20
64	8. Radiation and all other losses (by difference)	1,194	8.21
	Total losses	14,546	100.00

REFERENCES

Mechanical Engineers' Handbook, MARKS.
Finding and Stopping Waste in Modern Boiler Rooms, COCHRANE CORPORATION.
Bulletins Nos. 80, 40, and 135, BUREAU OF MINES.
Bulletin No. 116, UNIVERSITY OF ILLINOIS EXPERIMENT STATION.
Steam, BABCOCK and WILCOX Co.
Fundamentals of Combustion and Air Supply, AMERICAN ARCH Co.
Coal and its Combustion, COSGROVE.
Boiler Room Operation, Arts. 1–10, REPUBLIC FLOW METER Co.

REVIEW QUESTIONS AND PROBLEMS

1. Define: (a) combustion, (b) combustible.
2. Describe the changes which take place during combustion of a solid fuel.
3. Why is it sometimes more convenient to use the molecular form of chemical equation for combustion than the atomic form? Write Equation (45) in terms of the pound-mol.
4. Find the volume of carbon dioxide measured at 32 deg. fahr. and 14.7 lb. per sq. in. produced by completely burning 100 lb. of carbon. Convert this volume to that at 60 deg. fahr. and 30 in. mercury. What volume of oxygen was consumed?
5. A Montana coal "as fired" has the following ultimate analysis, expressed in per cent: sulphur, 7.69; hydrogen, 3.16; carbon, 44.16; nitrogen, 0.49; oxygen, 6.73; ash, 37.77. Find the theoretical amount of air required per pound of coal.

6. In Problem 5, what would be the weight of air required, provided 50 per cent excess air were used? What effect would this excess air have on the percentage of carbon dioxide in the flue gas? Check the value of 14.9 per cent CO_2 for 40 per cent excess air.

7. The ultimate analysis of a Kentucky coal "as fired," expressed in per cent, is as follows: sulphur, 1.22; hydrogen, 5.43; carbon, 77.37; nitrogen, 1.83; oxygen, 9.76; ash, 4.39. Find the weight of air theoretically required to burn a pound of the coal. Would this be sufficient under actual conditions?

8. What is the function of the air supplied for combustion?

9. Explain the meaning of all factors and symbols used in Equation (47).

10. What is a flue-gas analysis? How is it made, and what does it show regarding combustion?

11. Explain the principle of operation of the Hays CO_2 recorder.

12. During the test of a 508 horsepower B. and W. boiler, the following data were taken: weight of coal fired per hour, 21,954 lb.; weight of ashes per hour, 4087 lb.; carbon in the ash, 22.84 per cent.

Ultimate analysis is of coal "as fired," expressed in per cent:

Carbon.............	60.31	Nitrogen............	1.13
Hydrogen...........	4.06	Sulphur.............	5.23
Oxygen.............	6.71	Ash.................	22.56

Flue-gas analysis, expressed in per cent:

Carbon dioxide.......	10.69	Oxygen..............	8.23
Carbon monoxide.....	0.23	Nitrogen (by difference)	80.85

Find the weight of dry chimney gases per pound of coal fired. What must be added to the weight of the dry gases to obtain the total weight of chimney gases?

13. What is the effect upon the flue-gas analysis of air leaking into a boiler setting? How can this leakage be determined by means of flue-gas analyses?

14. What is shown by a boiler heat balance? Of what assistance is such a heat balance in correcting improper furnace conditions?

15. A rule, sometimes used, states that the air required to burn a pound of coal equals 7.65 lb. per 10,000 B.t.u. in the coal. Check the accuracy of this rule for a Pennsylvania coal which has a heat value of 14,688 B.t.u. and the following ultimate analysis: carbon, 83.43; hydrogen, 4.75; oxygen, 3.91; sulphur, 1.20; ash, 5.44.

16. Calculate the volume of the *mol* for oxygen at 60 deg. fahr., and 30 in. of mercury. Take data for oxygen from Table 4, page 84.

CHAPTER VIII

SMOKE PREVENTION, METHODS OF BURNING FUELS, FURNACES, AND REFRACTORIES

175. Foreword. — Within the past few years, great advances have been made in the methods used to prevent smoke in power plants burning coal. These improvements have been brought about by campaigns against smoke-producing plants, and by more stringent enforcement of city ordinances against smoke.

The better grades of fuel, which contain small amounts of volatile matter, may easily be burned without smoke; but with the poorer grades of fuel, high in volatile matter, the prevention of smoke is more difficult, and careful attention must be given to furnace conditions.

The actual loss of heat from smoke alone does not exceed $1\frac{3}{4}$ per cent of the heat value of the fuel. The presence of dense smoke, however, indicates poor furnace conditions, which may cause losses amounting to 20 per cent. Such losses materially reduce the efficiency and capacity of a boiler plant. On the other hand, a smokeless stack does not necessarily mean high efficiency in the boiler room; it may indicate too large an excess of air and the accompanying losses. A small amount of smoke may accompany the best of furnace conditions. In order that the boiler may be operated under the best possible conditions, it is necessary that the fireman have suitable instruments, such as flow meters for steam, air and water, draft gages, and CO_2 recorders, to guide him in the proper operation of the furnace. In some large high-pressure plants automatic control[1] of the furnace is resorted to in order to obtain satisfactory operation, and increase the operating efficiency.

176. Smoke and Its Cause. — Smoke consists of water vapor and gaseous products of combustion, colored with fine particles of carbon or soot, and unburned vapors of the tarry constituents of fuel.

Smoke is produced if the air supply is insufficient; if the air and gases are not intimately mixed; if the temperature is not high enough to ignite the carbon; or if there is not sufficient time to burn the gases completely, before they have been chilled below the temperature of ignition by contact with the cooler surfaces of the boiler.

To prevent smoke, the escape of the unburned smoke-making constituents must be prevented, and the conditions necessary for perfect combustion, Art. 152, must be maintained. The volume of the combustion

[1] Consult Chapter XIII, page 318, for description.

chamber must be sufficient to permit the proper mixing of the air and gases before they are cooled below the ignition temperature, and to assure their burning before reaching the boiler heating surface.

Coal that has a high volatile-matter content is not necessarily a smoke producer, the amount of smoke depending upon the constituents forming the volatile matter. A high hydrocarbon content in the volatile matter generally indicates that the coal will produce a large amount of smoke. A coal which has a large amount of tarry vapors will produce smoke, unless ample time is given in which to burn them.

A knowledge of firing methods is essential to a proper understanding of the methods employed to prevent smoke. The methods used to charge coal are (1) hand stoking, and (2) mechanical stoking.

FIG. 114. — Stationary Circular Grate.

177. Grates for Hand Stoking. — The purpose of the grate is to support the fuel bed, at the same time admitting air for combustion. It should be of such a form that it will be kept uniformly cool by the inflowing air, and should permit ashes to pass freely. The width of the air spaces should be as great as possible without permitting coal to pass through. This width depends upon the kind and size of coal used, and ranges from $\frac{3}{16}$ to $\frac{5}{16}$ of an inch. *The total area of the openings for air, when burning coal, varies from 30 to 50 per cent of the total grate area.*

The type of grate to be used with a hand-fired furnace is controlled by the type of boiler and kind of fuel used. With small vertical boilers, a **stationary circular grate,** Fig. 114, is used; it is made of a number of cast-iron pieces fitted to form a circle, and is supported by a suitable ring, Fig. 24, page 31. With large vertical boilers, **rocking** and **shaking grates** of the circular form are used, while horizontal boilers often use **stationary grates** with the grate surface formed by a number of grate bars, placed side by side and held in place by their own weight. Lugs at each end of the grate bars rest upon **bearer bars** fastened to the walls of the setting. The

common grate bar, Fig. 115, is made about 3 inches deep at its center, to give strength and thus prevent sagging under the weight of coal when hot. The width of each bar varies from $\tfrac{3}{4}$-inch at the top to $\tfrac{3}{8}$-inch at the bottom, to allow the ashes to drop through easily.

The tupper and herring-bone grate bar is made with side flanges, which stiffen the grate and reduce the liability of warping. Each grate bar is about 6 inches wide and has openings of the shape shown.

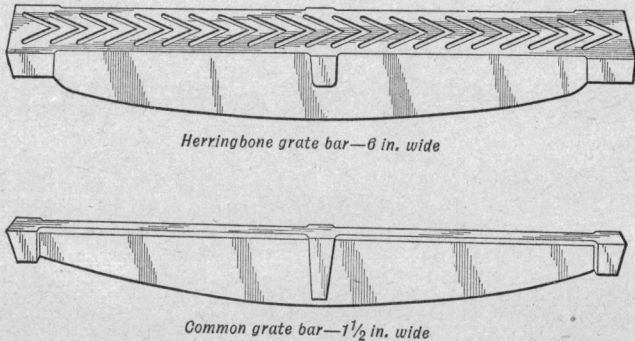

Fig. 115. — Types of Stationary Grate Bars.

The length of each grate bar is about 3 feet. The total length of the grate depends upon the type of fuel and size of boiler. For easy firing, the length should not be over 6 feet. The rear end of the grate is ordinarily from 3 to 5 inches lower than the front end. A solid plate known as the **dead plate** is often used, with coking coal, to support the front end of the grate bars and to hold the green fuel until the volatile gases are distilled off.

178. Shaking Grates. — With stationary grates, the fire is not easily cleaned, combustion is hindered and the fire is sluggish, unless the air spaces are kept free of clinker and ashes. Frequent cleaning is wasteful of coal, and requires the firing doors to be open. This admits a large quantity of excess air and consequently lowers the temperature of the furnace. The shaking grate, Fig. 116, permits removing the ash and refuse without opening the firing doors and also reduces the amount of labor required. At each end of the grate bars, which run at right angles to the length of the furnace, are trunnions resting in slots in the side **bearer bars.** Ordinarily the grate surface is 2 or 3 bars wide and a varying number of bars in depth. The side bearer bars are supported by a frame, which rests upon the ashpit floor. The rear and front grate bars are stationary, and the lower ends of the movable grate bars are joined in groups. Each group is connected to a lever attached to the operating lever. Movement of the lever tips the grate bars about the lugs resting in the bearer bars, and thus allows the loose ash to fall into the ashpit.

179. Tools Used to Handle Fires. — The tools required to handle the fire, in a furnace fired by hand, are the **hoe**, the **slice bar** and the **rake**. The hoe and the slice bar are used principally to clean the fire and break up the clinker; the rake is used to level the fuel bed. A **lazy bar** is often used to support the tools while working; it is laid from the hinge bracket to the latch lug, and the tools rest upon it.

180. Methods of Firing Coal by Hand. — Coal is charged into hand-fired furnaces by the **alternate, spreading, coking,** and **ribbon methods.**

When the *alternate method* is used, coal is fired on one side of the furnace while the other is burning brightly. The volatile gases that are distilled

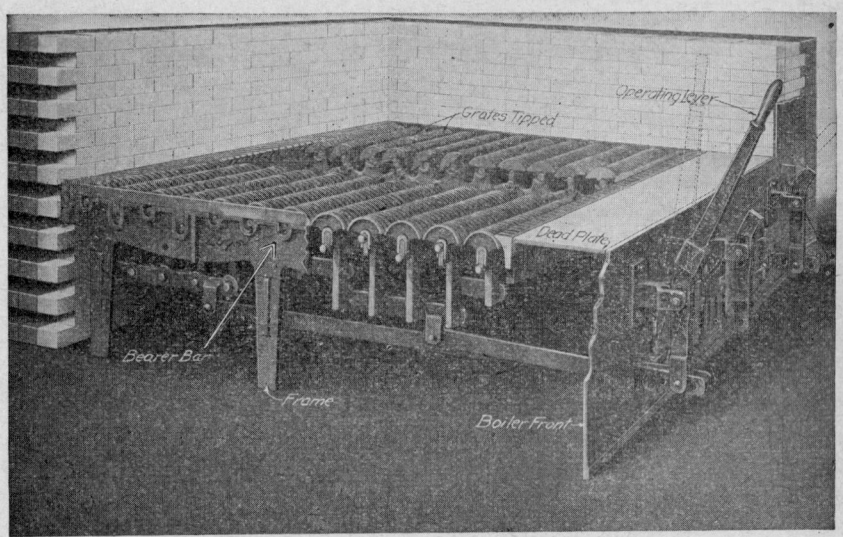

FIG. 116. — Hand-operated Shaking Grate.

off are mixed with air admitted over the fire and are burned in the presence of an intense heat from the bright side of the furnace.

With the *spreading method*, coal is spread in a thin layer over the grate at each firing. It is customary to commence at the bridge wall and work toward the firing door. This requires the frequent firing of small quantities of coal.

With the *coking method*, coal is charged on the dead plate or at the front of the fire, in considerable depth, and allowed to coke, after which the coke is pushed toward the bridge wall and spread evenly over the grate, before more coal is fired. The hot gases distilled during the coking period must pass over the burning fuel in the rear, and are burned.

By the *ribbon method*, coal is fired in narrow alternate strips, which extend the full length of the grate.

Of these four methods, tests have shown that the ribbon method of firing, with the coal fired in small amounts, gives the highest efficiency and practically no smoke; that the spreading method gives the lowest efficiency and produces the most smoke; and that the coking and alternate methods give nearly the same result and are better than the spreading method.

Coal is fired intermittently into hand-fired furnaces, and careful firing is required to prevent smoke. The results obtained depend upon the human element and generally are not consistent. The fires are liable to be carried too thick, with the result that the air supply is insufficient. As a general rule, a 9-inch fire is sufficiently thick, while tests by the BUREAU OF MINES show that *thinner fires* than this, *with the coal fired often and in small amounts, give the best results.*

181. Smoke Prevention, Hand-fired Furnaces. — The methods of hand firing which produce the least smoke approach the method of firing used with mechanical stokers. A modification of the shaking grate, known as a **hand-operated stoker,** is sometimes used effectively; the grate surface is formed as in an ordinary shaking grate, but is given a greater slope from front to rear, with a dump plate at the bottom of the slope and a coking plate at the top. This type of stoker has been improved by the addition of hopper feeds and power drives, and by using a combination arch and depression plate to introduce a mixture of preheated air and furnace gases over the fuel bed. It has the advantage over hand-fired grates in that the feeding of the fuel is more uniform, and in addition does not require frequent opening of fire-doors for feeding coal, thus preventing the admission of large amounts of excess air. It is adapted for low-load-factor low-rating plants, with good fuel, nut size or larger, and where the cost of fuel and the type of installation do not permit the expenditure of a larger investment.

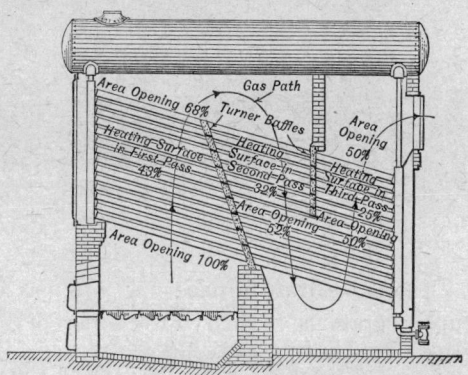

FIG. 117. — Inclined Baffle Walls.

The principal methods of smoke prevention in hand-fired furnaces are those which depend upon furnace construction to ensure thorough mixing of air and gases. Special baffle walls of the inclined type, Fig. 117, in which the cross-sectional area of the gas passages is reduced nearly proportional to the reduction in the volume of the gases as they cool, and furnaces of the Dutch-oven, Chicago-setting and Downdraft types are examples of the methods used.

The **Dutch-oven furnace,** Fig. 127, page 195, consists of an extension of the common type of setting, bringing the furnace out in front of the boiler.

The hot gases thus have a greater distance to travel before coming in contact with the cooler boiler and consequently have more time in which to burn. The **coking arch** above the grate absorbs heat from the burning coal, becomes white hot, radiates heat back upon the coal, and thus assists in securing better combustion. The chief defect of the Dutch oven is too rapid distillation of the volatile gases, without sufficient air supply to properly burn them.

The **Wing-wall Chicago setting** is used to burn Illinois bituminous coal. It differs from the ordinary boiler setting in that, directly back of the bridge wall, there is a wing-shaped firebrick wall, which changes the course of the furnace gases and produces better mixing of the air and gases. Like most hand-fired furnaces, it requires careful manipulation.

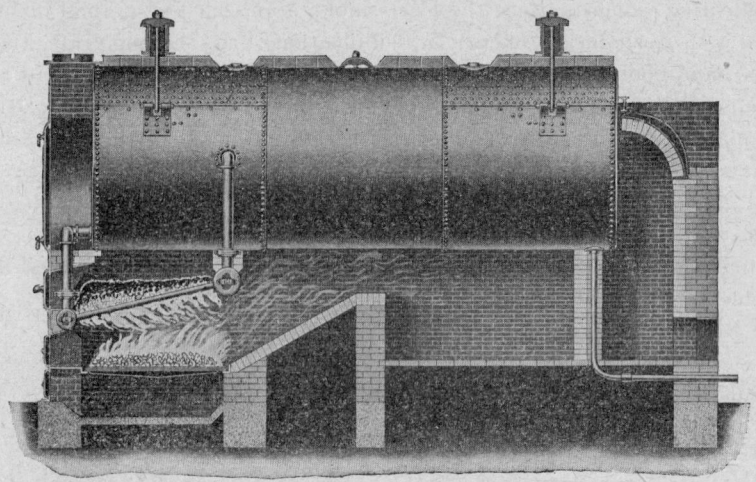

Fig. 118. — Vogt Downdraft Furnace.

The **Downdraft furnace,** Fig. 118, has an upper and a lower grate. The upper grate is formed by a series of pipes through which water from the boiler circulates, while the lower grate is formed by common grate-bars. Coal is thrown upon the upper grate, where it burns, and incandescent fuel, falling through to the lower grate, keeps a bright fire there. Air admitted above the upper grate mixes with the distilled gases from the coking coal and passes down through the coal on the upper grate. These gases are mixed with additional air admitted through the lower grate doors, and are completely burned by the heat from the incandescent fuel on the lower grate. *This gives practically smokeless combustion.*

Steam jets are often used for inducing air and mixing it with the gases. The jets are placed above the grate and at the front end of the furnace, and are operated during the firing period and for a short time after firing.

A form of setting known as the **Kilgour setting** has been used successfully for prevention of smoke. It makes use of steam jets located on the bridge wall, and draws heated air through a passage in the bridge wall at the time of firing. The steam, air, and distilled gases are then directed through a short passage lined with firebrick, which becomes nearly white hot under ordinary conditions of operation.

182. Smoke Prevention by Mechanical Firing Methods. — The best method of preventing smoke and increasing the efficiency in coal-fired furnaces is to use *mechanical stokers*. Their use presents the following **advantages:**

1. Reduces amount of labor required in the boiler room of large power plants. With stokers, a man can care for 5000 to 7000 boiler horsepower and, in extreme cases, as high as 14,000 boiler horsepower. With hand firing, a man is required for each 500 boiler horsepower.
2. Permits the burning of poorer grades of fuel.
3. Maintains better furnace conditions by feeding coal at a uniform rate.
4. Avoids excessive admission of air.
5. Saves labor of handling ashes and is self-cleaning.

The **disadvantages** are cost of operation and repairs, resulting from high furnace temperatures. This cost can, however, be decreased by judicious overstokering. As ordinarily operated, about 6 per cent of the steam generated is required to operate the stokers.

To obtain satisfactory results in the operation of stokers, intelligence and good judgment are required. Stokers are not a cure-all for smoke abatement. They are too often neglected, and therefore often fail to give satisfaction.

In general, the more simple the stoker the better are the results obtained. Stokers are in common use in all power plants of 500 boiler horsepower or more, and in many plants of 200 boiler horsepower (see page 220).

183. Classification of Stokers. — Stokers may be divided into three general classes:

1. Traveling, or chain grate.
2. Overfeed.
3. Underfeed.

The names of a few of the typical makes belonging to each class follow:

Chain grate { Natural draft or Forced draft.... { Babcock and Wilcox
Green
Harrington
Illinois
Westinghouse
Brady

188   SMOKE PREVENTION, FURNACES AND STOKERS

Overfeed
- Front-overfeed or inclined....
  - Roney
  - Wilkinson
- Side-overfeed or V..........
  - Murphy
  - Detroit

Underfeed
- Surface of fuel bed nearly flat
  - Jones
  - Type E
  - Westinghouse
- Surface of fuel bed inclined at a considerable angle........
  - Riley
  - Taylor
  - Westinghouse

Each type of stoker has its field of application, and cannot be used with all types of installations and fuels, but must be adapted to the fuel, rating and conditions under which the installation is being made.

184. Natural Draft, Traveling, or Chain Grate. — The grate surface of this type of stoker consists of an endless traveling chain made up of a large number of links, upon which the coal is fired at the front end and from which the ashes are discharged at the rear. The chain travels through the fur-

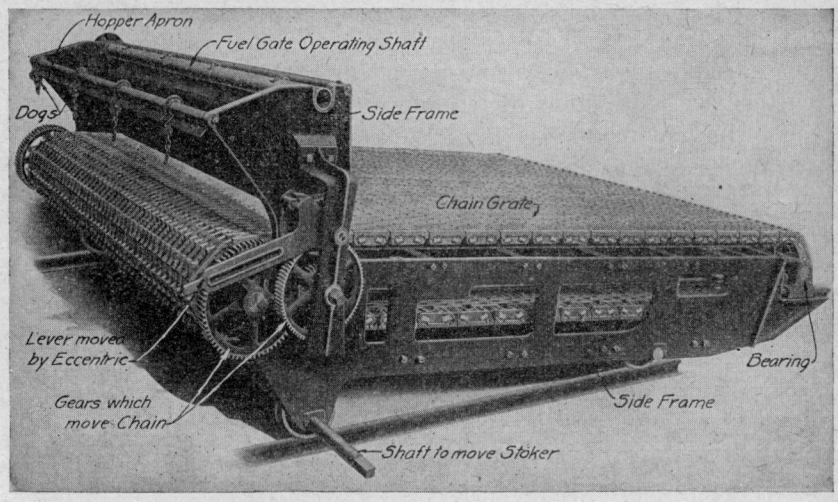

FIG. 119. — Green Type K Chain Grate Assembly.

nace at a uniform rate and does not agitate the fuel bed. As usually set, the surface of the grate is horizontal; some makes, however, have a slight incline toward the rear. The various makes differ mainly in points of design.

An assembly of a grate and frame is shown in Fig. 119, and a frame without the chain is illustrated in Fig. 120. The **frame** consists of cast-iron **side frames** joined by **steel cross girders** and shaped at the front end to form the sides of the coal hopper. The chains rest on top and bottom skids which are supported by the **cross frames,** thus producing a level grate surface and materially reducing siftings. The front **sprocket shaft** is supported on bearings fixed to the side frames, while the rear drum shaft bearings are made adjustable to permit changing the tension of the chain.

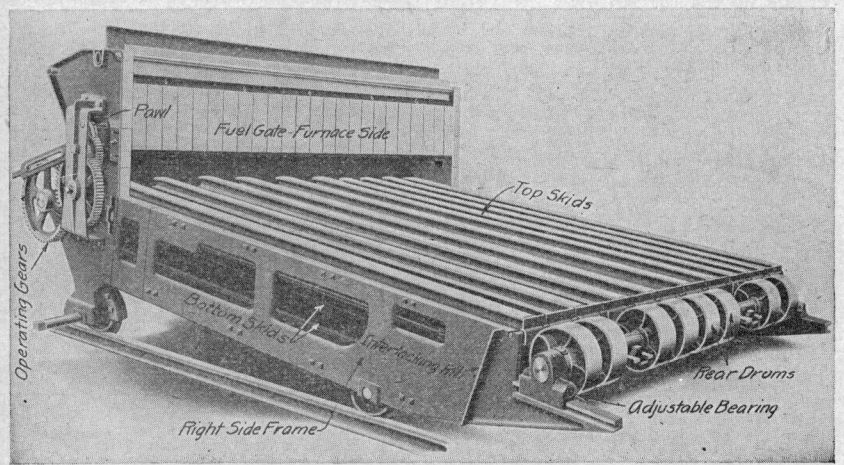

FIG. 120. — Assembly of Chain Grate Frame from Furnace End.

FIG. 121. — Green Chain Grate

The **links** of the chain are held together by oval bars, with the **driving links** riding over the sprocket teeth centrally and carrying all tension. The remaining links are slotted for easy removal and do no driving. A top and bottom view of the chain is shown in Fig. 121. It is driven through a system of steel **spur gears,** which drive the front sprocket shaft from both sides. The spur gears are driven by a **pawl** which is attached to an arm connected to an eccentric rod, having an **eccentric** located on a shaft at the front of the boiler or below the boiler room floor. The eccentric shaft is generally driven by a belt from a small steam engine or electric motor. The position of the eccentric rod on the arm is adjustable, to permit changing the rate of feed; the longer the arm the slower will be the feed.

Ventilated cooling plates are sometimes placed in the side frame above the top of the grate, to prevent overheating of the side frames. **Adjustable flanges** on the **ledge plates,** which are built into the side walls of the setting, fit against the side of the upper chain, preventing air leakage between the chain and walls and presenting a durable surface to the moving fuel.

The **front apron** of the hopper is made of steel plate and can be dropped to a horizontal position by releasing the **holding dogs.** The back wall of the hopper is formed by a **fuel gate,** which is adjustable in height. This gate has a firebrick lining on the fire side, and renewable **cast-iron shoes** on the hopper side. The depth of fuel on the grate is regulated by raising or lowering the gate by means of a worm and wheel. Leakage of air into the furnace through the rear is prevented by **side dampers** attached to the side walls. These dampers bear against projections on the side frame and make a tight joint at this point. A steel plate, Fig. 122, connecting the side frames, below the lower chain, prevents air leakage at the rear. The whole frame is mounted on wheels which rest on a track, thus permitting withdrawal from the furnace when making adjustments.

Coal is fed by gravity from the hopper, Fig. 122, to the moving chain grate. It passes under the fuel gate and is ignited, by the furnace fire, when under a **flat** or **sloping coking arch** which extends over the entire width of the furnace at the front end and *assists in the combustion of the fuel. The speed of the chain,* considering depth of fuel and draft, *is such that combustion will be complete by the time the chain passes the rear sprocket and the refuse falls to the ashpit below.* Average speeds vary from 1 to 6 inches per minute.

The fuel bed at the rear end of the grate is compressed by coming into contact with an overhanging bridge wall. When this wall is built of firebrick, incandescent ash adheres to it, the opening for ash discharge becomes closed, and the overhang burns off. To prevent this, and to reduce the ashpit loss, a **water back,** Fig. 122, consisting of a pipe or pipes spanning the rear of the furnace, is used. Water passing to the boiler circulates through the pipes and keeps them cool.

This type of stoker is commonly used in central stations under large boilers, and in industrial plants under boilers as small as 150-horsepower rating. It is adapted to free-burning coal that is high in volatile matter, but does not give satisfactory operation with the high-carbon coking coals of the Appalachian field. It successfully burns high-ash clinkering coals of the West and Middle West regions.

FORCED-DRAFT CHAIN-GRATE STOKER

When intelligently operated, the chain-grate stoker will produce little smoke, and will handle small sizes of coal successfully. *The type and shape of ignition or coking arch are important.*

For best operation, the depth of fuel bed should not be over 6 or 8 inches. Coal should not be permitted to burn in the ashpit, and the grate bars should be cold when re-entering the furnace. A combustion rate of 40 pounds of coal per square foot of grate surface per hour may be used with satisfactory results, and higher rates for the peak load may be obtained with increased draft and depth of fuel.

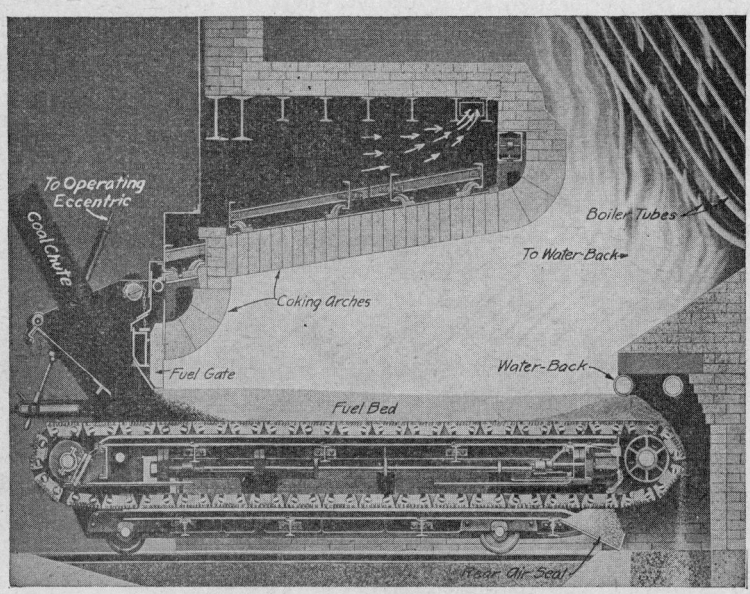

Fig. 122. — Green Chain Grate under Stirling Boiler.

185. Forced-draft Chain-grate Stoker. — This type of stoker has been gradually replacing the natural-draft chain grate because of its higher efficiency, greater flexibility, and only slightly greater total stoker and furnace cost. The Harrington forced-draft chain-grate stoker is shown in Fig. 123. Its general construction is similar to that of the natural-draft chain-grate stoker except that the entire active grate surface is divided into four or more compartments, each compartment being supplied with low velocity air through dampers controlled from the front of the stoker. The grate surface of this stoker consists of individual grate bars having interlocking edges mounted on transverse racks which are supported by driving chains in three or more places, depending upon the width of the stoker.

Combustion proceeds progressively from ignition at the front of the stoker to the ash which is discharged as formed at the rear of the stoker. Since the character of the fuel bed is constantly changing, the air pressure and volume are different at each stage of combustion, the proper air supply being provided by controlling the volume of the air in each of the compartments.

This type of stoker is adapted to a wide variety of coals and will burn anthracite, coke breeze, or mixtures of these coals with bituminous coal. Both front and rear arches are required.

With all coals, thorough air sealing of the ashpit is required, and the use of water backs is an advantage, especially with bituminous coals. **Water-cooled side walls** at the clinker line or carborundum brick, preferably air-cooled but not perforated, are a distinct operating help. Segregation of fine and coarse coal in the hopper is detrimental to good operation and a well-burned-out ash.

FIG. 123. — Harrington Forced-draft Chain-grate Stoker.

The use of draft gages and automatic furnace-draft regulators, Art. 297, page 316, is almost a necessity for the proper operation of forced-draft stokers.

186. Front-overfeed Stoker. — *This type of stoker has an inclined grate surface. Coal is fed on the grate at the top, and the ashes are discharged at the bottom.* Fig. 124 shows a typical stoker of this type, as applied to a Vogt water-tube boiler.

The Westinghouse-Roney front-overfeed stoker, shown in Fig. 125, has a fuel hopper located at the front of the boiler and extending the width of the furnace. **Rocking-grate bars**, Fig. 126, are inclined from front to rear and are supported at each end by bearer bars. The **main operating shaft** runs horizontally along the stoker front beneath the hopper. Keyed to this shaft is an eccentric which, by means of an arm cast integral with one-half of the eccentric strap, imparts an oscillatory motion to the agitator. This **agitator** swings freely on a hopper shaft, which has bearings at each end of the hopper. Connecting rods, by which the grate bars are rocked, are fastened to the agitator and the **rocker bar,** in which a projection on the grate bars rests. At each end of the hopper shaft, a **toothed sector** is keyed; it projects through a slot in the **pusher plate,** at the bottom of the hopper, and engages a series of **rack teeth** on the under side of the **pusher** which rests on the pusher plate and is operated by the toothed sector. To one of these sectors an offset arm is cast, the outer end overhanging the agitator. *A long bolt, attached by a swivel connection to the agitator, passes loosely through the offset arm of the sector, and, by means of a handwheel on the bolt, forms a means of adjusting the travel of the pusher,*

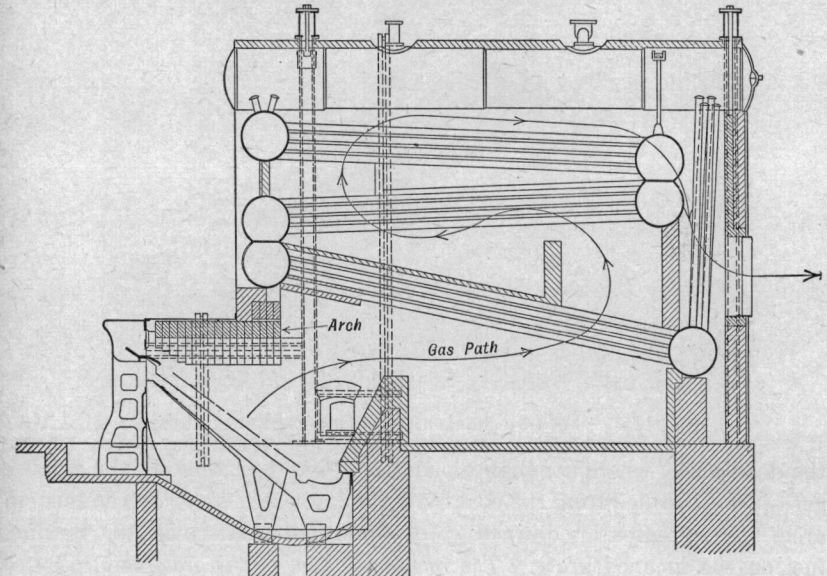

FIG. 124. — Westinghouse-Roney Stoker Applied to Vogt Water-tube Boiler.

by allowing lost motion between the handwheel and the sector. With the handwheel set up tightly, the agitator, through the hopper shaft and sectors, imparts maximum motion to the pusher. If the feed wheel is slacked off, so that there is lost motion equal to the arc described by the agitator, at the point where the bolt is attached, the pusher will not be moved.

A dump plate, made into several sections, for convenience in handling, is located at the bottom of the grate bars. At the top of the grate bars is a **dead plate** extending a short distance into the furnace, below the **combustion arch,** which extends over the width of the grate. A lug on each grate bar rests in a rocker bar, driven by the connecting rod attached to the agitator. *The rocker bar rocks the grate bars, on their trunnions in the side bearers, through an arc of 30 degrees, giving them first a hori-*

zontal and then an inclined position. The amount of motion given the rocker bar is regulated by changing the position of the sheath and lock nuts, to increase or decrease the lost motion.

When the pusher is drawn back by the sector, coal from the hopper settles down in front of it and, as it advances, the coal is pushed on to

Fig. 125. — Feed Mechanism, Westinghouse-Roney Stoker.

the dead plate, where it remains a sufficient length of time for the volatile gases to be distilled from the coal by the radiant heat from the combustion arch. These gases are burned in passing to the rear over the burning fuel on the inclined grate. *The rocking motion of the grate bars, assisted*

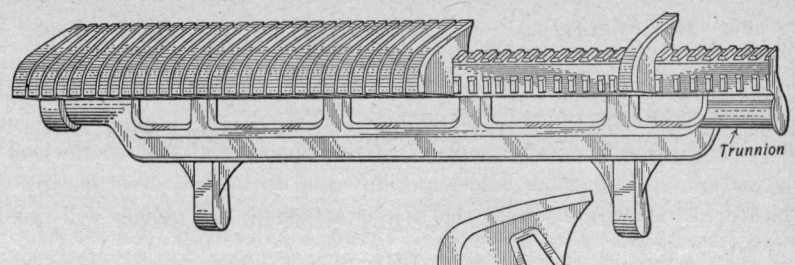

Fig. 126. — Section Grate Bar — Westinghouse-Roney Stoker.

by the force of gravity, causes the coal to move down the incline, and it is burned in traveling to the dump plate at the bottom, where the ashes collect and are dumped by hand as occasion demands. At the bottom of the incline a **guard plate** can be raised to hold the coal on the grate when lowering the dump plate.

SIDE-OVERFEED STOKERS 195

Air for combustion is supplied by natural draft from below the grates. Steam jets at the front of the combustion arch are sometimes used to assist in mixing the air and gases.

This type of stoker can be readily forced beyond its rated capacity to higher rates of combustion, and if installed in a properly designed furnace will not produce much smoke. It has given excellent results when used for loads such as are encountered in hospitals, schools and office buildings. It is adapted to low-ash high-grade coals, particularly of the semi-bituminous varieties. The rate of combustion can be as high as 40 pounds of coal per square foot of grate surface per hour, but for best operation it

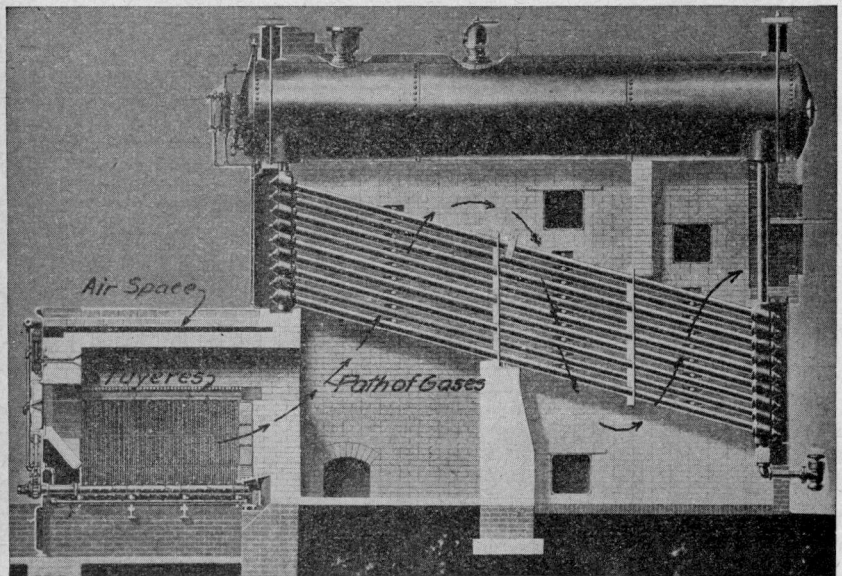

FIG. 127. — Murphy Stoker as applied to B. & W. Boiler.

should be about 20 pounds per square foot per hour. The depth of the fuel should not be over 7 inches.

187. Side-overfeed Stokers. — This stoker is generally located in a Dutch-oven furnace, as shown in Fig. 127. *The grate bars are arranged to form a V-shaped inclined fuel bed which is agitated by the movable grate bars.*

The Murphy stoker, Fig. 128, which is representative of this type, has two fuel hoppers, one on each side of the furnace, extending from front to rear. The inner wall of the hopper is formed by an **arch plate** supported front and rear by an upright frame. Ribs, forming passages through which heated air from the air space above the arch is admitted for combustion, are cast on the lower end of the arch plate. At the bottom of the hopper is a built-up iron and steel plate, called the coking plate. Below this is an air duct through which air circulates and keeps the coking

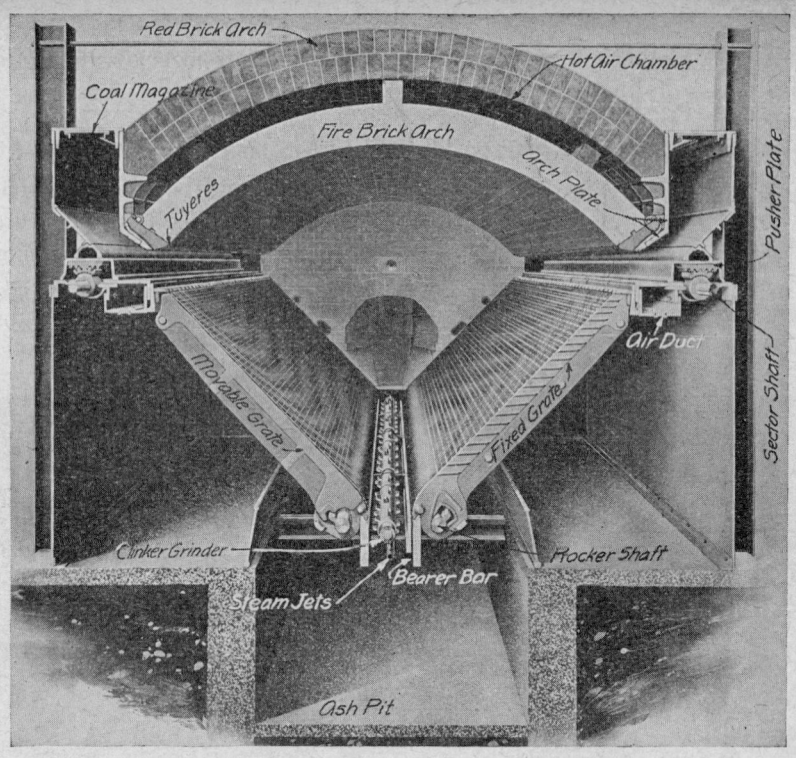

Fig. 128. — Murphy Stoker looking toward the Front End.

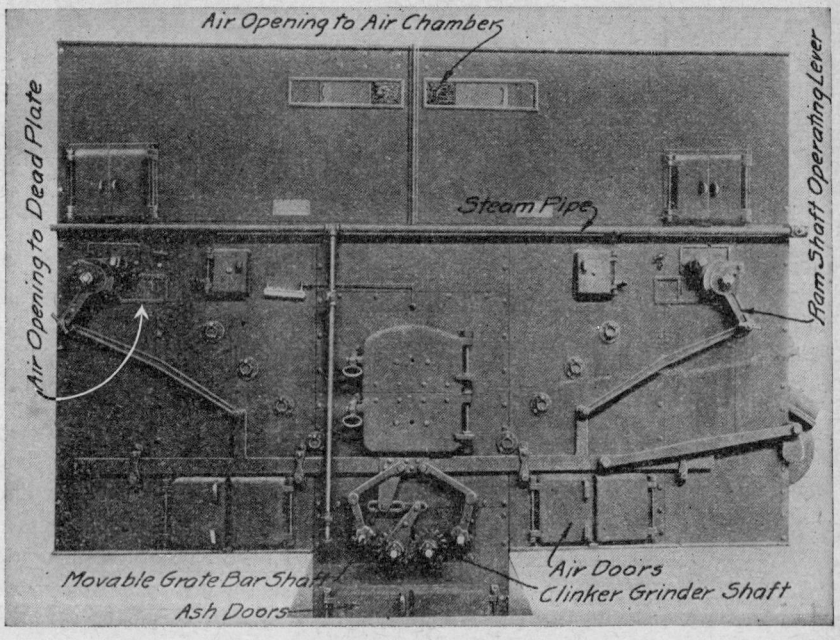

Fig. 129. — Front View of Murphy Stoker.

plate from becoming overheated. The pusher rests and slides upon the coking plate, as it is operated by a rack and sector. The sector is driven by a shaft, running below the hopper to the front of the furnace, Fig. 129, and is there connected to an operating lever.

The grate bars, Fig. 130, *are made in pairs, one fixed and one movable.* The fixed grate bars are ribbed to prevent the entrance of too much air from below the grate, and also to prevent excessive sifting of fine coal. Their upper ends rest against the coking plate and the lower ends are supported by the **grate bearer.** The movable grate bars are pivoted at their upper ends, and their lower ends are moved by the rocker shaft, first above and then below the stationary grate bars.

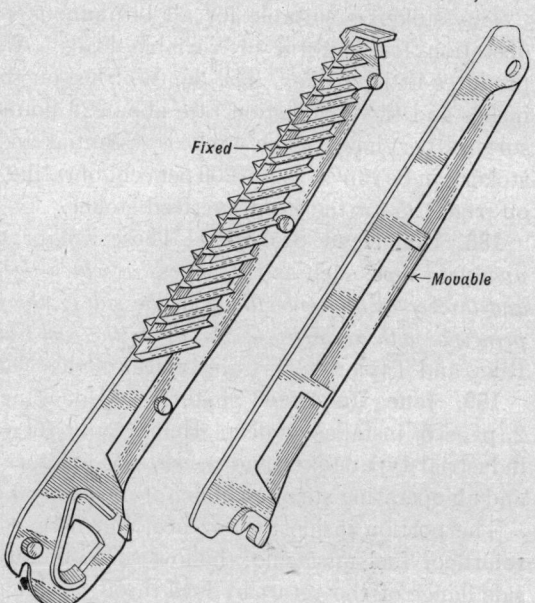

The grate bearer is located at the bottom of the V and, besides supporting the lower ends of the fixed grate bars, carries the **clinker grinder,** the form of which varies with the service it is to perform. The clinker grinder is driven by a reciprocating bar running across the stoker front. This bar also controls the feed of the fuel and movement of the grate bars, and is driven by an electric motor or a steam engine located at one side of the setting.

Fig. 130. — Grate Bars for Murphy Stoker.

A **triple air-cooled arch,** made in two parts and supported by the arch plate, extends over the whole grate. The lower part of the arch is made of firebrick and the upper arch of common brick. An air space, through which air is constantly circulating, separates these arches. Air is admitted to this air space through dampers on the front of the furnace, and then passes to the furnace through the passages formed by the ribs on the arch plate. The front and sides of the furnace are covered with plates of sheet steel.

Operation. — Coal is delivered to the hopper either mechanically or by hand, and is intermittently fed, by the **pusher,** from the hopper to the coking plate at the upper end of the grate bars. While it is on the coking plate, the volatile gases are driven off by radiant heat from the white-hot arch, and are immediately mixed with the heated air admitted through the arch plate ducts, forming a combustible gas which is burned as it passes toward the rear of the furnace. The fuel travels slowly down the inclined grates toward the clinker grinder, receiving the necessary amount of air through the grate to complete the burning process. By the time it reaches the clinker grinder, combustion is complete, and the ash and refuse are automatically discharged into the ashpit.

The speed at which the pusher pushes the coal to the grates can be regulated as desired. The movable grate bars have their greatest action at their lower end, and thus break up the fuel bed for free admission of air to that part of the furnace where it is most required.

This stoker is suitable for all bituminous coals and gives satisfactory operation for uniform or variable loads. When properly operated, it produces little smoke. The depth of fire on the grates should be about 6 inches and the combustion rate about 23 pounds per square foot of grate surface per hour. The arch is essential in eliminating smoke. This stoker can be run at 150 to 200 per cent of rating, and its field is as mentioned on page 195 for the front-overfeed stoker.

188. Underfeed Stokers. — These are of two types: *one is a pure underfeed type*, such as the Jones, *in which the fuel bed is slightly rounded, and thicker at the center than on the sides; the other combines the underfeed principle with a gravity feed and has the fuel bed inclined from front to rear,* Riley and Taylor stokers are typical of this class.

189. Jones Underfeed Stoker. — A Jones inclined stoker is shown in Fig. 2, page 6, installed under a Heine water-tube boiler. It is known as an industrial type stoker and consists of a retort located inside the furnace, and an operating steam cylinder or mechanical drive outside the furnace.

The portion inside the furnace, Fig. 131, consists of a **heavy cast-iron retort,** or fuel magazine; hollow **cast-iron tuyere boxes** fastened to the side flange of the retort by lugs through which a retaining rod passes; an **auxiliary pusher rod** driven by the main plunger; **pusher blocks** bolted to the auxiliary plunger rod; sloping combustion grates; and heavily **ribbed hand-operated dump plates,** which extend from the edge of the retort to the side wall of the furnace.

The portion outside the furnace, Fig. 131, consists of a **coal hopper;** a **plunger case** directly below the hopper and bolted to the retort at the boiler front; and a **plunger** connected through a system of spur gears to a constant speed motor, which drives the fuel feeding mechanism, and forced-draft fan.

Operation. — The fuel bed consists of three zones, the green coal at the bottom, the coking zone in the middle, and the incandescent fuel bed at the top. Coal is fed into the hopper by hand or by mechanical means, and drops in front of the intermittently operated plunger, which, by a quick thrust, forces the coal into the retort. The pusher blocks on the auxiliary plunger rod force the coal in the retort backward and upward, thus breaking up the fuel bed at each stroke. Air is supplied by a fan at about 3 inches of water pressure, the exact pressure depending on the thickness of the fuel bed. The air enters below the stoker, passes to the tuyere boxes and into the fuel bed below the fire. Thus the volatile hydrocarbons from the coking zone are thoroughly mixed with air before reaching the incandescent zone. The supply of air and fuel is regulated to meet the operating conditions. The ashes collect on the dead plate and are periodically dumped into the ashpit.

When larger capacity is desired, more than one retort is placed in the furnace, with a dead plate between each pair of retorts.

Fig. 131. — Riflex-drive Jones Underfeed Stoker.

This type of stoker is adapted to bituminous coals, but not to anthracite. It does not require as large a combustion space as the overfeed stoker, because the gases are burned in passing through the incandescent fuel bed. It can be applied in units from 75 horsepower, and upward. The thickness of the fire is about 15 inches.

In a recent development of the Jones stoker, the fuel bed and retort are given a slight incline from front to rear, with dump plates at the bottom of the incline. The ashes collect on the dump plates, which are dumped by hand from the side of the furnace.

190. Riley Underfeed Stokers. — This stoker is shown in Fig. 132 under a cross-drum water-tube boiler. It is a multiple-retort, inclined, underfeed stoker in which the coal is forced into the retorts of the furnace, Fig. 133, by a ram, from a point below that at which air is admitted.

That portion of the stoker which is outside the furnace is shown in Fig. 134. It consists of a **coal hopper** attached to the upper part of the **ram case**; **rams** driven by connecting rods from a crank shaft running the length of the stoker; an **enclosed speed-changing device** operated by a hand lever; a set of **change gears** connected to the crank shaft **by a worm** and **worm wheel;** a **change-gear shaft** operated by a steam engine or an electric motor; **levers** for changing stroke of underfeed side or grate bars; **angle irons** running the length of the furnace to support the ram case front and rear; and **pipe supports** upon which the angle irons rest.

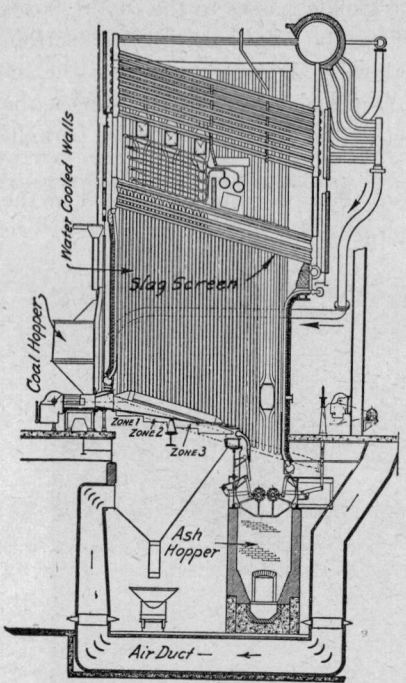

Fig. 132. — Riley Air-zoned Super Stoker.

The portion of the stoker which is inside the furnace, Fig. 133, consists of **underfeed reciprocating grate bars,** Fig. 135, fastened to the retort side bars by bolts; **side-bar shoes** to support the lower end of the side bar; **support rods** to carry the upper end and to give motion to the side bars; a **retort** formed between each two pairs of side bars and a metal bottom to which are attached sifting strips to prevent sifting of coal into the space beneath; **overfeed reciprocating grates** attached to and moving with the side bars; **rocker dump plates** attached to the lower end of the side-bar shoes and adjust-

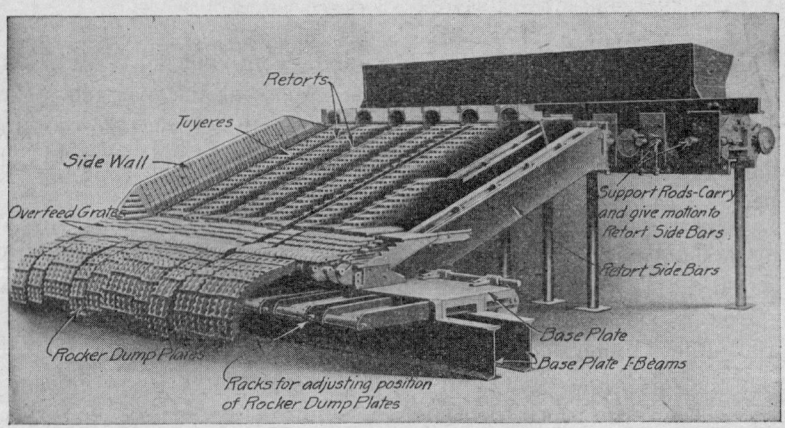

Fig. 133. — Perspective View of a 9-Retort Riley Underfeed Stoker.

able by a rack connected to a shaft operated by a handwheel located at one side of the setting; a **base plate** upon which the side-bar shoe moves; and **two I-beams** upon which rest the base plate and dump plate racks.

RILEY UNDERFEED STOKERS

Operation. — Coal is fed into the retorts from the hopper by plungers, Fig. 136, which push the coal forward and upward. At the same time, the side bars are pushed forward, giving the coal a movement in the

Fig. 134. — Riley Underfeed Stoker — Portion outside Furnace.

same direction. The overfeed reciprocating gates move the coal to the rocker dump plates, which continuously agitate, crush, and discharge the ash. The movement of the dump plates is adjustable, thus controlling the movement of the fuel bed and the dumping of refuse.

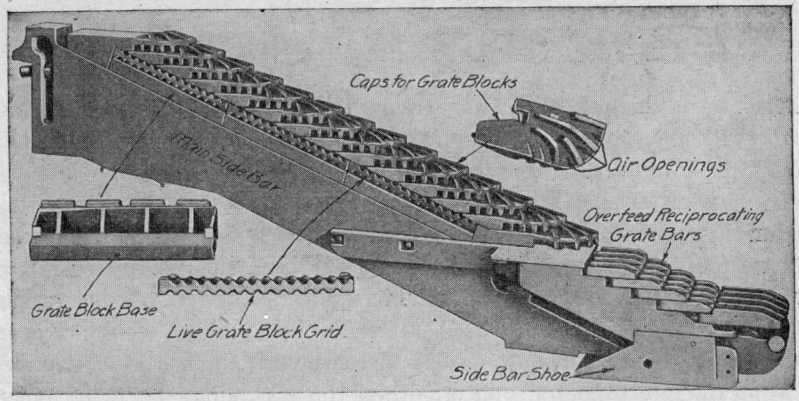

Fig. 135. — Reciprocating Grate Bar — Riley Stoker.

Distillation of the volatile gases occurs in the retorts, after which they are mixed with air discharged through openings, or tuyeres, under the grate blocks, which are carried by the moving side bars. The gases thus pass through an actual bed of burning coke and then through an incandescent fire zone.

Air for combustion is supplied, through air ducts, to the space below the stoker. A steel plate covers the front of this space and prevents the escape of air. From this space the air then passes into the furnace, through the tuyeres in the underfeed grates. The air to the overfeed grates is controlled by a damper.

This type of stoker is used mainly with boilers of large capacity, the size being determined by the number of retorts in the furnace. Each retort has a rating of about 100 boiler horsepower, when not forced. This stoker is capable of forcing the boiler 300 to 500 per cent of its rating.

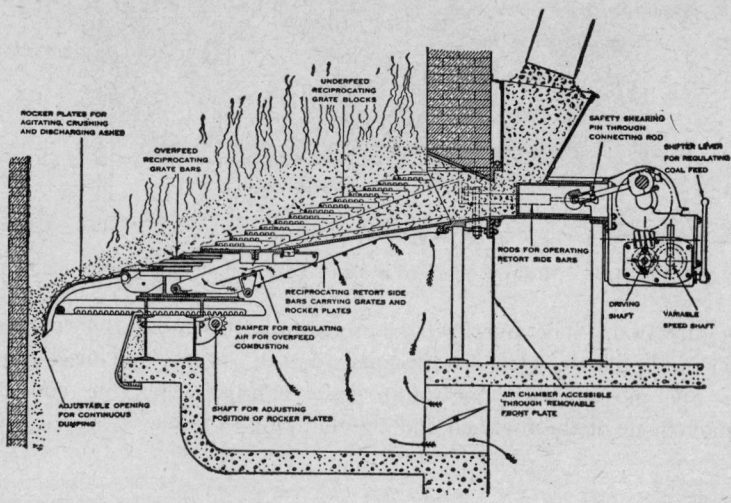

FIG. 136. — Section through Fuel Bed of Riley Stoker Fired Furnace.

Efficiencies as high as 80 per cent are obtainable when operating at normal loads without using air, preheaters or economizers; when forced, the efficiency is lowered slightly. All grades of free-burning bituminous coals can be used. The fuel bed varies from 2 to 3 feet in depth, and while ordinarily the maximum combustion rate should not exceed 50 pounds per square foot of grate surface, 70 pounds has been successfully burned in some installations. The main difference in the design of the various stokers of this class is at the discharge end. Some have a section of overfeed grates to assist in burning out the carbon in the ash after it passes beyond the retort section, while on others this is eliminated and the underfeed retorts discharge directly on to the dump plates. The overfeed section may be of the reciprocating type as shown in Fig. 136, or of the undulating link type, Fig. 137, page 203. For ash discharge, single and double dump plates, reciprocating apron, and single and double roll clinker grinders are used. The latter are usually located in a deep discharge pit, Fig. 137, to which air is admitted to burn out the carbon in the ashes. Water backs are used with nearly all types. In recent designs of large capacity models, the air supplied to the various portions of the fuel bed is carefully regulated by zones.

191. Taylor Underfeed Stoker. This was the pioneer stoker of the gravity underfeed type, and it is used extensively in large central stations. Like the Riley stoker, it has a feeding mechanism outside the setting, and a distributing portion inside the setting. The portion outside the setting consists of a hopper attached to a ram box in which is located a ram driven by a connecting rod through a planetary spur gear transmission which permits a speed variation of 50 per cent. The portion inside the furnace consists of stationary retorts, the bottom of which is formed by a series of distributing rams so arranged that the amount of fuel fed can be changed by the adjustments of the individual rams. The rams are driven by an attachment to the main ram located outside the setting.

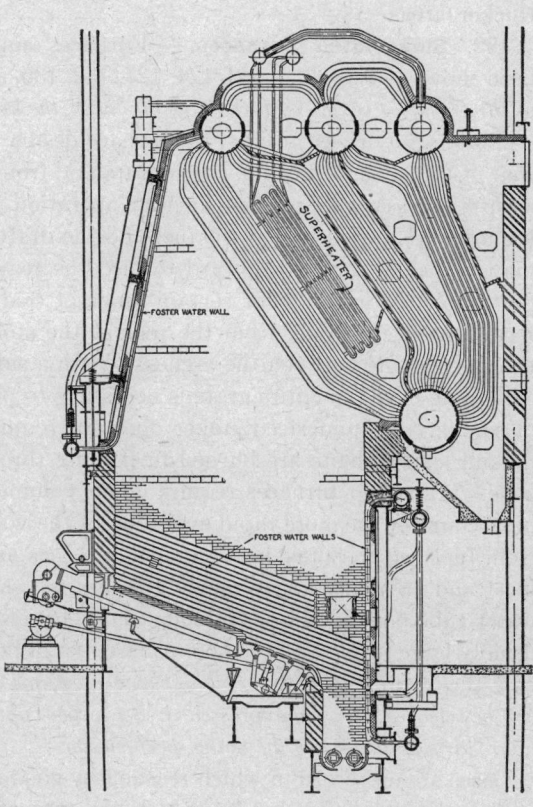

FIG. 137. — Westinghouse Underfeed Stoker.

At the bottom the coal is discharged to the dump plates which do not reciprocate and are dumped by either hand or power.

A 57 tuyere Taylor stoker is shown in Fig. 39, page 44, as installed under a 500,000 pounds per hour steam unit at the Delray Station of the Detroit Edison Co. It is installed in an extended type furnace, which has replaced the duplex arrangement formerly used for large capacity units. The furnace has four water-cooled walls, using bare metallic block on the sides and rear walls and insulated block on the front wall. The stoker is hydraulically operated with a connection to each distributing ram. Preheated air is supplied to a continuous chamber beneath the stoker by forced-draft fans. From this chamber the air flows to the fuel bed through the extension grate and tuyeres which are designed to convert the static pressure into velocity pressure.[1] The flow of preheated air is automatically controlled to maintain a uniform fire and a fuel bed of even thickness by balancing the air supply to each section of the stoker by using **air motors** that operate veins to reduce the air supply to sections where the fuel bed tends to become thin, and increase the supply to thicker areas.

192. Stoker-fired Furnaces. — Furnaces equipped with stokers have been shown in Figs. 2, 3, 122, 124, 127, 132, 136, and 137. *The general form of the furnaces varies with the type of stoker, the kind of fuel used, and the load conditions.* For best results, the furnace design must be coördinated with that of the boiler. For the chain-grate and front-overfeed stoker, a coking arch is necessary to give satisfactory operation. The height and shape of the arch are varied to suit the fuel and the draft. With the chain grate it may be flat or inclined upward toward the rear, and only extends over a portion of the grate. For certain kinds of coal, it is necessary to have a second arch extending from the rear of the stoker toward the front with the gas opening between the arches. With a side-overfeed stoker, an arch extending over the entire grate is necessary to prevent excessive formation of smoke. An underfeed stoker does not require an arch, as the smoke-making constituents are burned in passing through the incandescent fuel zone. All stoker furnaces require larger volume than hand-fired furnaces, on account of the more rapid evolution of the volatile gases. The distance from fuel bed to boiler heating surface varies and should be from 5 to 20 feet, and in many cases much greater, for satisfactory operation with water-tube boilers and bituminous coal. Recent installations have an unusually large furnace volume per square foot of heating surface; in a typical case this ratio is 0.42 to 1. *The furnace volume and the amount of heat that can be released, usually expressed in B.t.u. per cubic foot of furnace volume per hour, largely determine the boiler performance.*

Cases are on record in which reasonably good results have been obtained with stoker-fired furnaces, in which the **rate of heat release** approached 100,000 B.t.u. per cubic foot of furnace volume per hour. The rate of heat release, however, generally is between 25,000 and 50,000. Higher rates are usually accompanied with excessive fixed carbon losses. It is impossible to give any simple rule for furnace volume. However, it may be said that as furnace walls are protected better with air-cooled, water-cooled, or steam-cooled surfaces, and as more highly preheated air is used, the rate of heat release per unit of volume may increase without affecting materially the operating results. The principal limits are set by the furnace brickwork and the tendency to deposit fused ash on the boiler tubes nearest the furnace. The latter can be obviated by the use of an adequate slag-screen, Fig. 132, page 200, arrangement consisting of one or more rows of widely

---

[1] For explanation of these terms consult page 314.

spaced tubes which cool the molten fly ash before it strikes the closely spaced boiler tubes.

193. Stoker Comparisons. — Data compiled by R. J. S. Pigott, pertaining to the various types of stokers, are given in Table 19. The prices are those given in 1929.

TABLE 19. — STOKER DATA

Type of Stoker	Step and Slope Overfeed	V Overfeed	Chain Overfeed 150 hp.–1000 hp.	Gravity Underfeed 2000 hp.	Horizontal Retort Underfeed 600 hp.
Average price per rated boiler horsepower	$9.50	$9.50	$21.20–$9.70	$13.90	$11.50
Normal forcing ability in per cent of rating	190	175	325	300–450	300
Price per maximum horsepower developable	$3.60	$5.15	$6.60–$3.00	$3.10	$4.00
Maintenance per ton coal fired, in cents	10–12	11–14	6–14	5–10	8–14
Attendance in man-hours per active hour	0.45	0.45–0.50	0.20–0.30	0.10–0.20	0.30–0.40
Pounds coal per square foot grate surface (maximum)	35–38	35–42	45–48	60–75	50–65

194. Locomotive Stokers. — Stokers are used on locomotives to lighten the labor of the firemen and to give a greater capacity than that attained by hand firing. There are a number of makes of locomotive stokers, differing mainly in the method used to feed the coal into the firebox. A typical stoker, manufactured by the Locomotive Stoker Co., is shown in Fig. 138 (next page). It consists of:

1. A **conveyor** and **crushing system** for crushing and carrying the coal from the tender to the locomotive. The conveyor screw is operated by spur gears connected through a driving arm to a rack driven by the stoker engine. The speed of the conveyor shaft is regulated by the speed of the stoker engine, which is controlled by a hand-operated globe valve and operates at from 8 to 90 pounds per square inch pressure.

2. An **elevating system,** consisting of a transfer hopper located beneath the cab deck and containing an adjustable gate; and right and left elevator screws, driven by gears which mesh with a rack driven by the stoker engine, to raise the coal to the **distributors** set in the back boiler head.

3. A **distributing system,** which spreads the coal over the grate, using a low-pressure steam jet located in the back and bottom portion of the distributors. The grate is the same as those used for hand firing.

Operation. — The shovel sheet is provided with an opening, 18 inches wide, extending from the coal gates to the slope sheet of the feedwater tank. Coal passes through this opening to the trough beneath, and is carried by the conveyor screw through the crusher, where it is broken to a suitable size. Entering the transfer hopper, the coal is divided, equally or in the proportion desired, between the elevating screws, by a movable dividing rib. The elevating screws elevate the coal and drop it into tubes fitted with elbows which extend through the back head on each side of the fire door. Steam jets in the elbows constantly blow the coal through the tubes to the distributors located inside the firebox. These distributors deflect and spread the coal over the entire surface of the fire. A level, light fire is carried, and excellent combustion results, the temperature being 400 deg. fahr. to 500 deg. fahr. higher than that attained with hand firing.

195. Methods of Burning Gas. — The burning of a gaseous fuel is accomplished by passing it into burners, where it is mixed with the required amount of air for combustion. The type of burner used depends upon the kind of gas to be burned.

Blast-furnace gas burners are of two general types, those in which the air for combustion is admitted around the burner proper, and those in which the air is admitted through the burner. Provision is made for regulating the air and gas supply independently.

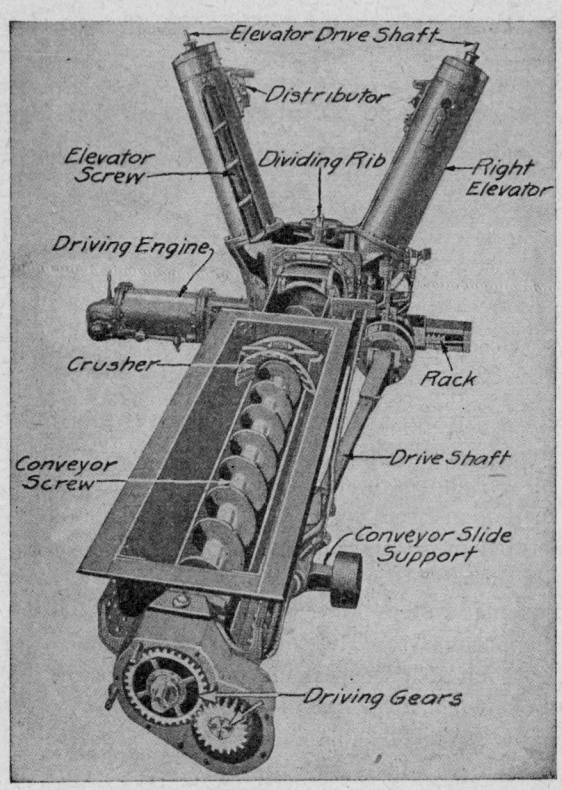

Fig. 138. — Rear View of Locomotive Duplex Stoker.

A gas opening of 0.8 square inch per rated horsepower will enable a boiler to develop its nominal rating with a gas pressure of 2 inches of water. The pressure is ordinarily about 6 to 8 inches of water, and thus the burner will be good for ordinary overloads. The size of the air openings varies from 0.75 to 0.85 of a square inch per rated horsepower. For good results the burners are slightly inclined downward toward the rear of the furnace.

For natural gas, a large number of small burners are used, each capable of handling 30 nominal rated horsepower. The use of a large number of burners obviates the danger of laneing or blow-pipe action, which would result if large burners were used. The gas pressure entering the burners is about 14 inches of water. The burners should give the gas and air a rotary motion to ensure proper mixture. With an excess of air below 15 per cent most gas burners back-fire or pulsate which loosens the furnace brickwork.

196. Furnaces for Gas Fuels. — The essential feature in gas-burning furnaces is sufficient combustion space to give the gases ample time to burn completely before striking the heating surfaces. For blast furnace gases, a furnace volume of 1.5 to 2 cubic feet per rated horsepower is satisfactory. For natural gas, a volume of 2 cubic feet gives good results. Gas furnaces ordinarily do not have a bridge wall.

197. Methods of Burning Oil Fuel. — Oil is burned under boilers by first atomizing or vaporizing it, so that it may be burned like gas. Burners of the following types are used to atomize the oil: (1) **Spray burners,** in which the oil is atomized by steam or compressed air; (2) **Mechanical burners,** in which the oil is atomized by submitting it to a high pressure and passing it through a small orifice.

The simplicity of the steam atomizer spray burner, the excellent economy of the better types, and the low oil pressure and temperature required, make it a favorite for stationary plants, where the loss of fresh water is not a vital consideration. In marine work, or in any case where it is necessary to save feedwater that otherwise would have to be added in the form of *"make up,"* either compressed air or mechanical means are used for atomizing.

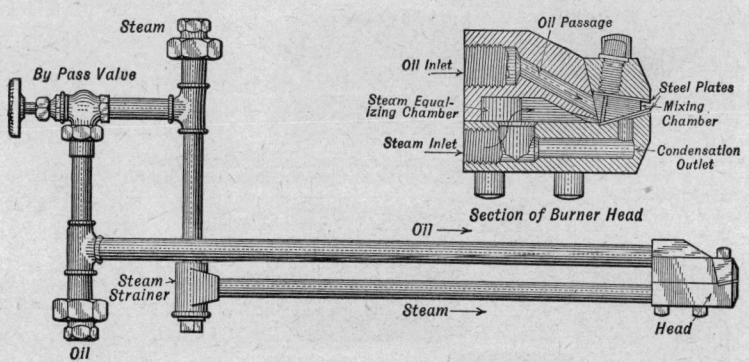

FIG. 139. — Hammel Steam Atomizing Oil Burner Complete.

Steam spray burners, as now used, may be divided into two classes: (1) Inside mixers in which the steam and oil come in contact within the burner and the mixture is atomized in passing through the orifice of the burner nozzle; (2) Outside mixers in which the steam flows through a narrow slot or horizontal row of small holes in the burner nozzle, the oil flowing through a similar opening above the steam orifice, and being picked up and atomized by the steam outside the burner.

The Hammel oil burner, Fig. 139, belongs to the inside-mixing flat-flame type. Steam, issuing from three small slots into the mixing chamber, cuts across the oil stream, and the energy of the steam is fully utilized in atom-

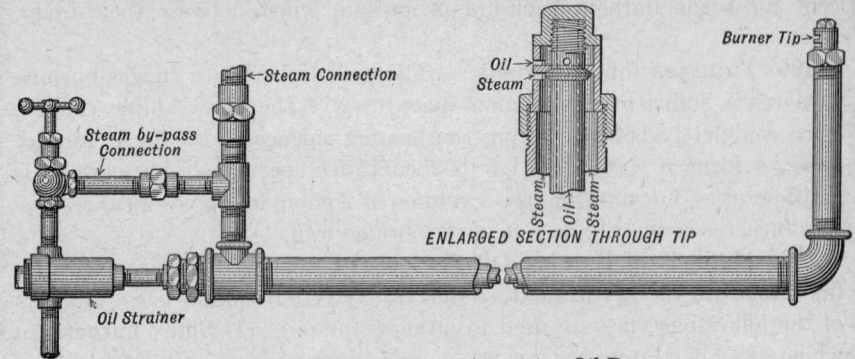

Fig. 140. — Peabody Steam Atomizing Oil Burner.

izing the heavy hydrocarbons and vaporizing the lighter ones. The mixture issues from the burner and ignites like a gas flame. The steam used amounts to about 2 per cent of the steam generated by the boiler. This burner will handle heavy or light oils with steam pressures from 25 to 80

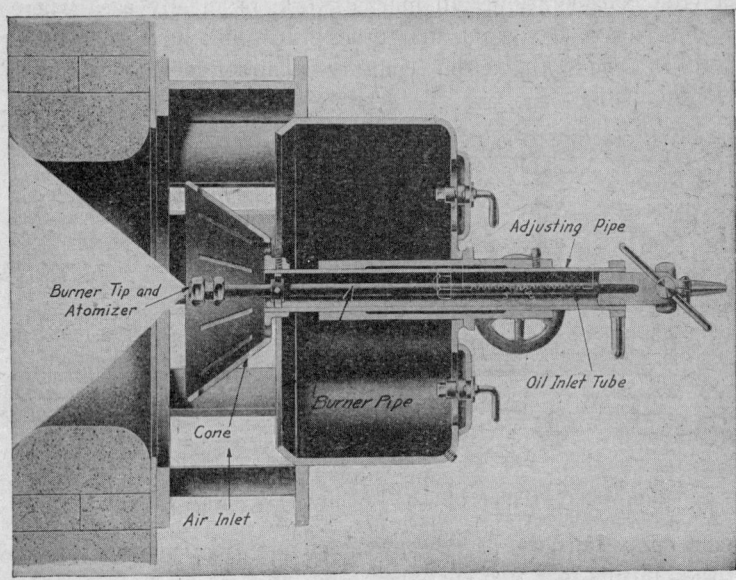

Fig. 141. — Dahl Mechanical Atomizing Burner.

pounds per square inch gage. The steam passage is continued into a condensation chamber which acts as a steam separator.

The Peabody outside-mixing type of oil burner, shown in Fig. 140, has given good satisfaction. The construction is clearly shown in the figure.

Mechanical burners, of which Figs. 141 and 143 are typical, are being used in many stationary as well as in marine-power plants. The most suc-

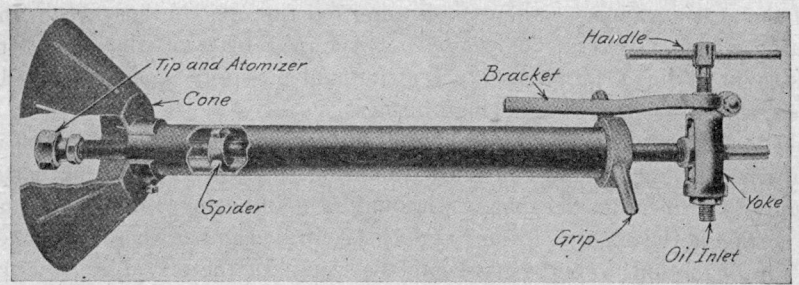

Fig. 142. — Assembly of Dahl Burner Parts.

cessful burners of this type have a short conical flame and give a whirling motion to the oil within the burner tip. The oil is delivered to the burner under a pressure that varies from 100 to 250 pounds per square inch and a

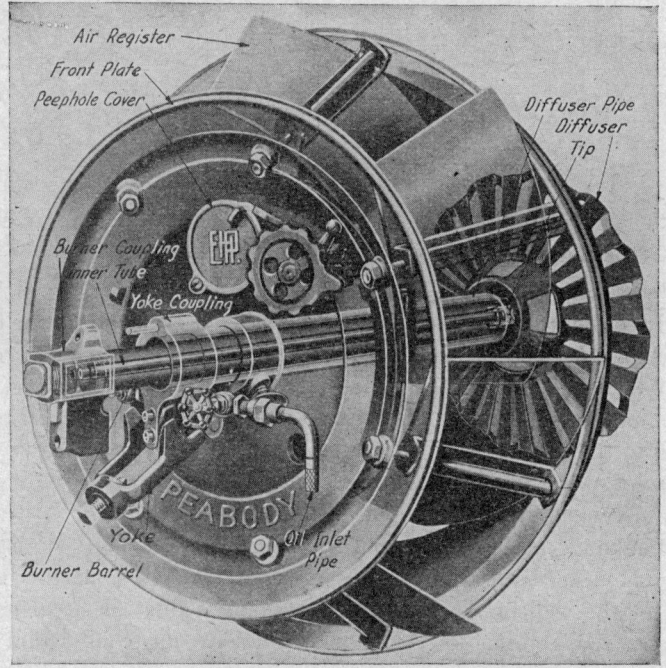

Fig. 143. — Peabody-Fisher Wide Range Oil Burner.

temperature that varies from 150 deg. fahr. to 200 deg. fahr. depending upon the range of oil and the make of the burner. A constant oil pressure is produced by a pump having a governor controlled by steam pressure.

These burners are located at the front of the furnace, and the air required for combustion is admitted around the burner through suitable adjustable shutters of various shapes. This is a better means of regulation than that of the steam atomizer, where air is admitted through checkerwork in the furnace floor (see Fig. 25, page 32). A forced blast of air is used to assist in giving a proper mixture of air and oil spray, and is especially necessary when operating at high ratings.

The Peabody-Fisher wide range oil burner, Fig. 143, is a widely used burner of the mechanical pressure atomizing type without moving parts. It gives a wide range in capacity without loss in atomizing power by causing a portion of the oil, after its entry into the central chamber where it assumes a rotary motion, to be by-passed to the section of the oil tank or service

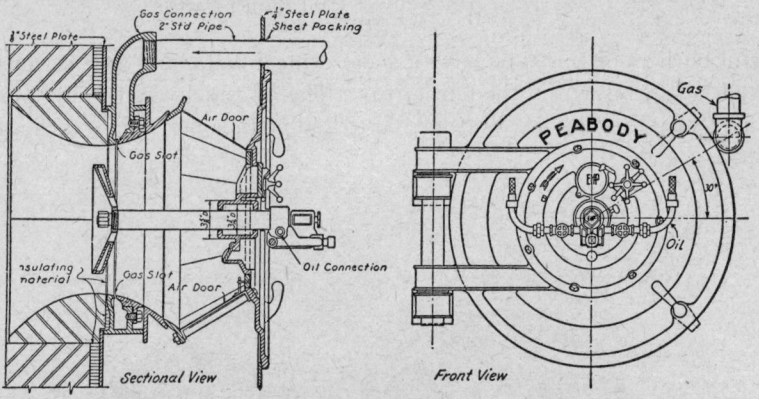

FIG. 144. — Peabody Combined Oil and Gas Burner.

tank. It consists of a *diffuser*, a *barrel* threaded at each end, to the furnace end of which a *burner tip* is fastened and to the other end of which a *coupling casting* is attached. The burner is attached, by a yoke coupling, to the diffuser pipe and the diffuser, the whole being surrounded by an air register, attached to the furnace front, containing curved veins which serve to give the air a rotary motion as it enters the register. The quantity of air and velocity are controlled by the area of the register openings.

The **Peabody combined oil and gas burner,** Fig. 144, consists of a typical oil burner into which gas is injected through a narrow annular opening in the wall of the casting surrounding a duct of circular section leading to the combustion chamber. The air for combustion enters through an air register where it is given a rotary motion, and the gas is forced into this rotating column of air from all sides, thus insuring an intimate mixture of gas and air.

198. Capacity of Oil Burners. — A good steam atomizing burner, properly located in a well-designed oil furnace, has a capacity of about 400 boiler horsepower, while recent installations of mechanical atomizing burners are developing 2000 horsepower per burner.

199. Oil Furnace. — *To burn oil successfully, the oil should be thoroughly atomized, and the furnace volume should be proportioned to give approximately one cubic foot of combustion space per horsepower developed, when operating under natural draft. With forced draft the volume can be reduced to 0.75 cubic feet, with the air delivered at a pressure of 1.5 inches of water. These figures are based on the furnace volume below the tubes.* With boilers operating at high ratings, with correspondingly high furnace temperatures, air-cooled walls are necessary to prevent the brickwork from melting. With a satisfactory burner the furnace arrangement and the method of introducing air for combustion are important factors.

The point at which the atomized oil is introduced depends upon the type of boiler under which the burner is installed. With horizontal return-

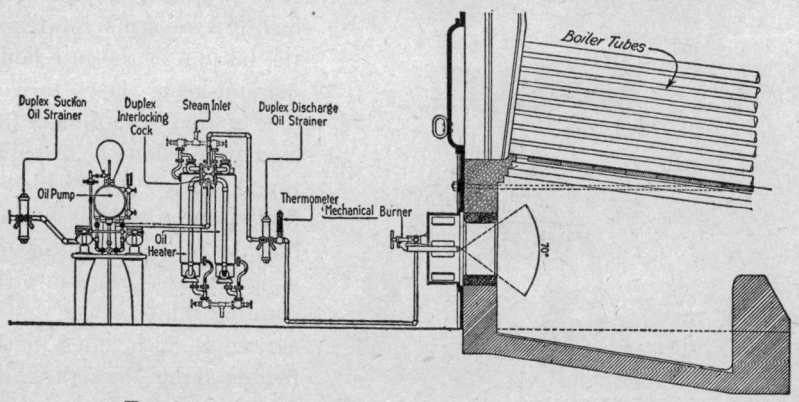

Fig. 145. — Koerting Mechanical Oil Burner Installation.

tubular, and Heine boilers, the burners, when steam is used for atomizing, are located at the front of the furnace, with the flame directed toward the rear. No bridge wall is used, and the path of the gases and the volume of the furnace are satisfactory for complete combustion. For Stirling and horizontal drum water-tube boilers the burner is located at the rear under the bridge wall, and the flame is directed toward the boiler front. The furnace thus increases in volume in the direction of the flame, and ensures free expansion and a thorough mixture of the oil and air. If the flames are directed against the tubes they are soon burned out, because of the blowpipe action of the burner. Air for combustion is admitted through a checkerwork of firebrick supported on the furnace floor, the openings in the checkerwork being near the burner discharge. The burners should be located so that the flames from the individual burners do not interfere or strike the side walls of the furnace. The burners are operated from the boiler front, and peep-holes are supplied through which the operator may watch the flame while he is adjusting it. Mechanical atomizing burners are generally located as in Fig. 145.

With oil as a fuel, the efficiency of the boiler is generally high, because combustion is more nearly perfect. Ordinarily, 14 per cent carbon dioxide can be attained, with only a trace of carbon monoxide.

200. Pulverized-fuel Burning. — Pulverized fuel is fed to the burner at a uniform rate either from an overhead storage bin through a feeder or directly from the mill. Feeders which produce uniform and regular feed, and at the same time aerate the coal are aids to combustion. The feeder, shown in Fig. 146, consists of a hopper attached to the pulverized coal bin and containing two shut-off gates. A distributor having two arms rotates at the bottom of the gate housing and keeps the entire volume of coal in motion. Immediately below the distributor is a fixed plate called a **fluffer apron,** which has an opening on one side through which the coal drops into the rotating fluffer wheel, and is carried to an opening in the **feeder apron** through which it drops to the **feeder wheel** located below. The feeder wheel rotates and carries the coal around the feeder apron to the open ports in the spacer, through which the coal drops into the discharge pipe where it is spread into a thin sheet by inclined baffles. It is then picked up by the primary air and carried to the burner, where it is mixed with sufficient secondary air for proper combustion.

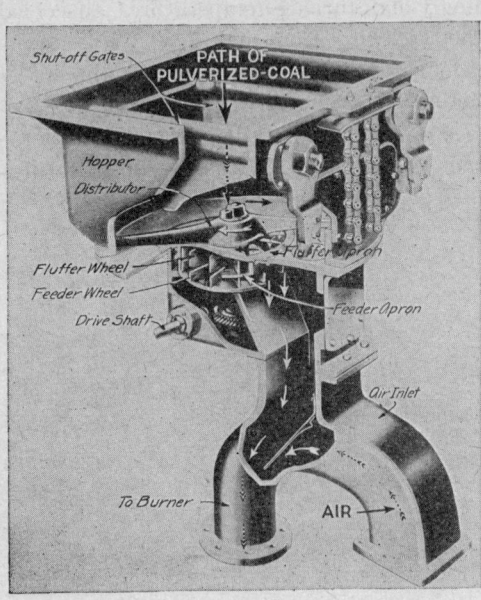

FIG. 146. — Bailey Feeder.

The burners used for pulverized fuel may be of either the horizontal flare type or the turbulent type. In the latter the coal and air are intimately mixed and agitated as they enter the furnace which speeds up combustion during the early stages, and increases the rate of evaporation which in some cases may be as high as 20 pounds of water per square foot of heating surface. Turbulence also decreases the volume of the furnace by making the entire volume of the furnace effective, and by increasing the frequency of contact between the solid particles of the fuel and the oxygen, thus giving better combustion. The function of the burner is to deliver the coal and air to the furnace in right proportions, each particle of coal having the air

PULVERIZED-FUEL BURNING

available shortly after it enters the furnace and enough turbulence to bring the molecules of oxygen in contact with fuel in quickest time. *The ideal burner is one by which the combustion of fuel is completed with a minimum of furnace volume and a reasonably low amount of excess air.* For best operation, the burners should be designed to meet the requirements of the type of furnace and the kind of fuel being burned.

A typical **horizontal turbulent type burner,** having a capacity of 7500 pounds of coal per hour and using an orifice, and interference principle for creating turbulence, is illustrated in Fig. 147. It is suitable for use with

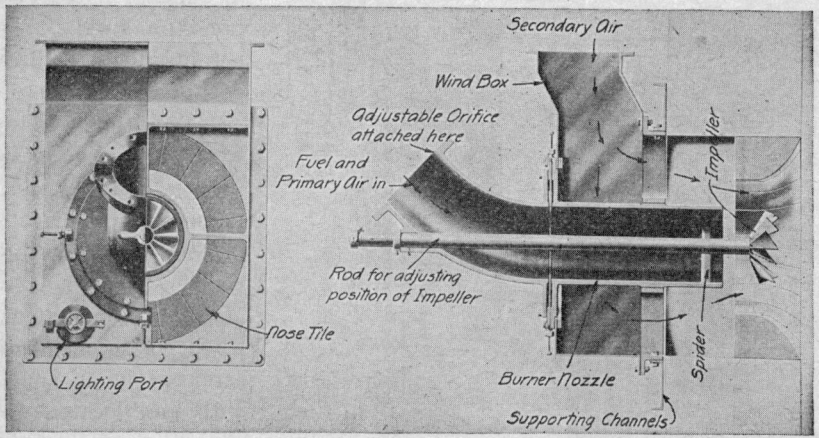

FIG. 147. — Bailey Horizontal Turbulent Type Pulverized Coal Burner.

solid refractory front walls, in which the secondary air for combustion is supplied by forced draft. Coal carrying primary air enters the burner nozzle at the outer flange of the elbow and the secondary air enters from the windbox. An adjustable orifice (not shown) attached to the outer flange provides a re-mixing device for the coal and primary air as it enters the burner and effects a uniform coal and air mixture by nullifying any segregation taking place in the coal and air piping leading to the burner. An **adjustable impeller** attached to a rod which is supported by a spider deflects the primary air and coal mixture, leaving the burner nozzle, into the secondary air stream. Refractory nose tile, built into the refractory furnace wall, form the orifice through which the secondary air passes on entering the furnace. The intimate mixing of coal and air is accomplished by the impeller which deflects the coal in the primary air stream. Eddy currents in the secondary air orifice restoration zone create the violent mixing and re-mixing effect, not only making oxygen available to every coal particle but creating a scrubbing effect and removing spent gases from the active combustion zone.

The coal, mixed with the proper amount of air for combustion, enters the combustion chamber in a finely divided state, the volatile portion being burned like a gas, and the solid portion while suspended in a mixture of gas and air. The burning should be complete before the coal comes into contact with the cooler surfaces of the boiler.

A burner designed for use with the Bailey water-cooled wall construction is illustrated in Fig. 148. It injects the primary air carrying the pulverized coal into the furnace horizontally through a long narrow vertical slot, located between the water-cooled tubes. The primary air is intimately mixed with the secondary air, upon entering the furnace, in such a manner that a short flame of rectangular cross-section is produced. This permits highly efficient use of combustion space and makes possible an economical furnace design with a minimum of flame impingement upon the walls. The secondary air enters through staggered ports which are arranged at an angle with and located on either side of the primary air jets.

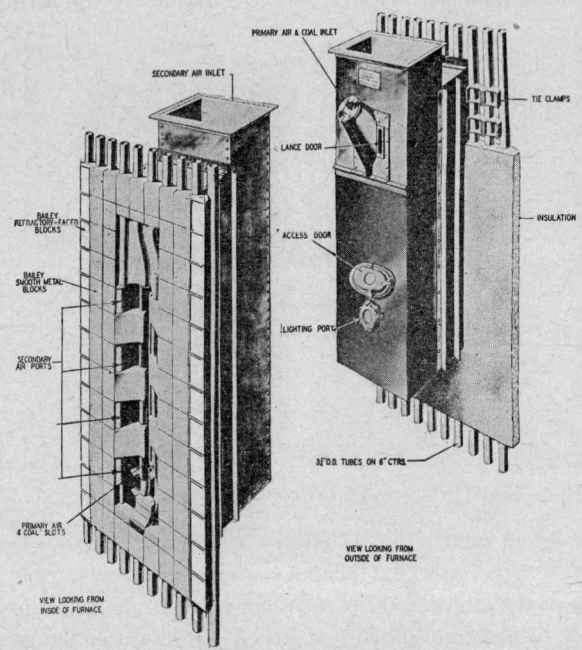

Fig. 148. — Calumet Burner.

In furnaces that are fired vertically a burner of the fantail type shown in Fig. 149 is generally used. The primary air and coal enter through the fan-shaped central portion of the burner, the secondary air being supplied through a hand or automatically operated damper which controls the space surrounding the primary air tube, or through numerous small ports in the front wall (see page 120).

About 30 per cent of the ashes resulting from the combustion are carried from the chimney; the remaining 70 per cent are deposited in the soot chamber, on the tubes, or in the ashpit. The ashes which collect in the ashpit will fuse and produce slag in an improperly designed furnace. In a medium-sized plant the ashes should be removed once a day, and the tubes should be blown once in eight hours.

The cost of installation and operation using pulverized fuel compares favorably with those using an underfeed stoker. The overall efficiencies are slightly higher for the same capacity, and existing plants have attained an overall efficiency of nearly 90 per cent. Plants using pulverized fuel do not have standby losses caused by change of load or banked fires.

201. Furnace Construction for Pulverized Fuel. — Furnace design is the most important single factor in the burning of pulverized coal. *The shape, size and temperature of the furnace must be adapted to the coal being burned, and to the type of boiler being used.* The furnace should be proportioned for the maximum rating of the boiler, and should be so arranged that the entering pulverized coal may be raised as rapidly as possible to the temperature at which it will begin to burn. This temperature varies with the chemical and physical properties of the coal, and in general is lower the

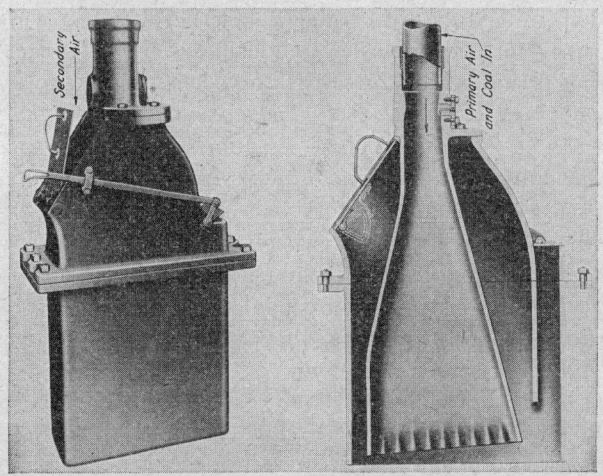

FIG. 149. — Lopulco Fantail Burner.

higher the volatile content, the lower the ash and moisture content, and the finer the pulverization. The greater the surface exposed to radiation from the wall and flame, and the lower the heat capacity of the entering mixture of coal and air, the more rapidly will this temperature be attained. Further, the more slowly the coal or mixture of coal and air is admitted, the less distance will it have traveled for the same amount of radiation it has received. To further increase the efficiency of combustion, the **ratio of the exposed tube surface to grate surface** is being increased, thus increasing the amount of heat absorbed by radiation, and in addition cooling the furnace walls. Present ratios are about 1.8 to 1.

Furnaces are frequently made large to decrease the velocity, which is ordinarily about 7 feet per second, and to increase the length of the gas travel, which is sometimes 60 feet vertical and 40 feet horizontal, thereby giving ample time for more complete oxidation of the small coal particles. The volume of a modern furnace is often 12,000 cubic feet, or 6.5 cubic feet per rated boiler horsepower. This may however be decreased by using a **water screen** consisting of water-cooled tubes laid across the top of the ashpit, and spaced far enough apart so that ash or slag can readily drop between them. This prevents slagging in the ashpit, and since these tubes are connected into the boiler circulation, they form a cooling zone in the lower part of the furnace which prevents the fusing of ash to the walls or bottom. The practical success of the water screen has made possible the elimination of the inefficient "air blanket cooling zone" sometimes resorted to.

Further decrease in the furnace volume can be made by using either air-cooled refractory walls, water-cooled walls or steam-cooled walls to replace solid brick walls. In many installations the side, front and rear walls are equipped with water tubes which, taken together with a water screen and the boiler tubes, form a furnace entirely enclosed by water-cooled surfaces and requiring preheated air for satisfactory operation. *The proportion of the entire wall area that is water cooled has a direct effect upon the rate of heat release that can be obtained.* This is sometimes called the "fraction cold" and varies in pulverized fuel fired furnaces from 0.15 to 0.90. The limit in the heat release for solid refractory walls is about 20,000 B.t.u. per cubic foot of furnace volume per hour, whereas with completely water-cooled walls it may be as high as 80,000 B.t.u. per cubic foot per hour.

Various types of construction are used in the bottom of the furnace, giving rise to the names **well type, hopper bottom, slag bottom,** Fig. 246, page 333, and **water-screened bottom.**

The *well type furnace,* commonly known as the Buffalo type, consists of a square refractory-faced water-cooled chamber having a specially designed flat refractory bottom on which the ash gathers in liquid form and is tapped off in that condition at convenient intervals. The size of this chamber is about 10 feet on the side and 6 feet deep, and is placed in the bottom of the combustion chamber. The burners are placed near the bottom of the well and tangential to the furnace in order to produce high turbulence.

The hopper bottom furnace, Fig. 99, page 145, is composed of water-cooled tubes covered with smooth cast-iron blocks. With this construction the ash may chill but does not adhere to the furnace bottom, or the ash may run into the ashpit in liquid form where it is quenched and broken up by a water spray for easy removal by means of sluice conveyors.

202. Furnace Refractories. — *The term* **refractories** *is applied to the fire-resistant material which is used in the construction of boiler-furnace walls, arches, baffles, and the bottom of the furnace.* It includes also the finely ground material used as mortar in furnace walls, plastic compounds and the so-called high temperature cements.

The modern tendency in boiler operation toward high overloads, and the call for high efficiency at these high ratings, has greatly increased the severity of the service which the furnace refractory is called upon to stand,

and has in many instances so greatly increased the cost of furnace maintenance that the problem of boiler furnace refractories is one of major importance.

The present need appears to be for refractories that will give a minimum of trouble from fusion, slag action, softening under load, spalling, and thermal expansion and contraction.

The choice of a suitable refractory is a difficult problem to solve because each kind of fuel, and each set of operating conditions calls for a different solution. A commercially satisfactory solution is obtained only when all the above factors are properly taken into account, and that refractory is chosen which will give the best compromise between many conflicting factors. The ultimate and only real test for refractories appears to be that of service under actual furnace conditions.

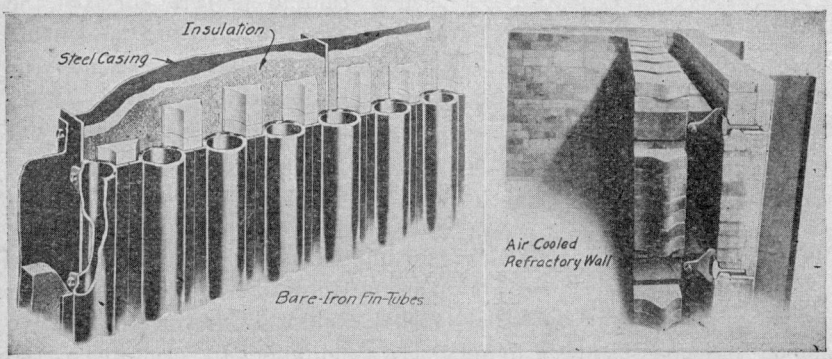

FIGS. 150 and 151. — Water-cooled and Air-cooled Furnace Walls.

The solution of the boiler furnace refractory problem undoubtedly lies in furnace design to avoid the deposition of ash on the surfaces, the introduction of air to deflect the gases, the introduction of water-cooled, air-cooled and steam-cooled furnace walls, and the protection of the brick walls by so placing them that they have the best opportunity to radiate to cold surfaces, such as the boiler tubes.

With **water-cooled furnace walls,** Fig. 132, page 200, the water in the walls forms part of the boiler circulation with the water flowing from the boiler drum through the down-take tubes to the lower water-wall header, then up through the tubes to the upper water-wall header, and finally to the steam drum by way of the up-take tubes. Several types of water-wall construction are being used, some of which have bare-iron fin tubes, Fig. 150 or plain boiler tubes, while others use tubes with the refractory or cast-iron covered faces, as illustrated in Fig. 152. The tubes forming the water walls are so proportioned that not over 50 per cent of the mixture discharged from the tube is steam when the tube is absorbing heat at its maximum rate. The steam generated in the tubes is ordinarily collected in the upper header and delivered to the main boiler drum. The temperature of the furnace is ordinarily lowered by using water walls as a result of the rapid absorption of heat by the bare walls, and in addition, the furnace and stoker maintenance is decreased. Experience seems to indicate that water-cooled walls have little or no effect on the speed of combustion.

Hollow air-cooled walls, Fig. 151, made of refractory materials give better results when pulverized fuel or fuel oil is employed than when stokers are the firing medium. This is due to the fact that in pulverized fuel and fuel oil firing, air infiltration is not as detrimental as in stoker firing, and that if the inner lining of the hollow walls, when stoker fired, is made sufficiently thin to properly air cool the refractories, then the necessity of cutting away clinker or the abrasive action of the moving fuel takes away the wall as quickly as the extreme thermal conditions would have affected a solid wall.

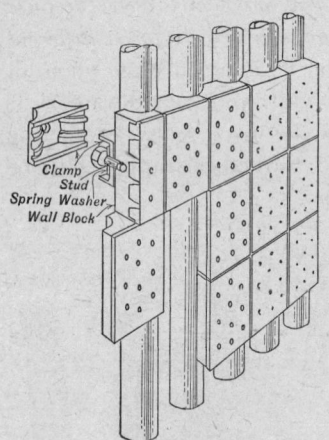

FIG. 152. — Refractory Block Covered Water Wall.

203. Methods of Smoke Determination. — No satisfactory method of determining the amount of smoke, either qualitatively or quantitatively, has yet come into use; and it cannot even be said that a reliable method of fixing the relative amount of smoke has been found. The condition of the atmosphere, the appearance of the background, and the personal equation all enter into the making of a determination. For qualitative smoke measurement, the

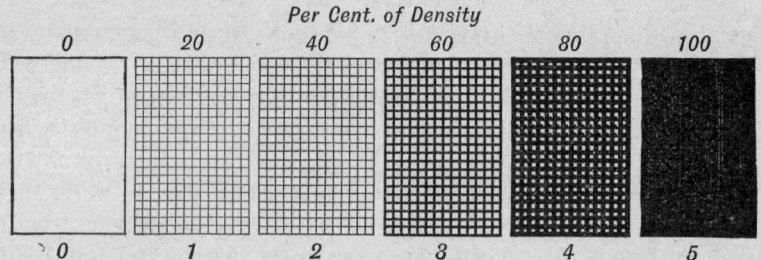

FIG. 153. — Ringelmann Chart for Grading the Density of Smoke.

Ringelmann chart, Fig. 153, is generally used. This consists of four large charts ruled with vertical and horizontal lines forming squares, and numbered from 1 to 4, No. 1 representing a smoke density of 20 per cent; No. 2, 40 per cent; No. 3, 60 per cent; and No. 4, 80 per cent. A white card is 0 per cent density and a black card 100 per cent. These charts, together with the white and black cards, are placed 50 feet from the observer, in line with the chimney. The lines on the chart disappear at this distance and the charts have a gray appearance. The observer glances rapidly from the chimney to the charts and judges which chart corresponds, in color, with the smoke. He then records his observation and the time, on a smoke record chart. Readings are taken at one-half or one minute intervals.

In some cities, notably Chicago, a **"smoke unit"** has been adopted as a method of estimating the total smoke emitted by a chimney. This unit is the equivalent of No. 1 smoke emitted for one minute. Thus, if a stack emits No. 2 smoke for ten minutes it would be emitting 20 smoke units.

REFERENCES

Finding and Stopping Waste in Modern Boiler Rooms, COCHRANE CORPORATION.
Steam Power Plant Engineering, GEBHARDT.
Other references as under Chapter VI.
Some Fundamental Considerations in Design of Boiler Furnaces, W. J. WOHLENBERG, and F. W. BROOKS.
Some Factors in Furnace Design for High Capacity, E. G. BAILEY, Fuels and Steam Power, Vol. 50, page 253.
Boiler Furnace Refractories, HIRSHFELD and CARTER, Fuels and Steam Power, Vol. 50.

REVIEW QUESTIONS AND PROBLEMS

1. Explain the cause of smoke, and state the factors necessary for its prevention.
2. Name the types of grates used with hand-fired furnaces.
3. What three methods are used to fire coal by hand?
4. With hand-fired furnaces, how may the amount of smoke be minimized?
5. What are the advantages of mechanical methods of firing coal?
6. Classify mechanical stokers.
7. Describe the following types of stokers: (a) chain grate, (b) Riley underfeed, (c) Murphy.
8. Find the amount of coal burned per hour on a 5 ft. by 12 ft. chain grate which moves 4 in. per min. and has a depth of fuel bed of 8 in. Average weight of coal on grate, 45 lb. per cu. ft.
9. What can be said regarding the volume of stoker-fired furnaces when compared with hand-fired furnaces?
10. Describe a powdered-coal-burning furnace.
11. Describe the construction of the Peabody oil, and gas burner.
12. Describe an oil-burning furnace.
13. Name the essential qualifications of refractories.
14. Describe the "slag" bottom furnace.
15. Explain the meaning of the "smoke unit," as used in Chicago.

CHAPTER IX

RATING, EFFICIENCY AND TESTING OF STEAM BOILERS

204. Boiler Rating and Capacity. — The question of a suitable unit of boiler capacity has been widely discussed, and efforts have been made to obtain a definite and rational basis for stating the capacity. The methods commonly used to rate steam boilers, until recently, has been, (1) *by boiler horsepower*, (2) *by amount of heating surface*, and (3) *by area of grate surface*. The Power Test Code of the A. S. M. E. now recommends that boiler output be expressed by: (1) the *heat output in steam* expressed in B.t.u. per hour; (2) *actual evaporation*, pounds of steam per hour at the observed steam pressure and quality or temperature, and observed feed-water temperature; and (3) *units of evaporation*.

While the last three methods are logical, and should be used, the term capacity, through long usage, has come to be expressed in **boiler horsepower** (*boiler hp.*). This term has been so generally accepted in the United States that it will probably be a long time before it can be entirely abandoned. *It has no definite relation to the term horsepower as applied to a steam engine*, since the boiler does not perform work as ordinarily considered. It does, however, designate the capacity of a boiler as regards its ability to evaporate a definite amount of water per hour under a given set of conditions, and is, consequently, a unit of evaporation, and not one of power.

At the Centennial Exposition in Philadelphia, in 1876, **a boiler horsepower** *was taken as the evaporation of* 30 *pounds of water per hour into steam, at a gage pressure of* 70 *pounds per square inch with a feedwater temperature of* 100 *deg. fahr.* This was the weight of steam used, at that time, by a simple steam engine in producing one horsepower when the steam pressure was 70 pounds per square inch, gage. This rating is known as the **Centennial rating**, and gave a standard for comparing boiler performances.

The **commercial rating** of a boiler, as adopted by the AMERICAN SOCIETY OF MECHANICAL ENGINEERS,[1] *is the evaporation of* 34.5 *pounds of water into steam, from and at* 212 *deg. fahr. per hour.* Expressed differently, a boiler, to develop one boiler horsepower, must convert 34.5 pounds of water into steam per hour, when the feedwater temperature is 212 deg. fahr. and the absolute pressure 14.7 pounds per square inch; that is, the

---

[1] The American Society of Electrical Engineers have proposed that the **myriawatt** be adopted as a standard for rating boilers. A myriawatt is the power equivalent of 10,000 watts, or 34,130 B.t.u. per hour.

boiler must supply the latent heat of steam at atmospheric pressure for each pound of water evaporated.

The heat required to produce a boiler horsepower is, therefore, **34.5 × 970.2, or 33,471.9 B.t.u. per hour.** The horsepower rating of a boiler can be found by multiplying the weight of water evaporated per hour by the heat required to evaporate a pound of the water and dividing this product by the heat corresponding to a boiler horsepower.

As steam may exist in a wet, dry, or superheated condition, the boiler horsepower may be found by using one of the following equations:

$$\text{Wet steam,} \qquad \text{Boiler hp.} = \frac{W(xL + h_f - h_{f_1})}{34.5 \times 970.2} \quad \ldots \ldots \quad (65)$$

$$\text{Dry steam,} \qquad \text{Boiler hp.} = \frac{W(h_g - h_{f_1})}{34.5 \times 970.2} \quad \ldots \ldots \ldots \quad (66)$$

$$\text{Superheated steam,} \quad \text{Boiler hp.} = \frac{W[h_g + C_p(t_s - t_v) - h_{f_1}]}{34.5 \times 970.2} \quad \ldots \quad (67)$$

in which

W = total weight of water evaporated per hour, lb.
x = quality of the steam.
L = latent heat of steam at absolute boiler pressure, B.t.u.
h_f = heat of liquid corresponding to absolute boiler pressure, B.t.u.
h_{f_1} = heat of liquid corresponding to feedwater temperature, B.t.u.

Note. — [$(t - 32)$ can be used for h_{f_1} without serious error.]

h_g = total heat per pound of dry steam at absolute boiler pressure, B.t.u.
t_v = temperature of saturated steam at absolute boiler pressure, deg. fahr.
t_s = temperature of superheated steam at absolute boiler pressure, deg. fahr.
c_p = specific heat of superheated steam at constant pressure. (See Fig. 73, page 102.)

Example 26. — During the test of a 150 hp. horizontal return-tubular boiler, the following data were taken:

Water evaporated per hour, 3030 lb.; steam pressure, 110 lb. per sq. in. gage; barometer 29.50 in. mercury; moisture in steam 2 per cent; feedwater temperature 162 deg. fahr. Find the boiler horsepower developed.

Solution. — Using Equation (65), because the steam is wet,

$$\text{Boiler hp.} = \frac{W(xL + h_f - h_{f_1})}{34.5 \times 970.2} = \frac{3030\,(0.98 \times 875.2 + 315.3 - 130)}{34.5 \times 970.2} = 94.4$$

$W = 3030$ lb.; $x = 0.98$
Absolute pressure in boiler, lb. per sq. in. = $29.50 \times 0.491 + 110 = 124.5$.
L corresponding to 124.5 lb. per sq. in. = 875.2 B.t.u.
h_f corresponding to 124.5 lb. per sq. in. = 315.3 B.t.u.
h_{f_1} corresponding to 162 deg. fahr. = $(t - 32) = 162 - 32 = 130$ B.t.u.

222   RATING, EFFICIENCY AND TESTING OF STEAM BOILERS

The **Unit of Evaporation,** $U.E.$, equals the absorption of 1000 $B.t.u.$ per hour, therefore the capacity of a boiler when expressed in units of evaporation equals the total heat output delivered by the boiler unit per hour divided by 1000.

205. Heating Surface, $H.S.$ — *Manufacturers base their ratings of boiler horsepower on a definite amount of water-heating surface per boiler horsepower.* Formerly this amount of H. S. was 10 square feet for water-tube boilers, 12 square feet for tubular boilers, and 8 square feet for Scotch marine boilers. It is now customary to rate all stationary boilers on a basis of 10 square feet of heating surface per boiler horsepower. This does not mean that the boiler should not develop more than its rating; but that, under ordinary operating conditions, 10 square feet of heating surface will evaporate the equivalent of 34.5 pounds of water into dry steam at 212 deg. fahr., when the feedwater temperature is 212 deg. fahr. This method of rating is arbitrary and is used for convenience only. It is known as the **manufacturer's rating** and is the **nominal rated capacity** of the boiler. Boilers in modern power plants are being operated at capacities undreamed of fifteen to twenty years ago. In some cases ratings as high as 600 per cent above rated capacity for short periods of time have been reported. Under normal operation, however, 150 to 400 per cent above the rated capacity is the common rule.

By this method, all boilers of equal heating surface are given the same rating. This is unfair to purchasers, as, of two boilers having equal heating surface, one may have its heating surface arranged more advantageously than the other and hence have greater capacity under the same conditions of operation.

Engineers ordinarily select boilers on a basis of heating surface sufficient to supply the demand, allowing a reasonable rate of evaporation per square foot of heating surface per hour.

206. Heating Surface of H. R. T. Boiler. — *The heating surface of a return-tubular boiler equals one-half the external area of the cylindrical shell, plus the inside area of all the tubes, plus two-thirds the area of the rear head, minus the combined external cross-sectional area of the tubes, all expressed in square feet.*

Example 27. — Find the rated boiler horsepower of a boiler 60 in. diam., 16 ft. long, and having 70 three-inch tubes.

Solution. — The heating surface is found by applying the above rule.
1. Area of shell = $\frac{1}{2}$ (3.14 × $\frac{60}{12}$ × 16)   = 125
2. [1]Area of all the tubes = 3.14 × $\frac{2.78}{12}$ × 16 × 70   = 817
3. Two-thirds area of rear head, less tube sections
$$= \tfrac{2}{3}[\tfrac{1}{4} \times 3.14 \times (\tfrac{60}{12})^2] - [70 \times \tfrac{1}{4} \times 3.14 \times (\tfrac{3}{12})^2] = \quad 9.5$$

Total H. S.   951.5

[1] For data on tubes consult Table 20, page 223.

Allowing 10 sq. ft. of heating surface per boiler horsepower, the rating equals approximately 95 boiler horsepower.

An approximate rule, which is sometimes sufficiently accurate to use in finding the total heating surface of this type of boiler, is to divide the area of the tubes by 0.85.

207. Heating Surface of a Water-tube Boiler. — For a water-tube boiler of the Heine type, *the heating surface equals one-half the area of the shell, plus the outside area of the tubes, plus the inner area of each water leg, minus twice the area of all the tubes.*

An approximate rule is to divide the area of the tubes by 0.90.

The *heating surface for a* **steam generating unit** consists of that portion of the surface of the heat transfer apparatus exposed, on one side to the gas or refractory being cooled, and on the other to the fluid being heated, measured on the side receiving heat, and shall be divided into the boiler heating surface including the boiler, water walls, water screen and water floor, superheating surface, economizer surface, reheater surface, and air-heater surface.

TABLE 20. — LAP-WELDED CHARCOAL-IRON BOILER TUBES

Diameter		Thickness, Inches	Wire Gage Number	Circumference		Transverse Areas			Length of Tube per Sq. Foot of	
External, Inches	Internal, Inches			External, Inches	Internal, Inches	External, Sq. Inches	Internal, Sq. Inches	Metal, Sq. Inches	Ex. Surf., Feet	Int. Surf., Feet
1	.856	.072	15	3.142	2.689	.785	.575	.21	3.819	4.462
1¼	1.106	.072	15	3.927	3.475	1.227	.961	.266	3.056	3.453
1½	1.334	.083	14	4.172	4.191	1.767	1.398	.369	2.547	2.863
1¾	1.56	.095	13	5.498	4.901	2.405	1.911	.494	2.183	2.448
2	1.81	.095	13	6.283	5.686	3.142	2.573	.569	1.909	2.11
2¼	2.06	.095	13	7.069	6.472	3.976	3.333	.643	1.698	1.854
2½	2.282	.109	12	7.854	7.169	4.909	4.09	.819	1.528	1.674
2¾	2.532	.109	12	8.639	7.954	5.94	5.035	.905	1.389	1.509
3	2.782	.109	12	9.425	8.74	7.069	6.079	.99	1.273	1.373
3¼	3.01	.12	11	10.21	9.456	8.296	7.116	1.18	1.175	1.26
3½	3.26	.12	11	10.996	10.241	9.621	8.347	1.274	1.091	1.172
3¾	3.51	.12	11	11.781	11.027	11.045	9.676	1.369	1.018	1.088
4	3.732	.134	10	12.566	11.724	12.566	10.939	1.627	.955	1.024
4½	4.232	.134	10	14.137	13.295	15.904	14.066	1.838	.849	.902
5	4.704	.148	9	15.708	14.778	19.635	17.379	2.250	.764	.812
6	5.67	.165	8	18.85	17.813	28.274	25.429	3.025	.637	.673
7	6.67	.165	8	21.991	20.954	38.485	34.942	3.543	.546	.573
8	7.67	.165	8	25.133	24.096	50.266	46.204	4.062	.477	.498
9	8.64	.18	7	28.274	27.143	63.617	58.629	4.988	.424	.442
10	9.594	.203	6	31.416	30.14	78.54	72.292	6.248	.382	.398

208. Grate Surface, *G. S.* — In rating boilers by grate surface, one-third of a square foot of grate surface is sometimes taken as equivalent to a boiler horsepower. This value varies, however, with the conditions under which a boiler is operated, and also differs with the method of firing.

The grate surface of a stoker-fired furnace is taken as the projected area within the four walls surrounding the furnace measured at the middle of the grate.

209. Ratio of Heating Surface to Grate Surface. — The ratio of heating surface to grate surface is a variable quantity. It varies with the type of boiler, and even among boilers of the same type. An average value for water-tube boilers is 40 to 1 and for fire-tube boilers, 50 to 1.

210. Equivalent Evaporation, W_e. — *The economy of a steam boiler is defined as the number of pounds of water evaporated per pound of coal used.* The heat required to evaporate one pound of water, from a feedwater temperature of 60 deg. fahr. into steam at 165 pounds per square inch absolute, is different from that necessary to evaporate one pound of water under other conditions; as for example, water at a temperature of 180 deg. fahr. into steam at a pressure of 150 pounds per square inch absolute. The number of pounds of water evaporated under the first condition represents a different quantity of heat from that required to evaporate the same number of pounds under the second condition. A comparison of boiler performances under these two sets of conditions would be without value; to make a just comparison, the boiler performance must be reduced to some standard condition. The standard adopted is called **equivalent evaporation from and at 212 deg. fahr.** *By this standard the water actually evaporated is expressed in terms of the amount of water which would be evaporated, if the pressure were* 14.7 *pounds per square inch absolute and the feedwater temperature* 212 *deg. fahr., by using a quantity of heat equal to that used under the actual conditions of pressure and temperature.* Each pound of equivalent evaporation represents the addition of 970.2 B.t.u., or the latent heat of evaporation at a pressure of 14.7 pounds absolute.

The equivalent evaporation may be expressed thus:

$$\text{Wet steam,} \qquad W_e = \frac{W(xL + h_f - h_{f_1})}{970.2} \quad \ldots \ldots \quad (68)$$

$$\text{Dry steam,} \qquad W_e = \frac{W(h_g - h_{f_1})}{970.2} \quad \ldots \ldots \ldots \quad (69)$$

$$\text{Superheated steam,} \quad W_e = \frac{W[h_g + C_p(t_s - t_v) - h_{f_1}]}{970.2} \quad \ldots \quad (70)$$

in which W = actual weight of water fed per hour per pound of fuel, lb.
W_e = equivalent weight of evaporation per hour, lb.
970.2 = latent heat corresponding to a temperature of 212 deg. fahr., B.t.u.

Example. 28. — Using data from Example 26, page 221, find the equivalent evaporation from and at 212 deg. fahr.

Solution. — Since the steam contains moisture, substitute the various values in Equation (68). The numerical values of all symbols are as in Example 26.

$$W_e = \frac{W(xL + h_f - h_{f_1})}{970.2} = 3030 \times \frac{(0.98 \times 875.2 + 315.3 - 130)}{970.2}$$
$$= 3030 \times 1.075 = 3257 \text{ lb. per hr.}$$

BOILER PERFORMANCE

211. Factor of Evaporation, F. — *The factor of evaporation is the factor by which the actual evaporation is multiplied to obtain the equivalent evaporation.* It may be defined as the number of pounds of water that should be evaporated from a feedwater temperature of 212 deg. fahr. into dry steam at 212 deg. fahr., by the expenditure of the same amount of heat as was used to evaporate one pound of feedwater under the actual conditions. For the various states in which steam may exist, the factor of evaporation may be calculated from the following equations:

$$\text{Wet steam,} \qquad F = \frac{xL + h_f - h_{f_1}}{970.2} \qquad (71)$$

$$\text{Dry steam,} \qquad F = \frac{h_g - h_{f_1}}{970.2} \qquad (72)$$

$$\text{Superheated steam,} \quad F = \frac{h_g + C_p(t_s - t_v) - h_{f_1}}{970.2} \qquad (73)$$

Example 29. — The following data were taken during the test of a cross-drum B. and W. boiler, fired by Riley stokers, at the Columbus Railway, Power and Light Co.: Steam pressure, 256.5 lb. per sq. in. abs.; temperature of steam, 614 deg. fahr.; temperature of feedwater entering economizer, 128 deg. fahr. Find the factor of evaporation, including economizer and superheater.

Solution. — The steam is superheated, and Equation (73) will therefore be used.

$$F = \frac{h_g + C_p(t_s - t_v) - h_{f_1}}{970.2} = \frac{1200.9 + 0.56(614 - 403.2) - 95.9}{970.2} = 1.26$$

h_g at 256.5 lb. per sq. in. abs. .... = 1200.9
t_v at 256.5 lb. per sq. in. abs. .... = 403.2 deg. fahr.
t_s .... = 614 deg. fahr.
C_p corresponding to temperature and pressure of superheated
    steam, from curve, page 102 .... = 0.56
h_{f_1} at 128 deg. fahr. .... = 95.9

212. Boiler Performance. — The performance of a boiler or steam generator may be expressed in terms of: (1) efficiency, (2) rate of combustion in B.t.u. per square foot of grate surface per hour for solid fuels, as fired, (3) rate of combustion in pounds of fuel, as fired, per hour per cubic foot of furnace volume, (4) heat transferred per square foot of heating surface per hour, and (5) heat developed per cubic foot of furnace volume.

Boiler efficiency equals the ratio of the **output** to the **input**.

The **output of a boiler,** in heat units, is calculated from the weight of water evaporated. The calculation may be made in either of two ways: (1) *The equivalent evaporation, W_e, may be multiplied by* 970.2; or (2) *the actual evaporation, W, may be multiplied by the heat actually used in evaporating one pound of water under the existing conditions.* That is, the output may equal 970.2 W_e; or $W(xL + h_f - h_{f_1})$ for wet steam, $W(h_g - h_{f_1})$ for dry steam, and $W[h_g + C_p(t_s - t_v) - h_{f_1}]$ for superheated steam.

The **input to a boiler** may be based on the high heat value of the fuel as fired; that is, the fuel that goes into the furnace door; or it may be based upon the heat evolved from the coal burned on the grate, not taking into account the heat value of the fuel which falls through the grate into the ashpit. The efficiency found by using the first value of input is the **overall efficiency of the boiler, furnace and grate,** and that found by using the second value is the **efficiency of the boiler and furnace,** excluding the grate. These efficiencies may be expressed as follows:

1. *Efficiency of boiler, furnace, and grate*

$$= \frac{\text{Weight of water per hour} \times \text{heat per lb. of steam}}{\text{Weight of fuel per hour} \times \text{heat per lb. of fuel}} = \frac{\text{Output}}{\text{Input}}$$

$$= \frac{W(xL + h_f - h_{f_1})}{W_c C}, \text{ for wet steam} \quad \ldots \ldots \ldots \ldots \quad (74)$$

2. *Efficiency of boiler and furnace*

$$= \frac{\text{Weight of water per hour} \times \text{heat per lb. of steam}}{\text{Weight of combustible per hour} \times \text{heat per lb. of combustible}}$$

$$= \frac{W(xL + h_f - h_{f_1})}{W_c C - W_a C_a}; \text{ for wet steam} \quad \ldots \ldots \ldots \ldots \quad (75)$$

in which W_c = weight of fuel per hour, lb.
C = heat value per pound of fuel, B.t.u.
W_a = weight of ash and refuse per hour, lb.
C_a = heat value per pound of ash and refuse. In practically all cases, the combustible in the ash is carbon, and C_a may be taken

$$= \frac{14{,}600 \times \text{per cent carbon in ash}}{100}.$$

Other symbols are as in Art. 204, page 221.

In Equation (74), the heat value per pound of steam should correspond to the condition of the steam; that is, wet, dry or superheated, and the weight of coal may be either "as fired" or "dry," and the heat value per pound should correspond with the condition of the coal used. The final result is, however, the same, in either case.

The year-round efficiency of boiler, furnace and grate ordinarily does not exceed 60 per cent.

Example 30. — Additional data from the boiler test of Example 29, page 225, are as follows: weight of coal as fired, 5460 lb. per hr.; actual evaporation, 36,680 lb. per hr.; heat value of coal as fired, 10,775 B.t.u.; weight of ashes, 965 lb. per hr.; per cent carbon in ash, 20.26. Find: (a) the overall efficiency of boiler plant, and (b) efficiency of boiler and furnace.

Solution. — (a) Using Equation 74.
Efficiency, boiler, furnace and grate

$$= \frac{W[h_g + C_p(t_s - t_v) - h_{f_1}]}{W_c C} = \frac{36{,}680 \times 1223}{5460 \times 10{,}775} = 0.76$$

$W = 36{,}680$ lb. per hr.; $W_c = 5460$ lb.; $C = 10{,}775$ B.t.u.
$[h_g + C_p(t_s - t_v) - h_{f_1}]$ from *Example* 29 = 1223.

(b) Efficiency, boiler and furnace

$$= \frac{W\,[h_g + C_p\,(t_s - t_v) - h_{f1}]}{W_c C - W_a C_a} = \frac{36{,}680 \times 1223}{5460 \times 10{,}775 - 965 \times 2964} = 0.803$$

All values as in (a), except $W_a = 965$ lb. per hr., and

$$C_a = \frac{14{,}600 \times 20.3}{100} = 2964 \text{ B.t.u.}$$

213. Effect of Capacity on Efficiency. — When a boiler is forced beyond its normal rated capacity, its efficiency is usually lowered. The decrease in efficiency will not be great until the boiler is 50 per cent or more overloaded. This drop in efficiency is caused by the inability of the boiler to absorb the excess heat produced when forcing the fires.

214. Steam Boiler Testing. — From time to time it is necessary to make checks on the performances of the boiler unit. This may be provided for by installing testing equipment arranged to obtain the necessary data during the regular operation of the boiler plant, or by special tests made for the following purposes:[1]

1. To determine the capacity and efficiency of the boiler, for comparison with standard or guaranteed results.
2. To determine the cause of superior or inferior results.
3. To compare different kinds of fuels.
4. To compare different conditions and methods of operation.
5. To determine the effect of changes in design or proportion of the boiler or furnace, upon capacity and efficiency.

A boiler test generally consists of the making of certain observations, which are briefly described in the present chapter. The person in charge of the test should have the aid of a sufficient number of assistants, so that he may be free to give his attention to any part of the work whenever and wherever it may be required. *He should make sure that the instruments and testing apparatus continually give reliable indications and that the readings are being correctly recorded.* He should also keep a close watch on the operation of the boiler under test, and see that the operating conditions determined upon are maintained, and that nothing occurs to vitiate the data.

The object of the test should be clearly kept in view at all times. *Accuracy and reliability must characterize the work from beginning to end.*

215. Preparation for the Test. — The dimensions of the principal parts of the apparatus which bear on the object in view should be determined. The general features of the boiler should be observed and sketches should be made to show unusual points of design. The physical conditions of all parts of the boiler and apparatus should be thoroughly examined and a record made of the conditions found, together with any points in the

---

[1] A considerable portion of the material in this chapter has been taken from the A. S. M. E. "Test Code for Stationary Steam-Generating Units."

operation which should affect the results. The brick walls and cleaning doors should be examined for air leaks, either by shutting the damper and observing the escaping smoke or by the candle flame test. Leakage of steam through blow-offs, drips, or any steam or water connections of the apparatus undergoing test should be prevented by blanking off; or satisfactory assurance should be obtained that there is no leakage either outward or inward. *This is a most important matter, and no assurance should be considered satisfactory which is not susceptible of absolute demonstration.*

For tests of maximum capacity and efficiency of the boiler the coal should be of some kind which is commercially recognized as a standard for the locality where the test is made.

In guarantee tests with fuel of a specified calorific value, or other characteristics, there should be a clear understanding as to the permissible variations from the specified characteristics.

Fig. 154. — Cochrane V-Notch Meter.

Tests made with oil or gas fuels accord with the rules for making a test with coal. The "running" method of starting and stopping is used, and the length of the test may be shortened.

Before the start of the test, the boiler should have been in operation a sufficient length of time to attain working temperatures and proper operating conditions throughout. For tests to determine efficiency, it is desirable to run preliminary tests to determine the most advantageous conditions of operation.

216. Apparatus and Instruments. — The weight of coal and ashes used on a test is generally obtained by using **platform scales,** which should be tested for accuracy by calibrating with standard weights.

The quantity of feedwater should be obtained by means of tanks and platform scales. A suitable arrangement for this purpose consists of one or more tanks, each resting on a platform scale, and elevated a sufficient distance above the floor to empty into a receiving tank below. The feed pump is connected to the lower tank. Each tank should be of sufficient size to give ample time for the proper weighing of the water between successive emptyings.

The feedwater is often measured by some form of **water-meter,** of which there are many commercial types. *The meter should be calibrated in* *place before and after test to ensure accuracy.* One typical water-meter, Fig. 154, uses a triangular weir over which the water flows, mounted in a suitable chamber made of either cast iron or steel plate with a recorder to register the head and flow, the readings of which can be checked by calculation using the height of water surface above the bottom of the weir as explained in Art. 437, page 465.

For measuring water flowing in pipes, the **Venturi meter**, Fig. 155, is well suited. This meter consists of a short constricted passage, or throat, joined to the full section of the pipe by conical converging or diverging sections. *The amount of water flowing is proportional to the difference in* **static**, *or radial, pressure at the sections indicated*, this pressure difference being obtained by a manometer. The meter is calibrated by weighing, and the amount of water flowing is found as follows:

$$Q = A_t V \qquad (76)$$

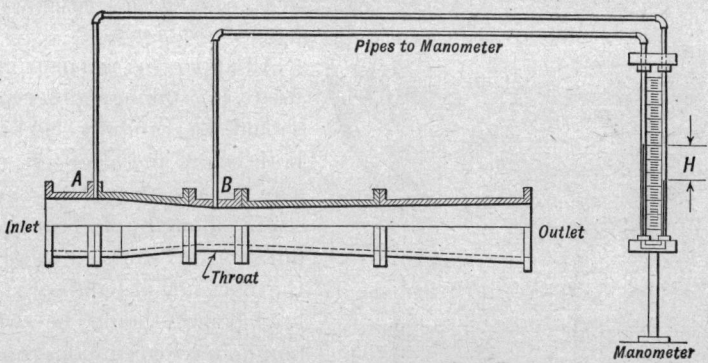

FIG. 155. — Venturi Meter.

in which Q = cubic feet of water discharged per second.
A_t = area of throat section, square feet.
V = velocity of water at throat, feet per second.
$= \dfrac{A_a}{\sqrt{A_a{}^2 - A_t{}^2}} \sqrt{2gH}$

in which A_a = area of upstream section, square feet.
H = difference in pressure, measured in feet of water by the manometer.[1]

The various pressures and temperatures are measured by pressure gages, draft gages, thermometers and pyrometers. These instruments are described in Chapter III.

The amount of moisture in the steam, when wet, should be determined by a throttling or separating calorimeter (see pages 102 and 105). When the

[1] Where there is water above the indicating liquid the actual head equals H as measured, multiplied by $(S - 1)$, where S is the specific gravity of the indicating liquid.

steam is superheated, its temperature should be determined by a thermometer inserted in a thermometer well attached to the superheating surface.

The analysis of the flue gases is generally determined by some form of Orsat apparatus, Art. 165, page 167. If momentary samples are obtained, the analysis should be made as frequently as possible, say every fifteen or thirty minutes, and the conditions of the furnace and firing should be noted when the sample is withdrawn.

The amount of air used for combustion is often measured, by an **anemometer**, Fig. 156, as the air enters the space beneath the grate. This instrument consists of a light vane which is revolved by the movement of the air, the number of revolutions being recorded by a system of gears and pointers. *The instrument measures the linear velocity only.*

All apparatus and instruments used in taking observations should be carefully calibrated, both before and after test, to ensure their accuracy.

217. Length of Test. — The duration of tests to determine the efficiency of hand- or stoker-fired boilers should be twenty-four hours of continuous running, or such time as may be required to burn a total of 250 pounds of coal per square foot of grate area.

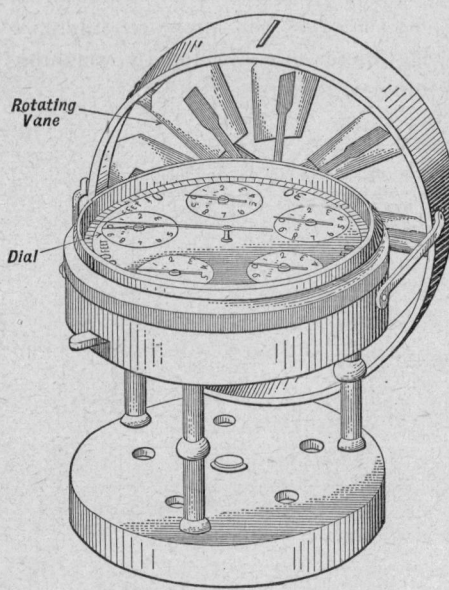

Fig. 156. — Anemometer.

In commercial tests where the service requires continuous operation night and day, with frequent shifts of firemen, the duration of the test, whether the boilers are hand fired or stoker fired, should be at least twenty-four hours. Likewise, in commercial tests, either of a single boiler or of a plant of several boilers, where the plant operates regularly a certain number of hours and during the rest of the day the fires are banked, the duration should not be less than twenty-four hours. The duration of tests to determine the maximum evaporative capacity of a boiler, without determining the efficiency, should not be less than three hours.

218. Starting and Stopping. — Combustion, fuel, draft, and temperature conditions, the water level, rate of feeding water, rate of steaming, and the steam pressure should be as nearly as possible the same at the end as at the beginning of the test.

If the water level is not the same at the end of the test as at the beginning a correction should be made by computation rather than by feeding additional water after the final readings are taken. For specific methods to be used in hand-firing and stoker-firing equipment "The Boiler Test Code of the A. S. M. E." should be consulted.

219. Records. — *A log of the data should be entered in notebooks or on blank sheets suitably prepared in advance.* This should be done in such manner that the test may be divided into hourly periods, or, if necessary, periods of less duration; the leading data may then be obtained for any period or periods, as desired, thereby showing the degree of uniformity obtained. A sample feedwater log is shown in Table 21.

TABLE 21. — FEEDWATER LOG, BOILER TRIAL NO. 1

Made at............
Date.............. By................
Boiler No.......... Fireman...........

Time	Water Delivered to Feed Tank Pounds	Temp. of Water in Tank °F.	Time	Water Delivered to Feed Tank Pounds	Temp. of Water in Tank °F.	Remarks
7:00	7855	196				Test began at 7:00 A.M.; March 26. Test closed at 6:00 P.M.; March 26.
8:30	3169	197				
9:00	3244	196				
9:40	2567	196				
10:00	5845	196				
11:00	5638	198				
12:01	6339	199				
1:00	6449	197				
2:03	9076	198				
3:45	3312	200				
4:03	3275	200				
4:52	2600	195				
5:04	4290	198				

The readings of the various instruments and apparatus concerned in the test, other than those showing quantities of consumption (such as fuel, water, and gas), should be taken at intervals not exceeding half an hour and entered in the log. Whenever the indications fluctuate, the intervals should be reduced, according to the extent of the fluctuation. In the case of smoke observations, for example, it is often necessary to take observations every minute, or still oftener, continuing these throughout the period covering the range of variations. *When it is essential that a number of instruments be read simultaneously, there should be an observer stationed at each one, and the readings should be taken at a signal from a time-keeper.*

Coal should be weighed and delivered to the firemen in portions sufficient for one hour's run, in order that the degree of uniformity of firing may be ascertained. An ample supply of coal should be maintained at all times, but the quantity on the floor at the end of each hour should be as small as practicable, so that it may be readily estimated and deducted from the total weight, to determine the hourly weight.

The records should be such that the amount of feedwater used during each hour can be ascertained, thus giving a check on the uniformity of evaporation.

Make a memorandum of every event connected with the progress of a test, however unnecessary it may appear at the time. A careful record should be made of the time of every such occurrence and the time of taking every weight and every observation. *For the purpose of identification, the signature of the observer, and the date, should be affixed to each log sheet or record.*

In the matter of weighing coal by the barrow-load, or weighing water by the tankful, which is required in many tests, a series of marks, or tallies, should never be trusted. The time each load is weighed or emptied should be recorded. *The weighing of coal should not be delegated to unreliable assistants, and, whenever practicable, one or more men should be assigned solely to this work. The same may be said with regard to the weighing of feedwater.*

220. Sampling and Drying Coal. — During the progress of the test, the coal should be regularly sampled for the purpose of analysis and determination of moisture, using the American Society of Testing Materials standard method of sampling, by which a gross sample is first obtained by taking a definite weight of coal from each portion delivered to the firemen. This sample is stored in a cool place and in a covered metal receptacle. When all the coal has been thus sampled the lumps are broken up, the whole quantity thoroughly mixed and finally reduced by the process of repeated quartering and crushing to a sample weighing about 5 pounds, the largest pieces being about the size of a piece of pea coal. This should then be placed in a suitable container and sealed in such a manner as to preclude tampering, and be preserved for subsequent determinations of moisture, calorific value and chemical combustion. When the sample lot of coal has been reduced by quartering to approximately 100 pounds, a 5-pound portion is withdrawn for the purpose of immediate moisture determination.

221. Ashes and Refuse. — The ashes and refuse, withdrawn from the furnace and ashpit during the progress of the test and at its close, should be weighed, if possible, in a dry state. If this material is wet, the amount of moisture should be ascertained and allowed for, a sample being taken and dried for this purpose. This sample may serve also for analysis and the determination of unburned carbon and fusing temperature.

Clinker should be weighed separately from the fine ash. The sample may be prepared as explained for coal, and should contain approximately the same proportions by weight of clinker and ash as for the whole test.

222. Smoke Observations. — In tests of bituminous coals, requiring a determination of the amount of smoke produced, observations should be made regularly throughout the trial at intervals of five minutes, or, if necessary, every minute. At the same time, the furnace and firing conditions should be noted.

223. Data and Results. — It is well to plot all the data on a chart, Fig. 157, which shows the degree of uniformity of operation at a glance. This should be done during the progress of the test.

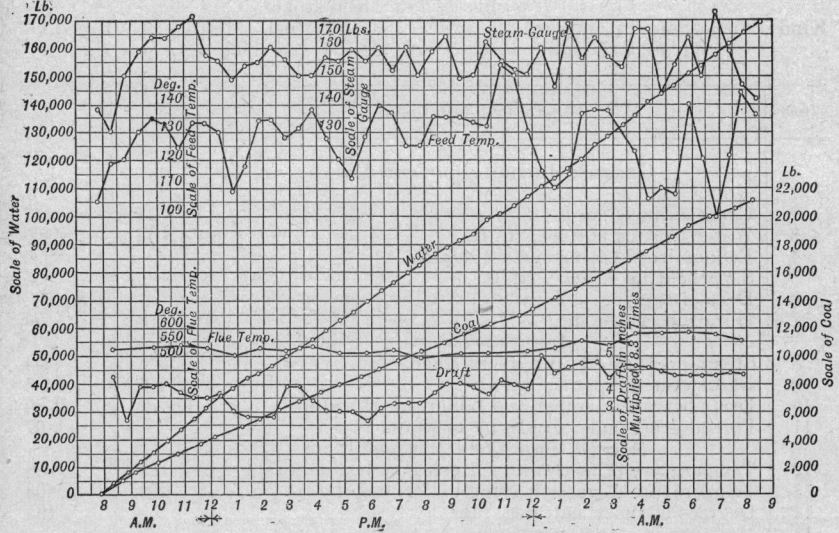

Fig. 157. — Boiler Data Chart.

224. Report of Test. — A report of the test should present all the leading facts bearing on the design, condition, and operation of the apparatus tested, together with such sketches and photographs as may be needed for a clear understanding of all points of the test. It should state the object and character of the test, the methods followed, the conditions maintained, and the conclusions reached, and should close with a tabular summary of the principal data and results. The form of a data-and-result sheet is given in Table 22, in which two tests are recorded. The first test is for a boiler which has neither superheater nor economizer; the second is for a boiler having both. Observed data are recorded in roman type, and calculated results in **bold-face type**.

TABLE 22. — DATA AND RESULTS OF A BOILER TEST

Test No. 1
Made by: Engineering Department.
Location: United States Naval Station, Key West, Fla.
Purpose of test: Efficiency and Capacity.

Type of boiler: Keeler water-tube.
Kind of furnace: Hand-fired.

Kind of fuel: Semi-bituminous. Run of the mine.

Test No. 2
Made by: Riley Stoker Company.
Location: Columbus Railway Light and Power Co., Columbus, Ohio.
Purpose of test: Overall efficiency of stokers, furnaces, boilers, and economizers.

Type of boiler: Cross-drum B. and W.
Kind of furnace: Riley 8-retort underfeed stoker.

Kind of fuel: Mixture of crushed Ohio No. 6 and slack.

Item	Name of Item with Units		
	GENERAL INFORMATION AND DATA		
1	No. of test...............	1	*2
2	Grate surface, sq. ft.....	45	117
3	Water heating surface, sq. ft....	2043	4400
4	Economizer heating surface, sq. ft.....	....	6290
5	Superheating surface, sq. ft.....	....	855
6	Date of test............	....	Aug. 23, '18
7	Duration of test, hr.....	8	24
	Average pressures		
8	Steam pressure, lb. per sq. in. gage.....	135.6	244
9	Barometer, lb. per sq. in. abs.....	14.6	14.5
10	Draft between damper and boiler, in. of water.....	0.78	0.89
11	Forced draft under grate, in. of water.....	....	1.97
	Average temperatures and quality		
12	Temperature of feedwater entering boiler, deg. fahr.	122	†238
13	Temperature of gases leaving boiler, deg. fahr.....	530	403
14	Degrees of superheat, deg. fahr.....	....	165
15	Moisture in steam, per cent.....	1.25	....
	Fuel		
16	Proximate analysis, as fired, per cent by weight.....		
	Moisture.....	1.20	8.13
	Volatile matter.....	21.54	31.55
	Fixed carbon.....	71.88	46.25
	Ash.....	5.38	14.07
17	Ultimate analysis, as fired, per cent by weight		
	Carbon.....	82.54	62.40
	Oxygen.....	4.99	16.16
	Sulphur.....	0.65	1.20
	Hydrogen.....	4.76	5.05
	Nitrogen.....	1.68	1.12
	Ash.....	5.38	14.07
18	Heat value per pound dry coal, by calorimeter, B.t.u.	14,807	11,730
19	Heat value per pound combustible, by calculation, B.t.u.....	15,650	13,690

* Calculations are for boiler including economizer and superheater.
† Temperature of water entering economizer, 128 deg. fahr.

TABLE 22. — DATA AND RESULTS OF A BOILER TEST. (*Continued*)

Item	Name of Item with Units	1	2
20	Weight of coal as fired, lb...............................	8,936	139,910
21	Weight of dry coal, lb....................................	8,830	128,500
22	Weight of ash and refuse, lb.............................	840	28,350
23	Combustible in ash, per cent.............................	.....	22.05
24	Weight of combustible burned, lb.........................	7,990	100,150
25	Weight of dry coal consumed per hour, lb.................	1,104	5355
26	Dry coal per sq. ft. of grate surface per hour, lb........	24.53	.....
27	Combustible burned per hour, lb..........................	.....	.....
28	Combustible burned per sq. ft. grate surface per hr., lb.	.....	.....
29	Analysis of flue gases, per cent by volume...............		
	Carbon dioxide...	.....	12.10
	Oxygen..	.....	6.93
	Carbon monoxide...	.....	0.19
	Nitrogen (by difference)................................	.....	80.78
	Water, total for test		
30	Total weight of water fed to boiler, lb..................	86,872	907,500
31	Total weight of water evaporated, corrected for moisture, lb.	85,786	.....
32	Factor of evaporation...................................	1.14	1.24
33	Equivalent water evaporated into dry steam from and at 212 deg. fahr., lb.	97,696	1,123,000
	Water, per hour		
34	Water evaporated per hour corrected for moisture, lb.....	10,723	37,810
35	Equivalent evaporation into dry steam per hour from and at 212 deg. fahr., lb.	12,212	46,800
36	Equivalent evaporation per hour from and at 212 deg. fahr. per sq. ft. of water heating surface, lb.	5.98	10.63
37	Horsepower developed, hp................................	356.8	1358
38	Builder's rated horsepower on basis of 10 sq. ft. per hour, hp.	204.0	444
39	Percentage of builder's rated horsepower developed.......	174.6	305
	Economic results		
40	Water apparently evaporated under actual conditions per pound of coal as fired, lb.	9.74	6.49
41	Equivalent evaporation from and at 212 deg. fahr. per pound of coal as fired, lb.	10.90	8.05
42	Equivalent evaporation from and at 212 deg. fahr. per pound of dry coal, lb.	11.06	8.75
43	Equivalent evaporation from and at 212 deg. fahr. per pound of combustible, lb.	12.22	11.20
	Efficiency results		
44	Efficiency of boiler, furnace and grate, per cent.........	72.8	72.2
45	Efficiency of boiler and furnace, per cent................	76.4	79.6

REFERENCES

Mechanical Equipment of Buildings, Vol. II, HARDING and WILLARD.
Steam, BABCOCK and WILCOX COMPANY.
Code on Definitions and Values, A. S. M. E.
Power Plant Testing, MOYER.
Mechanical Laboratory Methods of Testing Machines and Instruments, SMALLWOOD.
Experimental Engineering and Manual for Testing, CARPENTER and DIEDERICHS.
Test Code for Stationary Steam Boilers, A. S. M. E.
Test Code for Steam Generating Units, A. S. M. E.

236 RATING, EFFICIENCY AND TESTING OF STEAM BOILERS

REVIEW QUESTIONS AND PROBLEMS

1. What is the significance of the term "boiler horsepower"?

2. Define the A. S. M. E. standard for boiler horsepower. What is the relation it bears to the Centennial rating?

3. What is the standard amount of heating surface equivalent to one boiler horsepower?

4. The following data apply to Dillon return-tubular boilers:

	1	2	3	4	5
Length of tubes, ft...............	16	15	19	17	22
Length of boiler (overall)........	17	16' 2''	20' 2''	18' 8''	23' 10''
Diam. of shell, in................	36	54	60	84	96
No. of tubes.....................	32	62	82	150	262
Size of tubes, in.................	$2\frac{1}{2}$	3	3	$3\frac{1}{2}$	3

Find: (a) Square feet of heating surface in each boiler, and check with the approximate rule, Art. 206.

(b) Rated horsepower of each boiler.

5. What should be the grate area for each of the boilers in Problem 4?

6. The following data were obtained during the test of a Wickes vertical water-tube boiler: absolute steam pressure, 212.1 lb. per sq. in.; superheat, 162.2 deg. fahr.; temperature of feedwater, 198.8 deg. fahr.; water evaporated, 70,938 lb.; coal fired, 9193 lb.; total ash and refuse, 953 lb.; per cent combustible in ash, 8.68; heat value per pound of dry coal, 13,245 B.t.u.; moisture in coal fired, 4.42 per cent; duration of test, 10 hr. Find: (a) Factor of evaporation.

(b) Total equivalent evaporation, including superheater.

(c) Overall efficiency of plant.

(d) Efficiency of boiler and furnace combined.

7. During the test of a B. and W. cross-drum marine boiler used on the U.S.S. Battleship Wyoming, the following data were taken: steam pressure, 209.9 lb. per sq. in. gage; feedwater temperature, 168.6 deg. fahr.; barometer, 30.0 in. mercury; weight of oil used, 5943 lb.; quality of steam, 99.19 per cent; weight of water evaporated, 74,898 lb.; duration of test, 2 hr.; heat value per pound of oil, 19,291 B.t.u. Find: (a) Equivalent evaporation from and at 212 deg. fahr.

(b) Actual evaporation, pounds per hour.

(c) Heat output B.t.u. per hour and units of evaporation.

(d) Factor of evaporation.

(e) Boiler horsepower developed.

(f) Overall efficiency.

8. During the test of a 468 hp. boiler using pulverized coal, data were taken as follows: pressure in boiler, 167 lb. per sq. in. gage; barometer, 29.25 in. mercury; temperature of feedwater, 157.2 deg. fahr.; total weight of fuel, 47,775 lb.; total weight of water, 393,168; degrees superheat, 374.9 deg. fahr.; B.t.u. per pound fuel as fired, 10,779; duration of test, 24 hr. Find the items called for in Problem 7.

9. A B. and W. boiler, equipped with convection and radiant superheaters, under test conditions gave the following data: duration of test, 54.6 hr.; coal fired, 154,300 lb.; water actually evaporated, 1,561,170 lb.; B.t.u. per pound coal (as fired), 13,904; feedwater temperature, 207 deg. fahr.; steam pressure, 314 lb. per sq. in. abs.; temperature leaving convection superheater, 460 deg. fahr., and leaving radiant superheater, 632 deg. fahr.

Find: (a) Efficiency of unit, (b) efficiency of unit without radiant superheater, (c) per cent of total output absorbed by radiant superheater.

10. A boiler having 2000 sq. ft. of heating surface evaporated 218,860 lb. of water in 8 hr. Temperature of feedwater, 211 deg. fahr. Steam pressure, 125 lb. per sq. in. gage. Barometer, 30 in. mercury. Quality of steam = 1.

Find: (a) The per cent of builders rating developed, (b) heating surface per horsepower developed, and (c) the weight of water evaporated per hour per square foot of heating surface.

11. What are the purposes for which tests are made on steam boilers?

12. Describe the preparations necessary to make a steam boiler test.

13. Name the conditions which should exist at start and finish of a steam boiler test.

14. How may the amount of air used during a boiler test be measured? Where this is not feasible, what method is used?

15. Prepare a data sheet suitable for the data pertaining to the weight of coal.

16. What data are shown on the boiler data chart?

17. Calculate all items which are printed in bold-face type for test No. 2, page 234.

18. Using the analysis for test No. 2, make a heat balance for the boiler, based on coal as fired.

19. During the test of a Babcock and Wilcox water-tube boiler at the Pennsylvania Navy Yard, the data were: water evaporated per hour, 13,552 lb.; oil used per hour, 871 lb.; quality of steam, 0.999; heat value of oil per lb., 19,525 B.t.u.; steam pressure, 300 lb. per sq. in. gage; barometer, 30.12 in. mercury; feedwater temperature at boiler, 199 deg. fahr. Find overall efficiency of the plant and equivalent evaporation per pound of oil. How does the equivalent evaporation compare with that ordinarily obtained per pound of coal?

20. During a boiler test, the flue-gas analysis gave carbon dioxide, 15.10; oxygen, 3.09; carbon monoxide, 0.00; nitrogen by difference, 81.81. Find the weight of dry chimney gases per pound coal.

21. In Problem 20, the sum of the carbon dioxide and oxygen is 18.19 per cent. Give reasons why it is not 21 per cent. (See Art. 167, page 170.)

22. During an 8-hr. test of a Heine-Horizontal water-tube boiler, fired by 3 Coen pulverized fuel burners, and having 5090 sq. ft. of heating surface, the following data were taken: water evaporated, 188,374 lb.; wet coke fired, 18,900 lb.; volume of furnace, 1236 sq. ft.; superheater heating surface, 174 sq. ft.; pressure of steam, 180 lb. abs.; steam temperature, 471 deg. fahr.; feedwater temperature, 208 deg. fahr.; heating value, 14,400 B.t.u. as fired. Find: (a) Actual evaporation from and at 212 deg. fahr.

(b) Actual evaporation per square foot of heating surface.

(c) Actual evaporation per pound of fuel as fired.

(d) Horsepower developed.

(e) Rate of combustion in lb. of fuel as fired per hr. per cu. ft. of furnace volume.

(f) Heat developed per cubic foot of furnace volume. How does this compare with present practice?

(g) Efficiency of boiler, furnace and superheater.

CHAPTER X

PIPE SYSTEMS, PIPE, VALVES, AND PIPE ACCESSORIES

225. Foreword. — For the efficient and economical operation of a power plant, the layout of the piping system is of prime importance, and the care with which this is arranged will have its effect continuously on the operation of the plant. The points that should be considered in a piping system are: (1) continuity of service and provision for extension, (2) economical size and sufficient strength, (3) expansion, (4) drainage, (5) support, (6) insulation.

226. Division of Piping Systems. — The piping of a power plant may be conveniently considered under the following divisions:

1. *High-pressure piping.* — This includes the piping connecting the boilers with the engines, turbines, and steam pumps. It includes the boiler leads, main steam header and auxiliary header, engine and turbine leads, connections to auxiliaries and low-pressure traps.

2. *Low-pressure piping.* — This includes all atmospheric exhaust lines, connections to feedwater heaters and exhaust steam heating.

3. *Vacuum piping.* — Includes all exhaust connections between the prime mover and condenser.

4. *Feedwater piping.* — Includes all connections to and from the water end of feed pumps and injectors and the feed lines of the boilers.

5. *Blow-off piping.*

6. *High- and low-pressure drainage systems.* — Includes drips, traps, and seals for the return of condensation either to the feedwater heater or, direct to the boilers.

227. High-pressure Piping Systems. — The systems of piping employed for the main steam piping are: (1) single-header, (2) spider, (3) loop, or ring, and (4) unit.

Of these systems the one to be used depends upon the size of the plant and the character of the load. In plants having several boilers and engines, or turbines, the leads are connected to a common header which forms a flexible tie between the boilers and engines. This permits cutting any boiler into or out of service and re-distributing the load among the remaining boilers.

The **single-header system,** which is largely used in small and medium-sized plants, is illustrated in Fig. 158, with the boilers and engines arranged back to back. It has a low first cost and is simple and easy to ex-

HIGH-PRESSURE PIPING SYSTEMS 239

tend. Two stop valves are located between the header and each battery of boilers, to permit cutting out any section when necessary. Several methods of providing for expansion are shown.

The **spider system** is an adaptation of the single-header system in which the boiler leads are connected to a short header, long enough to contain

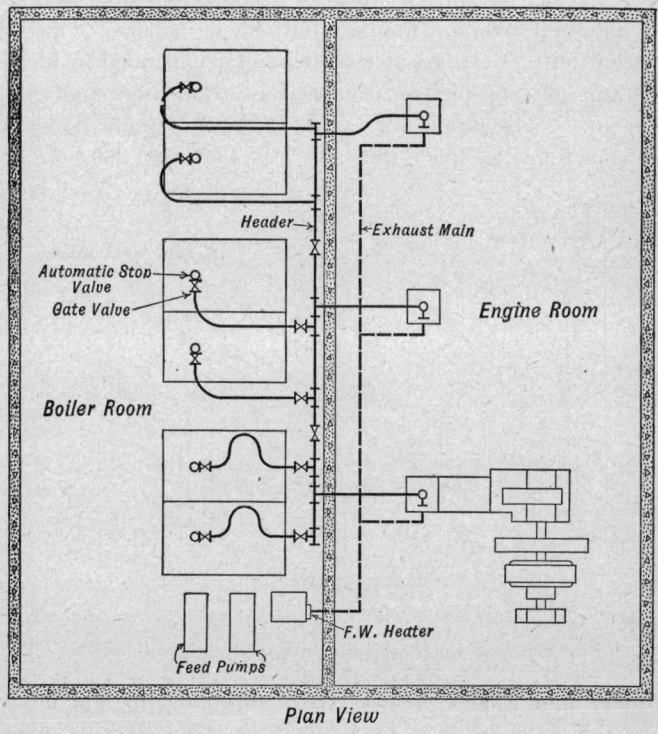

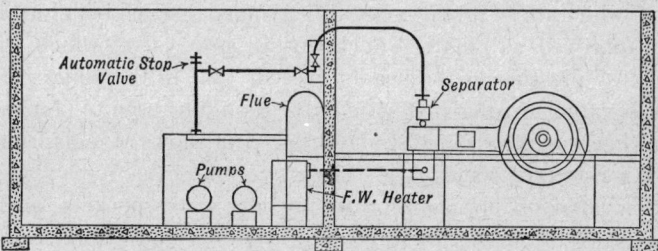

Fig. 158. — Single Header Pipe System.

the necessary valves and fittings. The short header reduces the danger of shutdown, reduces condensation, and is easily extended.

The **loop**, or **ring, system** has piping in which the header is arranged in the form of a loop, and hence requires a greater length of pipe than the single header and the spider systems. The valves in the header should be so located as to permit cutting any section out of service and supplying steam from the remaining boilers. This system is not in general use, because of the high cost. It cannot be easily extended and has a large number of joints from which leakage may occur.

The **unit system** uses a single header, and each prime mover is supplied by its own battery of boilers, with the units cross connected to permit throwing any unit over on another battery of boilers. This system is used in large central stations and is frequently expanded to include separate feedwater heaters, pumps, economizers, condensers and chimney for each main unit. The piping for a modern high-pressure boiler unit and turbine is shown in Fig. 159. (See also Fig. 184, page 260.)

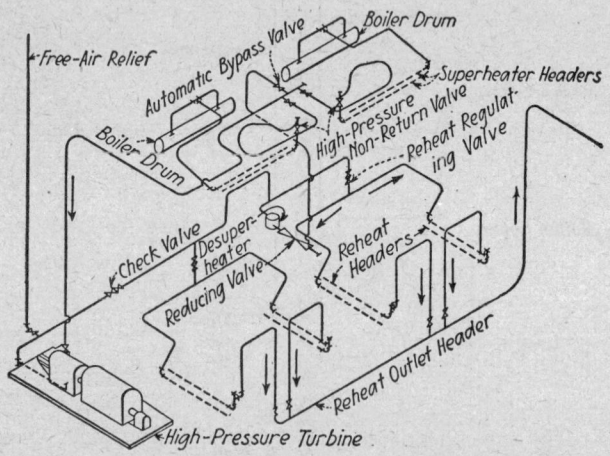

Fig. 159. — Piping for High-pressure Boiler and Turbine.

228. Boiler and Engine Leads. — An approved method of arranging the piping for boiler leads is shown in Fig. 160. By this arrangement pockets in which water might collect are avoided. Each boiler lead should have an approved automatic, quick-closing stop valve, which will close whenever the pressure in the header exceeds that in the boiler. In addition to this valve, a rising-stem gate valve should be used to permit closing the lead when making repairs. The rising stem shows at a glance whether the valve is closed or open.

Valves located on horizontal runs prevent water pockets, and this is the approved location of gate valves for all boiler and engine leads. The slope of the piping should be in the direction of steam flow, with the header connection at the upper side. Expansion in engine and boiler leads is provided for by long radius bends. The leads should be supported from above by hangers.

A typical engine lead is also shown in Fig. 160. It has a rising-stem gate valve and discharges into a receiver separator located just above the engine throttle valve. By this arrangement, known as the **triple swing** connection, the necessity of springing the bends into place is prevented, because a swing adjustment can be made on three planes. It must have *a horizontal and two vertical joints* on one end of the connection.

229. Size of Leads and Headers. — Boiler and engine leads are ordinarily made one size smaller than the boiler nozzle or flange on an engine or turbine. The velocity of steam in the pipe is limited by good practice to 6000 feet per minute. Velocities as high as 10,000 to 12,000 feet per minute are often used with turbines when the steam flow is constant, while 9000 feet per minute is a maximum for steam-engine piping.

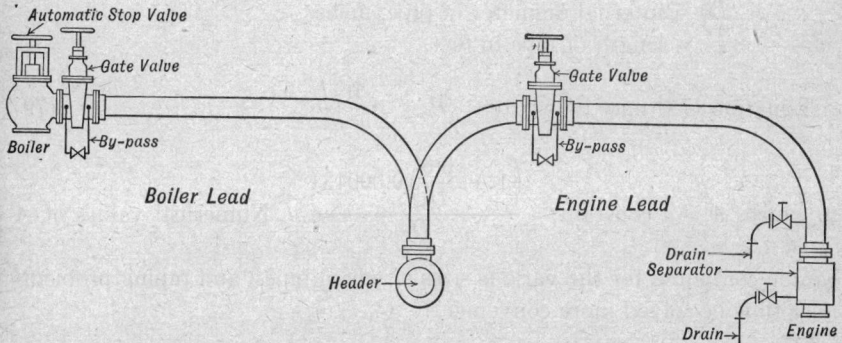

Fig. 160. — Boiler and Engine Leads.

The size of a steam pipe, based on an allowable velocity, can be found by using the following equation:

$$A = \text{area of pipe in square inches} = \frac{144 \times W \times v_g}{V} \quad \ldots \quad (77)$$

in which W = equivalent weight of steam flowing, pounds per minute.

v_g = specific volume of steam at absolute pressure in pipe, cubic feet.

V = velocity of steam, feet per minute.

Example 31. — Using the data given below, find the diameter of steam pipe required to keep the velocity of the steam to 6000 ft. per min.

Solution. — Using Equation (77),

$$\text{Area in sq. in.} = \frac{144 \times W \times v_g}{V} = \frac{144 \times 260 \times 2.830}{6000} = 17.75 \text{ sq. in.}$$

W = 260 lb. per min.
V = 6000 ft. per min.
v_g = 2.830 cu. ft. per lb. at 160 lb. per sq. in. abs., Table 7, page 96.

$$\text{Diameter of pipe} = d \sqrt{\frac{17.75 \times 4}{3.14}} = \sqrt{22.60} = 4.76 \text{ in.}$$

Note: A 5-in. pipe would be used, as it would be the nearest commercial size.

In practice, the limiting factor, in determining the size of steam piping, is the allowable pressure drop, since the permissible velocities are governed by this drop. There are various equations for computing the drop in pressure. **Babcock's equation** appears to give as satisfactory results as any:

$$P = 0.000131 \left(1 + \frac{3.6}{D_1}\right) \frac{W^2 L}{d D_1^5} \quad \ldots \ldots \quad (78)$$

in which P = the difference in pressure between the two ends of the pipe in pounds per square inch.
W = weight of steam passing, pounds per minute.
d = mean density of steam, pounds per cubic feet.
D_1 = internal diameter of pipe, inches.
L = length of pipe in feet.

Equation (78) may be written: $P = A \dfrac{W^2 L}{d} \quad \ldots \ldots \ldots \quad (79)$

in which A = a constant = $\dfrac{\left(1 + \dfrac{3.6}{D_1}\right) 0.000131}{D_1^5}$. Numerical values of A can be computed for the various sizes of steam pipes, and piping problems may thus be solved more conveniently.

Example 32. — A 5-in. steam pipe line has a length of 300 ft. and delivers 15,000 lb. of steam per hr., at a steam pressure of 165 lb. per sq. in. abs. Find the loss in pressure in the pipe line.

Solution. — Using Equation (79) with proper substitutions

$$P = \frac{A W^2 L}{d} = 0.000,000,07 \times \frac{250^2 \times 300}{0.363} = 3.62 \text{ lb. per sq. in.}$$

A for a 5-in. pipe = $0.000,000,07$

$W = \dfrac{15,000}{60} = 250$ lb. per min.; $L = 300$ ft.

d = from steam table = $\dfrac{1}{v_g} = \dfrac{1}{2.748} = 0.363$ lb. per cu. ft.

The *steam header* for a single header system should be made equal in size to the largest boiler lead. If the engine and boiler are located at opposite ends of the header, the size may be computed by using a velocity of 8000 feet per minute for the rate of steam flow. The header should generally be carried on rigid supports and anchored to a support at the center of the header and the *gate valves used between each battery of boilers should be of the rising stem type.* An 18-inch O. D. forged and welded steam header, suitable for a pressure of 350 pounds and a temperature of 700 deg. fahr. and as used in the Cahokia Station of the Union Electric Light and Power Company, is shown in Fig. 161.

230. Exhaust Piping. — The exhaust piping connecting each engine or turbine with the exhaust steam main should have a gate valve to isolate the unit when making repairs. The use of pipe bends is not general.

All piping should be short and direct, with a minimum number of joints which should be made tight to prevent leakage of air when attached to a condenser.

The size of exhaust piping is generally based on a velocity of 8000 to 9000 feet per minute for non-condensing operation; velocities as high as 20,000

FIG. 161. — Forged Steel Header.

feet per minute are permissible for condensing plants because the loss caused by friction is low.

231. Feedwater Piping. — The piping used for feedwater lines should be made of extra heavy steel or brass pipe, to guard against corrosion. Flanges made of cast steel give better service than cast-iron flanges. A feed, check, and stop valve are generally used for each feedwater con-

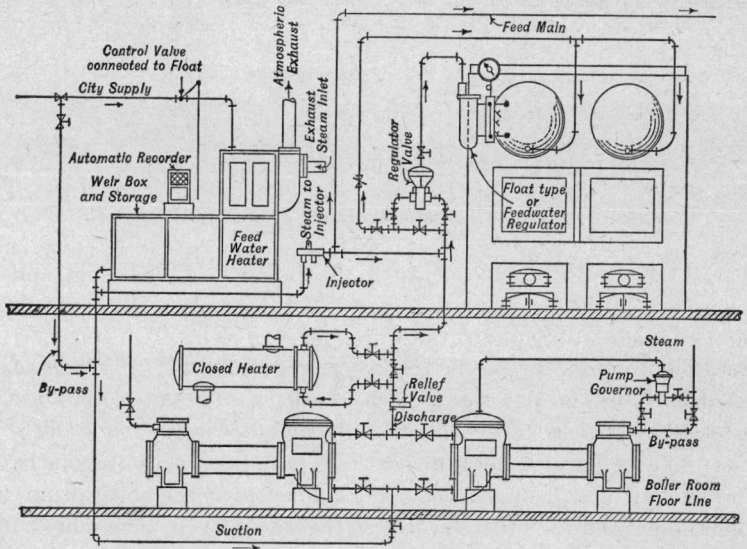

FIG. 162. — Non-condensing Feedwater Piping with Feed Pumps Installed in Duplicate.

nection to the boiler. The stop valve used should be of the globe pattern, as the gate valve cannot be used to regulate the flow, and pulsations from the feed pump cause it to clatter. The complete feedwater piping for a non-condensing plant is shown in Fig. 162. In general, all equipment should be by-passed to permit the making of repairs without closing down the plant.

The flow of feedwater may be controlled by hand-operated valves or by a feedwater regulator which automatically maintains the water level in the boiler as nearly constant as possible. These regulators are not always positive in their action, and most engineers prefer to rely on hand regulation. With high pressures regulators appear to be a necessity.

232. Feedwater Regulators. — Feedwater regulators are generally of the float, thermostatic, or expansion type, and ordinarily operate in conjunction with a governor attached to the feedwater pumps. The regulator moves a valve located in the feedwater pipe and thus increases or diminishes the opening through which the feedwater flows. This action decreases or

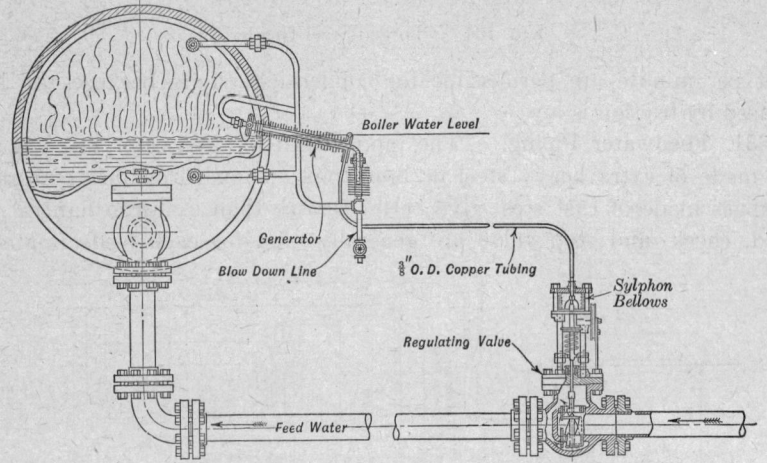

Fig. 163. — Bailey Thermo-Hydraulic Feedwater Regulator.

increases the pressure against which the pump is discharging, and the pump governor then changes the speed of the pump to feed the proper amount of water to maintain the water level.

The **Bailey thermo-hydraulic feedwater regulator,** Fig. 163, *depends for its operation upon the fact that a given weight of water, when converted into steam, occupies a much greater volume than that of the water from which it was formed.* The regulator **generator** consists of two tubes, the outer one having fins for cooling, and the inner one being connected to the boiler drum, with one end above and the other end below the water level. The inner tube is subjected to boiler pressure and contains both steam and water, the water level being the same as that in the boiler drum. The annular space between the inner and outer generator tubes, the copper tubing connecting the generator and regulating valve of the feedwater line, and the metal bellows or **sylphon** of the regulating valve form a closed system, which is filled with water when the regulating valve is closed and the sylphon contracted. In operation, heat from the steam in the upper portion of the inner tube causes the water surrounding that portion of the tube containing steam to rise in temperature until a portion of it flashes into steam. This produces an increased pressure in the generator to force water out of the generator until its water level is the same as that in the inner generator tube. The water forced out of the generator passes into the sylphon on the **regulating valve,** expanding it, compressing the valve spring and opening the regulating valve a proportional amount. A pressure of one pound will move the valve 1/16th of an inch. With a dropping off in the steam demand, the boiler water level tends to rise, and cold water from the water storage leg rises into the inner generator tube. This cold water plus radiation condenses the steam in the annular space thereby reducing the generator pressure

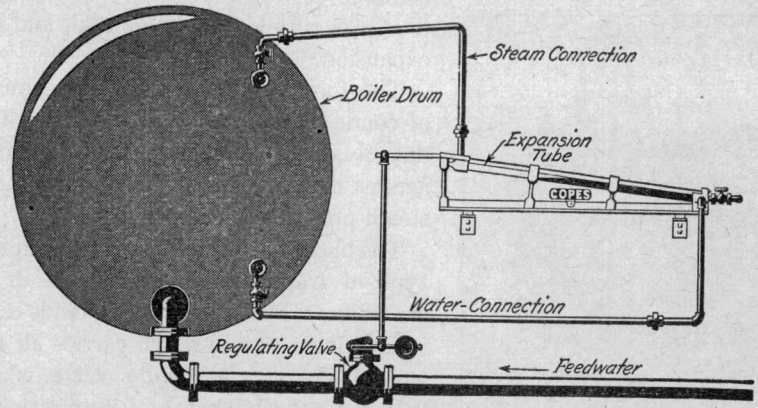

FIG. 164. — Copes Expansion Type Feedwater Regulator.

until the water levels are again equal. This reduction in volume of steam and water in the closed system, permits the regulating valve spring to force water back to the generator, and partially close the regulating valve. Feedwater regulators of this type are sometimes arranged with a vertical and inclined thermostat so connected that the feedwater level is controlled by a combination of steam flow and water level movement.

The **Copes regulator,** Fig. 164, is a simple type of regulator depending for its operation upon the expansion and contraction of an inclined tube. This tube is connected to the steam and water spaces of the boiler, in such a manner that when the water is at its lowest level it contains only steam and then has its maximum length. As the water level rises, water rises in the tube, which then contracts, the length of the tube depending upon the level to which it is filled with water. The tube is connected to a balanced valve in the feedwater line by a system of levers which move the valve as the height of the water level changes. With low water level, the tube is relatively long and the valve nearly open, thus giving maximum rate of flow; as the water rises, the tube is shortened and the rate of flow decreased.

233. High-pressure Drip Piping. — This piping automatically returns the steam condensed in the headers, steam separators, or other high-pressure piping to the boiler or feedwater heater, by means of **steam traps, pumps,** or the **steam loop.** The size of header drips is usually from $\frac{3}{4}$- to 1-inch pipe, and that of throttle and engine drips $\frac{1}{2}$-inch pipe. *A check valve should be located in each drip when two or more drips are connected to the same trap.* In plants using superheated steam, the drips are often wasted.

234. Steam Traps. — *Steam traps collect the water of condensation from steam apparatus, and automatically discharge it to a tank or hot well, with minimum loss of steam.* There are many types of steam traps, the most common of which are: (1) bucket, (2) float, (3) tilting, (4) differential, and (5) expansion.

The force of gravity causes the water of condensation to flow to the trap. Ordinarily, a steam trap will discharge against a pressure 5 pounds less than the steam pressure in the trap.

The **bucket trap,** Fig. 165, is the simplest type of trap, and is intermittent in its operation. It consists of an outside casting having a cover, which carries all the working parts. A **needle valve** is attached to the **bucket,** and its position is determined by the position of the bucket.

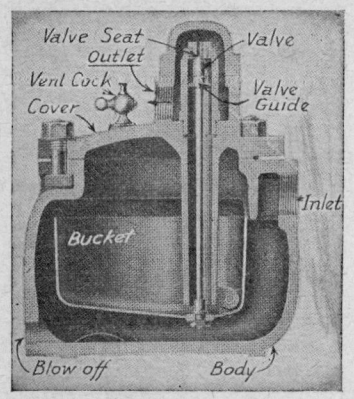

FIG. 165. — Bucket Steam Trap.

Water of condensation and steam enter the trap at the right and pass into the body of the trap. The water gradually rises, thus raising the bucket and closing the discharge valve. This valve remains closed until water rises in the body of the casting and overflows into the bucket in sufficient quantity to make the weight of the bucket, plus the water, great enough to sink the bucket and open the discharge valve wide. Steam pressure, acting upon the water within the bucket, then forces the water up through the sleeve which surrounds the valve and out at the left. This action continues until enough water is forced from the bucket to lighten the bucket sufficiently to permit it to rise and close the valve again. A **vent cock** in the cover permits removal of air in case the trap becomes air bound.

A **float trap** is shown in Fig. 166, with all parts labeled. Water and steam enter at the inlet, and any sediment in the water is removed by the strainer. The water rises in the chamber until it seals the valve with 3 inches or more of water, the height of which is indicated by a gage glass. The float is then raised by the rising water, and the discharge valve is opened. The rate of discharge depends upon the position of the float, which is controlled by the amount of water entering. The valve is under water at all times. A small cam is provided to permit by-passing the trap when necessary. The float is **counter-weighted** to permit using a smaller float. An **air valve** at the top is provided for removing air from the trap, which will not work when air collects in the chamber. This type of trap gives a continuous discharge; but if it is not well made, leakage of steam is liable to occur. The high velocity of discharge also wears the valve seat.

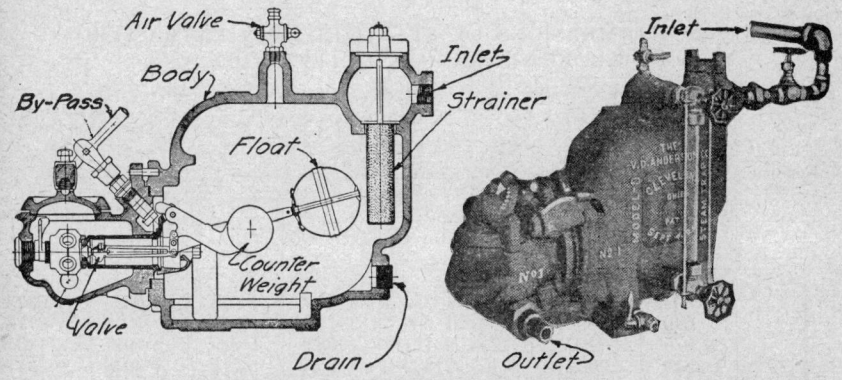

Fig. 166. — Anderson Float Trap.

The **tilting trap** is essentially a steel tank swung on a brass trunnion between two supports which are secured to a cast-iron bed plate. The tank is normally held horizontal by means of a weighted lever. When sufficient water has entered the tank to overcome the action of the lever and weight, the tank tilts backward. *This opens the discharge valve, and steam pressure forces the water from the tank until the weight on the lever again returns the tank to a horizontal position.* The weight is adjusted to allow sufficient water to remain in the tank to keep the outlet always water-sealed.

235. Return Traps. — The traps described above are of the **non-return type,** as the water is not returned to the boiler. There are traps that are located above the boiler level; when the trap is full, live steam at boiler pressure is admitted automatically to the trap. The pressure in the boiler and trap is thus equalized, and the water flows from the trap to the boiler by gravity. This type of trap is known as a **return trap.**

Steam traps are usually rated by the weight of water they will discharge per hour.

236. Commercial Piping. — Commercial pipe used in power plants should conform to standard specifications, such as those of the **American Standards Association** and the **American Society of Testing Materials.** It is made of the following materials: (1) wrought iron, (2) mild steel, (3) cast iron, and (4) brass.

Cast-iron pipe is used mainly for water and sewage systems; wrought-iron and steel pipe is used principally for steam, air and oil. *The size of pipe is stated in nominal inside diameter, up to and including 12-inch pipe. Above 12 inches, the size is based on the outside diameter, the thickness of metal must be always given, and the pipe is known as* **O. D. pipe.** The grades of steel and wrought-iron pipe in general use are (1) **standard,** (2) **extra strong,** and (3) **double extra strong.** Table 23 gives the principal dimensions and areas of standard and extra-heavy pipe. *The outside*

TABLE 23. — DIMENSIONS OF STANDARD AND EXTRA-STRONG WROUGHT-IRON AND STEEL PIPE

Nominal Size	Diameter			Circumference			Internal Transverse Area	
	External Standard and Extra Heavy	Internal		External Standard and Extra Heavy	Internal		Standard	Extra Heavy
		Standard	Extra Heavy		Standard	Extra Heavy		
1/8	0.405	0.269	0.215	1.272	0.848	0.675	0.0573	0.0363
1/4	.540	.364	.302	1.696	1.144	.949	.1041	.0716
3/8	.675	.493	.423	2.121	1.552	1.329	.1917	.1405
1/2	.840	.622	.546	2.639	1.957	1.715	.3048	.2341
3/4	1.050	.824	.742	3.299	2.589	2.331	.5333	.4324
1	1.315	1.049	.957	4.131	3.292	3.007	.8626	.7193
1¼	1.660	1.380	1.278	5.215	4.335	4.015	1.496	1.287
1½	1.900	1.610	1.500	5.969	5.061	4.712	2.038	1.767
2	2.375	2.067	1.939	7.461	6.494	6.092	3.356	2.953
2½	2.875	2.469	2.323	9.032	7.753	7.298	4.784	4.238
3	3.500	3.068	2.900	10.996	9.636	9.111	7.388	6.605
3½	4.000	3.548	3.364	12.566	11.146	10.568	9.887	8.888
4	4.500	4.026	3.826	14.137	12.648	12.020	12.730	11.497
4½	5.000	4.506	4.290	15.708	14.162	13.477	15.961	14.454
5	5.563	5.047	4.813	17.477	15.849	15.121	19.990	18.194
6	6.625	6.065	5.761	20.813	19.054	18.099	28.888	26.067
7	7.625	7.023	6.625	23.955	22.063	20.813	38.738	34.472
8	8.625	7.981	7.625	27.096	25.076	23.955	50.040	45.664
9	9.625	8.941	8.625	30.238	28.089	27.096	62.776	58.426
10	10.750	10.020	9.750	33.772	31.477	30.631	78.839	74.662
11	11.750	11.000	10.750	36.914	34.558	33.772	95.033	90.763
12	12.750	12.000	11.750	40.055	37.700	36.914	113.098	108.43

NOTE. — Dimensions are normal and are in inches.

diameter of standard, extra-heavy, and double extra-heavy pipe is the same for any given size; the inside diameters are different. Most piping comes in random lengths of from 16 to 20 feet and may or may not be threaded, the threading being done by means of **dies.** The type of die required for wrought iron does not give satisfactory threads when used on steel pipe, because it tears the threads.

The standard system of pipe threads is the **Briggs system.** The thread is given a taper of ¾-inch per foot, thus making it possible to secure a tight joint when screwing the pipe into a fitting.

237. Expansion. — *Whenever water, steam, or gas at high temperatures is to be conveyed by piping, proper provision must be made for expansion and flexibility.* With increased steam pressures and temperatures, the stresses and reactions caused by thermal expansion of piping has caused increased attention to be given to the details and methods of designing. The movement between pipe lines may be taken up by **expansion joints, swing joints,** and **pipe bends.**

238. Expansion Joints. — The slip type of expansion joint, Fig. 167, may be a single or a double slip joint. It consists of a main casting, or

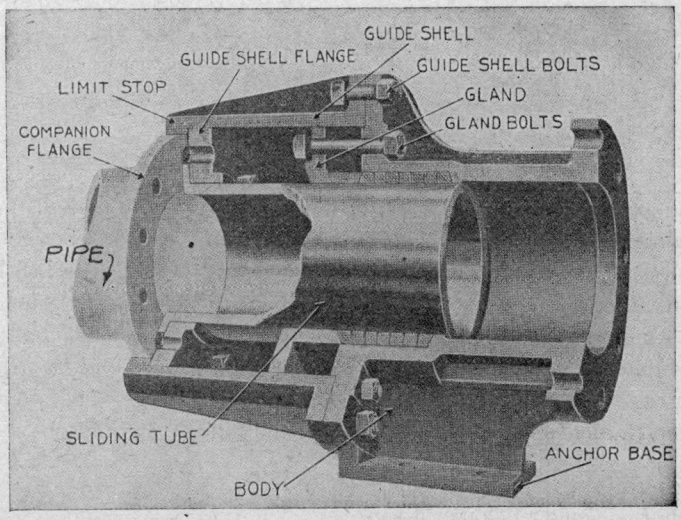

FIG. 167. — Howard Guided Expansion Joint.

body, which is anchored, a bronze sliding tube which fits into this main casting and a **gland and packing** placed around the tube. The gland is adjusted to prevent leakage, by gland bolts. The pipe line must be held in line, in order that this joint may work properly. Expansion and contraction move the sleeve in the main anchored casting. The number of expansion joints installed in a pipe line depends upon (1) the amount and direction of expansion and (2) the amount of expansion permitted by each joint.

In addition to the slip joint, a **corrugated copper expansion joint** is frequently used. It consists of a cylinder of corrugated copper held between two flanges. When used for high pressures the corrugations are reinforced with steel rings.

239. Pipe Bends. — Several typical expansion bends are shown in Fig. 168. The amount of expansion provided for by each bend depends upon the radius of the bend and the thickness of the pipe. Increased flexibility can be obtained when using pipe bends by creasing the metal on the inside of the bend. These **creased bends** permit the use of thinner walled pipe and bends of shorter radii. The bend is placed at a suitable point in the pipe line and installed so that it will not act as a dam to obstruct steam or water flow. *Particular attention must be given to the proper drainage of the steam line when pipe bends are used.* The bend, when cold,

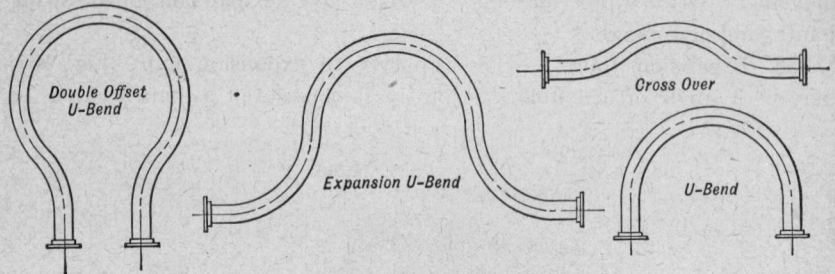

Fig. 168. — Typical Pipe Bends.

is customarily placed under initial tension equal to about one-half the total expansion; this decreases the amount of stress on the joint when hot. As an example of the amount of expansion of a typical bend: a 10-inch **U-pipe bend** will allow 3.2 inches expansion when bent to a radius of 80 inches.

240. Amount of Expansion. — *The amount of expansion in piping depends upon the material from which the piping is made and the difference in temperature between the pipe when hot and when cold.* The amount of expansion in inches may be calculated by the following equation:

$$\text{Expansion in inches}[1] = L \times C \times (t_1 - t_2) \times 12 \quad \ldots \quad (80)$$

in which L = length of pipe in feet.

C = coefficient of linear expansion per inch of length per deg. fahr.
  = .00001111 for bronze.
  = .0000068 for wrought iron.
  = .0000067 for steel.
  = .0000065 for cast iron.

t_1 = final temperature, deg. fahr.

t_2 = initial temperature, deg. fahr.

Example 33. — A steel steam pipe 293 ft. long carries steam at 190 lb. per sq. in. abs. pressure and superheated 125 deg. fahr. Find the theoretical amount of expansion on being heated from a room temperature of 80 deg. fahr.

Solution. — From Equation (80)

$$\text{Expansion in inches} = L \times C \times (t_1 - t_2) \times 12$$
$$= 293 \times 0.000{,}006{,}7 \times (502.6 - 80) \times 12$$
$$= 9.95 \text{ in.}$$

L = 293 ft.; C = 0.000,006,7; t_2 = 80 deg. fahr.

t_1 = 377.6 + 125 = 502.6 deg. fahr., where 377.6 is found from the Steam Table at 190 lb. abs. pressure.

[1] See also Equation for expansion in "Piping Handbook" by Walker and Crocker.

AMOUNT OF EXPANSION

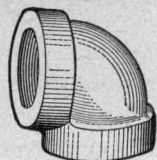
Straight Size Elbow

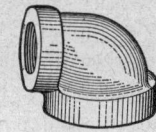
Reducing Elbow

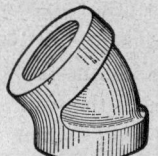
45° Elbow

Elbows

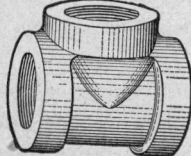
Straight Size

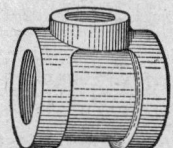
Reducing on Outlet

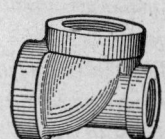
Reducing on Run

Tees

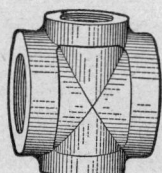
Crosses

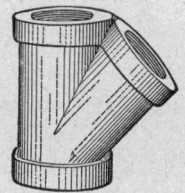
Y Branch or Lateral

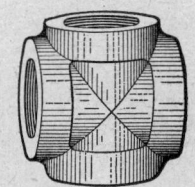
Crosses

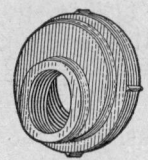
Reducer

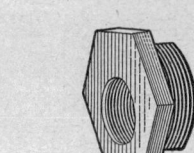
Bushing

Eccentric Fittings

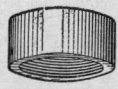
Cap, Plain
$\frac{3}{4}''$ and smaller

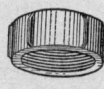
Cap, Ribbed
$1''$ and $1\frac{1}{4}''$ sizes

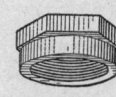
Cap, Octagon Head
$1\frac{1}{2}''$ and larger

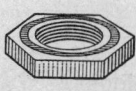
Lock Nut

Reducer

Caps

Plug

Counter Sunk Plug

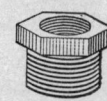
Bushing

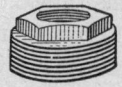
Bushing

Faced Bushing

Plugs

Fig. 169. — Cast-iron Fittings.

The actual expansion is somewhat less than the theoretical amount of expansion. It was formerly thought that the amount of expansion was about one-half of the theoretical, because the exterior surface does not reach the full steam temperature. It has, however, been demonstrated that, when the surface is well covered, the actual expansion is nearly the same as the theoretical expansion.

241. Pipe Fittings. — Commercial fittings used to join the various lengths of pipe are made in a variety of forms. *Screwed, or welded, fittings are generally used in sizes up to $3\frac{1}{2}$ inches, and flanged, or welded, fittings for larger sizes.*

The material from which fittings are made depends upon the type of service as regards pressure and temperature. Cast-iron, malleable-iron, steel, steel alloy, and brass fittings are common. Cast-iron fittings were formerly classed as: (1) **low-pressure fittings,** for pressures around 25 pounds per square inch; (2) **standard fittings** for 125 pounds per square inch; and (3) **extra-heavy fittings** for pressures of 250 pounds per square inch. It is, however, now common practice to refer to fittings as *"American Standard" followed by the name of the material and the safe working steam pressure*, thus: **American Standard cast-iron fitting for 250 pounds W. S. P.**

242. Screwed Fittings. — The various screwed fittings are illustrated in Fig. 169. **Nipples,** or short pieces of threaded pipe, **couplings, elbows, return bends, T's, crosses, laterals, Y-branches, bushings,** and **reducing fittings** are used to join pipe lengths.

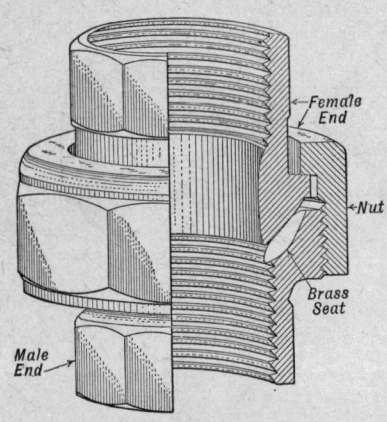

Fig. 170. — Screwed Union.

Plugs and **caps** are used to close the ends of fittings and pipes. Reducing bushings and fittings may be tapped eccentrically to permit free drainage.

Unions, Fig. 170, are used to provide a means of disconnecting a pipe line. They are seldom used above 4 inches nominal diameter. The parts of the union are the **nut, male end,** and **female end.** The seat is usually made of brass to prevent rusting.

Malleable iron and brass fittings have a round instead of a flat bead. Fittings having the thread on the outside are called **male fittings,** and those with the thread inside, **female fittings.**

In designating reducing fittings having more than two openings, it is customary to state first the run beginning at the larger end, and to end with the side outlet. Tees having a side outlet larger than the size of the run are known as **bullhead** tees.

FLANGED JOINTS AND FITTINGS

243. Flanged Joints and Fittings. — Pipe sizes above 6 inches are joined with flanges, which are made of cast iron or steel and are attached to the pipe in such a way that either a screwed, lap or welded flange joint is

FIG. 171. — Flanged Steel Fittings.

formed. *The grades of flanges as adopted by the American Society of Mechanical Engineers are as given in Art. 241, for fittings.*

Several flange fittings suitable for a working steam pressure (**W. S. P.**) of 150 pounds and a temperature of 750 deg. fahr. are shown in Fig. 171. A reducing flange is specified by stating: *first*, the smaller pipe and *next* the outside diameter of the flange, as a 10-inch by 19-inch flange. This means that a 10-inch line is to be connected to a 12-inch line, since a 12-inch standard flange has a 19-inch diameter.

The screwed joint made by screwing the pipe into the flange is satisfactory for medium working pressures, in sizes up to 12 inches. The **Vanstone** or **lap joint,** Fig. 172, is the best type of commercial joint for high pressure and temperature; it is dependable and requires no attention besides occasionally replacing a gasket. The pipe is rolled over against the flange at

FIG. 172. — Vanstone Joint.

right angles to the axis of the pipe. The lap is then faced on front and edge, and acts as a bearing for the gasket when making the joint, while the flanges being loose act as swivel collars to hold the pipe together. This flange is adaptable to all classes of service, for pressure up to 1000 pounds per square inch or above. A modi-

FIG. 173. — Ball Joint.

fication of the Vanstone joint, known as the **Sargol**, has the flanges smooth faced with the projecting lips welded together after erection. With this construction a gasket is not required. **Welded** flanges are made of forged steel welded to the pipe and afterward faced to a true face. In installations requiring a flexible connection **ball joints,** Fig. 173, are used.

244. Methods of Facing Flanges. — In order to make a tight joint the faces of the flanges are machined according to the American Standard, Fig. 174, as follows: (1) raised face, (2) tongue and groove, and (3) male and female.

The raised face is made by raising the face of the flange between the bore and inside of the bolt holes from $\frac{1}{32}$- to $\frac{1}{16}$-inch above the remainder of the flange. It makes a satisfactory joint, and is the most common type of joint for medium steam pressures. A ring gasket is used between the flanges, or it may be machined true for a ground joint.

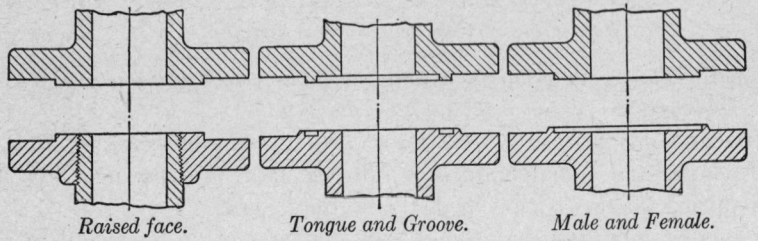

Raised face.　　　Tongue and Groove.　　　Male and Female.

Fig. 174. — Standard American Flange Facings.

The tongue and groove facing is made by machining a circular ring on one flange and a corresponding recess on the other. For ammonia and hydraulic pipe lines there is no better joint.

The large male and female is standard for stock material. It is adapted to high pressures, as it prevents the gasket from blowing out.

An undesirable feature of facings such as the male and female and tongue and groove is that the units must be shifted longitudinally when the joint is assembled or disconnected.

The finish of the contact surfaces varies with the ideas of the designer. In some a smooth surface is required and in others using gaskets spiral and concentric grooves are used. **Serrated joints,** having 16 grooves per inch, and **phonographic joints,** having 32 grooves per inch, have given satisfactory service under high pressures.

245. Welded Joints. — The practice of joining the various lengths of pipe by welding, using either the oxyacetylene or electric arc process, is gradually playing a more and more important part in the field of pipe fabrication. Welded pipe, when properly constructed, may be made to develop the full strength of the pipe material. *The results obtained, however, depend to a large extent on the skill of the welder.* The various advantages which result from the use of welded piping are: (1) elimination of screwed or flanged joints, which simplifies design and layout; (2) lower material and operating cost; (3) less weight to support; (4) continuity of insulation as a result of the elimination of flanges with resulting decrease of radiation losses; and (5) saving of time in erection where special and odd-shaped fittings are necessary.

Several common types of welds are shown in Fig. 175. The "butt weld" is generally used for making extra long lengths of piping required for headers, or where elimination of flanges in the pipe line is desirable. A modification of this joint in which a sleeve is placed over the butt weld,

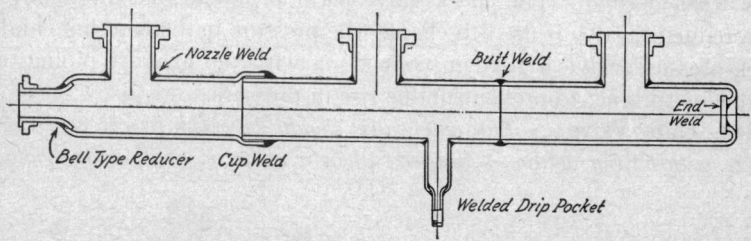

Fig. 175. — Illustrating various Types of Pipe Welds.

each end of the sleeve welded is known as a " sleeve or cup weld." It is used where unusual stresses occur as in pipe bends or where a reinforced type of weld is specified. This type of weld is required by standard specifications of the Power Piping Society to reinforce butt welds on all lines of pipe $\frac{1}{2}$-inch or more in thickness.

246. Gaskets. — For steam and water lines, where the pressure does not exceed 160 pounds per square inch, **wire insertion gaskets,** $\frac{1}{16}$-inch thick, give satisfactory service. For high pressures and superheated steam, **corrugated steel gaskets,** extending the full width of the flange inside the bolt holes, are satisfactory. **Asbestos-filled corrugated copper gaskets,** Fig. 176, are used for medium steam pressures, and special **sheet-rubber packing** for low steam pressures.

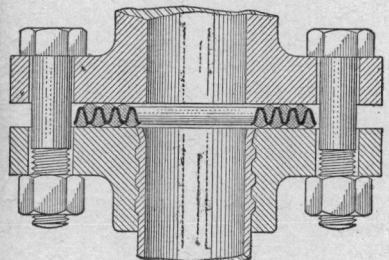

Fig. 176. — Corrugated Copper Gasket.

Canvas-insertion black-rubber gaskets may be used for low-pressure water lines, and wire-insertion rubber gaskets for high-pressure water lines.

When gaskets are used on oil lines, they must be made of special material to withstand the action of the oil.

247. Valves. — The valves used commercially may have either flanged or screwed ends. When flanged, the bolt spacing corresponds to the American Standard for flanged fittings. *The material used in small-sized valves, up to 3 inches, is usually brass; for large-sized valves, either cast iron or cast or forged steel, is used. The seats, disks, and spindles of all valves used with steam or water are made of brass. When used with acids or ammonia, valves are made of iron or steel throughout.* Valves are made for the same class of service as given for fittings, page 252. All valves above **6** inches should be by-passed.

248. Types of Valves. — The more common types of valves are **gate, globe, angle, automatic stop valves, check, reducing** and **back-pressure** valves. The first four are known as stop valves and are used to stop the flow when desired. The check valve permits flow in one direction only. The reducing valve is used to change the pressure in a steam line, and the back-pressure valve is used in connection with the exhaust piping of an engine or turbine, to prevent undue rise in pressure.

249. Gate Valve. — *The gate valve should be used in all cases except where a throttling action is desired; then a globe or angle valve should be*

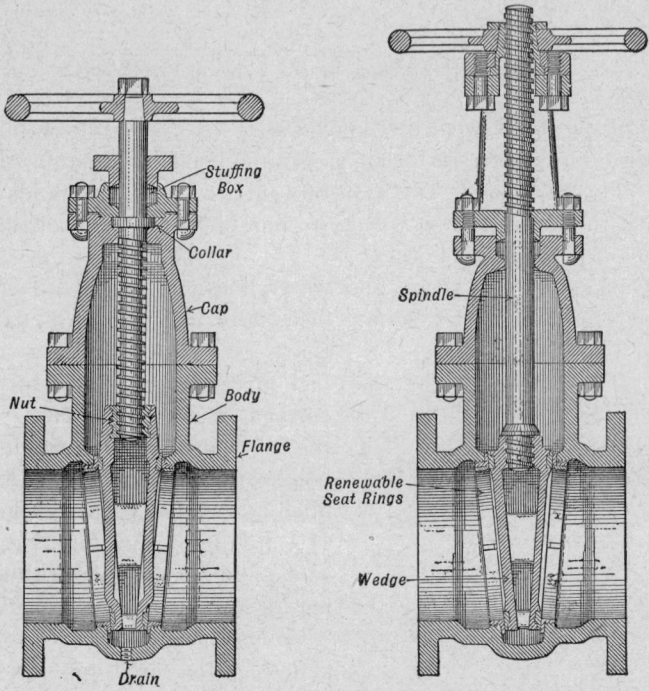

Fig. 177. — Non-rising and Rising Spindle Gate Valves with Iron Body and Brass Mountings.

used. This type of valve gives an unobstructed passage to the flow of gas or fluid.

The valve consists of a body, bonnet, wedge, seat, spindle or stem, stuffing box, and handwheel. The body, bonnet, and wedge are generally made of cast iron, and the seat and spindle of brass or bronze, to prevent corrosion. The spindle may be of the rising or non-rising type, as shown in Fig. 177. The **rising spindle** type has the wedge attached directly to the spindle, which is threaded into a yoke; while in the **non-rising spindle** type the spindle screws into the wedge as the wheel is revolved, in which case it is not apparent to the eye whether the valve is closed or open. The wedge may be solid or split, and should have a slight taper, so that it will close tight when pushed down by the spindle. The top of the wedge is made to seat against the bonnet when wide open, which permits packing the stuffing box with the valve open.

Small sizes up to 3 inches are made of brass and usually have screwed ends, Fig. 178. The bonnet, which is screwed to the body by a **union nut,** carries a stuffing box through which the spindle rises and to which the gland is held by a nut. A wheel is attached to the top of the spindle, and a shoulder and hub on the lower end of the spindle turn in a split wedge which seats against the tapered faces. The top of the shoulder is finished and, when open, bears against the bottom of the bonnet and permits packing the valve under pressure.

250. Globe and Angle Valves. — Large sizes of these valves are made of cast iron with brass mountings. The construction is similar to that of the gate valve, except that the seat is horizontal, and the valve disk is attached to the end of the spindle, which is of the rising type, Fig. 179. A yoke guides the spindle. This valve should be placed in a horizontal steam line with the spindle horizontal, to prevent the collection of water at the horizontal valve seat. *When closed, the steam pressure should always be beneath the disk.* The spindle is guided at its lower end to seat the disk squarely. *The disk is made of rubber composition for cold water, babbitt metal for hot water, copper or bronze for high-pressure steam, and nickel for superheated steam.* A recent valve of this type has a cylindrical piston or plunger, sliding vertically through two special packing rings, separated by a bronze cage, in which are ports through which the steam flows when the valve is open.

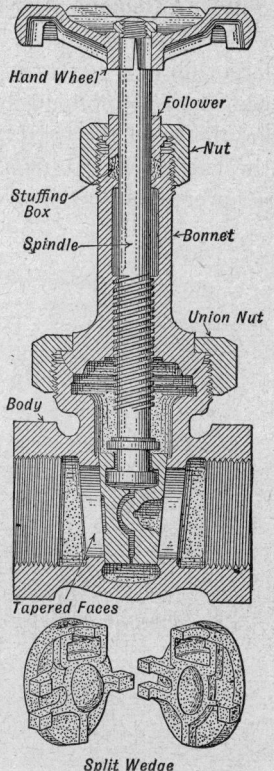

Fig. 178. — Brass Gate Valve with Union Bonnet and Split Wedge.

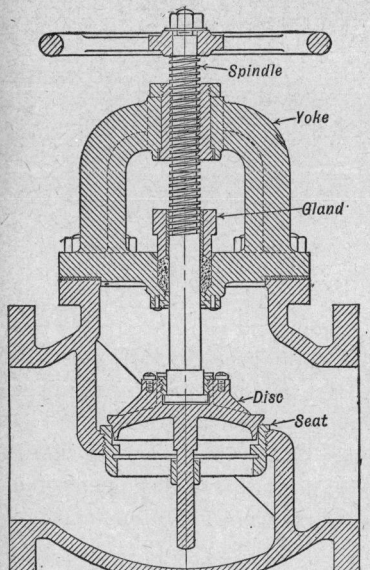

Fig. 179. — Large Size Globe Valve.

Small sizes up to 4 inches are made of brass with screwed ends, Fig. 180. The disk is made removable, so that it may be easily replaced.

251. Automatic Non-return Stop Valve. — The Foster non-return stop valve, shown in Fig. 181, is used to equalize the pressure between the different units of two or more boilers which are arranged in battery and connected to a common header. In case an accident lowers the pressure in any boiler of the battery, the valve disk, which slides in a guide fitted in the valve body, automatically closes, because the pressure below the disk is lower than on the header side. The disk remains closed as long as the

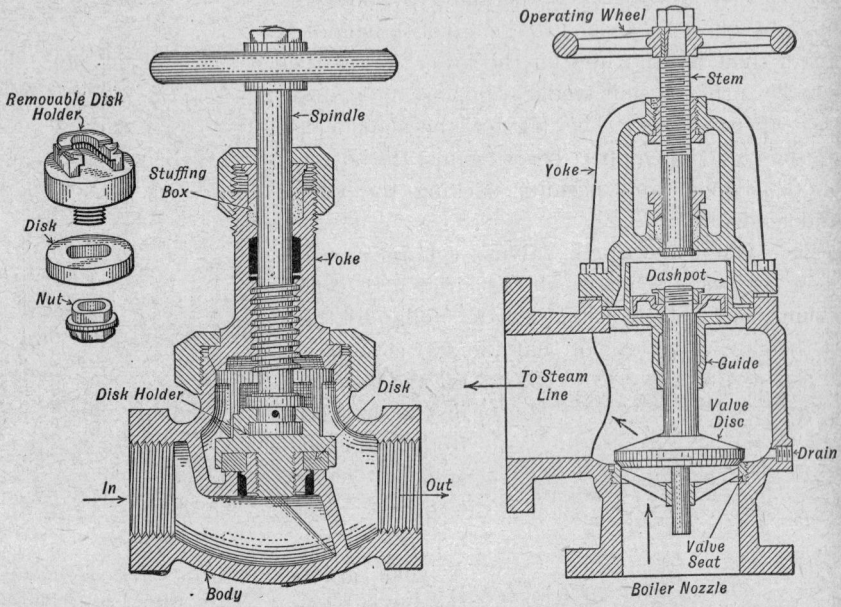

Fig. 180. — Powell Brass Globe Valve showing Method used to hold Disk.

Fig. 181. — Non-return Stop Valve.

pressure in the boiler is below that in the header and can only be opened with pressure on the boiler side. The valve should not stick, chatter, or hammer while performing its function. A dash pot is fitted to the upper end of the disk spindle to prevent chattering or violent movement. The outside stem is not attached to the valve disk, but the valve can be closed like any stop valve by screwing the handwheel down. The disk opens on a difference of pressure of one pound. Valves of this type are often arranged to close by steam pressure when desired.

252. Reducing Valve. — This valve is used where it is desired to reduce the steam pressure, as for use in a heating system. Fig. 182 illustrates the construction of a weight-and-lever type of reducing valve, consisting of a body connected into the pipe line. In the body is a balanced disk valve attached to a spindle, to the lower end of which a diaphragm is attached; just above the diaphragm is a lever, which is pivoted on an extension of the valve body. High-pressure steam enters from the left and passes through the balanced valve, which is normally held open by a weight located on the lever. The low-pressure side of the valve is connected by a pipe to the chamber below the diaphragm, in order to have the low-pressure steam act on the under side of the diaphragm. When the steam pressure acting on the diaphragm builds up sufficiently to lift the weight, the spindle is raised, the balanced valve partially closed, and the pressure reduced to a value that will just balance the weight.

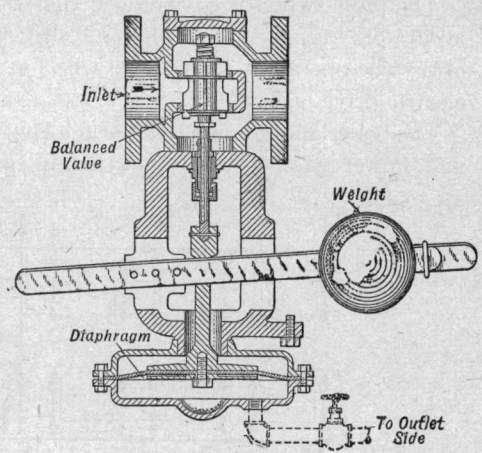

Fig. 182. — Kieley Reducing Valve.

The position of the weight determines the lower pressure.

253. Automatically Controlled Valves. — The increasing use of high pressure and superheat in modern power plant practice has made it necessary to have a safe and positive means of controlling the operation of the steam lines and headers. This is now being done by using solenoid, electric motor, air motor, and hydraulic motor-controlled valves, which provide safety in cases of emergency, as well as rapid and convenient control under normal operating conditions. This type of control makes it possible to operate all valves from a central control board on which red and green lights show the exact position of every valve throughout the plant. One type of motor controlled valve is shown in Fig. 183, and consists of driving motor, reduction gears, limit switch mechanism, and a pawl device, the whole being mounted in a single case. The driving motor is connected to a system of combination worm and planetary gearing for the necessary reduction of speed.

Fig. 183. — C-H Valve Control Unit.

The normal speed of the motor is 4000 revolutions per minute, while that of the slow-shaft is 100 revolutions per minute.

The limit switch is so arranged that the amount of valve travel can be accurately adjusted to eliminate drift or overtravel. The valve can always be operated by the hand wheel at any time without disturbing the electric drive or control.

The boiler and piping layout of the Hell Gate Station of the United Light and Power Co., in which electrically operated valves are employed, is

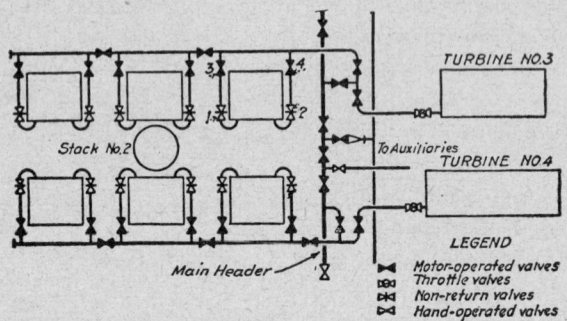

Fig. 184. — Boiler Piping Layout.

shown in Fig. 184, the layout being so arranged that the boiler-lead valves can be closed separately, and the main and branch-header valves can be used for sectionalizing.

254. Pipe Coverings. — *Coverings made from heat insulating material, such as magnesia or asbestos fiber, should be placed on all air, water and steam piping, valves and fittings, in order to prevent the loss from condensed steam as a result of radiation from uncovered steam lines.* For instance, with a 10-inch pipe and the following conditions in a plant, the loss per square foot of bare pipe in dollars amounts to $62.80 per month; steam pressure 125 pounds and temperature 350 deg. fahr.; air temperature 70 deg. fahr.; boiler efficiency 70 per cent and heat value of the coal 13,000 B.t.u. The amount of heat saved, by using pipe covering, may often be as high as 85 per cent of the heat lost from uncovered pipe. The economical thickness of the covering depends on the external temperature and the cost of steam. The main steam line in the Hell Gate Station is covered with 4 inches of 85 per cent magnesia. Cold-water pipes are often covered, to prevent sweating in hot weather.

255. Pipe Hangers, Rollers, and Supports. — Pipe hangers, Fig. 185, are used to relieve fittings of strain and keep a pipe line from sagging. They are ordinarily attached to the framework of the building, and are free to swing as the pipe expands. In addition to hangers, pipes are supported on rollers which permit the pipe to expand and contract, and which are carried by a roller frame attached to a bracket located on the wall or frame of the building. When it is necessary to support pipes at a considerable distance from the floor, a pipe column is used.

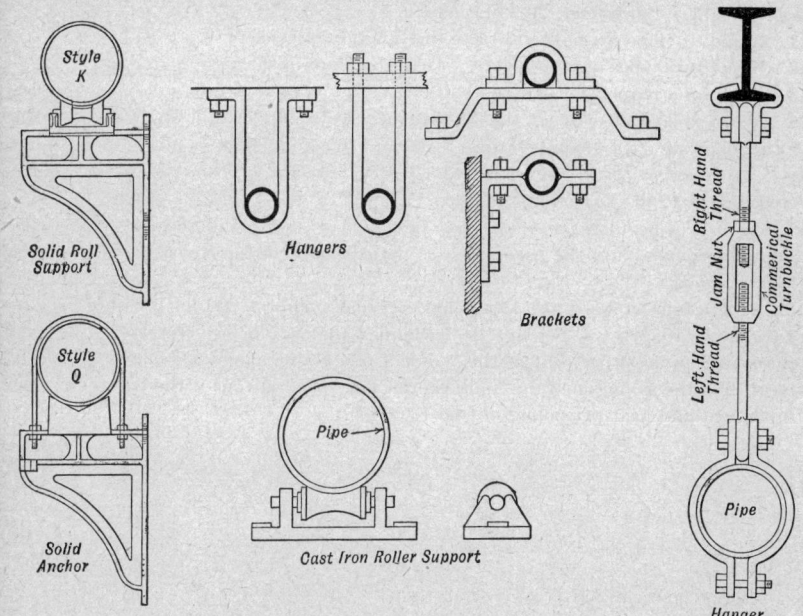

FIG. 185. — Typical Pipe Hangers, Brackets and Roller with Frame.

REFERENCES

Steam Power Plant Engineering, GEBHARDT.
Mechanical Equipment of Buildings, Vols. I, and II, HARDING and WILLARD.
Mechanical Engineers Handbook, MARKS.
Piping Handbook, WALKER and CROCKER.
Handbook on Piping, SVENSON.
Power Plant Engineering, FERNALD and ORROK.
Steam, BABCOCK and WILCOX Co.

REVIEW QUESTIONS AND PROBLEMS

1. Name five requirements which a well-designed pipe line should satisfy.
2. Sketch the layout of a single-header system of piping.
3. Sketch a proper arrangement of piping for a boiler lead.
4. A 1700 i.hp. engine, using 12.5 lb. of steam per i.hp. per hr. has a supply pipe 40 ft. long from the header. Average steam pressure, 160.3 lb. per sq. in. gage; average amount of superheat, 40 deg. fahr. Pressure drop from header to engine, 1.0 lb. What should be the diameter of the supply pipe?
5. Describe a typical feedwater piping system, using sketch.
6. What is a steam trap? Name four types and describe one type mentioned.
7. What is a return trap?
8. Name the grades of steam piping.
9. Name three methods of providing for the expansion of a steam line.

262   PIPE SYSTEMS, PIPE, VALVES, AND PIPE ACCESSORIES

10. A 5-in. steel pipe line, 600 ft. long, carries steam at an average pressure of 155.3 lb. gage, and average superheat in the line, 50 deg. fahr. Temperature before turning on steam, 50 deg. fahr. Find the allowance that must be made for expansion if the pipe is straight. Barometer, 29.92 in. mercury.

11. Name ten pipe fittings, and state the function of each.

12. Name two types of stop valves. Describe a non-return stop valve.

13. Describe a reducing valve and state why such a valve is used.

14. A pipe-insulation test at the Milwaukee Electric Light and Railway Company gave the following data: temperatures — steam, 365.2 deg. fahr., room, 83.9 deg. fahr.; areas — covered, 58.53 sq. ft., bare, 13.3 sq. ft.; weight of condensate per hour — covered pipe, 4.42 lb., bare, 14.03 lb.

Find: (a) The heat loss per square foot, per hour, per degree difference in temperature for both the covered and the bare pipe, (b) the efficiency of the covering which was 85 per cent magnesia.

15. A bare 3-in. steam pipe, 1 ft. long, carrying steam at 200 lb. pressure will lose approximately 1052 B.t.u. per hr. It is claimed that a 2 in. covering of 85 per cent magnesia will save 90 per cent of this. For a pipe 80 ft. long and operating 7200 hr. per year, find the saving in B.t.u. and also the coal saved in tons if the boiler efficiency is 70 per cent and heat per pound of coal 13,500 B.t.u.

CHAPTER XI

HEAT SAVING EQUIPMENT — FEEDWATER HEATERS, ECONOMIZERS, SOOT BLOWERS, AND STATION HEAT BALANCING

256. Foreword. — In addition to the heat-saving equipment discussed in Chapter V, the efficiency of the boiler plant can be increased by increasing the temperature of the boiler feedwater using feedwater heaters. *The heating may be done by live steam from the boiler, exhaust steam from the engine or turbine, or steam extracted, i.e. bled, from one or more stages of the main turbine.* When using live steam, the heater is often called a **purifier.** The resulting gain in efficiency of the boiler is about 1 per cent for each 10 to 12 deg. fahr. increase in the feedwater temperature when operating at medium boiler pressures, the greatest gain occurring when the feedwater is heated to approximately the temperature of saturated steam in the boiler. Incidentally, increasing the temperature of the feedwater reduces strains caused by difference in temperature in the parts of the boiler, prolongs the life of the boiler, and increases its steaming capacity. With large heaters the storage of water is an additional advantage, since a considerable quantity of hot water may be ready to supply a sudden demand.

A further increase in efficiency may be made by reclaiming excess heat from the flue gases by either an **economizer** or **air preheater.** This is increasingly important because of the trend to higher boiler ratings and the heating of feedwater by stage bleeding in steam turbine plants. The increased temperatures of the flue gases, which are the result of high boiler ratings, cannot be fully utilized in an economizer, this makes profitable the further recovery of heat from the flue gases by using an air preheater. The maximum gain is obtained when the air for combustion is heated to a temperature approaching that at which the flue gases leave the boiler.

257. Classification of Feedwater Heaters. — In addition to the classification by source of heat, as mentioned in the previous article feedwater heaters may be classified according to, (1) method of transmission of heat (contact or conduction), (2) pressure of steam, (3) arrangement for flow of the steam.

(1) If steam comes directly into contact with the water, the heater is an **open-type heater.** If heat is transmitted through walls which separate the steam from the water, the heater is of the **closed type.**

(2) Closed heaters, through which the exhaust steam from an engine passes on its way to the condenser, are called **primary heaters,** because

cold feedwater sometimes receives heat first in this way. Steam exhausted at or near atmospheric pressure is sometimes used to heat water in open or closed heaters, which are called **secondary, or atmospheric, heaters.**

(3) The heater is of the **induced type** when only that portion of the steam which is induced enters the heater, but all the steam entering is condensed, thus lowering the pressure slightly and drawing more steam in. It is called a **through heater** when all of the steam is forced through the heater and only a part of the steam is condensed, the remainder being discharged.

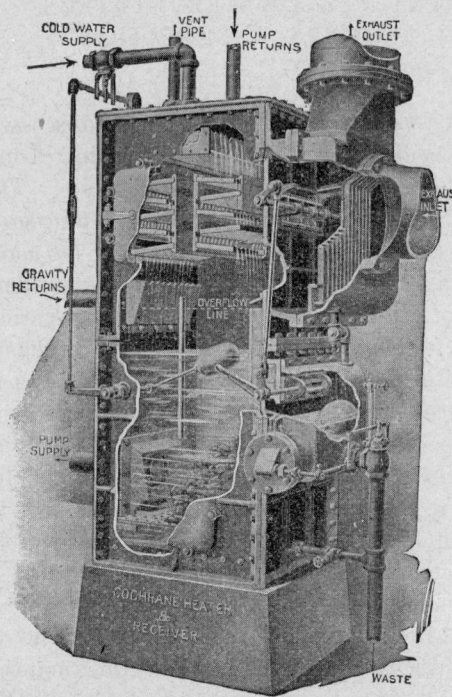

Fig. 186. — Cochrane Open Type Steam Stack Cut-out Feedwater Heater.

258. Open Feedwater Heater. — A section of a Cochrane open feedwater heater and receiver, with part of the shell cut away to give a view of the interior, is shown in Fig. 186. It is made of cast-iron plates bolted together, is provided with doors to allow cleaning, and is intended for pressures as high as 10 pounds per square inch.

This heater is called a "steam stack cut-out" heater and is arranged for use when more exhaust steam is available than is required for heating the feedwater.

Part of the steam enters the heater through an opening controlled by a **cut-out valve,** the remainder passing directly to the exhaust outlet. A large oil-separator, in the exhaust inlet to the heater, removes the oil from the exhaust steam.

Cold water enters a reservoir at the top and overflows to a series of **corrugated copper trays,** or pans, and thence to the filter compartment below. While falling from the reservoir and pans, the water is so subdivided into sprays that it offers a large amount of surface to the entering steam, which is immediately condensed by direct contact. The quantity of cold water entering the heater is controlled by a **float** in the tank, which operates an inlet valve in the supply pipe.

The hot water passes down through the **coke filter** and enters the compartment at the left, where a section of the partition is shown cut away to show the outlet to the feed pump. The heater is flooded to the overflow level, and thus floating impurities pass off into the **float trap** at the right and are discharged to the waste pipe, by action of the trap. The discharge from the oil-separator is also removed by this trap.

A considerable amount of the soft scale-making material in the water is deposited on the trays and may be removed by opening the cast-iron doors of the heater and taking out the trays.

This heater is located on the suction side of the boiler feed pump and *serves as a* receiving tank or *hot well*, to which hot condensate is returned by gravity or by means of a pump. A **vent pipe** is provided to discharge

FIG. 187. — Hoppes Horizontal Open Type Feedwater Heater.

the gases that are released and to ensure low pressure in the heater. Since the amount of steam mixed with the vented gases may exceed 1 per cent of the weight of water heated, the heat in this steam is often recovered by condensing it in a surface type **vent condenser**.

The **Hoppes horizontal open feedwater heater,** Fig. 187, illustrates the pan type of heater. The water entering through the supply pipe is controlled by a float valve. It overflows from pan to pan, receiving heat from the exhaust steam, and is drawn away by the pump, which is generally placed below it. The exhaust steam passes over an oil-separator before entering the space between the trays. After considerable scale has accumulated the heater head can be removed and the trays taken out and cleaned. A small crane is provided to assist in swinging the head out of the way and back into position.

Open heaters may also be of either the **deaërating type**[1] or the **jet type**.

[1] See Art. 271, page 282.

The former are similar to the standard type of open heaters, but are arranged to heat the water to full steam temperature, and to eliminate the dissolved gases, including oxygen and carbon dioxide. The latter are similar to the jet condenser, page 560, and are used where large capacity is desired.

259. Closed Heater. — This type of heater is connected into the feed line between the pump and the boiler. Such heaters are classified, according to the position and direction of movement of the steam and water, as (1) water-tube, or (2) steam-tube, either type having parallel currents or counter currents; that is, with steam and water moving in the same direction or in opposite directions. In **single-flow** water-tube heaters the water flows in one direction only; in a **multi-flow** heater the water passes back and forth in its passage through the heater tubes. A **coil heater** has one or more coils of pipe carrying the water.

A vertical closed heater of the multi-flow water-tube type, cut away at the top and bottom to show its construction, is illustrated in Fig. 188. The shell is made of cast iron and the tubes of corrugated copper, expanded in the tube heads and supported by a thimble. The corrugations increase the heating surface, and, by agitating the water in its passage, tend to bring cooler particles of the water stream against the heating surface, and thus increase the efficiency of heat transfer. The irregular movement of the water and the movement of the tube by expansion and contraction also tend to prevent deposit of scale on the tubes. The upper tube head is made **floating** to allow expansion and contraction of the tubes with change in temperature. Exhaust steam enters against a baffle plate, used to protect the tubes from impact of the steam, and is distributed to the top and bottom. Condensed steam is discharged through the **drain pipe**

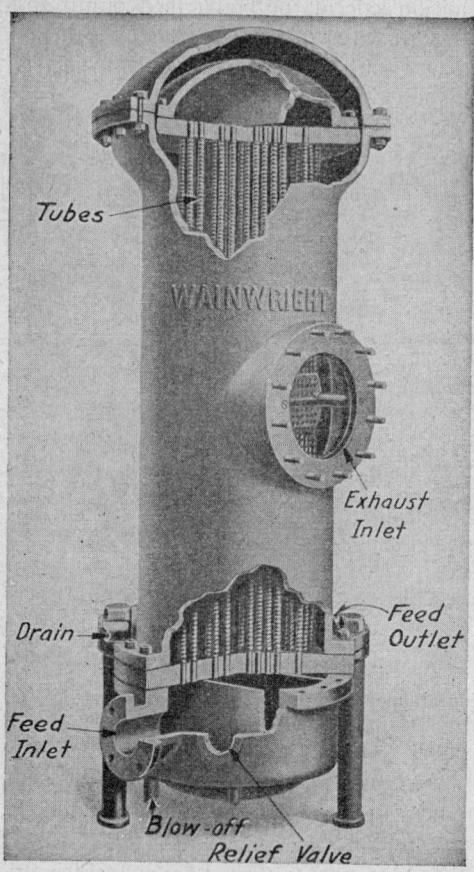

Fig. 188. — Wainwright Closed Feedwater Heater.

shown at the left of the figure. Feedwater enters near the bottom, passes to the left top compartment, thence to the center bottom and right top compartments, and down to the feed outlet. There are two **partition plates** at the bottom and one at the top to direct the flow. A blow-off connection is shown and, as this heater is under boiler pressure or slightly higher, when the pump is working, provision is made for connecting a relief or safety valve.

A spiral coil is sometimes used, in straight brass tubes, for agitation of the water. This type of tube is expanded into one tube sheet and pro-

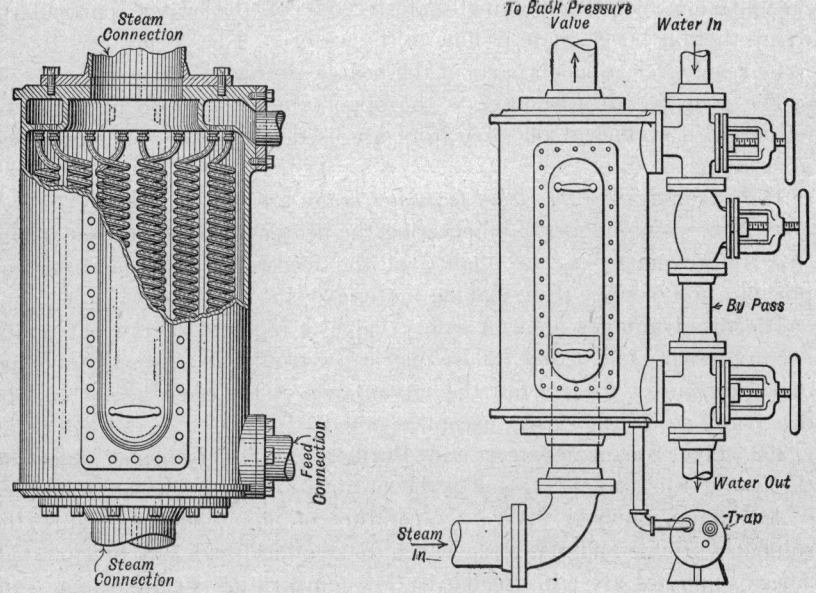

FIG. 189. — Reilly Multi-coil Exhaust Steam Heater and Piping Connections for through Type of Heater.

vided with a deep packing box and screwed gland in the other tube sheet, to allow for expansion and contraction.

260. Coil Tube Heater. — A multi-coil heater is illustrated in Fig. 189, in which a Reilly through-type of closed heater, with part of the cast-iron shell removed, shows the coils and their connections to the manifolds. The coils are interchangeable and are connected to the headers at top and bottom by metal screwed union joints. Expansion is taken care of by the form of the coils.

The arrangement of piping for a through-coil heater is also shown in Fig. 189. The pressure of the steam passing through the heater is kept constant by a weighted back-pressure valve. A trap is connected as shown, to receive the steam condensed in the heater.

261. Advantages and Disadvantages Claimed for Open and Closed Types of Feedwater Heaters. — Advantages *of the open type feedwater heater* are:

1. Direct contact between steam and water which gives efficient heat transmission and requires less steam to heat the water through a given temperature range.
2. A saving in water because the condensed steam can be used instead of an equal quantity of cold water.
3. Improvement in the quality of the feedwater since the temperature in the heater is sufficient to liberate gases and air in the feedwater and precipitate some scale-making material out of the water, particularly where there is bicarbonate of lime in the water.
4. Easy to clean as all parts of the heater are readily accessible.

The principal **disadvantage** is the opportunity for oil to get into the feedwater if inefficient oil separators are used, or if care is not exercised in cleaning the filter.

The **advantages** *of the closed feedwater heater* are:

1. There is no danger of oil entering the heater from the exhaust steam.
2. Improvement in the quality of the feedwater resulting from the precipitation of some scale-making material.

The **disadvantages** are: A safety valve is required to prevent undue rise in pressure, and a tube failure may cause trouble because the tubes are under pressure. In general, the advantages of the open heater are the disadvantages of the closed, except as noted.

262. Live Steam Heaters and Purifiers. — The principal object in the use of this type of heater is the purification of the water from hard scale-making material. At a temperature of about 300 deg. fahr., the sulphates of lime and magnesia, which make a hard scale, are precipitated. These sulphates are not affected by low-temperature steam. Some gain in efficiency in heating feedwater by the use of live steam heaters has been shown, an explanation for which was offered in *Steam*[1] by the suggestion that heat may be transferred more readily at considerable temperature differences.

The live steam purifier is ordinarily of the pan type and in construction resembles the Hoppes heater (Fig. 187, page 265). It is, however, operated at boiler pressure instead of at exhaust pressure. Water is delivered to the top pan and runs over the sides to the pans below; it then enters the boiler by gravity, because the location of the heater is above the water level. Air and gas expelled from the water must be vented to some high-pressure steam-pipe line, to prevent the heater from becoming air bound.

263. Fuel Saved by Heating Feedwater. — For the sake of economy, exhaust steam should be used for heating feedwater, wherever possible.

[1] Babcock and Wilcox Co.

The heat in the exhaust of an engine or turbine is a large proportion of the heat supplied. All the heat, except that used in work and lost by friction, radiation and leakage, can be returned to the boiler in the feedwater. If an engine using steam at 150 pounds per square inch gage pressure exhaust into a heater at atmospheric pressure, and the feedwater enters the heater at 50 deg. fahr. and leaves at 210 deg. fahr. the exhaust steam furnishes 159.93 B.t.u., which is about 13.5 per cent or nearly one-seventh of the total heat, 1176.8 B.t.u., required to evaporate one pound of water under the conditions. Thus the capacity of the boiler plant is increased without increasing the number of boilers; or it will be possible to furnish the same amount of steam with fewer boilers and less coal.

264. Fuel Economizers. — The use of an economizer results in a gain in the overall efficiency of the plant which varies from 4 to 20 per cent, depending on the operating conditions. Its use is of particular value in stations carrying a periodic or regular overload because it practically increases the heating surface and saves the heat that would otherwise pass to the stack at such times. The economizer serves as a feedwater purifier like any closed water-tube heater, and also reduces by a small amount the soot otherwise discharged with the gases. Since there is some reduction in draft caused by using an economizer, it is customary to use mechanical methods to obtain the necessary draft.

The **Green fuel economizer,** shown in Fig. 190, consists of a series of vertical cast-iron pipes 9 to 12 feet long and $4\frac{9}{16}$ inches external diameter, arranged in **parallel rows** across the flue and connected into cast-iron headers at top and bottom. The bottom headers are connected by flanged joints at the front side to bottom branch distributing pipes. The top headers are similarly connected at the rear to the top branch collecting pipes. The distributing and collecting pipes, for the groups of vertical pipes making up the economizer, are connected by U-shaped, cast-iron, flanged fittings to allow for expansion and contraction. The tubes have metal-to-metal taper fits into the headers at both ends and are forced into place by hydraulic pressure. Opposite the top of each tube end in the top header is a hand-hole, closed by a lid having a conical seat and held in place by a yoke and bolt. The lids are removed through one opening which is made larger than the rest.

Feedwater enters the bottom distributing pipe at the end nearest the point where the gases leave the economizer and, after passing through headers and pipes, leaves the top collecting pipe at the end where the gases enter, thus ensuring the hottest water for boiler feed. A safety valve is located on a flanged nozzle on the top branch pipe, next to the exit connection to the boiler-feed line. A bottom blow-off valve is connected to the bottom branch pipe to discharge sediment.

Scrapers, for removing the soot which falls to a space below the pipes, are continuously moved up and down the pipes by chains connected to a

270 HEAT SAVING EQUIPMENT

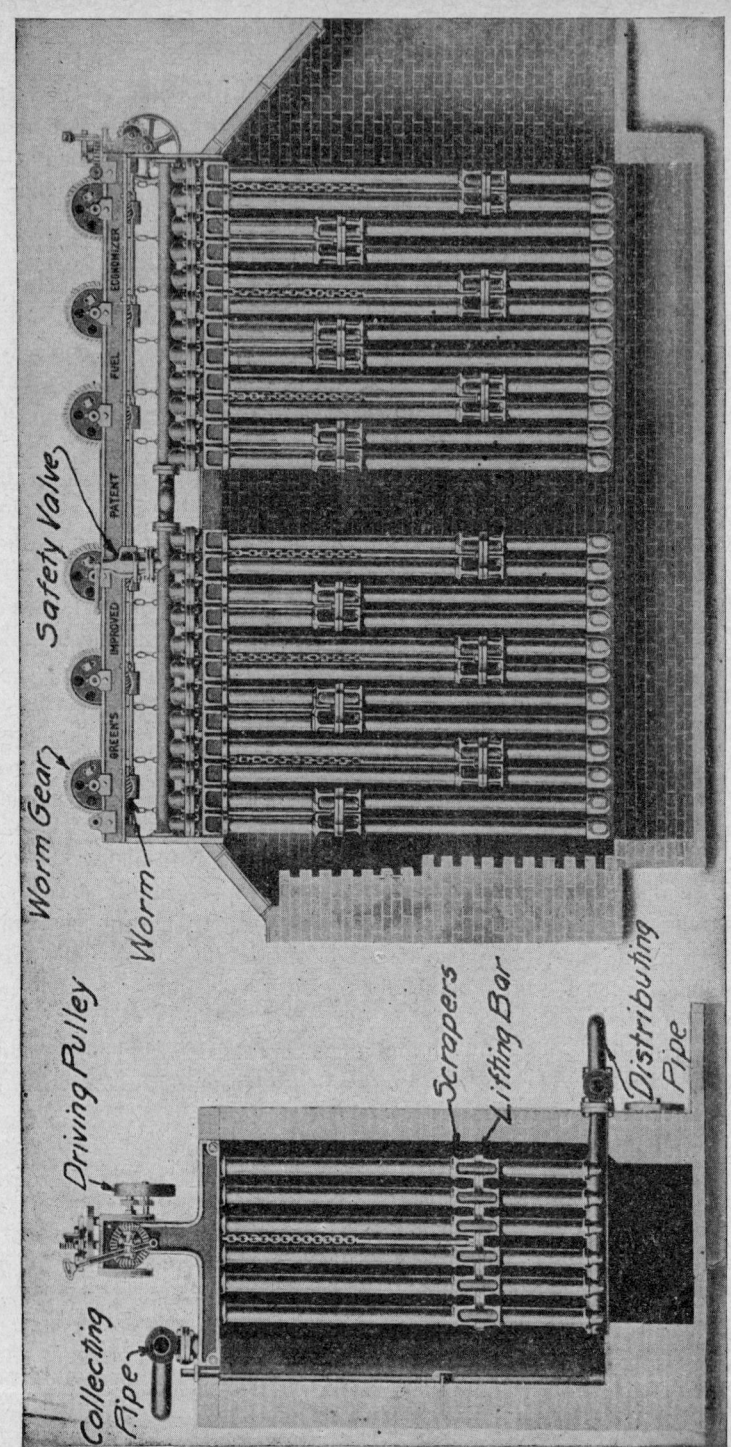

Fig. 190. — Side and End View of Green Fuel Economizer.

power-driven mechanism located above the economizer. The scrapers for two pipe sections are carried by a **lifting bar** and are held in place on the down stroke by a guard bolted to the bar. A chain from the lifting bar passes up between the headers, over a grooved pulley and down to a similar lifting bar and set of scrapers on the two adjoining tube sections. Thus, the weight of one set of pipe scrapers balances the weight of the other set.

The **reversing mechanism** has a large bevel gear fastened to the shaft running lengthwise above the economizer. This operates the chain pulleys by means of worms and worm gears. The direction of rotation of the shaft is reversed by a lever which is thrown first to one side and then to the other by the action of fingers attached to an idler wheel, which is

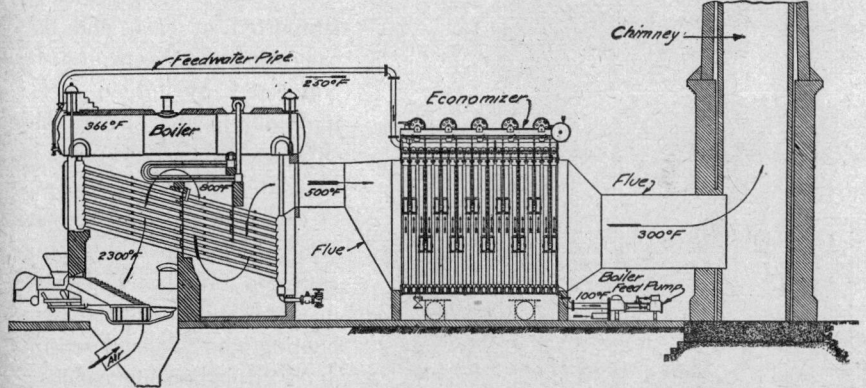

FIG. 191. — Section Drawing of Green Fuel Economizer Installation.

operated by a worm gear and wheel from the shaft. The lever slides a toothed clutch keyed on the cross shaft, making the small bevel gears, located at each end of the cross shaft, alternately drivers and idlers.

The economizer is enclosed in a brick setting or, in more modern construction, in an asbestos-lined sheet-metal casing. A **by-pass** with dampers is provided for the gases, to allow for cleaning and repair, and a by-pass is also arranged in the feedwater piping, as for any heater. An illustration of a typical economizer installation is shown in Fig. 191, together with the temperature change and arrangement of feedwater piping.

The life of this type of economizer is sometimes lengthened by using a multi-stage centrifugal turbine-driven feed pump from the first stage of which the water passes to the economizer at low pressure and returns to the remaining stages of the pump to receive an increase in pressure sufficient to enter the boiler. The economizer just described is ordinarily used to serve a group of boilers, and is adapted to low pressures.

The majority of economizers now being used are of either the **unit type,** in which a single economizer handles the gases and feedwater of one boiler only, or the **integral type** in which the economizer forms an integral part of the boiler, replacing some of the less efficient heating surface of the boiler proper. Both of these economizers are of the steel-tube construction which is necessary for pressures from 400 to 1200 pounds per square inch. The use of the steel-tube economizer has been made possible by removing the air, oxygen, and carbon dioxide from the feedwater in deaërating feedwater heaters and before it enters the economizer.

265. Foster Steel-tube Economizer. — A phantom view of this economizer is shown in Fig. 192, and consists of steel heating elements supported at each end by cast-iron tube sheets, and so connected by forged steel manifolds located outside the tube sheets that the water circulates through the tubes of each row in series. The elements are surrounded by cast-iron protecting sleeves in order to prevent overheating, and are enclosed in a cast-iron housing consisting of inside and outside plates with insulating material placed between the plates. The bottom of the housing consists of a chamber to which the flue connection is made and below which are cast-iron pans arranged

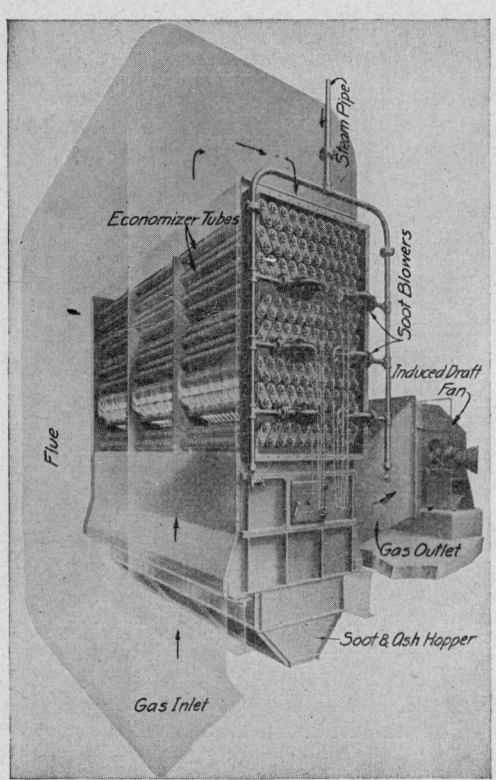

Fig. 192. — Steel-tube Economizer.

to collect the flue dust deposited by the soot blowers, and to catch and discharge the water, which may be used for cleaning the heating surface, through suitable drains placed in the pan bottom.

The location of the economizer relative to the fan and the boiler flue, and also the location of the soot blower nozzles, is clearly shown in the illustration.

The **integral** or **steaming** economizer is shown in Fig. 193, and consists essentially of an upper and a lower drum connected by suitable tubes, the general construction of which is similar to that of any bent tube type of boiler.

266. Power Station Heat Balance and Extraction Feedwater Heating. — Recent improvement in the efficiency of steam power plants has resulted

HEAT BALANCE AND EXTRACTION FEEDWATER HEATING

through a study of **heat balancing**; *that is, of so choosing, proportioning and arranging auxiliary equipment that the plant processes will conform closely to the requirements of an efficient heat cycle, as defined by thermodynamic theory.*

The question of heat balancing is intimately connected with the method of driving the auxiliaries and the method of heating the feedwater. Stations using steam extracted from the main turbine for feedwater heating rely mainly on motor-driven auxiliaries. In some cases the important auxiliaries which will insure the station remaining in operation are supplied from a separate source of power, such as a **house turbine**, or such auxiliaries may have duplicate drives consisting of motors and simple, rugged, but comparatively efficient turbines. The house turbine, when used, is often of sufficient size to supply all the exhaust steam above that furnished by the turbine-driven feed pumps, exciters and other auxiliaries, but present practice is to carry a smaller load on the house turbine and obtain a major portion of the steam for heating the feedwater from one or more stages of the main turbine unit.

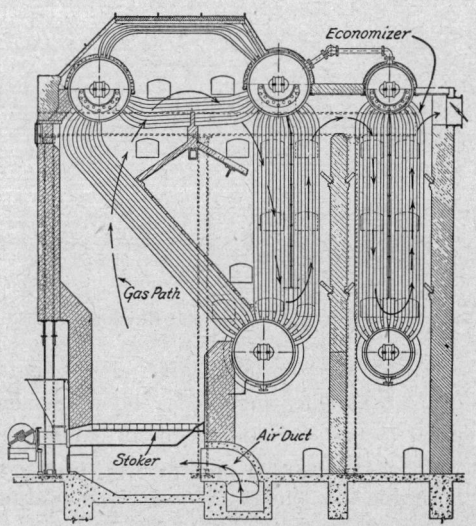

Fig. 193. — Integral Economizer.

It was pointed out in Chapter V that the *Rankine Cycle* in which steam is expanded through a prime mover, condensed and the condensate returned direct to the boiler is an irreversible cycle, and therefore not the most efficient means of converting heat into mechanical energy. The efficiency of this cycle is improved from 4 to 6 per cent by heating the feedwater with pressure-expanded steam extracted or "bled" from a steam turbine. The use of high-steam pressures increases the gain from this type of feedwater heating. The number of stages in which the heating takes place varies from 2 to 4. It is probable that as the units become larger, and coal more expensive, the number of stages at which it will be economical to bleed from the turbine will increase.

If three or more stages are used, the heaters are usually arranged so that the hot-well pump sends the condensate through the tubes of the low-pressure heaters into a surge tank in order to allow for the increase in volume of the feedwater as the result of the heating, or into a deaërator which drives out the dissolved gases and thereby protects the piping, economizers and boilers against oxygen corrosion. From here the condensate flows directly to the boiler feed pump which in turn forces it through the high-pressure heaters into the boiler.

When the steam is extracted from the main unit, it is essential that there be no possibility of a flow of steam from heater to turbine in case the pressure at the bleeding point falls below the temperature of evaporation of the water in the heater. With an open heater it is necessary to install automatic closing valves, in order to prevent the turbine running away at light or no load as the result of steam entering from the heater.

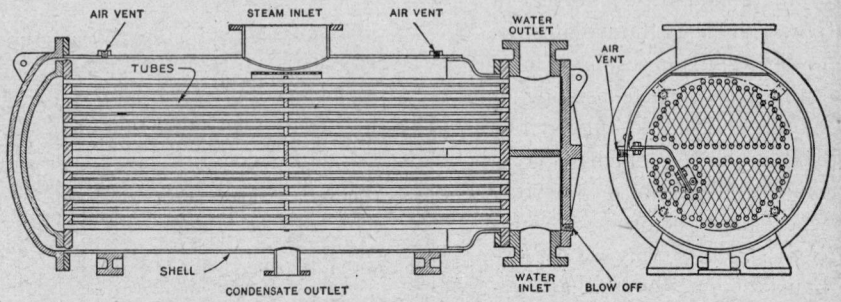

FIG. 194. — Croll-Reynolds Closed Type Extraction Heater.

Either an open heater or a closed heater may be used with this method of feedwater heating. It is, however, common practice in the majority of cases to use the closed type of extraction heaters, Fig. 194, in which the water is heated very close to the saturated steam temperatures. It is often desirable that one of the heaters should be of the open type in order to deaërate the water. The possibility of air absorption by the water in the system is almost completely eliminated if the water is heated to and maintained at 210 deg. fahr.

A typical heat balance diagram for the Long Beach Station of the Southern California Edison Co. is shown in Fig. 195, and as indicated, steam is extracted from the 5th, 10th, 14th, and 18th stages of the turbine for feedwater heating. The condensate from the four vertical condensers collects in the condenser hot-wells and flows by gravity to the condensate pump which forces it through two vertical low-pressure heaters and the evaporator condenser to the suction side of the boiler feed pump which forces the water through two high-pressure horizontal heaters into the boiler. The make-up water, which amounts to about $\frac{1}{2}$ of 1 per cent of the steam required by the turbine, is treated in zeolite water softeners, (Art. 277, page 288), previous to entering the evaporator (Art. 280, page 291).

The heat balance is often improved by recovering the steam from the final stages of the steam jet air pumps, the generator air cooler, and the vapor vented from the extraction heaters. Three schemes of venting are in common use: *one* is to return the non-condensible gases to a lower stage of the turbine; *another* is to vent each heater to the next lower heater, and the lowest to the condenser or a separate air pump; and the *last method* is to vent each heater separately to the condenser or to a separate air pump.

267. Preheated Air and Air Preheaters. — It was shown in Chapter VII that the greatest source of loss with steam boiler operation is the quantity of heat carried up the stack by the flue gases. With the increase in pressures and ratings at which the modern boilers are operated, the stack temperatures are increased; and in order to obtain better overall efficiencies, air preheaters are used to recover a portion of this loss. While economizers were formerly used for this purpose, the practice of preheating

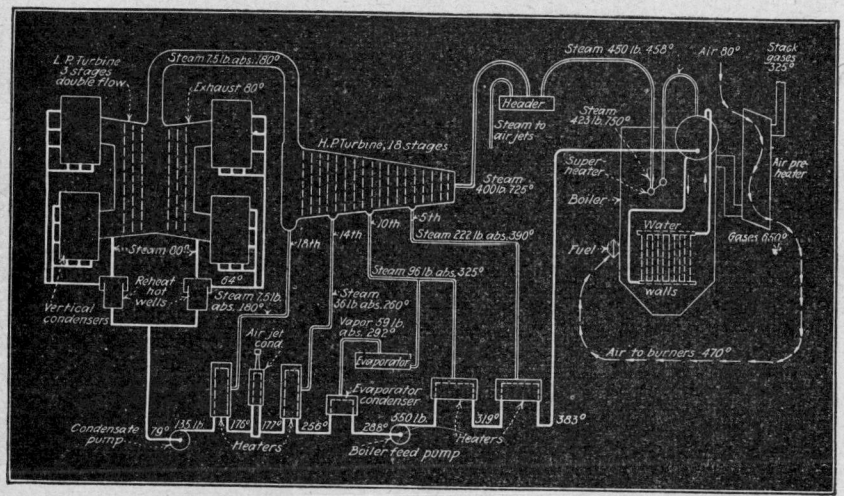

Fig. 195. — Heat Balance Diagram — Long Beach Station.

the boiler feedwater by using steam bled from the main turbines has limited the usefulness of the economizer as an agent in reducing the flue gas temperature.

The **advantages** to be obtained from the use of air preheaters, in addition to the saving of heat in the flue gases, are: (1) *improved combustion*, (2) *use of poorer grades of fuel, and* (3) *increase of the boiler capacity*. When using preheated air with solid fuels the combustion starts quicker, is more uniform, is more nearly completed within the furnace; less soot and smoke are produced; and less combustible is lost in the ash. With pulverized coal, preheated combustion air is even more beneficial than in stoker-fired furnaces, as a higher preheat temperature can be carried, in many cases 500 deg. fahr., and the amount of excess air can be reduced.

Under certain conditions of operation and with fuels which contain large amounts of moisture and sulphur preheaters are subject to considerable corrosion at the cold end. Cleaning is ordinarily provided for by using steam blown soot blowers.

Air preheaters are of either the **convection** or **regenerative** type, the convection type being further subdivided into **plate** and **tubular** preheaters.

The *plate-type* preheater is widely used, with the various makes differing mainly in details of construction. The preheater shown in Fig. 196 con-

FIG. 196. — Sturtevant Plate Type Air Preheater.

sists of a number of flat chambers formed by the sides of the air passages. They are assembled and properly spaced in a structural steel frame with all the inlet and outlet channels brought together on one face. The chambers are surrounded with a casing lined with high-grade insulating material and so designed that an inlet and outlet passage for the gases can be attached at the top and bottom. The chambers divide the air and gas into alternate thin streams flowing in opposite directions and separated by the thin walls of the chambers, the air receiving its heat from the hot gases

PREHEATED AIR AND AIR PREHEATERS 277

as they rise through the passages between the chambers. The gas and air passages are installed vertically which facilitates self-cleaning as there is sufficient vibration of the plates to dislodge any adhering material. The scouring action of the cinders entrained in the flue gases assists in keeping the surfaces clean.

The *tubular type* of preheater consists of tubes through which the flue gas passes with suitable arrangements for causing the air to flow over the external surfaces of the tubes. The air ordinarily makes several passes

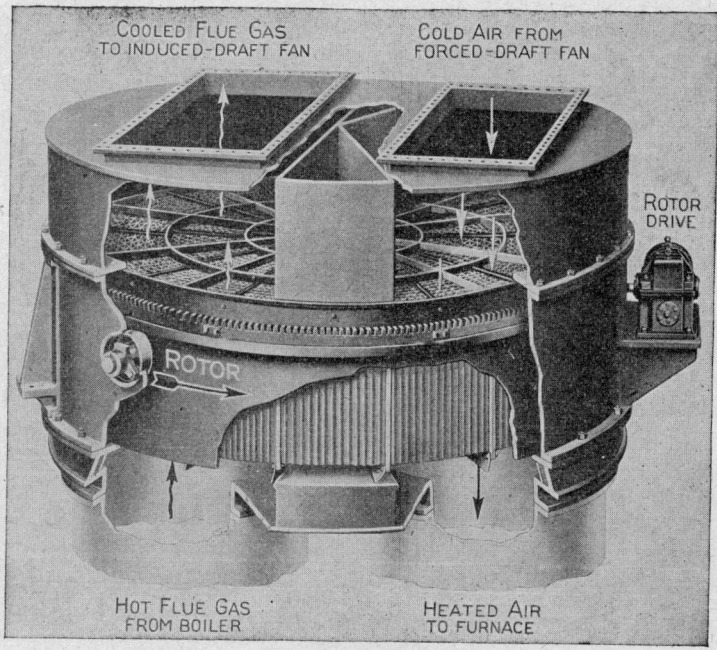

FIG. 197. — Regenerative Type Air Preheater.

across the tubes and is generally arranged so the gas and air flow in opposite directions.

The **Ljungström Air Preheater,** Fig. 197, is of the *regenerative type* and consists of a housing divided into three compartments. The middle one contains a slow-moving rotor, and the upper and lower ones are divided by vertical partitions which confine the flue gas to one side of the apparatus and the air to the other.

The rotor contains the heating surface which is formed into a cellular or honeycombed structure with short parallel vertical passages.

Flue gas from the boiler enters the lower left semi-circular chamber, flows continuously upward through the rotor passages into the upper left chamber and is drawn to the stack by the induced-draft fan.

Air from the boiler room is forced continuously into the upper right chamber by the forced-draft fan, is forced downward through the passages in the rotor, and discharged from the lower right chamber into a duct leading to the boiler furnace.

The small parallel streams of hot flue gas flowing through the rotor give up heat to the metal heating surface which acts as a storage reservoir, and as the rotor rotates carries the heated metal into the path of the cool air in the opposite side of the preheater. At this point, the thin streams of comparatively cold air traveling through the rotor passages absorb the heat from the metal and the metal mass cools.

The uniform slow rotor motion at a maximum speed of four revolutions per minute makes the process continuous. The heated portion of the metal mass is constantly coming into the path of the cool air, and at the same time, the cooled portion is continually returning into the path of the hot flue gas to be reheated. Every revolution of the rotor produces a complete cycle of heat exchange. An even flow of both gas and air is maintained through the rotor by forced- and induced-draft fans which may be independent and located in the ducts leading to the preheater, or integral with the preheater.

268. Soot Removal and Soot Blower. — Soot is an amorphous carbon, and as such acts as an insulator when deposited on heating surfaces. Its removal therefore saves heat. In small plants scrapers and brushes, attached to rods made of small pipe, are used to remove soot and light incrustations of ash. In large plants washing or some form of soot blower is used for this purpose.

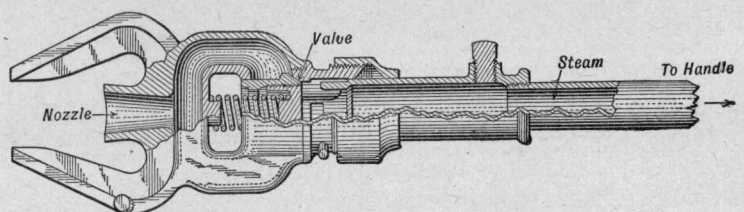

Fig. 198. — Hand-operated Soot Blower Head.

Several types of soot blowers, differing principally in the construction of the blower head, operate by using steam for power. Some use the steam blast only, while in others the steam jet induces a volume of air to assist in blowing out the soot. The latter form has greater cleaning power. An open-ended steam pipe, called a steam-lance, is often used by hand to remove soot, but it is uneconomical in the use of steam. In the efficient soot blower steam is expanded in a nozzle located in the blower head, Fig. 198, and attains a high velocity, thus drawing in a large volume of air. The combined blast of steam and air sweeps the tube effectively.

A permanent installation of steam jets, Fig. 192, page 272, for the removal of soot, is commonly used in important plants, since the gain in economy more than warrants the expense. The mechanical device is more likely to be used frequently, as it is comparatively easy and agreeable to operate, while cleaning the tubes with a hand blower is a disagreeable job likely to be neglected.

When applied to water-tube boilers the mechanical soot cleaner consists essentially of a cleaning element, which may be stationary or movable. A typical soot cleaner as applied to a B. and W. type boiler consists of one or more elements for each pass of the boiler setting. Each element is located above the tubes and at right angles to them, and has steam nozzles arranged to direct the steam, on the tubes, along diagonal paths at 60 degrees to the horizontal. Steam is supplied to the elements by pipes having automatic valves. Each element is mounted on bearings and is provided with a chain sprocket located outside the setting, so that the jets of steam can be made to sweep over the tubes and thus clean them more effectively.

REFERENCES

Feedwater Heaters and Purifiers, J. C. W. GRETH of Wm. B. Scaife & Sons Co.
Power, see index of each volume.
Mechanical Engineers Handbook, MARKS.
Steam, BABCOCK and WILCOX Co.
Steam Power Plant Engineering, GEBHARDT.

REVIEW QUESTIONS AND PROBLEMS

1. What is the primary purpose of heating boiler feedwater?
2. Name three methods of heating feedwater and explain the essential differences in the methods.
3. Describe the construction and operation of an open type of feedwater heater.
4. Describe the construction of a modern steel-tube economizer. What types are used?
5. Describe the heat balancing set-up for unit No. 3 of the Long Beach Power Station.
6. The test of a Heine cross-drum boiler, operating at 287 lb. per sq. in. abs. pressure and 263 deg. fahr. superheat, showed that the feedwater temperature was increased from 157 to 283 deg. fahr. by using an economizer. Find the per cent saving resulting from the use of the economizer.
7. (*a*) Name the advantages resulting from the use of air preheaters. (*b*) Name two types of air preheaters.
8. What is soot and what methods are used to remove it from boiler-tube surfaces?

CHAPTER XII

BOILER FEEDWATER CONDITIONING, SCALE REMOVAL, EVAPORATORS, DEAËRATORS, AND TUBE CLEANERS

269. Foreword. — An essential requirement in boiler plant operation, particularly in those plants where high pressures and extremely high ratings are used, is a source of pure water for boiler feeding purposes. Water obtained from natural sources, such as rain water, water from creeks, rivers, lakes, wells and the sea, is never free from impurity, the relative degree of purity being in the order named. Falling rain absorbs carbon dioxide, air, and other gases, and in the earth becomes impregnated with mineral compounds, loaded with organic matter and mud in suspension, and often becomes defiled by acid wastes from industrial plants.

If impure water is used as boiler feed it is obvious that, as the water evaporates, and is removed from the boiler in the form of steam, the scale-making impurities and mineral salts will remain and will collect at the bottom as a sludge, or, depending upon their nature, will form a scale on the interior surfaces. With the high pressures and ratings now being used, the formation of scale cannot be permitted without endangering continuity of operation. The cost of tube replacements in water walls, water screens, or in the boiler proper, and the losses incurred through disarrangement in plant operation, caused by the outage of large steam-generating units, render imperative exact water conditioning that can obviate these losses. High ratings at the same time necessitate segregation from the steam of greater amounts of boiler water in less time. Under these conditions, maintenance of clean surfaces throughout the boiler and production of dry steam are increasingly important.

Operating experience with the 1300 pound pressure boiler unit at the Lakeside Station of the Milwaukee Electric Light and Power Company indicates that some form of feedwater treatment is imperative even though ideal evaporated make-up water conditions are maintained. Scale-forming materials are so greatly less soluble at high pressures, and corrosion so much more active that experience with moderate pressure boilers is definitely inapplicable to high-pressure units.

Boiler feedwater treatment has become a scientific chemical procedure that aims to: (1) prevent scale formation; (2) eliminate corrosion in all water and steam spaces; (3) prevent caustic embrittlement; and (4) prevent priming, foaming, and wet steam.

270. Boiler Scale and its Effects. — *Boiler scale is a solid deposit, or incrustation, caused by the action of heat and the concentration of impurities* *held in solution and suspension in the water.* The concentration causes crystallization of the salt out of the solution and deposits on the metal. *The principal scale-making materials are the sulphates, nitrates, and chlorides of lime and magnesium, which make a hard scale, and the carbonates of lime and magnesium, which make a softer scale.* The latter are precipitated at temperatures around 210 deg. fahr. A boiler tube in which scale has been allowed to accumulate and also the appearance of a typical scale formation, are shown in Fig. 199. The presence of scale on boiler plate prevents the rapid transmission of heat to the water, and may cause the overheating of tubes, flues, or shell, with disastrous effects. Reports of boiler insurance companies show that a large proportion of the damage caused each year in steam boilers is caused by boiler scale. It is also a direct item of cost in that the fuel requirements are greater, tube maintenance is excessive,

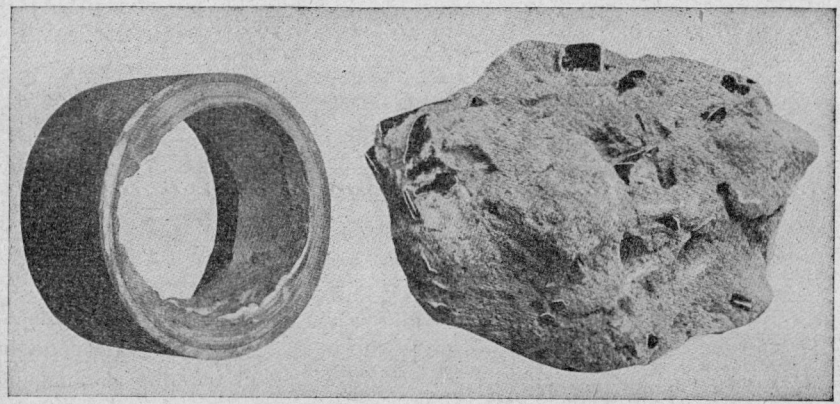

Fig. 199. — Boiler Tube showing Scale and a Typical Scale Formation.

the boiler is out of service more than it should be, heat losses as a result of blowdowns are high, and labor charges are increased.

271. Corrosion. — Corrosion, or wasting away, is caused by **electrolytic action** of metal in contact with water, and is especially active in the presence of acid and dissolved oxygen. *If the oxygen can be eliminated from the water, corrosion may be practically prevented.* The presence of carbon dioxide, since it causes acidity, always implies a tendency toward corrosion; but if oxygen is not present also, the possibility of corrosion is slight. **Pitting** is due to local corrosion as the result of galvanic action between impurities, mill scale, and rust at points on the surface of the metal.

Other factors being the same, the corrosive effect of water depends upon its active acidity *or hydrogen ion concentration, i.e., the actual concentration of hydrogen ions in a given volume of the solution.* This is according to the electrolytic theory of dissociation which states that all liquids, having water as a constituent, contain free H- and OH-ions. In case these ions are equal the solution is called neutral. When the H-ions predominate it is said to be acid, and with the OH-ions in excess, alkaline.

The concentration of H-and OH-ions in pure distilled water are each equal to 1/10,000,000 or 10^{-7}. The term pH is now generally used to express hydrogen ion concentration, its value being the logarithm of the reciprocal of the H-ion concentration. The numbers on this scale designate the fraction of a gram of hydrogen present in each liter of water. *Neutral solutions* have a pH value of 7, acid solutions less than 7, and alkaline solutions greater than 7. A solution having a pH value of 5 is ten times as acid as one with a pH of 6. Analogously, a solution of pH 4 is ten times as acid as one of pH 5. A pH of 4, therefore, indicates an acidity 100 times as great as one of 6. A similar relationship holds on the alkaline side of the scale. *It has been found that by maintaining a pH value of 10, internal corrosion will be greatly improved, if not entirely eliminated.* The determination of the pH value for water is extremely simple, and gives an easy method for determining the acidity of boiler water.

The *elimination of corrosion* in heaters, feedwater lines, economizers, and boilers in addition to *requiring a low concentration of oxygen in the water requires the maintenance of definite conditions of hydroxide alkalinity*, using either sulphates, phosphates, tannates or acetates to maintain the desired concentration of hydroxide alkalinity. Ordinarily 85.5 parts per million of caustic alkalinity will eliminate corrosion.

An example of the effect of corrosion is given in an article presented at the annual meeting of the American Society of Mechanical Enginners, December, 1915. In twelve water-tube boilers at the Frontvale plant of the Southern Pacific Company, one-third of the tubes required replacing in eighteen months after the plant had been placed in operation.

Slabs of zinc securely fastened to the stays of marine boilers are used to prevent corrosion of the boiler metal. The zinc requires renewal, as it is gradually wasted by galvanic action.

As indicated above the amount of corrosion is directly proportional to the free oxygen content of the water. To remove oxygen from the water as it enters a hot-water system, either a **deactivator** or a **deaërator** may be used. *The deactivator removes only oxygen using a chemical process.* It consists of a vessel containing many sheets of iron chemically treated so that rapid corrosion takes place in the presence of oxygen. The oxygen is thus consumed in corroding these cheap plates instead of the expensive piping system. This process is not extensively used at present.

The deaërator removes oxygen and all dissolved gases by a physical process. A deaërating heater of the direct-contact type is shown diagrammatically in Fig. 200, and consists of a heating chamber for heating the water, a separating chamber for deaërating it, an auxiliary steam jet between the heating and deaërating chambers to inject into the water sufficient steam to compensate for the temperature drop in the heater, and a vent condenser to recover the heat in the vented mixture. Water drawn from the supply line passes through the **vent condenser** and is heated in the heater to the condenser temperature by the steam which has previously been used for deaëration. The heated water next flows to the separator where it comes into contact with the auxiliary steam jet, and then enters the separating chamber at substantially the temperature of the steam. Here it cascades from trays and is agitated in the presence of the steam. The steam sweeps through the separating chamber, deaërates the water by liberating the dissolved gases, such as oxygen, nitrogen and carbon dioxide, and then flows to the heater. The water drops into the lower part of the separating or deaërating chamber, and is pumped to the boiler or economizer.

When deaërators of the so-called flash type, in which the gases are freed by explosive boiling of the feedwater at the low pressure in the deaërator, are used with bled steam there is a loss in the useful power output and an impairing of the cycle efficiency as a result of the **degradation of heat** of the bled steam.

A well-vented open feedwater heater will often give a degree of oxygen removal sufficient to prevent corrosion.

The employment of protective coatings, such as apexior, to eliminate corrosion in tubes and drums of boilers

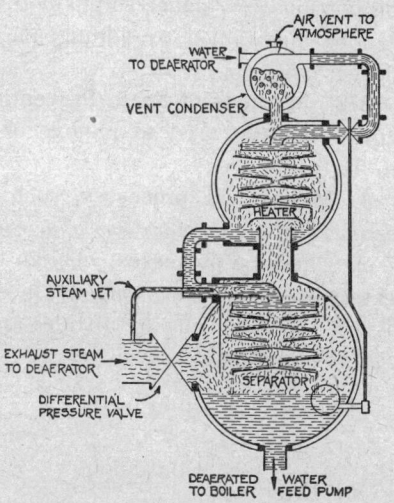

Fig. 200. — Elliott Deaërator.

is being greatly advanced and is of value, even though the feedwater is thoroughly deaërated and the proper hydrogen ion concentration maintained. Apexior is a gray-black liquid which is applied, to the clean dry surfaces to be protected, by either a hand- or power-operated brush. As soon as applied, the spirit vehicle starts to evaporate and in a few hours there is left a smooth, hard, continuous, protective coating on the metal surface.

Caustic embrittlement, a form of corrosive attack, manifests itself by the formation of cracks in the steel plate and suggests that the metal has undergone a complete change in its properties, from a ductile material to one almost entirely lacking in this characteristic. The exact cause of embrittlement of boiler plate is as yet not fully known. It is, however, known that embrittlement is not found in boilers where: (1) the ratio of sodium sulphate to sodium carbonate alkalinity is greater than 2 to 1 for pressure below 250 pounds per square inch, (2) boilers are calked on the inside to prevent high concentrations of caustic between the seams. The late Prof. S. W. Parr attributes embrittlement to nascent hydrogen, and at high temperatures to hydrolysis of sodium hydroxide.

Water-softening procedure (Art. 276) usually introduces into the feedwater either sodium bicarbonate or sodium carbonate, which, under the high temperature in a boiler, break down to caustic soda. Investigation has disclosed that the presence of uncontrolled caustic in the boiler water leads to embrittlement. Further, it has been determined that to minimize the effect of caustic in the boiler, a definite ratio between the sulphate content and the alkalinity must be maintained. A safe sulphate ratio[1] has been established for each working pressure, and with increasing pressures proportionately higher sulphate ratios are required. The sulphate ratio can be maintained by adding magnesium sulphate or sulphuric acid in proper amounts.

272. Methods of Scale Prevention and Water Softening. — The principal processes used to remove or neutralize the scale-making material in feedwater are:

1. **Mechanical processes,** such as: *deaëration, filtration, evaporation deconcentration by blowdown and sedimentation.*

2. **Chemical processes,** wherein the reaction takes place either inside or external to the boiler itself, the most common of which are:

(a) *Internal* using hydroxides and carbonates, aluminates, phosphates, and foam preventives.

(b) *External*
- Intermittent and Continuous } cold, using { Lime and soda ash / Lime and barium / Lime and sodium aluminate
- Continuous — hot, using { Lime and soda ash / Lime and sodium aluminate
- Continuous — cold — Zeolite or base exchange

The process to be used depends upon the amount and quality of the feedwater. **Filters,** which may be of either the pressure or gravity type, using granulated quartz or other sand as a bed through which the water is passed, may be used to remove material carried in mechanical suspension by the feedwater. Where there is a feedwater heater, a large part of the carbonates of lime and magnesium is precipitated out of the water by the heat. In small installations having a fair quality of feedwater, the use of a boiler compound directly in the boiler is common. *Chemical treatment using either the precipitation process or the zeolite process, is used in most large boiler plants.* The precipitation process involves the addition of certain reagents to prevent the formation of scale and to minimize priming and foaming within the boiler. It uses either the hot or cold process, the former being used extensively since the chemical reactions are greatly

[1] See A. S. M. E. Code for Construction of Boilers.

accelerated by heat and the equipment occupies a much smaller space than the cold process.

273. Boiler Compounds. — Internal treatment uses boiler compounds which usually consist of carbonate of soda, to which caustic soda and phosphate of soda are sometimes added. Tannin and starchy materials are occasionally added, with the purpose of coating the particles of incrusting materials, holding them in suspension to prevent the formation of a solid mass. This method adds impurities to the water in the form of precipitates and insoluble impurities and necessitates the periodical blowdown of the boiler to prevent the accumulation of sludge and an occasional opening and washing out to prevent foaming. The amount of compound to be used depends on the quantity of water evaporated, as well as its quality. The compound is sometimes pumped in continuously with the feedwater by a small pump made especially for that purpose, or forced in periodically by an injector or by use of a by-pass in the feed line. There is a tendency to look upon this method of treatment as a "cure-all" for any water difficulties. This is a mistake, but when proper care is taken to suit the compound to the water in use, the results secured are fairly effective in the prevention of scale. To be most effective, the compound, in addition to effecting the precipitation of scale-forming impurities, should also render the precipitated material non-adherent by mechanical action.

The **Navy Standard Boiler Compound** has been used successfully in a district having one of the worst waters used by locomotives on the Southern Pacific Lines and under conditions in which other treatment was unsuccessful. This compound is composed principally of sodium carbonate but also contains tri-sodium phosphate, dextrine and a tannin compound. The tannic acid and starch are added to prevent the formation of scale, the action being to hold the impurities in suspension in a **colloidal**, or jelly-like, state. The tri-sodium phosphate prevents the rise of the surface tension of the solution and consequent priming caused by the impurities in the water and by the application of the other ingredients in the compound. In using this compound, a sufficient quantity must be added to each boiler to render the alkaline strength of the water in the boiler 3 per cent of the normal or above.

Kerosene and **graphite** are sometimes used in small quantities to loosen scale. They should be used with caution, especially where there is a chance for dislodged scale to accumulate and cause overheating of the metal.

274. Analysis of Feedwater. — The only satisfactory way to deal with the feedwater problem in taking water from an untried source is to begin with an analysis of the water. Companies which manufacture heaters, water-treatment apparatus, and boiler compounds maintain departments for that purpose; or the work may be done by a recognized chemist having suitable experience and equipment. It often happens that a water which appears to be suitable, on a casual examination, may carry corrosive elements, such as acid from mill waste, or bad scale-making material. The analysis of a well water and a river water are given in Table 24, which shows the incrusting or scale-making constituents in grains per gallon.

TABLE 24. — ANALYSES OF A BOILER FEEDWATER

Name of Substance	Well Water Grains per U. S. Gallon		River Water Grains per U. S. Gallon	
	Raw	After Heating	Raw	After Heating
Calcium carbonate	18.39	2.50	3.51	1.26
Calcium sulphate	....	0.98	0.54	0.42
Magnesium nitrate	0.89	0.27	....	....
Magnesium sulphate	7.80	2.31	2.52	1.90
Magnesium chloride	2.13	1.13	....	....
Magnesium carbonate	1.97	....	....	....
Silica	0.85	0.15	0.75	0.60
Iron oxide and alumina	....	....	0.50	0.25
Sodium chloride	0.33	0.08	0.81	0.68
Volatile and organic	1.55	1.10	2.73	3.23
Total solids	33.91	8.52	11.36	8.34
Suspended matter	0.15	0.35	....	....
Free carbonic acid	0.99	0.06	1.00	....
Incrusting solids	33.58	8.44	10.55	7.66
Non-incrusting solids	0.33	0.08	0.81	0.68

275. Effect of Heating Feedwater. — The result of heat alone on feedwater is shown by the column adjoining the analysis of the raw water in Table 24. The water was passed successively through an open heater and a live-steam purifier. It is evident, that when purification by heat alone is used, the larger the percentage of incrusting solids, the larger is the percentage of removal effected. In both cases, the diluting effect of the condensed exhaust steam was present in the second analysis, possibly amounting to over 25 per cent, or more than one-quarter, of the reduction noted in the solids. Time is an important factor in the extent of purification in the heater; to remove *temporary hardness* by boiling at atmospheric pressure, a period of about thirty minutes is required for precipitation of the carbonates of lime and magnesia to the limit of solubility. As the feedwater does not generally remain as long as this in the heater, the full effect is not realized.

276. Water-softening Apparatus. — Where considerable quantities of water are required for boiler purposes, and the supply is poor in quality, *external systems* known as water-softening plants have come into use. The method of treatment is both **mechanical** and **chemical,** and the process may be **intermittent** or **continuous.** These systems eliminate the scale-forming constituents, as insoluble solids, prior to admission of the water into the boiler, by using a lime reagent to convert free carbon dioxide and bicarbonate to the normal carbonate, with the precipitation of calcium and magnesium. When non-carbonate hardness, such as sulphates and chlorides, is present, soda ash is used. In the Scaife intermittent system of water purification, the apparatus, Fig. 201, consists of two or more **treating and settling tanks,** a **chemical mixing tank** with pump or injector, **filter tanks,** and a motor to drive the **mixing paddles** in the treating tanks. While one treating tank is being filled, treated and settled, the other tank is supplying treated water.

The exact quantities of the reagents, **lime** and **soda ash,** are weighed to correspond to the quantity of water to be treated, and are mixed with

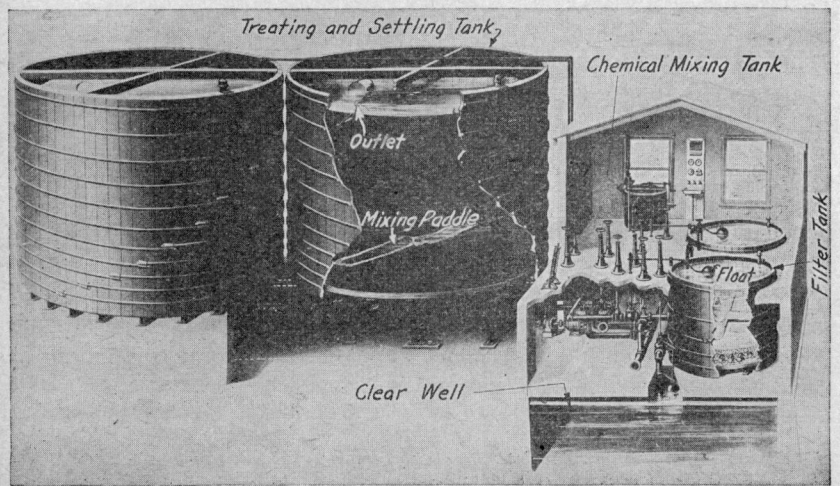

FIG. 201. — We-Fu-Go Water-softening Plant.

water in the chemical mixing tank. The mixture is then pumped into the tank under treatment. The lime unites with the bicarbonates of lime and magnesium, which produce temporary hardness in the water, and forms the insoluble carbonates of lime and hydrate of magnesium. These appear as a sludge at the bottom of the tank. The action of the paddles mixes the reagents and stirs up the sludge which has been allowed to remain from a preceding purification.

The sludge assists in setting the newly formed precipitate. After the mixing has stopped and sufficient time has been allowed for the precipitate to settle, the softened water is drawn out through a floating outlet pipe, from a point near the surface where it is cleanest. It passes to the filter tank, where the water level is controlled by a float and valve, and then to the supply tank, or clear well, for the boiler feed pumps, or to an open feedwater heater.

In the **continuous type** of apparatus, the water passes through compartments in which the reagents are continuously added during the stirring process, and then to a settling compartment and through a sand filter to the boiler feed line. In the "hot" continuous process, the action takes place more rapidly than in the cold process, and less storage capacity is required.

The chemical reaction equations for the removal of temporary hardness, together with the names of the components that enter and result from the combination, are as follows:

$$Ca(HCO_3)_2 + Ca(OH)_2 = 2 CaCO_3 + 2 H_2O$$
Calcium bicarbonate + lime hydrate = calcium carbonate + water

$$Mg(HCO_3)_2 + CaO = MgCO_3 + CaCO_3 + H_2O$$
Magnesium bicarbonate + lime = magnesium carbonate + calcium carbonate + water

$$MgCO_3 + Ca(OH)_2 = Mg(OH)_2 + CaCO_3$$
Magnesium carbonate + lime hydrate = magnesium hydrate + calcium carbonate

Magnesium chloride is also removed, and the resulting reaction is:

$$MgCl_2 + Ca(OH)_2 = Mg(OH)_2 + CaCl_2$$
Magnesium chloride + lime hydrate = magnesium hydrate + calcium chloride

The sulphates, nitrates, and chlorides of lime and magnesia cause *permanent hardness* and dense scale in boilers. The soda ash, or sodium carbonate, precipitates insoluble calcium carbonate and also forms sodium sulphate. It also acts on the calcium chloride and makes calcium carbonate and sodium chloride. The sulphate and chloride of sodium are soluble. The reaction equations are as follows:

$$CaSO_4 + Na_2CO_3 = CaCO_3 + Na_2SO_4$$
Calcium sulphate + soda ash = calcium carbonate + sodium sulphate

$$CaCl_2 + Na_2CO_3 = CaCO_3 + 2 NaCl$$
Calcium chloride + soda ash = calcium carbonate + sodium chloride.

With a high sulphate concentration, barium is often used as the reagent. The barium sulphate formed being practically insoluble, decreases the soluble solids going to the boiler. In many instances, especially in the cold process work, coagulants, such as alum or sodium aluminate are used to hasten the sedimentation of the insoluble products of the reaction by forming a gelatinous substance or "flocs."

277. Zeolite Process. — This method is known as the base exchange method of water softening, and uses a natural or artificial material called **zeolite**, made up largely of sodium compounds. *It has the property of removing the scale-forming calcium and magnesium from the water being treated and replacing them by sodium which is not scale-forming.* The reaction can be represented thus:

$$\text{Na zeolite} + \begin{Bmatrix} Ca \\ Mg \end{Bmatrix} \text{salts} = \begin{Bmatrix} Ca \\ Mg \end{Bmatrix} \text{zeolite} + \text{Na salts}$$

The zeolite softening systems consist of one or more closed steel tanks partly filled with zeolite material and a separate tank for making a saturated salt solution. The untreated water is passed through the zeolite tank and emerges with zero hardness. After the zeolite is exhausted, or has given up its available sodium, a solution of common salt is passed through the zeolite which removes the calcium and magnesium and replaces them with sodium, thus returning the zeolite to its original condition.

The best results in softening water for boilers by the zeolite process are obtained with waters which are clear, practically free from iron, and low in carbonate hardness. A raw water high in carbonates of calcium and magnesium results in a softened water high in sodium carbonate (soda ash) which is often objectionable as boiler feedwater because it is a frequent cause of foaming. In such a case all the water is first put through a continuous tank in which it is subjected to the lime treatment, after which it passes through a filter and then to the zeolite softener. Water from natural sources usually contains suspended matter, organic matter and iron which should be properly coagulated and filtered before being softened by the zeolite process.

The zeolite process often gives a low ratio of sodium sulphate to sodium carbonate alkalinity, which tests indicate may be one of the causes of caustic embrittlement. Under these conditions, a combination of the lime-soda and zeolite processes may be used effectively, or the water may be treated with sulphuric acid and some of the carbonate thereby changed to sodium sulphate.

278. Hot-process Water-softener. — The Cochrane water-softener, Fig. 202, supplies a chemical treatment to feedwater that has been heated to 210 deg. fahr. or higher. The effect of heat upon chemical reactions is to hasten them; and the rapid deposit of scale-making material is favored by increasing the temperature of the water, since this material descends more quickly in a medium of less density and viscosity. Less of the scale material being held in solution in hot water than in cold water, less will pass along to the boiler.

This softener is a combined open-feedwater heater and lime-soda softener, with the feedwater heater and oil-separator mounted above the softener in which chemical treatment and sedimentation take place. The hot water, freed from carbon dioxide, air, and other gases in the heater, is mixed with reagents from the automatic chemical proportioner, and falls into the tank below, where the scale material is deposited. The hot, treated water passes up into an inverted funnel, out to a sand filter, and thence to the feed pump and boilers. The sludge is blown out of the sedimentation tank by opening the valve at the bottom of the cone. The process is continuous. Where a part of the supply to the heater, such as condensate, does not require chemical treatment, a compartment is provided in the heater from which the heated water is supplied directly to the feed pump.

279. Foaming and Priming. — *Foam* in boilers consists of unbroken bubbles of steam, and is caused by concentration of soluble and insoluble salts, together with other impurities such as organic matter which are carried in suspension and thus render difficult the free escape of the steam bubbles as they rise to the surface of the water. As the impurities become more concentrated, foaming increases. Scum, which may be caused by oil, vegetable matter, or sewage collecting on the surface of the water, is a frequent cause of foaming, because the bubbles of steam are prevented from breaking as they arise to the surface. If the water contains an alkali, it will change any vegetable or animal oil mixed with it into soap, which forms suds and causes foaming. Excessive foaming causes water to be

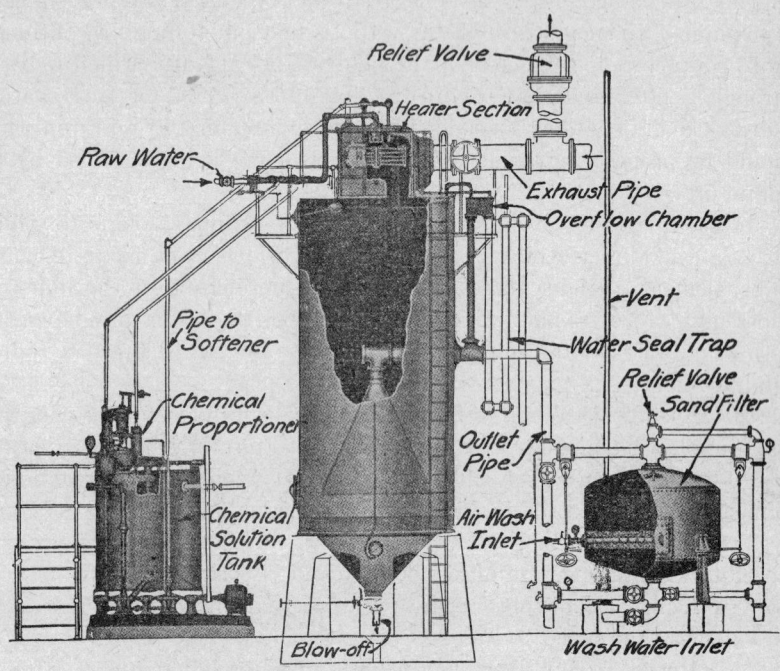

FIG. 202. — Cochrane Hot Process Water-softener.

carried into the steam pipe in sufficient quantities to become dangerous, because it is carried along by the rapidly moving steam and may rupture the piping. Foaming is generally attributed to sodium and potassium salts. It is less in high-pressure boilers because of the higher density of the steam bubbles. It may be prevented by using pure feedwater, and may be controlled by: (1) blowing down from 2 to 8 per cent of the boiler feedwater and admitting fresh water to hold the concentration below the point at which foaming occurs, or (2) using deconcentration equipment, Fig. 203, which involves the definite chemical control of reactions within the boiler to produce the maximum quantity of suspended solids or sludge, and the continuous circulation of a portion of the boiler water to a settling tank where the suspended solids deposit and from which the clarified water returns to the boiler.

It is difficult to blow down intermittently such amounts as will keep the concentration between certain limits and yet avoid excessive waste of heat. This waste may be avoided by using **continuous boiler blowdown** by which the heat in the blowdown water is recovered by passing feedwater through some form of heat exchanger in which the heat contained in the blowdown is given up to the feedwater.

Priming, or **carry-over,** consists of water carried from the boiler by the steam. It may be in the form of a mist, or, if the boiler is priming badly,

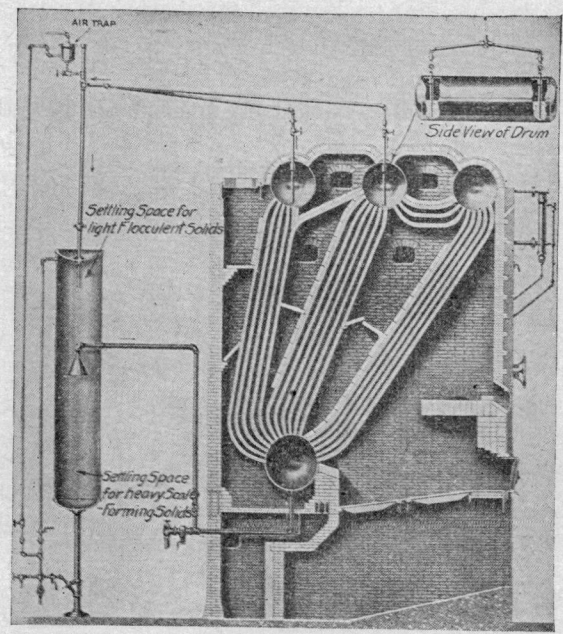

FIG. 203. — Elgin Deconcentrator.

the water may leave the boiler in belches. It is due to several causes: the rapid disengagement of steam from too small an area of water surface, the method of operation of boiler and engine, a high water level, overloading, and impurities in the water. PARSONS, in *Steam Boilers,* states that to avoid excessive priming the velocity of disengagement should not exceed 2.5 feet per second.

280. Evaporators for Distilled Make-up Feedwater. — The use of distilled water for boiler make-up affords such obvious advantages in keeping boilers free from scale, and from the corrosive action of acids and free oxygen that it has become common practice in modern power plants. Not only are the dissolved solids removed from the raw water by distillation, but the entrained air also. When using high pressures with some grades of raw water chemical treatment ahead of the evaporator has been found profitable, the zeolite method of treatment being generally used. It is often necessary to treat the condensate plus the make-up water with sodium phosphate or soda ash in order to maintain a positive alkalinity. The cost of evaporation will, in many cases, be more than offset by the decrease in operating and maintenance costs resulting from the use of pure water.

As part of the water evaporates, that remaining becomes saturated with impurities so it is necessary to blowdown the evaporator, either continuously or intermittently, in order to maintain efficient evaporation. In the case of fresh water, the blowdown amounts to 4 or 5 per cent of the amount evaporated. The heat used for evaporation may be obtained by using live, exhaust or "bled" steam, and the evaporator is usually introduced as a step in the station heat balance.

Two types of evaporator systems are in common use; the **pressure system,** in which either live, exhaust or "bled" steam is supplied at a pressure considerably above that of the atmosphere, and the **vacuum system,** in which the pressure is at or below atmospheric. Evaporator plants are sometimes classified as *"low heat"* or *"high heat"* level. In the "low heat level" type, evaporation takes place at low temperatures and pressures, and in the "high heat level" type, evaporation takes place at high temperatures and pressures. The vapor from the former is usually condensed in an open heater by the boiler feedwater, before it enters the boiler heater, and at a pressure slightly above atmospheric, while the vapor from the latter is ordinarily condensed in a special surface type heater, using boiler feedwater from the highest temperature level heater as the condensing medium — the heat lost in the evaporating process thereby being recovered by boiler feedwater used as circulating water in the evaporating condenser. (See Fig. 195, page 275.)

Three types of evaporators are used: (1) the **film type** in which a thin film of water passes over the surface of pipes filled with steam; (2) the **flash type** in which the vapor flashes from water injected into a chamber under a partial vacuum, and (3) the **submerged type** in which evaporation occurs in much the same way as it does in a fire-tube boiler.

A submerged type shell evaporator is shown in Fig. 204. It consists of a closed cast-iron shell containing several removable evaporating sections having two manifolds connected with bent tubes. Steam enters the top of the front manifold, passes through the bent tubes to the rear manifold, returns to the front manifold and leaves at the bottom as water. The tubes expand and contract and thus break up any scale which forms. The locations of steam inlet, vapor outlet, and blow-off are marked on the illustration. A crown baffle located above the tubes is used to prevent water being carried to the vapor outlet.

Depending upon operating conditions in the plant, a single evaporator may be used or two or more evaporators used in series, the arrangements being called **single, double** or **multiple effect.** In the single effect, the vapor driven off from the water is led directly to a condenser, the condensate being delivered to the hot well. This condenser may be that serving the main unit but is usually a separate condenser using condensate as a condensing medium. With multiple effect, the vapor from the first effect evaporator is condensed in following evaporators and so on. The vapor from the last effect being condensed either in a closed type condenser or in an open heater.

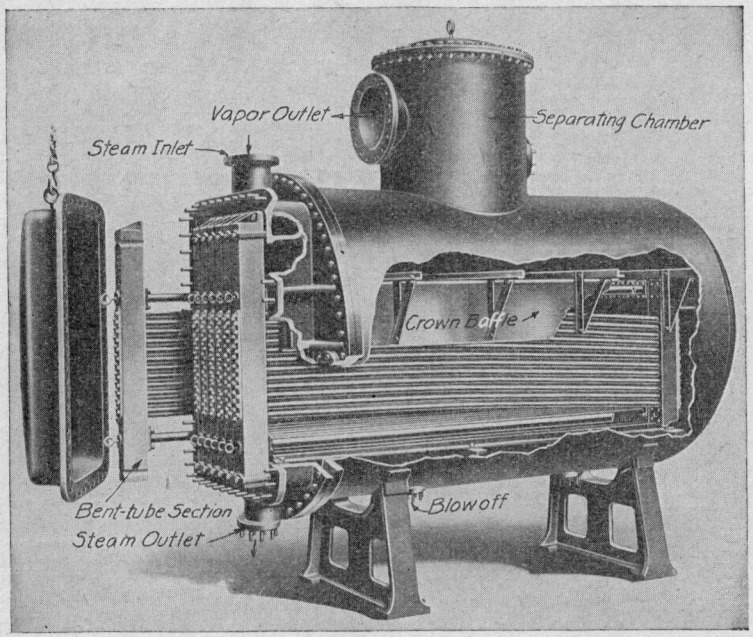

Fig. 204. — Griscom-Russell Submerged Type Bent-tube Evaporator.

281. Scale Removal. — Although a considerable amount of scale-making material may be removed from the feedwater by using heaters, and by methods of water treatment, nevertheless, scale is deposited in boilers, and some mechanical means is required to remove it.

In accessible parts of drums or shells, the scale is removed by using a hammer and a blunt chisel, while for boiler tubes some type of tube cleaner is used; the tubes of water-tube boilers are freed from scale by passing through them a cleaner head carrying a tool rotated by a motor driven by air, steam, water, or electricity, the loosened scale being carried out by the steam, air, or water used to drive the motor.

A **pneumatic cleaner,** using air at 50 to 75 pounds per square inch gage pressure, is shown in Fig. 205. Air is supplied through **armored hose**

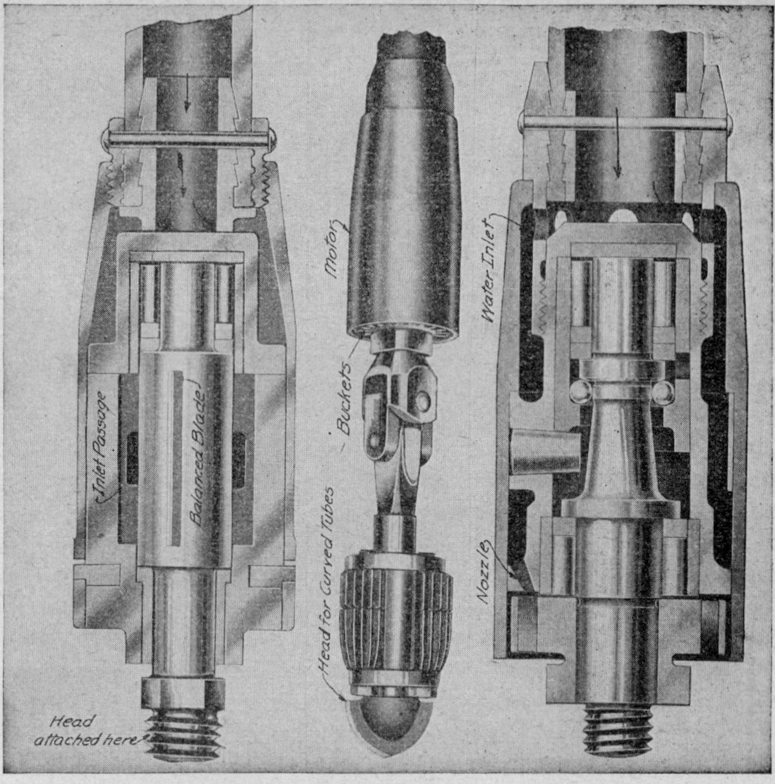

294 BOILER FEEDWATER CONDITIONING

Fig. 205. — Pneumatic and Water-driven Tube Cleaners for Water-tubes.

to the motor, which consists of a cylindrical casing in which revolves a shaft carrying the cleaning tool. The shaft is slotted parallel with its axis, to carry a sliding steel blade against which the air presses and thus revolves the shaft. Fig. 206 shows the casing, blades, and inlet and exhaust ports. Oil for lubricating the motor is carried through the hose from a sight-feed oiler attached to the air-pressure pipe. Steam may be used in this cleaner, instead of air.

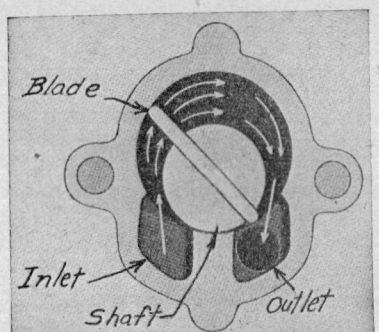

Fig. 206. — Section of Pneumatic Cleaner showing Operating Blade.

A tube cleaner driven by a water turbine is also shown in Fig. 205. A pressure of 100 to 125 pounds per square inch gage is desirable for rapid work. The turbine wheel is at the open end of the hardened steel casing, and drives the rotating shaft which carries the cleaning tool. A steel shrouding band over the ends of the

SCALE REMOVAL 295

turbine buckets prevents loss of water radially and thus increases the efficiency.

The most important part of a mechanical cleaner is the **cutting head,** several types of which are shown in Fig. 207. The pins of the straight chilled-iron cutter wheels are centered in forged steel blocks that move

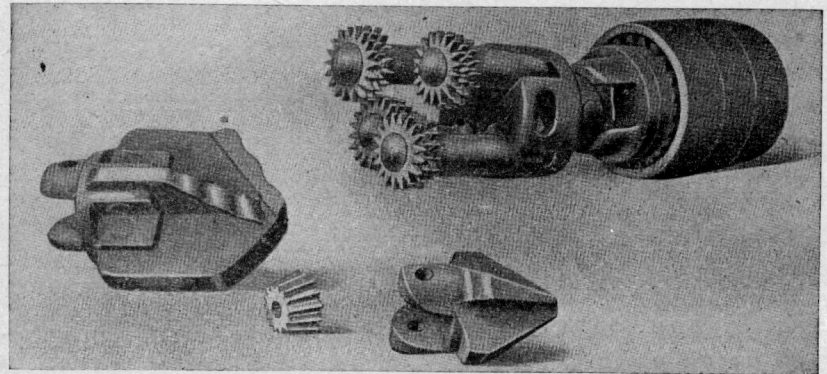

FIG. 207. — Cutting Heads for Scale Removers.

freely in slots in the supporting disks. The head is often driven by a pulley, operated from a motor or engine by a belt, and the fragments of scale are carried away by water supplied to the boiler tube through a hose connection.

Electrically driven motors, are located outside the tubes and drive the cutter head by a straight shaft, or flexible jointed connections to allow for bends in the tubes.

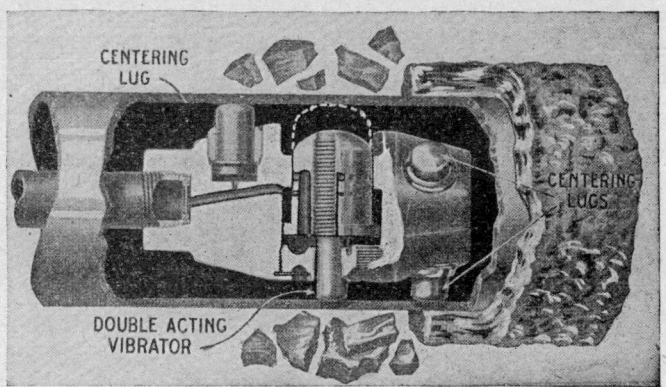

FIG. 208. — Vibrating Type of Fire-tube Scale Remover.

For cleaning scale from the outside of fire tubes, the cleaner head may be like those previously described, except that a swinging percussion tool or knocker head is attached to the rotating shaft; or the head may carry a vibrating plunger, Fig. 208. A large number of rapid, light blows are delivered against the inside of the tube, and the scale is cracked off as a result of the vibrations. Compressed air or steam should be used as the motive power instead of water, in cleaning fire tubes, because it is generally inconvenient to dispose of the water in fire-tube boilers.

REFERENCES

Catalogues of WM. B. SCAIFE and SONS Co.
Catalogues of COCHRANE CORPORATION.
Hydrogen Ion Concentration, W. A. TAYLOR, Vol. LXV, Water Works.
Boiler Feedwater Purification, S. T. POWELL.
See also references, Chapter IX.

REVIEW QUESTIONS

1. Name three methods of purifying feedwater.
2. Describe the construction and operation of an intermittent water-softening plant.
3. By what means is scale, which has accumulated on the surfaces of tubes, removed?
4. Describe the construction of the apparatus used to remove scale from a water tube.
5. Describe the method of removing soot with a permanent soot-removal apparatus.
6. Explain what is meant by pH as applied to boiler feedwater. Which has the greater alkalinity, a feedwater having a pH of 8, or a pH of 9?
7. Explain how water is softened when the zeolite process is used.
8. Describe the bent-tube evaporator, and explain what is meant by a multiple effect evaporator.

CHAPTER XIII

DRAFT AND METHODS OF PRODUCING DRAFT

282. Foreword. — The steaming rate of a boiler depends upon the rate at which coal can be burned. This in turn depends upon the difference in static pressure[1] available to produce a flow of air, through the fuel bed on the grate, to the chimney. This difference in static pressure is known as **draft** and is required to overcome various obstacles which retard the flow of air. Draft is ordinarily measured by the use of a draft gage, and **its intensity is expressed as head in inches of water,** at some standard temperature. A multiple-scale draft gage having as an operating mechanism a slack leather diaphragm and a phosphor-bronze cantilever spring attached to each pointer is shown in Fig. 209.

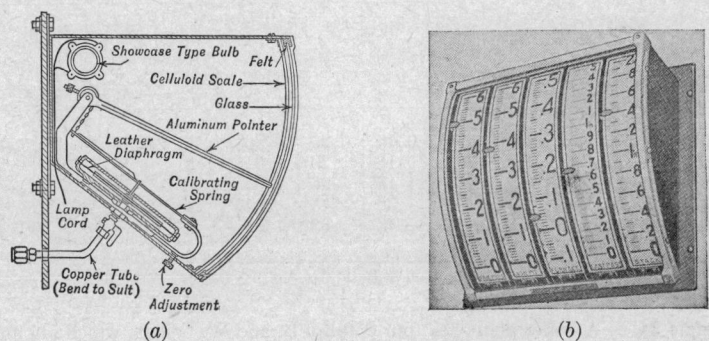

FIG. 209. — Hays Multiple-scale Draft Gage.

The draft necessary for a given plant depends upon two factors: the pressure loss caused by retarding the gas flow by the breeching, boiler damper, baffle walls, tubes, superheaters, economizers, and preheaters; and the depth and kind of fuel on the grates.

The draft required to overcome the friction, caused by the flow of gases in the flue and breeching, is small. It amounts to approximately 0.13 of an inch of water per 100 foot length of flue and is increased 0.05 of an inch of water for each right-angle bend in the flue.

To overcome the resistance offered to the flow of gases by the baffle walls and boiler tubes, the amount of draft required varies with the rate at which the boiler is working, the type of boiler, and type of baffling used. Table 25 gives the draft required for a few typical boilers at normal rating.

[1] See Art. 294, page 313.

TABLE 25. — LOSS OF DRAFT IN BOILERS

Type of Boiler	Inches of Water
Horizontal return-tubular	0.25 to 0.30
Babcock and Wilcox	.20 to .35
Heine	.49
Stirling	.51
Wickes vertical tubular	.43
Cahall vertical tubular	.45

The pressure required to force the air needed for combustion through the fuel bed on the grate is one of the most important factors in determining the necessary draft. This pressure varies with the kind of fuel and the rate of burning, as shown in Table 26.

TABLE 26. — DRAFT BETWEEN FURNACE AND ASHPIT TO BURN COAL*

Kind of Coal	Coal Burned per Square Foot of Grate per Hour, Lb.						
	15	20	25	30	35	40	45
	Force of Draft Inches Water						
Bituminous, Ill., Ind., Kan	0.14	0.20	0.26	0.33	0.40	0.48	0.57
Bituminous, Ala., Ky., Pa., Tenn.	.16	.23	.31	.40	.49	.60	.72
Semi-bituminous	.18	.26	.35	.45	.57	.71	.87
Anthracite Pea	.30	.45	.64	.88	1.23	...	...
Anthracite Buckwheat No. 1	.43	.68	1.00	1.50	...	...	...

* From Mechanical Equipment of Buildings, Vol. II, Harding and Willard.

Example 34. — A boiler plant has four 250-hp. B. and W. boilers, which at times are operated at 200 per cent of their rated capacity. A flue 50 ft. long and having two right-angle bends connects the boiler to the chimney. Semi-bituminous coal is burned at a rate of 15 lb. per sq. ft. per hr. under normal operation. Temperature of gases 550 deg. fahr., outside temperature 70 deg. fahr. Determine the draft necessary to overcome these resistances.

Solution. — At 200 per cent rating the boiler would burn 30 lb. of coal per sq. ft. of grate surface, and the friction losses through the grate and boiler would be double those given in Tables 25 and 26, consequently

Loss through grate, Table 26 = 0.45 in.
Loss through boiler, Table 25 = 0.40 "
Loss in flue = $\frac{1}{2} \times 0.13$ = 0.07 "
Loss in bends = 2×0.05 = 0.10 "

Total loss, exclusive of stack loss = 1.02 in. water

The addition of an economizer will increase the necessary draft by about 0.30 of an inch of water, a superheater will add about 0.15 of an inch and an air preheater another 0.30 of an inch to the total amount of draft required without their use. The total draft should be sufficient to operate the boiler at its maximum rating and ordinarily varies from 1 inch to 18 inches.

There are two kinds of draft: (1) natural draft, produced by chimneys; and (2) mechanical draft, produced by fans.

283. Natural Draft. — *Draft, as produced by a chimney, is caused by the difference in weight between a column of hot gases inside the chimney and the weight of a column of cold gases, of equal height, outside the chimney.* Chimney draft varies

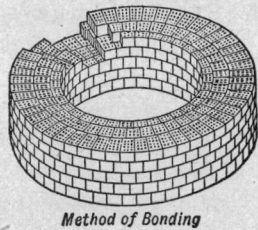

Method of Bonding

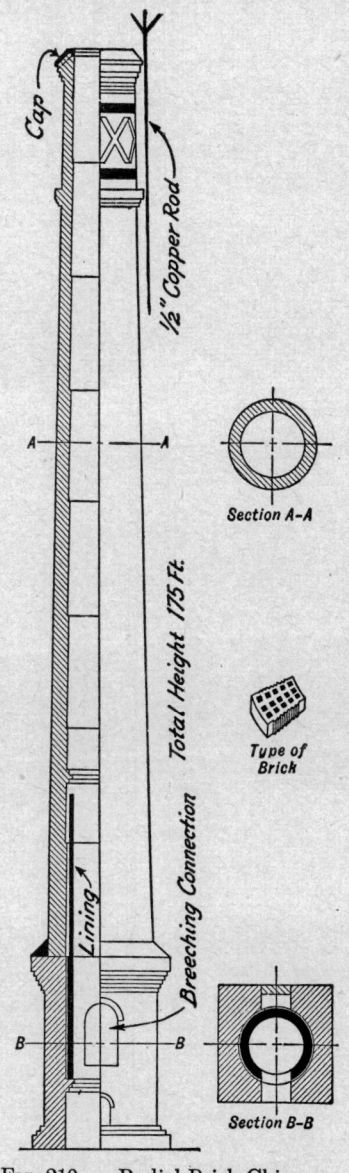

Fig. 210. — Radial Brick Chimney.

with climatic conditions, and is less when the outside temperature is high. The most common types of chimneys are made of: (1) brick, (2) steel, and (3) concrete.

284. Radial Brick Chimney. — This type of brick chimney has virtually replaced the common brick chimney. It is circular in section and is made of moulded radial perforated brick, which are curved to conform to the inside and outside curvature of the chimney. The perforations aid in securing good quality of brick during burning and serve to increase the bonding action between bricks. The bricks vary in size and are laid as illustrated in Fig. 210, each row of brick being covered with a thin layer of cement. The small dead-air spaces, thus formed in each layer of brick, provide an excellent insulation and, by keeping the chimney gases hot, aid in producing good draft. The sides of radial brick chimneys usually taper 2

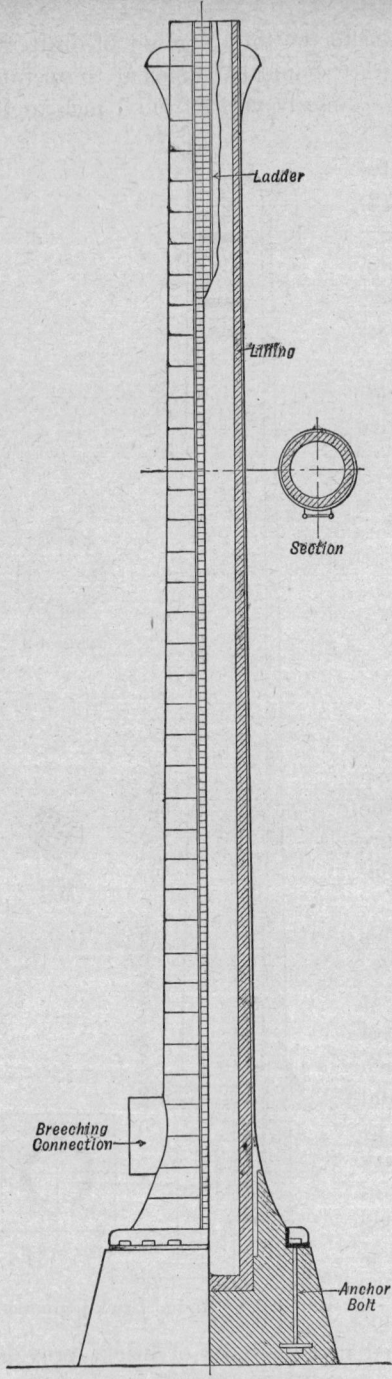

Fig. 211. — Self Supporting Steel Chimney.

feet in 100 feet of height. An inner firebrick lining usually extends to a height of 35 or 50 feet above the flue connection. At the top the chimney is **corbeled** and covered with a cast-iron or terra-cotta **coping** or cap to protect the top from the action of the weather and to hold it in position. The chimney usually rests upon a concrete foundation distributed over sufficient area to prevent settling or tipping, and a **clean-out door** and **flue connection** are provided near the base.

285. Steel Chimneys. — Steel chimneys, or stacks, are either self-supporting or guyed. The **self-supporting** steel chimney, shown in Fig. 211, is generally made of a series of rings having single-riveted lap joints. The steel plate varies from $\frac{3}{16}$- to $\frac{1}{2}$-inch in thickness, depending upon the size of the chimney. Cylindrical rings of two different diameters may be used in one chimney, the wider alternating with the narrower; or, all the rings may be of the same diameter and slightly conical, to allow the top of each ring to fit inside the bottom of the ring next above. The various rings are fastened together by a single row of rivets, and the bottom ring is flared to form a base which is held to the foundation by anchor bolts passing through lugs on the base. The steel structure above the boiler is often used to support the stack, in order to reduce the ground space required.

This type of chimney depends for its stability on the anchor bolts and the foundation. It is ordinarily lined to prevent excessive radiation and to protect the inside from the corrosive action of the gases; the first 50 feet of the lining is ordinarily made of firebrick and the remainder of common red brick. The lining is usually of uniform thickness, set in contact with the steel and thoroughly grouted to prevent rapid depreciation. It is sometimes made in independent sections resting on brackets riveted to the steel plate. The stack is reinforced at the opening to which the breeching connection is attached and is protected from lightning by several platinum tipped copper joints connected to a ground plate by a cable.

The **guyed** steel chimney may be supported on a foundation, or directly by the boiler shell, as shown in Fig. 2, page 4, and when thus supported, it is short and limited in weight. The pressure of the wind is generally resisted by three steel cables, the upper ends of which are fastened to an angle or T-band attached at about two-thirds the height of the chimney, and the lower ends to the ground or building at such a distance from the chimney that the guys make an angle of about 60 degrees with the vertical.

286. Concrete Chimney. — The concrete chimney, Fig. 212, is finding favor with many engineers, and is being used quite extensively. The sides of the chimney are made either straight or tapered. When they are straight, the thickness of the wall at the top is about 6 inches, and when tapered about 4 inches, with a taper of 4 inches per 100 feet. Since concrete is strong in compression and weak in tension, vertical steel rods are used to reinforce the concrete, in order to resist the tensile strains resulting from the pressure of the wind. Horizontal rings of steel are used to take up stresses caused by temperature changes, and to assist in holding the vertical rods. The walls are monolithic, and air leakage is thus prevented. Concrete chimneys should ordinarily be lined with hard-burned brick for about 50 feet of their height, and the space between the lining and the outside wall should be covered with a concrete cap.

287. Comparison of Chimneys. — The life of brick and concrete chimneys is longer than that of steel chimneys, the former retaining their usefulness for about fifty years, and the latter from five to fifteen years, depending upon the care taken to prevent corrosion. Because of cracks and imperfect bonding of the bricks, the brick chimney may have considerable air leakage, which the concrete and steel chimneys do not have.

The concrete chimney is light in weight, requires a smaller space than either steel or brick chimneys, has great resisting power, and can be rapidly constructed, 6 feet per day being an average rate. If unlined, it may have an excessive loss of heat by radiation, which tends to lower the draft produced.

The steel chimney can be constructed at a more rapid rate and has less weight than either the brick or the concrete chimney. It, however, re-

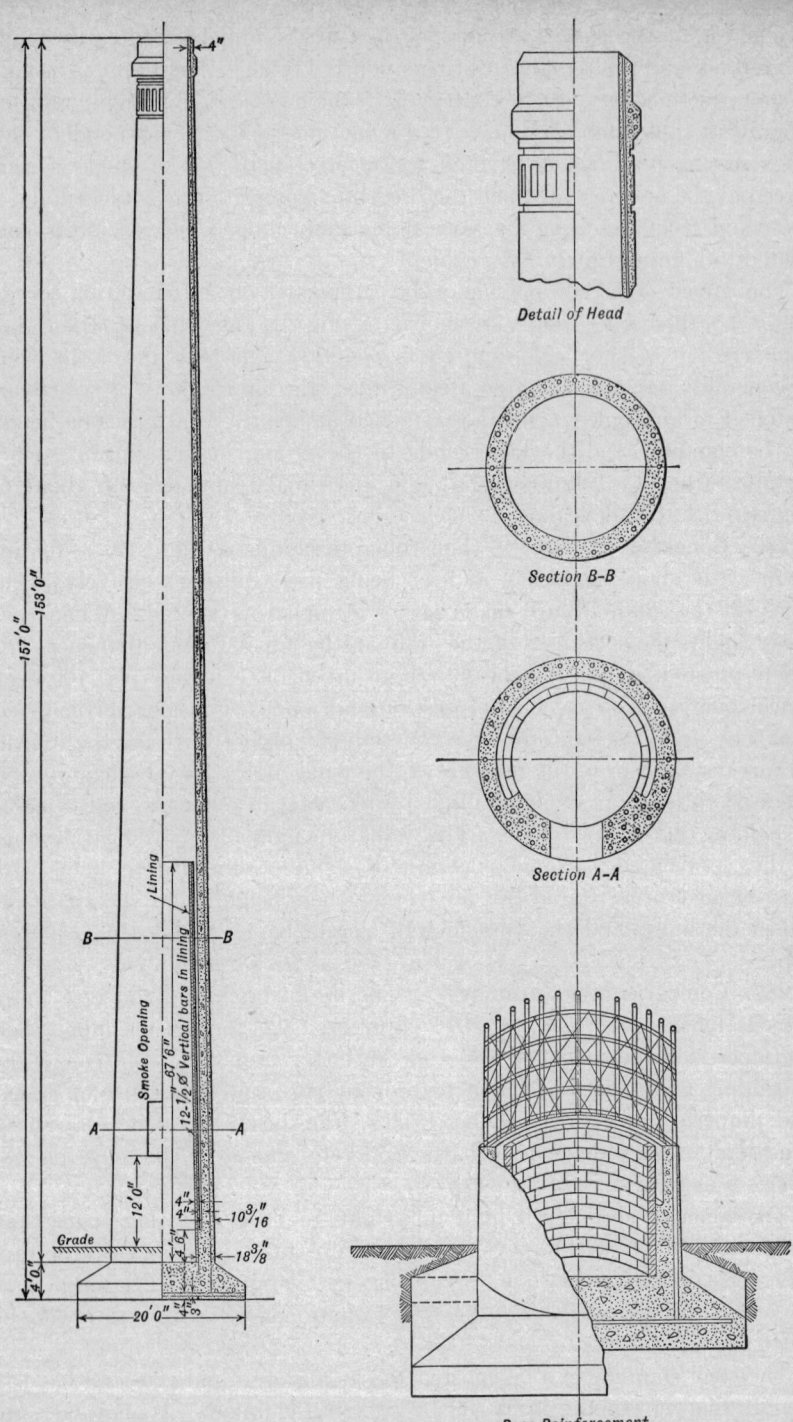

Fig. 212. — Weber "Coniform" Concrete Chimney.

quires frequent painting to prevent corrosion, which is rapid in air that contains salt water, as when near the ocean.

The cost of the brick chimney is the highest, and that of the unlined steel stack the lowest. The variation in the cost of concrete chimneys, based on costs in 1917, was from $2000 for a chimney having an inside diameter of 4 feet and a height of 105 feet, to $18,000 for a chimney with an inside diameter of 16 feet and a height of 258 feet.

288. Determination of Chimney Size. — There are two steps in determining the size of a chimney: *first*, finding the net area required to pass the maximum volume of flue gases which will be discharged by the boilers when operating at maximum rating, and *second*, finding the requisite height above the grates to give a draft somewhat in excess of the sum of the required furnace draft plus all draft losses.

The *net area* of the chimney is ordinarily based on the volume of the gases to be discharged at the maximum rating and an assumed velocity[1] in the chimney. This net area determines *the* **capacity** *of the chimney, that is, the number of cubic feet of gases the chimney will discharge per hour*. The capacity determines the number of pounds of coal a plant will burn and it is influenced by the ratio of the height of the chimney to the diameter. In general, the diameter should be about 8 per cent of the height. The *volume of the furnace gases* is determined by the weight of coal that can be burned per square foot of grate surface and the weight of air per pound of fuel. For average conditions, using coal in a hand-fired boiler operating at 150 per cent of rating, 120 pounds of gas per boiler horsepower are produced, and when using oil 55 pounds of gas per boiler horsepower with 50 per cent excess air.

The *height of the chimney* required to produce a given draft is obtained by calculation, using the weight of gas inside and the weight of air outside the chimney, and varies with the elevation above sea level. To determine the amount of draft produced by a chimney, consider a chimney filled with hot gases and closed at the bottom by a horizontal diaphragm. The pressure of the air, at the top of the chimney, produced by the atmosphere above that level, is the same on the gases inside and the air outside the chimney. *The pressure on the upper side of the diaphragm* is, therefore, the sum of the air pressure at the top of the chimney plus the pressure produced by the column of hot air inside the chimney. *The pressure on the under side of the diaphragm* is the pressure of air at the top of the chimney plus the pressure produced by a column of cold air equal to the height of the chimney. Since the air inside the chimney has less weight per cubic foot than the air outside, there is a difference in pressure on the two sides of the diaphragm equal to the *draft produced by the chimney*. This may be written as an equation, thus:

---

[1] This velocity is usually taken from 20 to 30 feet per second.

$$D = \frac{12H}{K}(d_c - d_h). \quad\ldots\ldots\ldots\ldots (81)$$

in which D = intensity of draft in inches of water.

K = density of water in U-tube, pounds per cubic foot.
 = 62.3 for a temperature of 70 deg. fahr.

H = the height of the chimney above the grate level, feet.

d_c = density of the cold air outside the chimney, pounds per cubic foot.

d_h = density of the hot gases inside the chimney, pounds per cubic foot.

With the diaphragm removed, as in an actual chimney, this difference in pressure produces a movement of gases from the bottom to the top of the chimney. When one leg of a U-tube partly filled with water is connected to the inside of the chimney and the other leg open to the air, the level of the water in the two legs will be displaced, as shown in Fig. 65, page 76. The difference in level between the surface of water in the two legs is the draft, measured in inches of water.

Equation (81) may be put in a more convenient form by substituting, for the values d_c and d_h, terms containing the corresponding absolute temperatures. This can be done by using the combined law of gases, $PV = WRT$, in which P is the absolute pressure in pounds per square foot; V = volume in cubic feet; W = weight in pounds; R = a constant = 53.37 for air; and T = absolute temperature, deg. fahr.

By writing $V = 1$, and P = pressure in pounds per square foot at sea level = $14.7 \times 144 = 2116.8$, in the above equation, W will equal the density, or weight per cubic foot, at atmospheric pressure; and for conditions outside and inside the chimney

$$W = d_c = \frac{39.7}{T_c}, \quad \text{and} \quad W = d_h = \frac{39.7}{T_h}$$

in which T_c and T_h equal respectively the absolute temperature of the air outside and the gases inside the chimney.

Substituting these values of d_c and d_h in Equation (81) and assuming the density of the gases to be the same as that of air at the same temperature, there results:

$$D = 7.64 H \left(\frac{1}{T_c} - \frac{1}{T_h}\right) \quad\ldots\ldots\ldots\ldots (82)$$

in which

D = the total draft produced by the chimney in inches of water.

H = the height of the chimney above the grate in feet.

T_c = absolute temperature of outside air, deg. fahr.

T_h = absolute temperature of inside gases, deg. fahr.

EMPIRICAL CHIMNEY FORMULAE

When the gas temperature is taken as it leaves the boiler, the **available draft** is taken as 0.80 of D in Equation (82), to allow for friction losses in the chimney itself, and for the cooling of the gases in the flue and chimney. The draft lost in the chimney, h_c, may, however, be calculated from the equation:

$$h_c = \frac{fW^2CH}{A^3} \qquad \qquad (83)$$

in which W = weight of gases, pounds per second.

f = 0.0015 for steel, and 0.002 for brick at 600 deg. fahr.
C = perimeter of chimney, feet.
H = height of chimney, feet.
A = cross-sectional area of chimney, sq. ft.

Equation (82) is suitable for all altitudes up to an elevation of 1500 feet, above which the height for a given draft should be increased inversely as the barometric pressures.

The amount of draft produced by a chimney is seldom above 1.5 inches. This limits the height of the chimney to about 250 feet, and for installations serving 700 to 800 boiler horsepower, the height is ordinarily limited to a minimum of 150 feet. A given height of chimney when used for oil is suitable for double the rating that it could serve when operated with coal. Chimneys for high buildings are made small in order to increase the friction and thus reduce the available draft.

Example 35. — A chimney 100 ft. high is filled with gases at 550 deg. fahr., when the outside air temperature is 70 deg. fahr. Find the available draft produced.

Solution. — Using Equation (82) and making proper substitutions,

$$D = 7.64H \left(\frac{1}{T_c} - \frac{1}{T_h} \right) = 7.64 \times 100 \left(\tfrac{1}{530} - \tfrac{1}{1010} \right) = 0.76 \text{ in. of water.}$$

$H = 100$ ft.; $T_h = 550 + 460 = 1010$ deg. fahr.; $T_c = 70 + 460 = 530$ deg. fahr.

The available draft $= 0.80 \times 0.76 = 0.61$, or the friction loss in the chimney could be calculated and subtracted from D above.

For the data given in Example 34, a chimney 168 ft. high would be required, but would probably be made 175 ft. high.

289. Empirical Chimney Formulae.[1] — Various empirical equations are used to check the height and area of chimneys. These should be used with caution, since conditions have so changed as regards the kind of boilers, combustion rates and use of stokers. When used, the results obtained from these equations should be checked by taking into account all the factors which have been previously discussed. The most common of these equations is that proposed by William Kent, and is as follows:

$$\text{Boiler horsepower} = 3.33\ E\ \sqrt{H} \qquad \qquad (84)$$

[1] For other equations, see Gebhardt's Steam Power Engineering.

in which E = effective area of a chimney in square feet. For circular chimneys, Kent considered a dead air space 2 inches wide extending around the inside of the chimney.
$= A - 0.60 \sqrt{A}$, in which A = actual area of chimney, square feet.
H = height of chimney in feet.

Kent's equation assumes the burning of 5 pounds of coal per boiler horsepower per hour, and that 24 pounds of air is used per pound of coal. For other rates of combustion Equation (84) should be multiplied by the ratio of 5 to the new rate of combustion.

Example 36. — Find the necessary area to discharge the gases formed in a boiler plant of 981 boiler hp. burning coal at a rate of 5 lb. of coal per boiler hp. Height of chimney to be 200 ft.
Solution. — Using Equation (84)

$$E = \frac{\text{Boiler hp.}}{3.33 \sqrt{H}} = \frac{981}{3.33 \sqrt{200}} = 20.83 \text{ sq. ft.}$$

Boiler hp. = 981, H = 200 ft., E = effective area.

The effective area $20.83 = A - 0.6 \sqrt{A}$.
Solving for A the desired area is 23.76 sq. ft. and the corresponding inside diam. 66 in.

290. Mechanical Draft. — In order to reduce the necessary height of chimney and at the same time obtain a draft that is independent of weather conditions and is easily and positively controlled, fans and blowers are used. Two systems, known respectively as induced draft and forced draft, are used to produce draft by mechanical means. In the **induced draft system** the pressure over the fuel bed is reduced below that of the atmosphere, by means of a fan located between the boiler setting and chimney, and air passes through the fuel bed because of this difference in pressure. In the **forced draft system,** air under pressure is forced either through the fuel bed on the stoker or grates, or the pulverized fuel burners, or the oil burners, into the furnace, and the gases are removed from the furnace by either a chimney or induced draft fans. In each of these systems, the power required to operate the fan varies from 4 to 6 per cent of the boiler capacity, and for forced draft rarely exceeds 7 kilowatts per ton of coal. The induced draft fan must be of larger capacity, in the ratio of approximately 2 to 1, than the forced draft installation because of the increased volume resulting from the hot gases handled. A combination of forced and induced draft, commonly known as **balanced draft,** is sometimes used in which the pressure above the fire is maintained approximately atmospheric.

291. Induced and Forced Draft System. — In the majority of power plants a combined forced and induced draft system is used, with forced draft supplying the air for burning the fuel, and the induced draft removing the gases from the furnace. An installation of this type is shown in Fig. 213, in which the induced draft fan overcomes the draft losses through the boiler, economizer, air preheater and connecting flues, and the forced draft fan overcomes the resistance of the air preheater on the inlet side and the resistance of the forced-draft chain-grate stoker on the discharge side. The induced draft fans are generally installed in duplicate, and are located on a suitable foundation directly above the boilers. In small installations, **steam jets,** which are simple, cheap, and easily applied, are

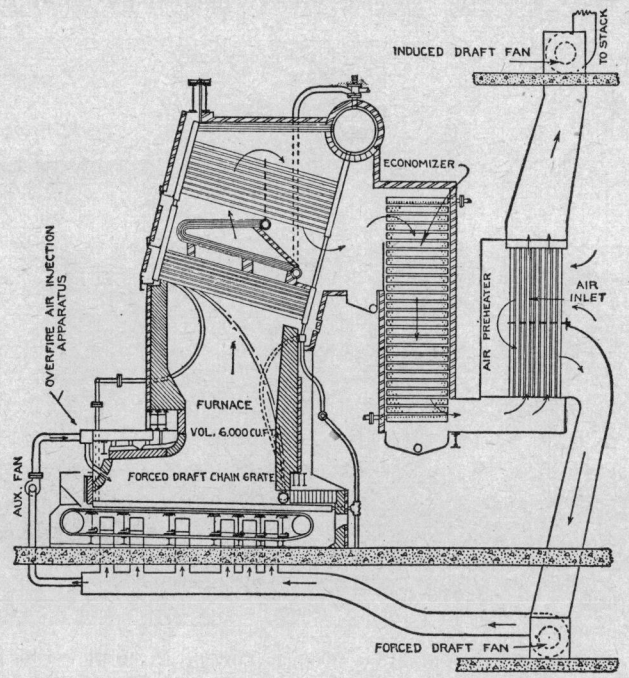

Fig. 213. — Forced and Induced-draft Installation.

sometimes used as an aspirator to remove the gases from the furnace. Forced draft is required by all underfeed stokers carrying a thick fuel bed where the draft for satisfactory operation usually varies from 4 to 18 inches of water with the fans installed either above the boiler, or in the basement below the furnaces. With either system, the fans may be driven by either a small vertical engine, a steam turbine, or an electric motor, depending upon the conditions under which the installation is made and are usually arranged to control the fans automatically.

The **Prat, or Evasé, system** of producing induced draft is used extensively in Europe, and its use is increasing in the United States. A recent adaptation of this system, known as a **Thermix, or streamline stack,** Fig. 214, combines in a single unit the fan, breeching connections, and the various interrelated parts which make up the draft-producing apparatus as a whole. The most common type of this stack consists of a double inlet induced draft fan built into the base of the circular stack which flares outward from the fan connection at its base, giving a shape like an inverted frustrum of a cone. The sides of the stack are tapered about seven degrees, to produce the least possible resistance to the passage of the gases, and to act as a diffuser to diminish the energy loss which would result from dis-

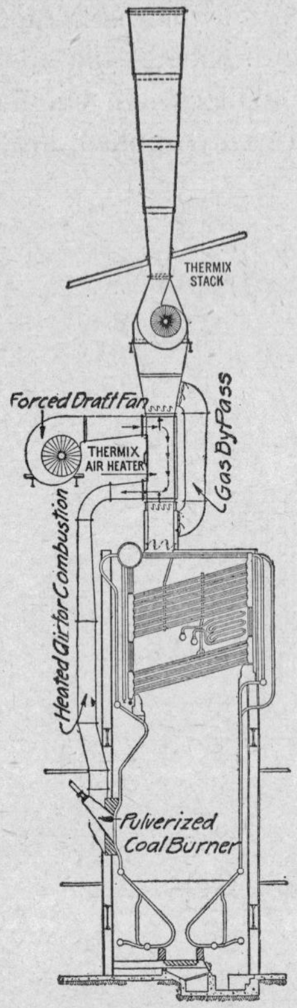

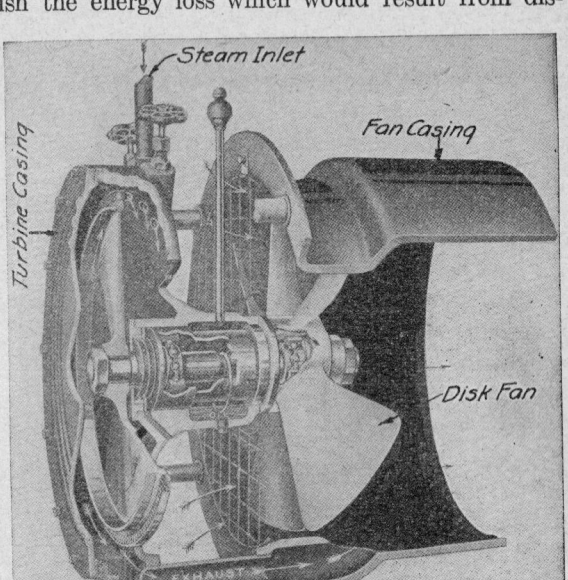

Fig. 214. — Thermix Draft System.

Fig. 215. — Coppus Turbine-driven Blower for producing Forced Draft.

charging the gases at high velocity into the air. The system is arranged either to have the fan handle the entire amount of gas, or by changing the location of a damper provided, may operate with nearly all natural draft. There are several different arrangements of this stack in addition to the one described above, designed for special purposes and to give various drafts.

A type of forced draft installation, which is used for both stationary and marine service, consists of a small turbine-driven blower, Fig. 215. When used for stationary service, the blower is placed at an opening in the side wall of each boiler setting; and when used for marine service is generally supported by a steel framework attached to the frame of the ship and above the firing room. This type of blower is well suited to the forced draft system, since it can be operated at high speeds and is easily controlled.

BALANCED DRAFT SYSTEM

For marine work, two systems of forced draft are used, namely, the closed ashpit and the closed boiler room. In the **closed ashpit system**, air is supplied by a suitable fan, through a duct, to the ashpit of the boiler, which is made air-tight. In this system the draft should be shut off when firing, to prevent forcing fire into the boiler room. With the **closed boiler room** or **Howden system** the fan is located either in the boiler room compartment, or at a suitable location outside the boiler room, and connected to the boiler room by suitable ducts as shown in Fig. 7, page 15. The boiler room is then made tight against air leakage and the ashpits are open. When this method is used, the boiler tubes are sometimes injured by a rush

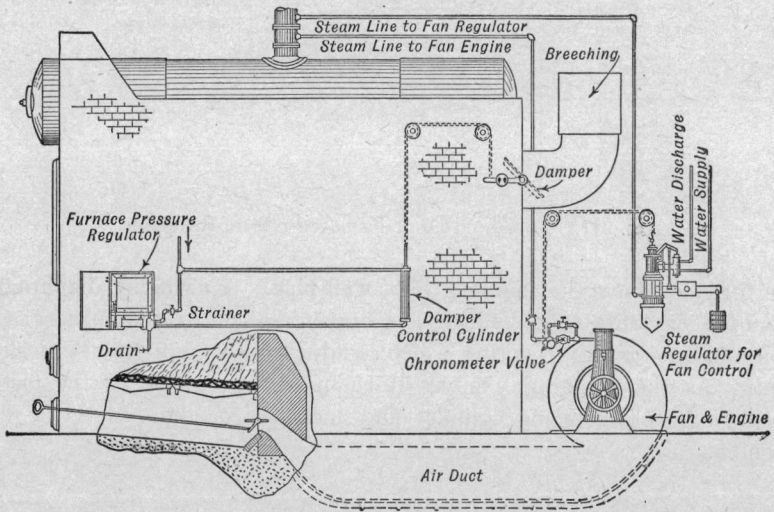

Fig. 216. — Diagram showing Balanced Draft with Steam Controlled Fan and Damper Regulator.

of cold air through the firing doors during firing. This difficulty is not encountered when oil burners are used.

292. Balanced Draft System. — A typical **balanced draft system** of combustion control is shown in Fig. 216, with the various operating parts named. With this system the draft over the fire can be maintained within 0.02 of an inch of water of that desired. The system operates as follows: when an increase of load occurs, the pressure will drop in the steam header; the steam regulator will then speed up the fan, thus forcing more air through the fuel bed and increasing the rate of combustion. This will produce more gas and tend to decrease the furnace draft. The furnace pressure regulator then increases the damper opening sufficiently to remove the gases at the new rate and thus maintain the predetermined draft. With stoker installations, a compensating stoker can control the rate of fuel fed to meet the demand.

The **steam regulator**, Fig. 216, is a refined boiler damper regulator, which controls a reducing valve in the steam line to the forced draft fan and thereby controls its speed. The **furnace pressure regulator**, Fig. 217, is mounted on the side wall of the furnace and connected with

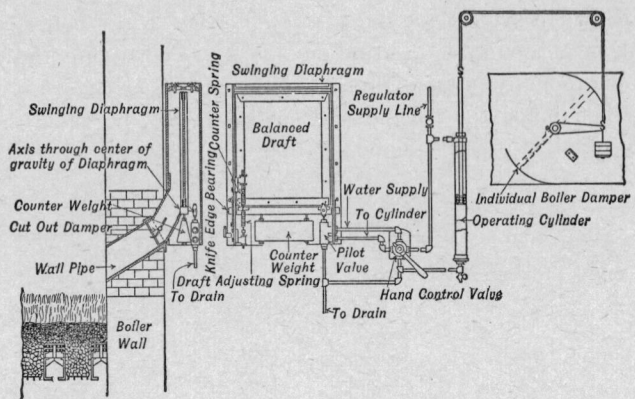

Fig. 217. — Balanced Draft Furnace Pressure Regulator.

the furnace, above the fuel bed, by a **wall pipe**. A **swinging diaphragm**, mounted on **knife-edge bearings**, is drawn inward by the furnace suction and outward by a spring which is adjusted to maintain the desired draft. As the diaphragm moves it changes the position of the piston in the damper-operating cylinder and increases or diminishes the flue opening.

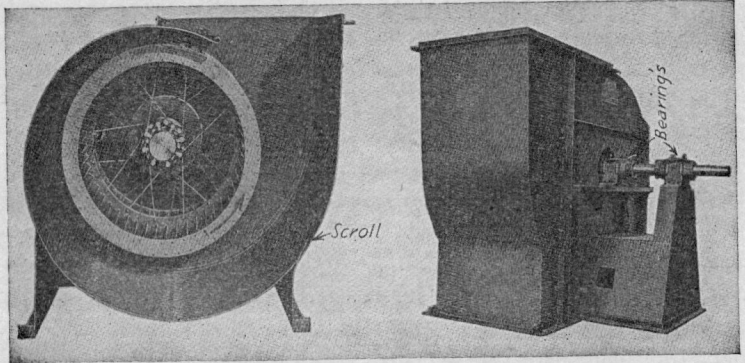

Fig. 218. — Internal and External View of Bayley Flexiform Fan.

293. Fans Used to Produce Draft. — Fans used for the production of artificial draft in modern furnace installations are of either the **centrifugal type** or the **propeller type**. In the former, the air is drawn in axially from one or both directions and discharged tangentially; and in the latter, the air is discharged axially by using propeller type blades. Centrifugal fans are used in the majority of cases and are generally designated by the type of the rotating element used, the most common of which are: (1) *steel plate,* (2) *multi-vane,* (3) *conoidal,* and (4) *radial flow.* The multi-vane fan is the most efficient of the first three types. As generally constructed, a fan, Fig. 218, consists of a **wheel** or **impeller,** to which plates or vanes are attached, and which rotates within a sheet metal **casing** or **housing** having a scroll shape to permit the air to leave the impeller with the minimum of disturbance. The scroll may be built with a *full* or *three-quarter* housing, and may be arranged for *top* or *bottom horizontal discharge,* or *top* or *bottom vertical discharge.*[1] The full housing fan has the entire housing above the foundation or floor, and the three-quarter housing fan has one-quarter of the enclosure surrounding the wheel

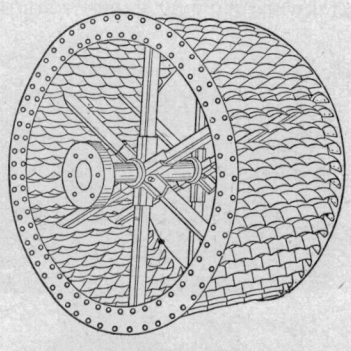

Multi-vane.

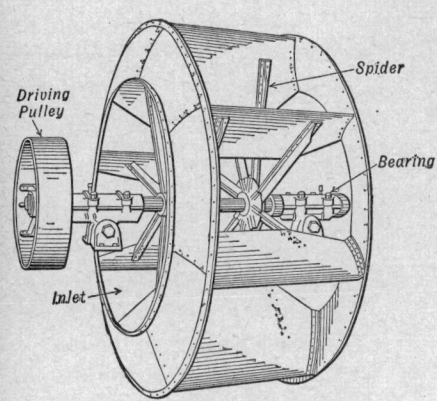

Plate or Paddle-wheel.

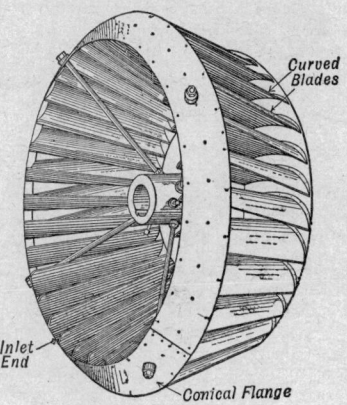

Conoidal.

Fig. 219. — Typical Fan Impellers.

formed by the foundation. When used for induced draft, the inlet to the housing is generally from one side and is called **single inlet,** but when used for forced draft the inlet may be from both sides and it is called **double inlet.**

The impeller or spider is supported on a shaft, which rotates in bearings generally ring oiled and water cooled when used for induced draft, and carries at its circumference blades which may be of any of the forms shown in Fig. 219. The curvature of these blades, with respect to the direction

---

[1] For other arrangements consult manufacturers' catalogs.

of rotation, is a vital factor in determining the **dynamic pressure** producing capacity with respect to the peripheral speed. When the blades curve forward, the air leaves the tips at a velocity higher than that of the impeller, with consequent higher pressure of discharge. Conversely, when the blades incline backward, the air leaves the tips at a lower speed than that of the impeller and with a lower pressure of discharge. An impeller with backward curving vanes is adapted to direct high-speed drive and in order to prevent vibration must be given a running balance test. The impeller may be arranged to overhang the bearing, when it is called an **overhung impeller,** or it may be supported from both sides. In operation, air is drawn into the inlet as the impeller revolves, is delivered at the circumference of the impeller, and flows along the casing to the outlet at a pressure equal to the centrifugal pressure generated by the revolving impeller.

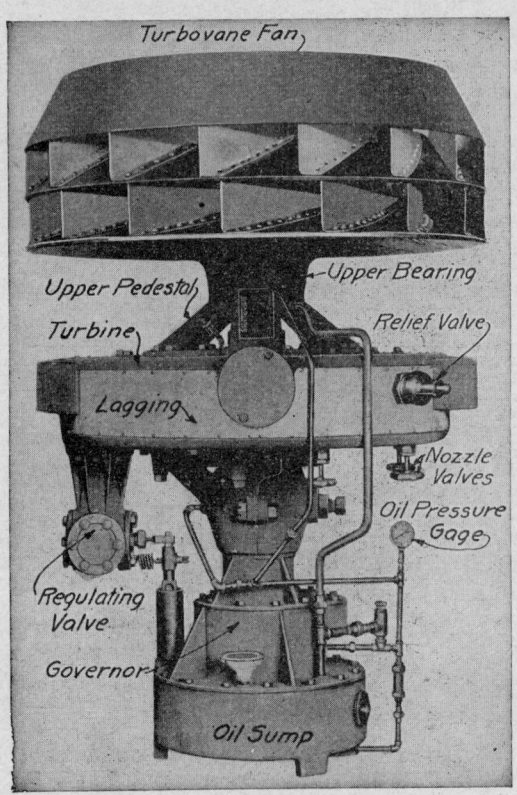

FIG. 220. — Turbo-vane Turbine-driven Fan.

The radial flow type of fan, Fig. 220, has an impeller so constructed that it causes a radial flow of air, and thereby increases the efficiency of the fan.

For induced draft installations, the volume delivered by the fan may be regulated by an ordinary blade or louvre damper placed either in the discharge or inlet of the fan which runs at constant speed, or the volume may be controlled by varying the speed of the prime mover, using a slip-ring motor. Recently a fan, Fig. 221, has been developed to run at constant speed and the volume effectively controlled by changing the relative position of vanes, located at the inlet of the fan, and which are provided to reduce the inlet losses and give a better distribution of gas in the inlet box. The vanes are opened and closed by overlapping each other, similar to the action of the shutter leaves of a camera. This is done by a link motion actuated by a worm gear.

An induced draft fan arranged to produce draft and at the same time collect cinders is shown in Fig. 222. It makes use of the sudden changes in air direction and velocity, that occur in all **paddle-wheel type fans,** to separate the cinders from the gases. The cinders and flue gases enter at the inlet of a paddle-wheel fan, the blades of which are provided with channels to catch and direct the cinders into the chambers in the sides of the casing.

294. Elementary Fan Theory. — Fans when in operation deliver air against the **static pressure** or resistance of the system, and in addition

Fig. 221. — Sturtevant Vane-control Fan.

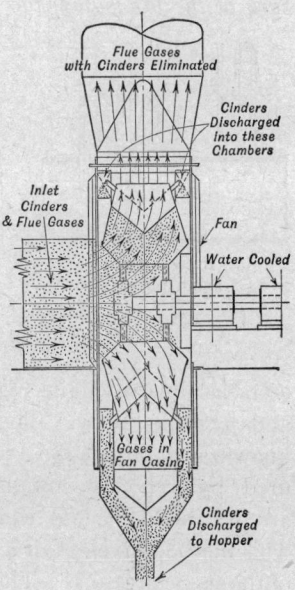

Fig. 222. — Sturtevant Cinder-vane Fan.

impart to the air a velocity at the fan outlet. This velocity is dependent on the quantity of the air and the area of the fan outlet. The **velocity pressure,** expressed in inches of water, corresponding to this velocity, may be determined from the following equation:

$$v = \sqrt{2gh} \quad \ldots \ldots \ldots \ldots \quad (85)$$

in which v = velocity in feet per second.
  g = acceleration of gravity, 32.2 feet per second per second.
  h = head in feet of fluid flowing.

By substituting in Equation (85) the value of g, and expressing h in terms of the gas density, and the average velocity pressure in inches of water, a simpler form of equation results. Thus:

$$V_a = 60 \sqrt{\frac{2 \times 32.2 \times 62.3 \times p_{av}}{12 \times d}}$$

$$= 1096 \sqrt{\frac{p_{av}}{d}} \quad \ldots \ldots \ldots \ldots \ldots \quad (86)$$

in which V_a = average velocity, feet per minute.

p_{av} = pressure drop producing velocity, inches of water.

d = density of gas flowing, pounds per cubic foot.

62.3 = density of water at 70 deg. fahr., pounds per cubic foot.

The **total or dynamic** *pressure is the algebraic sum of the velocity pressure at the fan outlet, the static discharge pressure, and the suction or inlet static pressure, and is the pressure upon which the performance and efficiency of the fan is usually based.*

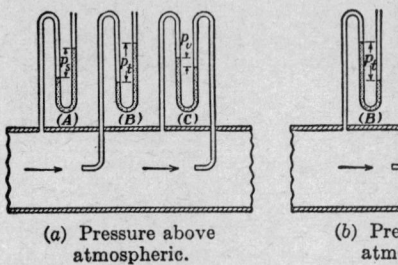

(a) Pressure above atmospheric.　　(b) Pressure below atmospheric.

FIG. 223. — Static, Total, and Velocity Pressures Illustrated.

The *total, static* and *velocity pressures* existing in ducts through which air or gas is moving, are measured by U-shaped water manometers as illustrated in Fig. 223 and are denoted by the symbols p_t, p_s, and p_v respectively. The form of tube shown at B and C, Fig. 223a is known as a Pitot tube. The *static pressure* in a duct is commonly measured by a hollow annular ring, known as a **piezometer ring,** passing around the duct and connected to the duct by a series of small tubes. This method gives a fair average of the static pressure. The *velocity pressure* is usually obtained by using a commercial form of Pitot tube, Fig. 224, which combines in a single instrument both static and total pressure readings, the velocity pressure being found by subtracting the static from the total pressure. Measurements of this kind should be made in a straight section of

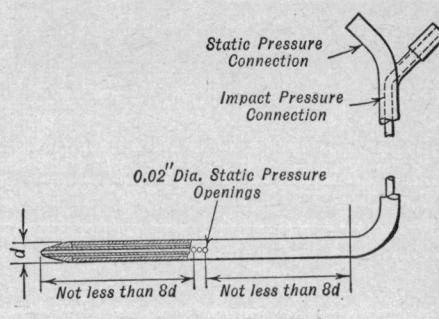

FIG. 224. — Pitot Tube.

pipe at least 20 duct diameters long. For accurate results there should be a length of duct of at least 10 duct diameters each side of the Pitot tube location. Since the velocity varies across the diameter of the duct, it is necessary to take readings at various locations across the duct and then average the readings algebraically. This average is used in all calculations. Ordinarily a single reading taken at the center of a circular duct varies between 0.91 and 0.93 of the average value.

The amount of power required to move air by a fan, generally called **air-horsepower**, may be found by using the following equation:

$$\text{Fan air-horsepower, a.hp.} = \frac{62.3 \times Q \times p_t}{33,000 \times 12} = \frac{Qp_t}{6356} \quad \ldots \quad (87)$$

in which Q = quantity of air flowing, cubic feet per minute.

p_t = total fluid pressure, inches of water.

The quantity of fluid (Q) flowing is obtained by multiplying the average velocity in feet per minute, as found from Equation (86), by the cross-sectional area of the duct in square feet.

The **mechanical efficiency of a fan** equals the ratio of the air-horsepower to the horsepower input, required for driving, at the fan shaft.

Example 37. — Determine the horsepower necessary to deliver 1000 cu. ft. of air per min. to a boiler at a pressure of 1.6 in. of water. Mechanical efficiency of the fan, 55 per cent.

Solution. — Using Equation (87)

$$\text{Air horsepower} = \frac{Qp_t}{6356}$$
$$= \frac{1000 \times 1.6}{6356} = 0.25$$
$$\text{Horsepower input} = \frac{0.25}{0.55} = 0.45$$
$$Q = 1000 \text{ cu. ft.}; \quad P_t = 1.5 \text{ in.}; \quad \text{efficiency} = 0.55$$

295. Fan Characteristics and Performance. — Characteristic curves, Fig. 225, ordinarily show the variation in the horsepower, static, and total pressure, the total and static efficiency and the blast area for changes in capacity. The **static efficiency** equals the mechanical efficiency multiplied by the ratio of the static pressure to the total pressure, and the **blast area** is the maximum effective discharge area, which an enclosed fan of this type may have, and still maintain the pressure equivalent of the tip velocity of the blades. This area is ordinarily about one-third the projected area of the wheel, i.e., $D \times W \div 3$, where D = the diameter and W the width.

Performance data are contained in tables, obtained from manufacturers, and usually give the *speed in r.p.m.*, the *brake horsepower*, and *capacity* in cubic feet per minute for static pressures within the range of the fan. Variation in the speed of the fan affects the values given in the tables as follows: the *capacity* varies directly as the speed ratio, the *static pressure* as the square of the speed ratio, and the *power* required to drive the fan as the cube of the speed ratio. These tables are used in selecting fans to meet the required conditions.

For forced-draft installations a fan with backward curved blades is usually selected, since the horsepower required to drive this type of fan reaches a maximum and then declines with further load, giving a non-overloading characteristic. Hence, if a motor is selected of a capacity sufficient for this maximum horsepower, it is impossible to overload it, no matter how the fuel bed resistance is cut down or how high the volume delivery goes. With oil and pulverized fuel firing, this is not so essential, as there is no varying fuel bed resistance to cause excessive fluctuations in air requirements.

For induced draft work, the choice generally lies between the straight blade fan and the forward curved type of design, the backward curved type generally being unsuited because it has a tendency to accumulate soot in the hollows of the blades.

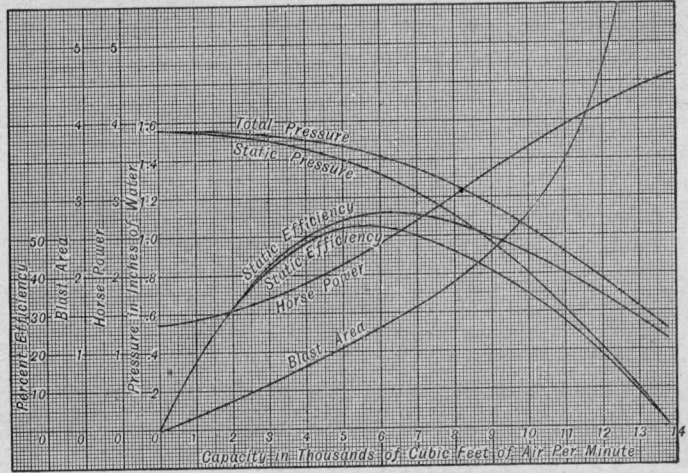

Fig. 225. — Fan Characteristic Curves — Buffalo Cinder Fan.

296. Breeching Connections. — Adjacent to the chimney and connecting the chimney to the boilers, there is usually a smoke flue or breeching. This breeching may be circular or rectangular in section, and is made of steel plates, lined to reduce the loss of heat. Its area is generally 10 to 15 per cent greater than the area of the chimney, with the branch connections entering the flue in such a manner that the gas stream is disturbed as little as possible. Several forms of breechings are shown in Fig. 226.

Breechings for large installations must be stiffened in order to prevent failure. A method which is extensively used for this purpose is shown in Fig. 227, and consists of a V-shaped expansion stiffener, so built into the breeching that it provides for expansion both vertically and horizontally. The outside of this breeching is ordinarily covered with a wire and asbestos covering to prevent excessive loss of heat.

297. Combustion Control. — The use of the right type of combustion control results in a considerable increase in boiler economy, a decrease in

COMBUSTION CONTROL

furnace maintenance, in some cases lower labor costs, and an increased generating capacity of the boiler plant itself.

The earlier forms of combustion control consisted of a damper regulator operated by water pressure, steam pressure or electricity, which supplied

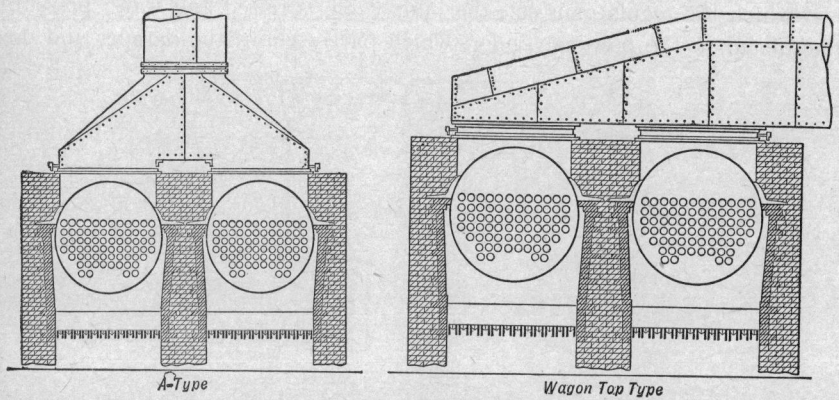

FIG. 226. — Typical Breeching Shapes.

more or *less* air in proportion to the demand for steam, leaving it for the firemen to keep the fuel bed in the desired condition to utilize this air whenever it entered the furnace. A typical hydraulic type of damper regulator is shown in Fig. 228 in which the damper is normally held open by the

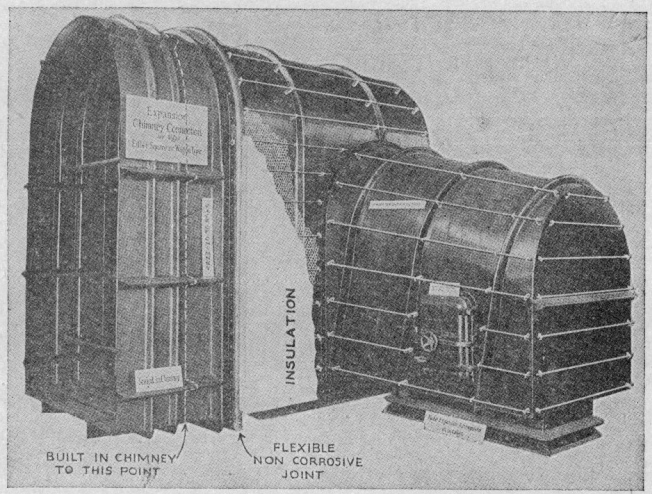

FIG. 227. — Connery's Improved Breeching.

counter-weight on the lever arm. This is the position for maximum draft. Steam pressure from the boiler acts on the upper side of the phosphor-bronze diaphragm in the steam chamber and is balanced by the counter-weight on the lever. The pilot valve is connected to this lever and controls the admission and exhaust of the water under pressure, to and from the power cylinder, which is proportioned to give a pressure seven times the water pressure. The power-cylinder piston rod is connected to the damper lever by a chain or rigid rod. With increase in steam pressure above that at which the regulator is set, the pilot valve is raised and water pressure admitted to the power cylinder, which partly closes the damper and de-

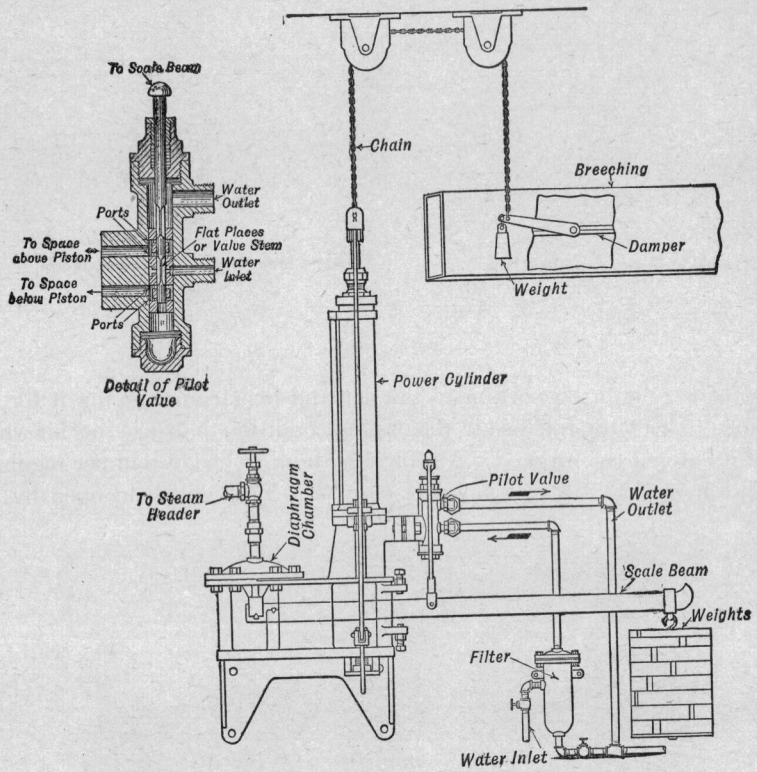

Fig. 228. — R–K Double Acting Hydraulic Damper Regulator.

creases the draft. With a fall in pressure, the counter-weight moves the pilot valve to discharge the operating fluid from the power cylinder and increases the opening of the damper.

The steam-operated regulator has steam pressure acting directly on the power-cylinder piston, which moves against a spring of the proper strength. The piston rod is connected directly to the damper by suitable means.

As mechanical stokers replaced hand firing and better oil and gas burners became available, **automatic combustion control systems** were developed. *The primary feature of these systems is that the fuel and air supply is automatically varied in proportion to all boiler loads in order to meet a fixed set of conditions.*

The **Bailey combustion control system** automatically changes the quantity of fuel and air, admitted to the furnaces, of all boilers in a plant, in order to satisfy the steam requirements as indicated by changes in the steam pressure. The air or fuel supply to each individual boiler is re-adjusted automatically to maintain the proper relation between them for best combustion efficiency. *This is accomplished by using the steam flow-air flow relation of the Bailey boiler meter as an indication of the amount of excess air being used for combustion.* If the boilers are equipped with forced, as well as induced draft, the forced draft is re-adjusted automatically

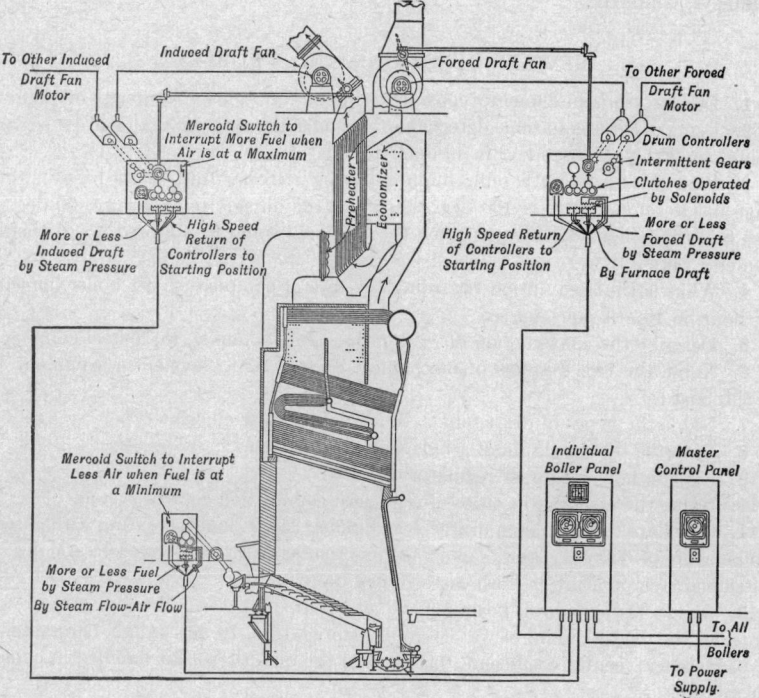

FIG. 229. — Bailey Meter Control at Edgar Station of Boston Edison Company.

to maintain a balanced furnace draft or a furnace draft which may vary slightly with boiler rating if desired.

The application of Bailey meter control to a stoker fired boiler, equipped with an economizer, air preheater, and forced and induced draft fans is shown by the diagrammatic layout, Fig. 229. When the load on the station increases, the steam pressure acting through a master steam-pressure contactor changes the speed of the stokers, forced draft fans, and induced draft fans, on all boilers, connected with the control system, distributing the load changes equally among the various boilers.

The control drives are designed so that the rates of supply of air and fuel are varied in substantially the correct proportion. The slight readjustments necessary in the fuel supply to maintain the correct fuel-air ratio are made automatically by the **steam flow-air flow contactor**, while the air supplied to the fuel bed by forced draft is governed by the furnace draft. In this way all three conditions for proper operation are satisfied.

REFERENCES

Mechanical Equipment of Buildings, Vol. II, HARDING and WILLARD.
Steam Power Plant Engineering, GEBHARDT.
Chimneys, CHRISTIE.

REVIEW QUESTIONS AND PROBLEMS

1. Deduce, from fundamental considerations, an equation for the height of a chimney.
2. Using Kent's equation, determine the boiler horsepower that will be served by a chimney 200 ft. high and 12 ft. in diameter.
3. Find the height of a mill chimney having 1.16 in. draft at its base, if the average inside temperature is 400 deg. fahr., and the outside temperature, 60 deg. fahr. Also find the boiler horsepower, which the chimney would be adapted for, if the inside diameter were 9 ft.
4. What is the assumption regarding the rate of combustion per boiler horsepower per hour, in Kent's equation?
5. Describe the construction of (a) a radial brick chimney, (b) a steel chimney.
6. Name the two systems of mechanical draft. Which system is commonly used on shipboard?
7. Name the types of draft fans. Which is the most efficient type?
8. Describe the operation of a balanced draft system.
9. Explain how a damper regulator operates.
10. Name the essential features of a modern combustion control system.
11. Calculate the change in draft for a chimney 250 ft. high operating with gases at a temperature of 600 deg. fahr., should the gas temperature be lowered to 400 deg. fahr. Outside air temperature in both cases 70 deg. fahr.
12. Air at 80 deg. fahr. is flowing in a duct 4 ft. in diameter. Velocity pressure at the center of the duct, 0.30 in. of water; room temperature, 70 deg. fahr.; barometer, 29.4 in. of mercury; center coefficient, 0.90. Find the quantity of air handled in cubic feet per minute.
13. Calculate the diameter, in feet, for a chimney 200 ft. high to serve 1200 boiler hp. Use Kent's equation.
14. A certain power plant has a turbo-vane fan, which handles 193,000 cu. ft. of air per min., at a pressure of 7.15 in. water. Assume efficiency, as 60 per cent. Find the power required to drive the fan.
15. A fan having a mechanical efficiency of 65 per cent delivers 21,000 cu. ft. per min. against a static pressure of 2.5 in. Air delivered through a duct 4 ft. sq. Density of air, 0.065. Find: (a) air horsepower of the fan; (b) horsepower required for driving.

CHAPTER XIV

COAL AND ASH HANDLING EQUIPMENT

298. Foreword. — In small power plants, coal and ashes are moved by hand. Coal is shoveled into a wheel-barrow or small truck at the storage pile, moved to the front of the boiler and dumped on the floor or fired directly into the furnace from the barrow. From the floor it is shoveled into the furnace or hopper of the stoker, according to the method of firing employed. The ashes are raked from beneath the grates to the front of the boiler and shoveled into wheel-barrows or dump cars. The cost of these operations varies, and, in the case of coal, is generally about 50 cents per ton; in the case of ashes, the cost is difficult to determine, because the men performing this work also perform other duties.

FIG. 230. — Coal Tower — Richmond Station, Philadelphia.

In large power plants, the expense of handling coal and ashes by the above method would be excessive, and, therefore, coal handling equipment is used. The raw coal is ordinarily received by rail or by water. If by rail in hopper bottom cars, the coal is discharged into a track hopper, Fig. 2, page 6, or a rotary car dumper is used to dump the coal into the hopper. From here it is moved by a conveyor to a crusher, and after passing through the crusher is elevated to overhead raw coal bunkers from which it falls by gravity into the hoppers of the stokers, or to pulverizing mills, depending on the method of firing used. When received by water, traveling or stationary coal towers, Fig. 230, are used to unload the colliers. The cost of handling a ton of coal, using a conveyor, varies from 4 to 20 cents per ton, and the cost of handling ashes varies from 5 to 50 cents per ton.

In general, coal and ashes should not be handled by the same conveyor. When they are handled separately the conveyor system is more flexible and has a longer life. Coal is graphitic and causes small wear of the conveyor parts, while ashes are abrasive and cause excessive wear of the conveyor machinery. Furthermore, the ashes may be hot or wet. When hot they produce distortion and rapid disintegration of the conveyor buckets; when wet they produce corrosion, which results from the formation of sulphuric acid.

In connection with the use of pulverized coal, the elimination of heavy ash discharge into the atmosphere is a problem which is assuming great importance. In some instances this has become a decided civic problem and almost prohibitive legislation has been passed regulating its control. Much effort has been made to collect the pulverized coal ash from large central stations by using electrical precipitation, gas washing and mechanical collection in baffle and cyclone collectors.

299. Classification of Coal Conveyors. — Conveyors used for moving coal may be classified as follows: (1) *scraper, or flight*, (2) *bucket*, (3) *belt*, (4) *apron*, (5) *grab bucket*, and (6) *telpherage systems*.

300. Scraper, or Flight Conveyor. — This type of conveyor is used extensively for conveying coal horizontally and for inclinations up to 45 degrees. It is a simple method, but is subject to maximum wear, because the material is pushed and not carried. Its capacity may run as high as 100 tons per hour.

It consists of one or two strands of chain, to which are attached steel scrapers, or flights, which scrape the coal through a trough having the same shape as the flight. The coal is discharged from the trough through gate-controlled openings in the bottom of the trough. There are several types of scraper conveyors, the most common of which are: (1) the plain scraper, Fig. 231, in which the flights are suspended from a chain and drag along the trough; (2) the suspended flight, in which the flights are supported by a cross bar having wearing shoes at each end, the wearing shoes move along a track and the flight does not touch the trough at any point; (3) the roller flight, in which rollers replace the shoes of the suspended flight conveyor, is the most common type.

301. V-Bucket Elevators, and Conveyors. — The V-bucket elevator consists of steel V-buckets rigidly fastened to a continuous steel chain, which runs over upper and lower sprockets. The buckets are equally

V-BUCKET ELEVATORS, AND CONVEYORS 323

spaced on the chain, and receive their load by dipping into a coal pocket, or **boot**, at the lower end of the system. The material elevated may be discharged by centrifugal force at the top of the elevator, when the direction of motion is reversed (**centrifugal discharge**), or by the action of a pair of

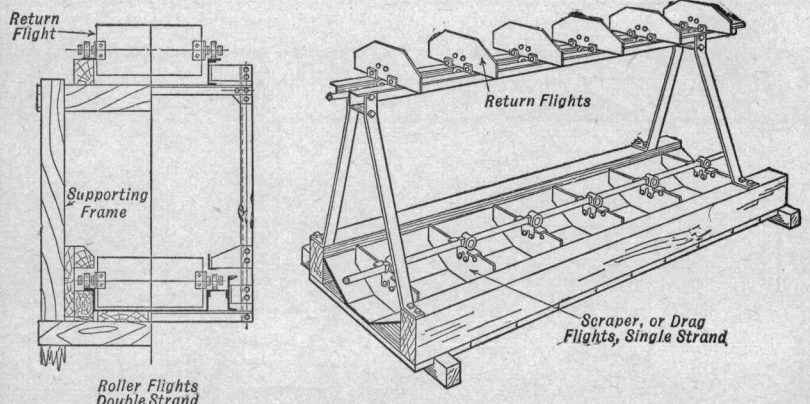

FIG. 231. — Flight Conveyor.

idler sprockets, which draw back the buckets, on the discharge side, at the top (**positive**, or **perfect**, **discharge**). In each case the material is positively discharged.

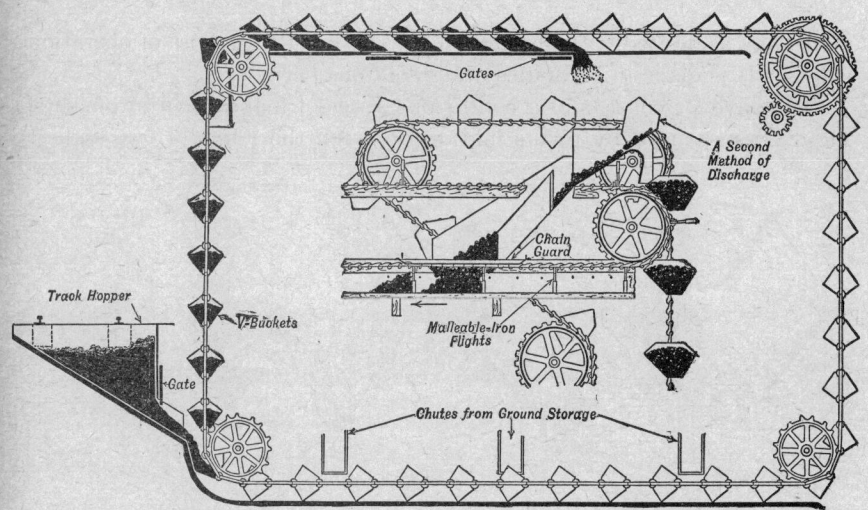

FIG. 232. — Diagram showing Operation of V-bucket Conveyor and Elevator.

The V-bucket elevator and conveyor, Fig. 232, has the buckets attached to a chain running over sprockets. The buckets lift the coal vertically and act as a drag conveyor on horizontal runs, each bucket pushing its half-spilled load ahead of it in a trough of suitable shape. The discharge, on horizontal runs, is through openings controlled by hand-operated gates. The capacity is ordinarily limited to 90 tons per hour.

302. Pivoted Bucket Conveyor. — This type of conveyor, Fig. 233, carries its load in buckets suspended from a pivot shaft, and can convey

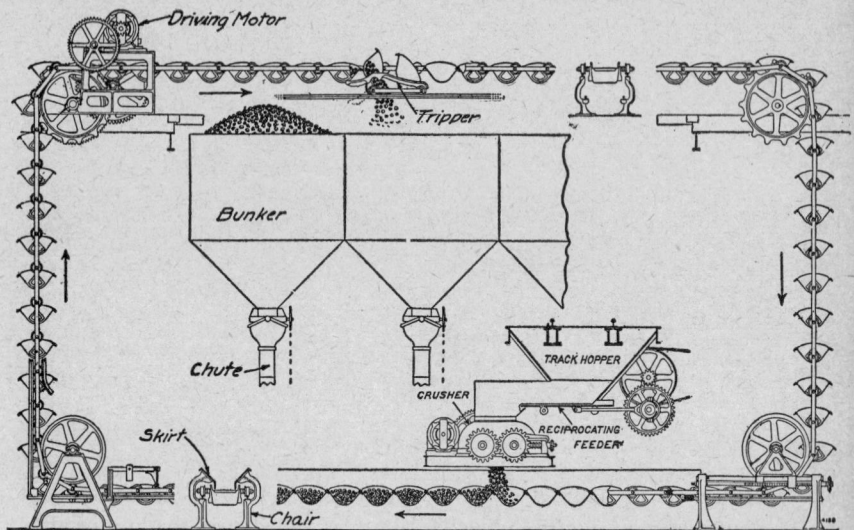

Fig. 233. — Diagram showing Operation of the Peck Pivoted Bucket and Conveyor.

or elevate, as desired. Its initial cost is high, but the expense of operation is low. It is made in capacities up to 350 tons per hour.

The conveyor consists of a continuous series of malleable iron buckets, Fig. 234, suspended by pivots midway between the joints of two endless

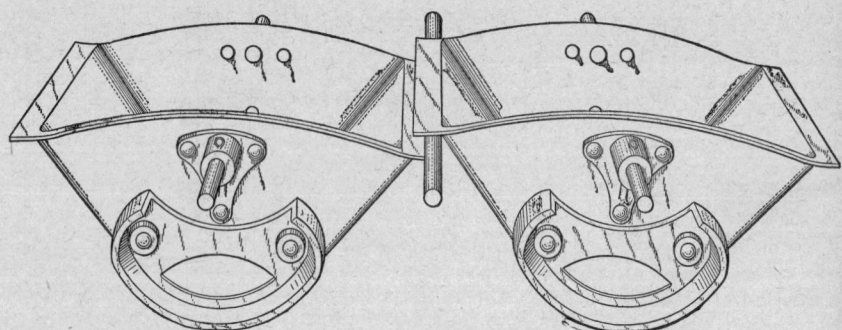

Fig. 234. — Buckets of the Jeffrey Pivoted Bucket Conveyor.

chains, which are driven by a motor located at some convenient point, generally at the top of a vertical rise. The buckets maintain their position by gravity and, when traveling horizontally, are supported by a pair of wheels located at each joint of the chains.

BELT CONVEYOR

The conveyor is loaded by passing below a coal crusher. A **skirt**, supported by castings, overlaps the buckets at the point of loading and permits continuous loading without spilling of the coal. The coal is dis-

FIG. 235. — Jeffrey Belt Conveyor and Tripper above the Coal Bunkers.

charged into the bunkers by a stationary or movable tripping device, which engages the cams located on the sides of the buckets and tilts the buckets sufficiently to discharge the coal.

303. Belt Conveyor. — The belt conveyor, Fig. 235, gives a flexible means of conveying coal. It is limited, in the inclination at which coal may be elevated, to about 20 degrees. The driving mechanism for the belt conveyor is simple, and may be located at any point along the length of the belt. Belts 1000 feet long are not uncommon. Five hundred tons per hour can be delivered by this type of conveyor.

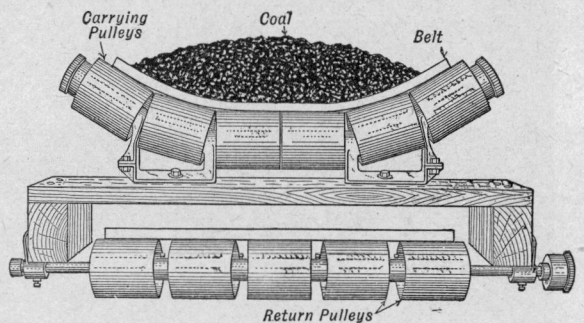

FIG. 236. — Pulley Arrangement of Robins Belt Conveyor.

The conveyor consists of an endless belt traveling over pulleys arranged

326 COAL AND ASH HANDLING EQUIPMENT

to give the carrying side of the belt the shape of a trough, as shown in Fig. 236, the return side being supported on flat idler pulleys. The trough-

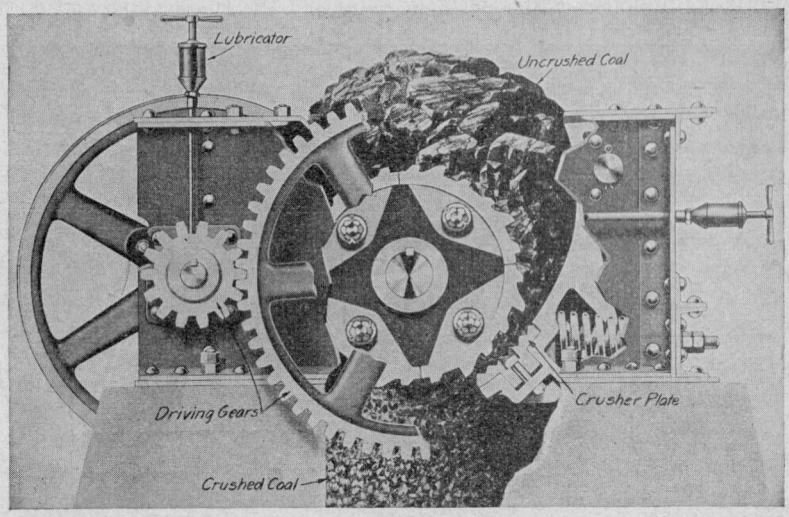

Fig. 237. — Pennsylvania Roll-type Crusher.

Fig. 238. — Bradford Coal Breaker.

ing pulleys are spaced from 3 to 6 feet between centers. The coal carried is discharged by movable or fixed trippers, which consist of two pulleys, one above and slightly in advance of the other. The belt runs over the upper and under the lower one, thus causing the coal to fall into a chute which discharges into the storage bunker. A revolving brush is generally placed at the return end of the conveyor to keep the belt clean.

304. Coal Crusher. — When *run-of-the-mine* coal is to be used in a stoker, it must be reduced in size by passing through a crusher, before it is stored in overhead bins or bunkers. The types of crushers used are: *high-speed hammer and ring mills, roll crushers, and breakers.* Fig. 237 shows a typical roll crusher, to which coal is generally fed continuously

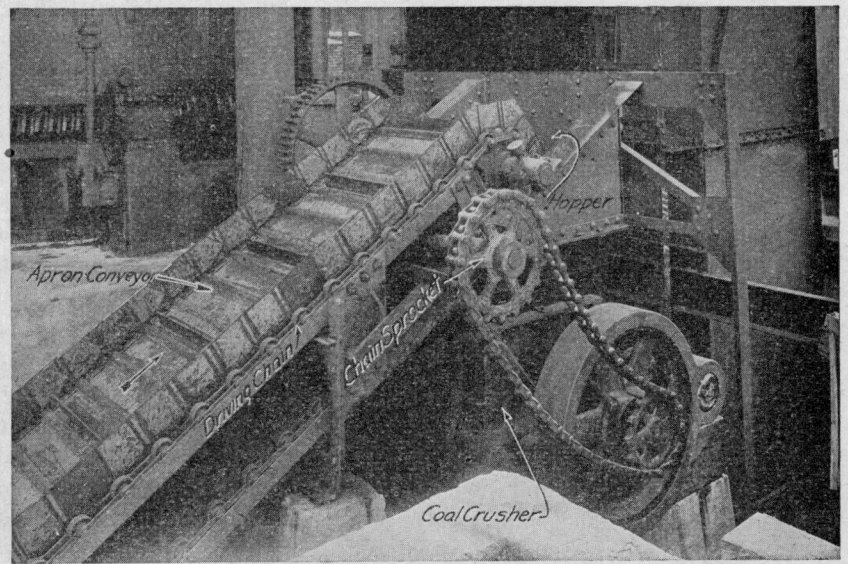

Fig. 239. — Jeffrey Apron Feeder carrying Coal from Track Hopper to Crusher.

by a reciprocating pusher or apron feeder. The crusher drum shaft is driven from the flywheel shaft by spur gearing, and the crusher drum carries teeth which grasp the coal and draw it between the drum and **crusher plate,** thus reducing the lumps to small size. The crusher plate is grooved for the larger teeth, which are required to grasp and break the large lumps into sizes suitable for the smaller teeth to still further reduce in size.

The crusher is ordinarily located in a pit below the track hopper, from which the coal is delivered by gravity to some form of apron conveyor or reciprocating feeder, which regulates the supply of coal to the crusher, and prevents choking in an ordinary two-roll crusher.

The Bradford Breaker, Fig. 238, is in common use in large central stations, and can crush from 25 to 600 tons of coal per hour. The coal is crushed, sized and cleaned in one operation. It enters the slowly revolving drum at one end, where it is caught up by special lifting shelves attached to the screen plates. As each shelf approaches the top its load of coal is automatically dumped, and striking the heavy steel screen plates, is broken down. The coal made fine enough by this dropping, along with a small percentage of the crushed impurities, immediately escapes through the screen plates into storage. The lifting operation is repeated until the remaining uncrushed coal is reduced.

The **apron feeder**, Fig. 239, consists of overlapping metal slats riveted to two strands of roller chain traveling slowly on tracks, and is used especially when the coal must be lifted in transit or carried some distance to the crusher. The inclination at which this feeder will handle coal is 30 degrees, and only end discharge is possible.

305. Coal Weighing Hoppers and Coal Valves. — Some suitable method of weighing coal is necessary in order that satisfactory records of the quantity of coal burned may be kept. In power plants having overhead bunkers the coal is weighed in stationary or traveling hoppers, Fig. 240, which have a scale attachment. Coal from the bunker is delivered into the coal-weighing hopper, its weight is recorded, and the coal is then discharged into the hoppers of the stokers through a chute having a flexible joint which permits swinging through several degrees either side of the central position of the chute, at the lower end of which a **coal valve** is used. The hopper mechanism is sometimes arranged to discharge 100 pounds of coal at one time.

Coal valves are made in a variety of shapes and sizes. The simple slide valve and the simplex and duplex rotating valves are used to draw coal from overhead bunkers, and the flap and rotating valves to draw coal from the side of a bin. The simplex valve has a single rotating jaw moved by a lever, and the duplex valve two rotating jaws, which are moved simultaneously by a common actuating lever. The coal is thus discharged centrally even with a partially opened valve. Both these types cut through the coal without jamming and are easily operated.

The flap valve is the most simple form and consists merely of an iron flap hinged to the bottom of the chute. The flap is lowered to discharge coal and raised to stop the flow.

306. Coal Handling from Ground Storage. — The amount of coal that can be stored in overhead bunkers is limited to a few days' supply, especially in large plants. For instance, the Commonwealth Edison Company in one of its Chicago stations uses 3000 tons of coal in twenty-four hours. In such cases the remaining supply, sufficient for several months, is stored in large outdoor piles. The above mentioned company stores 300,000 tons in these piles, from which coal is delivered to overhead bunkers as needed. When the storage space is near the water front, coal is delivered in barges, and grab buckets and telpherage systems, such as shown in Fig. 240, are used to remove the coal from the barges and deliver it to the storage piles.

COAL HANDLING FROM GROUND STORAGE 329

Cars of coal are generally loaded and unloaded by means of a long-radius steam- or electric-driven crane, Fig. 241, having self-filling buckets. Such cranes are capable of moving from 40 to 250 tons of coal per hour,

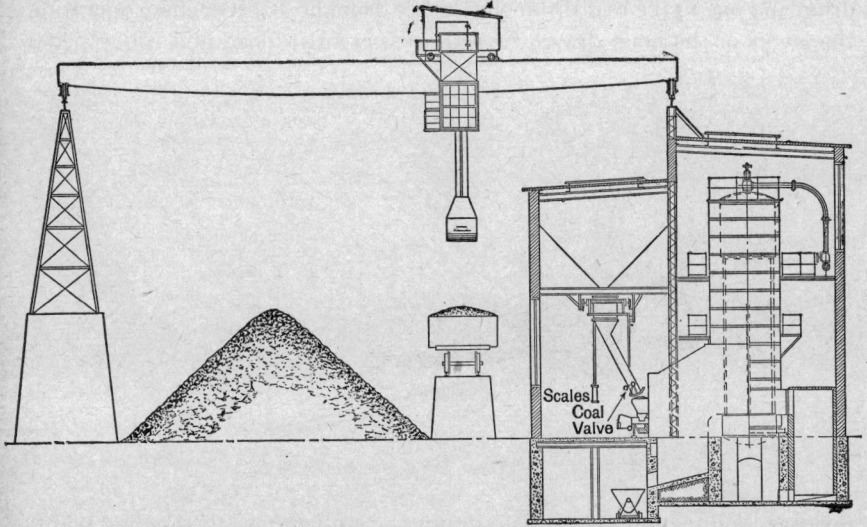

FIG. 240. — Coaling Wickes Boiler House with Overhead Electric Traveling Crane.

according to the size of the buckets. Coal can be handled at a small cost per ton by this method.

The Power Drag Scraper System, Fig. 242, is extensively used for distributing coal to storage from an initial pile formed by discharging the coal

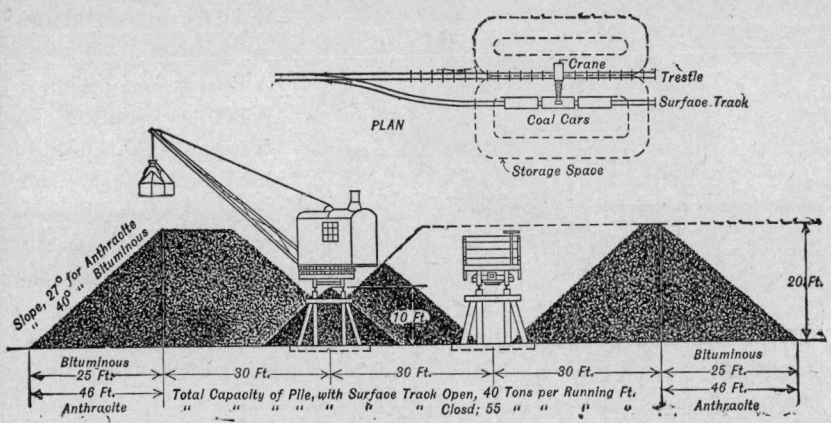

FIG. 241. — Storing Coal using a Locomotive Crane and Parallel Tracks.

from an elevator, belt conveyor, skip joist, pivoted bucket conveyor, grab bucket or any of the accepted types of coal conveying and elevating machinery. It consists of a scraper, Fig. 243, which is drawn forward or backward by means of cables attached to the front and rear of the bucket, thus distributing the coal in layers over the entire storage area. These cables pass through a series of blocks which guide them, one to the front drum and one to the rear drum of a double drum hoist. One man operating the levers of the hoist draws the scraper forward to load and convey, and

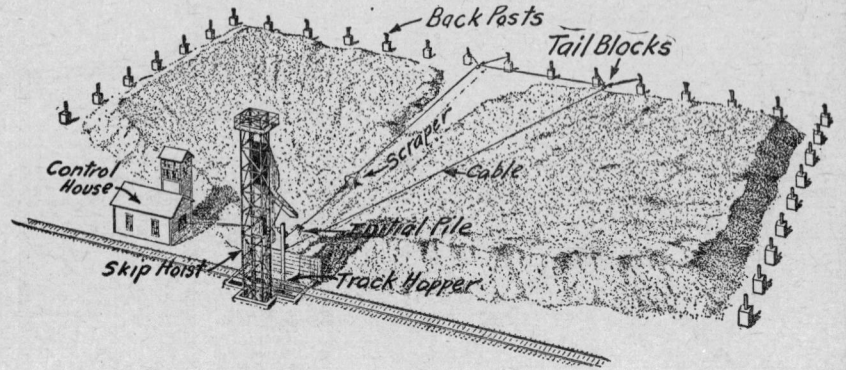

FIG. 242. — Power Drag Scraper System.

reverses its direction of travel to dump and return it to the loading point. The blocks which guide the operating cables to the hoist drums are attached either to a permanent post or mast located back of the initial pile or point to which the coal is to be reclaimed, or to a tail tower moving on a track. The line of operation of the scraper is changed by moving the mast or tail tower. Scrapers of this type are built for any capacity up to 1000 tons per hour.

FIG. 243. — Sauerman Crescent Drag Scraper.

For small plants a portable V-bucket or belt conveyor is used to load trucks or carts from storage. A small gasoline engine is usually the means of driving the conveyor.

307. Pulverized Coal Handling. — When using the storage or bin system with pulverized coal, the coal is conveyed by either a screw conveyor, a compressed air transport system, or a pumping system. The latter system is shown in Fig. 244, and consists of a motor-driven pump; a source of compressed air for fluidizing the material; a conduit or delivery line; either manually operated or electro-pneumatically remote-controlled distributing valves and electric bin-level indicators. The pulverized coal from the pulverized mills is fed direct to the pump and is carried to the discharge end of the screw where the mass is aerated by a small amount of compressed air admitted through a series of small holes, and at a pressure of from 5 to 75 pounds per square inch, depending upon the material, the distance and the elevation. The nature of the fuel is thus changed from a compact mass into a semi-fluid, in which state it is transported through a black steel pipe and distributing valves to the individual service bins. The distributing valves are actuated by an air line by-passing the pump, and are of the disc multiple discharge port type. The valve disc inside the casing is carried on a spindle on the outer end of which is mounted a mercuroid switch. The switch tilts from side to side in synchronism with

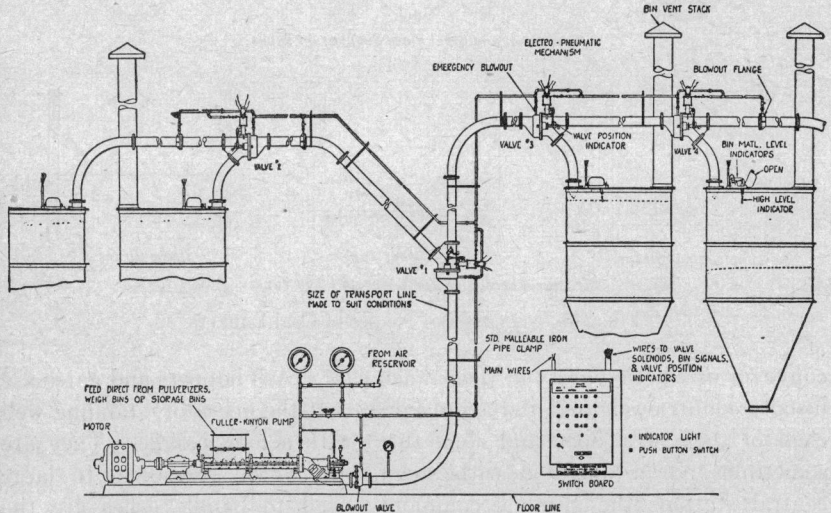

FIG. 244. — Fuller-Kinyon Pulverized Coal Conveying System.

the valve disc, and in each extreme position it closes an electric circuit, which lights an electric light on the switchboard indicating to the operator whether the flow is to the main or to the branch line.

Each bin is provided with a bin level indicator which gives a visible or audible signal when the bin is empty or full. The signal is ordinarily located on a signal or switchboard which can be located at any convenient point, thus centralizing the control of the system. This system may have a capacity of 100 tons per hour, and is being used successfully for distance as great as 3600 feet.

308. Overhead Coal Bins, or Bunkers. — The general shape of overhead bunkers is shown in Fig. 245. They are constructed of $\frac{1}{4}$- to $\frac{3}{8}$-inch steel plate and structural shapes, or of reinforced concrete. The **suspension or catenary type of bunker** is the cheapest form, because the plates are only subjected to tension and do not require structural members, except for the ends. The shape of this type of bunker is that of a parabola. To permit self-cleaning, the sides are inclined at an angle of 45 degrees and the bottom is made with a hopper.

309. Ash Handling. — In small plants where ash-handling machinery is not installed, the ash and clinker are hoed out, shoveled into a wheelbarrow and wheeled to the nearest dump available. When ash hoppers are provided, the ash falls by gravity into the hoppers and may be removed by a

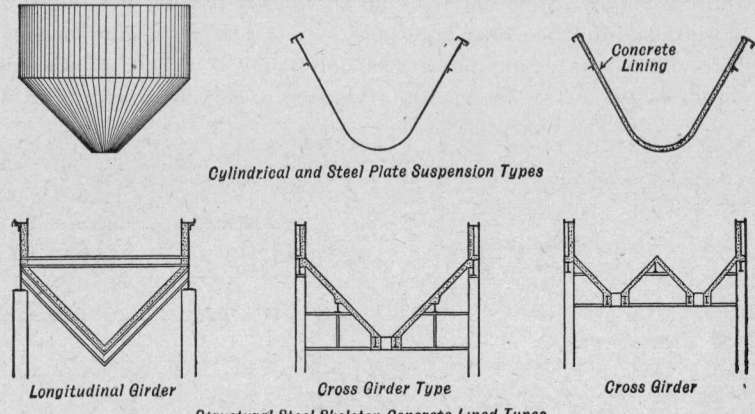

Fig. 245. — Types and Shapes of Coal Bunkers.

conveyor or wheeled away in special ash cars. Ash hoppers and gates are usually counter-weighted and arranged so that the operator, standing well clear of them, can open and close them with a long handle. They are sometimes power operated with steam, air, or oil pressure. In large central stations the systems commonly used to handle ashes are the **mechanical conveyor, steam or vacuum conveyor, hydraulic system,** and the **gravity system.** The types of mechanical conveyors used are generally of the *V-bucket, skip hoist, drag chain,* and *pivoted* or *overlapping bucket.* The latter is extensively used since it carries instead of pushes the material. For this service the equipment is generally more rugged than that used in handling coal.

Ash from pulverized fuel furnaces presents many problems, since the temperature of the ash when it gets to the bottom of the furnace must be below the fusing point if slagging and clinkering are to be prevented. Water screens are often used, the ash dropping through them into hoppers, from which it is removed in the dry state. In some cases the lower part of the furnace, Fig. 246, is especially adapted for accumulating the ash in a molten condition and tapping it out intermittently. Experience with horizontal turbulent burners in the lower part of the furnace indicates that there is no difficulty in keeping ash having a fusing temperature as high as 2500 or 2600 deg. fahr. in a satisfactory molten condition with the boiler operating at any reasonable rating.

ASH HANDLING 333

In addition to this ash, it is estimated in pulverized fuel operation that from 12 to 25 per cent of the ash in the coal will pass out through the stack in flocculent form unless it is entrapped, for which several methods are

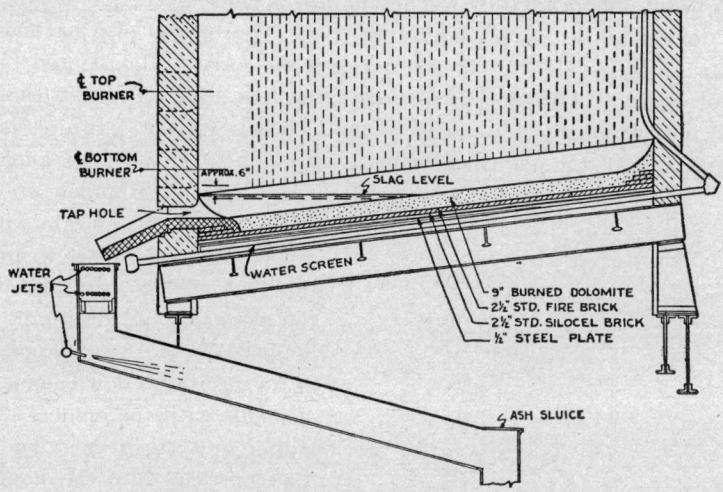

Fig. 246. — Slag Bottom Furnace — Toronto Station.

Fig. 247. — Sturtevant Cinder Trap.

used. In some cases a special type of induced draft fan, designed to catch cinders and ash, Fig. 222, page 313, is used. Other ash-catchers employ a cyclone or centrifugal separator connected with the stack, the gas being forced into it by the induced draft fan. At the Trenton Channel Plant a Cottrell electrical precipitator is used, and the flue gas is passed between oppositely charged electrodes that produce a strong electrostatic field, the ash being precipitated by the action of this field.

One device, known as a cinder trap, Fig. 247, consists of a battery of elements set vertically, arranged in staggered rows, and mounted on a frame or casing. This unit is placed in the flue above a special shaped hopper, into which the lower end of the elements project. In operation the main body of the gas passes directly through the cinder trap, and the gases containing cinders are divided into vertical strata which make an abrupt change in direction, the cinders are trapped in the elements and drop into the hopper below. A small portion of the gases pass down along the elements with the cinders, thus aiding gravity in carrying the dust particles into the hopper. This portion returns to the main flue through the opening at the rear of the elements. A special arrangement of baffles in the hopper prevents the cinders being carried through with the gases into the main flue.

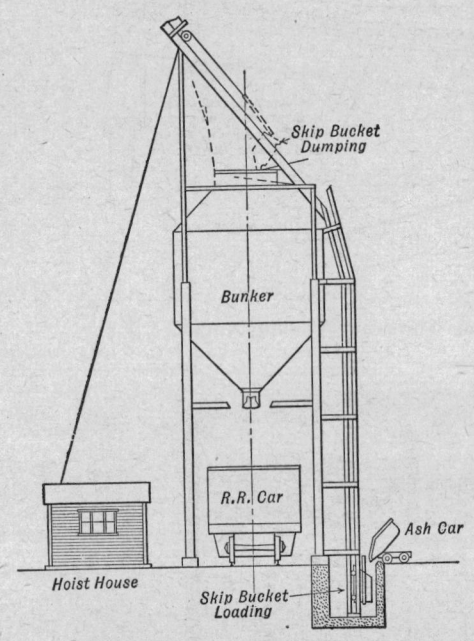

Fig. 248. — Skip Hoist.

In addition to these methods, gas-washers or scrubbers are often used ahead of the induced draft fan. They are of the same type generally used for cleaning air for generators, and eliminate from 90 to 95 per cent of the dust where using from 500 to 800 gallons of water per minute.

310. Skip Hoist. — The skip hoist, Fig. 248, is simple and has few parts at which wear and corrosion can occur. It consists of a bucket running on inclined or vertical tracks, and hoisted by means of a steel cable attached to a winding machine. The bucket consists of a rectangular steel box, open at the top and fitted with guide rollers and hoist bale. In operation, the ashes are delivered into the **skip** by a dump car, and the skip is hoisted by cable and electric motor. At the top the skip is turned upside down and the ashes discharged, after which the skip automatically returns to the bottom for another load.

311. Steam-jet Ash-conveyor. — The steam-jet conveyor, Fig. 249, consists of lengths of hard extra-heavy cast-iron pipe, held together by means of bolts which pass through lugs cast on each pipe. Openings into the pipe are made in front of each ashpit, and the ashes are raked into these openings. Expansion joints are used to provide for expansion, and rollers or hangers to support the pipe.

Steam is introduced through one or more nozzles located at an elbow, causing a rapid movement of air through the pipe toward the discharge end. The moving air carries with it the ashes entering the line at the intakes. The air inlet end of the pipe has a protected hood to prevent clogging of the inlet, and in the discharge pipe provision is made for a water spray to prevent dust. The discharge pipe opens into a **baffle box,**

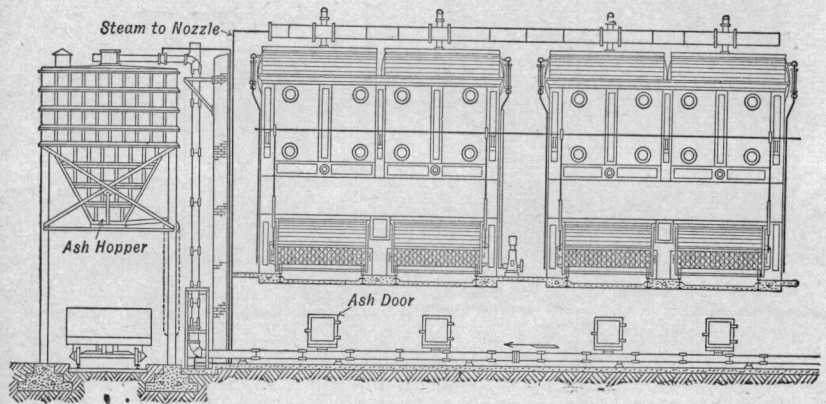

FIG. 249. — Green Steam Jet Ash Conveyor.

in which the ashes pack and thus eliminate wear, because the discharging ash strikes against ashes. This box also prevents packing of ashes, by breaking the force at which the ashes enter the ash bunker. The elbows where the pipe turns are subject to rapid wear, and are lined with renewable iron inserts which are easily replaced.

312. Drag-chain Conveyor. — The drag-chain conveyor consists of a single-strand grit-proof chain about 7 inches wide, running very slowly in a cast-iron trough. The chain rides on the ashes and does not wear the trough. The cost is low and the space occupied is small.

313. Hydraulic System for Ash Handling. — The sluicing and pumping of ash by hydraulic methods are being used to a considerable extent. One method known as the Hydrojet, Fig. 250, is a totally enclosed system for the removal and transportation of ashes in liquid form. The ashes drop from the furnace to a hopper where it is quenched by sprays, and, as often as is necessary, is discharged at a rate of about one ton per minute into a high velocity water stream, by which the ash is carried to a centrally located concrete bin or sump, or to adjacent lowlands. The ashes can be pumped from the sump, or removed by a grab bucket into an overhead storage bin from which they are later unloaded to a railroad car or auto truck.

314. Ash Bunkers. — Ash bunkers resemble coal bunkers in shape and construction. The material coming in contact with the hot, moist ashes must be such that it will not corrode. Steel, wood, concrete, and cast

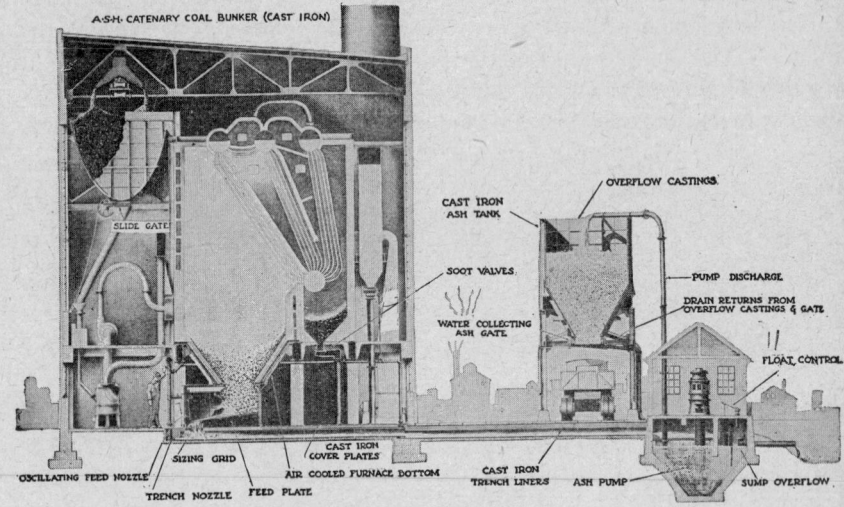

Fig. 250. — Hydraulic Ash Handling.

iron are used. When the hopper is in the power house, it generally has a side inlet with a flap valve. A chute delivers the ashes into whatever conveyance is provided to carry them away.

REFERENCES

Mechanical Equipment of Buildings, HARDING and WILLARD.
Power Plant Engineering, FERNALD and ORROK.
Catalogues of the following Companies: LINK BELT, JEFFREY, LAMSON, STEPHENS ADAMSON.
Power Plant Engineering, GEBHARDT.

REVIEW QUESTIONS

1. What is the principal reason for the adoption of mechanical methods of handling coal and ashes?
2. Classify coal-conveyors and describe the construction of a belt conveyor.
3. Describe the coal- and ash-handling equipment in some typical power plant.
4. Name two methods used to handle ashes. Describe one of these methods.
5. Describe the operation of a coal-crusher.
6. Describe the hydraulic system of handling ashes.
7. Describe the construction of a slag-bottom furnace.
8. Discuss the question of handling the ash from pulverized fuel furnaces.

CHAPTER XV

RECIPROCATING STEAM ENGINE — SIMPLE ENGINES

315. Foreword. — The **reciprocating steam engine** is the most widely known prime mover, and although its field of usefulness has been encroached upon, in recent years, by the steam turbine and gas engine, it is still used extensively. *The steam engine is superior to the turbine for work requiring variable speed, slow rotative speeds and large starting torque, while the turbine has superseded the engine for large central-station units and for auxiliaries requiring high rotative speed.* The high-speed turbine, when used with efficient reduction gearing, has many advantages for low-speed drives, and is rapidly replacing the steam engine in that field. Considering only efficiency in the use of heat, the Diesel type of oil engine is superior to the steam engine; and the turbine is more economical in space requirements. However, the reciprocating steam engine has been enabled to hold its place as a prime mover, by an increase in economy obtained by improved design; that is, by utilizing the uniflow principle in the action of the steam, and balanced poppet valves for high temperatures and pressures. In general, the modern steam engine, in sizes up to 2500 horsepower, is more economical than the turbine, and with the increase of steam pressures the steam engine is suitable for high-back pressure operation at better steam economy than small-sized turbines, and in addition is well adapted for bleeder service.

316. Classification. — Steam engines may be classified in the following ways:

1. By valve gear
   - Slide valve
     - "D"
     - Balanced
     - Multi-ported
     - Piston
   - Corliss valve
     - Drop cut-off
     - Positively operated cut-off
   - Poppet valve

2. By position of the axis of the cylinder
   - Vertical
   - Inclined
   - Horizontal

3. By number of cylinders in which steam expands
   - Single expansion, or simple engine
   - Multiple expansion
     - Compound
     - Triple
     - Quadruple

338 RECIPROCATING STEAM ENGINE — SIMPLE ENGINES

4. By use
   - Stationary
   - Portable
   - Locomotive
   - Marine
   - Hoisting

5. By rotative speed
   - Low
   - Medium
   - High

6. By ratio of stroke to diameter
   - Long stroke
   - Short stroke

7. By method of exhausting steam
   - Condensing
   - Non-condensing

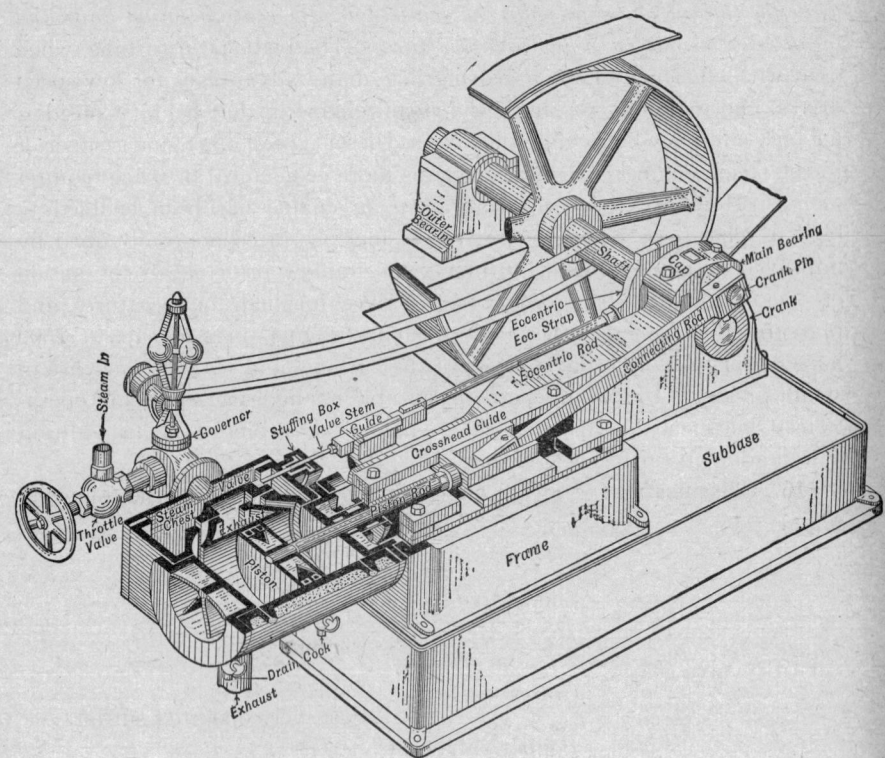

Fig. 251. — "D" Slide Valve, Simple Engine.

317. Parts of a Steam Engine. — A simple form of reciprocating steam engine is shown in Fig. 251. Its parts may be separated into the following groups:

1. **Stationary parts:** frame, bed or base, cylinder, cylinder heads, steam-chest cover, bearings, stuffing-box.

2. **Moving parts:** piston, piston rod, crosshead, connecting rod, crank, shaft, and flywheel, all of which are moved by the action of the steam.

3. **Valve gear,** which controls the distribution of steam and consists of eccentric rod, valve rod guide, valve stem or rod, and valve.

318. Function of Engine Parts. — The frame supports and holds the moving parts in proper relative position and gives rigidity to the various moving members. It may rest directly upon the foundation or upon a **cast-iron bed plate** on the foundation. The **bearings** support the shaft and are attached to one end of the engine frame or are supported on a separate foundation. The bearing which is made integral with or rests on the frame is known as the **main bearing,** or **pillow block,** while that resting on a **pedestal** and supporting the outer end of the shaft is called the **outboard,** or **pedestal bearing.** That part of the shaft which turns in the bearing is called a **journal.** The **cylinder** is bolted to the frame at the end opposite the bearings, and forms a chamber in which the piston moves under action of the steam. The ends of the cylinder are closed by **cylinder heads,** and the joint between the cylinder and head is made tight by a **gasket** (a thin piece of asbestos or rubber) or by grinding the surfaces with fine emery to produce a smooth surface, in which case the joint is metal to metal and is called a **ground joint.** The outer surface of the cylinder is covered with **non-conducting material,** to prevent excessive loss of heat by radiation.

The **piston** transmits the pressure of the steam to the crosshead, to which it is attached by the piston rod. At the point where the piston rod passes through the cylinder head, a **stuffing box** filled with packing prevents loss of steam. The **crosshead** slides back and forth, in a straight line, between guides formed in the engine frame. These **guides** make a sliding surface for the crosshead and keep the piston rod from bending. One end of the **connecting rod** is attached to the crosshead and the other to the crank pin, which is fastened to the outer end of the crank and is moved in a circle about the center of the shaft. Hence the connecting rod changes the straight line motion of the crosshead into rotating motion of the crank, which is keyed or pressed on to the shaft to which the flywheel is fastened. The **key** is a rectangular piece of metal which fits into corresponding grooves in two parts of a machine and prevents their relative rotary motion.

The **valve** is located in the **steam chest,** which is ordinarily a part of the cylinder casting and is connected to the cylinder by ports or **passages** made in the casting. It is moved back and forth upon its seat by an **eccentric,** so located on the crank shaft with reference to the crank that it moves the valve to admit or discharge steam to or from the cylinder at the proper time. The **valve rod,** or **valve stem,** connects the valve and the **valve rod guide,** which is used to prevent bending of the valve **rod.** The **eccentric rod** connects the valve rod guide and the eccentric.

The **flywheel** has a heavy rim which absorbs energy when the supply of energy is in excess of the demand and gives up energy when the supply is not equal to the demand. It thus prevents rapid fluctuations in speed during a revolution.

The **governor** maintains the speed of the engine nearly constant, by controlling the amount or the pressure of the steam supplied to the engine.

The parts of the engine that move backward and forward in a straight line, such as the piston, piston rod, crosshead, valve, and valve rod, are known as **reciprocating parts.** The parts that rotate about an axis, such as the shaft, crank, flywheel, and eccentric, are known as **rotating parts.**

The **size of a steam engine** is ordinarily given by stating, *first*, the diameter of the cylinder in inches; *second*, the length of stroke in inches; and *third*, the number of revolutions, as 7 inches by 10 inches — 300 r.p.m.

319. Engine Nomenclature. — The following terms, some of which do not appear to have any logical basis, are applied to steam engines:

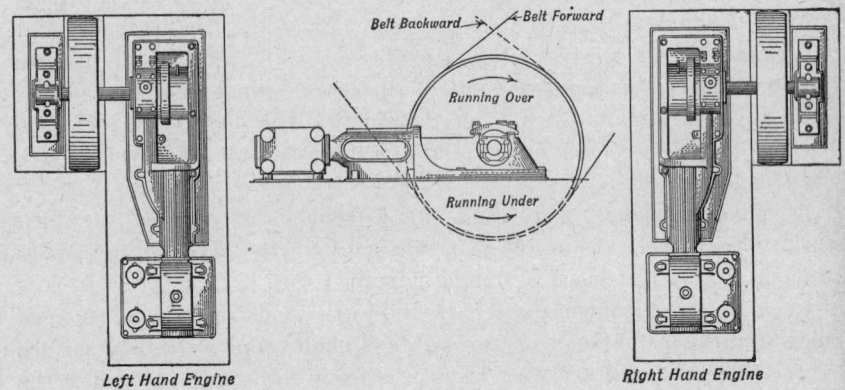

Fig. 252. — Illustration of Several Engine Terms.

Running over is a term applied to the action of an engine when the top of the flywheel revolves away from the cylinder end. **Running under** is the reverse of running over, the top of the flywheel revolving toward the cylinder end. A **right-hand engine** is one in which the flywheel and valve gear are located on the right-hand side of the steam cylinder, as seen by a person standing at the cylinder end and looking toward the shaft. The flywheel is located to the left of the cylinder on a **left-hand engine.** These terms are illustrated in Fig. 252. The **head end,** or front end, of an engine is the end of the cylinder farthest from the crank. The **crank end,** or back end, is the end of the cylinder that is nearest to the crank and flywheel. The **stroke** is the distance traveled by the engine piston in passing from the head end to the crank end or vice versa. The **crank**

ENGINE NOMENCLATURE

throw is equal to the length of the crank, or one-half the stroke. The **forward stroke,** or out stroke, is made while the piston passes from the head to the crank end. The **return stroke,** or back stroke, is made while the piston travels from the crank to the head end. A **long-stroke** engine is an engine whose stroke is long as compared with the diameter of the cylinder. A **short-stroke** engine has a stroke that is equal to or slightly less than the diameter of the cylinder. **Dead center** is the point at the end of the stroke where the center lines of the connecting rod and crank are in the same straight line. There are two dead centers, one at each end of the cylinder, called the **head-end dead center** and the **crank-end dead center.** **Piston displacement** is the volume through which the piston sweeps in traveling from one dead-center position to the other. *Numerically it equals the net piston area times the length of stroke.* **Clearance** may be either **mechanical** or **volumetric.** The former is the linear distance between the end of the piston and the nearest cylinder head, when the crank pin is on dead center. *Volumetric clearance is the volume of the space included between the piston and the cylinder head, when the crank pin is on dead center, and includes the volume of the steam port for the end considered. Volumetric clearance is usually expressed in percentage of the piston displacement.* A cylinder having a clearance of 6 per cent would have a clearance volume equal to 6 per cent of the piston displacement.

Example 38. — Compute the volumetric clearance for the head end of a Corliss engine that has a cylinder diameter of 8 in., a stroke of 24 in., and a clearance of 5 per cent.

Solution. — Clearance volume, cu. in. = area of piston in sq. in. $\times$ stroke in inches $\times$ per cent clearance

$$= \tfrac{1}{4} \pi d^2 \times 24 \times 0.05$$
$$= \tfrac{1}{4} \times 3.14 \times 64 \times 24 \times 0.05 = 60.3 \text{ cu. in.}$$

The term **speed,** as applied to engines, refers to **rotative speed,** and is expressed in revolutions per minute. Speeds above 250 r.p.m. are called *high speeds;* from 150 r.p.m. to 250 r.p.m. *medium speeds*, and below 150 r.p.m. *slow speeds.* High-speed engines usually have a short stroke and low-speed engines have a long stroke, compared to the diameter of the cylinder. Piston speeds, expressed in feet per minute, range from 500 to 700. The piston speed is obtained from the formula:

$$S = 2\,LN \quad \dots \dots \dots \dots \quad (88)$$

in which S = piston speed, feet per minute.
L = length of stroke, feet.
N = revolutions per minute.

Example 39. — The piston speed, in ft. per min. of a 7 in. by 10 in. — 300 r.p.m. Buckeye engine would be

$$S = 2\,LN = 2 \times \tfrac{10}{12} \times 300 = 500 \text{ ft. per min.}$$

320. Engine Frames. — The form of the engine frame is determined by the type of engine and the purpose for which it is to be used. The frame is made of cast iron, and should have the metal so distributed that the center of gravity of the frame is as near the foundation as possible.

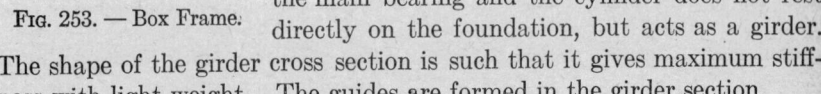

The **box type of frame,** Fig. 253, largely used on high-speed engines, has a rectangular base that rests directly on the bed plate or foundation. The lower crosshead guide is machined as a part of the engine frame, and the upper guide is detachable. The cylinder overhangs one end of the frame and the opposite end carries two bearings, which are separated by a web that stiffens the frame. The recess formed by this web and the side pieces serves to collect any oil that drips from the bearings, and thus keeps it from the floor and foundation. In some types, this chamber is used as an oil reservoir into which the crank dips at each revolution.

The **girder frame,** Fig. 254, is used for Corliss engines doing light work. It is called a girder frame because the part of the frame that connects the main bearing and the cylinder does not rest directly on the foundation, but acts as a girder.

Fig. 253. — Box Frame.

The shape of the girder cross section is such that it gives maximum stiffness with light weight. The guides are formed in the girder section.

The **heavy duty,** or **rolling mill, frame,** Fig. 255, has a low center line and heavy construction. The pillow block and guide are made in one casting, thus giving strength and rigidity. The crosshead guides are

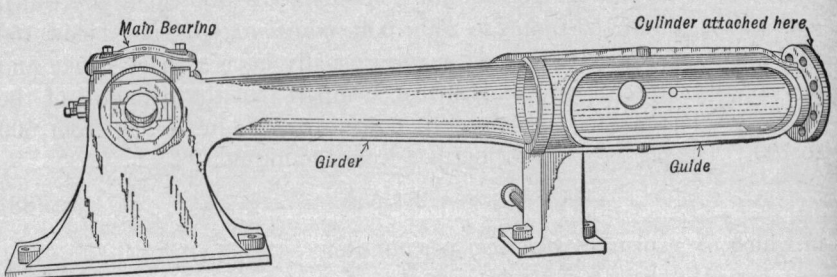

Fig. 254. — Girder Frame — Crank Side.

bored, and the frame is carried under the crank, forming an integral crank case in which oil from the bearings and crosshead guides collects. The oil is then drained from the lowest part of the frame. This frame is of pleasing design and affords continuous contact with the foundation.

CYLINDERS

A type of frame, Fig. 357, page 446, known as the **Tangye frame** is extensively used. The construction is graceful and adapted to heavy duty, since the frame has a stiff back behind the guides.

The frame used for vertical engines generally has the shape of the letter A and is known as an **A-frame**. This frame is usually made with an upper and a lower part, the lower part having a rectangular shape and resting directly on the foundation or on a bedplate. A web, cast across the bottom, serves to catch oil. The upper part rests upon the lower, and forms the guides and a support for the cylinder. The vertical frame used on marine engines, Fig. 310, consists of columns, made of cast iron, cast steel, or steel forgings and shaped like an inverted Y. These columns

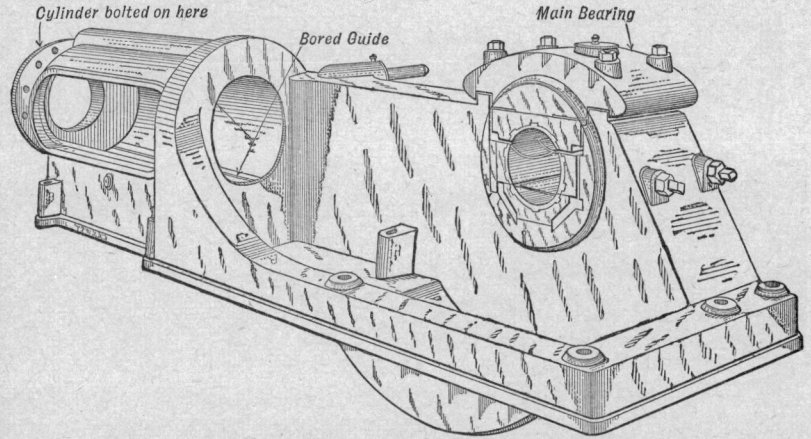

FIG. 255. — Rolling Mill Frame — Crank Side.

support the cylinder and serve as supports for the crosshead guides. The front and rear columns are often tied together by rods which give sufficient resistance to withstand the severe stresses.

321. Cylinders. — The cylinder and steam chest, Fig. 256, are generally made of cast iron in a single casting, with connecting passages or ports. The body of the cylinder is a shell of uniform thickness, with flanges at each end, to which the cylinder heads are bolted. The diameter of the central portion of the cylinder is called the **bore** and that of the enlarged part at each end, the **counter-bore**. The length of the bore is such that the **piston** ring will overtravel the edge of the bore slightly at each end, and thus prevent the wearing of shoulders in the bore. The counter-bore is made from $\frac{1}{8}$- to $\frac{1}{4}$-inch larger in diameter than the bore, to permit reboring when the cylinder becomes worn.

The heads are flat pieces, having shoulders which fit into the counter-bore. The head-end head is sometimes recessed to permit the piston to travel closer to the cylinder head, thus reducing the clearance. The crank-end head carries a cylindrical stuffing box through which the piston rod passes. The stuffing box is filled with fibrous or metallic packing which is compressed by a **stuffing box gland** held by studs and nuts.

The part of the casting that forms the steam chest and steam passages is more complicated than the cylinder proper. The **valve seat,** or surface upon which the valve slides, has a rectangular form, and is raised above the surface of the valve chest and machined to a smooth surface. The passages for steam extend from the valve seat to each end of the cylinder; in Fig. 256, these passages are long, and thus increase the clearance. The top of the valve chest is covered with a flat cover plate bolted to the valve or steam chest.

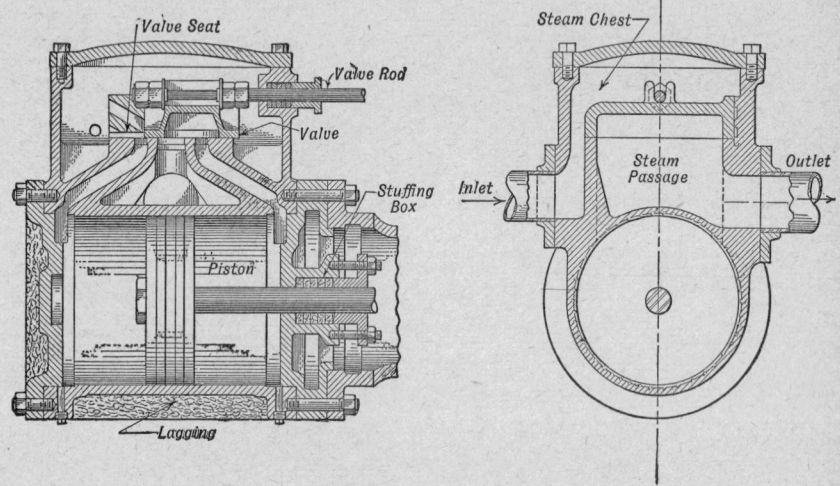

Fig. 256. — Sectional View of Steam Cylinder.

The cylinder is given a covering of planished steel, to improve its appearance, and the space between the covering and the cylinder is filled with asbestos, magnesia, or other insulating material, to prevent loss of heat by radiation. The crank end of the cylinder is machined to fit a corresponding machined part on the frame, and is held in position on the frame by a ring of bolts.

An outside view of a **Corliss engine cylinder** is shown in Fig. 257a, and the appearance of the cylinder after being insulated and covered with planished metal is shown in Fig. 257b. This cylinder is much heavier than the slide-valve cylinder; and rests directly on the bedplate foundation and is bolted to the frame. The valve seats are circular in section, and extend across the cylinder casting at each end. This permits the use of short steam passages and reduces the clearance. The steam chest is at the top and is separated from the cylinder bore by the cylinder wall, thus serving to **steam-jacket** the upper part of the cylinder. The exhaust-steam passage is at the bottom of the casting and is separated from the cylinder wall by an air space, which prevents heat loss from the cylinder to the exhaust passage. In some types of Corliss engines the cylinder is made with a cylindrical barrel, and the valve openings are placed in the cylinder heads, which are bolted to the cylinder. This makes a simple cylinder casting.

322. Cylinder Head. — The cylinder head is generally made, like that shown in Fig. 320, of a heavy casting stiffened with ribs. At the point where the steam passages enter the cylinder, the surface of the cylinder head is relieved.

323. Stuffing Box Packing. — The stuffing box used with saturated steam is generally filled with **fibrous packing** impregnated with graphite, Fig. 258. Leakage of steam past the packing is prevented by tightening the nuts on the gland thus compressing the packing against the rod. A brass bushing is usually placed in the

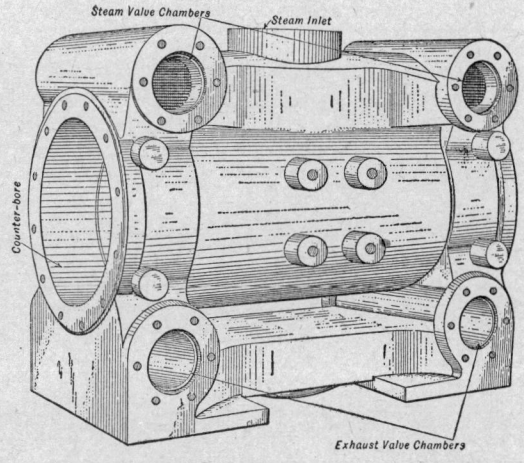

FIG. 257a. — Corliss Cylinder — Outside View.

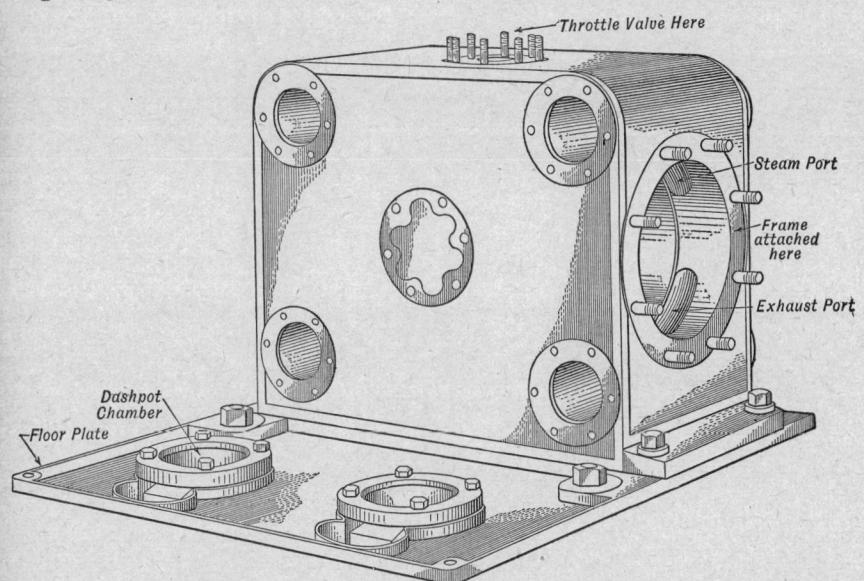

FIG. 257b. — Corliss Cylinder, Floor Plate and Lagging.

bottom of the stuffing box to prevent the packing from being forced into the cylinder. Fibrous packing often becomes hard and scores the piston rod, because of scanty lubrication. It is not satisfactory for use with superheated steam, because the high temperatures destroy the packing.

Metallic packing, Fig. 259, is used with superheated steam. It has

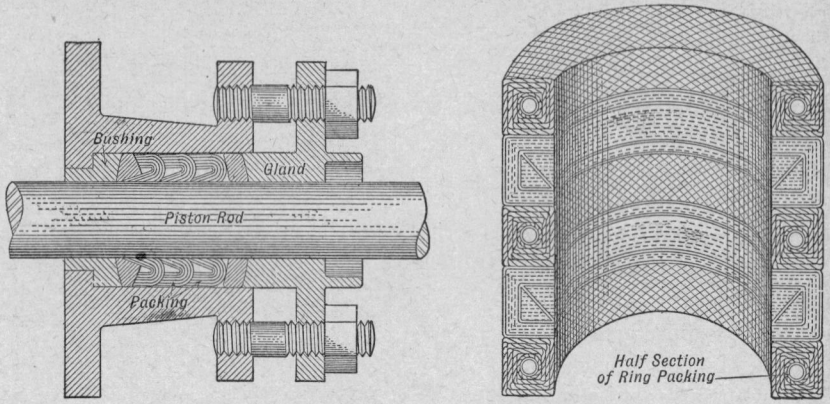

FIG. 258. — Stuffing Box and Several Types of Ring Packing.

long life, has proved satisfactory in service, automatically adjusts itself as wear takes place, and is sufficiently flexible to allow the piston rod to move out of alignment.

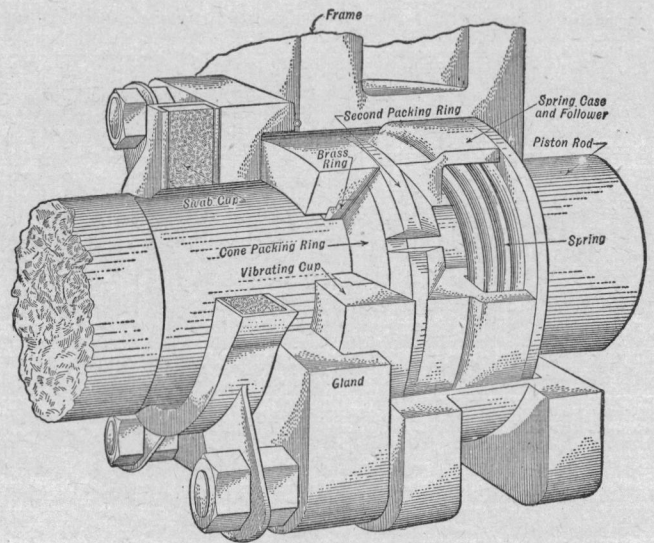

FIG. 259. — Sullivan Metallic Piston Rod Packing.

A special gland holds the packing in place, and a small copper wire is placed between the gland and the stuffing box to prevent leakage. The packing consists of a number of beveled rings which are split and fit closely together. The **vibrating cup** surrounds a beveled brass inner ring, made in halves which bear against the piston rod. The **inner packing ring** has a double bevel, one side of which fits the vibrating cup, while the other side fits the **outer packing ring**. The **spring case** fits over the outer packing ring and retains the spring, which holds the various rings in proper position. The **spring** also puts some pressure on the packing, this pressure being increased by the steam pressure acting against the inner end of the spring cage. An **oil-soaked swab cup,** attached to the gland bolts, provides sufficient lubrication.

324. Pistons. — The piston should be of simple construction, of light weight, especially for horizontal cylinders, and strong enough to stand the pressure of the steam without deformation. A piston for a horizontal engine should have a wide face to distribute its weight over a larger surface and thus reduce wear.

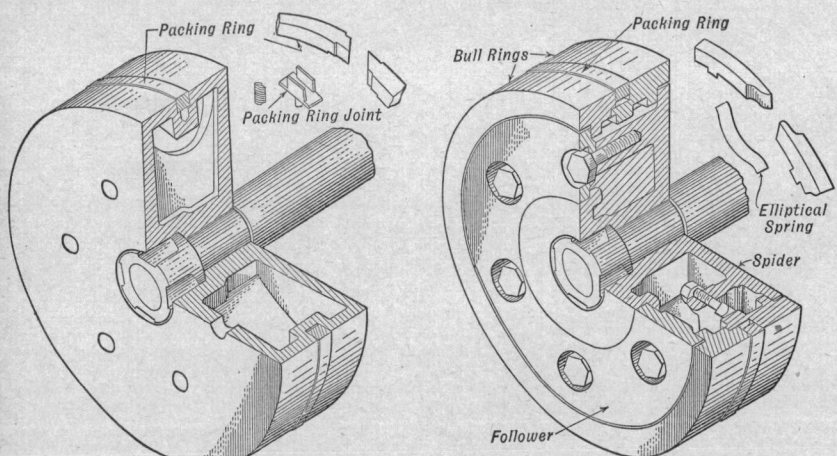

FIG. 260. — Solid Piston with Single Packing Ring.

FIG. 261. — Built-up Piston, with Narrow Follower and Elliptical Spring.

The **solid type** of piston, Fig. 260, is quite commonly used. It consists of a hollow cylindrical casting made with flat surfaces of uniform thickness on both sides. A hub, to which the piston rod is attached, is cast at the center of the piston. The hollow part of the casting is divided into several small compartments, by webs which strengthen the end surfaces. Holes are cast in each compartment to permit the removal of the core sand and are plugged after it is removed.

The piston rod is machined to fit into the hub of the piston and is held in place by a countersunk nut screwed on the rod. The nut pulls the hub tight up against a shoulder on the rod, which is often made a straight forced fit into the piston.

Leakage of steam past the piston is prevented by a single packing, or **piston ring,** placed in a groove machined in the circumference of the piston.

The ring is made from a piece of cast iron with its outer surface and sides finished. It has a diameter slightly larger than the bore of the cylinder, and is generally made thicker on one side than the other, with a diagonal cut at its thinnest section. The ring is sprung into the groove in the piston and, being larger than the bore of the piston, bears lightly against the walls of the cylinder when sprung into place. Leakage of steam through the diagonal cut is sometimes prevented by a tongued clip.

A **built-up piston,** Fig. 261, is much used on Corliss engines. The body of the piston is a flanged spider to which the piston rod is attached. Two **bull-rings,** or **junk-rings,** form the sliding face of the piston, and a single

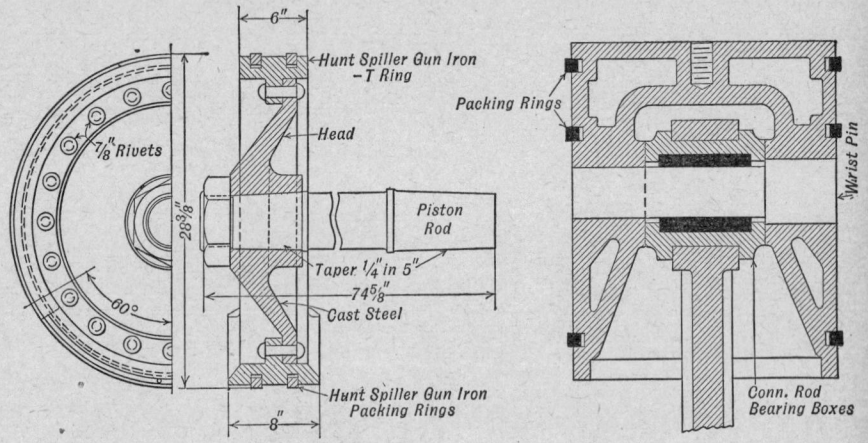

FIG. 262. — Locomotive Piston.

FIG. 263. — Trunk Piston — Sectional View.

packing ring is placed in the opening between the bull-rings. A **follower plate,** attached to the spider by bolts, holds the bull-ring in position. The position of the bull-rings can be adjusted for wear by set screws fastened to the spider. Elliptical springs are placed at various points between the bull-ring and packing ring. Packing rings are often made in sections, with elliptical springs placed under each section to hold them out against the cylinder bore.

One form of **locomotive piston,** shown in Fig. 262, is made from a simple T-section of cast steel, and has two packing rings. Pistons generally have more than one packing ring, with the rings so placed that the joints do not come in the same line. Leakage past the rings is thus prevented without the use of tongued clips. The piston rod is fastened to the piston by a taper fit, and a nut, placed on the end of the piston rod, holds the piston from coming loose. The nut is generally pinned to prevent its working loose. Locomotive pistons are often made broader at the bottom, as in the illustration, to increase the wearing surface.

On some types of steam engines a trunk piston, Fig. 263, is used. Steam can only act on one side of this piston, which is made long because it performs the function of the crosshead. One end of the piston is open, to admit the connecting rod, and bosses are cast on the inside to support the wrist pin.

325. Piston Rod. — The piston rod is circular in cross section and is made of a good grade of steel, with the crosshead end threaded and the piston end machined to fit a similar machined surface in the piston.

326. Crossheads. — The crosshead is attached to the piston rod by a thread and nut, and to the connecting rod by a crosshead pin, or **wrist pin**, which is carried by the body of the crosshead and which holds one end of the connecting rod. During one half of each revolution the crosshead pulls the connecting rod and during the other half it pushes the connecting rod. This causes a pressure on the guides, which always acts downward when the engine is " running over."

The most common types of crossheads are the " wing," " block," and " slipper." The **" wing,"** or **locomotive type**, Fig. 264, consists of two side blocks, or wings, united by a yoke. The piston rod is screwed into the yoke and provided with a lock nut placed on the piston rod and backed up against the yoke, to prevent the piston rod from working loose in the crosshead, since vibration or removal of the load might

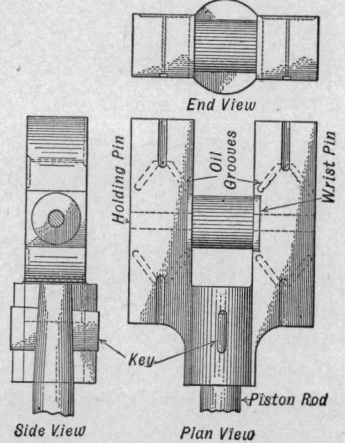

FIG. 264. — Wing Type Crosshead.

loosen the fit between the rod and crosshead and permit the rod to turn out. The pressure of the piston is carried by a steel wrist pin with a flange on each end that fits into recesses in the wings. A small steel pin passes through the wings and the wrist pin, and holds the wrist pin from being lifted vertically. The wings slide between upper and lower guides, which can be raised or lowered by **shims** to provide for wear. The sliding surfaces of the wings are generally made with holes that are filled with babbitt metal to reduce friction. **Babbitt metal** is an alloy of copper, tin, and zinc or antimony, having anti-friction qualities and a tin content exceeding 50 per cent.

The " **block**," or **Corliss crosshead**, Fig. 265, has a heavy cast-iron body into which the piston rod screws. The body is made hollow to admit the end of the connecting rod. The crosshead pin has tapered ends, where it fits into the body, and is held in position by a washer. An **upper** and a **lower shoe,** or slipper, are fastened to the body by side bolts. The surfaces of the shoe and body, where they come together, are made on an incline, and nuts and bolts attached to the body permit adjustment of the shoes along the inclined surface. The sliding surfaces of the shoes are babbitted.

The **slipper crosshead,** Fig. 266, is used extensively on marine and locomotive engines. The crosshead body and slipper are made in one casting, with the piston rod attached to the body above or below the

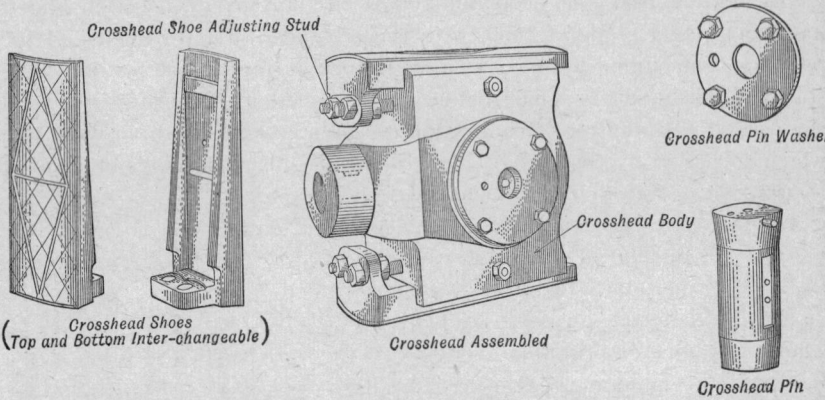

FIG. 265. — Corliss Type Crosshead.

slipper. The sliding surface of the slipper is flat and broad. The guide has a flat planed surface on the engine frame, with adjustable side pieces which fit on top of each side of the slipper and hold the slipper in position. In some types of locomotive crossheads, the guide is a square rod and the crosshead surrounds the guide.

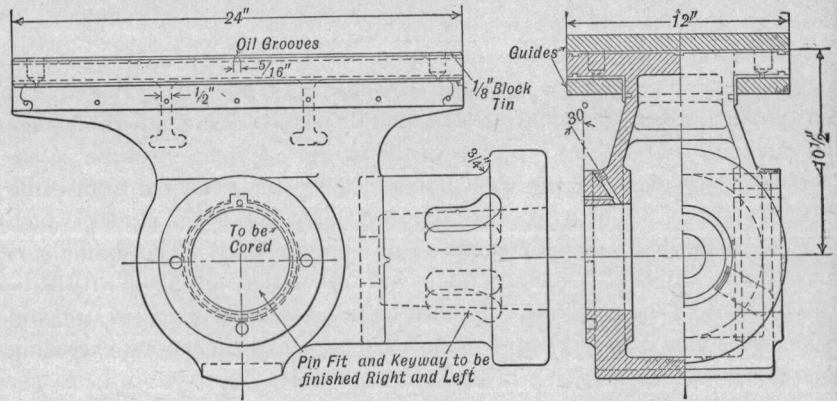

FIG. 266. — Slipper Type Crosshead for Locomotive.

The shape of the sliding surfaces may be flat, circular or triangular. The circular form is the most common and is easily adjusted and machined.

327. Connecting Rods. — The connecting rod is usually made of a good grade of steel, and may have a circular, rectangular or I-section. Corliss and marine engines generally have round rods, and high-speed engines have rectangular rods. Circular rods are largest at the center and taper toward each end. Rectangular rods have the largest section at the crank-pin end.

Common types of connecting-rod ends are the solid, strap, and marine end. The **solid-end rod**, Fig. 267, can only be used with a projecting, or overhung crank pin. It has enlarged and flattened ends, which are slotted to hold the **boxes**, or **brasses**, in which the crank pin and wrist pin fit. The boxes do not come quite together at the center. This permits adjustment for wear by wedges and bolts placed one inside and one outside, at their respective pins. With the wedges thus located adjustment for wear can be made without changing the length of the connecting rod. The wedges are threaded, and turning the adjusting bolt changes the position of the wedge at right angles to the pin, while side flanges machined on the boxes prevent movement parallel to the axis of the pin.

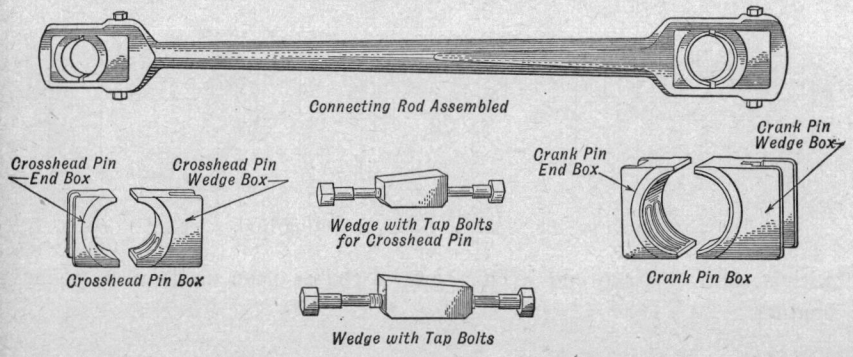

Fig. 267. — Solid End Connecting Rod.

In the **strap-end rod**, Fig. 268, the brasses are held to the enlarged end of the rod by a strap that passes around them and is held to the rod end by two through bolts. The adjusting wedge is similar to that of the solid-end rod.

The adjusting wedge and through bolts are sometimes omitted and a **gib** and **key** used. The gib and key pass through a hole cut in the end of the rod, and the inner brass bears directly against the rod. The gib prevents the end of the strap from spreading. Adjustment is made by driving the key into the rod, and a set screw prevents the key from coming out while the engine is running.

The **marine-end rod**, Fig. 269, has the upper end forked, each branch of the fork having a bearing, or **eye**, and connections for the crosshead pins. Pins used at the crosshead end of this type of rod are called **gudgeon pins**. The lower, or crank, end is enlarged, and has an upper and a lower part made of brass and babbitted. Two through bolts hold these parts to the enlarged end of the rod, and **lock nuts** keep the bolts from coming loose while running. **Thin pieces of metal, called shims,** are placed between the upper and lower halves of the crank-pin boxes, and adjustment is made by removing or adding shims. The upper end is sometimes made

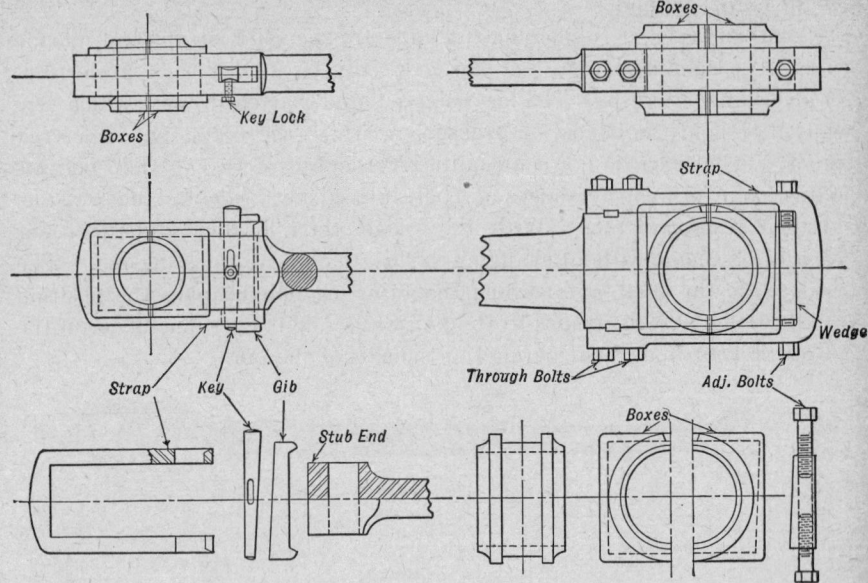

Fig. 268. — Strap End Connecting Rod.

with a solid or strap end. This type of rod is often used on stationary engines.

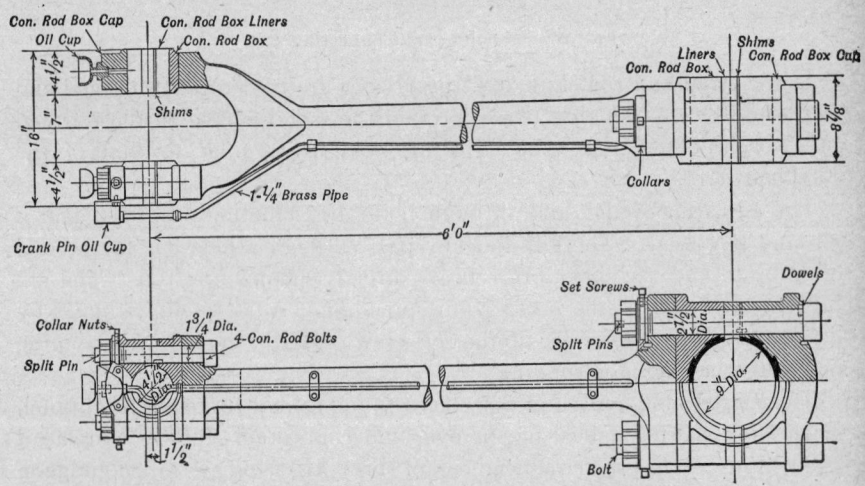

Fig. 269. — Marine Type Connecting Rod.

328. Cranks. — The most common types of cranks, Figs. 357, 311, and 270, are the disk crank, the center crank, and the overhung crank, the first being used to the greatest extent on modern engines. It is generally made heavier on the side opposite the crank pin, to counterbalance the shaking forces produced by the crank and the reciprocating parts of the engine, and thus make the engine run more smoothly. The overhung crank is attached to the end of the shaft, and is often a disk crank; the center crank has the crank pin fastened between two crank disks. Cranks are ordinarily made of cast iron.

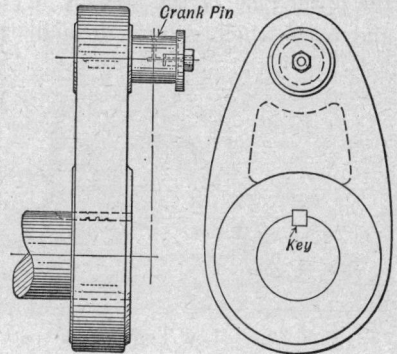

The marine engine crank shaft, Fig. 271, has the crank and the crank shaft formed in one piece. It is commonly made in sections, which are bolted together to make a continuous shaft, each section consisting of two crank webs, or disks, a crank pin, and a piece of shaft at each end, which has flanges for coupling. The crank

FIG. 270. — Overhung Crank with Crank Pin.

pins are arranged to have the cranks come at 90 or 120 degrees. This gives more even running and distributes the stresses more equally. The entire crank shaft is often forged in one piece, and the crank pins and shaft made hollow to decrease weight and increase strength. The locomotive crank is formed as a part of the drive wheel.

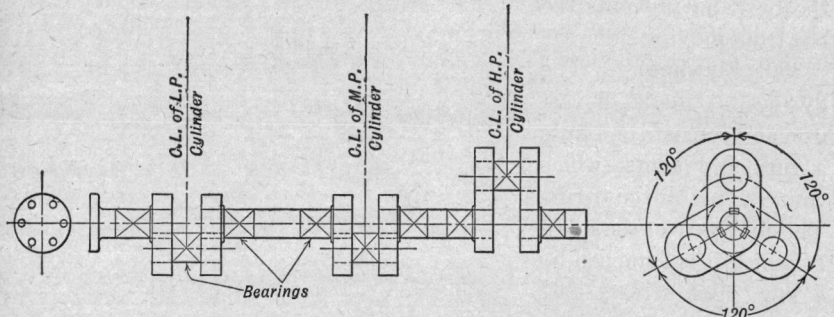

FIG. 271. — Marine Crank Shaft Assembly.

329. Bearings. — The simplest form of bearing is the two-part bearing, Fig. 272, generally used for the outboard bearing. It is made of cast iron and consists of an upper and a lower half bolted together and lined with babbitt metal to form the bearing surface. The babbitt lining is ordinarily grooved to facilitate the flow of oil to the bearing surfaces. The lower half is bolted to the foundation. When used as an outboard bearing, it serves to support the weight of the flywheel and keeps the outer end of the shaft in alignment.

The main bearing, or **pillow block**, Fig. 273, is made heavier than the outboard bearing to take the thrust from the piston. It consists of a lower part, or **housing,** which may be made separately or as a part of the main frame, and a **bearing cap.** The bearing is formed by three or four cast-iron boxes, which are babbitted and are held in proper position by the lower part of the housing. The boxes are adjusted by inside wedges and bolts which extend through the walls of the housing to the outside.

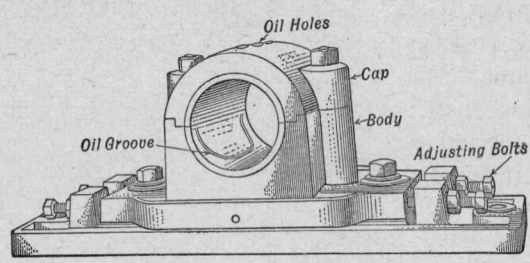

Fig. 272. — Low Type of Two-part Bearing.

Bolts, which pass through the bearing cap, hold the upper box in position. The lower box is raised or lowered by adding or removing shims from between the box and housing.

The bearing, often called a **journal box,** used on the locomotive axle is shown in Fig. 274. The box is made of cast iron and has a brass or bronze bushing which covers only the upper part of the journal. It is held from rotating by the bearing box. The pressure is always downward on this bearing, and only one bushing is necessary. The lower part of the box forms a cellar, which is filled with oily waste, and also prevents the journal from leaving the box. The locomotive frame fits between projections on the side of the box and prevents the box from moving sidewise.

330. Flywheel. — The flywheel is made of cast iron and consists of a **hub,** a **rim,** and **arms,** which connect the hub and rim. The hub fits over the shaft, to which it is attached by a key. **Set screws** pass through the hub and bear on the key, to prevent axial movement of the flywheel. The hub, rim, and arms may be made in one piece, or they may be made in a number of parts and bolted together. To maintain constant speed, the flywheel should be heavy.

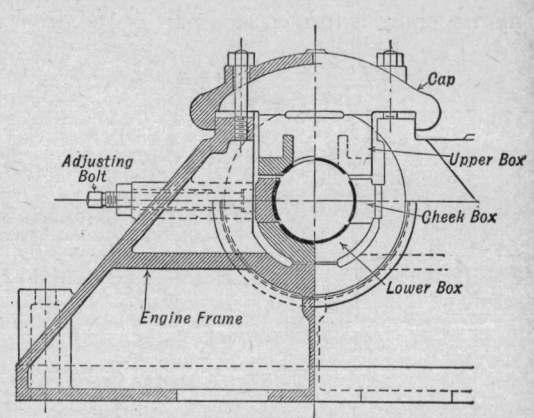

Fig. 273. — Main Bearing, or Pillow Block.

Fig. 357, page 446, shows a flywheel made of two pieces, with **through bolts** holding the parts of the hub together, and **stud links,** which fit holes made in the rim, holding the rims together. The length of the links between shoulders, before being put in place, is made shorter than the length between the shoulders in the rim, by a few thousandths of an inch. The link is heated and placed in position; it shrinks when cooling and holds the parts firmly together.

331. Foundation. — In order that an engine may remain level and in satisfactory running condition it is necessary that it rest upon a suitable foundation which may be constructed of concrete, brick, or stone. Concrete is the best material for the part below the ground while hard brick with a granite cap is satisfactory for the part above the ground. When brick or stone are used, they should be laid in a good quality of Portland cement and clean sharp sand. When concrete is used a satisfactory mixture is made of 1 part cement, 2 parts sharp sand, and 4 parts gravel or crushed stone.

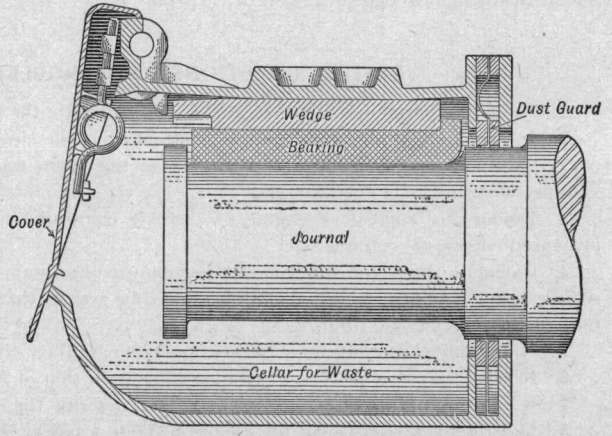

Fig. 274. — Journal Box for Locomotive — Sectional View.

The shape of foundations varies, wide and shallow foundations being generally preferable to deep and narrow foundations. Straight lines give a more pleasing appearance than curved lines. That part of the foundation which supports the bearings should be tied to the main foundation.

The area over which the foundation should extend is determined by the character of the soil. Foundations built on soils having a low bearing resistance such as clay require a greater area than those built on soils having a high bearing resistance, such as granite rock.

Foundations built on soils of low bearing resistance are generally made in two parts: a sub-foundation and a main foundation. The sub-foundation extends over a greater area than the main foundation, and when made of concrete, its surface is left rough to give a good bond with the main foundation.

Soils which are moist and have low bearing values are generally prepared for the sub-foundation by driving piles, spaced about 3 feet apart, to hardpan or bedrock. The piles are made of red pine, oak, beech, steel, or reinforced concrete, trimmed to an even height and covered with a layer of concrete not less than 24 inches thick, which is often reinforced with steel. The soil directly around the foundation should be well drained.

The weight acting on the soil consists of the weight of the engine plus

356   RECIPROCATING STEAM ENGINE — SIMPLE ENGINES

the weight of the foundation. In general, the weight of the foundation is from 4 to 5 times the weight of the engine.

332. Valve Mechanism. — The various types of valve gears, with their operating mechanism, are described in connection with the engines upon which they are used, Chapters XVI and XVII.

REFERENCES

Steam Power, HIRSHFELD and ULBRICHT.
Steam Engines, SHEALY.
Steam Power Plant Engineering, GEBHARDT.
Steam Bled Engines, page 972, Vol. 67, POWER.

REVIEW QUESTIONS AND PROBLEMS

1. Name the parts of a simple steam engine, and state the function of each part.
2. In a manufacturer's catalogue, it is stated that the clearance of the engine is 8 per cent. Explain what is meant. If the size of the engine was 7 in. by 10 in. find the volumetric clearance of the head end.
3. Define: (a) right-hand engine, (b) stroke, (c) dead center, (d) high-speed and low-speed engine, (e) running over.
4. Calculate the piston speed in feet per minute of an 8 in. by 24 in. Harris Corliss engine, running at 110 r.p.m. Would this speed be greater than a 7 in. by 9 in. Harrisburg engine making 300 r.p.m.?
5. Name three types of engine frames, and compare them with regard to strength.
6. In what way does a Corliss cylinder differ from that of Fig. 256?
7. Name three types of connecting rods, and describe the construction of each.
8. Describe how adjustment for wear is made in a box type of crosshead.
9. Describe the construction of a four-part bearing.
10. State the function of a piston ring, and explain how it performs this function.

CHAPTER XVI

SLIDE-VALVE ENGINES, VALVE DIAGRAMS, AND SLIDE-VALVE SETTING

333. Foreword. — Slide-valve engines are made in many forms and differ mainly in the type of valve and the method of governing. Most slide-valve engines use a single valve, which may be flat or cylindrical, to control the flow of steam. A few engines use two valves, and some use four valves.

As classified in Art. 316, the most common types of slide-valve engines are: " D " slide-valve, balanced, multi-ported and piston.

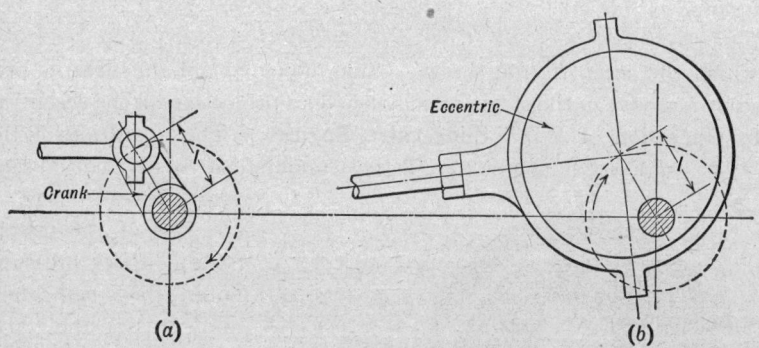

Fig. 275. — Analogy Between Crank and Eccentric.

334. Plain " D " Slide-valve Engine. — The plain " D " slide-valve engine is the simplest of the slide-valve engines, and takes its name from the fact that the section of the valve resembles the letter D. The **valve gear** consists of the valve, valve rod, valve-rod guide, eccentric rod, and eccentric, in which the eccentric is direct-connected to the valve.

The eccentric is essentially a short crank, which moves the valve back and forth on its seat. Consider a short crank of length l, Fig. 275a. If the crank pin is enlarged to include the shaft, as in Fig. 275b, an eccentric is formed. The distance between the center of the eccentric and the center of the shaft is the **eccentricity,** or throw of the eccentric, and is equal to l, of Fig. 275a.

The eccentric, Fig. 276, is a circular disk, or **sheave,** of metal containing a hole, which is not in the center of the disk, through which the crank shaft passes. A set screw generally secures the eccentric to the shaft and thus permits changing the position of the eccentric by loosening the set screw.

An **eccentric strap**, Fig. 277, made in two parts, surrounds the eccentric. These two parts are held together by through bolts having lock nuts to prevent the bolts coming loose. One part of the strap has a hub, or yoke,

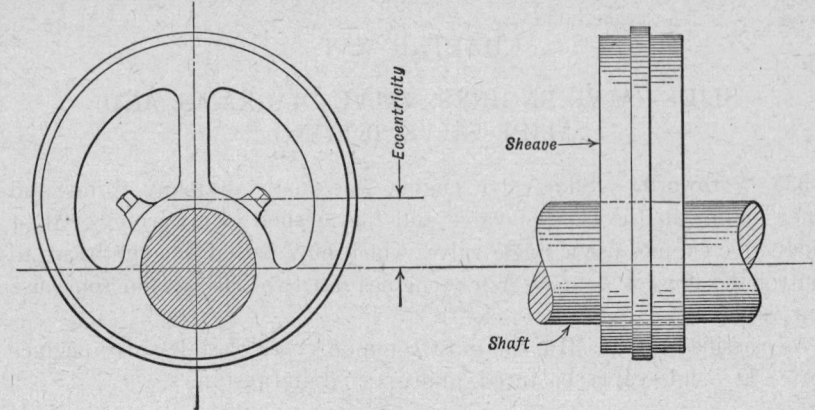

Fig. 276. — Eccentric.

into which the eccentric rod screws. Side movement of the strap is prevented by a groove in the strap, which fits over a projection on the eccentric.

335. Operation of "D" Slide-valve Engine. — The operation of the plain "D" slide-valve engine may be best understood by first considering what takes place on one side of the piston. Steam, under pressure, enters the steam chest and presses the valve tightly against its seat.

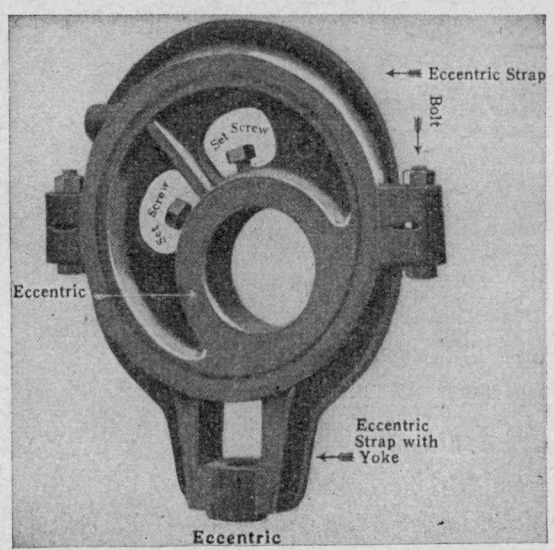

Fig. 277. — Eccentric and Eccentric Strap.

The eccentric is so located on the crank shaft, ahead of the crank in this case, that it will move the valve in the proper direction to admit steam to the cylinder, just before the piston has reached the end of the stroke. Fig. 278 shows the valve in this position, which is called the **point of admission**. The valve and piston are traveling in opposite directions, and any further movement of the valve to the left admits steam behind the piston. The pressure of the steam, acting on the piston, pushes it to the left and produces rotation of the crank and crank shaft, to which the eccentric is attached. The valve is thus moved and the steam passage opened wider. Steam continues to enter behind the moving piston, until the eccentric has moved the valve to its extreme right-hand position and

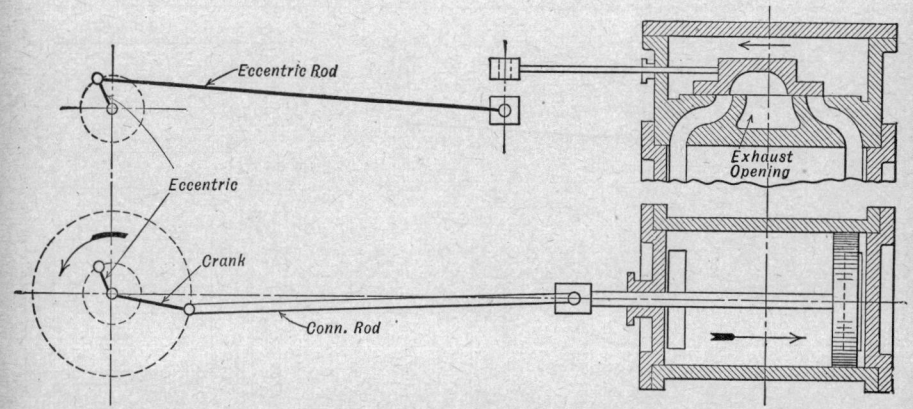

FIG. 278. — Positions of Crank and Eccentric at Point of Head End Admission.

returned it to the position shown in Fig. 279. **Cut-off** then occurs, and steam can no longer enter the cylinder. The valve and piston are moving in opposite directions, and each is traveling opposite to its direction of motion at admission. The steam in the cylinder at cut-off **expands,** that is, increases in volume, and performs work as the piston moves to the left. *As the volume increases the pressure falls.* When the valve and piston reach

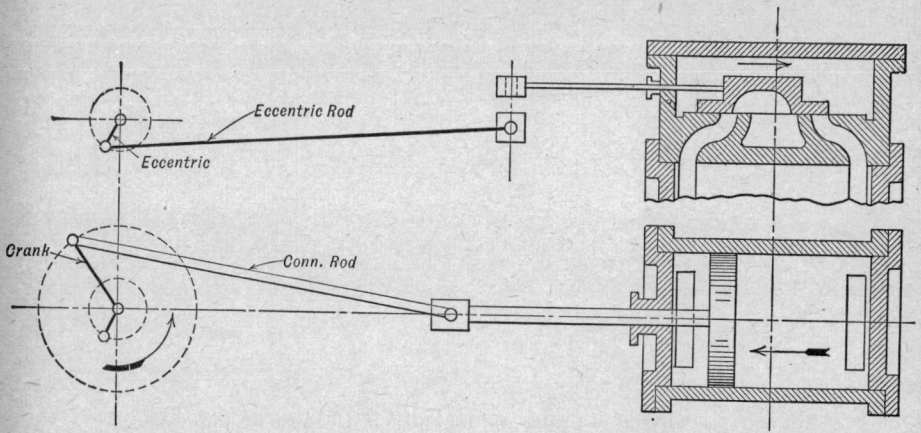

FIG. 279. — Positions of Crank and Eccentric at Point of Head End Cut-off.

the position shown in Fig. 280, **release** occurs, the valve opening the passage from the cylinder to the exhaust pipe. The exhaust steam passes from the cylinder through the steam passage and the cavity on the under side of the valve to the exhaust opening. At release, the valve and piston are traveling in the same relative direction as at cut-off. Upon further movement of the valve, the exhaust opening is increased, and the pressure in the cylinder falls nearly to the pressure in the exhaust pipe. Shortly after release occurs, the direction of the motion of the position is reversed.

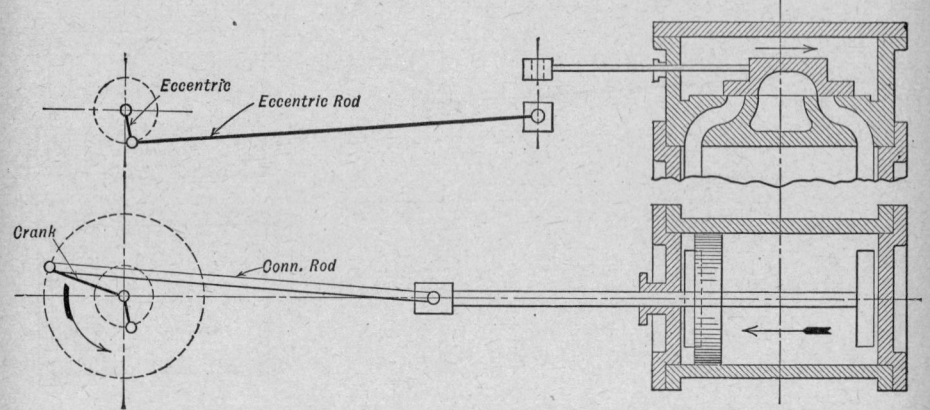

Fig. 280. — Positions of Crank and Eccentric at Point of Head End Release.

It now moves to the right and pushes the steam remaining in the head-end side of the cylinder out through the exhaust passage. When the crank has nearly reached the head-end dead center, and the direction of valve motion has been reversed by the eccentric, the valve and piston occupy the positions shown in Fig. 281. This is the **point of compression,** and

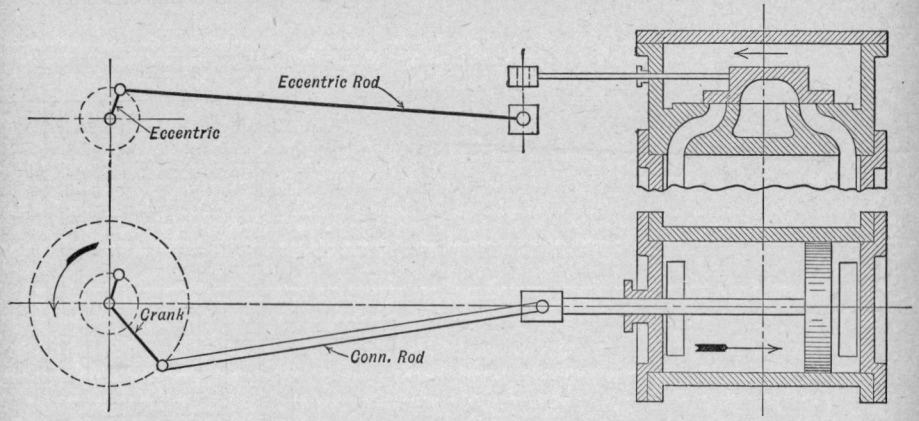

Fig. 281. — Positions of Crank and Eccentric at Head End Compression.

the exhaust passage is closed by the valve. The valve and piston are then traveling in opposite directions.

A **cycle of events** is performed by an engine when it passes through a series of operations and returns to its starting position. That part of the *steam-engine cycle* between the points of admission and cut-off is known as **admission**; between the point of cut-off and the point of release is called **expansion**; between the points of release and compression is **exhaust**; and between the points of compression and admission, **compression**. When admission and expansion are occurring in the head-end, exhaust and compression are occurring in the crank end, and *vice versa*.

An engine like the one described, which has steam entering alternately on one side of the piston and then on the other, is called a **double-acting** engine. Most steam engines are double-acting.

If steam acts only on one side of the piston, the engine is known as a **single-acting** engine.

The cycle of events for an engine may be studied by a diagram, Fig. 282, known as an **indicator diagram.** *The vertical height of the diagram*

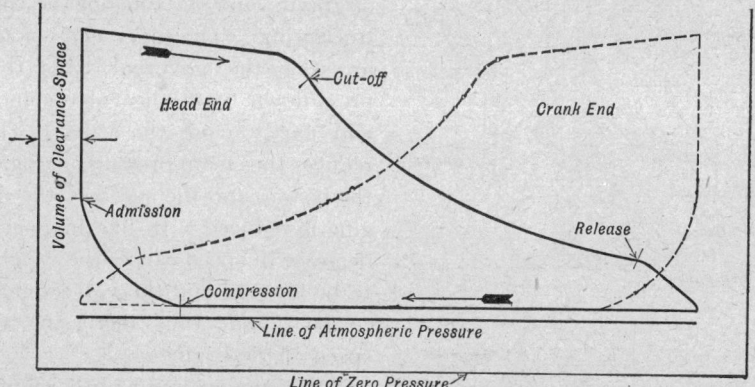

FIG. 282. — Indicator Diagrams.

represents the pressure acting on the piston, to some scale. Horizontal distances represent the volume of the piston displacement at any point in the stroke. The arrows show the direction in which the piston travels. A line representing atmospheric pressure is drawn on the diagram, for reference, to show how the pressure varies as compared with the pressure of the atmosphere.

336. Governor. — The governor used on the " D " slide-valve engine is a **throttling governor.** *It controls the speed by reducing or increasing the steam pressure, depending upon the load to be carried.* The governor operates a throttle valve placed in the main steam pipe, just before the pipe enters the steam chest.

One type of throttle governor generally called a flyball governor is shown in Fig. 283. The valve has an upper and a lower disk. Steam at the pressure in the main steam line acts on top of the upper and on the bottom of the lower disk. The same steam pressure acts on the outlet side of both disks. The valve is, therefore, **balanced,** and moves easily. The **valve spindle,** which passes through a **sleeve,** bears against the upper ends of the **governor arms.**

A **bracket** attached to the upper part of the valve body carries a horizontal bearing for the belt pulley shaft, a vertical bearing for the sleeve, and a chamber to which the adjusting lever is attached. The arms supporting the governor balls are pivoted to the upper parts of the sleeve and revolve with it. A **bevel gear** is keyed to the lower part of the sleeve and meshes with a similar gear on the belt pulley shaft. A **belt** connects the **belt pulley** to the engine shaft.

The governor operates as follows: With an increase in the speed of the engine, the speed of rotation of the governor balls is increased. Centrifugal force causes them to move outward and upward against the force of gravity and the force of the **control spring**. The inner ends of the arms of the governor balls thus press down upon the valve spindle and partly close the valve. This reduces the steam pressure acting on the piston, and the speed of the engine is reduced. In like manner, a decrease in speed causes the weights to be lowered and the valve spindle to be raised, thus increasing the speed of the engine.

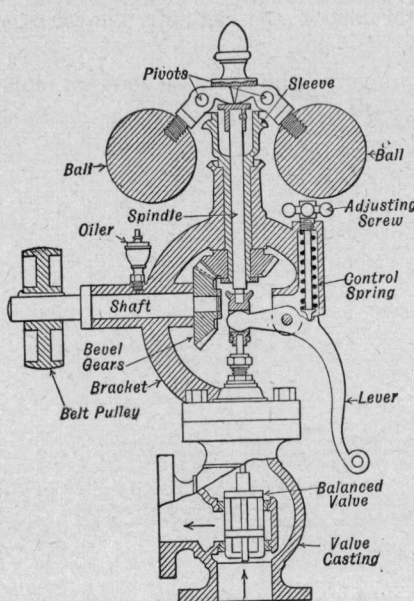

Fig. 283. — Flyball Throttling Governor.

For any given speed of the engine, the balls will take a position at which their centrifugal force just balances the force of gravity and the tension of the spring. If the engine speed changes, the weights will rise or fall until the steam pressure is adjusted to suit the load to be carried.

The speed at which the engine will run can be changed by turning the adjusting screw and thus changing the tension of the control spring. An increase in tension of the spring will increase the speed of the engine and *vice versa*.

A governor, to give satisfactory speed regulation, must have stability; otherwise it would not be able to control the speed, and its movements would be irregular and uncertain.

337. Stability. — The throttle governor already described is **stable**; that is, there is a definite position of the weights for any definite speed. If the speed changes the weights assume a new position corresponding to the new speed. This changes the position of the throttle valve and brings the speed of the engine back to the speed at which the governor is set to run.

338. Sensitiveness and Hunting. — A governor is **sensitive** when a small variation in speed will cause it to move from one extreme of its position to the other extreme. Naturally the more sensitive the governor is, the less stable it will be; but both of these qualities are essential. A governor should not be too sensitive, however, or it will **hunt;** that is, swing first to one extreme and then to the other extreme of its travel, thus making the speed first too high and then too low. Hunting is overcome by using dashpots attached to the governor in such a manner that its motion is damped.

The **dashpot** is a small cylinder having a tight-fitting piston in which there are one or more small holes. The cylinder is filled with heavy oil, which requires time to pass through the holes from one side of the piston to the other and any momentary change of the governor is thus prevented.

339. Terms Applied to Slide Valves. — There are many terms applicable to all types of valve gears, which are more clearly understood when studied with the slide-valve engine.

Valve travel is the distance moved by the valve from one extreme of its motion to the other. When the eccentric is connected to the valve without an intervening **rocker arm,** page 366, the *valve travel equals twice the eccentricity.*

Mid-position is the position occupied by the valve when it is halfway between the extreme positions of its motion. *The position of a valve is described by giving its displacement from mid-position.* In most engines, the eccentric is nearly vertical when the valve is in mid-position, or **central.**

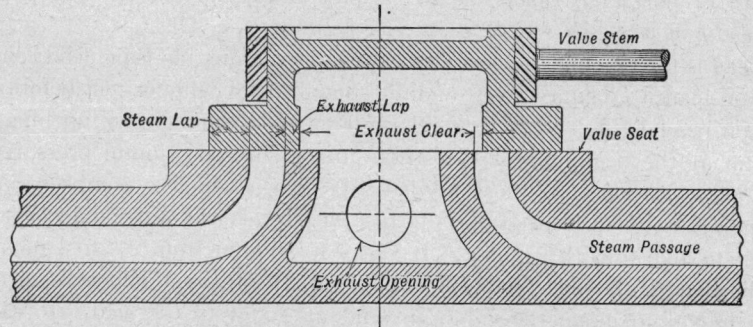

Fig. 284. — Valve Central Showing Laps.

Lap is the distance the edges of the valve overlap the corresponding edges of the port when the valve is in mid-position, as shown in Fig. 284. When applied to the inside edges of the valve and port, it is called **inside lap** and when applied to the outside edges, **outside lap.** When steam is admitted to the engine past the outer edge of the valve, the *outside lap* is the **steam lap,** and the *inside lap* is the **exhaust lap.** Steam is often admitted from the inside edge of the valve; the inside lap is then the **steam lap** and the outside lap the **exhaust lap.** *Steam lap may be defined as the distance the valve must move from mid-position to allow admission* *of steam, and exhaust lap as the distance the valve must move from mid-position to allow steam to escape from the cylinder.*

The steam lap is always positive or zero, while the exhaust lap may be positive, negative or zero. When the exhaust lap is negative it is called **exhaust clearance,** and is used to give better distribution of steam. The **angle of lap** is the angle turned through by the eccentric to move the valve a distance equal to the lap.

By referring to Figs. 278 to 281, it will be seen that:

1. When the engine is at admission, the valve has moved from mid-position an amount equal to the steam lap and is moving in a direction to uncover the port.
2. When the engine is at cut-off, the valve is in the same position as at admission but moving in a direction to cover the port.
3. When the engine is at release, the valve is displaced an amount equal to the exhaust lap and moving to uncover the port.
4. When the engine is at compression, the valve is in the same position as at release on that end, but moving to cover the port.

Port opening is the amount the port is open to steam at any instant, and *equals the valve displacement minus the steam lap.* **Maximum port opening** is the amount of port opening at the extreme travel of the valve. This term should not be confused with width of port.

Overtravel is the amount the maximum port opening differs from the width of port, and *equals the eccentricity minus the sum of the lap plus the width of port. It may be positive or negative.*

Lead is the amount the port is open when the crank pin is on dead center. **Steam lead** is given a valve to admit steam to the cylinder just before the piston reaches the end of its stroke. This gives an adequate opening for steam at the beginning of the stroke and assures maximum pressure on the piston when it is most needed. It also assists the compression in bringing the reciprocating parts to rest without shock, and should be sufficient to give smooth running. It varies in amount from $\frac{1}{32}$- to $\frac{1}{4}$-inch.

The **angle of lead** equals the angle, Fig. 285, through which the eccentric moves to move the valve an amount equal to the lead. It varies from 2 degrees to 8 degrees, depending upon the size and speed of the engine.

340. Angle of Advance. — Referring to Fig. 285, it is seen that *a direct-connected eccentric must be moved ahead of the crank* 90 *degrees* + (*the angle of lap* + *the angle of lead*) to have steam admitted before the piston has reached dead center. The amount by which this angle exceeds 90 degrees is the angle of advance. *It is the angle through which the center line of the eccentric moves to displace the valve from mid-position, an amount equal to the steam lap plus the steam lead.* The Greek letter delta (δ) is used to designate this angle.

341. The Effect of Lap. — For a valve without lap to admit steam to the cylinder, the eccentric must be 90 degrees ahead of the crank in the direction the engine is to run. Admission would then occur with the crank on one dead center, and cut-off would not occur until the crank had reached the other dead center. Steam would be admitted for the entire length of stroke, and there would be no expansion of the steam. Such an engine would require an excessive amount of steam for each horsepower developed.

Steam lap is given a valve to prevent this waste of steam, and to permit expansion of the steam by bringing cut-off before the end of the stroke.

If there were no exhaust lap, the events of release and compression would occur with the valve in mid-position and with the piston at the end of its stroke, and there would be no compression. To have the exhaust edge of the valve close the port before the crank pin has reached its dead-center position, exhaust lap is given. Compression traps steam in the clearance space and forms a steam cushion, which assists in bringing the reciprocating parts of the engine to rest without shock and makes the engine run more quietly. Zero exhaust lap and negative exhaust lap are given to valves of high-speed engines having a large angle of advance.

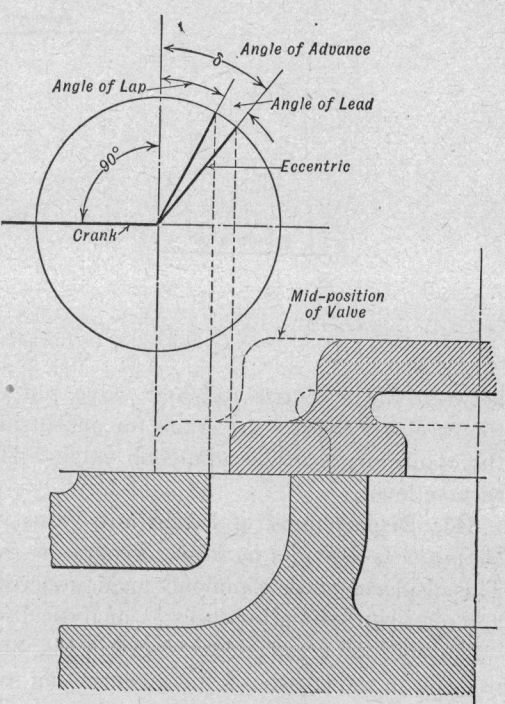

FIG. 285. — Diagram Showing Angle of Lap, and Angle of Lead.

342. Relative Position of Crank and Eccentric. — It has been seen that a plain "D" slide valve, connected to the eccentric without an intervening rocker, and having outside admission, has the eccentric ahead of the crank by 90 degrees plus the angle of advance. For a direct-connected slide valve having inside admission, the eccentric would follow the crank by 90 degrees minus the angle of advance, to have the valve open the port; otherwise, the valve would be closing the port. In general, the eccentric is located, with respect to the crank, in a position to open the proper steam port when the crank pin moves from dead center in the direction in which the engine is to run.

If a **rocker arm**, Fig. 286b, is placed between the eccentric rod and the valve rod, and has a pivot as shown, the rocker arm increases the movement of the valve for a given eccentricity, but does not change the relative position of the eccentric and crank. If the pivot of the rocker arm comes

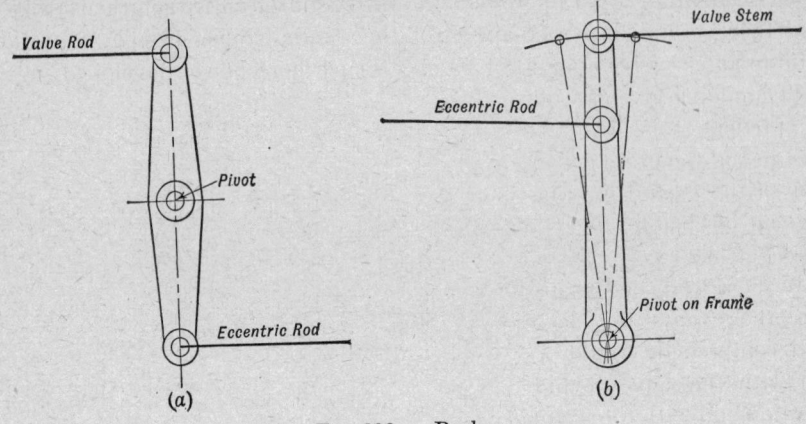

FIG. 286. — Rockers.

between the eccentric rod and valve rod connection, as in Fig. 286a, the eccentric follows the crank for an outside admission valve and leads the crank for an inside admission valve. This type of rocker is called a **reverse lever**.

343. Displacement of Piston and Valve. — *The linear displacement of the piston is described by stating its position from either dead-center position.* This displacement is commonly given in percentage of the length of stroke, and may be found graphically or analytically.

Consider the arrangement shown in Fig. 287, where the crank pin works in a slot in the crosshead and a connecting rod is not used. The displacement of the piston for any crank position would equal the horizontal distance traveled by the crank pin or $(OD - OV)$ where V is the foot of the perpendicular, from the crank position to the center line of the piston. From this it is seen that the position of the piston may be found by studying the position of

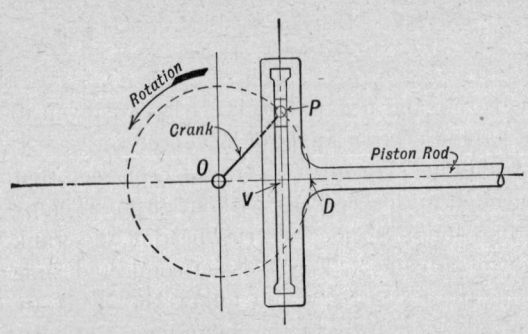

FIG. 287. — Connecting Rod of Infinite Length.

the crank pin. This arrangement is equivalent to a connecting rod of infinite length, and the motion given the crosshead and piston is known as **harmonic motion**.

The ordinary connecting rod is relatively short when compared with the length of the crank; the ratio of the length of the connecting rod to the crank usually ranges from $4\frac{1}{2}$ to 6 for stationary steam engines, and up to $7\frac{1}{2}$ on locomotives. This makes an appreciable angle between the connecting rod and the center line of the cylinder at certain crank positions. For any crank position, the piston will thus be at a different point in the stroke from that at which it would be with a connecting rod of infinite length.

In Fig. 288, the circle represents the path traveled by the crank pin during one revolution; the diameter of this circle therefore represents the stroke.

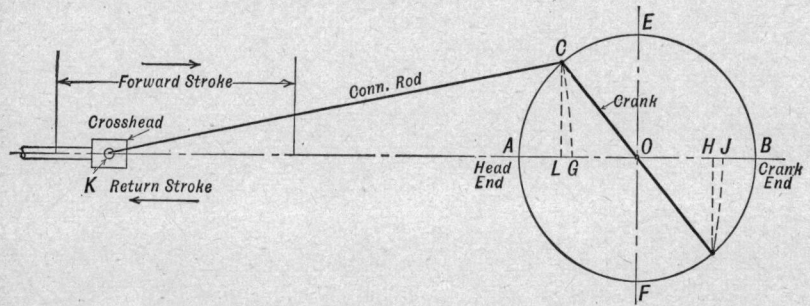

FIG. 288. — Diagram Showing Effect of Angularity upon Movement of the Piston.

When the crank is in any position, OC, the corresponding position of the piston may be located by taking a radius equal to the length of the connecting rod and, with K as a center, cutting the path of the crank pin and engine center line with an arc drawn through point C. The position of the piston is represented by the point G. Its position for a connecting rod of infinite length is shown by the point L. The actual position of the piston for the forward stroke is displaced from its position for an infinite connecting rod by the amount LG; for the return stroke this difference is HJ. *For the forward stroke, a finite connecting rod gives a greater movement of the piston for a given position of the crank pin than is shown by the length AL. For the return stroke, a finite connecting rod gives a smaller movement of the piston than is shown by the length BH.* This effect is caused by the **angularity,** or angular position of the connecting rod. The effect is greater for crank pin positions corresponding to the mid-position of the crosshead, and is also greater when the length of connecting rod is short compared with the length of the crank.

Since the eccentric, eccentric rod, valve rod, and valve have a motion similar to that of crank, connecting rod, piston rod, and piston, the position of the valve may also be found by using a circle having a radius equal to the eccentricity. The diameter of the circle will represent the valve travel. In considering movement of the valve, the angularity of the eccentric rod is neglected, as its length compared with the eccentricity is large, and the valve movement is measured from mid-position, or the vertical axis of the circle.

Since the position of the eccentric and the valve displacement may be represented by using a circle having a radius equal to the eccentricity, and the position of the crank and the piston displacement may also be represented by using a circle with a radius equal to the length of the crank these circles may be superimposed and the movement of the valve, with reference to the piston may be shown by two concentric circles, Fig. 289. *The scale to which the eccentric circle is drawn should be made large, for accuracy of measurement.* The crank position is first drawn at any desired

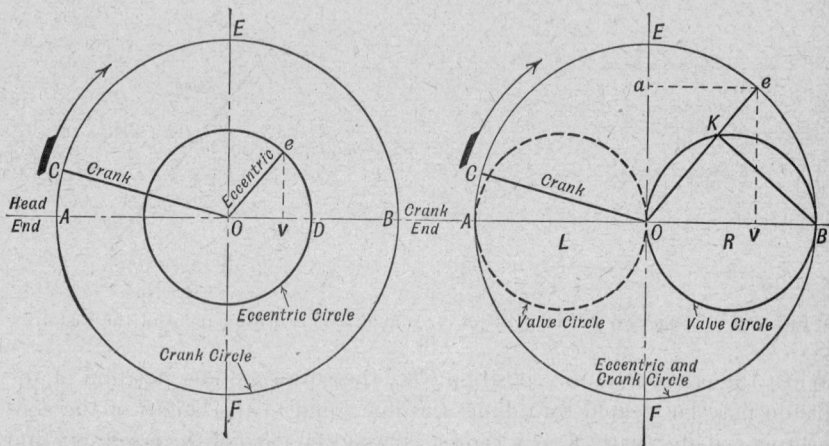

Fig. 289. — Crank Circle and Eccentric Circle Superimposed.

Fig. 290. — Development of Zeuner Diagram.

point, and the position of the eccentric located by laying off the angle COe, by which the eccentric is separated from the crank. The above construction must be repeated for each crank position. A more convenient method of studying the valve displacement for all positions of the crank is a graphical construction, termed a **valve diagram**.

The most common forms of valve diagrams are:

Valve ellipse..........applicable to any valve.

Zeuner diagram
Reuleaux diagram } applicable to valves that have harmonic motion.
Bilgram diagram

The **Zeuner** and **Bilgram diagrams** are in common use, and engineers should be familiar with both. *The valve diagram must be interpreted by intelligent reference to the actual mechanism to which it is applied.* A person should be able, when looking at a diagram, to picture in his mind just what the valve, valve gear, piston, and crank are doing at any given time.

THE ZEUNER DIAGRAM

344. Valve Ellipse. — This diagram is a closed curve made up of points which represent the displacement of the valve for each corresponding position of the piston. It is drawn by finding the displacement of the valve for each piston position and laying off the displacement, at the corresponding piston position, perpendicular to a center line, which represents the mid-position of the valve. Displacements to the right of mid-position are usually laid off above this center line and those to the left below. A curve drawn through the points thus found will be the valve ellipse. Lines drawn parallel to the center line and separated from it by a distance equal to the steam lap and exhaust lap will cut the ellipse at points corresponding to admission, cut-off, release and compression. A valve ellipse for a Corliss valve engine is shown in Fig. 330, page 406.

345. The Zeuner Diagram. — Referring to Fig. 289, OV is the amount the valve has traveled to the right of mid-position with the crank at C, for a valve having outside admission and the engine running over. Since the angular displacement of the crank does not depend upon its length, the crank circle may be omitted, and the position of the crank and eccentric represented upon the eccentric circle, as shown in Fig. 290. *The diameter of the eccentric circle then represents the piston travel to one scale and the valve travel to another scale.*

For any crank position, OC, the corresponding position of the eccentric for clockwise rotations of the crank is Oe, and the displacement of the valve is ae. If a perpendicular, KB, is drawn from B to the eccentric position, Oe, a triangle, OKB, is formed which is equal to triangle Oae. Therefore $OK = ae$, the displacement of the valve. By taking other positions of the crank from A to B it will be found that K will always lie upon the circumference of a circle R, called the **valve circle** and having the eccentricity, OB, as a diameter. The displacement of the valve to the right of mid-position will equal the chord OK cut from the circle R by any eccentric position. For displacements to the left of mid-position there will be a similar valve circle L, shown dotted in the left semicircle.

As the relative position of the crank pin and valve are of greater importance than that of the eccentric and valve, the valve circles, R and L, may be revolved against the direction of rotation until the eccentric, Oe, falls upon the line OC, as in Fig. 291. *The chords now cut from the circle R, by any crank position, will represent the displacements of the valve to the right of its mid-position and the chords cut from L will represent the displacements to the left of mid-position.* **This is the fundamental principle of the Zeuner diagram.** The triangle OKB in Fig. 290 has been turned back through an angle (90 degrees $+$ δ), and the line OB in Fig. 290 now occupies the position OP in Fig. 291. Angle BOE equals 90 degrees and therefore EOP equals the angle of advance.

It has been previously shown that the valve displacement at admission and cut-off is equal to the steam lap. The position of the crank at cut-off and admission can therefore be located on the diagram by drawing an arc of a circle with O as a center, Fig. 292, and a radius, OL, equal to the steam lap. This arc will cut the valve circle R at the points H and J, and lines OD and OG drawn through these points will represent the position of the crank at admission and cut-off respectively, since the valve displacement OH and OJ is, in each case, equal to the steam lap.

The lead of the valve, which is the amount of port opening with the crank pin on dead center, is represented by the length LM, because the port opening is equal to the displacement of the valve minus the steam lap, and OA is the dead-center position of the crank, OM the valve displace-

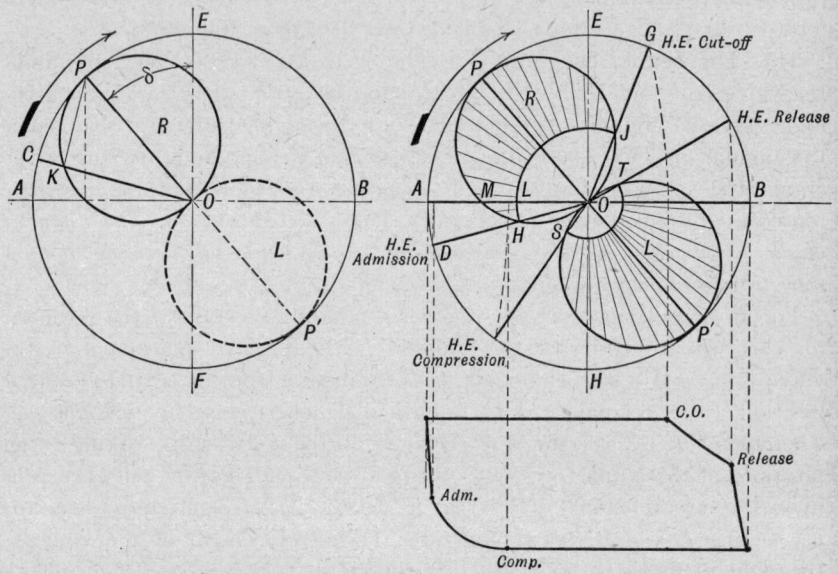

Fig. 291. — Development of the Zeuner Diagram.

Fig. 292. — Zeuner Diagram Complete for Head End.

ment, and OL the steam lap for this crank position. The amount of steam port opening for any crank position equals the valve displacement minus the steam lap, or the distance cut from any crank position by the valve circle and the lap circle, as shown by the chords of circle R. The port opening increases from H to P, where it is a maximum, and then decreases from P to J. The port is closed to live steam from J to H. The crank position corresponding to mid-position of the valve is perpendicular to the line OP drawn through the center O, of the eccentric circle.

Since release and compression occur when the valve displacement is equal to the exhaust lap, an arc drawn to cut the circle L at S and T will locate the position of the crank at the point of compression and release. If the valve has exhaust clearance instead of lap, the exhaust lap arc ST would be drawn on the circle R instead of on the circle L, because with

USEFUL CHARACTERISTICS OF THE ZEUNER DIAGRAM 371

exhaust clearance the valve displacement is to the right, instead of to the left, of mid-position for head-end release and compression.

The approximate shape of the indicator diagram, for the above conditions, is as shown in Fig. 292. The points of admission, cut-off, compression, and release are projected to the center line AB, with a radius equal to the length of the connecting rod, drawn to the same scale as the piston stroke AB. The pressure at admission and compression is assumed, and all points projected vertically to the admission and exhaust lines and the indicator diagram, drawn as shown.

The valve diagram, Fig. 292, shows only the events that occur for the head end of the cylinder. The same valve diagram may, however, be used to show the events occurring on the crank end of the cylinder, by

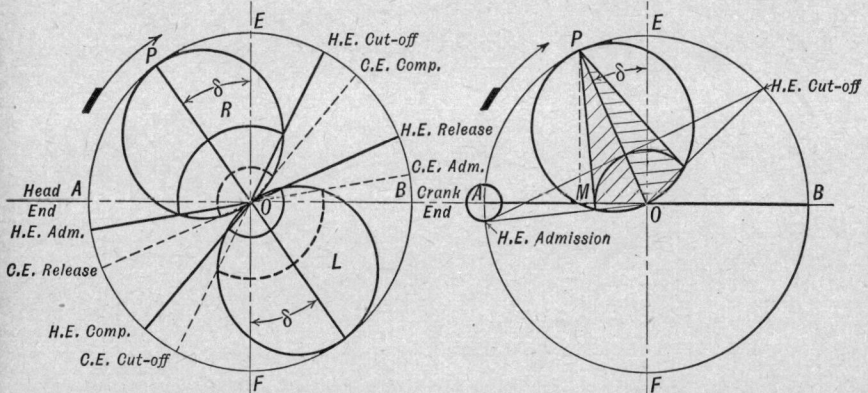

FIG. 293. — Zeuner Diagram for both Head and Crank End.

FIG. 294. — Useful Characteristics of Zeuner Diagram.

drawing the crank-end steam lap arc upon the valve circle L, and the crank-end exhaust lap arc upon the valve circle R, as has been done in Fig. 293. This gives a complete diagram, which shows the events for each end of the cylinder. Head-end events are shown with **full lines** and crank-end events with **dotted lines.**

The important points to remember when constructing the Zeuner diagram are:

1. The angle of advance (δ) is laid off from the axis that is perpendicular to the travel of the piston, in a direction opposite to that in which the crank turns.
2. The chord cut from the valve circle by any crank position shows the displacement of the valve from mid-position.

346. Useful Characteristics of the Zeuner Diagram. — When the *Zeuner* diagram is drawn with the radius of the crank pin circle equal to the eccentricity, it has the following characteristics:

372   SLIDE-VALVE ENGINES, VALVE DIAGRAMS AND SETTING

1. A perpendicular dropped from the end of the eccentric position, Fig. 294, always intersects the center line through O, at a distance from O equal to the steam lap plus the lead. This is shown by the right-angled triangle OPM, where OM equals the steam lap plus the lead.

2. A perpendicular dropped from P to the crank position for admission and cut-off intersects the valve circle at the same point as the steam-lap arc, Fig. 294. This is also true of the exhaust lap and the crank positions for compression and release.

3. A circle with its center at the dead-center position, OA, of the crank and a radius equal to the lead will be tangent to the crank position at admission. Also, a line connecting the ends of the crank positions at cut-off and admission will be tangent to the lap circle and perpendicular to the line OP.

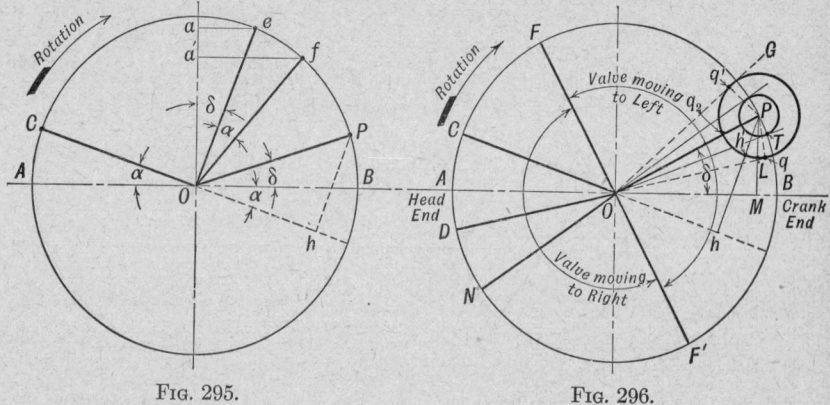

Fig. 295.   Fig. 296.

Figs. 295 and 296. — Fundamental Idea of Bilgram Diagram.

347. The Bilgram Diagram. — Draw a circle with a radius equal to the eccentricity, and a pair of axes as shown in Fig. 295. When the crank is in position OA, the eccentric center line is at Oe. The valve is displaced by an amount which is equal to the steam lap plus the lead and equals ae.

If the crank moves to a new position OC, through any angle α, the eccentric will move to Of, and the displacement of the valve will be $a'f$.

Lay off the angle of advance, δ, from the horizontal axis AB, in a direction opposite to that in which the crank rotates, and call the point on the valve circle, at the end of the eccentric, P. From this point drop a perpendicular Ph upon the crank position OC or OC extended. The length of this perpendicular is the displacement of the valve from mid-position for the crank position OC, because triangle $a'Of$ equals triangle POh and $a'f$ equals Ph. *The fundamental idea of the Bilgram diagram is, therefore, that the length of a perpendicular dropped from P to any crank position, or crank position extended, is the displacement of the valve from mid-position.*

The steam-lap circle is represented on the Bilgram diagram by a circle with P as a center and a radius equal to the steam lap, Fig. 296. The exhaust-lap circle is represented by a circle, with P as a center and a radius equal to the exhaust lap. *A negative exhaust lap is shown by a dotted circle.*

The crank position for head-end admission is at OD (the extremity of OD is tangent to the head-end steam-lap circle) because the displacement of the valve to the right of mid-position is equal to the steam lap. For the dead-center position, OA, the valve is displaced by an amount PM, which is equal to the steam lap plus the lead, and the port opening is equal to LM, which equals the lead.

For any other position of the crank, OC, the valve is displaced from mid-position by an amount Ph, and the port opening equals hh'. When the crank has moved to OF, perpendicular to OP, the valve has moved to its extreme right-hand position, and from crank position OF to OF' the valve is moving to the left.

When the crank reaches the position OG, tangent to the steam-lap circle, the valve is displaced to the left a distance equal to the steam lap, and the crank is at the crank position for head-end cut-off. At OP the valve is in mid-position.

As the crank moves from OF to OF', the valve is moving to the left, and at T is displaced by an amount equal to the head-end exhaust lap. OT is, therefore, the crank position for head-end release.

At the crank position OF'', the valve is at its extreme left-hand position.

Compression for the head end occurs when the crank is at the position ON, because the valve is displaced to the left by the amount of head-end exhaust lap.

A similar diagram can be drawn for the crank end, by repeating this construction with OP extended to cut the valve circle 180 degrees from its position for the head end.

348. Facts Shown by Valve Diagrams. — As the steam lap, exhaust lap and eccentricity are constant for any slide-valve engine, the only change that can be made is in the angle of advance. *An increase in the angle of advance increases the lead and makes all events occur earlier.*

With a constant eccentricity and angle of advance, an increase in the steam lap means a later admission and earlier cut-off. This means a larger valve and increased friction between the valve face and its seat, which is undesirable.

The cut-off for a plain slide valve, having a single eccentric, is limited to half-stroke or later. It is desirable to have as much expansion as possible, and this can only be obtained by increasing the angle of advance, which makes all events occur earlier. Early cut-off is accompanied by small port opening, the effect of which may be counteracted by using multi-ported valves. The valve diagram shows that the angle turned through by the crank during compression equals the angle turned through by the crank during expansion. Therefore to make cut-off occur earlier, by changing the angle of advance, makes compression occur earlier. An early compression is generally, undesirable as it means a loss of power, except on a high-speed engine, where an early compression is often considered desirable to bring the reciprocating parts to rest without shock.

Example 40. — Given: Steam lap $\frac{1}{4}$ in., exhaust lap $\frac{1}{8}$ in., valve travel 2 in., lead $\frac{1}{8}$ in., no rocker arm, and engine to run over. Find the angle of advance, and the position of the piston at head-end release, compression, admission, and cut-off. Take connecting rod five times the length of the crank.

Solution by Zeuner diagram. — Draw two axes, AB and EF, at right angles as in Fig. 297.

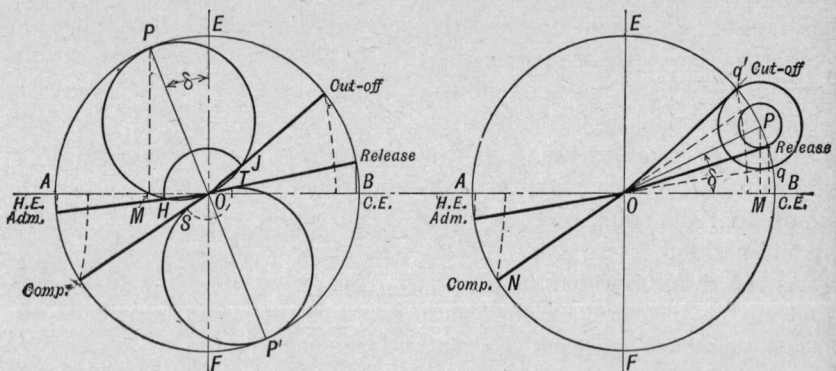

FIG. 297. — Solution of Problem 39 Using Zeuner Diagram.

FIG. 298. — Solution of Problem 39 Using Bilgram Diagram.

With O as a center, draw the eccentric circle P, with the eccentricity, or one-half the valve travel, as a radius. (*Take twice the size, for accuracy.*)

From O measure off on OA a distance OM, equal to the steam lap plus the lead. (*Double scale because eccentricity is double scale.*)

Erect a perpendicular at M, cutting the eccentric circle at P; then EOP is the angle of advance.

Upon OP draw the valve circle.

Extend OP to P', and upon OP' draw a second valve circle.

With O as a center, draw a circle having a radius equal to the steam lap, cutting the valve circle on OP at H and J. Draw crank position through H for admission and through J for cut-off.

With O as a center, draw a second circle with a radius equal to the exhaust lap and cutting the valve circle on OP' at S and T. From O through T, draw the crank position for release, and through S the crank position for compression.

With a radius equal to the length of the connecting rod, swing these points to the line AB, and measure each from the dead-center position for the end considered. The scale of measurement used is that to which the eccentric circle represents the stroke.

Solution by the Bilgram diagram. — Draw axes AB, EF, and eccentric circle, Fig. 298, as for Zeuner diagram.

To the right of EF lay off a distance PM, equal to the steam lap plus the lead, perpendicular to AB and cutting the eccentric circle. The angle POM is the angle of advance.

With P as a center, draw two concentric circles with radii equal to the steam lap and exhaust respectively.

Draw the crank position for head-end admission tangent to the head-end steam-lap circle at q, and for head-end cut-off at q'.

Draw the crank position for head-end release, tangent to the exhaust-lap circle on the side below OP, and for head-end compression, tangent to the exhaust-lap circle above OP, and extend through O to N.

The remainder of the solution is the same as that based upon the Zeuner diagram.

349. Automatic High-speed Engine. — A typical automatic high-speed engine is shown in Fig. 299. The frame is of the box type, open at the top, and rests upon a cast-iron sub-base. A light, nickel-plated oil shield

Fig. 299. — Automatic High-speed Engine.

covers the top opening. The frame is for a center crank and is so made, between the bearings, that it forms an oil reservoir into which the crank pin dips at each revolution.

The valve for the engine shown in the illustration is of the **flat-slide-valve** type and is balanced. The ordinary " D " slide valve has full pressure acting upon its outer surface and pushing its face against the valve seat. The steam pressure in the ports covered by the valve face reduces the pressure holding the valve to its seat, but the resultant pressure is sufficient to produce a heavy friction load upon the valve gear that moves the valve. As there is 60 to 90 per cent less pressure upon the **balanced valve** than upon the ordinary valve, less force is required to move it, and thus the wear is reduced. In one form of balanced valve, used on the Ball engine and shown in Fig. 300, the valve is thin and flat, and slides between a **pressure plate** and the valve seat. The pressure plate rests upon **distance pieces** at the side of the valve, and keeps the steam pressure from the top of the valve. Springs placed between the steam-chest cover and the pressure plate hold the plate against the distance pieces. Short bolts center the pressure plate and hold it from moving with the valve. This valve is made with ports through it at each end. Steam can pass to the cylinder through the ports and around the edge of the valve, as shown by the arrows. Such a valve is **double ported;** that is, it has two openings for the simultaneous passage of steam. A double-ported valve halves the travel of the valve for a given port opening.

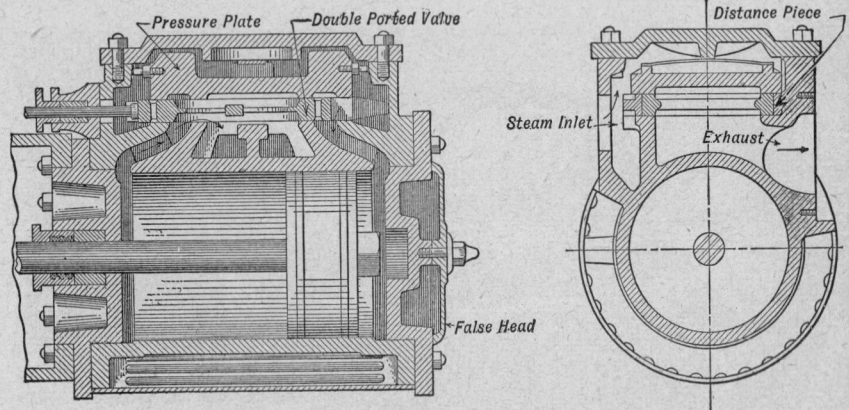

Fig. 300. — Cylinder Showing Sweet Balanced Valve.

The piston valve shown in Fig. 311, page 389, is commonly used on the high-speed engine. It is a cylindrical slide valve fitted to slide back and forth on a cylindrical seat in the steam chest, and may be single or double ported. As it is generally made, steam is admitted to the cylinder from the inside of the valve and exhausted past the outer edge. In the illustration the valve seat is formed by a thin liner made with slots through which the steam passes. Piston valves are perfectly balanced and are light in proportion to their size. They increase the clearance volume, because of the additional space which surrounds the valve; they make a complicated cylinder casting, and, when wear occurs, they cause increased leakage of steam. To prevent leakage resulting from wear, packing rings are often used.

The valve gear shown in Fig. 299 differs from the plain " D " slide-valve gear previously described, in that it has a rocker arm between the valve rod and the eccentric rod, so arranged that the eccentric, in this case a pin with a spherical end mounted on an arm attached to the governor arm, follows the crank instead of leading the crank.

A **shaft governor** controls the speed, regulating the amount of steam entering the cylinder, by changing the point of cut-off. The governor shown in the illustration has a **weighted eccentric arm** supported on the flywheel by a hardened steel **roller bearing** and connected to a **leaf spring** by a **link**. Hardened steel pins and bushings are used to attach the weighted link to the governor arm and spring. *The pull of the spring and the size of the weights determine the speed at which the engine is to run.* To decrease the speed weights are added, and to increase the speed weights are removed.

350. Shaft Governors. — There are many makes of shaft governors, but the underlying principle of all is covered by the types to be described.

The shaft governor is fastened to the flywheel and rotates with it in a plane perpendicular to the shaft. The position taken by the governor depends on centrifugal and inertia forces, acting in opposition to the pull of one or more springs attached to the wheel on which the governor is mounted.

In all forms of shaft governors, the eccentric is under absolute control of the governor. The eccentric is fastened to the governor in such a manner that any movement of the latter causes a displacement of the eccentric center. Such an eccentric is called a **swinging** or **shifting eccentric,** and in operation changes the point of cut-off, thus proportioning the supply of steam to the load.

The movement of the governor changes the point in the **stroke** at which cut-off occurs by:

1. Changing the angle of advance.
2. Changing the eccentricity.
3. Changing both the angle of advance and the eccentricity at the same time.

The last method is generally employed with single-valve engines.

Shaft governors may be classified according to the predominating force acting to move the governor as: (1) **centrifugal** governors, or (2) **inertia** governors.

It should be remembered that in all centrifugal governors there are some inertia forces, and that in all inertia governors some centrifugal forces acting on the movable parts.

351. Centrifugal Shaft Governor. — The Armstrong governor, Fig. 301, is a centrifugal governor which swings the eccentric in such a manner that the eccentricity and angle of advance are changed at the same time. The eccentric is not attached directly to the shaft; it is pivoted on the flywheel and has a curved slot through which the shaft passes. An arm attached to the eccentric is connected to the governor weight, which is fastened to the end of the leaf spring. This spring is firmly fastened to an arm of the flywheel, with bolts. The eccentric is so located that, with the governor weight clear in, cut-off will occur at its latest point.

For the position of the crank pin shown, the eccentric center is at e. An increase in the speed of the engine causes the governor weight to swing away from the shaft center against the force of the spring. This action pulls the end of the eccentric arm and revolves the eccentric around its supporting pin. The effect is to increase the angle of advance and decrease the eccentricity.

As the eccentric is revolved about the pin by the governor, because of an increase in load, the eccentric center will travel in an arc ee', with a center at the pivot; at the same time, the angle of advance will increase, as for instance from Coe to Coe', and the eccentricity will decrease from oe

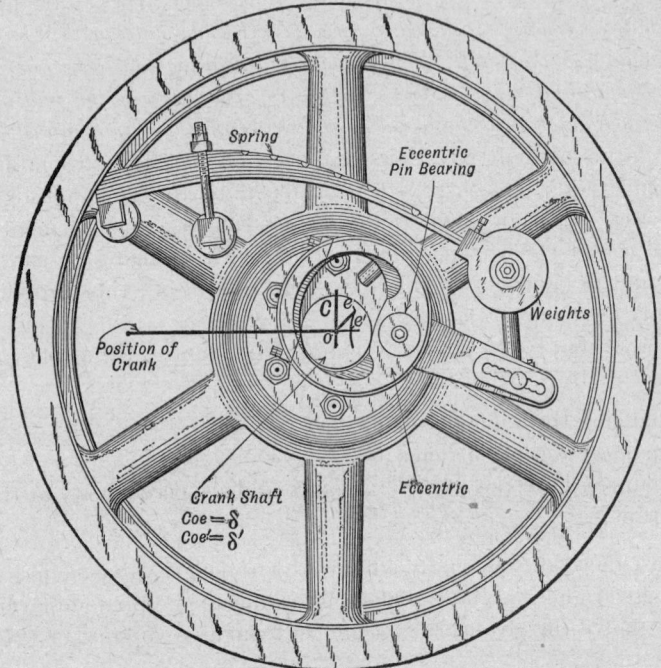

FIG. 301. — Armstrong Centrifugal Governor.

to oe'. *When the eccentric center moves in an arc, as shown, the lead of the valve decreases as the speed increases.*

Centrifugal governors are sometimes arranged to move the eccentric center in a straight line across the shaft. The lead then remains constant as the eccentric changes, and the eccentric is called a **shifting eccentric.** An example is the Fitchburg governor.

An illustration of the Zeuner diagram applied to the swinging eccentric is shown in Fig. 302, in which the point of cut-off changes as the eccentricity and angle of advance change. The diagram is lettered to correspond to Fig. 292, page 370.

In the Zeuner diagram, the point of eccentric support, R, is shown 180 degrees from its actual position, R'. On the Bilgram diagram this center would be on OC and would be 90 degrees from its actual position, R'. Since the lap does not change while the eccentricity and valve travel change, the steam-lap circle, drawn with O as a center, will cut the valve circles drawn upon the various eccentricities at the points of admission and cut-off. With the eccentric at e, the cut-off is at G and the lead is LM. When the eccentric is at e', cut-off is at g, earlier than at G, and the lead has decreased nearly to zero.

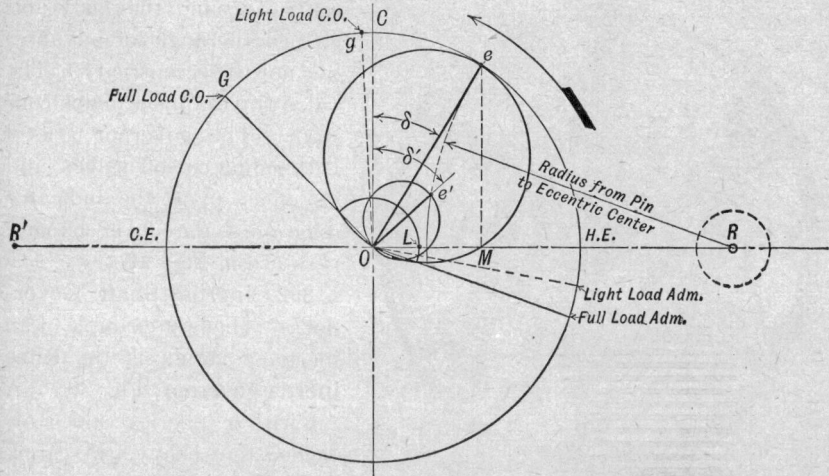

Fig. 302. — Zeuner Diagram Applied to Swinging Eccentric.

It should be noted that the position of the pin supporting the eccentric is important. As has been shown, the lead for a swinging eccentric changes as the eccentric center changes from e to e'. An increase in the length of the eccentric arm will keep the lead more nearly constant, which is desirable in order to obtain as much cushioning effect as possible. The lead will always change with this method of moving the eccentric. The amount should, however, be kept as small as is practicable.

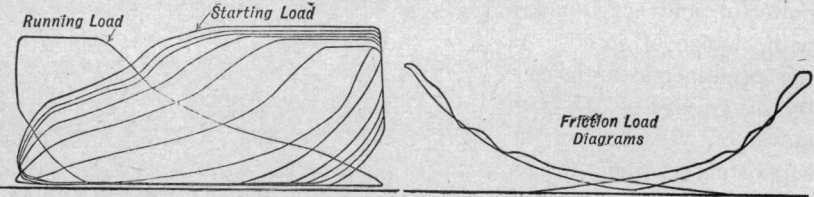

Fig. 303. — Indicator Diagram showing Action of Shaft Governor for a 14 in. by 10 in. Engine.

The effect upon the indicator diagram of governing by the swinging eccentric is shown in Fig. 303. This shows that the compression period becomes greater as the cut-off becomes shorter. The compression should be constant to obtain smooth running. On high-speed engines, the changes in compression have small effect because of the large clearance space used.

On an automatic high-speed engine, the valves must be balanced in order that they may be moved easily; otherwise the sensitiveness of the governor will be decreased.

The **Buckeye governor**, Fig. 304, is a centrifugal governor which has coil springs instead of leaf springs. The governor weights are so arranged that they turn the eccentric around the shaft and change the angle of advance, but not the eccentricity. The valve travel thus remains constant. This governor is used with **riding cut-off valves**, and its effect upon the indicator diagram, as the cut-off changes, is shown in Fig. 305.

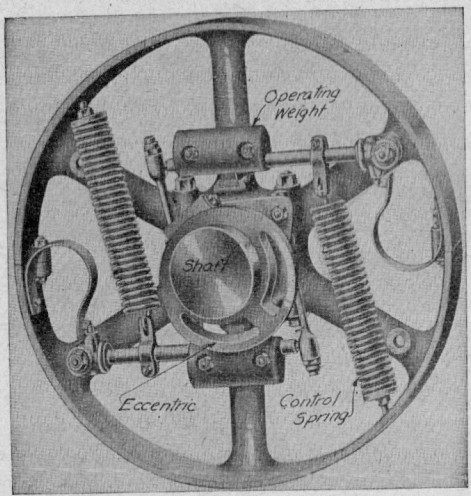

Fig. 304. — Buckeye Centrifugal Governor.

352. Inertia Shaft Governor. — The best example of an inertia governor is the **Rites inertia governor**, Fig. 306. A pin with a spherical end is attached to a heavy arm, carrying a weight at each end, and pivoted to a pin on the flywheel. The pin replaces the swinging eccentric of the Armstrong governor and substitutes for the large sliding surfaces of the eccentric a connection having a small amount of friction and a wider variation of control over the valve motion. The location of the pin is such that the valve motion is the same as for the swinging eccentric, Fig. 301. A coil spring, fastened to the swinging arm and the rim of the flywheel, is the controlling force.

The inertia governor is powerful, and acts quickly with change of speed. As the speed increases, the center of gravity of the arm moves outward around the supporting pin under the action of the centrifugal force C, and against the action of the spring. This movement shifts the center of the eccentric from E toward e, and changes the cut-off by reducing the eccentricity and increasing the angle of advance.

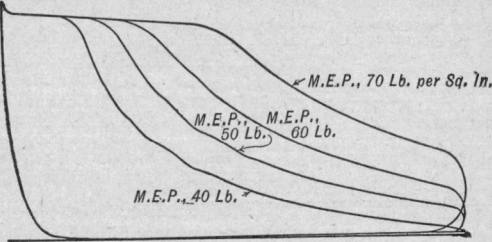

Fig. 305. — Diagram showing Action of Buckeye Governor.

Superposed upon this action is the inertia effect of the arm; that is, the tendency of the weights to keep moving at a constant speed when the speed of the flywheel changes. If the load on the engine is reduced, the speed of the flywheel will increase, but the governor arm will tend to rotate at the same speed, because of its inertia. The governor arm will thus lag behind the flywheel. This action moves the eccentric pin nearer to the center of the engine shaft, which decreases the eccentricity and makes the cut-off occur earlier, thus bringing the speed back to normal. The position thus assumed will later be maintained by the centrifugal force, if the new speed is maintained. The particular advantage of the inertia effect is in the rapidity of its action. The Rites inertia governor is simple, but, like the Armstrong governor, does not give a constant lead. It is also unbalanced, as it is pivoted away from the shaft center. For low speeds, this lack of balance affects the point of cut-off and becomes noticeable as a jerk in the action of the governor, which may cause the arm to swing through its whole range every second or third revolution. A drag or brake spring is attached to the flywheel, in such a manner that it bears against one of the weights with sufficient force to prevent this sudden swing,

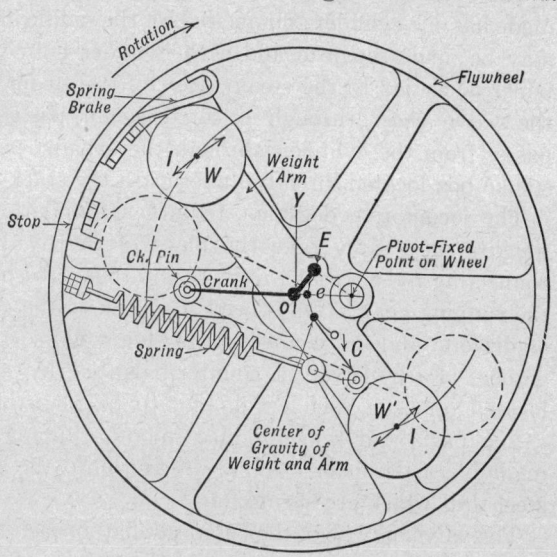

FIG. 306. — Rites Inertia Governor.

but not to prevent the satisfactory operation of the governor. To prevent the arm from swinging too far, a stop is placed on the flywheel rim.

To secure a balanced inertia governor, two parallel arms, arranged on opposite sides of the shaft center, are sometimes used. This permits running at slow speeds.

353. Locomotive. — The locomotive engine is of the plain slide-valve type, having a complete engine, with cylinders and valve gear, located on each side of the locomotive. The cranks are arranged 90 degrees apart, which is necessary to permit starting when one side is on dead center.

The frame, Fig. 6, page 12, differs from the usual type of engine frame in that it is not a solid casting, but is formed by wrought iron or steel **side rails**, which are held together by **cross ties**, or girders. There is a rail on each side to which the cylinders are bolted and which runs the entire length of the locomotive. The cylinder castings are bolted together at the front of the frame to form a **saddle**, or curved portion, upon which the smoke box rests. The smoke box is bolted to the saddle and forms an airtight connection. The crosshead guides may have one, two, or four bars and are supported at the front by **guide blocks** fastened to the rear head of the cylinder and at the rear by a **guide yoke** which extends across the frame. The guide yoke supports the guides on both sides and is stiffened by a **brace** connected to the guide and riveted to the boiler shell.

Pedestal legs, Fig. 309, are formed in each side rail to hold the bearing housings in position. One leg is made at an angle, and a **wedge** is placed between the housing and leg to permit adjustment of the bearing.

The cylinder casting is bolted to the front of the frame and as generally made has one cylinder and one-half of the saddle in one piece, but the saddle may be made separate and each cylinder bolted to it. Steam from the boiler is carried to the steam inlet by the steam pipe, and then passes to the valve chest through a cored steam passage. The exhaust steam passes from the cylinder, through the exhaust passage, to a **nozzle** in the smoke box located directly underneath the stack.

The locomotive does not require a foundation. The wheels and the weights they carry take the place of the flywheel. The wheels, axles, connecting rods, parallel rods, piston rods, crossheads, and frame compose the **running gear**. The wheels attached to the connecting rod are known as **drivers**, and are connected by **side rods**, so that they will revolve together. Each driver is counter-balanced by a crescent-shaped weight located opposite each crank pin, to balance the revolving and reciprocating parts and thus produce smooth running. The wheel frames are made of cast-iron or steel, are forced onto the shaft and carry hardened steel tires which are shrunk on.

The valve may be of the balanced, flat, or piston type, Fig. 311, page 389, and as usually made have an exhaust clearance of from $\frac{1}{8}$- to $\frac{3}{16}$-inch, instead of an exhaust lap. This reduces the back pressure on the engine, since the exhaust port is thus held open for a longer period.

The locomotive valve gear should be easy to operate, and should be so made that the engine can be reversed at will. The principal locomotive valve gears used in American practice, together with typical makes, are given below:

1. Link motions — Stephenson, Gooch

2. Radial valve gears
   - Walschaert, Baker, Young, Southern — Modern Types
   - Joy, Marshall, Hackworth — Old Types

STEPHENSON LINK MOTION

354. Stephenson Link Motion. — In Fig. 307, *e* represents the position of the center of the eccentric, with reference to the crank, for a plain slide valve having a fixed eccentric, which is connected to the valve rod without an intervening rocker and which has an outside admission. Such an engine can be reversed by turning the eccentric around the shaft until the center of the eccentric is at *e'*, directly across the shaft from *e*. To reverse the engine by this method would not be practical for locomotive, marine, hoisting, traction, and rolling mill engines, which have to be reversed frequently.

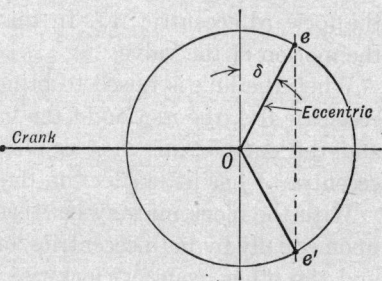

The Stephenson link motion, Fig. 308, has two fixed eccentrics, *A* and *B*, each having its own eccentric rod. The eccentrics are located essentially as the eccentrics would be in Fig. 307. The ends of the eccentric rods are connected by a **slotted link,** suspended by a **link hanger rod** from a **bell crank,** which is pivoted to a shaft carried by the frame and is operated by the reversing lever. A **saddle block** attached to the back of the link supports a pin to which the lower end of the hanger rod is attached. A slide block which fits the slot in the link, and is free to move in it, is con-

Fig. 307. — Diagram showing Positions of Eccentric when Reversing Rotation of Engine.

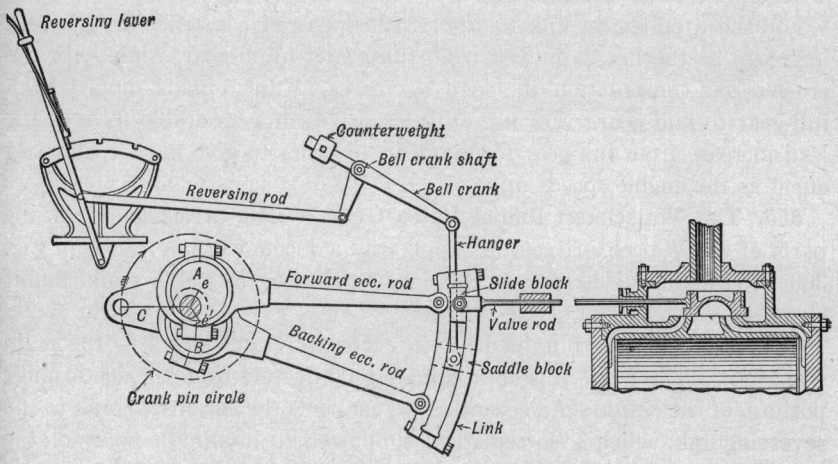

Fig. 308. — Stephenson Link Motion.

nected to the end of the valve rod. The slide block and valve rod remain stationary during movement of the link by the bell crank.

The eccentric rods may be open, as in Fig. 308, or crossed. For an outside admission valve and no reversing rocker arm, the link motion is open-rod construction, if the rods are open when the crank is in the dead-center position on the side of the shaft away from the valve. Under the same conditions, the valve gear is of the crossed-rod construction if the rods are crossed when the crank is in the dead-center position on the side away from the valve gear.

With the reversing lever in the position shown, the valve is moved by the forward eccentric A. In this case eccentric B has small effect upon the motion of the valve.

When the link is raised to bring the valve rod in line with the backing eccentric rod, the motion of the valve comes mainly from the eccentric B, and the engine runs in a counter-clockwise direction. In this case the eccentric A has little effect on the motion of the valve.

With the block midway between two ends of the link, the valve is acted upon equally by both eccentrics, one tending to produce clockwise rotation, and the other, counter-clockwise rotation; therefore, the engine will not run in either direction.

The valve mechanism is said to be in **mid-gear** when the block is in the middle of the link, and to be in **full-gear** when the block is at its extreme position near the end of the link. There are two full-gear positions, one called **full-gear forward** and the other, **full-gear reverse.**

With the link in full-gear position, the full steam pressure acts upon the piston for nearly the entire stroke, and the latest cut-off is obtained. This is the condition at starting a locomotive pulling a load. As the link is brought toward mid-gear position, the cut-off becomes earlier.

For the Stephenson link motion, with open-rod construction, the lead increases as the link is moved from full-gear to mid-gear, while with the crossed-rod construction the lead decreases as the link is moved from full-gear to mid-gear. For use on locomotives, it is desirable to have the lead increase from full-gear to mid-gear, in order to give more cushioning effect as the engine speeds up.

355. The Walschaert Radial Valve Gear. — The arrangement of the parts of the Walschaert gear, as applied to a locomotive, is shown in Fig. 309. With the mechanism in the position shown, the valve would admit steam to the head end of the cylinder.

The eccentricity is obtained by an **eccentric crank** keyed to the main crank pin. The point E is approximately 90 degrees from the dead-center position of the crank. An eccentric rod connects the eccentric crank to the **reversing link,** which is slotted and is supported, at its middle point, on the frame of the engine by a **trunnion.** The eccentric rod is attached to an arm that projects from the bottom of the link, in order to bring the eccentric rod as near the horizontal as possible. This arm is shaped to correct the distorted motion of the valve, caused by the angularity of the eccentric rod. A **radius rod** connects the link with the **lap-and-lead lever,** or **combination lever,** and thus transmits the movement of the link to the valve.

THE WALSCHAERT RADIAL VALVE GEAR

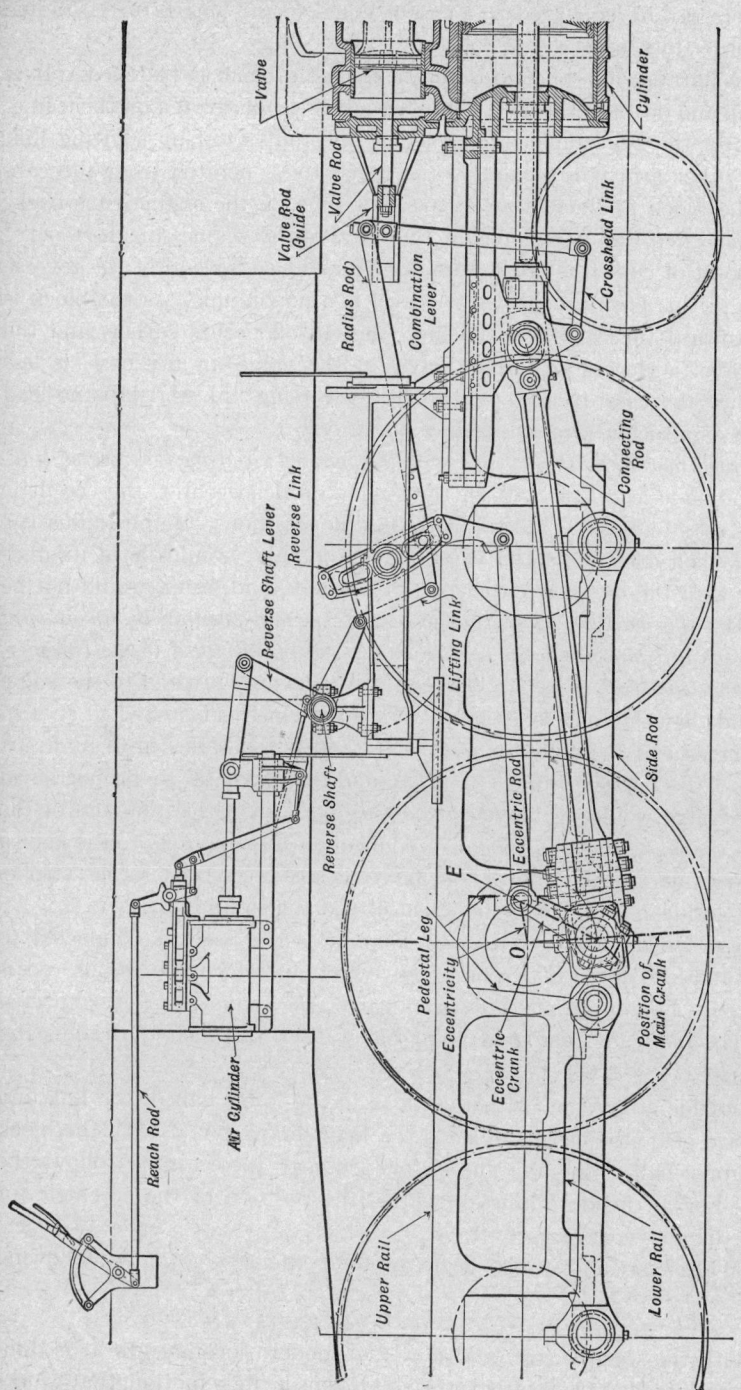

Fig. 309. — Arrangement of Walschaert Valve Gear for Chicago, Burlington and Quincy Pacific (4-6-2) Locomotive.

The reversing link is curved to a radius equal to the length of the radius rod, in order to keep the lead constant.

To the link end of the radius rod, a **movable block** is fastened. It is moved up and down in the slotted link by a bell crank pivoted on the frame. One arm of the bell crank is attached to the radius rod by a lifting link and the other arm is attached to the reach rod, operated from the cab. When the block is above the reverse link fulcrum the engine runs backward, and when the block is below the fulcrum the engine runs forward.

The point of cut-off is latest with the block at either end of the reverse link, for in this position the valve travel is a maximum. As the block is moved toward the center of the link, the cut-off occurs earlier, and the valve travel is decreased. The travel of the valve, in any case, is less than twice the eccentricity, because the reversing link and lap-and-lead lever act as a rocker arm and reduce the travel.

The valve motion is the resultant of the motion given by the reverse link, and the lap-and-lead lever. With the engine on dead center, the eccentric crank is so set that the link is in its middle position. If the radius rod were attached directly to the valve stem, the valve would be in its midposition with the block at either end of the link, and steam would not be admitted. *To have the valve displaced from its mid-position by an amount equal to the lap plus the lead, and thus have steam admitted to the cylinder, the lap-and-lead lever is used.* One end of the lever is pivoted to the valve stem, and the other end is attached to a link, which is fastened to an arm on the crosshead. The outer end of the valve rod slides in a guide attached to the engine frame. The lap-and-lead lever is so proportioned that the valve would be moved a distance equal to twice the sum of the lap plus the lead, when the point of connection to the radius rod is kept a stationary fulcrum and the piston moved a distance equal to the stroke. The lap-and-lead lever gives the effect of the angle of advance.

For a valve having *outside admission*, the valve stem is connected to the lap-and-lead lever at a point above that at which the radius rod is connected. For an *inside admission* valve, the valve stem is connected to the lap-and-lead lever at a point below that at which the radius rod is connected, Fig. 309.

For outside admission, with the block in the lower half of the link and in forward gear, the eccentric leads the main crank pin. With the block in the upper half of the link and in forward gear, the eccentric follows the crank. For an inside admission valve, the position of the eccentric for each of the above cases is reversed.

The Walschaert and similar gears are used extensively, for the following reasons:

1. With the weight, size, and power of modern locomotives, it is difficult to design a satisfactory Stephenson link motion that works

between the sides of the frame. The radial gears, being outside the frame, overcome this difficulty.
2. The straps and eccentrics of the Stephenson link motion are heavy and wide. They are located underneath the frame where they are subject to dust, and rapid wear results. This, together with the wear in the rockers and transmission bars, results in a large amount of lost motion. The radial gears do not have eccentrics, and as only hardened pins are used, they are more easily kept in repair.
3. The position of the radial gears, outside the frame, makes them more accessible for proper lubrication.
4. The frame can be made stronger by bracing, because the valve gear is not inside the frame.

The locomotive does not require a governor. The speed is regulated by changing the pressure of the steam admitted to the cylinder by the throttle valve, or by changing the point of cut-off by means of the valve gear.

356. Marine Steam Engine. — A marine engine of the latest type is shown in Fig. 310. These engines are seldom made with a single cylinder, but are made compound and multiple-expansion, the steam passing in succession from one cylinder to one or more larger cylinders. The most common types are compound, triple-expansion, and quadruple-expansion engines, and are made in sizes up to 6000 horsepower.

The frame consists of inverted columns resting on a heavy bed-plate, which has six cross girders by which the main bearings are supported. Two-part babbitted bearings are used, with the lower part of the bearing housing cast as a part of the bed-plate.

The valves are balanced piston and balanced flat valves, actuated by a Stephenson link motion which permits reversing the direction of rotation of the engine. A small steam engine attached to the main frame operates the valve gear when reversing. The valves are attached to balancing pistons, located in cylinders above the valves, to counter-balance the weight and inertia of the valves and valve gear, when conditions require. Crossed eccentric rods are generally used to permit stopping of the engine by setting the link at mid-gear. With open rods and the link at mid-gear, the engine might not stop. All parts of the valve gear are made heavy, and the moving parts are often babbitted.

A governor is not required, as the resistance of the propeller increases, with the velocity, sufficiently to control the speed. For low speeds the engine is governed by means of the throttle valve.

The thrust, or push, of the propeller is taken by a **thrust bearing** and transferred to the frames of the ship. This bearing is at the left of the illustration in Fig. 310. It is a modern type of **horseshoe-collar thrust**

388 SLIDE-VALVE ENGINES, VALVE DIAGRAMS AND SETTING

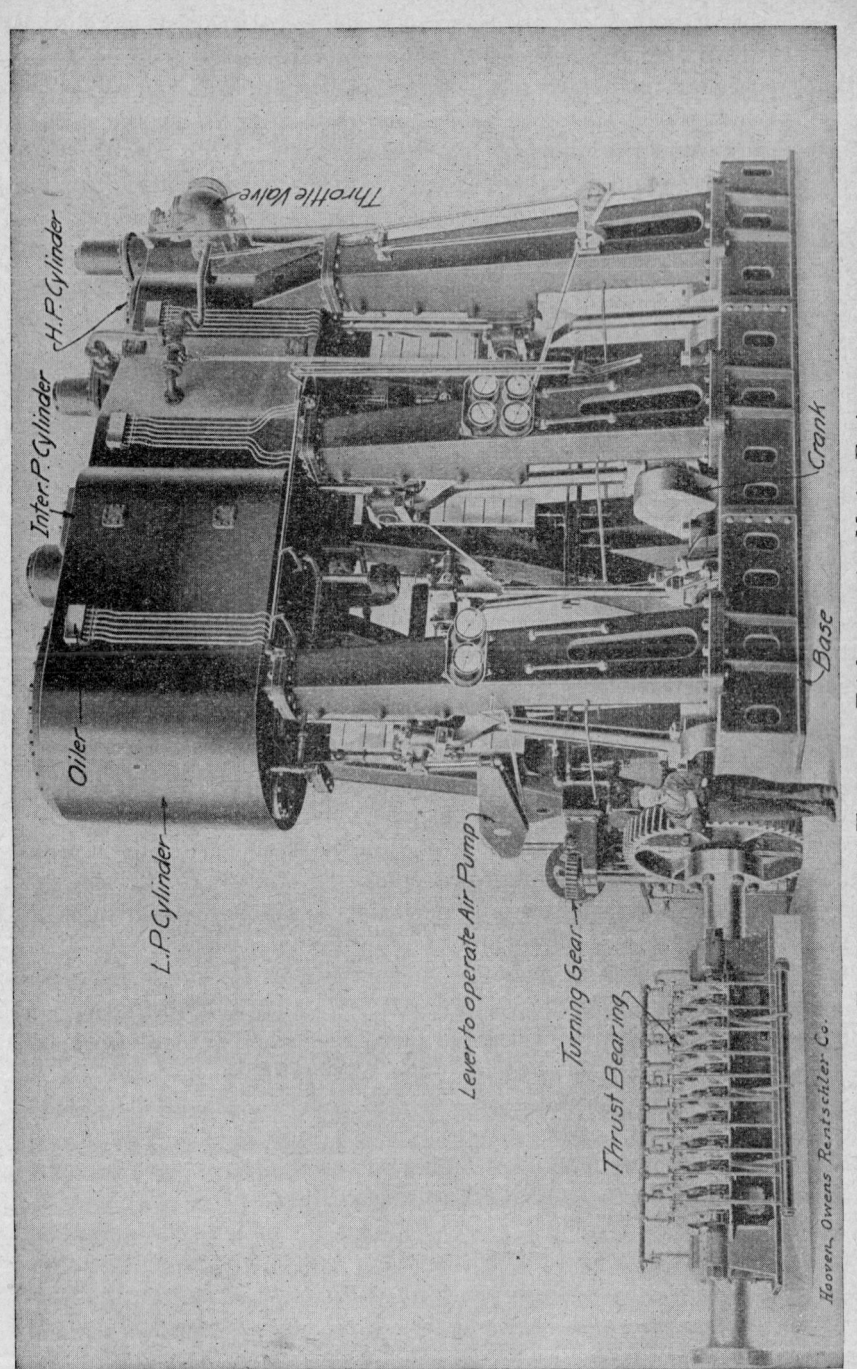

Fig. 310. — 2500 Horsepower Triple-expansion Marine Engine.

MARINE STEAM ENGINE

bearing, having a series of rings or **collars** made on the shaft and a series of separate thrust yokes, which fit between the shaft collars. These thrust yokes are provided with ears, or lugs, which bear against nuts on side rods attached to the bearing casing. This arrangement permits each thrust yoke to be adjusted separately for wear.

The thrust from the propeller is transferred from the faces of the shaft rings to the thrust yokes, thence through the lugs and nuts to the side rods, which transmit the thrust to the bearing casing bolted to the frames of the ship. Each thrust yoke has an oil chamber at the top, from which the oil is siphoned by wicking into the oil pipe, and in addition the lower part of the bearing housing is made rectangular and filled with oil, into which the shaft collars dip.

The **turning gear** used to turn the main shaft, when the main engine is not operating, is shown to the left of the engine. It consists of a large **worm wheel** on the main shaft, geared to a **worm.** On some engines the worm is operated by a small steam engine, and on others by hand; a lever with **pawl** and **ratchet** being used in the latter case, to operate the worm.

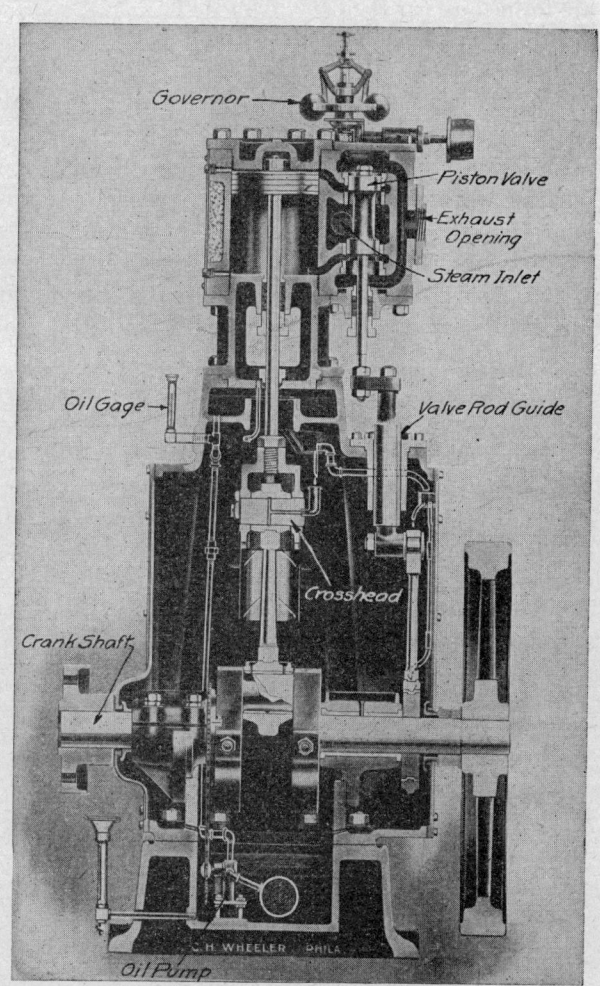

FIG. 311. — Small Vertical Steam Engine.

An **air pump** is attached to the side of the low-pressure cylinder frame, and is driven by the low-pressure crosshead through a lever connection. It is used to maintain the vacuum on the engine when operating condensing.

The eccentric is made of cast-iron and keyed to the shaft, and the strap is also made of cast-iron and babbitted. The eccentric rod, to which the forked valve-rod end is bolted, is of forged steel.

The crosshead may be of the single or double slipper type, and the guides are formed as a part of the main frame.

357. Small Vertical Slide-valve Engine. — This type of engine is used to drive stokers, fans, and other auxiliaries. The engine shown in Fig. 311 has an enclosed frame with balanced piston valve. The frame is of the A-type, completely enclosed. The guides are formed as a part of the main frame, which is supported by a bed-plate that makes a reservoir for oil. Between the cylinder and the top of the frame, is a distance piece which prevents water that leaks past the piston-rod stuffing box from getting into the enclosed frame. All moving parts are completely enclosed.

358. Special Forms of Slide Valves. — In addition to the slide valves already described, the following types are used on slide-valve engines:

1. Double and multi-ported.
2. Riding cut-off.

359. Double and Multi-ported Valves. — The plain slide valve gives a restricted port opening at the start of the stroke. To overcome this, and to reduce the movement of the valve necessary for a given port opening, **multi-ported** valves are used. Fig. 312 shows a **double-ported** valve used on marine engines. The valve is surrounded by steam, and has two steam chambers that run clear through the valve and are filled with live steam. The valve seat has two ports for each end, which merge into a passage connecting with the cylinder. The two

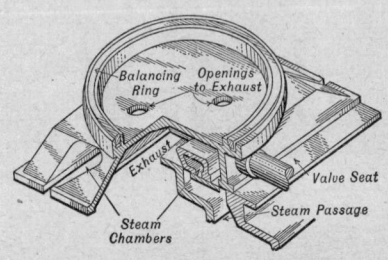

Fig. 312. — Marine Double-ported Slide Valves.

ports on each end are covered with valve feet which are duplicates and, as the valve moves, both ports are opened at the same time, thus increasing the size of the passage for steam for any movement of the valve.

A double-ported valve used on locomotives, and known as the **Allen trick valve**, is shown in Fig. 313. It resembles a "D" slide valve but has a steam passage cored in it. The valve is so proportioned that when it is in the position shown steam will enter the cylinder, past the outer edge of the valve and also from the crank end

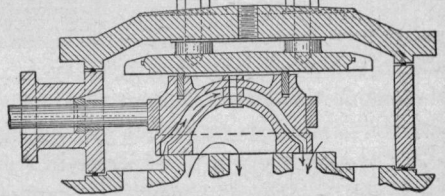

Fig. 313. — Allen-Richardson Balanced Slide Valve — Double-ported.

through the cored passage. This gives a more rapid opening and closing of the port.

A multi-ported valve known as the **gridiron** valve is used on multi-valve engines. It consists of a flat plate having a number of parallel rectangular passages through it. The valve seat has an equal number of parallel rectangular ports, which are simultaneously opened and closed by the action of the valve. When this type of valve is used, a separate valve or valves must be used for the inlet and exhaust.

360. Riding Cut-off Valves. — It was shown in Art. 348, page 373, that, with a single valve controlling all events of the engine, a change in the cut-off, by changing the angle of advance, also changed the point at which compression, release, and admission occurred, an earlier cut-off being accompanied by an earlier compression.

High compression on a slow-running engine is not desirable, and to provide a means of changing the cut-off without altering the point at which release, compression, and admission occur, a **riding cut-off valve** is used.

A simple form of this valve is shown in Fig. 314. A **main valve** which controls the point of admission, compression, and release, slides upon the valve seat. It is essentially a plain slide valve having steam passages near the ends. Steam enters the cylinder through these passages, instead of past the outside edge of the valve, and a riding cut-off valve slides upon the back of the main valve and controls the point of cut-off. It is a flat valve with a projection to which the riding cut-off valve stem is attached.

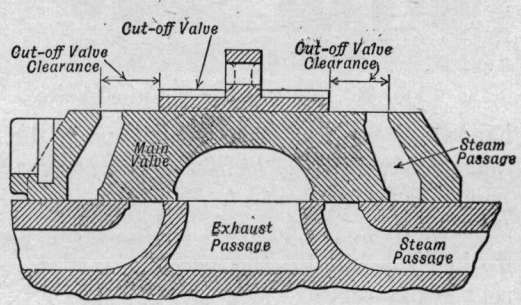

FIG. 314. — Riding Cut-off Valve.

The main and riding cut-off valves are controlled by separate eccentrics. The eccentric for the main valve is fixed and set to bring cut-off at about three-quarter stroke. The riding valve eccentric is loose upon the shaft, and is generally controlled by a shaft governor which controls the speed by changing the point of cut-off. The riding valve and its eccentric are so designed and adjusted that at admission the cut-off valve is not obstructing the port but at cut-off the block is moved to cover the port before the main valve has reached its cut-off position. Having closed the valve port for either end, the riding valve must keep it closed until the main port is closed by the main valve, or a second admission of live steam will occur.

When the riding valve is located centrally on the back of the main valve, it fails to cover the ports in the main valve by an amount indicated, and called **cut-off valve clearance,** or **negative lap.** The riding valve must be displaced relative to the main valve, by a distance equal to this clearance, in order to close the port to steam.

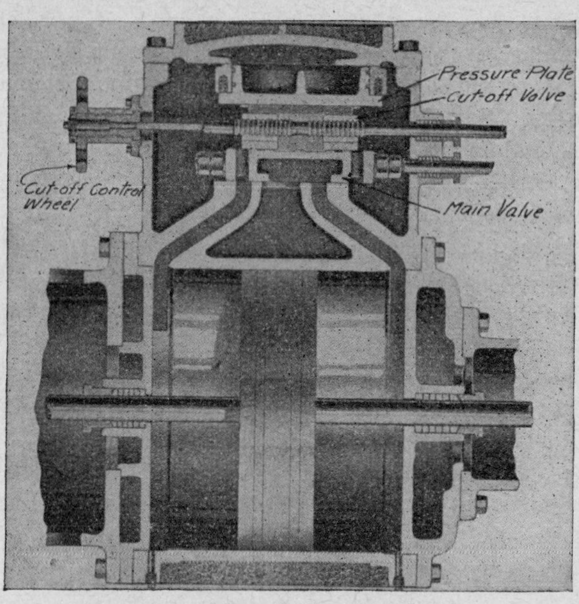

361. The Meyer Valve. — A common type of the Meyer valve, Fig. 315, is used on air compressors and pumps where the load is practically constant. The riding valve is made in two blocks, one of which is threaded right-handed and the other left-handed. The valve rod has right- and left-hand threads upon which the blocks fit. It extends through the steam chest, and a handwheel is placed on a squared end of the rod. *Turning the handwheel changes the position of the blocks, thus altering the clearance and changing the point of cut-off without altering the other events.* The wheel is marked to indicate the percentage of stroke at which cut-off is occurring.

Fig. 315. — Meyer Riding Cut-off Valve.

With a Meyer valve the speed is controlled by a throttle governor. The action of the Meyer valve is made clearer by studying the valve diagram. In the Zeuner diagram, Fig. 316, the valve circle, M, is for the main valve and is drawn as for any slide valve, with an angle of advance (δ). The auxiliary or cut-off valve circle, A, is laid off with YOP as the angle of advance, the angle between the crank and auxiliary eccentric being BOP. This angle is generally 180 degrees.

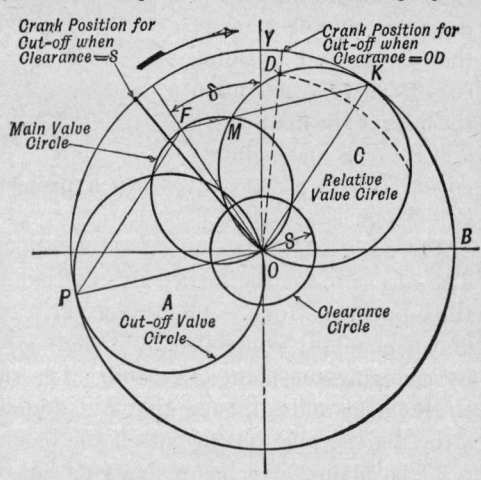

Fig. 316. — Zeuner Diagram applied to Meyer Valve for one Position of Cut-off Blocks.

Since both valves are moved by eccentrics, their relative positions with respect to each other may be represented by a third valve circle drawn with a diameter OK, equal to PF. The position of the crank at cut-off for any amount of clearance may be located on the valve circle, C, by drawing it through any point, D, where OD equals the clearance.

To find cut-off for the opposite end of the cylinder, a second circle of the same size may be drawn on OK extended. For other events the diagram may be completed, as for a slide valve having outside admission, by using the main valve circle.

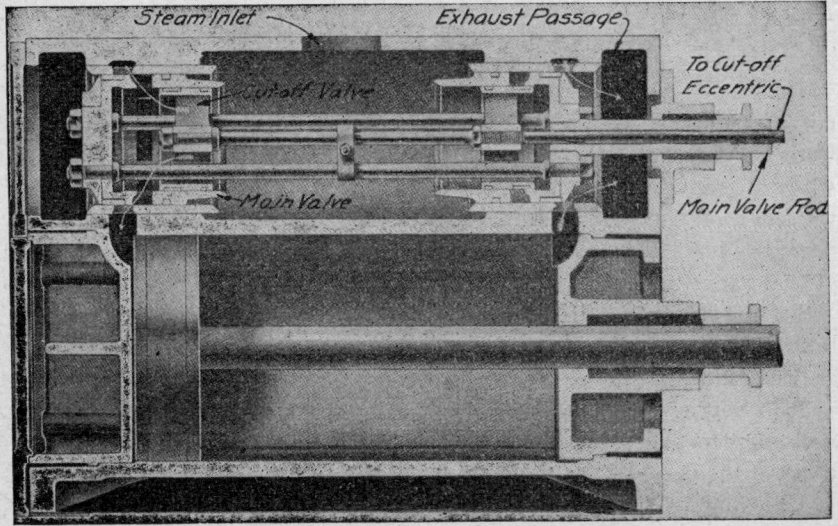

Fig. 317. — Buckeye Type Riding Cut-off Valve.

The **Buckeye engine** uses a valve, Fig. 317, of the **Polonceau** type. The main valve, which controls admission, release, and compression, is a steam-tight box, within which the riding cut-off valve moves. The cut-off valve controls the cut-off and is under control of the governor. These valves are so connected through a compound rocker that uniform travel of the cut-off valve, with respect to the main valve, is obtained for all cut-offs.

When drawing the Zeuner valve diagram for the Buckeye riding cut-off valve gear, the relative valve circle coincides with the cut-off valve circle.

362. Slide Valve Setting. — Most valve gears are provided with adjustments for changing the position of the valve. The making of these adjustments is known as **valve setting**.

The valve must be carefully set to obtain an economical steam distribution and smooth running of the engine. For economical distribution of steam, the cut-off at each end of the cylinder should be equal, and for smooth running the lead at each end should be equal, in order to give equal cushioning effect at each end of the stroke. Valves are therefore set for **equal leads** or equal **cut-offs**. With the ordinary construction of slide valves, the steam laps on each end are equal. When so made, a valve set for equal leads will not give equal cut-offs, because of the angularity of the connecting rod, and for the same reason, a valve set for equal cut-offs will not give equal leads. Valves are ordinarily set for equal leads.

In setting a slide valve, the laps and eccentricity are fixed in amount, and adjustments that can be made are:

1. Moving the valve on its stem.
2. Changing the position of the eccentric.

Moving the valve on the stem has the effect of changing the laps. When it is moved bodily along on the seat, the result is as though the steam lap were increased and the exhaust lap were decreased for the end toward which it is moved, and vice versa for the opposite end. Hence, changing the length of the valve rod increases the lead on one end of the valve and decreases it on the other. Lengthening the valve rod increases the lead at the crank end and decreases it at the head end.

Shifting or turning the eccentric around the shaft increases or decreases both leads, depending upon whether the angle of advance is increased or decreased.

363. Setting the Crank on Dead Center. — In setting valves it is necessary to place the crank on dead center. This must be accurately done, or the valve will not be set in its proper position relative to the piston.

The **trammel method** is generally used to place the engine on center. The **tram** is a rod pointed at both ends with one end bent to form a right angle.

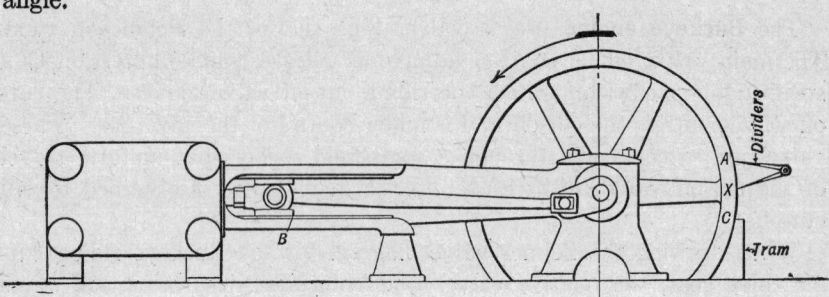

FIG. 318. — Method of Marking Guide and of Using Tram and Dividers.

The flywheel is turned until the crosshead is within 2 or 3 inches of the end of the stroke, and a mark is made on a chalked surface on both guide and crosshead, as shown at *B*, Fig. 318. Whenever these marks are brought together, the crosshead will be in its original position. The straight end of the tram is then placed on a fixed mark on the floor or frame,

SETTING A PLAIN SLIDE VALVE FOR EQUAL CUT-OFFS

and a mark is made with the bent end of the tram on the rim of the flywheel, as shown at C. The crank pin is then turned past the dead-center position until the marks on crosshead and guide again coincide and a second mark is made with the tram on the rim of the flywheel as shown at A in Fig. 318. The crank is now as far past dead center as it originally was ahead. A point X is located mid-way between A and C, and a mark is made at X. With the tram resting on the permanent mark on the floor, the engine is turned until the bent end of the tram rests squarely on the line at X. The engine is now exactly on dead center. The dead center for each end should be found in the above manner. The engine should always be turned, to take up any lost motion, in the same direction as when running, and thus have the same brasses in contact with the crank pin; otherwise the setting may not be correct.

364. Setting a Plain Slide Valve for Equal Leads. — The methods employed to set valves are:

1. By removing the valve-chest cover and locating the valve by measurement.

2. By taking indicator diagrams and adjusting the valve gear until the desired diagrams are obtained.

365. Setting a Plain Slide Valve for Equal Leads, by Measurement. — The valve is first adjusted to give equal travel each side of mid-position, or to have equal preliminary leads. The setting by the second method is made as follows:

1. With the crank on dead center, the angle between crank and eccentric is adjusted until the valve opens the port leading to the cylinder, by a slight amount. The width of the opening should be measured and recorded as a preliminary lead on that end. Suppose, for example, that it is $\frac{1}{8}$-inch. The engine should now be placed on the opposite dead center, and the port opening on that end measured and recorded as the preliminary lead on that end. Suppose it to be $\frac{1}{16}$-inch; there is, then, $\frac{1}{16}$-inch difference in the lead on the two ends. To make the leads equal in amount, the valve must be moved on its stem, or the length of the valve rod changed by an amount equal to half the difference. The valve should be moved toward the port having the larger opening.

2. With the leads equal, but not necessarily correct in amount, the required lead is given by placing the crank on head-end dead center and, after loosening the eccentric, moving it around the shaft, in the direction the engine is to run, until the desired lead is obtained. The engine is then turned to the other dead center and the lead on the crank end checked. The leads should now be nearly equal. The head-end lead is generally made slightly less than the crank-end lead, because of the distortion caused by the angularity of the connecting rod.

366. Setting a Plain Slide Valve for Equal Cut-offs. — 1. The valve is first set to travel equal distances each side of mid-position, as explained under setting for equal leads. 2. The eccentric is then adjusted to give cut-off when the piston has moved equal distances from dead center for each end. This is done by marking the limits of the stroke on the guide and setting the crosshead at the percentage of stroke at which cut-off is to occur. The eccentric is moved on the shaft in the direction the engine is to run, until it can be seen that the valve is just closing the steam port for the end from which the piston is moving. The eccentric is then fastened on the shaft and the engine turned over until the crosshead has moved the same distance from the other dead center and the valve should be just closing the port. If the setting is not correct the difference in measurements should be halved and a correction made for one-half by moving the eccentric on the shaft, and for the other half by moving the valve on the stem. This operation should be repeated until the required setting is attained.

Setting a Slide Valve by Means of a Tram. — Valves are often set by means of a tram and tram-marks on the valve stem. By this method the valve need not be seen and the steam chest cover is not removed. The tram-marks on the stem are obtained as follows: a center punch mark is made on some fixed part of the engine, and the straight part of the tram placed in it. The valve is then moved until its edge is at the head-end edge of port and a punch mark made on the valve stem into which the bent end of tram will fall. A similar punch mark is made on the valve rod when the edge of the valve is at the edge of the crank-end port. Knowing the position of the valve as indicated by these marks, the valve can be quickly adjusted.

For valve setting by indicator, see Art. 396, page 431.

367. Setting Special Types of Slide Valves. — For valves controlled by shaft governors, the length of the valve stem and position of the eccentric are found as for the plain slide-valve engine. The eccentric position is determined by turning the whole governor wheel and eccentric until the desired setting is obtained. This is done with the governor giving latest cut-off. For an engine having reversing devices, the setting is made as in the case of valves driven by a fixed eccentric, except that the valve must be set for both full-gear forward and reverse.

For riding cut-off valves, the main valve is set for equal leads, exactly as for a single slide valve. The riding valve is set to give equal cut-offs at some point of the stroke. When rocker arms are used, care must be taken to turn the eccentric in a direction to open the port with the crank pin on dead center. The cut-off is generally equalized at the middle of the governor range.

On a shaft-governor engine, the weights should be blocked out against the action of the springs, to the position for earliest cut-off, thus making sure that the governor can cut-off sufficiently near the beginning of the stroke to control the engine.

REFERENCES

American Steam Engine, HAWLEY.
Elements of Mechanism, JAMES and DOLE.
Steam Engine, SHEALY.
Cyclopedia of Engineering.
Valve Gears, FESSENDEN.
Steam Power Plant Engineering, GEBHARDT.
Bulletin No. 1018, AMERICAN LOCOMOTIVE Co.
Practical Marine Engineering, DYSON.

REVIEW QUESTIONS AND PROBLEMS

1. Name the parts of the eccentric connection to the valve rod, and define eccentricity.

2. Describe the operation of a "D" slide-valve engine.

3. What type of governor is used with the plain "D" slide valve? How does it control the speed?

4. Define: (a) valve travel, (b) mid-position of valve, (c) lap, (d) lead, (e) port opening.

5. How is the displacement, of the piston and valve respectively, stated?

6. Why is a valve diagram used? Describe the Zeuner diagram and state its fundamental principle.

7. A $6\frac{1}{2}$ in. by 10 in. engine with a connecting rod 30 in. long has a valve travel of 3 in., a steam lap = $\frac{3}{4}$ in., a lead = $\frac{1}{8}$ in., an exhaust lap = $\frac{3}{16}$ in. Use a 6 in. circle and find, by the Zeuner and Bilgram diagram, the distance from the end of the stroke at which the following occur for each end of the cylinder: (a) cut-off, (b) release, (c) compression.

8. Given: admission at $\frac{1}{120}$th. before start of stroke, cut-off at $\frac{3}{4}$ stroke, valve travel, $4\frac{1}{4}$ in. Find the steam lap, lead and angle of advance, using the Zeuner and Bilgram diagram. Neglect the effect of the angularity of the connecting rod.

9. What is meant by an automatic high-speed engine? What method of governing is used on such an engine?

10. Describe the operation of the Rites inertia governor.

11. Describe the construction of the Walschaert valve gear.

12. Describe the construction of a marine thrust bearing of the collar type. Why is it used?

13. Describe the construction and operation of a Meyer valve.

14. Find the radius of suspension and the coördinates of the point of suspension, for a swinging eccentric on a 7 in. by $10\frac{1}{2}$ in. engine having a steam lap = $\frac{5}{8}$ in.; cut-offs at 1, $3\frac{1}{16}$ and 7 in. on stroke; lead at the corresponding cut-offs, $-0.035, 0.00, 0.055$ in. Neglect the effect of the angularity of the connecting rod.

15. Explain the meaning of the term "relative valve circle."

16. State the methods used to set a slide valve.

17. Explain the trammel method of finding dead center.

18. Explain the method of setting a slide valve for equal leads.

19. In setting a slide valve on a 6 in. by 6 in. engine, it was found upon trial that with the proper lead on one end, the opposite end gave $\frac{1}{8}$-in. too great a lead. State what should be done to obtain correct setting and give reasons for your answer.

20. A slide-valve engine has a reverse rocker between the valve and the eccentric rod. What should be the location of the eccentric center with respect to the crank, to have the engine run over?

21. What is the advantage of using trams in setting a slide valve?

22. State the procedure in setting (a) a slide valve controlled by a shaft governor, (b) a riding cut-off valve.

CHAPTER XVII

MULTI-VALVE AND UNAFLOW ENGINES

368. Foreword. — It has been shown in Chapter XVI that, when a single valve controls all events of the engine, the opening of the steam port is restricted at admission. This causes throttling, or a drop in pressure, between the valve chest and the cylinder. Engines having single valves have large clearances because of the long steam passages, and the exhaust steam in passing from the cylinder cools the steam passages and cylinder

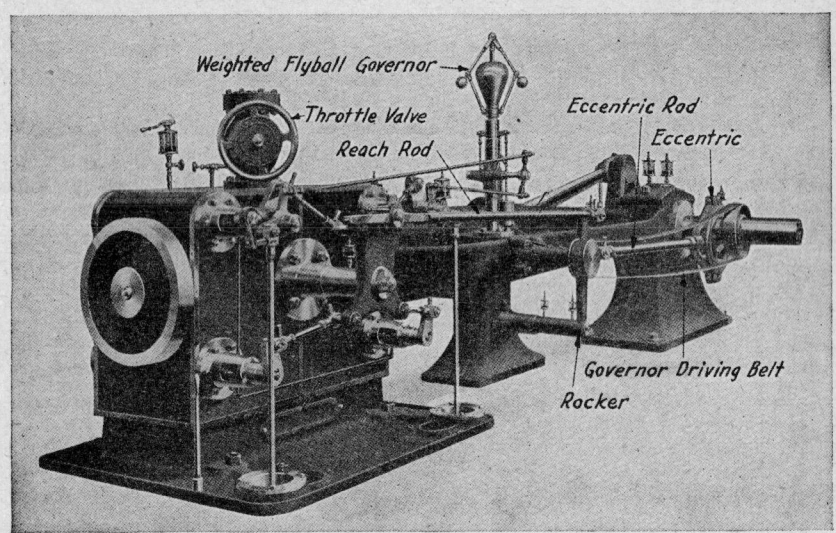

FIG. 319. — Corliss Engine — Valve Gear Side.

walls. When live steam again comes in contact with the cooled walls, some of it is condensed and, therefore, more steam is required to perform a given amount of work.

By the use of engines having four valves — two for admission and two for exhaust — the steam passages may be shortened and the clearance space reduced, because the valves may be located close to the ends of the cylinder. The use of separate valves for admission and cut-off permits the separate adjustment of each valve, and also reduces the initial condensation of steam.

369. Corliss Engine. — All parts of this engine, Fig. 319, with the exception of the valves, valve gear, and method of governing, were described in Chapter XV. The Corliss valve gear, invented by Geo. H. Corliss in 1850, has been principally instrumental in reducing the amount of steam, per horsepower, used by large reciprocating engines; and this has been the predominating type of valve gear used, since that time, on important stationary engines in the United States, England, and France. It consists essentially of cylindrical valves, which are given a semi-rotary motion by a

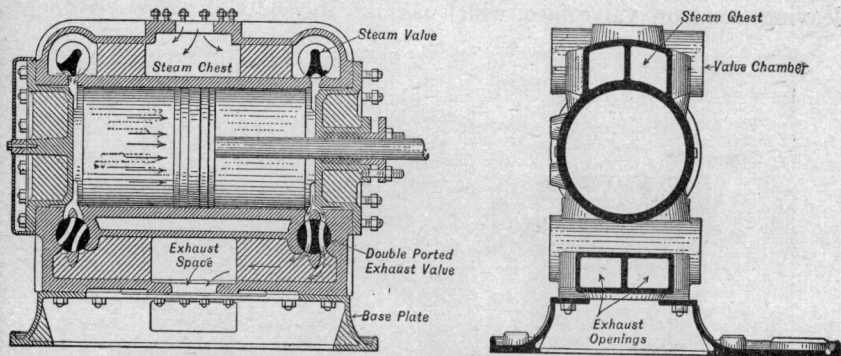

FIG. 320. — Corliss Cylinder showing Location of Valves.

releasing mechanism, under control of a governor. Valve gears using cylindrical valves and a non-releasing gear, which gives the same effect as the original Corliss releasing gear, are also termed Corliss gears. The Corliss engine is essentially a slow-speed engine, since satisfactory operation of the ordinary Corliss valve gear limits the speed to 100 or 125 revolutions per minute.

370. Classification of Corliss Valve Gears. — Corliss valve gears may be classified as: (1) *standard releasing*, (2) *long range cut-off*, (3) *high speed*, and (4) *non-releasing, or positively operated*.

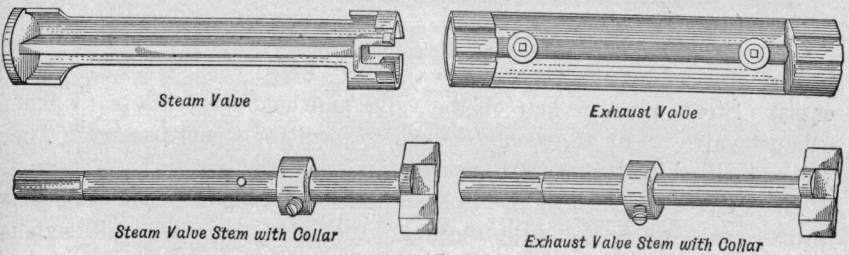

FIG. 321. — Steam and Exhaust Valves of a Corliss Engine.

371. Corliss Valves. — These valves are cylindrical, cast-iron pieces, extending across the cylinder, and are usually located at the four corners of the cylinder casting, Fig. 320. Engines which have large cylinders often have the valves in the cylinder heads, to simplify the cylinder casting. The **steam valves** of the horizontal engine, Fig. 321, are located at the top and control admission and cut-off; the **exhaust valves**, Fig. 321, are located at the bottom and control exhaust and compression. The valves may be single-ported, double-ported, or multi-ported. By the use of more than one port, the movement of the valve necessary for a given port opening is decreased.

The ends of the valves are made cylindrical, to form a surface which supports the valve. The middle portion of the steam valve is cut away, leaving a narrow valve face, which permits the valve to be moved more

FIG. 322. — Standard Releasing Corliss Valve Gear.

easily. Steam pressure acts on the valve and holds it on its seat. The exhaust valve is cut away only enough to form the steam passage. The valves oscillate back and forth, to open and close the ports.

The valve stem is made of brass and has a rectangular end which fits into a corresponding slot cut in the end of the valve. It extends through a stuffing box in the front bonnet, which carries a bracket to support the valve operating mechanism, and to form a bearing for the outer end of the valve stem. A collar on the valve stem takes the end thrust of the valve.

372. Standard Releasing Corliss Valve Gear. — In the standard valve-gear mechanism, shown in Fig. 322, the valves are moved by arms and rods attached to the **wrist plate**, which is pivoted on the cylinder casting. The **steam rods** are attached to the wrist plate in such a manner that the valves, while opening, move rapidly and cover a large proportion of their

STANDARD RELEASING CORLISS VALVE GEAR

total travel. The wrist plate receives its motion from the eccentric through the eccentric rod and the **reach rod.** A **rocker arm** is placed between the reach rod and eccentric rod to reduce the length of the eccentric rod and to increase the movement of the wrist plate for a given movement of the eccentric.

The exhaust valves are attached permanently to the wrist plate by **exhaust arms** and adjustable **exhaust-valve rods.** They are thus positively operated and receive, from the wrist plate, an oscillating motion, which is small while the valves are closed.

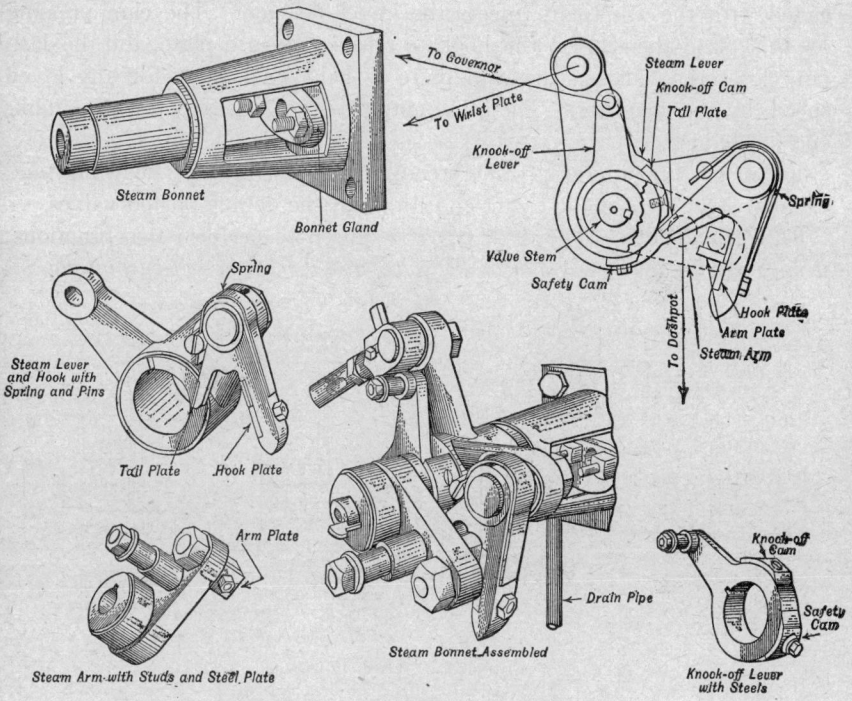

Fig. 323. — Corliss Detaching Valve Gear:

A **trip mechanism,** shown assembled and disassembled in Fig. 323, controls the operation of the steam valves, since they are not attached directly to the wrist plate. The valves are opened by movement of the wrist plate, released by the **knock-off cam,** and closed by the **dash pot.** The valve stem is keyed to the steam-valve arm, to which the **dash-pot rod** is attached. Mounted to turn on the **bonnet bracket,** in which the valve stem turns, are the **steam lever** and the **governor or knock-off lever.** These levers are not connected directly to the valve stem. One arm of the steam lever is attached to the wrist plate by an adjustable steam-valve rod; to the other arm of the steam lever is pivoted the **steam-valve latch,** or hook, which turns against the action of a flat spring. The **knock-off cam,** which lies in the plane of the valve hook, carries an arm, connected to the **governor** by a **governor rod** and **lever**. With the wrist plate, Fig. 322, at the extreme left of its travel, the hook is forced in by the flat spring and engages the **arm plate**, or block, on the head-end steam-valve arm. As the eccentric travels to the right, it rotates the steam lever and lifts the steam-valve arm, which rotates the steam valve on its seat and opens the steam port, and at the same time lifts the dash pot. The movement of the valve is rapid at this point, to prevent throttling of the steam pressure. The valve is held open by the hook, until the **tail plate**, which is not engaged with the arm plate, meets the knock-off cam. The cam pressing on this plate then forces the hook to release the arm plate, and the dash pot closes the valve. The cam plate is held stationary, for any given speed, by the governor. The same action is then repeated by the crank-end mechanism.

For the cylinder in Fig. 320, steam is flowing into the head end and exhaust steam out of the crank end, through the double-ported valves.

373. Dash Pot. — The dash pot is required to perform two functions: *it must exert a downward pull on the valve, and it must be self-cushioning.*

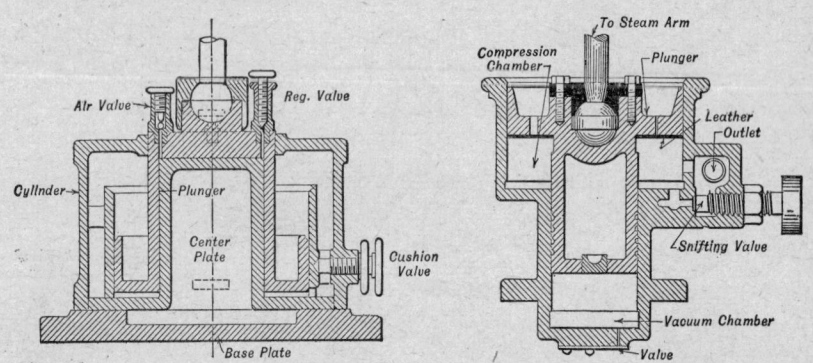

FIG. 324. — Dash Pots.

There are numerous types and arrangements of dash pots, two of which are shown in Fig. 324, with the various parts named. The outer casting, or cylinder, is generally fastened to the bed-plate by bolts. The **plunger** fits into the cylinder, and at its upper end is attached to the steam-valve arm by the dash-pot rod. The joint between the plunger and the dash-pot rod is a **ball and socket joint**, adjustable for wear. The lower part of the casting into which the lower end of the plunger fits forms the **vacuum chamber**, at the bottom of which is a small hole covered by a **flat spring** or **snifting valve**. The upper part of the cylinder and casting is enlarged to form a **compression chamber** for cushioning the plunger. This chamber is connected to a **cushion valve** which regulates the amount of cushioning.

The plunger is raised by the steam-valve latch, and a partial vacuum is thus formed below it. When raised, the plunger has the pressure of the atmosphere acting on its upper surface and a lower pressure, depending upon the degree of vacuum, acting on the bottom part of the plunger. The unbalanced pressure closes the valve suddenly when the latch is released by the knock-off cam. As the plunger descends, air is caught in the **cushion space** and escapes slowly through the **cushion-regulating valve.** This brings the dash pot to rest quietly, and without shock. Any air in the vacuum chamber is forced out through the opening at the bottom, and the vacuum is thus maintained. In some recent types of dash pots, a spring is used to cushion the plunger.

374. Governor. — The governor used on most Corliss engines having a releasing valve gear, is the **Watt,** or **pendulum governor,** Fig. 325. The governor balls are pivoted to the top of a spindle by ball arms. The spindle is driven through bevel gears, one gear being attached to the lower end of the spindle and the other to the pulley shaft and pulley which is driven by a belt from the engine shaft. The spindle of the latest types of governors revolves at two or three times the speed of the engine shaft. This permits smaller weights to be used and increases the sensitiveness of the governor. The spindle revolves in a vertical column attached to the engine frame, and, at its upper end, is surrounded by a movable sleeve to which a yoke connected to the ball arms is attached.

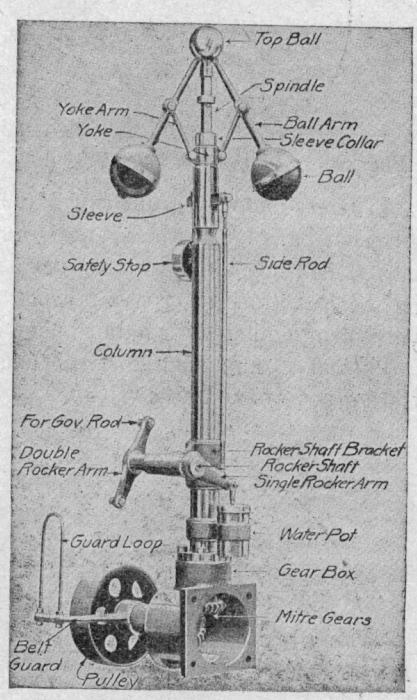

FIG. 325. — Flyball, or Pendulum, Governor.

The sleeve is connected by a **side rod** to a **double rocker arm** to which the governor rods are connected. It rises and falls with change in speed and thus determines the position of the knock-off cams and consequently the point of cut-off. A movable collar is located on the spindle to prevent the balls from rising too high, and a dash pot is attached to the side rod to prevent sudden fluctuations in the action of the governor.

A recent form of Corliss governor, made by the Allis Chalmers Co., has heavy weights acting against a spring. The weights and springs replace the balls and the force of gravity of the Watt governor. This makes a more compact, but a more complicated form of governor. All rotating parts are enclosed in a stationary case which is filled with oil.

404 MULTI-VALVE AND UNAFLOW ENGINES

Loaded governors, Fig. 319, page 398, that have a heavy weight mounted on the spindle sleeve, are used extensively. This type of governor is more powerful than one without the weight, and it can be operated at a lower speed without loss of sensitiveness.

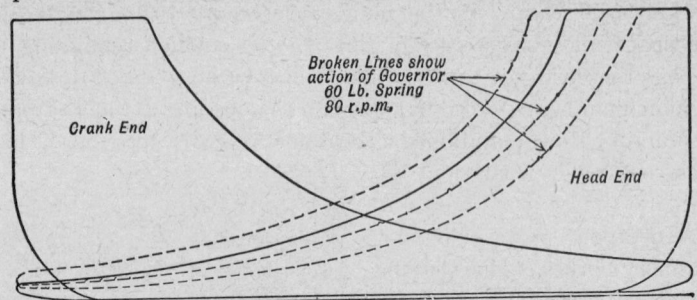

FIG. 326. — Indicator Diagrams from 16" × 38", Single Eccentric Corliss Engine.

A **safety stop** is ordinarily attached to the governor. Should the governor balls fall to their lowest position for any reason, the safety stop moves the cam-plate lever into such a position that the **safety cams** located on the cam-plate lever will prevent the latch plate from picking up the valve arm, and steam will not be admitted to the cylinder.

Typical Corliss diagrams, showing the effect of the operation of the Corliss governor, are shown in Fig. 326. The diagrams show the rapid operation of the valves at admission and at cut-off.

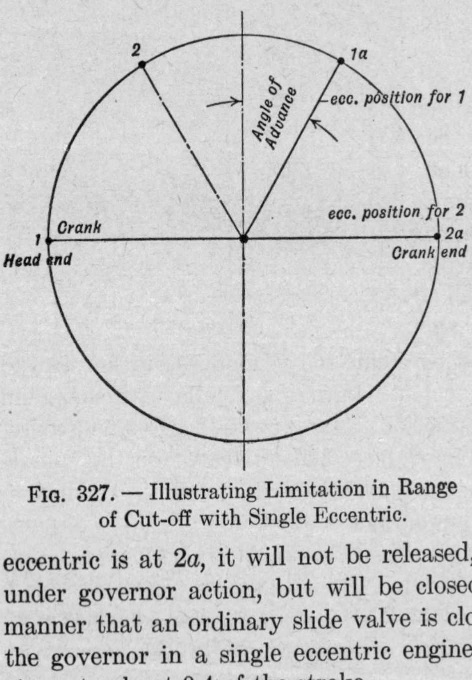

375. Limitation in Range of Cut-off for Single Eccentric Corliss Valve Gear. — Referring to Fig. 327; with the crank on head-end dead center, the eccentric will be at $1a$. The latest point of cut-off will occur when the crank is at 2, because the eccentric has then advanced to its extreme position $2a$, and any further movement will cause the eccentric to travel to the left. If the latch arm has not been released by the governor cam before the eccentric is at $2a$, it will not be released, and the valve will not cut-off under governor action, but will be closed by the eccentric in the same manner that an ordinary slide valve is closed. This limits the cut-off by the governor in a single eccentric engine, with the angle of advance as given, to about 0.4 of the stroke.

FIG. 327. — Illustrating Limitation in Range of Cut-off with Single Eccentric.

By retarding the eccentric to less than 90 degrees, cut-off by the governor can be made later. This, however, destroys the advantage of the Corliss gear, namely, quick opening of valves, because with the crank at dead center the eccentric is also nearer dead center and the speed of opening of the valves is reduced. Retarding the eccentric in this manner also effects the operation of the exhaust valves, since they are operated by the same eccentric.

By using two eccentrics, one for the admission valves and one for the exhaust valves, the cut-off eccentric may be set to give any cut-off up to seven-eighths stroke, and the exhaust valve eccentric may be set to give the proper compression and release.

376. Nordberg Long-range Corliss Valve Gear. — The essential difference between this gear, Fig. 328, and the single eccentric Corliss valve gear is that the hook is positively thrown in and out of engagement with the **valve arm**, by means of a lever which terminates in a roller resting in an **oscillating cam** attached to the valve arm. *This cam has two concentric circular slots joined by a transition slot. When the roller passes from one slot to the other, the hook is released and cut-off occurs.* The cam is oscillated by the **knock-off rod** attached to the

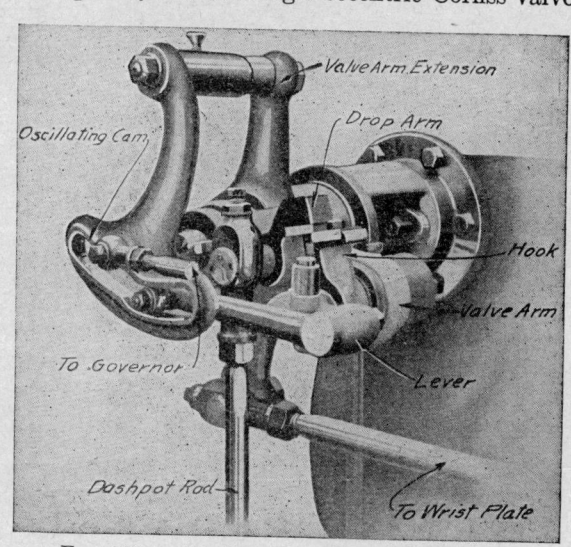

Fig. 328. — Nordberg Long-range Valve Gear.

governor, Fig. 329, and receives its motion from a separate cut-off eccentric, the throw of which corresponds to the length of each circular slot in the cam. The position of the cam and cut-off is determined by the governor and the cut-off eccentric. This gear permits higher speeds, because only one eccentric is used, and gives a range of cut-off up to 0.8 of the stroke. The movement of the valve, given by this gear, is shown by the valve ellipse, Fig. 330.

377. Nordberg High-speed Corliss Valve Gear. — The ordinary releasing Corliss gear is limited in speed because of the inability of the releasing mechanism to operate satisfactorily under high speeds. The Nordberg high-speed valve gear is adapted to speeds as high as 250 r.p.m. The essential difference between this type and the long-range cut-off gear is that the trip mechanism is symmetrical with respect to the central plane,

406 MULTI-VALVE AND UNAFLOW ENGINES

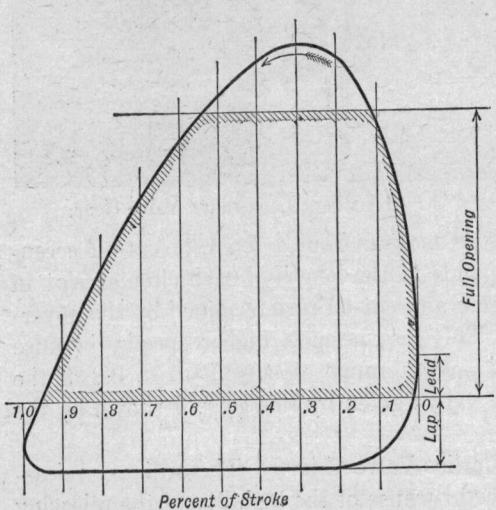

Fig. 329. — Nordberg Engine with Long-range Valve Gear.

Fig. 330. — Valve Ellipse for Admission Valve operated by Nordberg Long-range Valve Gear.

and does not have heavy overhanging parts. The valve stem and arm are supported by two bearings which support the stem on each side of the hook. The valves are made four-ported, to reduce the travel of the valve mechanism, and the dash pot is mounted on top of the valve bonnet.

378. Positively Operated Corliss Gears. — Engines having this type of gear are sometimes called **four-valve engines.** The typical Corliss valve motion is attained by using a system of links, proportioned to give the valve the desired motion. The pins on which the links oscillate are made of hardened steel. In the **Chuse non-releasing**

gear, shown in Fig. 331, the exhaust valves are driven by one eccentric and the steam valves by another. The links forming this gear are so proportioned that at the point of opening and closing, the valves travel at a high rate of speed, but at the end of travel there is practically no motion. This permits quick valve opening and closing with practically no reversal stresses on the working parts.

The location of the pins of the wrist plate and those of the valve cranks is such that a **toggle motion** is formed. This mechanism gives the valves a

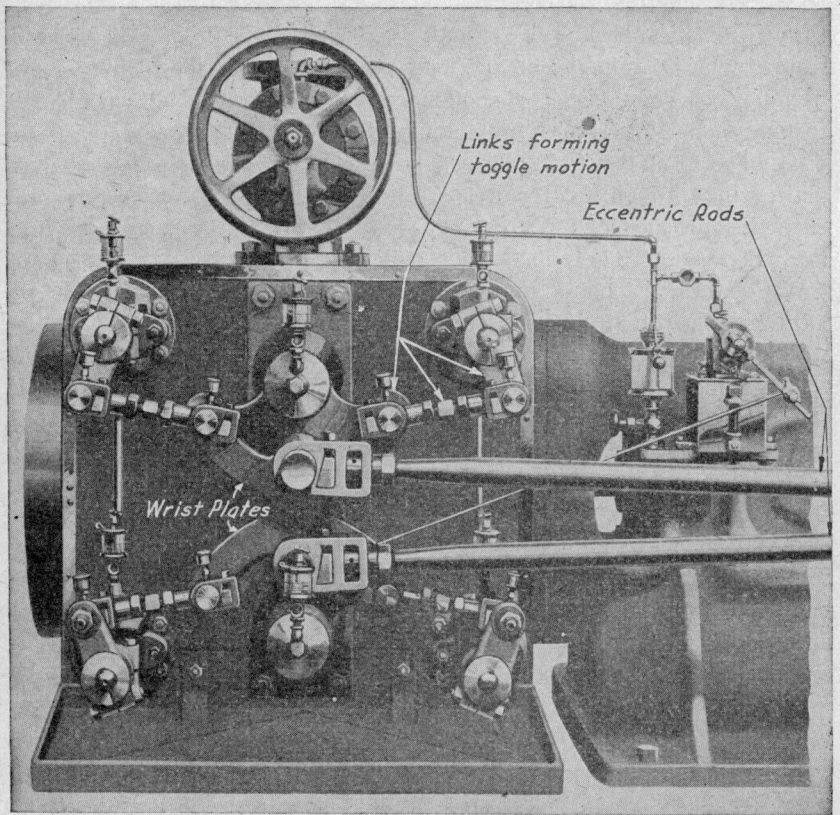

Fig. 331. — Chuse Non-releasing Corliss Valve Gear.

distinct pause during the period of greatest unbalanced pressure, starting the motion at a slow rate of speed, accelerating to a maximum at the instant of opening and closing. As a result the valves open with the same rate of speed as they close, thus improving the steam distribution and the resultant efficiency.

379. Corliss Valve Setting. — The principles that apply to the setting of the plain " D " slide-valve engine are applicable to setting Corliss valves. The valves of four-valve engines may be set by essentially the same method as that used for Corliss valves. The method used in setting Corliss valves is as follows:

1. The head- and crank-end dead-center positions are determined, as explained in Art. 363, page 394.

2. The bonnets covering the rear end of the valves are removed and the marks on each valve and seat verified. The mark on the end of each valve locates the position of the working, or opening edge, of the valve, and a corresponding mark on each seat locates the opening edge of the port.

3. The length of the eccentric rod is adjusted to make the rocker swing through equal angles on each side of its vertical position. This may be tested by using a **plumb bob** and marking each position on the floor. The dash-pot rods should be short enough to escape being caught and bent when turning the engine " over " in making this test.

4. The reach rod is next adjusted, if it is adjustable, to make the wrist plate swing through equal angles each side of its vertical position, when the engine is turned over. Marks are usually found on the hub, supporting the wrist plate. The outside marks locate the extreme positions of the travel of the wrist plate, and a mark midway between the outside marks locates the vertical position. On the hub of the wrist plate a single line will be found. When this line coincides with the central line on the hub, the wrist plate is **vertical, or central**. This position corresponds to that of the slide valve in mid-position, and it is the position of the wrist plate when measuring the laps.

5. The steam arms are hooked up, the wrist plate placed central, and the steam valves given the proper amount of lap, by adjusting the lengths of the steam valve rods. The amount of lap for the admission valves varies from $\frac{3}{16}$-inch for an 8-inch cylinder to $\frac{3}{8}$-inch for a 36-inch cylinder. The best method of measuring the amount of lap is by using a pair of dividers and a scale. The exhaust valve rods are next adjusted to be line-and-line, or to have a small clearance.

6. The crank is now placed on head-end dead center, the steam valves hooked up, and, with the wrist plate connected to the reach rod, the eccentric is moved *around the shaft in the direction the engine is to run*, until the steam valve nearest the piston shows the desired lead, which ordinarily varies from $\frac{1}{32}$-inch to $\frac{1}{16}$-inch for ordinary sized engines. The eccentric is next fastened on the shaft and the engine shaft turned in the direction it is to run, to the other dead center and the lead noted. If the leads are not the same, the connection between the wrist plate and eccentric is shortened or lengthened slightly or the length of the steam valve rod changed to give the desired lead. This affects the steam lap, but not seriously.

7. The length of the dash-pot rod is adjusted to give the proper hook clearance, by turning the eccentric; first, to one extreme of its travel and lengthening or shortening the dash-pot rod to have the arm plate clear the

hook by about $\frac{1}{16}$-inch, and then, to the other extreme and making the same adjustment.

8. The governor rods are adjusted to give equal cut-off at each end, as follows: block the governor halfway between its up and down positions

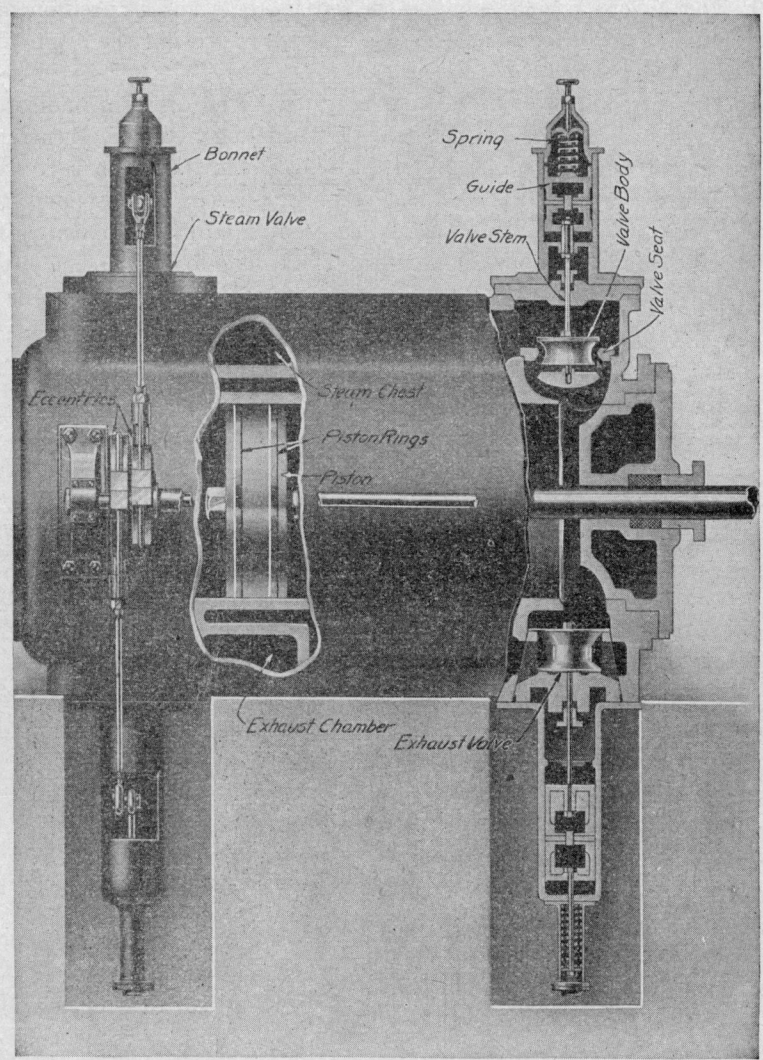

Fig. 332. — Poppet-valve Engine Cylinder.

and see if the rocker arm to which the governor rods are attached is about at right angles to a line midway between the governor rods. Place the piston at one-quarter stroke and raise the governor slowly until the valve is released by the latch. Block the governor in this position and turn engine over, to see if cut-off occurs at same point for the other end.

If it does not, shorten or lengthen the governor rod for that end to make the latch release the valve. Block the governor in a new position and see if cut-off is equal on each end. *A few trials may be needed to strike an average.* If desired, the governor may be blocked in its running position.

9. The governor balls are now placed in their lowest position and safety stop or cam adjusted, as the case may be, to prevent the hook from picking up the arm plate.

10. The correctness of the settings is tested by means of an indicator, Art. 396, page 431. For double eccentric Corliss engines, the eccentric is placed central with no lap on any of the valves, and the steam valves set as for a single eccentric. The exhaust eccentric is set by placing the piston at seven-eighths stroke and advancing the eccentric until the exhaust valve, at the end the piston is approaching, is just beginning to close. The piston on the opposite end is next placed at seven-eighths stroke, and the process repeated for the other exhaust valve.

Fig. 333. — Lentz Poppet-valve Compound Engine.

If there is any doubt as to the direction in which to move the eccentrics, remember that the *eccentric should always be set to have steam valve opening when crank starts from dead center, in direction engine is to run,* and that the exhaust valve at same end should be closing.

380. Poppet-valve Engine. — The poppet-valve engine is used to a greater extent in Europe than in the United States. The Lentz poppet-valve engine was developed in 1899. Since that time many improvements have been made and it is now used extensively. *Poppet valves are adapted to high speeds and can be used with superheated steam.*

A common type of poppet-valve engine cylinder, shown in Fig. 332, has two steam and two exhaust valves, each valve consisting of two cast-iron disks joined by a cylindrical body. The faces of the upper and lower disks rest on seats in the valve cage when the valve is closed, and the form of the valve makes it nearly balanced. The valve remains stationary when closed. The valve stem passes through a stuffing box in the valve cage, and is attached to a valve-stem guide which operates against a spring at the top of the bonnet. The valve cage and bonnet are removable and are bolted to the top and bottom at each end of the cylinder. Steam passages connect the valve seat and the cylinder.

The Lentz poppet-valve gear is shown in Fig. 333 and the admission-valve gear in more detail in Fig. 334. The valve stem extends from the valve through a bearing that serves as a stuffing box. In order to prevent loss of steam, the valve stem has grooves cut around its circumference where it slides in the bearing.

A spring which closes the valve presses on the upper end of the valve stem and bears against an **oscillating cam.** The cam is moved by an eccentric located on a **lay shaft,** which runs the length of the engine and is driven by bevel gears from the main shaft.[1]

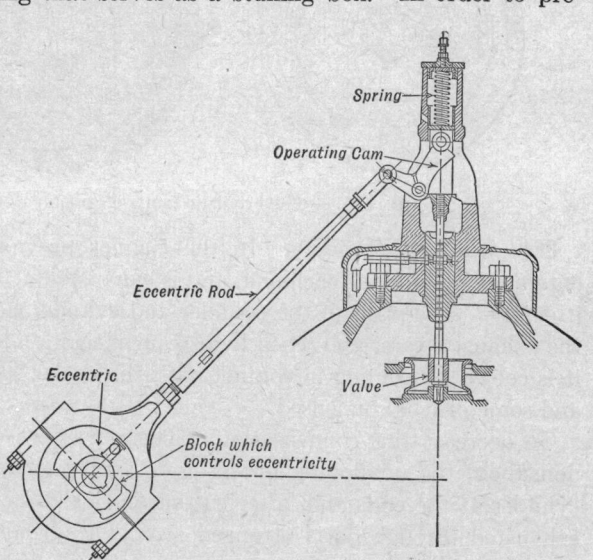

FIG. 334. — Admission Valve Gear of Lentz Engine.

Each valve has a separate eccentric. The exhaust-valve eccentrics are fixed on the lay shaft to open and close the exhaust valves at the proper time. The steam-valve eccentrics are slotted, and are under the control of the governor, which regulates the speed by changing the point of cut-off.

The small centrifugal governor, Fig. 335, is mounted on a sleeve which rotates on the lay shaft. The governor sleeve carries a pin, connected to the eccentric in such a way that it moves the eccentric on a block, which is fixed on the governor shaft, thus changing the eccentricity, and consequently the length of time the valve remains open.

A poppet-valve engine made by the Nordberg Engine Co. differs from that previously described, in that the cylinder proper is a straight barrel

---

[1] In the latest type of this engine, the lay shaft is omitted, the admission valves are operated by an eccentric controlled by a shaft governor, and the exhaust valves by a fixed eccentric.

without valve passages. The valves, together with the steam and exhaust passages, are located in the cylinder heads, which are bolted to each end of the barrel. The valves are positively opened and closed, without the aid of a spring, by a cam which acts on a roll plate pivoted to the bonnet and attached to the valve stem. The operation of this engine is similar to that of the Lentz engine.

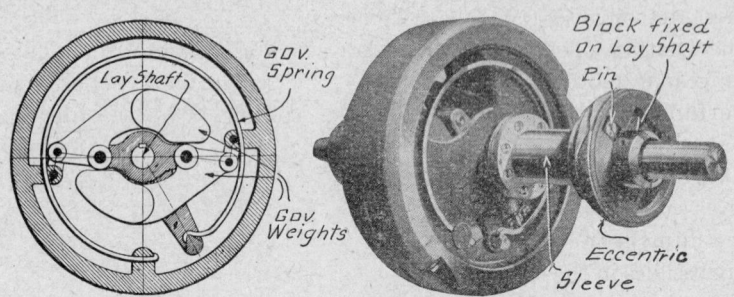

FIG. 335. — Governor of Lentz Poppet-valve Engine.

381. Unaflow Engine.[1] — In the engines previously described, steam enters the cylinder at each end, and is exhausted at the same end at which it enters. The walls of the cylinder and cylinder head are thus cooled by the exhaust steam, and when live steam is again admitted to the cylinder it is cooled by coming in contact with the cooler surfaces of the cylinder, and some of it is condensed.

To decrease this condensation of steam which is known as **initial condensation,** the unaflow engine is used. In this engine steam enters the cylinder at the ends and, after cut-off and expansion have taken place, is exhausted through ports arranged around the center of the cylinder and uncovered by the piston near the end of the stroke. The steam thus flows in one direction through the cylinder, and the cylinder walls and heads are not cooled by the exhaust steam.

The cylinder of the unaflow engine, Fig. 336, is made slightly different from the cylinders already considered, in order to accommodate the valves, and the cylinder heads at each end are made with a passage for live steam to prevent radiation loss from the cylinder head. There are two steam poppet valves, operated as for the poppet-valve engine shown in Fig. 332. The exhaust valves are omitted, and their function is performed by the piston, which is made long for that purpose. The exhaust ports have a large area which permits the steam to escape rapidly, as the exhaust is only open for a short time. The exhaust passage is formed by a ring cast around the cylinder and connected to the exhaust pipe at the bottom.

The European unaflow engine is generally operated condensing; that is, the exhaust pressure is below the pressure of the atmosphere. Compres-

[1] This type of engine is also called *uniflow*.

sion, beginning as soon as the piston covers the exhaust ports on its return stroke, takes place during 90 per cent of the stroke. When the engine is operating condensing, the final pressure will reach the initial steam-pipe pressure, but, when it is operated non-condensing, will rise above steam-pipe pressure with ordinary clearances. To prevent this rise in pressure when operating non-condensing, the following methods are employed:

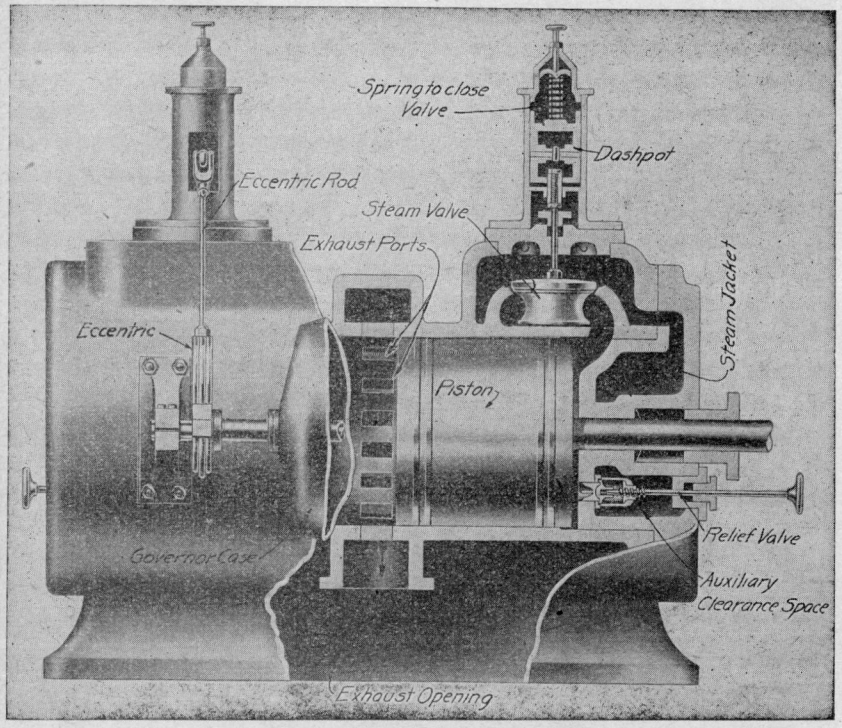

FIG. 336. — Unaflow Engine Cylinder.

1. Increasing the clearance by using clearance pockets in cylinder heads, controlled by hand-operated relief valves as illustrated in Fig. 336, or by concaving the ends of the piston.
2. Using automatic relief valves.
3. Using exhaust ports in the piston.
4. Using auxiliary exhaust ports to delay compression.

The additional clearance necessary varies with the steam pressure and for low steam pressures may be as high as 20 per cent.

The Skinner unaflow engine shown in Fig. 337 is a high-speed engine which runs at speeds ranging from 250 to 275 r.p.m. When operating non-condensing, an excessive pressure during compression is prevented by using auxiliary exhaust ports, located near each end of the bore of the cylinder, as shown in Fig. 338, where the piston and double-beat poppet valves are in such a position that *exhaust* is taking place at the head end and

Fig. 337. — Skinner Unaflow Engine—Valve Gear Side.

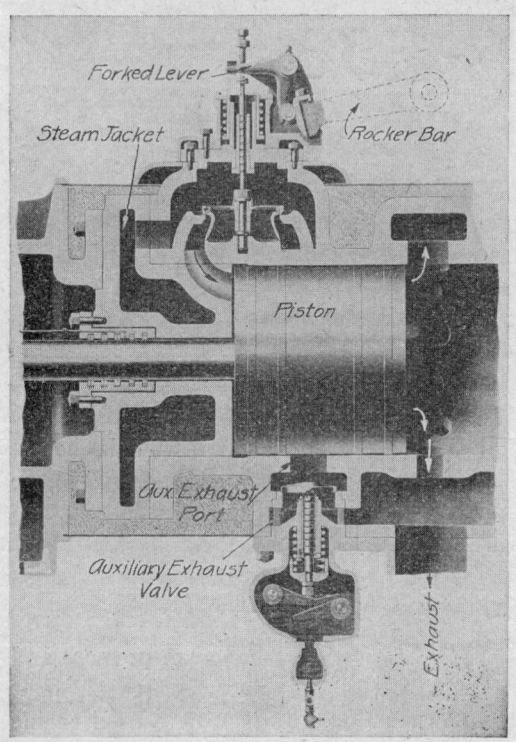

Fig. 338. — Sectional View of Cylinder.

admission at the crank end.

The steam valve gear consists of two lifter bellcranks operated by a rocker bar located in a case containing oil and moved by a rocker and reach rod connection to the inertia type governor. One end of each lifter bellcrank supports a hardened steel roller, which at all times bears against the cam on the rocker bar, and the other end makes contact with a set screw at the upper part of the valve stem for all positions before that of cut-off, after which there is a small clearance in order to permit the valve to close tightly. The inlet valves are thus opened by the cams on the rocker bar and are closed by a spring located in the valve bonnet and surrounding the valve stem, which is

UNAFLOW ENGINE

grooved and made a close fit in the sleeve, in which it slides, in order to prevent leakage of steam at this point. The position of the rocker is controlled by the governor, which changes the amount of movement and position of the cams, thereby changing the point of cut-off to meet the load requirements.

The **auxiliary exhaust valves** are designed to open and close under no difference in pressure, and are operated by a cam driven by a fixed eccentric located on the opposite side of the engine from the steam-valve gear. These valves are of the poppet type and are so arranged that the engine may be operated either condensing or non-condensing, without loss of efficiency, by means of an automatic disengaging device.

When operating condensing, the auxiliary exhaust valves are not used; when operating non-condensing, the auxiliary exhaust valves are automatically opened by the device shown in Fig. 339. An **idler** is located on a shaft which is free to move axially under the action of a spring located in a pocket connected to the exhaust pipe.

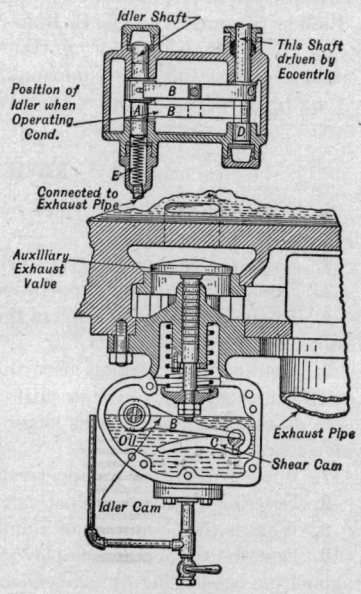

Fig. 339 — Auxiliary Exhaust Valve, Automatic Operating Device.

A **shear cam,** operated by the engine valve gear, through a connection to the shear cam shaft, raises the exhaust valve when in the position shown. The spring around the valve stem has just enough tension to insure quick closing at high speeds.

Under vacuum, the tension in the spring on the idler shaft is overcome, the idler is moved out of register with the cam driven by the valve gear, and the exhaust valve remains closed. Should the vacuum be lost, the spring moves the idler between the valve stem and cam, the auxiliary exhaust begins to function, and the engine automatically operates non-condensing. A typical indicator diagram is shown in Fig. 340.

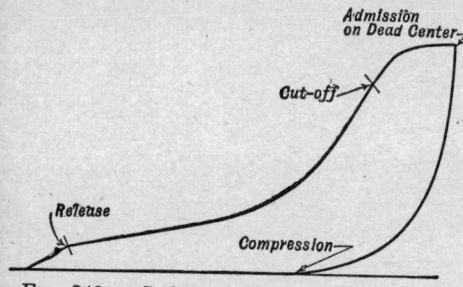

Fig. 340. — Indicator Diagram from Skinner Unaflow Engine.

The unaflow engine manufactured by the Nordberg Mfg. Co. uses a long-range high-speed Corliss valve gear for the admission valves, and additional clearance is obtained by using clearance pockets.

REFERENCES

Steam Power, HIRSHFELD and ULBRICHT.
American Steam Engines, HAWLEY.
Journal of the Institute of Mechanical Engineers, London, October 1920, Paper on the Uniflow Steam Engine, E. B. BERRY.
Steam Power Plant Engineering, GEBHARDT.
Catalogues of the following companies: NORDBERG ENGINE Co., SKINNER ENGINE Co., ERIE IRON WORKS, MURRAY IRON WORKS, A. L. IDE and SONS.

REVIEW QUESTIONS

1. What advantage is obtained by using four valves in place of one, to control the distribution of steam?
2. Name the various types of Corliss valve gears.
3. Explain the method of operation of the standard Corliss valve gear.
4. To about what proportion of the stroke is the cut-off limited on a single eccentric Corliss engine?
5. Mention two methods used to increase the proportion of the stroke at which cut-off can occur on a single eccentric Corliss engine.
6. Explain how the Corliss long-range valve gear differs from the standard Corliss valve gear.
7. What is meant by a poppet-valve gear?
8. Describe the Lentz poppet-valve gear.
9. What is the advantage of a unaflow engine over a double-flow engine?
10. Describe the method employed to change the operation of the Skinner unaflow engine from condensing to non-condensing.

CHAPTER XVIII

STEAM ENGINE INDICATOR AND ITS APPLICATIONS
STEAM ENGINE EFFICIENCIES AND LOSSES

382. Foreword. — A steam engine **indicator** is an instrument which records graphically the variation in the pressure of the steam, occurring in the cylinder of an engine during the stroke of the piston. This graphical record, when made for one revolution of the engine, is a closed curve, called an indicator diagram. It is formed by (1) a horizontal movement of the paper in exact correspondence with the movement of the piston, and (2) a vertical movement of the pencil in exact ratio to the pressure exerted in the cylinder of the engine. *The length of the diagram, therefore, represents the length of the stroke to a reduced scale, and its height at any point represents the pressure on the piston at the corresponding point in the stroke.* The indicator diagram may be used for the following purposes:

1. To furnish data for calculating the power developed by an engine.
2. To supply information regarding the accuracy of the setting of the valves.
3. To estimate the theoretical amount of steam used by the engine.

383. Description of the Steam Indicator. — There are many makes of indicators which differ mainly in details of construction, the essential operating parts being nearly the same in all. Steam engine indicators may be classified as: (1) *inside spring*, (2) *outside spring*, and (3) *continuous*. The essential parts of the steam indicator are as follows:

1. A piston arranged to move in a cylinder and upon which the steam pressure acts.
2. A pencil motion, made of a number of small links, arranged to move the outer end of a pencil arm in a direction parallel to the piston, and to amplify its movements.
3. A piston rod connecting the piston and the pencil motion.
4. A spring, interposed between the piston and pencil motion, to measure the pressure in pounds per square inch.
5. A drum, which can be rotated about its vertical axis and which carries the paper upon which the graphical record is made.

The indicator is ordinarily attached to a short length of pipe leading into the clearance space of the cylinder. Steam from the cylinder enters below the piston, compresses the spring and moves the pencil arm, which stands at a height proportional to the steam pressure in the engine cylinder. The drum, upon which the indicator card[1] is held, is rotated by a suitable connection to the crosshead. As the rotation occurs, the pencil is pressed against the indicator card and a line is drawn, which is produced by a combination of these two movements and which forms the diagram.

384. Crosby Inside Spring Indicator. — A sectional view of this instrument is shown in Fig. 341. The cylinder in which the piston slides is made of an alloy suited to the varying temperatures to which it is subjected. It is held between an upper and a lower casting, from which it is separated by

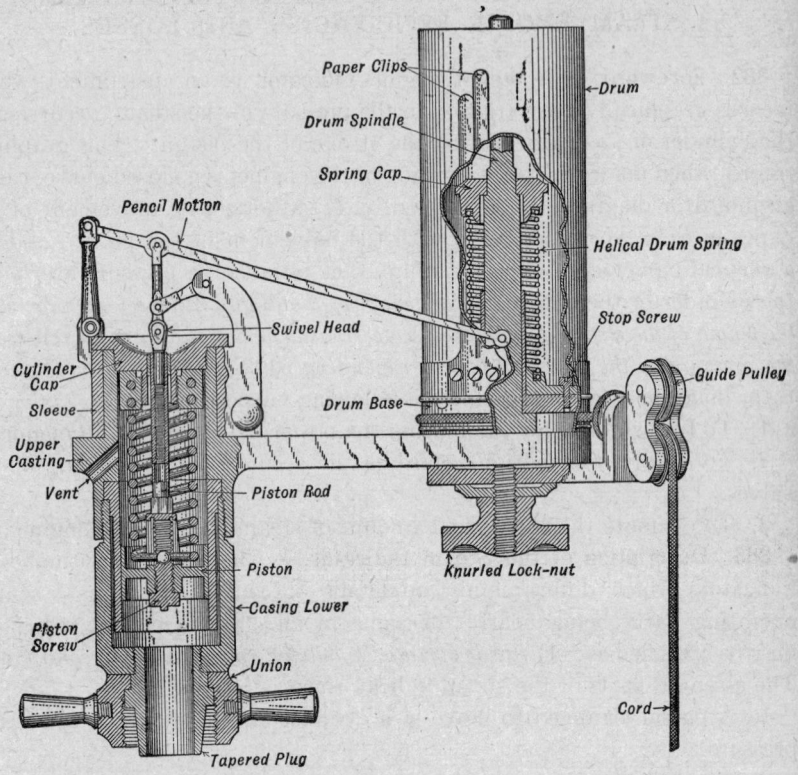

Fig. 341. — Sectional View of Crosby Inside Spring Steam Indicator.

a space, forming a steam jacket for the cylinder. The upper casting carries an arm, to the outer end of which the drum is attached, and its lower end is threaded inside to screw on the top part of the bottom casting. A series of vent holes in this casting serve to maintain atmospheric pressure above the piston. A cap screws into the upper part of this casting and holds the sleeve carrying the pencil motion in place; in the center of the cap is a hole fitted with a hardened-steel bushing which guides the upper end of the piston rod. The lower casting carries a tapered sleeve and nut, at its lower end, by which the indicator is attached to the indicator cock.

[1] The term *card* is here used to denote the paper upon which the diagram is drawn.

CROSBY INSIDE SPRING INDICATOR

The piston has an area of ½ square inch, and is a hardened cylindrical tool-steel shell having a transverse web near its center, with a hub projecting above and below the web. The part of the hub above the web is threaded inside to receive the lower end of the hollow steel piston rod, and has a longitudinal slot into which fits a cross-wire on the lower end of the spring. The part of the hub below the web is threaded for a hexagonal-headed screw, having a concave bearing in its upper end. The lower end of the piston rod is threaded to screw into the hub of the piston, and above the threaded part is a shoulder having a circular channel, on its under side, which fits over a machined portion of the hub and prevents it from spreading. The upper end of the piston rod is threaded inside and screws on a threaded rod attached to the pencil motion. This permits changing the position of the pencil point by turning the piston rod with the attached spring. The fundamental principle of the pencil motion is that of a **pantograph parallel motion.**

The spring, Fig. 342, is made of a single piece of spring-steel wire wound from the middle into a double coil, and fastened to a metal head threaded inside to screw on the inner threaded portion of the cap. The lower end of the spring has a cross-wire, midway on which is staked a small **bead.** When in position this bead is held between the lower end of the piston rod and the upper end of the piston screw. By this connection a ball and socket joint is formed, which prevents binding of the piston.

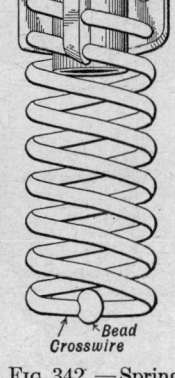

Fig. 342.—Spring for Crosby Indicator.

Indicator springs are made of various sizes of wire, to give a satisfactory height of diagram when used with different steam pressures. A number, known as the scale of the spring, is stamped on one of the flanges of the spring. *This number represents the pressure in pounds per square inch which would compress or elongate the spring sufficiently to move the pencil point one inch.* The most common scales of springs used in steam indicators are: 4, 8, 12, 16, 20, 30, 40, 50, 60, 80, 100, 120, 150, and 180.

The scale of the spring should be such that the diagram will be about 1¾ inches high.

The **drum** is a hollow, thin metal cylinder, closed at one end. Its open end fits tightly over a hub which turns on a bearing formed by the **drum spindle.** The lower end of the spindle is screwed into the supporting arm, with its upper end projecting through a hole in the top of the drum. A guide pulley bracket, held in place by a knurled lock-nut, fits over the screw part of the spindle below the supporting arm. The drum is rotated against the action of the drum spring by a cord wound around the hub of the drum. By changing the position of a small screw in the hub, the instrument can be prepared for use on either right- or left-hand engines.

The drum spring is helical with its lower end attached to the hub, and the upper end to a cap with a square hole which fits over a squared shoulder on the drum spindle. The tension of the drum spring can be changed, as required for the speed, by lifting the cap and turning in the proper direction. The higher the speed the greater the tension required.

A small handle, adjustable by a screw against a stop, is attached to the sleeve carrying the pencil motion, and the pressure of the pencil against the paper on the drum can thus be regulated. *It should be sufficient to give a clear line with minimum friction on the paper.*

Method of attaching spring. — To place a spring in the instrument the cap is unscrewed, and the pencil motion, sleeve, piston, and piston rod are lifted from the cylinder. The piston rod is unscrewed from the piston and pencil motion, and the piston-screw backed part way out. The piston rod is inserted into a hollow wrench provided for this purpose, and the rod and wrench inserted into the spring, until the bead on the cross-wire rests on the concave end

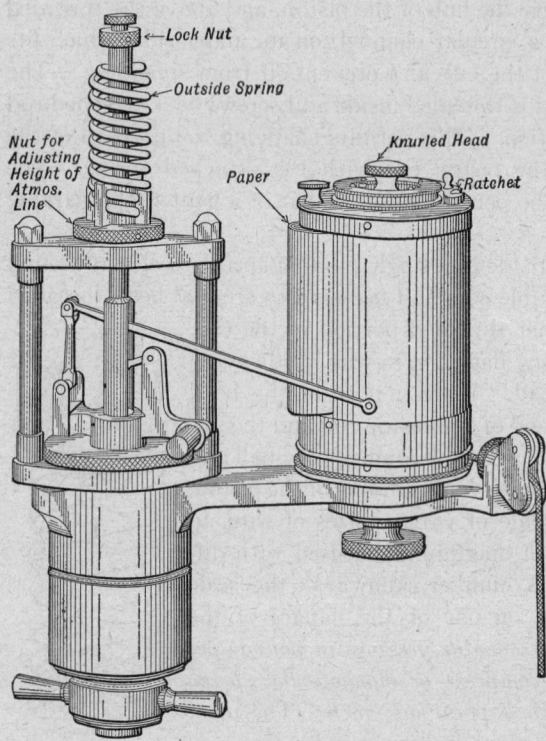

Fig. 343. — Crosby Outside Spring Indicator for Taking Continuous Diagrams.

of the rod. The piston is then screwed on the rod as far as it will go, or until the upper end of the hub is brought against the bottom of the channel. The piston-screw is tightened and the piston rod screwed on the threaded part of the swivel head of the pencil motion, until the head of the spring is firmly fastened to the cap. The cap is then released and the turning continued until the top of the piston rod is flush with the shoulder at the swivel head. *In this position the atmospheric line should be at the proper height. If it is desired to change its position, the piston and cap are turned clockwise to lower, and counter-clockwise to raise the atmospheric line. One turn changes the position of the pencil point $\frac{1}{8}$-inch.* The piston and its connecting parts are now placed in the cylinder, and the cap

INDICATOR PIPING AND COCKS 421

screwed down. The cap should always be examined before steam is turned on, to make sure it is properly screwed in position.

385. Outside-spring Indicator. — The new Crosby outside-spring indicator, Fig. 343, is illustrative of this type of instrument. The spring is outside the cylinder and is always at room temperature. The piston is made narrow and is, in form, the central portion of a sphere. It has an area of one square inch and the indicator springs are consequently made of larger-sized wire than those for inside-spring indicators. The spring is easily attached to an extension of the piston rod which extends above the spring supporting frame, by unscrewing the adjusting screw and lock-washer and screwing the threaded portion of the spring to the spring support, with the bead on the cross-wire in the slotted upper end of piston rod extension. With this done, the small nut at the top is screwed against the bead and locked. The atmospheric line is changed by changing the position of the **lock-washer,** against which the lower end of the spring rests when in position.

386. Continuous-diagram Indicator. — To obtain continuous records over an extended period of time, an indicator must be used that will permit the taking of diagrams in rapid succession. Fig. 343 shows such a device attached to the drum of a Crosby indicator. A roll of paper 12 feet long and 2 inches wide is located on a spindle within an opening in the shell of the drum. The paper passes from this roll around the drum and inward

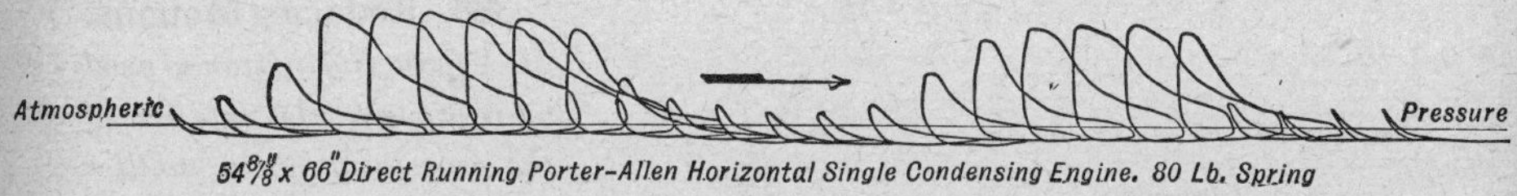

Atmospheric   *Pressure*

$54\tfrac{3}{8}'' \times 66''$ *Direct Running Porter-Allen Horizontal Single Condensing Engine. 80 Lb. Spring*

FIG. 344. — Continuous Indicator Diagram.

to a central cylinder, concentric with the shell of the drum, to which it is attached. Upon the top of the drum is a **ratchet wheel** which automatically unwinds a small length of paper from the roll and winds it on the inner cylinder, thus giving a series of diagrams which overlap each other slightly, as shown in Fig. 344. A **knurled head,** loosely attached to the drum spindle at the top, can be adjusted to move the paper around the drum by different amounts as desired, thus providing a method of controlling the distance between diagrams.

387. Indicator Piping and Cocks. — The indicator is connected to the cylinder by an indicator cock, Fig. 345a, and short $\tfrac{1}{2}$-inch pipe. *The piping connections should be as short and direct as possible, with few bends, to reduce condensation and loss of pressure to a minimum.* The **straightway indicator cock** consists of a casting threaded at both ends and having a circular passage from end to end. At right angles to this passage is a **conical plug** having a wooden handle. The upper end of the passage in the casting is tapered to form a tight joint with the tapered portion of the indicator coupling. With the handle of the plug parallel to the pipe, a straight passage is made to the cylinder. When the handle is at right angles to the pipe, the passage to the engine is closed, and the indicator is connected to the atmosphere by a small hole at the side of the casting.

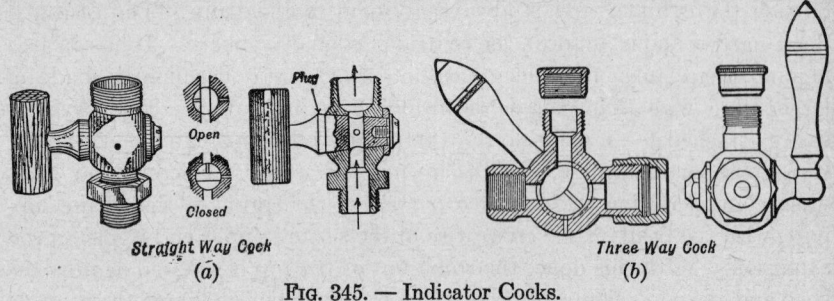

Fig. 345. — Indicator Cocks.

A **three-way cock,** Fig. 345b, is sometimes used with a single indicator for both ends of the cylinder; it is better, however, to have a separate indicator for each end of the cylinder. With the handle of the three-way cock vertical and toward the indicator, the passage between the indicator and cylinder is closed. When swung to the right, the left connection to the cylinder is open. In general, the opening is to the end opposite that toward which the handle is swung in opening.

388. Reducing Motions.
— *A reducing motion is used to reproduce the motion of the crosshead on a smaller scale.* The drum of the indicator is attached to the reducing motion by a cord, which should be as short as possible, and thus it is made to move with the crosshead. *The number of inches of the travel of the piston represented by one inch in length on the diagram is called the* **ratio of reduction.** To be satisfactory, the reducing motion should be proportioned to give a length of diagram ranging from $2\frac{1}{2}$ to 4 inches. The reducing motion should be accurate, all its joints should be free of looseness, and its parts should be light and rigid.

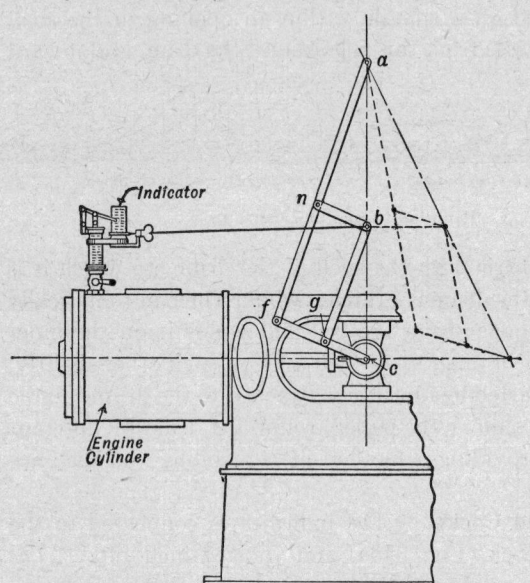

Fig. 346. — Pantograph Reducing Motion.

The **pantograph reducing motion,** Fig. 346, is well suited for slow moving engines. The indicator cord must be attached to the device at a point lying on the broken line *ac*, connecting the fixed point and the crosshead connection. *The ratio of reduction is then af divided by nf.* The four-bar linkage, *fnbg*, should be an exact parallelogram, and the chord to the drum should be parallel to the center line of the engine when leaving the linkage, or the reduction will not be in the above ratio.

The **reducing wheel**, Fig. 347, is portable and easily manipulated. The indicator is attached directly to the wheel casting, which in turn is attached to the indicator cock. The cord or string from the indicator passes around a small pulley, driven from the crosshead by a string wound around the larger, or driving, pulley. The movement of the driving pulley is communicated to the drum-string pulley through a set of spur gears. *The ratio of reduction is the number of teeth in the driven gear divided by the number of teeth in the driving gear multiplied by the diameter of the driving pulley divided by the diameter of the driven pulley.* This ratio may be changed by changing the pulley diameters or the size of the respective gears. A spring located in the spring case keeps the driving string taut.

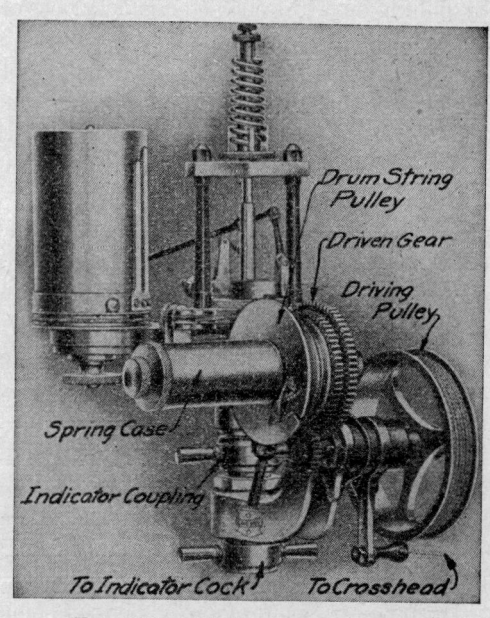

FIG. 347. — Crosby Reducing Wheel.

389. Accuracy of Indicator Diagrams. — It is possible to obtain accurate diagrams only when the indicator satisfies the following conditions:

1. Equal changes in pressure should be accompanied by equal movements of the pencil point. This requires an accurate pencil motion and spring.

2. Equal distances on the diagram should correspond to equal distances traveled by the piston. This requires a true cylindrical drum turning on its true geometric axis, a cord that does not stretch, and an accurate reducing motion.

390. Method of Taking Diagrams. — A satisfactory method of taking indicator diagrams is as follows:

1. The indicator cocks are opened and the steam is permitted to blow out any dirt, after which the indicator is attached and the length of the indicator cord adjusted to prevent the indicator drum striking the drum

stop-pin at either end of the stroke. The indicator cord should always leave the reducing motion parallel to the center line of the engine cylinder, and the hook on the cord should be so attached that it will not slip and yet will be easily loosened.

2. The tension of the drum is adjusted for the speed and the paper placed on the drum by folding one end over about $\frac{1}{4}$-inch, and hooking this bend over the longer of the paper clips; the paper is then looped around the drum and under the shorter clip, and with the forefinger and thumb the card is pulled down on the drum, at the same time it is kept tight against the drum by pulling outward on the ends held between the thumb and finger. The paper may have a prepared or a plain surface. The prepared paper requires a **brass stylus,** which will make a black line on the **zinc oxide** coating of the paper, while the plain paper requires only the lead of a pencil. The coated side of the prepared paper should be outward.

3. The pencil is adjusted against the paper to give a fine clear outline without undue friction on the paper.

4. The indicator cord is attached to reducing motion and the indicator cock partially opened to permit steam to blow through the cock into the atmosphere, thus removing water from the pipe connections to the indicator.

5. The indicator cock is next opened wide and the pencil point is pressed against the paper. If the pencil is held against the drum for too long a time, several diagrams will be obtained, in case the load is varying.

6. The indicator cock is now closed and the pencil pressed against the paper, to draw the atmospheric line.

391. Indicated Horsepower, i.hp. — *The indicated horsepower is the power developed in the steam cylinder, as determined from the indicator diagram.*

Horsepower, as previously defined, equals the product of a force in pounds multiplied by the velocity in feet per minute at which the force is moving, divided by 33,000. Written as an equation,

$$Horsepower = \frac{F \times V}{33,000} \quad \cdots \cdots \cdots \quad (89)$$

The force acting on the piston of a steam engine changes from point to point. *Its mean value in pounds per square inch is obtained from the indicator diagram, by dividing the area of the diagram by its length and multiplying by the scale of the indicator spring.* This pressure is called the **mean effective pressure,** or **m.e.p.** The **total force** equals the mean effective pressure, P, in pounds per square inch multiplied by the net area of the piston, A, in square inches, against which the pressure acts. *It should be noted that the area of the crank end of the piston is less than that of the head end by the area of the piston rod.*

The velocity of the piston in feet per minute equals the length of the

FINDING THE AREA OF AN INDICATOR DIAGRAM

stroke in feet, L, multiplied by the number of working strokes per minute, N. Substituting these values of force and velocity in Equation (89), there results

$$\text{Indicated horsepower, i.hp.} = \frac{PLAN}{33{,}000} \quad \ldots \ldots \quad (90)$$

For a double-acting engine, the total power equals the sum of the indicated horsepower for each end, or

$$\text{Total i.hp.} = \frac{P_h L A_h N}{33{,}000} + \frac{P_c L A_c N}{33{,}000} \quad \ldots \ldots \quad (91)$$

in which P_h and A_h refer to the head end and P_c and A_c to the crank end. In these equations for indicated horsepower, $\frac{LA}{33{,}000}$ is constant for a given engine and is known as the **engine constant**. When making a large number of calculations, the engine constant *should be computed for each end of the cylinder.* The indicated horsepower is then found by multiplying the engine constant by the mean effective pressure and the number of revolutions per minute.

It should be noted that the actual mean effective pressure is not the average pressure acting on one side of the piston. It is the average pressure acting on one side of the piston less that acting on the opposite side.

Example 41. — The following data were taken from a 12 in. by 24 in. Allis-Chalmers Corliss engine running at 102.9 r.p.m.: diameter of piston rod $2\frac{3}{16}$ in.; scale of indicator spring, 80 lb.; area of head-end diagram, 2.04 sq. in.; area of crank-end diagram, 1.85 sq. in.; length of each diagram, 3.76 in.

Find: (*a*) Engine constant for each end, (*b*) total indicated horsepower.

Solution. — (*a*) Head-end constant .... $= \frac{L_h A_h}{33{,}000} = \frac{24 \times 113.10}{12 \times 33{,}000} = 0.00685$

Crank-end constant ... $= \frac{L_c A_c}{33{,}000} = \frac{24 \times (113.10 - 3.76)}{12 \times 33{,}000} = 0.00664$

(*b*) Head-end i.hp. .... $=$ Head-end constant $\times N \times P$
$= 0.00685 \times 102.9 \times 43.40 = 30.5$

Mean effective pressure $= \dfrac{\text{area of diagram} \times \text{scale of spring}}{\text{length of diagram}}$

$= \dfrac{2.04 \times 80}{3.76} = 43.40$

Crank-end i.hp. .... $= 0.00664 \times 102.9 \times 39.40 = 26.8$

Crank-end m.e.p. .... $= \dfrac{1.85}{3.76} \times 80 = 39.40$

Total i.hp. .... $= 30.5 + 26.8 = 57.3$.

392. Method of Finding the Area of an Indicator Diagram with a Planimeter. — The **planimeter**, Fig. 348, is an instrument used to obtain the areas of irregular figures. It has a **guiding arm** joined to a **tracing arm** by a hinged joint. The guiding arm has a point used as a pivot and held in a fixed position by a small weight. The tracing arm carries a **tracing**

point, which is moved over the outline of the figure to be measured. Fixed to the tracing arm is a small calibrated wheel, which revolves as the arm is moved. This wheel is divided into ten equal parts, each part representing one square inch. A complete revolution of the measuring wheel measures an area of 10 square inches. The area of the figure equals the difference in

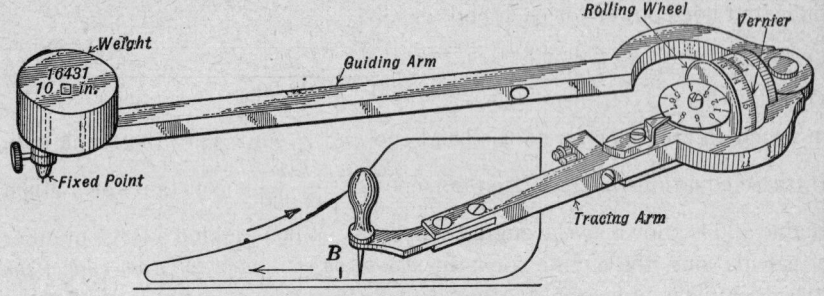

Fig. 348. — Fixed Arm Polar Planimeter.

the readings on the rolling wheel and vernier taken at the beginning and end of tracing around the figure.

The **vernier,** Fig. 349, is an auxiliary scale placed alongside the main scale to make it possible to secure an accurate reading of the hundredths place, which otherwise could only be estimated. The vernier scale is generally equal to the length of nine divisions on the main scale, and is divided into ten equal parts. Each vernier scale division is therefore one-tenth of a scale sub-division shorter than a sub-division of the main scale. To read, locate the scale mark before the index, and then look forward until a line of the vernier coincides with a line on the scale. Record the vernier reading as the last digit. The reading shown is 3.44.

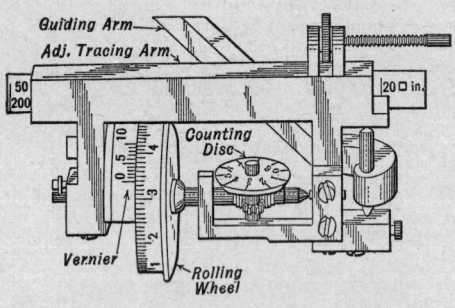

Fig. 349. — Rolling Wheel Showing Vernier.

The planimeter shown in Fig. 348 has a fixed tracing arm. Planimeters often have adjustable tracing arms which must be set to the proper scale before using. The area, as given by the planimeter, of a figure drawn to scale, must be corrected for the scale to which the figure is drawn, to obtain the actual area.

The indicator card, on which the diagram to be measured is drawn, should be fastened, by thumb tacks, to a piece of paper having a smooth unglazed surface for the wheel to roll upon; and the fixed point on the guiding arm should then be located so that the wheel will not come in contact with the edge of the card. The best location for the fixed point is above the card and with the arms making approximately 90 degrees with each other when the tracing point is midway between its extreme right and left positions on the figure.

A clean-cut mark at some point B, Fig. 348, is first made on the diagram. The tracing point is placed on this mark and the reading on the wheel and vernier taken. The tracing point is then moved **clockwise** over the outline of the diagram to the starting point and a second reading taken. The difference in these two readings is the area in square inches. *Care should be taken to start and stop at the same point.*

The length of the indicator diagram is obtained by drawing fine lines perpendicular to the atmospheric line from the extreme ends of the indicator diagram, and measuring the length between these lines on the atmospheric line. The area, as found above, divided by the length and multiplied by the scale of spring equals the *mean effective pressure.*

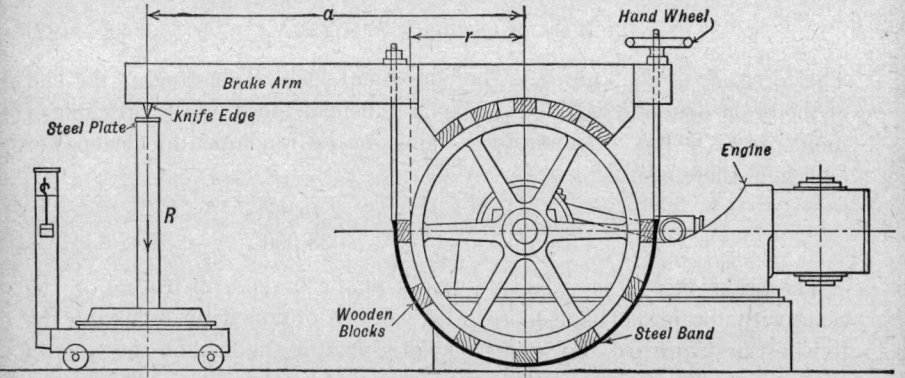

FIG. 350. — Prony Brake applied to Steam Engine.

When a planimeter is not available, the mean effective pressure may be determined by finding the average height of the diagram in inches as obtained by *averaging at least* 20 *ordinates measured between the lines of the forward and return strokes, and then multiplying this average by the scale of the spring.* If there are points on the expansion line below the back pressure line, the corresponding heights must be subtracted instead of added.

393. Brake Horsepower. — *The brake horsepower is the power delivered by the crank shaft of the engine, as determined by a brake mounted on the flywheel or brake drum. The power is absorbed by the brake as friction and appears as heat.*

A typical **Prony brake**, shown in Fig. 350, consists of an arm, one end of which bears on a standard that rests on a scale. The opposite end is shaped to fit on the flywheel or brake drum, and supports an adjustable band, which passes around the pulley. The amount of friction, and consequently the load on the engine, is controlled by tightening or loosening the band.

The power absorbed by a brake is obtained by using the fundamental equation

$$Horsepower = \frac{force \times velocity\ per\ minute}{33,000}$$

Let r = radius of the brake drum measured in feet.
N = revolutions of engine shaft per minute.
F = resistance at circumference of wheel of radius r, lb.
R = gross load on scale, lb.
w = tare weight on scale, lb.
W = net load on scale, lb. = $R - w$.
a = length of brake arm in ft. = horizontal distance between center of brake-drum shaft and the point of support at scale.

The work absorbed in friction equals the force times the distance. Expressed as an equation,

$$Work\ of\ friction = F \times 2\pi rN \quad \ldots \ldots \quad (92)$$

The force, F, is not known. The equivalent load, W, acting at the end of the brake arm of length a, may be used instead, since by the principle of moments $Fr = Wa$. Placing these values in the fundamental horsepower equation, there results

$$Brake\ horsepower,\ b.hp. = \frac{2\pi aWN}{33,000} \quad \ldots \ldots \quad (93)$$

The **brake tare** is the effective weight of the brake arm resting on the scale with the brake band loose. Its value is obtained by revolving the flywheel first forward and then backward, reading the load on the scale in each case. The speed in each direction should be the same. The weight on the scale when running forward is the weight of the arm acting on the scale plus friction; and when running backward, the weight of arm acting on the scale minus friction. Adding these weights eliminates the friction factor and gives twice the tare weight. For example; the weight when running forward is 38 pounds and backward 34 pounds; the tare weight then equals 38 plus 34 divided by 2, or 36 pounds. *The* **net scale load** *is found by subtracting the tare weight from the gross scale reading.*

Example 42. — An 8 in. by 12 in. steam engine is fitted with a Prony brake having an arm 54 in. long. During a test, the gross weight on the scale was 120 lb., the tare weight was 20 lb., and the revolutions per minute 221. Find the brake horsepower.

Solution. — Using Equation (93)

$$B.hp. = \frac{2\pi aWN}{33,000} = \frac{2 \times 3.14 \times 4.5 \times 100 \times 221}{33,000} = 18.9$$

$$a = \tfrac{54}{12} = 4.5\ \text{ft.};\ W = 120 - 20 = 100\ \text{lb.};\ N = 221$$

Friction Horsepower. — The friction horsepower represents the **power lost in friction** and is found by subtracting the brake horsepower from the indicated horsepower.

394. Mechancial Efficiency. — This efficiency is the ratio of the power output of the engine, as measured by the brake, to the power developed by the steam in the cylinder of the engine, as obtained from the indicator diagrams. Expressed as an equation,

$$Mechanical\ efficiency = \frac{b.hp.}{i.hp.} \quad \ldots \ldots \ldots (94)$$

Example 43. — Using the data from Example 42, in which the indicated horsepower was 21.7, find the mechanical efficiency of the engine.

$$Mechanical\ efficiency = \frac{b.hp.}{i.hp.} = \frac{18.9}{21.7} = 0.87\ or\ 87\ per\ cent.$$

395. Weight of Steam used, as Shown by the Indicator Diagram. — When the diagram is used to calculate the theoretical amount of steam used, the results obtained are only approximate, since water may be mixed with the steam at entrance, the steam condenses during admission, and

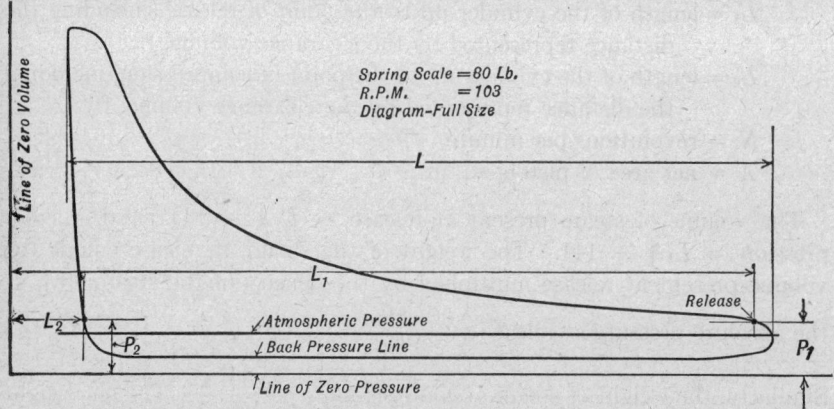

FIG. 351. — Calculation of Theoretical Steam Rate using Indicator Diagram.

steam leaks past the valves. This calculation was formerly included in the calculated results of an engine test, but it has been omitted in the recent *A. S. M. E. Code* because it is not essential to a performance test of an engine, and also because incorrect conclusions may be drawn from it, unless thoroughly understood. It has, however, sufficient theoretical interest to be discussed here.

Reference is made to the indicator diagram shown in Fig. 351. Lines representing **zero volume** and **zero pressure** are drawn upon the indicator card and are laid off to the proper scale. The zero-volume line is located by multiplying the length L of the diagram by the per cent clearance and laying it off as indicated, and the zero-pressure line is located by measuring to the proper scale, below the atmospheric line, a distance corresponding to the pressure of the atmosphere. At the point of cut-off or at release, there is a certain weight of steam represented by the volume of the cylinder up to that point, including the clearance volume. At the point of compression there is a weight of steam represented by the volume enclosed between the piston and cylinder head. This weight is known as **cushion steam,** and its amount is found by multiplying the volume of the cylinder at the point of compression by the density of the steam corresponding to the absolute pressure at that point. The difference between the weight of steam at cut-off or release and at compression is the **cylinder feed.** It represents the weight of steam supplied per stroke.

To compute the **cylinder feed** as found from the diagram, it is necessary to choose the points of compression and release. The point of release is used instead of the point of cut-off since at that point the amount of re-evaporation, on expansion, is included in the volume, and the net result is nearer to the actual steam used than it would be if the volume and weight were taken at cut-off.

Let L = length of stroke, ft.
L_1 = length of the cylinder up to the point of release, including the distance represented by the clearance volume, ft.
L_2 = length of the cylinder up to the point of compression, including the distance represented by the clearance volume, ft.
N = revolutions per minute.
A = net area of piston, sq. in.

The volume of steam present at release = $L_1 A \div 144$: and at compression = $L_2 A \div 144$. The weight of the steam at release equals the volume present at release multiplied by the density of the steam, W_1, at the absolute pressure at release = $\dfrac{L_1 A W_1}{144}$.

Similarly, the weight of steam at compression = $\dfrac{L_2 A W_2}{144}$. The net weight of steam supplied per revolution = $\dfrac{L_1 A W_1}{144} - \dfrac{L_2 A W_2}{144} = \dfrac{A}{144}(L_1 W_1 - L_2 W_2)$.

The weight of steam for N revolutions per minute = $\dfrac{AN}{144}(L_1 W_1 - L_2 W_2)$.

The *weight of steam used per indicated horsepower per hour* is, therefore,

$$\dfrac{\dfrac{60\,AN\,(L_1 W_1 - L_2 W_2)}{144}}{\dfrac{PLAN}{33{,}000}} = \dfrac{13{,}750}{P}\left(\dfrac{(L_1 W_1 - L_2 W_2)}{L}\right) \quad \ldots \quad (95)$$

The values corresponding to L, L_1, and L_2 may be taken from the indicator diagram, since lengths on the diagram are proportional to the length of stroke.

Example 44. — The following data were taken from a 12 in. by 24 in. Corliss engine running at 102.9 r.p.m.; clearance of head end, 7.89 per cent; clearance of crank end 7.40 per cent.

Head-end diagram data: $L_1 = 4.03$ in., $L_2 = 0.38$ in., $P_1 = 22.4$ lb. per sq. in. abs., $P_2 = 28$ lb. per sq. in. abs., $L = 3.87$ in., $W_1 = 0.055$ lb. per cu. ft., $W_2 = 0.068$ lb. per cu. ft., m.e.p. $= 42.8$ lb. per sq. in. Find the theoretical weight of steam per indicated horsepower per hour, from the diagram.

Solution. — Substituting in Equation (95)

Weight of steam per i.hp. per hr. $= \dfrac{13{,}750}{\text{m.e.p.}} \dfrac{(W_1 L_1 - W_2 L_2)}{L}$

$= \dfrac{13{,}750}{42.8} \left(\dfrac{0.055 \times 4.03 - 0.068 \times 0.38}{3.87} \right) = 16.20$ lb.

For the crank end, by a similar calculation, the weight of steam per i.hp. $= 13.30$ lb.

Weight of steam per i.hp. per hr. for whole engine $= \dfrac{16.20 + 13.30}{2} = 14.75$ lb.

396. Valve Setting with an Indicator. — In setting valves by means of an indicator, the diagrams taken from both ends of the cylinder are made to show a proper distribution of steam and to be as nearly alike as possible, except for a slightly later cut-off on the crank end to allow for the reduction of piston area by the piston rod. For a slide-valve engine, this is done by adjusting the length of valve stem and changing the position of the eccentric until the diagrams appear satisfactory.

It should be remembered that the information furnished by the indicator diagram, regarding the setting of the valves, is purely a matter of inference.

It is often difficult to determine the point at which an event on the diagram occurs, because the curves run into each other gradually, without clearly defining the point of separation of the curves. The point can be easily located by producing both curves along their regular trend until the curves meet; the point of tangency is the desired point.

397. Theoretical Indicator Diagram. — The theoretical indicator diagram, shown in Fig. 352, by dotted lines, for non-condensing operation, differs from the actual diagram, which is shown by

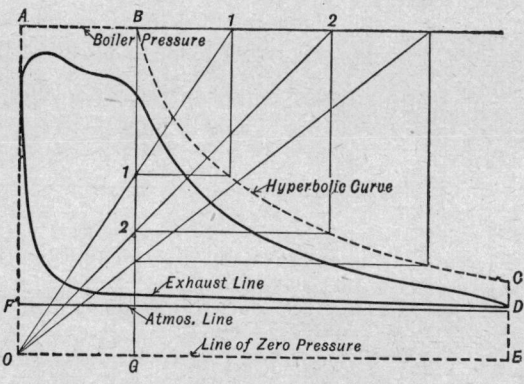

FIG. 352. — Theoretical Indicator Diagram.

solid heavy lines. An engine capable of giving the theoretical diagram would operate without clearance or compression; the valves would open and close instantaneously; the steam would enter the cylinder without drop in pressure during admission, and at the pressure in the boiler; the exhaust pressure would be that of the atmosphere for a non-condensing engine, and of the condenser for a condensing engine; and the expansion would be according to the equation $PV = $ a constant. This equation

432 STEAM ENGINE INDICATOR, ENGINE EFFICIENCIES AND LOSSES

is used because it is easier to construct than the adiabatic curve, to which the actual expansion curve, doubtless, more nearly coincides.

The expansion curve, or **rectangular hyperbola**, may be constructed by drawing *BG* perpendicular to the atmospheric line through the point of cut-off, and also a horizontal line through the same point. From *O*, the point of intersection of the line of zero pressure and zero volume, diagonal lines are drawn cutting each of these lines, as at 1-1 and 2-2, and from these points of intersection are drawn horizontal and vertical lines. The point of intersection, *K*, locates one point on the curve. By drawing a number of diagonal lines, a sufficient number of points may be located to draw the curve.

398. Theoretical Mean Effective Pressure. — In the discussion of the actual steam indicator diagram, it was shown that the mean ordinate equaled the area divided by the length of the diagram. Applying the same reasoning to Fig. 352, there results

$$\text{Mean ordinate} = \frac{\text{Area } ABCEO - FDEO}{\text{Length } OE} \quad \ldots \ldots \quad (96)$$

$$\text{Area } ABCEO = \text{Area } ABGO + \text{area } BCEG$$

$$= OA \times OG + P_1 V_1 \log_e \frac{V_2}{V_1}.$$

Let $OA = P_1$ = initial absolute pressure, lb. per sq. in.
$OG = V_1$ = volume of cylinder at cut-off, cu. ft.
$CE = P_2$ = absolute pressure at release, lb. per sq. in.
$OE = V_2$ = volume of cylinder at release, cu. ft.
$DE = P_3$ = absolute exhaust pressure, lb. per sq. in.

Substituting these symbols in Equation (96), there results

$$\text{Theoretical m.e.p.} = \frac{P_1 V_1 + P_1 V_1 \log_e \frac{V_2}{V_1} - P_3 V_2}{V_2}$$

Dividing this equation by V_1 and representing $\frac{V_2}{V_1}$ by r, the expression for the mean effective pressure becomes

$$\text{Theoretical m.e.p.} = \frac{P_1 (1 + \log_e r)}{r} - P_3 \quad \ldots \ldots \quad (97)$$

The ratio $\frac{V_2}{V_1} = r$ is known as the **ratio of expansion** and is the ratio of the volume of steam at the end of the stroke to the volume at cut-off. Since the area of the piston is constant, $\frac{V_1}{V_2} = \frac{L_1}{L_2}$, and the reciprocal of r is the fraction of the stroke completed at cut-off. With a cut-off occurring at one-quarter of the stroke, the number of expansions r would be $(1 \div \frac{1}{4})$, or 4.

A comparison of an actual and a theoretical diagram for the same conditions show that the actual diagram is smaller than the theoretical, because of imperfections in the working of the actual engine. During admission there is a loss in pressure caused by **throttling** of the steam when passing through the restricted openings and also by **initial condensation**. The valves do not open and close instantaneously, and a loss results, as shown by the rounded corners at cut-off, release and compression. Release occurs before the piston has reached the end of the stroke, and compression starts before the piston has completed the exhaust stroke.

Since the actual diagram is smaller than the theoretical, the **probable m.e.p.** may be obtained from the **theoretical m.e.p.** by multiplying it by a **diagram factor,** which is equal to the actual m.e.p. divided by the theoretical m.e.p. It varies for different engines, as shown in Table 27.

TABLE 27.— DIAGRAM FACTORS FOR SIMPLE ENGINES*

Type of Engine	Diagram Factor Per Cent
Simple, slide valve	55 to 90
Simple, Corliss	85 to 90
Compound, slide valve	55 to 80
Compound, Corliss	75 to 85
Triple expansion	55 to 70

* From "Heat Power Engineering," Hirshfeld and Barnard.

Example 45. — An 8 in. by 24 in. Corliss engine runs at 110 r.p.m. Initial steam pressure 100 lb. per sq. in. gage; back pressure 14.7 lb. per sq. in. abs.; cut-off at one-quarter stroke. Find (a) the theoretical m.e.p., (b) the probable m.e.p. if the diagram factor is 85 per cent, and (c) the probable indicated horsepower.

Solution. — Using Equation (97),

(a) Theoretical m.e.p. $= \dfrac{P_1 (1 + \log_e r)}{r} - P_3$

$= \dfrac{114.7 (1 + 1.38)}{4} - 14.7 = 53.6$ lb. per sq. in.

$P_1 = 100 + 14.7 = 114.7$ lb. per sq. in. abs.; $r = 4$
$\log_e r = 2.3 \log_{10} r = 2.3 \log_{10} 4 = 2.3 \times 0.6020 = 1.38$
$P_3 = 14.7$ lb. per sq. in. abs.

(b) Probable m.e.p. $= 53.6 \times$ diagram factor $= 53.6 \times 0.85 = 45.4$ lb. per sq. in.

(c) Probable horsepower $= \dfrac{PLAN}{33,000} \times 2 = \dfrac{45.4 \times 2 \times 50.27 \times 110 \times 2}{33,000} = 30.4$

$L = 24$ in. $= 2$ ft.; $N = 110$; $A = \frac{1}{4} \pi d^2 = \frac{1}{4} \times 3.14 \times 64 = 50.27$ sq. in.

399. Rating of Engines. — It is customary for manufacturers to rate steam engines on a basis of the indicated horsepower, when the cut-off occurs at one-quarter stroke and with a specified steam pressure at the throttle valve.

400. Performance of Steam Engines. — The performance of steam engines is customarily stated in the following ways:

1. *Steam rate*, pounds per hour or per i.hp.-hour.
2. *Heat rate*, B.t.u. per i.hp.-hr. or per i.hp.-minute.
3. *Thermal efficiency*, per cent.
4. *Rankine-cycle ratio*, per cent.
5. *Mechanical efficiency*, per cent.
6. *Duty*, for pumping engines (consult Art. 502, page 553).

401. Steam Rate.[1] — This is commonly stated as the weight of steam per unit of power under actual conditions, uncorrected for moisture or superheat. The steam rate is of small value when comparing engines, because

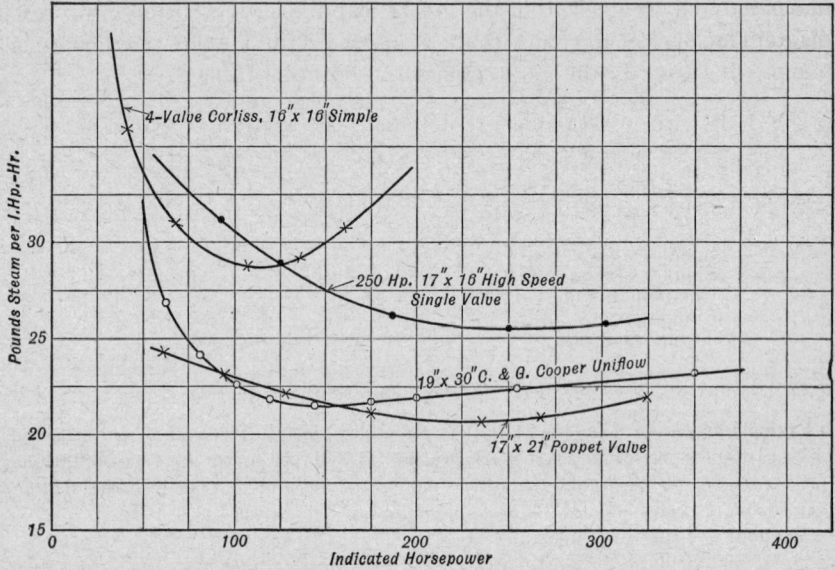

FIG. 353. — Typical Economy Curves — Non-Condensing Operation, Simple Engines.

of variations in the initial steam pressure, quality of steam, and back pressure. For small variations of these quantities, it is customary to correct the steam rate to conditions taken as standard. This should not be done when the variation in these quantities is over a few per cent. The variation in the steam rate of a few typical engines, with the power output, is shown in Fig. 353. When the total weight of steam used per hour, by an engine having a throttle governor, is plotted against the indicated horsepower, it will show a straight line, known as **Willans line**.

402. Heat Rate. — This affords the best method of comparing the performance of engines. The *heat rate of an engine per hour is found by multiplying the weight of steam used in pounds per hour by the difference between the total heat in one pound of steam at the average absolute pressure and quality found in the steam pipe near the throttle valve, and the heat in one*

---

[1] Also called **water rate**.

ENGINE EFFICIENCY, OR RANKINE–CYCLE RATIO

pound of water at the temperature of saturated steam at the average absolute pressure existing in the exhaust pipe near the cylinder.

Example 46. — The weight of steam used by an 8 in. by 18 in. Murray Corliss engine, during a test, was 780 lb. per hr.; steam pressure, 120 lb. per sq. in. gage; exhaust pressure, 0.10 lb. per sq. in. gage; barometer, 14.5 lb. per sq. in.; quality, 0.977. Find (a) the heat rate per hour, (b) the heat rate or consumption per i.hp. when the i.hp. equaled 26.4;

Solution. — (a) Heat rate = weight of steam per hour $\times$ ($x_1 L_1 + h_{f_1} - h_{f_2}$) in which x_1 = quality of steam = 0.977.

L_1 = latent heat of steam at initial abs. pressure = 870.25 B.t.u.
h_{f_1} = heat of liquid at initial abs. pressure = 321.47 B.t.u.
h_{f_2} = heat of liquid at abs. exhaust pressure = 180 B.t.u.

Absolute pressure at throttle = (120 + 14.5) = 134.5 lb. per sq. in.
Absolute pressure at exhaust = (0.10 + 14.5) = 14.6 lb. per sq. in.
Heat rate = 780 (0.977 $\times$ 870.25 + 321.47 − 180) = 773,526 B.t.u.

(b) Heat rate per i.hp. per hr. $= \dfrac{773{,}526}{26.4} = 29{,}292$ B.t.u.

403. Thermal Efficiency.[1] — *The thermal efficiency of an engine without reheat or extraction is the ratio of the heat converted into useful work to the heat supplied, measured above the temperature corresponding to the pressure of the exhaust steam.* It is commonly based on the steam used per i.hp.-hour.

The heat converted into useful work equals the heat equivalent of an i.hp. per hour or 2545 B.t.u. per hour, and since the heat supplied equals $w(h_1 - h_{f_2})$,

$$\text{The thermal efficiency} = \frac{2545}{w(h_1 - h_{f_2})} \quad \ldots \ldots \quad (98)$$

in which w = pounds of steam as supplied, per i.hp.-hour.
$\quad h_1$ = total heat above 32 deg. fahr. per pound of steam, at initial conditions prevailing before the throttle valve.
$\quad h_{f_2}$ = heat of liquid above 32 deg. fahr. in one pound of water, at the temperature of saturated steam at exhaust pressure.

Example 47. — Using the data given under Example 46 above, find the thermal efficiency.

Solution. — By substituting proper values in Equation (98)

$$\text{Thermal efficiency} = \frac{2545}{29.5\,(1171.7 - 180)}$$
$$= \frac{2545}{29{,}292} = 0.087, \text{ or } 8.7 \text{ per cent}$$

404. Engine Efficiency, or Rankine-cycle Ratio. — This efficiency is sometimes called **potential efficiency** or **efficiency ratio**. *It shows the extent to which the performance of the actual engine, expressed in heat units, approaches the heat available, for an ideal engine working on a* **Rankine cycle,** Fig. 354, which is the accepted standard for comparing engine cycles. This

---

[1] See A. S. M. E. Test Code on Definitions and Values.

cycle consists of (*a*) admission at constant pressure and temperature, (*b*) adiabatic expansion after cut-off to the back pressure, (*c*) exhaust at constant pressure and temperature, (*d*) return to the boiler of an equivalent amount of feedwater, at the pressure and temperature of the exhaust steam.

To obtain an indicator diagram or cycle of this kind, the walls of the cylinder and the piston would have to be non-conductors of heat; the action of the valves would be instantaneous; the steam passages would be of sufficient area to prevent wire-drawing; the engine would operate without clearance; there would be no heat leakage and no friction so that all stages of the cycle are ideally perfect; consequently all of the energy taken from the steam would be converted into useful work.

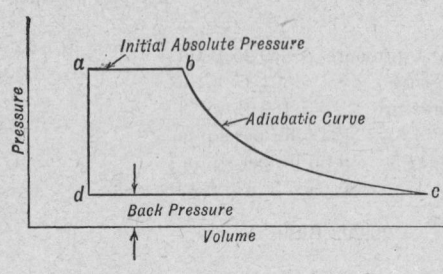

FIG. 354. — Diagram showing the Rankine Cycle.

The efficiency of the simple *Rankine cycle is the ratio of the heat available for conversion into work to the heat supplied*, and may be expressed as an equation as follows:

$$Rankine\text{-}cycle\ efficiency = \frac{Heat\ available}{Heat\ supplied} = \frac{h_1 - h_2}{h_1 - h_{f_2}} \quad . \quad . \quad . \quad (99)$$

in which h_1 = total heat above 32 deg. fahr. in one pound of steam at the absolute throttle pressure, B.t.u.

h_{f_2} = heat of liquid at the temperature corresponding to the exhaust pressure, B.t.u.

h_2 = total heat above 32 deg. fahr. in one pound of steam after adiabatic expansion to the absolute exhaust pressure, B.t.u.

The value of h_2 equals $x_2 L_2 + h_{f_2}$, and it may be found from the Steam Table, provided the quality at exhaust (x_2) is known. This quality may be found from the Mollier diagram, Fig. 81, page 112, or from the relation $\frac{x_1 L_1}{T_1} + s_{f_1} = \frac{x_2 L_2}{T_2} + s_{f_2}$, since with an adiabatic expansion there is no heat change and the entropy at cut-off equals the entropy at release. In this equation, x = quality, L = latent heat, s_f = entropy of liquid and T = absolute temperature. All these values, with the exception of x_2, may be found from the steam tables, as explained in Chapter IV, pages 97 to 101. Values of the efficiency of the Rankine cycle are given in Table 28 for different conditions of steam at inlet and exhaust. These values represent the maximum attainable thermal efficiency. For further study along this line, consult " Steam Power Plant Engineering " by GEBHARDT.

HEAT LOSSES IN THE STEAM ENGINE

$$\text{The engine efficiency referred to i.hp.} = \frac{\text{Thermal efficiency of the actual engine}}{\text{Thermal efficiency of the ideal engine}}$$

$$= \frac{2545}{w\,(h_1 - h_{f_2})} \div \frac{h_1 - h_2}{h_1 - h_{f_2}}$$

$$= \frac{2545}{w\,(h_1 - h_2)} \quad \ldots \ldots \ldots \ldots \quad (100)$$

in which the symbols are as defined in Arts. 403 and 404.

TABLE 28. — RANKINE–CYCLE EFFICIENCIES AND THEORETICAL WATER RATES

Initial Pressure Lb. per Sq. In. Abs.	Dry and Saturated Steam				Superheated Steam		
	*Condensing		† Non-condensing		Superheat ° F.	Condensing	Non-Condensing
	Rankine Efficiency	Water-rate	Rankine Efficiency	Water-rate		Rankine Efficiency	Rankine Efficiency
50	29.48	8.98	8.98	28.51	.....	.....	.....
100	28.47	7.85	13.88	18.22	272.2	29.8	15.4
150	30.60	7.26	16.65	15.08	.....	.....	.....
200	31.88	6.94	18.60	13.44	218.1	32.9	19.7
250	32.93	6.70	20.05	12.42	.....	.....	.....
300	33.76	6.52	21.22	11.71	182.5	34.5	22.0
400	35.10	6.25	23.07	10.73	155.2	36.1	23.7
500	36.06	6.07	24.46	10.10	132.7	36.7	25.0
600	36.84	5.94	25.57	9.66	113.4	37.3	26.0

* Back pressure ½ pound per sq. in., abs. † Back pressure 14.7 lb. per sq. in., abs.

Example 48. — Data as given in Example 46, page 435. Find the Rankine-cycle ratio.

Solution. — Using Equation (100) and data as given,

$$\text{Rankine-cycle ratio} = \frac{2545}{w\,(h_1 - h_2)} = \frac{2545}{29.5\,(1171.7 - 1005.1)} = 0.512$$

$w = 29.5$ lb. per i.hp.-hr.
$h_1 = xL_1 + h_{f_1} = 0.977 \times 870.25 + 321.47 = 1171.7$ B.t.u.
$h_2 = x_2 L_2 + h_{f_2} = 0.85 \times 970.7 + 180 = 1005.1$ B.t.u.

$$x_2 = \frac{\frac{x_1 L_1}{T_1} + \theta_1 - \theta_2}{L_2 \div T_2} = \frac{0.977 \times 1.075 + 0.503 - 0.3134}{1.447} = 0.85$$

Values of the Rankine-cycle ratio are given in Table 29 for a few typical steam engines, in order to give an idea of the value of this ratio for modern types of steam engines.

The data necessary to compute the various steam-engine efficiencies are obtained by making tests as explained in Chapter XXI.

405. Heat Losses in the Steam Engine. — It has been shown under engine efficiencies, that the actual steam engine has a lower efficiency than the ideal. This is caused by losses the most important of which are:

1. Cylinder condensation.
2. Leakage of steam.
3. Clearance volume.
4. Incomplete expansion.
5. Wire-drawing.
6. Friction of the engine parts.
7. Moisture in the entering steam.
8. Heat discharged in exhaust steam.
9. Radiation and other small losses.

Some of these losses are preventable, while others are inherent in the machine and cannot be prevented.

* TABLE 29. — VALUES OF RANKINE–CYCLE RATIO FOR TYPICAL STEAM ENGINES

Type of Engine	Ind. hp.	Size	Operating Conditions			Rankine Cycle Ratio	Water Rate
			Initial Pressure Lb. per Sq. In. Gage	Back Pressure Lb. per Sq. In. Abs.	Superheat ° F.		
Locomotive single valve...	975	22 × 28	196.0	atmos.	....	55.6	23.4
Ames...........	248	17 × 16	100.0	atmos.	....	65.9	26.0
Buffalo Forge..	86	12 × 12	80.0	3.0	....	40.4	27.5
Corliss.........	237	21.6 × 43.3	103.5	atmos.	....	79.5	11.7
Nordberg Poppet Valve.....	123	15 × 18	130.0	atmos.	....	80.0	13.4
Erie City Lentz	248	19 × 21 simple	133.0	atmos.	....	87.5	16.1
Worthington Pump Engine	648	Triple Exp.	146.8	0.78	87	72.5	10.0
McIntosh & Seymour......	2202	29 × 60 × 56	158.0	2.30	92	79.8	11.2

* From Tables 81 and 83 " Steam Power Plant Engineering," Gebhardt.

Cylinder Condensation. — This loss is the result of hot steam coming into contact with cooler surfaces, which absorb heat and produce condensation. The difference in the weight of steam as calculated from the indicator diagram and that actually used is caused mainly by cylinder condensation. This loss is the largest loss in a steam engine and varies from 16 to 20 per cent of the total loss. It is increased by using large clearances, and high ratios of expansion, because the volume of steam entering is then small in proportion to the volume of the cylinder and consequently a greater proportion of it is condensed.

Cylinder condensation may be reduced by using compound engines, unaflow and poppet-valve engines, by decreasing the clearance volume, increasing speed, decreasing ratio of expansion by making cut-off occur later, and by using superheated steam.

Leakage and Clearance. — The amount of leakage past the piston and valves is difficult to determine, but tests have shown that it may amount to from 4 to 20 per cent. The greater the wear of the parts upon which the steam acts, the greater is the leakage. It is greater with saturated steam than with superheated steam. This loss is generally included as a part of the condensation loss.

The loss caused by clearance is mainly that resulting from increased condensation. The greater the proportion of the clearance volume to the total cylinder volume, the greater is the loss. The clearance volume of engines varies from 1 per cent for large engines having short steam passages to around 12 per cent for small high-speed engines. Clearance affects the economy because of its effect on the ratio of expansion. Shortening the cut-off increases the ratio of the clearance space to the volume of the steam supplied, and hence increases the loss.

Loss Caused by Incomplete Expansion and Compression. — This loss is unpreventable, because the loss which results from having release occur before the expansion is complete is more than offset by reduced cylinder condensation since the cylinder walls are maintained at a higher temperature.

Compression is desirable to produce a smooth-running engine and, for a certain ratio of compression volume to clearance volume, means a saving in the amount of steam used.

Loss from Wire-drawing. — This loss is principally a loss in power resulting from the reduction in the pressure acting on the piston. The reduction in pressure is caused by restricted openings for the passage of steam. It is greatest in engines using single valves to control the distribution of steam. The use of multi-valve engines, such as Corliss and poppet valve, prevents this loss.

Loss from Friction. — The friction loss represents a loss in power, and varies from 4 to 20 per cent. It is larger when running under load than when under no load, because of the greater pressure on guides and bearings.

Loss from Moisture. — Unless a separator is used, the steam which condenses in the steam pipe passes through the engine as inert matter and reduces the work performed per pound of water and steam discharged. As this does not represent an actual loss, the performance of the engine is generally based on dry steam.

Loss from Heat Discharged in Exhaust Steam. — This loss is large and represents from 75 to 95 per cent of the heat supplied. Its amount depends upon the pressure against which the engine exhausts, an increase in the back pressure increasing the loss. In computing the quantity of heat in exhaust steam, the quality of the steam at exhaust must be known. This is a variable quantity, but tests have shown that its average value is approximately 90 per cent. Its value can be closely estimated by the following method:

$$\text{Heat in the exhaust steam, } h_2 = h_1 - h_r - \frac{2545}{w}$$

in which h_1 = total heat per pound of entering steam, B.t.u.

h_r = heat loss per pound of steam caused by radiation, B.t.u., may be assumed as 1 per cent of h_1 for most practical cases.

w = steam rate per indicated horsepower.

$h_2 = x_2 L_2 + h_{f_2}$

Substituting this value of h_2 in the above equation, there results

$$x_2 = \frac{h_1 - h_r - h_{f_2} - \frac{2545}{w}}{L_2} \quad \ldots \ldots \ldots (101)$$

Example 49. — The high-pressure cylinder of a triple-expansion engine develops 45.3 hp.; steam, dry and saturated; throttle pressure, 135.5 lb. per sq. in. abs.; exhaust pressure, 42.0 lb. per sq. in. abs.; weight of steam used per hr., 1626 lb. Find the quality of the exhaust steam.

Solution. — Using Equation (101),

$$x_2 = \frac{h_1 - h_r - h_{f_2} - \frac{2545}{w}}{L_2} = \frac{1191.8 - 11.91 - 238.95 - \frac{2545}{35.9}}{931.2} = 0.935$$

h_1 at 135.5 lb. = 1191.8 B.t.u.; $h_r = h_1 \times .01 = 11.91$ B.t.u.; h_{f_2} at 42 lb. = 238.95 B.t.u.; L_2 at 42 lb. = 931.2 B.t.u.; $w = \frac{1626}{45.3} = 35.9$ lb. per i.hp. per hr.

406. Methods of improving Economy in the Steam Engine. — The following methods are used to increase the economy of steam engines:

1. Increasing boiler pressure and rotative speed.
2. Decreasing back pressure (Chapter XXIV).
3. Superheating the steam.
4. Steam-jacketing the cylinders (in some cases).
5. Compounding and using reheating receivers between cylinders (Chapter XX).
6. Using unaflow engines.

By **increasing the boiler pressure,** with other conditions remaining the same, a higher theoretical efficiency is obtained. The limit of the increase, at the present time, appears to be about 600 pounds per square inch. Leakage of steam and cylinder construction are the limiting factors. Present practice ranges from 90 pounds per square inch for low-pressure engines to 175 for unaflow condensing engines, 210 for locomotives, and 250 for marine engines.

An **increase in the rotative speed** of an engine does not always result in increased economy, much depending upon the type of the engine. High-speed engines, as a class, have the advantage of being more compact for a given power, are simple in construction and are relatively low in first cost. They are, however, subject to quite rapid depreciation and, unless specially balanced, to excessive vibration, and are less economical in steam consumption.

The **saving effected by using superheated steam** is mainly a saving resulting from decreased initial condensation. The amount of steam used is reduced approximately 1 per cent for each 10 deg. fahr. increase in superheat, the limit of the increase in superheat being about 670 deg. fahr. Superheated steam requires special valves, such as piston and poppet valves; if other valves are used, trouble will be experienced from warping and sticking of the valves.

The **use of steam jackets** generally reduces cylinder condensation. Their use is questionable, however, except for low-speed engines, working under steady loads, such as pumping engines.

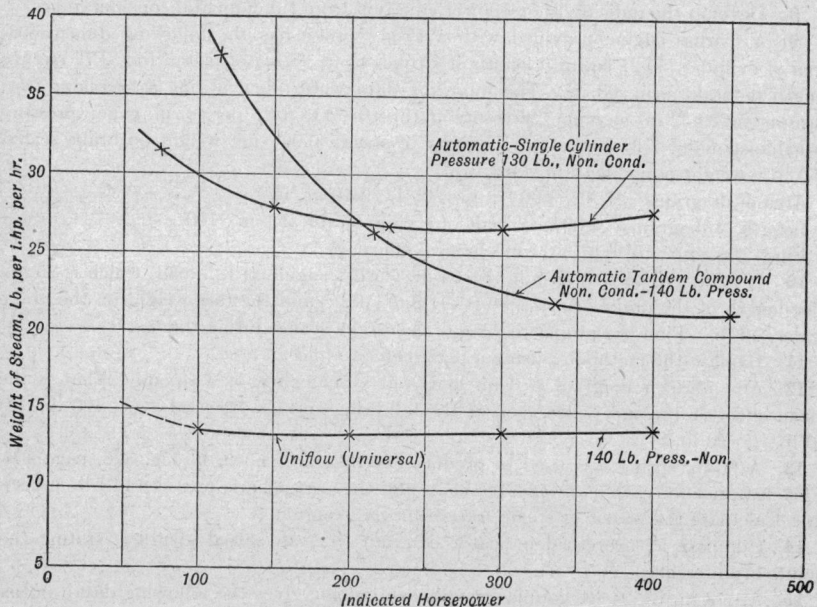

FIG. 355. — Steam Rate Curve of Unaflow Engine compared with Two other Types of Engines.

The **use of a unaflow engine,** in which the admission valves and cylinder walls are not cooled by the exhaust steam, results in increased economy for all loads, because of reduced cylinder condensation. The economy of the unaflow engine is equal to that of a compound engine, and, because of its flat economy curve, Fig. 355, it can be used over a wide range of loads. Economy curves for a simple and a compound engine are also plotted in Fig. 355, to give an idea of the difference in economy at light and heavy loads.

REFERENCES

Steam Power Plant Engineering, GEBHARDT.
Power Plant Testing, MOYER.
Thermodynamics, MOYER, CALDERWOOD and POTTER.
Elements of Heat Power Engineering, HIRSHFELD and BARNARD.
Thermodynamics, EMSWILER.

REVIEW QUESTIONS AND PROBLEMS

1. Describe the construction of a steam-engine indicator.
2. Give three uses of the steam-engine indicator.
3. State the proper size of indicator springs for steam pressures of 100, 150, and 200 lb. per sq. in.
4. Explain the method of placing a spring in the Crosby inside spring indicator.
5. What is the purpose of a reducing motion? Name two types.
6. A reducing motion like that shown in Fig. 346, page 422, is attached to a 6 in. by 6 in. engine. Find the length of bar fn to give a diagram $2\frac{1}{2}$ in. long, if af equals 18 in.
7. Describe the proper method of taking an indicator diagram.
8. Develop the indicated-horsepower equation from fundamental considerations.
9. A Corliss engine provided with a Prony brake has the following dimensions: area of cylinder, 50.27 sq. in.; length of stroke, 18 in.; area of piston rod, 2.77 sq. in.; length of brake arm, 63 in. The following data were taken during a two-hour test: barometer, 29.52 in. mercury; pressure at throttle, 118.5 lb. per sq. in. gage; pressure in exhaust main, 1 lb. per sq. in.; quality of steam, 0.99; net weight on brake scales, 250 lb.; weight of condensate, 1463 lb.

Area of diagrams: H. E., 3.00 sq. in.; C. E., 3.10 sq. in.
Length of diagrams: 4.00 in.; scale of spring, 80 lb.; r.p.m., 100.
Find: (a) i.hp., (b) b.hp., (c) mechanical efficiency.

10. It is desired to test an 8 in. by 24 in. Corliss engine at full load, which is 25 hp. The length of the brake arm is 7.86 ft.; r.p.m., 110; and the tare weight on the brake scales, 40 lb. Find the proper reading of the brake scales during the test.
11. Explain the method of using a planimeter to find an area.
12. An indicator diagram is 3 in. long and has an area of 2 sq. in. What is the mean effective pressure when each of the following springs are used: (a) 160 lb., (b) 60 lb., (c) 20 lb.?
13. A 60-lb. spring was used to produce the diagram shown in Fig. 326, page 404. The clearance for each end was 6 per cent, and the atmospheric pressure 14.7 lb. per sq. in. Calculate the water, or steam, rate from the diagram.
14. Compare a theoretical indicator diagram with an actual diagram, stating the points of difference.
15. An 18 in. by 24 in. double-acting steam engine gives the following data: initial steam pressure, 110 lb. per sq. in. gage; quality of steam, dry; exhaust pressure, 2 lb. per sq. in. gage; barometer, 28.5 in. mercury; cut-off, one-quarter stroke; r.p.m., 120; diagram factor, 0.80.

Find: (a) theoretical m.e.p., (b) probable m.e.p., (c) probable i.hp.

16. Using the data of Problem 9, find: (a) the thermal efficiency, (b) Rankine-cycle ratio.
17. An 8. in. by 12 in. engine uses 500 lb. of steam per hr. Steam pressure, 110.3 lb. per sq. in. gage; quality 0.99; indicated horsepower, 21.5. Engine operating non-condensing with pressure of atmosphere, 14.7 lb. per sq. in.

Find: (a) thermal efficiency, (b) Rankine-cycle ratio.

18. Name the sources of heat loss in a steam engine. Which of these losses is the greatest?
19. Name three methods of increasing steam-engine economy.

CHAPTER XIX

COMPOUND AND MULTIPLE-EXPANSION ENGINES

407. Foreword. — Compound and multiple-expansion engines use steam at high pressure. After expanding and doing work in the first, or high-pressure, cylinder, the steam passes to one or more larger cylinders, where it does more work by expanding to a low pressure.

The principal object of compounding is to reduce the amount of steam used per horsepower per hour, by reducing the loss resulting from condensation of the entering steam. If the expansion of the steam takes place in two or more cylinders, the difference in pressure and, therefore, the difference in temperature, in each cylinder is less than it would be if the whole expansion occurred in one cylinder. Consequently, the initial condensation of steam on the walls of the cylinder is reduced.

The loss of heat by radiation, from compound engines, is somewhat larger than for a simple engine, on account of the larger surface of the cylinders; and, in addition, the friction losses may be greater because of the larger number of moving parts.

If the gain by decreased condensation so far exceeds the loss by increased radiation and friction, as to allow for increased cost, the use of the compound engine is justified. This generally occurs when the initial steam pressure is over 100 pounds per square inch. Compound engines usually have the additional advantage of more uniform turning effort and smaller sizes of the separate parts.

The **nominal size** of a compound engine is stated in terms of diameters and the stroke, beginning with the smaller cylinder. Thus the size of an engine having high- and low-pressure cylinders 15 and 24 inches in diameter respectively, and a stroke 48 inches long is written, $15 \times 24 \times 48$ inches.

408. Classification. — Compound engines are classified in the following ways:

1. By position of cylinders — Center lines — vertical; horizontal or inclined; horizontal and vertical

2. By arrangement of cylinders —
   - Axis in same center line; tandem compound
   - Center lines parallel — duplex compound; cross compound
   - Center lines at right angles; angle compound

3. By number of cylinders
- Two cylinders
  - duplex
  - tandem
  - cross compound
  - angle compound
- Three cylinders
  - triple, expansion having high, intermediate, and low pressure cylinders
- Four cylinders
  - quadruple expansion, having high, first and second intermediate, and low-pressure cylinders

If two cylinders of a four-cylinder engine are low-pressure cylinders, the engine is called a four-cylinder, triple-expansion engine. Three-cylinder compound engines, like the inclined engines used on large side-wheel steamers, have one high-pressure and two low-pressure cylinders, one on each side of the high-pressure cylinder. A vertical compound engine is sometimes called " a fore-and-aft " compound, from its use on shipboard. An angle-compound engine has a horizontal high-pressure cylinder and a vertical low-pressure cylinder, the connecting rods bearing on the same crank pin.

409. Receivers. — If the cranks are at 90 degrees with each other, as in a cross-compound engine, a considerable part of the exhaust stroke of the high-pressure cylinder is still to be completed when cut-off comes in the low-pressure cylinder. For this reason, a receiver or storage vessel is provided to receive steam from the high-pressure cylinder and to furnish it to the low-pressure cylinder, as required, when the high-pressure cylinder is not exhausting. The volume of the receiver is ordinarily about twice the volume of the low-pressure cylinder, and the pressure is such that the work performed in each cylinder is nearly equal.

If the engine has a single crank, as in the tandem arrangement, or if the cranks are at 180 degrees, steam passing from the high-pressure cylinder enters the low-pressure cylinder up to cut-off. This cut-off is made late, and steam then remaining to be exhausted from the high-pressure cylinder is compressed, up to the end of the stroke, in the pipe connections between the cylinders. Such engines are called non-receiver compounds, or " Woolf " engines.

Receivers may be used between the cylinders on triple- and quadruple-expansion engines. They often contain reheating coils into which live steam is passed for the purpose of adding heat to the working steam.

410. Tandem-compound Engines. — The tandem-compound engine, Fig. 356, has two cylinders, high and low pressure, arranged on the same center line. The piston rod of the high-pressure cylinder is attached to the low-pressure piston. The power developed in the high-pressure cylinder, plus that developed in the low-pressure cylinder, is transmitted

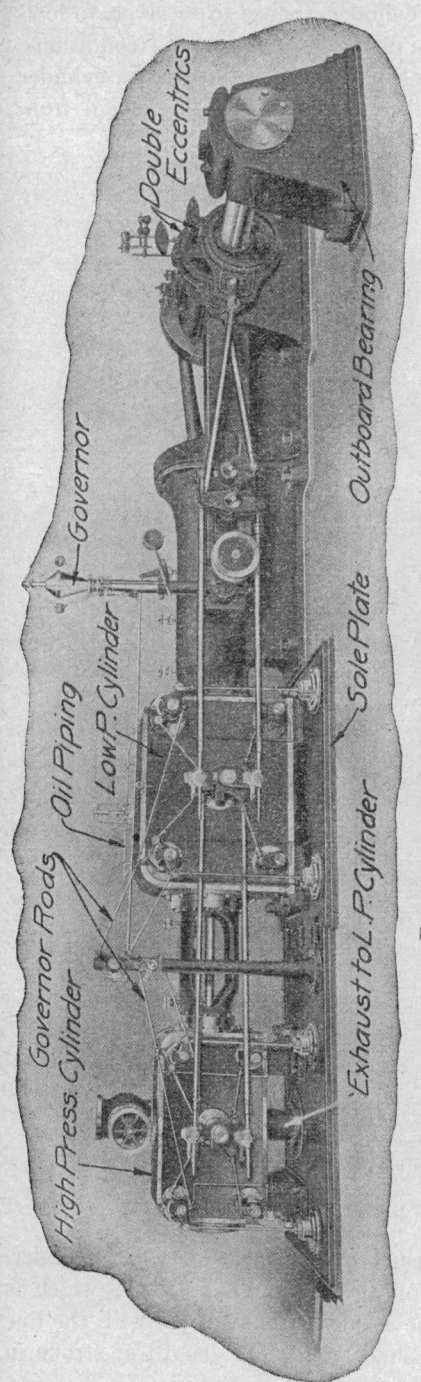

Fig. 356.—Tandem Compound Corliss Engine.

by the low-pressure piston rod, crosshead, connecting rod, and crank to the shaft and flywheel.

Steam is admitted first to the high-pressure cylinder, where it performs work. During the admission and expansion period for the head end, steam is exhausted from the crank end through a short pipe connection to the proper side of the low-pressure piston to move it in the same direction that the high-pressure piston is moving. Thus, the back pressure in the high-pressure cylinder is the forward pressure in the opposite end of the low-pressure cylinder, except for a small loss in pressure which occurs in passing through the ports and pipe connections.

This type of engine occupies less floor space than other compound engines of equal capacity, but the cylinder nearest the shaft is less readily accessible for repairs.

411. Cross-compound Engines. — The cylinders of a cross-compound engine, Fig. 357, are placed side by side, the space between them allowing room for the valve gear, flywheel, and generator, in case one is used. Each cylinder has its piston, piston rod, crosshead, crank, and valve gear, which are proportioned in accordance with the work to be done in the cylinder. The cranks are of the overhung type and generally are set at an angle of 90 degrees to each other.

This engine, with cranks at right angles, is usually provided with a receiver. The path of the steam is from steam supply to high-pressure cylinder, thence to receiver, to low-pressure cylinder, and to exhaust pipe under normal running conditions. When the engine is started, if the crank pin of the high-pressure cylinder is at dead center, a by-pass in the pipe connections allows steam from

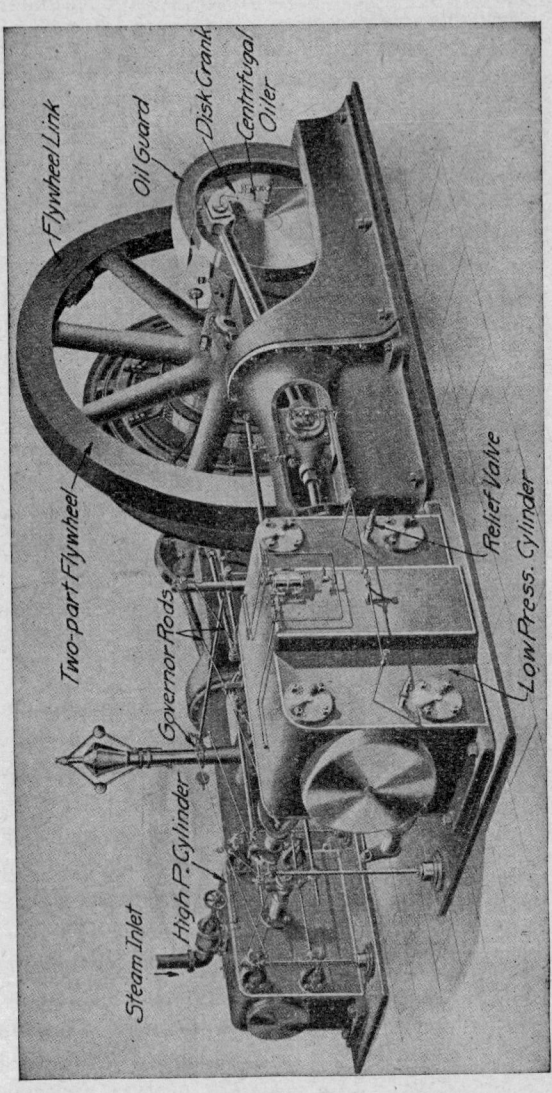

Fig. 357. — Heavy Duty Cross-compound Corliss Engine — Tangye Type.

the supply line to be admitted directly into the low-pressure cylinder. Thus the steam acts to advantage on the low-pressure piston, which is near the middle of its stroke while the crank is at 90 degrees with the line of centers. As in the tandem-compound engine, the length of stroke in both cylinders is made the same.

CYLINDER RATIO AND EXPANSION RATIO 447

The space required for the cross-compound engine is larger than that occupied by the tandem engine of the same capacity. The parts which transmit the power may be made lighter for the same total output, since the power is exerted through two separate sets of members. The turning force on the crank shaft is steadier, and, for the same uniformity in rotational effort, the flywheel may consequently be lighter than for a tandem engine.

Indicator diagrams from a 28 × 60 × 60 inch cross-compound engine are shown in Fig. 358.

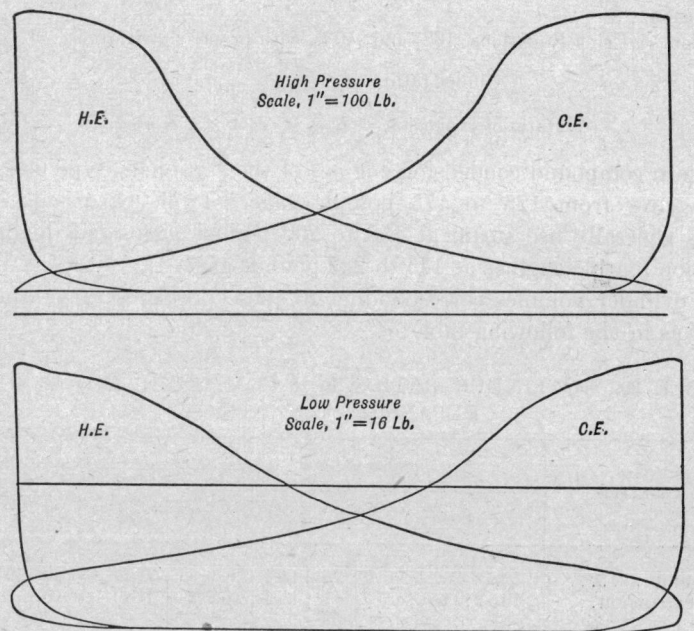

FIG. 358. — Indicator Diagrams from 28″ × 60″ × 60″ Cross-compound Engine.

412. Cylinder Ratio and Expansion Ratio. — *The cylinder ratio for multiple-expansion engines is the ratio of the piston displacement of the low-pressure cylinder to that of the high-pressure cylinder.* Since the stroke is made the same for all the cylinders, the volumes are to each other as the areas, or as the squares of the diameters.

For a compound engine, if C = cylinder ratio, L = stroke, d = diameter of the high-pressure cylinder, and D = diameter of low-pressure cylinder,

$$C = \frac{L \pi/4 \, D^2}{L \pi/4 \, d^2} = \frac{D^2}{d^2} \quad \ldots \ldots \ldots \quad (102)$$

The ratio of expansion, or number of expansions of steam, R, for a compound engine, is the ratio of the volume of the low-pressure cylinder to the volume in the high-pressure cylinder at cut-off.

If r = number of expansions in the high-pressure cylinder, then $\frac{1}{r} \times L \times \frac{\pi d^2}{4}$ = volume at cut-off in high-pressure cylinder, and

$$R = L\frac{\pi D^2}{4} \div \left[\frac{1}{r} \times L \times \frac{\pi d^2}{4}\right] = r\frac{D^2}{d^2} = rC \quad \ldots \quad (103)$$

No account of clearance volume is taken in the above ratios.

Example 50. — A cross-compound engine has the following dimensions: high-pressure cylinder, 20 in. diam.; low-pressure cylinder, 40 in. diam.; stroke, 30 in. Cut-off in high-pressure cylinder at one-quarter stroke. Find the cylinder ratio and the ratio of expansion.

Solution. — Using Equations (102) and (103), with proper substitutions,

$$\text{Cylinder ratio} = C = \frac{D^2}{d^2} = \frac{1600}{400} = 4$$

$$\text{Ratio of expansion} = R = rC = 4 \times 4 = 16.$$

Modern compound condensing engines of the stationary type use steam at pressures from 125 to 175 pounds gage. Triple-expansion marine engines generally use steam at 150 to 200 pounds gage, and quadruple-expansion marine engines at 175 to 225 pounds gage.

The cylinder volumes corresponding to these pressures range approximately as in the following table:

TABLE 30. — CYLINDER RATIOS FOR COMPOUND AND MULTI-EXPANSION ENGINES

Kind of Engine	Range of Ratio		Order of Cylinders in which Ratio is expressed
	From	To	
Compound..........	1 to 4.2	1 to 4.6	High to low
Triple expansion.....	1 to 2.3 to 6	1 to 2.5 to 7.8	High to intermediate to low
Quadruple expansion.	1 to 1.8 to 3.6 to 7.8	....	High to 1st intermediate to 2nd intermediate to low

Non-condensing engines using pressures from 100 pounds to 125 pounds have cylinder ratios from 2.5 to 3.25.

The tendency is toward much larger cylinder ratios for compound engines, tests having shown good economy with ratios as high as 7.0 to 1.0.

Cut-off in the high-pressure cylinder of stationary engines ranges from 0.25 to 0.4 stroke, under normal load conditions, and from 0.55 to 0.75 stroke for marine engines.

Triple-expansion engines are used in stationary practice, where the load is steady, as in pumping stations.

413. Indicated Horsepower. — *The horsepower of a compound or multiple-expansion engine is the sum of the horsepower developed simultaneously*

RATED HORSEPOWER 449

in the cylinders. The power for each cylinder may be found from the indicator diagrams, as explained in Art. 391, page 424.

Example 51. — A $20 \times 40 \times 30$ in. vertical cross-compound engine, with both piston rods 4 in. in diameter, running at 145 r.p.m., gave data from indicator diagrams as follows: Scale of indicator springs; high pressure, 80 lb.; low pressure, 16 lb.

INDICATOR DIAGRAMS: *high pressure*, H.E. area = 2.36 sq. in.; length, 3.95 in.; m.e.p., 47.80; C.E. area = 2.89 sq. in.; length, 3.95 in.; m.e.p., 58.53; *low pressure*, H.E. area = 2.72 sq. in.; length, 4.05 in.; m.e.p., 10.74; C.E. end area = 2.93; length, 4.05; m.e.p., 11.57. PISTON AREAS: *high pressure*, H.E. = 314.16 sq. in., C.E. = 301.59 sq. in.; *low pressure*, H.E. = 1256.64 sq. in., C.E. = 1244.07 sq. in. Find the i.hp. for the engine.

Solution. — Using Equation (91), page 425, and the data as given, the horsepower of the engine is found as follows:

High-pressure cylinder
$$\text{H.E. i.hp.} = \frac{PLAN}{33,000} = \frac{47.80 \times 2.5 \times 314.16 \times 145}{33,000} = 165$$
$$\text{C.E. i.hp.} = \frac{PLAN}{33,000} = \frac{58.53 \times 2.5 \times 301.59 \times 145}{33,000} = \underline{194}$$
$$359$$

Low-pressure cylinder
$$\text{H.E. i.hp.} = \frac{PLAN}{33,000} = \frac{10.74 \times 2.5 \times 1256.6 \times 145}{33,000} = 148$$
$$\text{C.E. i.hp.} = \frac{PLAN}{33,000} = \frac{11.57 \times 2.5 \times 1244 \times 145}{33,000} = \underline{158}$$
$$306$$

Total horsepower for the engine.......................... 665

414. Rated Horsepower. — The probable, or rated, horsepower can be calculated from the dimensions of the engine and the stated conditions, by considering that all the power could be developed in the low-pressure cylinder, except for practical reasons. In the actual engine, the same weight of steam, on the average, is exhausted per stroke from the low-pressure cylinder as is admitted to the high-pressure cylinder, and consequently the work may be figured as though it were all done in the low-pressure cylinder, with the total number of expansions equal to R. A suitable diagram factor is applied, to find the mean effective pressure. Thus, if

N = revolutions per minute.
P_1 = initial pressure, lb. per sq. in., absolute.
P_3 = back pressure for low-pressure cylinder, lb. per sq. in., absolute.
f = diagram factor.
C = cylinder ratio, and $R = Cr$, where r = number of expansions in high-pressure cylinder.

$$\text{Probable M.e.p.} = f\left(\frac{P_1(1 + \log_e R)}{R} - P_3\right) \text{lb. per sq. in.}$$

Rated horsepower (double-acting compound engine)

$$= \frac{2 \times \text{m.e.p.} \times L \times \text{Area of } L.P. \text{ cylinder (sq. in.)} \times N}{33,000} \quad (104)$$

The rated horsepower for a stationary engine is generally calculated with cut-off in the high-pressure cylinder at one-quarter stroke.

415. Diagram Factors for Compound Engines. — Slide-valve engines, 0.55 to 0.80; Corliss engines, 0.75 to 0.85; triple-expansion engines, 0.55 to 0.70.

416. Governing Compound Engines. — The compound engine is generally governed (1) by controlling the length of admission of steam into the high-pressure cylinder, with cut-off on the low-pressure cylinder fixed in one position, or (2) by simultaneous control of cut-off in both cylinders.

The governor may be of the loaded-pendulum type, or of the centrifugal-inertia type. The total work done by the engine is proportional to the amount of steam admitted into the high-pressure cylinder. *The distribution of work between the cylinders is controlled by the cut-off in the low-pressure cylinder and is generally such that the total work is equally distributed between the cylinders.*

In case (1), with fixed low-pressure cut-off, the work is evenly divided for only one set of conditions. In case (2), the distribution of work is controlled by low-pressure cut-off which is under control of the governor at all loads. It may be adjusted by hand while the engine is running. Early cut-off in the low-pressure cylinder increases the receiver pressure, since a smaller volume of steam is withdrawn than for a late cut-off. This increases the back pressure in the high-pressure cylinder and reduces the work done there, thus increasing the work done in the low-pressure cylinder if the total work remains the same. Therefore, in adjusting the valve gear to equalize the loads, if more work is to be done in the low-pressure cylinder the governor-rod yoke is adjusted to give earlier cut-off in that cylinder, thus increasing the receiver pressure. Later cut-off in the low-pressure cylinder reduces the receiver pressure, so that more work is done in the high- and less in the low-pressure cylinder. The receiver pressure for a condensing engine, with initial pressure 160-pound gage, and 27-inch vacuum, is about 11-pound gage.

417. Economy of Compound Engines. — Standard-type compound engines, having a single valve for each cylinder, generally require from 22 to 27 pounds of saturated steam per i.hp. per hour, non-condensing. It should be stated that compound engines are seldom run non-condensing and then usually for small sizes. With a standard vacuum of 26 inches of mercury, the water rate is decreased about 20 per cent.

For four-valve engines, the water rate with saturated steam may be from 17 to 22 pounds per i.hp. per hour, non-condensing. When operating condensing, the water rate may be 12 to 14 pounds per i.hp. per hour, depending upon the initial conditions of the steam and the vacuum.

For saturated steam, the triple-expansion Allis Corliss pumping engine at Chestnut Hill, Boston, has a record of 10.33 pounds of steam per i.hp. per hour, corresponding to 196 B.t.u. per minute.

Using steam at high pressure and superheat, compound condensing engines having poppet valves, unaflow engines, and locomobile engines have shown better economy — in some exceptional cases, as low as 6 to 8 pounds per i.hp.-hour.

REFERENCES

Steam Power Plant Engineering, GEBHARDT.
Mechanical Engineer's Handbook, MARKS.
Heat Power Engineering, HIRSHFELD and BARNARD.
Catalogues of Builders.

REVIEW QUESTIONS AND PROBLEMS

1. What is the principal reason for compounding steam engines?
2. How may compound engines be classified?
3. Why are receivers used?
4. Describe concisely the arrangement of the tandem-compound engine. What advantage has it over other compound engines?
5. What is "cylinder ratio," and what is ratio of expansion?
6. A double-acting cross-compound steam engine, $12 \times 22 \times 30$ in. has cut-off at quarter stroke in the high-pressure cylinder. What is the cylinder ratio? What is the ratio of expansion?
7. Using data in Problem 6, with initial steam pressure, 125.3 lb. gage; back pressure, 4 lb. abs.; r.p.m., 130; diagram factor, 0.75; find the probable horsepower.
8. Given the following data from a test on a $28 \times 52 \times 48$ in. Corliss cross-compound condensing engine, calculate the i.hp.:

Indicator diagrams: High-pressure, head-end area, 1.33 sq. in. length 3.89 in.; crank-end area 1.47 sq. in., length 3.89 in.; scale of spring, 80 lb. Low-pressure, head-end area, 1.46 sq. in., length 3.77 in.; crank-end area, 1.52 sq. in., length, 3.77 in.; scale of spring, 20 lb. Diameter of both piston rods, 6 in.; r.p.m., 82; at switchboard amperes, 600; volts, 615.

CHAPTER XX

METHODS OF LUBRICATION — ENGINE ACCESSORIES

418. Foreword. — The rubbing surfaces of metal bearings are never perfectly smooth; even when they are highly polished, small depressions and elevations can be seen under a microscope. Under pressure, these depressions and elevations interlock and resist the moving force, and the movement of such contact surfaces, relative to each other, produces friction and wear. Friction produces heat, which is a form of energy and in this case represents a loss of work, the amount of this loss varying from 4 to 20 per cent of the total power developed by the engine. The distribution of the friction loss in a 6 inch by 12 inch straight-line engine, having an unbalanced slide valve, is approximately as follows: main bearing, 35.4 per cent; piston and rod, 25 per cent; crank pin, 5.1 per cent; crosshead and wrist pin, 4.1 per cent; valve and rod, 26.4 per cent; eccentric strap, 4.0 per cent.

To reduce the loss caused by friction, a lubricant is placed between the surfaces. If the supply of lubricant is sufficient, an oil film is formed between the surfaces, and the resulting friction is merely fluid friction, or resistance of the molecules of the fluid to motion. The loss from fluid friction is much less than that resulting from poorly lubricated surfaces in which the oil film is not formed.

It would be impossible to operate machinery without lubrication, as the frictional heat developed would soon destroy the bearing surfaces.

The selection of the proper lubricant may mean a large increase in the economy of the engine. This selection is an important problem, which calls for a knowledge of the characteristics of lubricants and the best methods of applying them.

Lubricants used for engines may be classified as solid, such as graphite, semi-solid, such as heavy greases, and liquid. The most common liquid lubricant is mineral oil.

419. Characteristics of Oil. — An oil, to be efficient as a lubricant, should have the following characteristics:

1. Sufficient body, or combined capillarity and viscosity, to form an oil film between the bearing surfaces.
2. Least fluid friction compatible with sufficient body.
3. Freedom from tendency to decompose, or change in composition by gumming, on exposure to air or in use.

4. Absence of acids or properties liable to injure metal with which it comes in contact.

5. High temperature of vaporization, and low temperature of solidification.

6. Freedom from grit and foreign matter.

420. Methods of Lubrication. — The surfaces requiring lubrication in an engine may be divided into sliding and rotating surfaces. Flat sliding surfaces, such as the crosshead shoes, tend to scrape the oil from the guide, unless the corners are slightly rounded to prevent it; in general, such surfaces are imperfectly lubricated. Rotating surfaces, such as the journal in its bearing, tend to draw the oil in between the surfaces, because of the viscosity of the oil. Bearings are ordinarily made with a clearance amounting to 0.001 inch per inch of diameter, which permits the oil to form a film, when supplied in sufficient amounts. Bearing surfaces are generally grooved to assist in distributing the oil. Sliding surfaces, such as that of the piston in the cylinder, are more difficult to lubricate.

The methods used to lubricate steam-engine bearings may be classified as follows:

1. For individual bearings
   - Hand oiling
   - Drop-feed oiling
   - Ring or chain
   - Wiper
   - Grease cup
   - Oily waste

2. For group bearings
   - Splash system
   - Gravity circulating system

3. Cylinder lubrication
   - Hydrostatic lubricator
   - Mechanically operated lubricator

The methods listed in group 1 give imperfect lubrication, as the oil is supplied in limited amounts. Ring or chain oiling where applicable, is the best of these methods. Oil-bath and forced-feed methods give nearly perfect lubrication, since the oil is supplied in a sufficient amount to maintain an oil film.

Hand oiling is used for small bearings in which the rubbing speed is low, such as the moving parts of the valve gear. Oil is fed into an oil hole in the bearing cap, by an oil can having a flexible bottom. It then spreads and gradually works its way to the ends of the bearing. The oil hole is often covered with an oil cup having a spring-actuated cap which closes and keeps dust from getting into the bearing.

Drop-feed oiling consists in feeding a fairly regular supply of oil to the bearing. The devices used for drop-feed oiling are the wick-feed oiler, generally used to lubricate the main and thrust bearings of marine engines, and the sight-feed oiler.

The **wick-feed oiler** has an oil chamber covered by a cap, a siphon oil tube projecting above the oil level, and a wick made of strands of woolen yarn. The wick dips into the oil and has its longer end projecting into the tube. Oil is drawn from the oil chamber by capillary action and drops on the shaft.

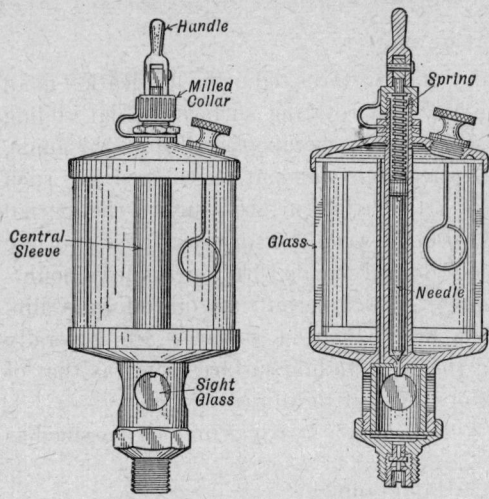

Fig. 359. — Sight-feed Oiler.

The **sight-feed oiler,** Fig. 359, is the most common type of oiler, and is easily regulated to feed the desired quantity of oil. The bottom of the **glass oil chamber,** which is used to show the height of oil, is made of brass and contains a conical seat at the bottom of a central sleeve. An **adjusting needle valve** screws into the top of the cap, which is provided with a hole for filling. By turning the **milled collar,** the position of the needle is changed with respect to the conical seat, and the rate of feed regulated. With the small handle attached to the top of the needle horizontal, a spring inside the tube pushes the needle against the conical seat and stops the flow of oil. When it is vertical the oil will feed as regulated. The rate of flow is observed through a sight glass placed in the base.

Sight-feed drop oilers are often arranged with multiple feeds, as shown in Fig. 382, page 480, in which there are seven oil feeders controlled by seven different needle valves located in oil tubes that deliver the oil to the various bearings.

The crank pins of overhung cranks are often oiled by a special arrangement, Fig. 360, using drop-feed oilers to regulate the rate of flow. The oil tube, bent to form a right angle, has one end attached to the crank pin and the other end enlarged to form a spherical chamber. The tube is of such length that the chamber is in line with the axis of the main shaft.

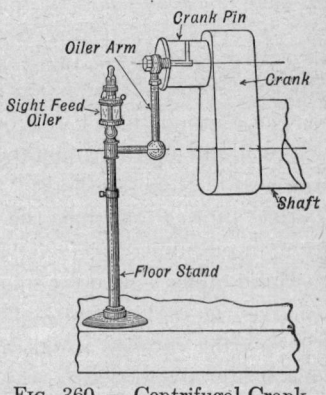

Fig. 360. — Centrifugal Crankpin Oiler.

Oil flows from the sight-feed drop oiler, usually fixed to a railing or standard, through a feed tube, and drops into the chamber, from which it is delivered to the crank pin by centrifugal force and enters the bearing through holes drilled in the crank pin, as shown.

CIRCULATION OILING SYSTEM

A **telescopic oiler** is used on many engines to lubricate the crosshead pin. A sight-feed oil cup delivers oil to a tube, which slides up and down in a tube attached to the crosshead. Oil flows from the oil cup through the tubes to the wrist pin, and then through drilled holes to the bearing.

An **oil wiper** is often used to lubricate the crosshead pin of slow-running engines. A wick, lubricated from a drop oil cup, hangs in the path of the wiper. The wiper strikes the wick and the oil thereon drops into the body of the oiler and flows to the wrist pin bearing, through holes drilled in the wrist pin.

A **grease cup**, Fig. 361, is used to lubricate eccentrics and other slow-moving parts. Grease is placed in the cup below the cup piston, where it is under a pressure produced by a light spring. When the bearing becomes sufficiently warm, some of the grease is melted and lubricates the bearing. When not in use, the small flanged nut raises the piston and removes the pressure from the grease.

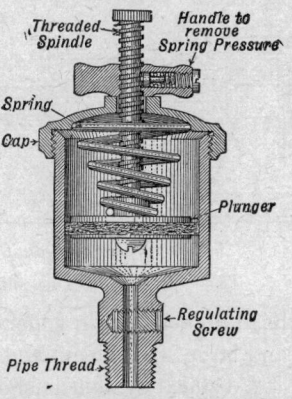

Fig. 361. — Grease Cup Oiler.

Ring and chain oiling are used to lubricate the main bearings of high-speed engines. The bearing housing forms an oil reservoir in which oil is maintained at a level sufficiently high to have the lower ends of the rings submerged. One or two rings or chains surround the shaft and dip into the oil. The rings revolve with the motion of the shaft and carry oil to the top of the shaft. The oil is then distributed to the bearing by the oil grooves. The surplus oil flows to the ends of the bearing and drops back into the oil reservoir, where it is sometimes cooled by water circulating through coils of piping.

Splash oiling is employed to lubricate a number of bearings in an enclosed casing. It is used extensively with small vertical and horizontal high-speed engines. The enclosed crank chamber, Fig. 362, contains oil at such a level that the crank pin dips into the oil at each revolution. The level of the oil is maintained constant by using an overflow connection. The crank disk, revolving inside the crank chamber, dips into the oil, and thus oil is picked up, and is thrown off the revolving rim by centrifugal force. Oil wells or **pockets**, cast on the inside of the casing, collect the oil and lead it through various channels, tubes, or troughs to the parts requiring lubrication. The main bearings and eccentrics, crank pin, crosshead pin, and crosshead guides are lunbricated either directly by means of oil spray or indirectly by means of the oil troughs.

421. Circulation Oiling System. — There are two **systems** embodying the oil circulation principle:
1. Non-pressure oil circulating system.
2. Pressure oil circulating system.

422. Non-pressure Oil Circulating System.

— By this system the oil delivered to the bearings is not under direct pressure. It is employed to lubricate automatically the main bearings, crank pins, crossheads, cross-

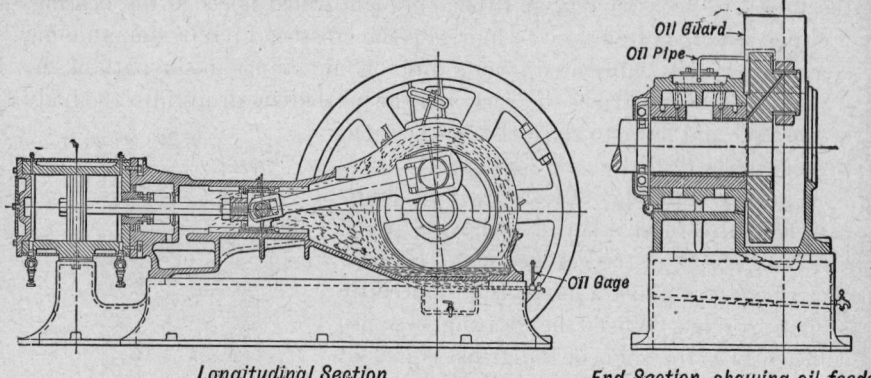

Longitudinal Section — End Section, showing oil feeds

FIG. 362. — Splash System of Oiling.

head guides; or, in general, most of the external moving parts in medium and large sized steam engines.

A typical system is shown in Fig. 363. Oil flows by gravity from the **supply tank,** through distributing pipes, to sight-feed glasses in each pipe

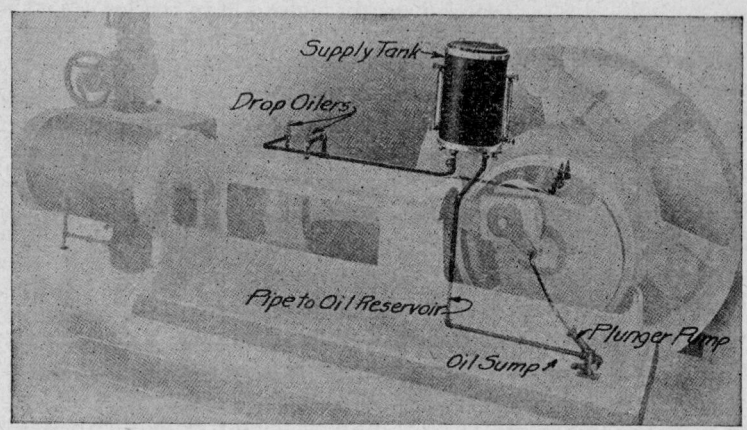

FIG. 363. — Non-pressure Oil Circulating System.

line, and the amount flowing to each bearing is regulated by needle valves at the sight-feed glasses. From the bearings, the oil drains through return oil pipes back into the **sump tank,** or lower part of the bed. An **oil pump,** driven by the engine, pumps the oil from the sump, either through an oil cooler or directly into the supply tank, which is provided with an overflow pipe to return the surplus oil to the sump tank.

423. Pressure Oil Circulation System.

— Oil is delivered by this system, as directly as possible, to the various bearing surfaces requiring lubrication.

It is used on enclosed types of steam engines and turbines. For description see Art. 476, page 527.

424. Lubrication of Internal Parts of the Steam Engine. — The internal parts of the steam engine, such as the valve, valve rod, piston, and piston rod, are exposed to steam temperature conditions. The valve and piston are not in view, and the condition of lubrication is not easily determined. For this reason, correct lubrication of these parts is more difficult than that of the external moving parts. It is accomplished by using an oil having a heavy body and known as **cylinder oil,** *which is introduced by a lubricator into the steam line at a point just above the stop valve. The oil is broken up and carried by the steam to the internal parts.* The following types of lubricators are used:

1. Hydrostatic lubricator.
2. Mechanically operated lubricator.

425. Hydrostatic Lubricator. — This lubricator, Fig. 364, is widely used because of its simplicity. It consists of a chamber having two connections to the steam pipe leading to the engine. The chamber is filled with oil and water the relative amounts of which are shown by a gage glass. At the bottom of the chamber is a drain cock for draining off the water before refilling the lubricator through the **filling plug.** *When refilling, the valves in both steam connections to the lubricator should be closed, to prevent oil from being blown out when the filling plug is removed, and causing possible injury.* The control, or water, valve at the base of the **head pipe** should also be closed; otherwise the water in the pipe will be drained out and the lubricator will not start feeding until the pipe again becomes filled with water. The head pipe contains a chamber, called a **condenser,** which exposes more surface to the air to condense the steam.

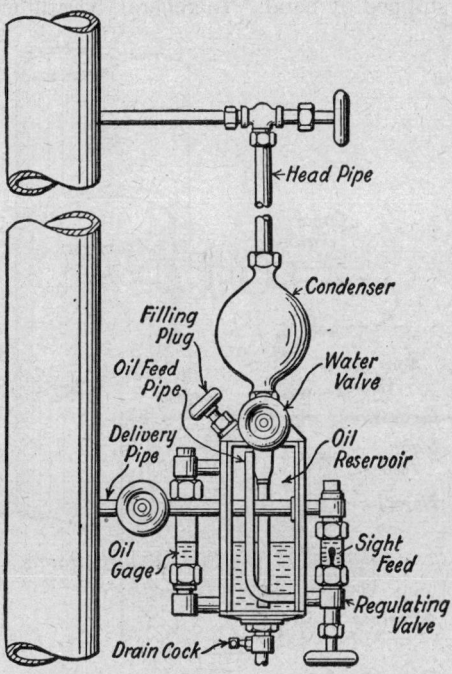

Fig. 364. — Hydrostatic Cylinder Lubricator.

Steam enters the condenser through the head pipe and is condensed. With the control valve open, as when feeding oil, the water passes through a pipe to the bottom of the oil chamber, displacing the oil, which leaves through the open end of the oil-feed pipe. In this oil-feed pipe is an adjusting valve to regulate the feed. The oil floats in drops through the water in the sight glass, and enters the steam pipe through the delivery pipe. Since the lubricator is connected to the steam line at two points, the steam pressure on the lubricator is balanced, *and the pressure to force the oil out of the chamber comes from the weight of water in the head pipe above the bottom of the oil in the lubricator.* The length of the head pipe should be at least 18 inches, to produce sufficient pressure to force the oil through the lubricator, and to provide sufficient condensing surface to keep the head pipe filled with water.

This lubricator is not automatic in action, but must be started and stopped by hand. In general, it is difficult to maintain a uniform cylinder feed with this type of lubricator. Three to five drops of oil passing the sight glass per minute is usually a sufficient amount.

426. Mechanically Operated Lubricator. — A mechanically operated lubricator gives a positive, automatic, uniform, and regular feed of cylinder oil. This type of lubricator is generally located in a fixed position on the engine cylinder, and is operated by some moving part of the engine. It therefore starts feeding as soon as the engine starts and stops feeding when the engine stops. There may be single or multiple delivery pipes, as demanded by the number of points requiring lubrication. Each delivery pipe should have sight-feed arrangements through which the quantity of oil being fed can be observed.

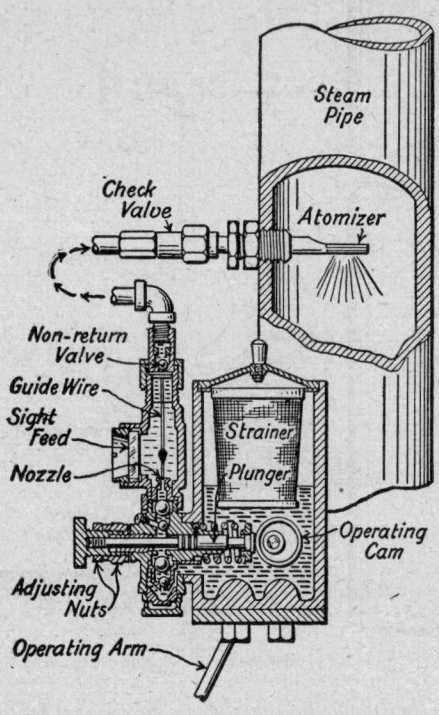

FIG. 365. — Positive Pressure Cylinder Lubricator.

One type of mechanically operated lubricator, shown in Fig. 365, has a **bull's-eye sight-feed arrangement** in the discharge pipe. It consists of an **oil reservoir** containing a **plunger pump** driven by a **cam** operated by the engine. The height of oil is indicated by an oil gage, not shown. At the top of the oil reservoir an **oil strainer** is provided to remove small particles which might clog the passages. The discharge pipe, near the steam line, contains a spring-actuated **check valve** to ensure that the discharge pipe is always kept full of oil and thus the oil supply starts as soon as the engine starts the pressure, because the pump opens the check valve.

When the engine starts, the plunger commences to travel back and forth. On the suction stroke of the plunger, oil flows past the two ball valves at the bottom. On the delivery stroke oil is discharged past the two upper ball valves and through a **nozzle.** Oil-drops form around the guide wire and rise through the water in the sight-feed. The oil then passes a non-return valve and is forced through the check valve to the **atomizer** in the steam pipe. The rate of feed is controlled by two adjusting nuts located below the sight-feed.

427. Effect of Oil in Exhaust Steam. — A part of the oil used to lubricate the internal parts of the steam engine passes out with the exhaust steam and is present in one of two forms: (1) oil in **suspension,** or (2) oil in **emulsion.** **Oil in suspension** is in the form of globules that can be removed from the steam by an exhaust-steam oil separator. The globules of oil which are not extracted from the steam in the separator in the exhaust pipe line will, in the case of condensing engines, mix with the condensed steam and pass to the hot well, where the greater portion will rise to the surface and can be skimmed off. Filters using sand, wood or wool may be used to retain the globules, in case the pump suction is not taken low enough in the hot well to avoid drawing in the " float " oil. **Oil in emulsion** consists of very minute particles of water coated with an oil film. The particles, which are so small that the exhaust-steam separator will remove only a portion of them, may be removed by chemical precipitation or electrical treatment.

A **splash guard,** made of sheet metal, is used to prevent oil from being thrown from the revolving crank pin and connecting rod end over the floor and walls of the building. The guard collects the escaping oil and drains it into the base of the frame. For a typical installation see Fig. 357, page 446.

428. Engine Accessories. — **A throttle valve,** which resembles the globe valve, Fig. 179, page 257, is placed near the engine cylinder to control the flow of steam to the engine. It consists of a cast-iron body having two flanges. One is bolted to the engine cylinder and the other to a flange on the main steam line.

The cylinders of steam engines are generally protected, by a relief valve or an explosion diaphragm, against dangerous rise in pressure caused by water from an undrained pipe or foaming boiler. The **relief valve** is a small safety valve which opens to let the water escape from the cylinder. The **explosion diaphragm** is a cast-iron disk having an accurately pre-determined section which will give way and release the water as soon as the pressure rises 50 per cent above initial steam pressure. Hand-operated valves are provided to shut off the diaphragms when necessary.

429. Steam Separator. — *The steam separator removes the entrained water, which collects at the bottom of a steam pipe and which is the result of moisture in the steam when it leaves the boiler and of condensation in the steam line.*

460   METHODS OF LUBRICATION — ENGINE ACCESSORIES

Unless this water is removed from the steam it may be carried into the cylinder, be caught between the piston and cylinder head, and result in breaking the cylinder head; if carried into a turbine it may pit and wear the blading. The danger from water in steam is increased when the velocity or pressure of the steam is increased. Steam turbines, reciprocating engines, steam pumps, or other equipment utilizing live steam should be protected by separators. The economy of engines and turbines is increased by decreasing the moisture content of the steam.

Steam separators may be classified as:

1. Reverse-current separators, in which the direction of flow is abruptly changed, usually through 180 degrees. The water in the steam, because of its greater weight and inertia, is thrown into the receiving vessel while the steam, being lighter, passes on.

2. Centrifugal-force separators, in which a rotary motion is imparted to the steam, and the water is thus separated from it.

3. Separators having baffle plates, placed in the path of the steam, to which water adheres and from which it then falls to a chamber below the plates.

4. Separators using screens in which the separation is brought about by mechanical filtration.

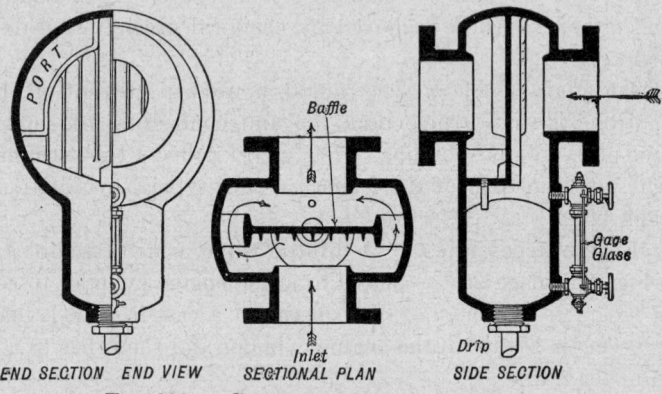

Fig. 366. — Cochrane Baffle-type Separator.

430. Separator with Baffle. — The baffle type of separator, shown in Fig. 366, consists of an upper and a lower chamber. A single solid baffle is located in the upper chamber, directly across the path of the steam. On the baffle are vertical ribs which prevent the steam current from scrubbing particles of water from the baffle. The steam, after striking the baffle, has its direction changed through 180 degrees and passes through a port at each side of the baffle. *The water separated from the steam runs into the lower chamber and is usually removed automatically by a float or bucket trap through a drip connection at the bottom.*

431. Oil Separators. — Oil permitted to remain in exhaust steam used for heating accumulates upon the heating surfaces of radiators and lowers their heat-transmitting power. It also forms a condensate unsuitable for boiler feed. Oil separators work upon the same general principle as steam separators, with changes in construction to adapt them to oil. The majority are compact and operate by splitting the steam into many small streams with frequently changing direction. The oil is trapped by baffle plates, so arranged that once the oil is separated it cannot be picked up again by the steam. The oil collects in a reservoir at the bottom of the separator and is removed as in the steam separator.

The **combined reverse-flow and centrifugal oil separator** is illustrated by the Stratton separator, Fig. 367. It consists of a vertical cast-iron cylinder with an internal central pipe extending from the outlet downward to about half the length of the separator. An annular space is thus formed between the cylinder and the central pipe. The current of entering steam is deflected by a curved partition and thrown tangentially to the side of the annular space. The velocity of the entering steam produces the centrifugal action, which throws the oil and water to the side, whence it runs down around the outer shell into the lower part of the cylinder. The steam follows the spiral course to the bottom of the central

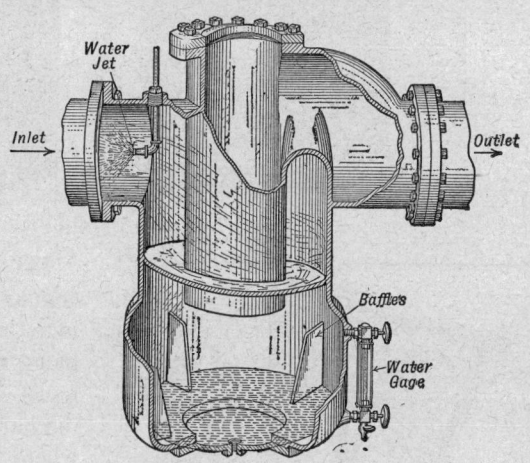

Fig. 367. — Stratton Oil Separator.

pipe and has its direction of flow changed through 180 degrees. Wings are placed in the bottom of the separator to break up the whirling motion of the water and allow it to settle quietly at the bottom. This separator uses a spray of water to condense any oil in the form of vapor before it enters the separator.

432. Oil Filter. — Oil which drains from bearings contains fine particles which should be removed before the oil is used again. For this purpose an oil filter, of which Fig. 368, is typical is used. It consists of a cylindrical tank within which are several compartments, one for pure oil, one for waste oil, and one for water. At the bottom of the water chamber is a coil to warm the oil when cold and thus hasten the action of the filter. Waste oil is poured in at the top and is passed through a thick layer of **waste** and into a pipe which leads to the bottom of the water chamber. The oil leaving this pipe is forced to spread out in a thin layer by two **baffle plates,** and is thus exposed to the action of the water which washes it out. The oil then rises through a second layer of waste into the pure oil chamber near the top of the filter and is ready to be used again.

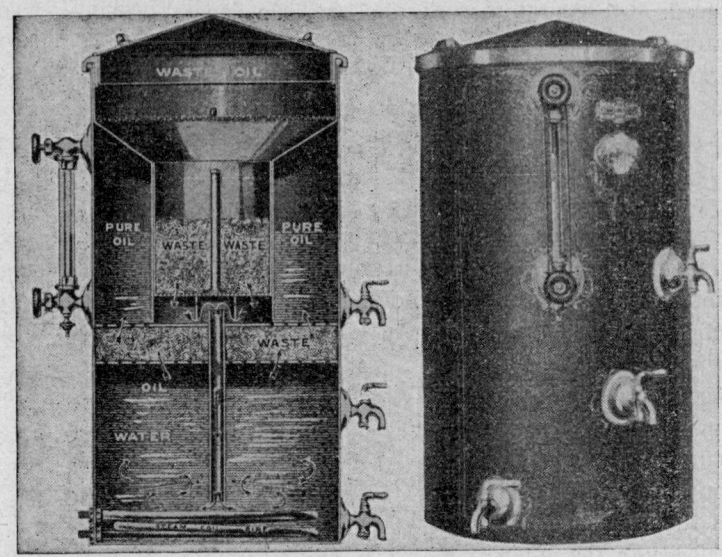

Fig. 368. — Oil Filter.

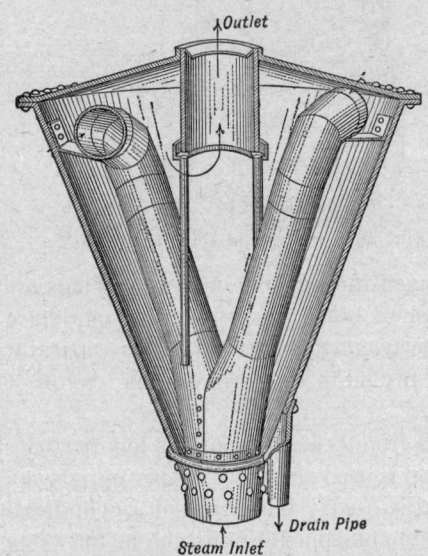

Fig. 369. — Centrifugal Type Exhaust Head.

433. Exhaust Heads. — When engines are operated non-condensing, the oil and water in the exhaust steam is removed by an exhaust head, before the steam is discharged into the atmosphere where the oil is likely to damage the exposed roofs and walls.

Exhaust heads are made of cast iron or sheet steel, the former being generally preferable. Figure 369 shows an exhaust head using the principle of centrifugal force and change in direction of the steam flow. The separated oil and water are removed by a drain pipe connected at the bottom of the chamber.

434. Back-pressure Valve. — This valve prevents excessive back pressure in the exhaust piping from engines, and is called a back-pressure valve in non-condensing plants and an **atmospheric relief valve** in condensing plants. There are many types of back-pressure valves, two of which are shown in Figs. 370 and 371. The Cochrane valve consists of several disks located on a deck plate which carries guide posts for the pressure plate. A dash pot is attached to the

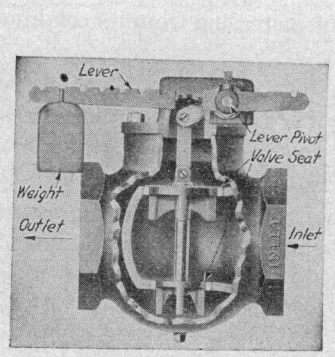

Fig. 370. — Davis Lever and Weight Back-pressure Valve.

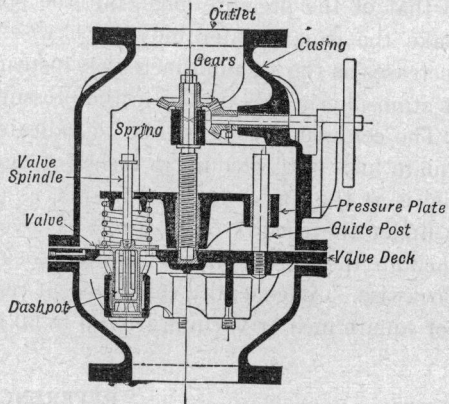

Fig. 371. — Cochrane Multiport Back-pressure Valve.

underside of the valve to prevent sudden closing. When the pressure in the exhaust piping becomes greater than the pressure of the atmosphere plus that produced by the springs, the valve disks rise and relieve the pressure. The opening pressure can be changed by changing the position of the pressure plate.

Fig. 372. — Vacuum and Compound Gages.

The Davis back-pressure valve has a weighted lever to regulate the opening pressure. The valve is of the double-disk type having wings to guide it properly. The disks are often provided with a dash pot to prevent pounding.

435. Compound and Vacuum Steam Gages. — These gages, Fig. 372, are used to indicate the pressure of the exhaust steam. The vacuum gage, of either the Bourdon tube or diaphragm type, is used for pressures below that of the atmosphere and is usually calibrated to read in inches of mercury. The construction of a Bourdon type of vacuum gage is similar to that of the pressure gage, but the position of the tube is reversed to make the gage pointer move clockwise. The area of the tube section decreases as the vacuum increases, instead of increasing from its condition at atmospheric pressure, as in the pressure gage.

The compound gage is used to indicate either a pressure in pounds per square inch or a vacuum in inches of mercury. The zero reading on the dial is at the top or at the left side, as in the figure. The needle pinion is under the control of a system of levers so arranged that when under vacuum the needle moves counter-clockwise, and when under pressure, clockwise. As generally calibrated, it reads pressure from 0 to 30 pounds per square inch or vacuum from 0 to 30 inches of mercury.

REFERENCES

Lubrication and Lost Work, THURSTON.
Lubrication Engineers' Handbook, BRATTLE.
Lubrication, VACUUM OIL CO.
Stationary Steam Engines, VACUUM OIL CO.
Steam Engines, SHEALY.

REVIEW QUESTIONS

1. What is the effect of friction in the moving parts of a steam engine, and how is this effect partially overcome?
2. Name three characteristics of a good lubricating oil.
3. Name four methods of lubricating engine bearings, and state which is the best and why?
4. Explain the operation of a hydrostatic lubricator. Why is such a lubricator used?
5. Describe the non-pressure oil-circulating system.
6. Name three engine accessories and state the use made of each.
7. Describe the operation of (a) a baffle steam separator, (b) centrifugal oil separator.
8. What is the purpose of an exhaust head?
9. How is oil prevented from being thrown on the walls and floor of the building by the revolving crank pin of an engine?
10. Describe the construction and operation of the Davis back-pressure valve.

CHAPTER XXI

STEAM ENGINE TESTING

436. Foreword. — Tests of steam engines are usually made for the following purposes:
1. To determine the weight of steam used by the engine per indicated horsepower per hour at various loads.
2. To determine the mechanical and thermal efficiencies.
3. To observe the operation of the engine under different running conditions.

The data obtained from these tests may be used for the same general purposes as those mentioned under Art. 214, page 227.

Apparatus and Instruments Necessary for a Test. — The customary tests made on a steam engine require the following instruments and apparatus to obtain the required observations:

1. Tanks and platform scales to weigh the water, or a suitable water-meter calibrated in place.
2. Graduated scales attached to the water glasses on the boilers, if the steam used is determined by measuring the boiler feedwater.
3. Condenser, to condense the steam in order to obtain its weight, or, in special cases, a steam-flow meter calibrated in position.
4. Thermometers, pressure and vacuum gages.
5. Steam calorimeters.
6. Barometer.
7. Steam-engine indicators.
8. Planimeters.
9. Tachometer, revolution-counter, or other speed-measuring device.
10. Friction brake, or dynamometer if available.

437. Water Measurement. — In making engine tests, it is often desired to measure the cooling water passing through the condenser. *With small condensers, this can be done by weighing; where the quantity of water flowing is large, some form of weir, orifice, nozzle, Venturi tube, or water-meter must be used.* These devices should be carefully calibrated in position.

A satisfactory type of **weir** is shown in Fig. 373. It consists of a 90-degree sharp-edged V-notch, cut in a bronze plate and so placed in a tank that the water to be measured passes over it. The amount of water passing can be computed by empirical formulae, such as Professor Thompson's formula, which is in general use,

$$Q = 2.544 \sqrt{H^5} \qquad \ldots \ldots \ldots \ldots (105)$$

in which Q = volume of water flowing, cubic feet per second.

H = head on the weir in feet of water.

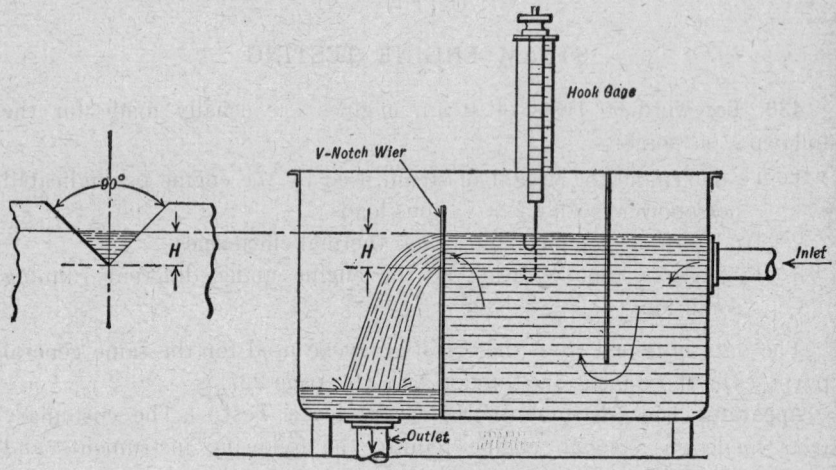

Fig. 373. — V-notch Weir and Hook Gage.

The **head** on the weir is the height shown in the drawing between the bottom of the V and the surface of the water. This height is obtained by a **hook gage** provided with a scale and vernier. A reading of the gage is first made when the hook is on the level with the bottom of the V-notch. With water flowing over the weir, a second reading is made, when the gage has been raised until the point of the hook begins to pierce the surface of the water. *The difference in these two readings gives the head, H.* These measurements must be made with the water nearly quiet; otherwise a correction must be made for the velocity at which the water approaches the weir.

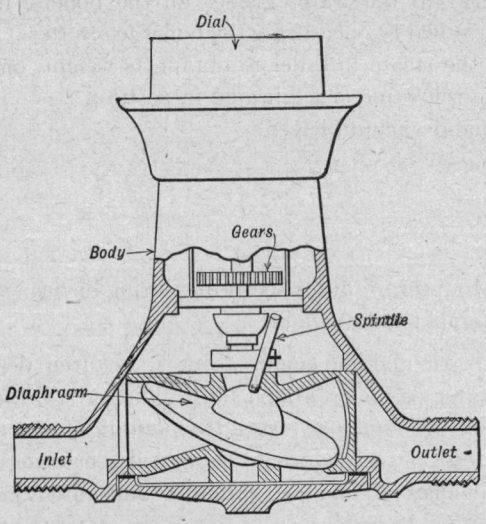

Fig. 374. — Pulsating Diaphragm Water Meter.

A typical water-meter, used to measure water under pressure, is shown in Fig. 374. It is known as the **pulsating-diaphragm meter** and consists of an inclined shaft attached to the **diaphragm**. This shaft travels around in contact with a small peg on a plate, which moves the counting mechanism through a system of gears. The frame has side chambers which are alternately filled and emptied, thus producing the "pulsating" motion that operates the recording mechanism. The diaphragm divides the measuring chamber into two compartments of equal volume, one of which is being filled while the other is being emptied. *Meters are generally calibrated to read in cubic feet of water flowing.*

438. Measurement of the Quantity of Steam. — The best method of measuring the steam used is by discharging the steam into a surface condenser and weighing the condensate. When this method is used, all leaks in the condenser should be repaired, or proper correction should be made

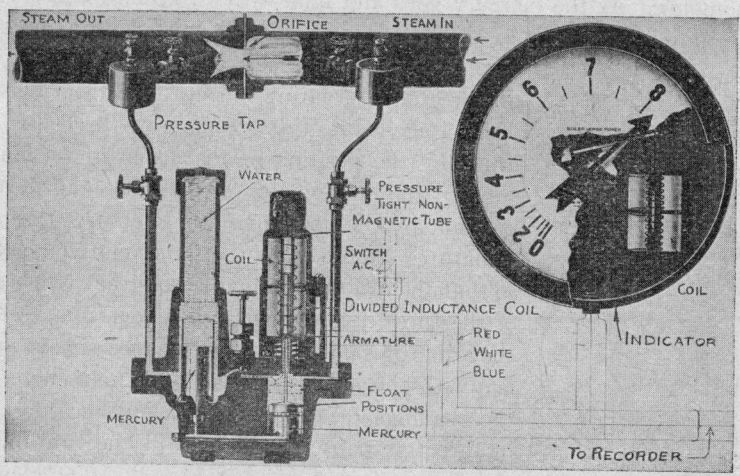

Fig. 375. — Brown Electric Steam-flow Meter.

for the amount of leakage. When a surface condenser is not available, the amount of steam used can be obtained by measuring the feedwater to the boiler supplying the engine. In this case, assurance must be had that all the steam from the boiler passes to the engine, and allowance must be made for all water wasted through steam traps or separators located in the steam pipe supplying the engine.

Where it is not possible to measure the amount of steam by either of the previously mentioned methods, a **steam-flow meter** connected into the steam pipe line may be used. The various forms of this instrument generally use an orifice, located at a suitable point in the steam line, to produce a pressure differential sufficient to operate either a mechanical or electrical recording device, which is ordinarily calibrated *to indicate the amount of steam flowing in pounds per hour or in boiler horsepower.*

The **Brown Electric Steam-flow Meter,** shown schematically in Fig. 375, uses the electric method of recording, and works upon the principle of a

self-balancing inductance bridge formed by 3 pairs of arms, each wound on a form made of moulded **bakelite** thus forming a divided inductance coil. One coil is located in the manometer and one each in the indicator and recorder.

The operating portion of the instrument consists of: a **manometer** connected by pressure taps to each side of the orifice in the steam line, a **float** riding on the surface of the right-hand leg of the manometer, and carrying a **non-magnetic rod** to the upper end of which a freely moving **magnetic armature** is attached, and a pressure-tight **non-magnetic tube** within which the armature moves.

When in operation, the difference in pressure produced by steam flowing through the orifice, is transmitted to the surfaces of the mercury in the manometer. As the mercury rises and falls with variation in the rate of steam flow, the float moves the armature up and down within the tube.

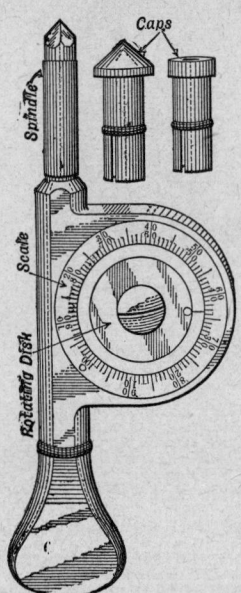

Fig. 376. — Hand Revolution Counter.

The divided inductance coil, being connected with a source of alternating current, the ratio between the voltage across the two parts of the coil is governed by the position of the armature within the tube. The manometer armature being held in position by the float upon the mercury, the other armatures are drawn to the same relative positions so that the ratios of voltage across the two sections of the three divided coils are made equal. Thus, the armatures in the instruments move in synchronism with the armature within the manometer and the readings of the indicator and recorder vary in direct response to changes in the rate of flow through the pipe.

439. Methods of Measuring Speed. — The **hand revolution counter,** Fig. 376, is much used for low speeds. It consists of a frame in which a spindle revolves. One end of the spindle carries a worm which meshes with a worm wheel driving a plate calibrated in revolutions; the other end carries a suitable point for connection to the shaft. The point is pressed against the shaft, and the revolutions are counted for one or two minutes. A **stop watch** should be used to observe the time.

The **stroke-counter** is suitable for low speeds. It has a frame on which are located a train of gears having dials with numbers attached to the gear spindles. The first gear is moved by a **ratchet wheel,** which in turn is moved by a lever with a fixed eccentric pin requiring a certain length of stroke. The outer end of the lever is attached to some reciprocating part of the engine and moves the ratchet wheel one tooth for each revolution. A **spring-pressed pawl** is provided, to prevent the ratchet wheel from moving backward and to hold it in the proper position to permit reading the numbers. The first disk reads units, the second tens, the third hundreds, and so on.

FIG. 377. — Electric Tachometer.

For high speeds, an **electric tachometer**, Fig. 377, a portable, geared hand tachometer, Fig. 378, or a chronometric tachometer is used. Tach-

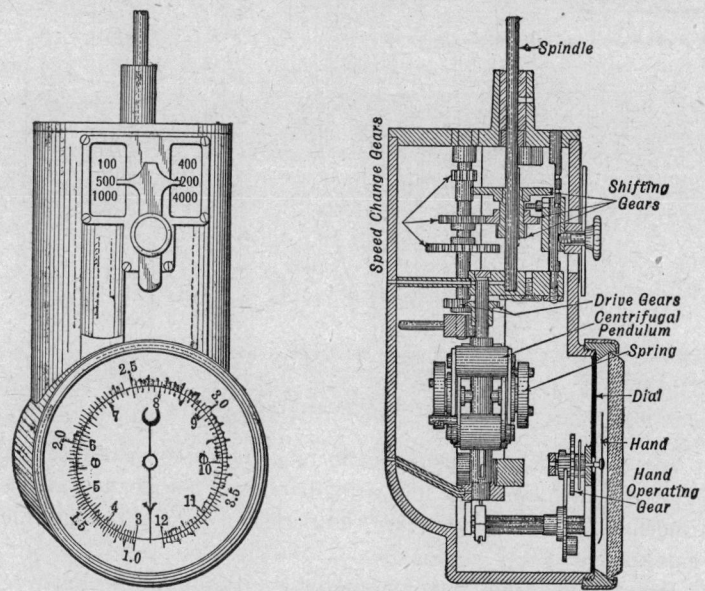

FIG. 378. — Centrifugal Geared Hand Tachometer.

ometers indicate instantaneous speed only and are suitable for speeds from six to several thousand r.p.m.

The electric tachometer has a permanent magnet, between the poles of which an **armature** revolves. An electromotive force, proportional to the speed, is set up, and is read by a voltmeter calibrated to read revolutions per minute.

The geared tachometer shown in Fig. 378, has a case in which are located the operating parts. The main spindle carries a set of gears which can be moved along the spindle by a thumb screw located on the front side of the instrument. Another shaft supports a second set of gears which drive the control portion of the instrument and thus move the indicating hand. The instrument can be set for different speed ranges, by moving the sliding gears to mesh with the various-sized gears on the stationary shaft. The operation of this instrument depends upon the change in centrifugal force, acting on the internal rotating parts, at different speeds.

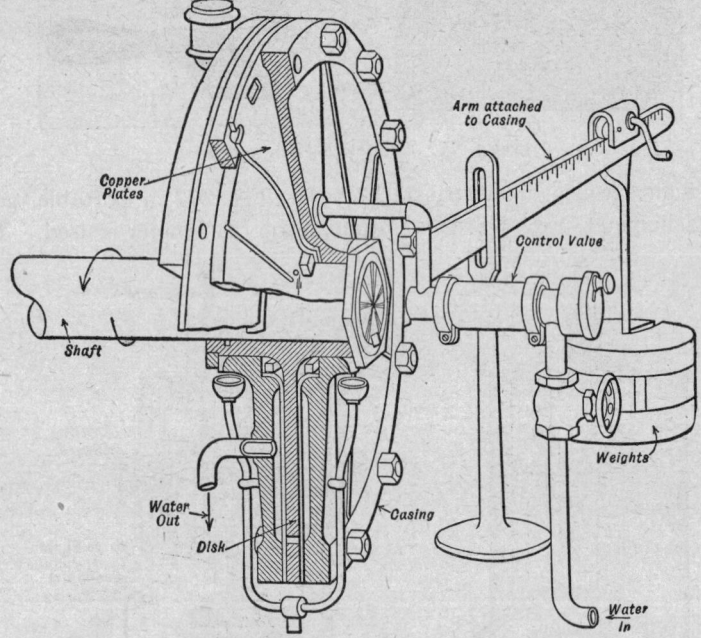

FIG. 379. — The Alden Water Brake.

The Von Sicklin-Elgin chronometric tachometer consists essentially of a set of gears, which move the indicating hand over a scale, and an accurate watch mechanism, which measures the period of time through which the hand moves.

440. Power-measuring Dynamometers. — Besides the Prony brake mentioned in Chapter XVIII, a form of brake known as a dynamometer is used to measure the power of large engines. Dynamometers may be either mechanical or electrical. A **water brake** of the former type is shown in Fig. 379. It has a disk attached to the power transmitting shaft, and revolves in oil between two copper plates. A casting, which is free to rotate, surrounds the disk and plates, and forms a chamber into which water under pressure is admitted. The pressure of the water forces the plates against the revolving disk, and the water carries away the heat resulting from friction. The friction produced tends to rotate the frame, but is prevented from doing so by weights hung on a calibrated arm attached to the casing. A special valve is provided to maintain a constant pressure of water on the brake, regardless of small pressure changes in the supply. Power is computed in the same way as for the Prony brake.

441. Preparation for the Test. — A careful examination should be made of the engine and the apparatus required in the test. The condition of the inside of the cylinder, valves, and valve seats should be ascertained, and notes made of all points which might affect the results. The cylinder and valves should be tested for leaks, and if a condenser is used its packing should be examined for air leaks. For this purpose a candle flame may be used to detect the leaks, or the piping and condenser may be filled with warm water under sufficient pressure to show the leaks if any exist.

The diameters of the engine cylinders should be obtained when cold, and the clearance obtained by finding the weight of water required to fill the clearance space. To determine the clearance by this method, the engine is set on dead center, and the clearance space filled from a quantity of water previously weighed. The water remaining when the clearance space is filled is subtracted from the original weight, to give the weight of water required. In converting this weight to volume, allowance must be made for the temperature of the water used. All air pockets should be vented to obtain accurate results. If there is leakage past the piston and valves, its amount should be allowed for by having the clearance space full and allowing it to settle for a certain length of time, then filling the clearance space again and obtaining the amount used to refill. The time required for filling, leakage, and refilling should be taken. The amount of water, W_1, required to fill the clearance space is then

$$W_1 = W - \frac{wT}{t + t_1} \quad \ldots \ldots \ldots \quad (106)$$

in which

W = weight of water poured into clearance space, first time.
w = weight of water to refill.
T = time to fill clearance space.
t = time for leakage.
t_1 = time to refill.

For large engines it is necessary to compute the clearance volume from working drawings of the engine.

If a surface condenser is used to determine the amount of steam used, it should be examined for leakage in the condenser tubes, by operating the condenser under vacuum with all steam shut off and observing the rate at which water is discharged by the condensate pump.

All instruments should be carefully calibrated before and after test.

The indicator springs should be calibrated under conditions as similar as possible to those prevailing in actual use. They should preferably be compared with a standard dead-weight tester, or a standard gage known to be correct. The calibration should be made for at least five equidistant points, and the arithmetical mean should be used as the average scale.

Platform scales should be compared with standard weights.

442. Conduct of Test. — The conditions under which the test is to be run should be maintained throughout.

The test should not be started until all the apparatus has been in operation for a sufficient length of time to be thoroughly warmed, and uniform conditions have been established. The time should then be noted and the observations started. The test should continue from three to ten hours, depending on conditions, and readings should be taken at least every fifteen minutes, or oftener if conditions fluctuate. Wide fluctuations should be recorded by recording instruments.

Each indicator card should be marked with the date, make and size of engine, number of card, time, scale of spring, and end of cylinder. One card of each set should be marked with the readings of the steam and vacuum gages.

A log of the test should be made, on which should be entered readings of steam and vacuum gages, thermometers, calorimeters, speed indicators, load-measuring devices, and steam used. These readings should be obtained at practically the same time the indicator diagrams are taken. The areas, lengths, mean effective pressures, and cut-offs shown by the diagrams should be placed on the log, and a set of representative indicator diagrams should be selected for inclusion in the record.

443. Calculation of Results. — *Before any calculations are made, the data should be examined carefully, and any observations which are obviously incorrect should be thrown out.* In making calculations involving the averages of a number of readings, the following methods may be used:

1. Primary averages in which all the readings of each instrument are averaged and the resulting averages used to calculate the final results.
2. Final averages in which the final result is calculated from each set of coincident observations. These final results are averaged as a grand average.

The choice between these two methods depends upon the degree of accuracy desired, and the type of formula involved. With a formula involving the sum or difference of first powers, either method is satisfactory; when the formula involves fractional powers or powers greater than

CALCULATION OF A HEAT BALANCE

the first, the method of final averages should be used to obtain the correct result.

Each item is calculated according to the methods explained in the previous chapters. The observations and the results should be tabulated in a form, such as shown in Table 33, page 474, in which roman figures indicate observed values and **bold-faced** type shows computed values.

To determine the point of cut-off from the diagram, the method shown in Fig. 380 is used. Through the point of maximum pressure during admission, a line is drawn parallel to the atmospheric line. Through a point on the expansion line where cut-off is complete, a **hyperbolic curve**, Art. 397, page 431, is drawn. The point at which the line through the point of maximum pressure intersects the hyperbolic curve is the point of **nominal cut-off**. The proportion of cut-off is found by dividing the length up to the point of cut-off, l, by the total length, L.

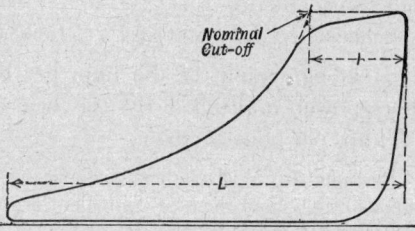

FIG. 380. — Method of Finding Point of Cut-off.

444. Calculation of a Heat Balance. — For some purposes it is often desired to make a heat balance for the engine, using the data obtained during the test. Such a heat balance is shown in Table **31**, and the method of calculating the various items is illustrated. The data used in this heat balance were taken from the form given in Table **32**.

TABLE 31. — HEAT BALANCE FOR AN 8-INCH BY 18-INCH CORLISS ENGINE

Name of Loss	B.t.u.	Per Cent
Heat supplied to engine per hour........................	913,926	100
Heat carried away by condensing water per hour.........	814,312	89.1
Heat equivalent of the i.hp. per hour....................	67,188	7.4
Heat lost by radiation, etc.............................	32,426	3.5

Heat supplied to the engine per hour equals the weight of steam used by the engine per hour multiplied by the quantity of heat per pound of steam above 32 deg. fahr. at the absolute pressure and quality in the steam main.

Example 52. — Weight of steam used, 780 lb. per hr.; quality equals 0.977; absolute steam pressure, 134.5 lb. per sq. in.; exhaust pressure, 15.6 lb. per sq. in. abs. Find the heat supplied per hour.

Solution. — Heat per pound $= xL + h_{f_1} = 0.977 \times 870.3 + 321.4 = 1171.7$ B.t.u. Heat supplied, B.t.u. per hour, equals $780 \times 1171.7 = 913,926$.

Heat absorbed by the circulating water per hour equals the weight of circulating water multiplied by the rise in temperature of the water in passing through the condenser. As an equation,

$$\text{Heat absorbed} = W_c(t_2 - t_1) \quad \ldots \ldots \quad (107)$$

in which W_c = weight of circulating water per hour, pounds.

t_1 = temperature at outlet, deg. fahr.

t_2 = temperature at inlet, deg. fahr.

Example 53. — W_c as observed during test given in Table 32, on this page, = 21,715 lb. per hr. t_1 = 55.5 deg. fahr.; t_2 = 93.0 deg. fahr. Find the heat absorbed by the circulating water per hour.

Solution. — Heat absorbed = 21,715 (93.0 − 55.5) = 814,312 B.t.u.

Heat equivalent of the i.hp. per hour equals the number of indicated horsepower multiplied by the heat equivalent of one horsepower-hour = i.hp. per hour × 2545.

Example 54. — The indicated horsepower, as shown by the engine test, Table 32, equals 26.4. Find the heat equivalent of the i.hp.

Solution. — Heat in B.t.u. = 26.4 × 2545 = 67,188.

TABLE 32. — DATA AND RESULTS OF RECIPROCATING
STEAM ENGINE TEST
A. S. M. E. Test Code, 1926

Item	Name of Item with Units	
	GENERAL INFORMATION, DESCRIPTION AND DIMENSIONS	
1	Date of test	Mar. 10, 1919
2	Builder of engine	Murray Iron Works
3	Object of test	Performance
4	Type of engine	Simple
5	Class of service	Mill
6	Auxiliaries	Steam driven
7	Type and make of condenser equipment	Wheeler, surface
8	Rated capacity of condenser	
9	Rated power of engine	25
10	Kind of valves	Corliss
11	Type of governor	Pendulum
12	Diameter of cylinder, in	8
13	Diameter of piston rod, in	$1\tfrac{7}{8}$
14	Stroke of piston, in	18
15	Clearance volume of head end, per cent	5.56
16	Clearance volume of crank end, per cent	5.10
17	Horsepower constant, head end	**0.00228**
18	Horsepower constant, crank end	**0.00216**
	TEST DATA AND RESULTS	
19	Duration of test, hr	3
	Average pressures	
20	Barometric pressure, in. of mercury	29.51
21	Barometric pressure, lb. per sq. in	**14.4**
22	Pressure in steam pipe near throttle, lb. per sq. in., gage	120
23	Absolute pressure corresponding to item 22, lb. per sq. in.	**134.4**
24	Pressure in exhaust pipe near engine, lb. per sq. in.	0.10
25	Absolute pressure corresponding to item 24, lb. per sq. in.	**14.5**

CALCULATION OF A HEAT BALANCE 475

TABLE 32. — DATA AND RESULTS OF RECIPROCATING STEAM ENGINE TEST — *Continued*
A. S. M. E. Test Code, 1926

Item	Name of Item with Units	
	Average temperatures	
26	Engine room temperature, deg. fahr..................	73
27	Temperature of saturated steam corresponding to pressure in exhaust pipe near engine, deg. fahr......	211.6
	Quality of Steam at throttle	
28	Per cent of moisture in steam, per cent..............	2.30
29	Quality...	0.977
	Total quantities	
30	Total steam consumed by engine, as measured, lb.....	2340
31	Total dry and saturated steam consumed, lb..........	2286
	Hourly quantities	
32	Steam consumed per hour as measured, lb............	780
33	Dry and saturated steam consumed per hour, lb.......	762
	Heat consumption	
34	Total heat above water at 32° F. per lb. of steam at throttle, B.t.u.....................................	1171.7
35	Heat of liquid at temperature of steam at exhaust pressure, B.t.u.....................................	180.0
36	Heat supplied per lb. of steam, B.t.u.................	1171.7
37	Heat consumed per hour, B.t.u.......................	773,526
38	Heat available for work per lb. of steam from adiabatic expansion between initial conditions and final pressure according to the Rankine cycle, B.t.u..........	167
	Indicator diagrams	
39	Nominal cut-off, per cent............................	19.6
40	Mean effective pressure (average), lb. per sq. in.......	......
	Speed	
41	Revolutions per minute, r.p.m.......................	102
	Power	
42	Indicated horsepower developed by whole engine, i.hp.	26.4
43	Brake horsepower developed by whole engine, b.hp..	25.6
44	Friction of engine, hp...............................	0.80
45	Mechanical efficiency, per cent......................	97
	Economy results	
46	Steam consumed per i.hp.-hr. as measured, lb.........	29.5
47	Dry and saturated steam consumed per i.hp.-hr., lb...	28.9
48	Steam consumed per b.hp.-hr. as measured, lb........	30.4
49	Dry and saturated steam consumed per b.hp.-hr., lb...	29.8
50	Heat consumed per i.hp.-hr..........................	29,292
51	Heat available according to Rankine cycle per i.hp.-hr., B.t.u...	4930
	Efficiency results	
52	Thermal efficiency referred to i.hp., per cent..........	8.7
53	Engine efficiency based on i.hp., per cent.............	51.3
54	Rankine cycle efficiency, per cent....................	14.2

Heat lost by radiation and other causes equals the heat supplied per hour minus the sum of the heat absorbed by the condenser per hour and the heat equivalent of the i.hp. per hour.

Example 55. — Using the data in the three previous examples, find the heat lost by radiation per hour.

Solution. — Heat supplied per hour, from Example 52 = 913,926. Sum of items in Examples 52 and 53 = 814,312 + 67,188 = 881,500. Heat lost by radiation, etc., including friction = 913,926 − 881,500 = 33,465 B.t.u.

REFERENCES

Power Test Code, A. S. M. E.
Power Plant Testing, MOYER.
Mechanical Laboratory Methods of Testing, SMALLWOOD.
Experimental Engineering, and Manual for Testing, CARPENTER and DIEDERICHS.

REVIEW QUESTIONS AND PROBLEMS

1. Name the apparatus required to make the observations necessary for a performance test of a steam engine.

2. State the purpose of each instrument named in Question 1.

3. Name three methods used to measure water, and describe one of them.

4. In testing a centrifugal pump, the water discharged was measured by a 90-deg. V-notch weir. The hook gage reading was 1.762 ft. at maximum discharge, zero reading, 1.4537 ft. Using Professor Thompson's formula, compute the flow in pounds of water, the observed temperature of which was 67 deg. fahr.

5. Describe one form of steam meter. Under what circumstances should it be used to obtain the weight of steam used by an engine?

6. Check items given in **bold-faced** type in Table 33.

7. Explain the meaning of items 39, 46, 52, and 53 in Table 33.

8. What is meant by "the method of averages," as applied to computing the results of a test.

9. Mention four methods of measuring speed. Which method should be used to give accurate readings at high speeds?

10. Explain the water method of finding the clearance of an engine.

CHAPTER XXII

STEAM TURBINES

445. Foreword. — A steam turbine is a steam engine having a rotating wheel or cylinder to which is fastened a series of buckets,[1] uniformly spaced, on its periphery. Steam from nozzles or guide passages is directed continuously against these buckets, thus causing rotation. Expansion of the steam in the nozzles or buckets converts its heat energy into energy of motion, and gives it a high velocity, which is expended on the moving buckets. *The difference in the various types of turbines is due to different methods of using the steam, depending upon the construction and arrangement of the nozzles, steam passages, and buckets.*

The steam turbine is essentially a high-speed machine. It is used to best advantage with **direct connection** to electric generators, centrifugal pumps and compressors; and with **geared connections** to marine propellers, rolling mills and machinery which should run at low speed.

Steam turbines range in capacity and speed from a few horsepower at 20,000 or 30,000 revolutions per minute, in the smaller, single-wheel turbines, to the modern 208,000 kw., triple-compound turbine, having one high-pressure and two low-pressure cylinders. Each cylinder having a rotor which runs at 1800 revolutions per minute.

The advantages claimed for the steam turbine are comparatively low initial cost, low expense for maintenance and attendance, small floor-space requirement, light foundations, large overload capacity, freedom from oil in the exhaust steam, absence of vibration, uniform velocity of rotation, and high efficiency over a wide range of load in large installations. The steam turbine can be built in units of much greater capacity than is practical with reciprocating steam engines.

446. Classification of Steam Turbines. — Turbines may be classified in the following ways:

1. By position of shaft
   { Horizontal
   { Vertical

2. By method of drive
   { Direct connected
   { Geared

3. By pressure of the entering steam
   { High pressure
   { Low pressure
   { Mixed pressure

---

[1] The terms **blade**, **bucket** and **vane** are used interchangeably by various manufacturers of steam turbines. Blade is generally applied to reaction turbines, and bucket to impulse turbines. The term bucket will be used in this chapter to avoid confusion.

4. According to pressure of exhaust
   - Non-condensing
   - Condensing
   - High back-pressure
   - Bleeder

5. In accordance with the action of the steam
   - Impulse
   - Reaction
   - Combined impulse and reaction

6. By method of subdividing the flow of energy in the buckets
   - Pressure stage
   - Velocity stage

7. By direction of flow of the steam with relation to the bucket wheel
   - Axial flow
   - Single flow
   - Double flow
   - Tangential flow
   - Radial flow

8. By number of cylinders used to expand the steam
   - Single cylinder
   - Compound cylinder
     - Cross
     - Tandem
   - Multiple cylinder

447. General Types of Steam Turbines. — In an **impulse type of turbine** *the expansion and consequent change in pressure of the steam occurs entirely within the nozzles, which direct the steam in jets against the moving buckets.*

In a **reaction type of turbine** *the steam is directed against the moving buckets by guide vanes or orifices, and the pressure of the steam changes as the steam expands through both stationary guide vanes and moving buckets.*

The forces of impulse and reaction are present in both these types, and the designation is made in accordance with the predominating effect. The meaning of these terms can be made clear by considering a jet of steam issuing from a stationary nozzle and impinging against a movable flat plate. The force causing the plate to move is due to impulse, because it is a force acting in a forward direction. The jet as it leaves the nozzle exerts a reaction, or backward, force upon the nozzle. As applied to commercial turbines, impulse and reaction are illustrated by a nozzle, Fig. 381b, discharging against a movable curved plate. The force causing motion of the plate is a combination of impulse and reaction, and is greater than the impulse or the reaction force acting alone. Buckets are thus made with a shape that reverses the jet of steam as much as is practical, thereby increasing the force acting on the bucket. The forms of buckets used in commercial impulse and reaction turbines are shown in Fig. 381a and c. The impulse element is better adapted to small steam volumes and the reaction to large steam volumes.

GENERAL TYPES OF STEAM TURBINES

In a **multi-pressure stage turbine** the total drop in pressure is divided into a number of small pressure drops, each of which is used as though in a separate turbine, and all the wheels of the turbine are carried on a common shaft. In the impulse type of pressure-stage turbine, the total drop in pressure of the steam, from inlet to exhaust outlet, is divided between two or more sets of impulse-type nozzles, each set having its row of moving buckets. The reaction turbine, from its construction, is a multi-pressure stage turbine, each stage being considered as made up of a single row of stationary vanes and its corresponding row of moving buckets. Consequently there are a larger number of stages than there are

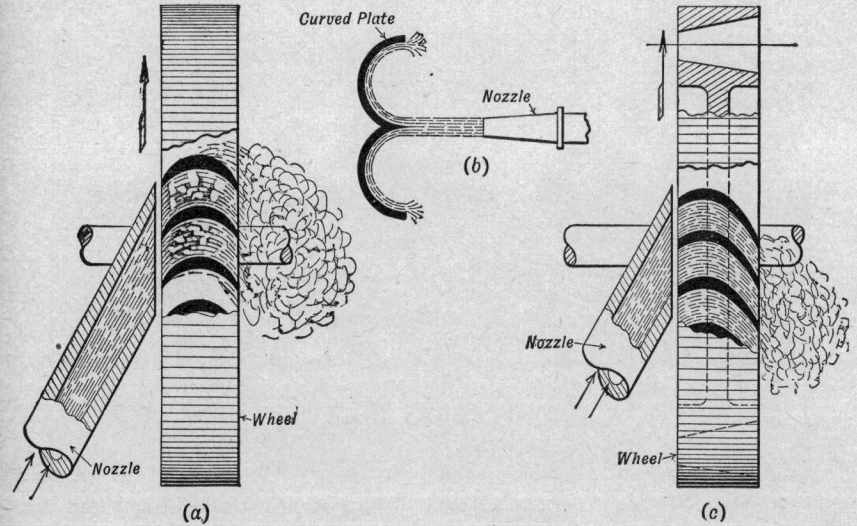

FIG. 381. — Action of Jet in Steam Turbines, and Commercial Forms of Impulse and Reaction Buckets.

for an impulse turbine, some reaction turbines having as high as 70 stages. *The multi-pressure stage construction is also called pressure compounding.*

In the **multi-velocity stage impulse turbine,** there are two or more rows of moving buckets to take up the velocity of the jets of steam, which is expanded from initial to final pressure in one set of nozzles. The steam is directed by the nozzles upon the first row of moving buckets. The steam leaving the first row of buckets is re-directed against a second row of moving buckets by the first row of stationary vanes, and so on through the various rows of buckets used to absorb the velocity.

The object of multi-staging is to reduce the rotative speed of the turbine shaft. When the entire expansion occurs in a single set of nozzles, as in the De Laval single-stage turbine, using a single row of buckets, the velocity attained by the jet of steam is between 3500 and 4000 feet per second, and to take the full energy out of the steam, the wheel velocity should be one-half the velocity of the steam jet. This speed is prohibitive, except with wheels having a small diameter,[1] on account of the strength of the materials required. For best economy, however, the corresponding velocity of the buckets is between 1200 and 1400 feet per second. By using multi-stage velocity turbines, the velocity of the buckets may be made less, since the initial velocity is taken up in several rows with a smaller amount per row. This permits using wheels of larger diameter with lower bucket speeds.

The same result is obtained by using pressure staging, since the pressure drop per stage is only a portion of the total, and the velocity to be taken up by each row of buckets is correspondingly reduced.

Fig. 382. — 35-kw. De Laval Impulse Steam Turbine with Reduction Gear.

448. Simple Impulse Steam Turbine. — A simple form of De Laval steam turbine representative of the elementary turbine, is shown in Figs. 382 and 383. It has a steel casing with three bearings supporting a flexible shaft, which carries a nickel-steel wheel having a single row of drop-forged steel buckets, around its periphery. A section of the rim of the wheel, Fig. 384, shows the buckets with bulb shanks forced into slots in the wheel. The upper ends of the buckets have projections, which fit closely together, forming a continuous ring around the circumference. On account of the high velocity of rotation, 20,000 to 30,000 r.p.m., the shaft is made light and flexible to permit the wheel and shaft to rotate about the center of mass and thus to prevent vibrations.

The bearings are made in halves, lined with anti-friction metal, reamed inside, and ground true outside. The outer bearing carries most of the weight on the shaft and has a spherical seat in the casing. This gives

---

[1] In the De Laval single-stage turbine the diameter of the wheel, to the center of the buckets, in a 5-hp. turbine, is 3.94 in., and the rotative speed is 30,000 r.p.m.

SIMPLE IMPULSE STEAM TURBINE

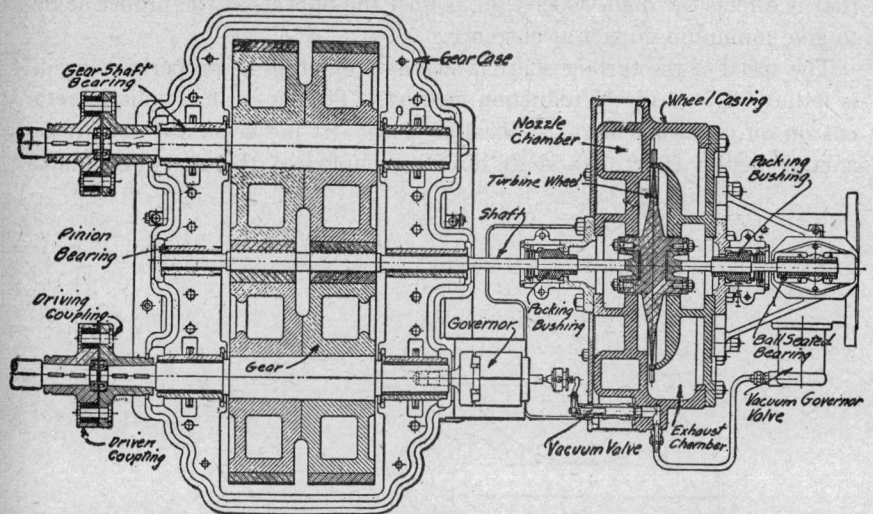

FIG. 383. — Horizontal Section of De Laval Double Geared Turbine.

flexibility and makes the bearing self-aligning. A helical spring, held in position by a cap, presses the middle bearing against its seat. The inner bearing is flexible and free to oscillate, and acts as a stuffing box to prevent leakage of steam from the casing. These bearings are oiled from a central reservoir.

Small disk wheels have holes through the center, and are forced on tapered sleeves shrunk on the shaft. The larger wheels are made solid for greater strength, and are bolted to flanges on the shaft. The thickness of the wheel section is increased from the rim to the center to aid in equalizing the stresses in the wheel. At the base of the rim, below the buckets, it is made thin enough to confine a break to that part of the wheel, in case of damage to the row of buckets.

Steam passing through the **double-beat poppet throttle valve,** Fig. 385, enters the **expanding nozzles,** in the steam chest, which are controlled by needle valves adjusted by hand. The number of nozzles depends on the size of the turbine, and ranges from 1 to 15. The cross section of the nozzles is generally round, though square or rectangular sections are sometimes used; their shape is such that the steam attains a maximum velocity during expansion. The nozzle is so located

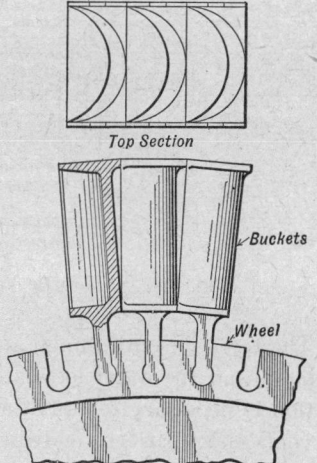

FIG. 384. — Buckets of De Laval Turbine and Method of attaching to Wheel.

that it directs the high-velocity jet against the bucket, at the proper angle to give minimum impact at entrance.

The speed of the turbine shaft, which is too high to be properly utilized, is reduced by means of **reduction gearing.** The pinion has **helical teeth** cut on an enlargement of the flexible shaft. It meshes with and drives a large helical gear on a shaft directly coupled to the generator shaft.

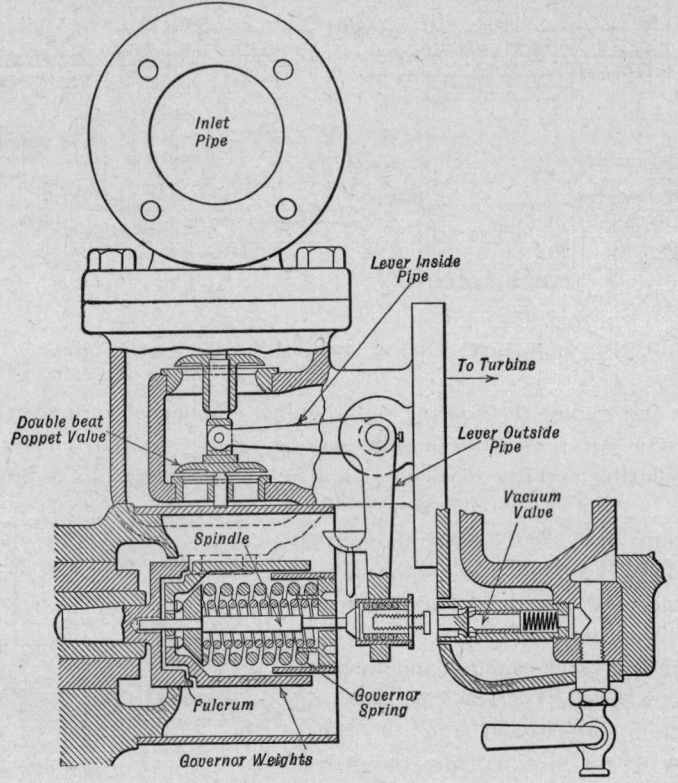

FIG. 385. — Throttle Valve and Governor De Laval 35-kw. Turbine.

The teeth on both pinion and gear are cut in two rows, one row cut right-hand and the other left-hand, at an angle of 45 degrees. This construction is necessary for quiet and efficient operation at high speeds and prevents end thrust, or endwise movement of the shaft.

The gear case is substantial and carries ample bearings for the pinion and gear shaft. The gear centers are of cast iron with steel rims shrunk on them. The teeth are cut on the rim with great accuracy and, if they are maintained in proper alignment with suitable lubrication, the wear after long periods of operation is slight, since the pressure on the teeth at the high speed of rotation is low.

The **governor** is a simple centrifugal throttling governor, Fig. 385, generally attached to the slow-speed shaft of the reduction gear. *It controls the speed by regulating the pressure of steam admitted to the nozzles.* The governor weights are cylindrical and surround a helical spring. This spring rests on a collar, against which pins in the governor weights bear. Knife-edges on the governor weights bear against a fulcrum on the main governor support. When the speed increases above normal, centrifugal force causes the governor weights to swing farther apart. This compresses the governor spring and moves the central spindle to the right, thus moving the bell-crank lever and operating the double-disk poppet admission valve to reduce the opening for steam, thus reducing the speed. The admission valve is normally held open by a spring attached, outside, to the bell crank. An adjusting nut screwed into the governor body controls the initial compression of the governor spring, and hence the speed of the turbine.

When the turbine is operated condensing, quick closing of the admission valve in case of emergency does not decrease the speed rapidly enough. Therefore, it is arranged so that further movement of the governor pins compresses a stiff spring in the stationary bell crank, and opens a vacuum valve. This admits air to the wheel chamber and decreases the speed by reducing the vacuum. This single-wheel turbine is built in sizes up to 600 horsepower, with peripheral speeds as high as 1300 feet per second.

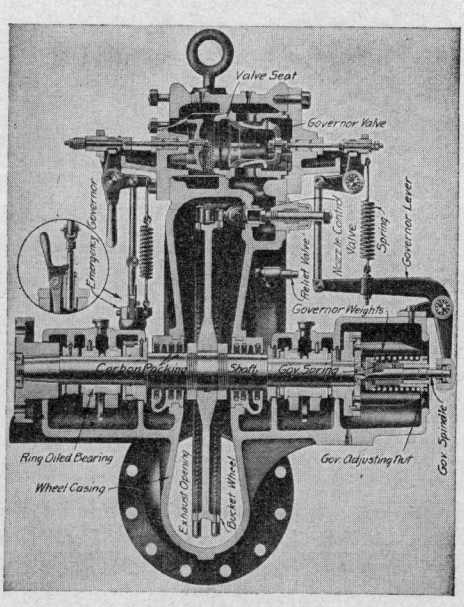

FIG. 386. — De Laval Velocity-stage Turbine.

The more recent types of De Laval turbines are of the velocity or pressure-stage construction and will be described as illustrative of these types.

449. De Laval Velocity-stage Impulse Turbine. — The velocity-stage De Laval turbine, Fig. 386, has two velocity stages, with the steam wholly expanded in one set of nozzles and the resulting velocity taken up in the rows of moving buckets, with the necessary stationary guide vanes for re-directing the steam upon the moving buckets. The forms of the nozzles are shown in Fig. 387. The buckets are similar to those described in the previous article, except that they are fastened to the wheel by a shank which passes through the flange of the wheel and is then riveted over. The shaft to which the bucket wheel is keyed is short and stiff, with a diameter sufficient to prevent vibration. It is supported at each end by a plain ring-oiled bearing. Thrust collars at the right-hand bearing prevent end play of the shaft. Leakage of steam around the shaft, from or into the casing, is prevented by **carbon ring packing**, Fig. 401a, page 492, supported in accurately machined housings and arranged for steam sealing when operated condensing. The speed is regulated by a throttle valve under the control of a centrifugal governor mounted directly on the turbine shaft.

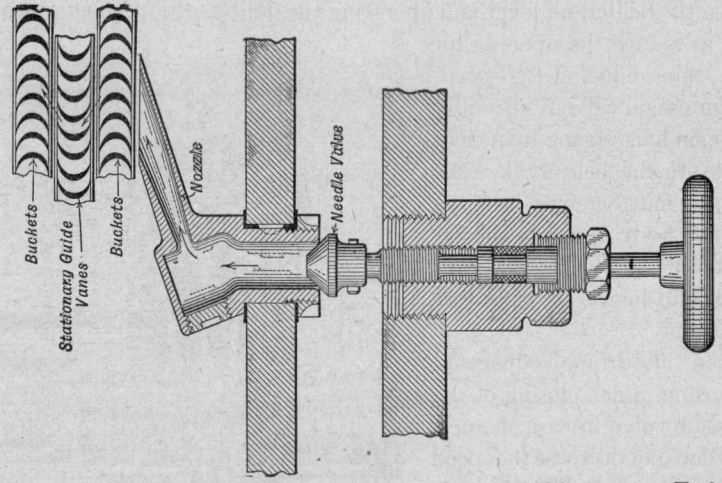

Fig. 387. — Nozzle and Bucket Arrangement of De Laval Velocity-stage Turbine.

450. Multi-pressure Stage Impulse Elements. — The impulse elements, consisting of sets of nozzles and the corresponding buckets, for medium- and large-sized pressure stage turbines, are of either the **Rateau** or **Curtis type**, Fig. 388. In the Rateau type a pressure stage is formed by expanding the steam in a single set of nozzles and taking its velocity up by a single row of buckets, the number of stages depending upon the capacity desired. In the Curtis type, the steam is expanded in a single row of nozzles to give it sufficient velocity to carry it through two rows of moving buckets and one row of stationary buckets, the nozzles and buckets forming one pressure stage.

Each type has its advantages. The Curtis type is especially suitable for high initial pressure. A combination of these types is used on most large turbines.

451. De Laval Pressure-stage Impulse Turbine. — This turbine, which is made in sizes from 50 to 15,000 horsepower, is illustrative of the Rateau type of construction, the general arrangement of the parts being shown in Fig. 389. A heavy shaft carries separate bucket wheels, each revolving in a separate chamber between diaphragms held by a cylindrical casing. The wheels are mounted on the shaft by means of tapered bushings which are drawn into position by a nut. Three bearings support the shaft, two of which are ring-oiled and the third a marine type of thrust bearing.

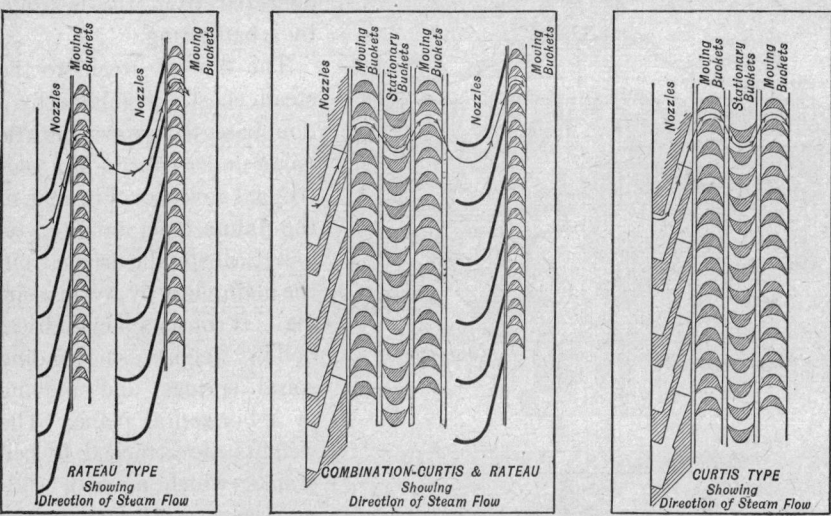

FIG. 388. — Rateau and Curtis Turbine Elements.

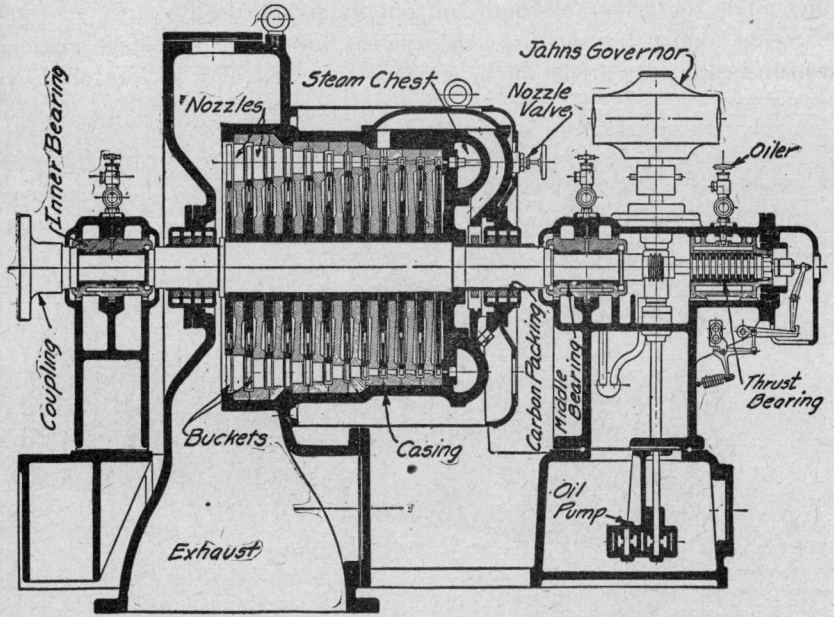

FIG. 389. — De Laval Pressure-stage Impulse Turbine.

To reduce the leakage of steam between stages and from the casing at the high-pressure end, removable **labyrinth packing rings**, Fig. 401b, page 492, carried by the diaphragm and casing, surround the rotating hub and shaft.

The space between the shaft and rings may be piped to a lower pressure stage or to the condenser. Condensation passing by the packing rings is deflected from the bearings by a **baffle ring.**

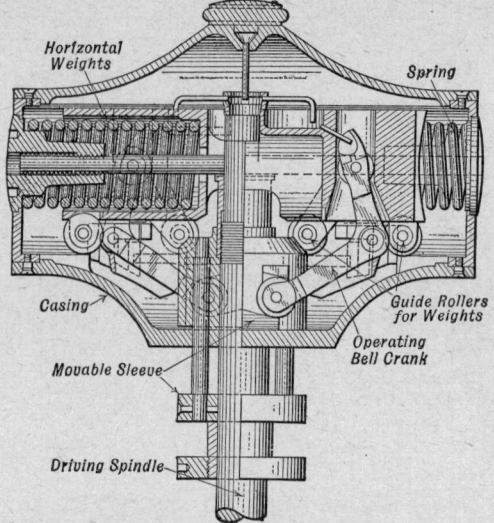

Fig. 390. — Jahns Governor.

The flow of steam to the steam chest is regulated by a double-seated poppet throttle valve under control of a centrifugal governor, Fig. 390, of the **Jahns type,** mounted on a vertical spindle driven from the main shaft by worm gearing. It consists of cylindrical hollow weights surrounding spiral springs and moving in a horizontal plane. The weights are attached to bell cranks which move a sliding sleeve connected to the throttle valve. Roller bearings guide the bell cranks both horizontally and vertically.

Steam admitted to the steam chest passes through the first-stage nozzles, which occupy only a part of the circumference and may be controlled by

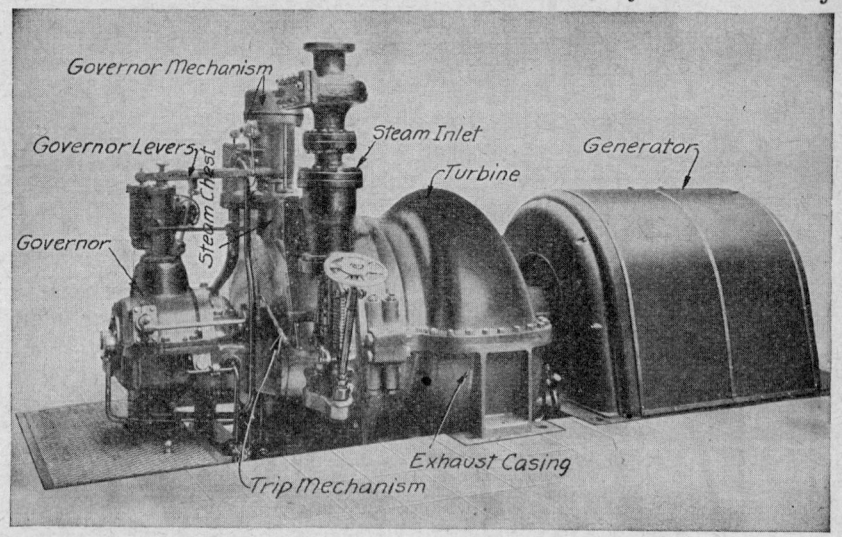

Fig. 391. — Curtis 7-stage Condensing Turbine — Outside View.

hand-operated valves. After leaving the first row of moving buckets, the movement of the steam is reversed in the **stationary guide vanes,** which are set in around the entire circumference of the **diaphragm.** As it leaves the guide vane, the steam is expanded again in short nozzles formed between the vanes in the diaphragm, and is thus re-directed against the second row of moving buckets. The path of the steam through the stages, each consisting of a row of nozzles and buckets, is continuous until the pressure has fallen to that of the steam at the exhaust connection. The height of the buckets increases as the exhaust end is approached, to allow for the increase in volume of the steam.

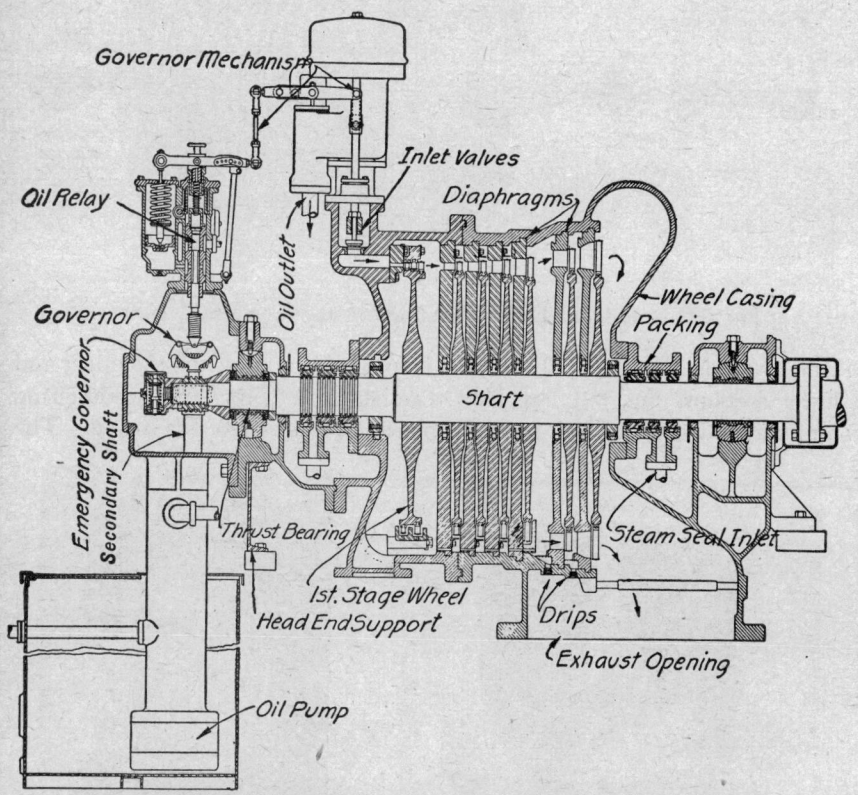

FIG. 392. — Curtis 7-stage Condensing Turbine — Sectional View.

452. Curtis Impulse Turbine. — Curtis turbines are of the pressure-stage impulse type, ranging from single-stage single-cylinder turbines of small power up to 20 stages in the larger 80,000-kilowatt units. The smaller turbines have Curtis multi-pressure staging, and the larger turbines have a combination of the Curtis and Rateau types. The 5000- to 10,000-kilowatt turbines were formerly made with vertical shafts, but in recent years horizontal construction has been used. Multi-cylinder turbines are also manufactured in capacities up to 208,000 kilowatts, with the cylinders arranged tandem, cross or vertical compound.

Outside and sectional views, Figs. 391 and 392, show the general construction of a medium-sized horizontal Curtis condensing turbine. The casing enclosing the wheels is made of cast iron, and carries the bearings

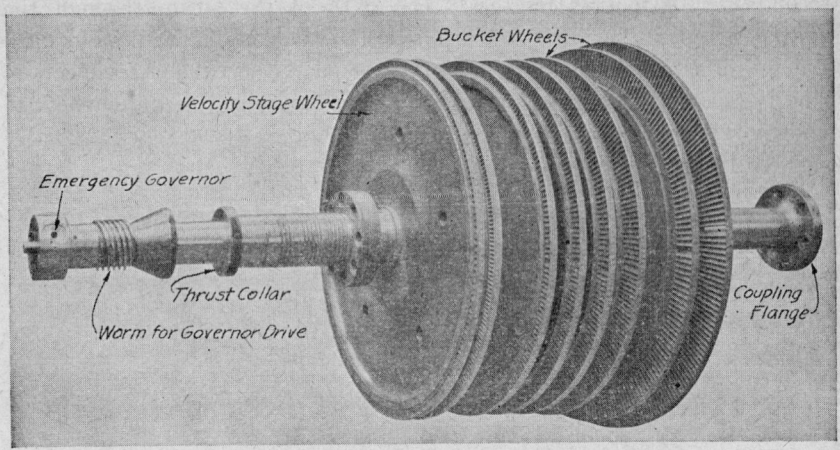

FIG. 393. — Rotating Element for Curtis 7-stage Condensing Turbine.

which support the shaft. The casing is split horizontally, and the upper and lower sections, Fig. 391, are held together with bolts. Non-conducting material, covered by sheet-metal, is placed over a part of the casing. The

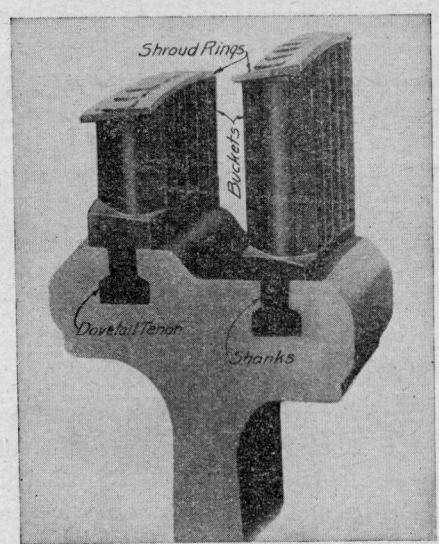 

FIG. 394a. — Method of Assembling Buckets — Curtis Turbine.   FIG. 394b. — Curtis Turbine Wheel showing Bucket Assembly.

location of the oil-operated valve gear, governor, governor shaft and steam connections is shown.

The rotating element, consisting of steel disks or wheels, mounted on a stiff shaft, is shown in Fig. 393. The first wheel carries two rows of buckets, arranged on the wheel as in Figs. 394a and b; the other wheels carry one row each. Each wheel runs in a separate chamber, formed by diaphragms supported between the wheels by the frame of the turbine.

The buckets are made of steel alloyed with nickel, vanadium, or chromium, the composition depending upon the strength required by the bucket sections. They are drop-forged to exact dimensions, and the **shanks** are then milled to form a **dovetail tenon,** which is inserted into a corresponding slot in the rim of the wheel. The tips of the buckets pass through slots in **shrouding rings,** and are riveted over to make a secure fastening, the method of attaching the buckets and shrouding being shown in Figs. 394a and b. The spacing, or distance between the buckets, is fixed by a projection on the base of the bucket. The opening made for insertion of the buckets into the slot is filled by a spacing block after the last bucket is in place. Buckets are now made with slots instead of tenons, Fig. 395. The stationary buckets are made like the moving buckets and are secured to segments which are bolted to the frame.

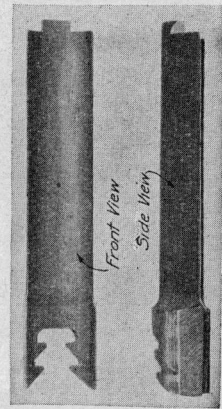

FIG. 395. — Impulse Buckets.

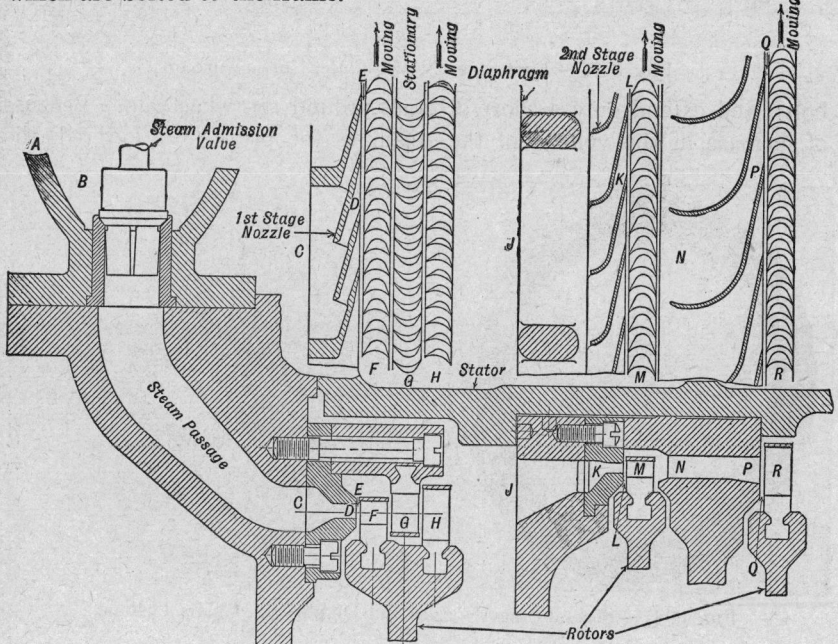

FIG. 396. — Diagrammatic Arrangement of Moving and Stationary Elements of Curtis Turbine.

490   STEAM TURBINES

The plan and elevation, Fig. 396, of sections through two stages of this turbine show the general arrangement of nozzles and buckets. The first-stage nozzles, Fig. 397, are formed in a plate attached to the casing by

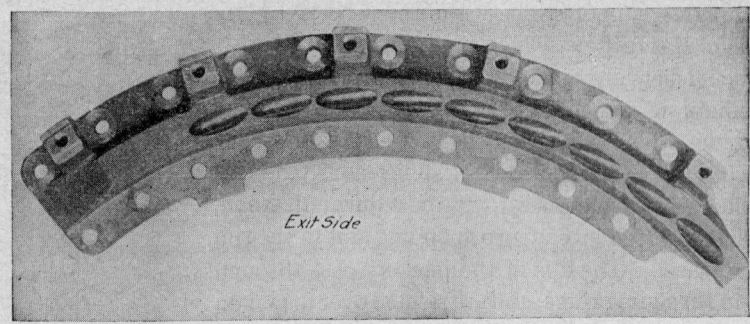

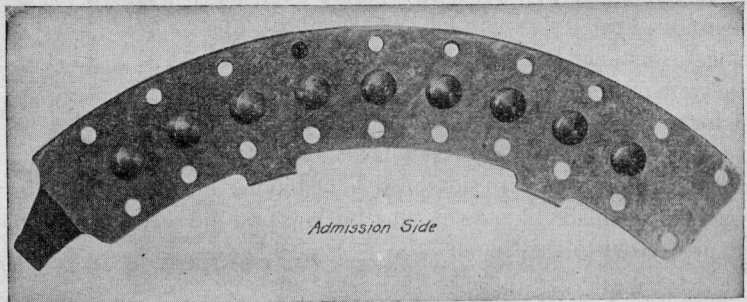

Fig. 397. — First Stage Nozzle Plate, Curtis Turbine.

bolts, and extend only a short distance around the wheel rim. Because of increase in the volume of the steam as the pressure is lowered, the

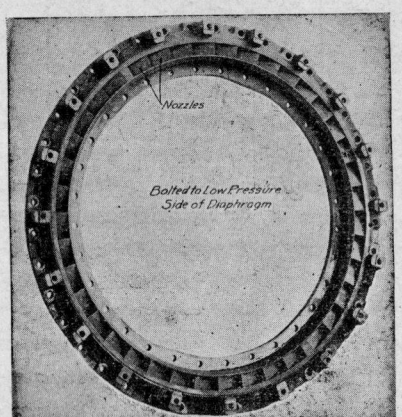

Fig. 398. — Second Stage Nozzles and Diaphragm, Curtis Turbine.

nozzle arc and the arc of the stationary reversing buckets, if used, are made progressively longer for the succeeding stages. In the lowest stage they extend entirely around the wheel. For this reason also, the height of the buckets is increased from stage to stage. The construction of the second-stage diaphragm and nozzles is shown in Fig. 398.

An increase in height on the exit side of a bucket is necessary to allow for decrease in velocity of the steam. This also accounts for the increase in height of the buckets in the same and following stages.

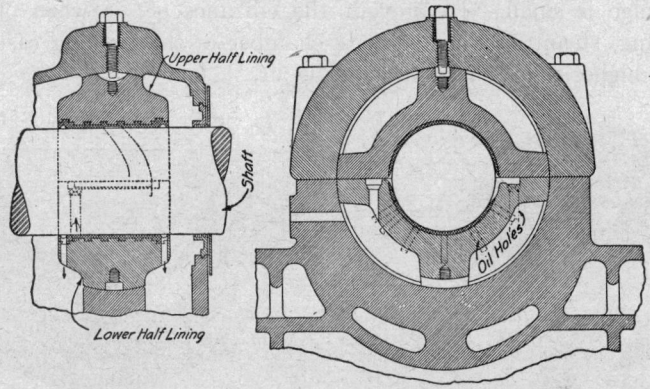

Fig. 399. — Curtis Turbine Bearing — Sectional Views.

The bearings have a shell divided horizontally into halves and lined with babbitt. A short length at the center of the shell is made spherical, to fit into a corresponding seat in the pedestal, for purposes of alignment. The bearing, Fig. 399, shows in detail the arrangement of the parts. The upper shell is prevented from rotating by a **screw bolt** extending into a slot in the upper half of the shell. Oil under pressure is fed through oil-holes to the lower half of the lining. The outer bearing ordinarily carries a **babbitt thrust bearing,** Fig. 400, which controls the position of the rotating parts, and consists of a collar keyed to the shaft of the turbine, and running between an inner and an outer thrust ring, having bearing faces. This view also shows the location of the emergency trip mechanism relative to the thrust bearing.

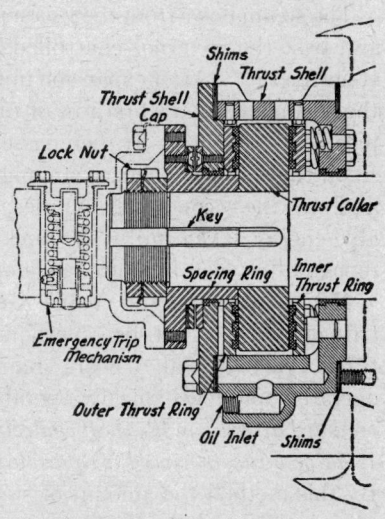

Fig. 400. — Babbitt Thrust Bearing.

Leakage of steam out of the high-pressure casing or between pressure stages, or leakage of air into the low-pressure end, is prevented by either carbon or labyrinth packing. Assemblies of **carbon** and **labyrinth packings** are shown in Figs. 401a and b. The carbon packing consists of several segments held between the partitions by a packing box, prevented from rotation and given a light bearing pressure on the shaft, by **flat or garter springs.** The labyrinth packing consists of metallic rings carried by the shaft between which are stationary rings supported by the frame, and held in position by garter springs. On account of the small clearance between the rings and the shaft and the long zigzag path for the escape of steam, the leakage is small. To prevent the entrance of air when operating condensing, steam is admitted to a chamber in the packing casing at a pressure higher than that of the outside air.

(a)

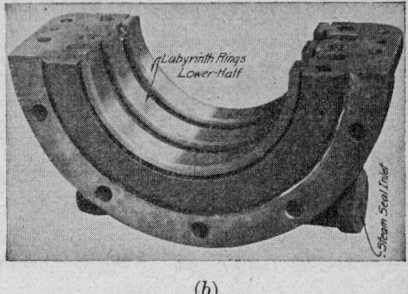
(b)

Fig. 401. — Carbon and Labyrinth Packing.

The steam flows from the steam pipe through a strainer in the steam chest, and past the governor-controlled admission valves, page 487, to the first-stage nozzles. After expansion in the nozzles, which results in high velocity, the steam strikes the first row of moving buckets on the rotor and gives up part of its velocity. It then passes to the stationary reversing buckets and so on through the velocity staging, to the second-stage diaphragm opening and into the second-stage nozzles, where it is again expanded and its available energy taken up by the buckets. This expansion and absorption are repeated for each pressure stage of the turbine.

453. Governors and Valve Gears for Curtis Turbines. — Both centrifugal and inertia type governors have been used on Curtis turbines, with three types of valve gears, namely, hydraulic or oil-operated, steam-operated, and mechanically operated. *In small units, a centrifugal governor mounted directly on the shaft controls the speed, by throttling the steam pressure. In large units, the speed is controlled by " cutting out nozzles " of the first stage.* By this method the amount of steam is varied, while the pressure is maintained constant, thereby keeping a constant, velocity in the nozzles and buckets.

454. Centrifugal Curtis Governor. — A sectional view, Fig. 402, of the centrifugal governor used with the hydraulic valve gear shows the weights, main spring, governor lever, or beam, and synchronizing springs, together with the other details. The weights, main spring, and connection rod or spindle are revolved by a vertical governor spindle, driven from the turbine shaft by a worm and worm gear. As the speed increases, the outward

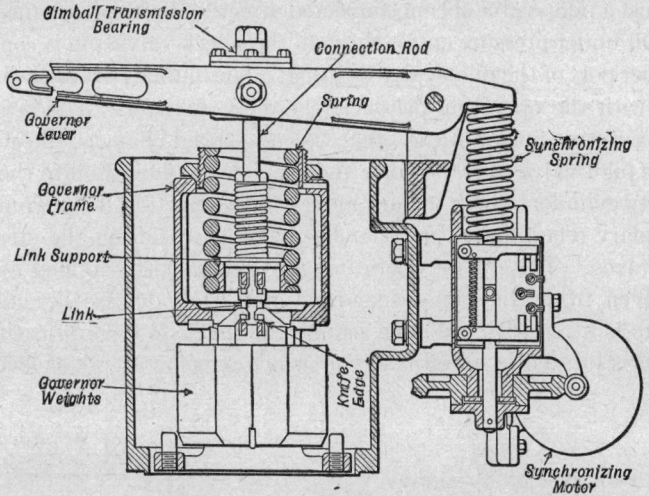

Fig. 402. — Centrifugal Governor for Curtis Horizontal Turbines.

movement of the weights extends the spring, and the motion is transmitted by the spindle to the governor lever, and thence through the valve gear to the admission valves.

The **auxiliary spring** is used to regulate the speeds through small limits, when synchronizing. Its tension is adjusted by either a handwheel, worm and worm gear, or a synchronizing motor.

455. Speed Governor and Primary Relay. — This governor and primary relay, Fig. 403, operates the governor lever through the agency of the oil pressure and spring-actuated piston, and the connecting rod. The governor is mounted on a secondary shaft driven from the main turbine shaft, page 487, by a worm gear. The forces in the governor are

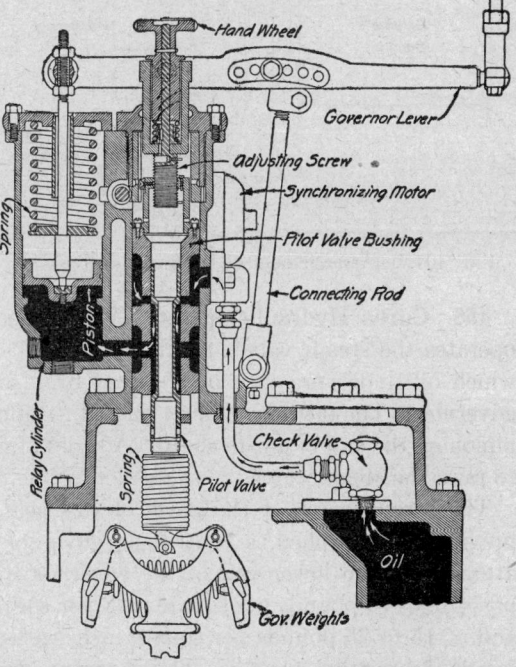

Fig. 403. — Speed Governor and Primary Relay.

balanced in order to minimize friction, and the governor is quick to respond to speed changes. The centrifugal weights are opposed by a coil spring, and bear against a pilot-valve spring connected directly to a small balanced pilot valve. Oil under pressure enters through the check valve and is conducted to the upper port of the pilot-valve bushing. The middle port of the bushing connects with the operating piston in the relay cylinder. The lower port serves as a return for oil in the tank. As the speed changes, the governor moves the pilot valve over the ports and controls the flow of oil to the piston in the relay cylinder, which in turn moves the connecting levers controlling the secondary relay apparatus, thereby opening or closing the steam admission valves. The oil for operating the governor is supplied by an oil pump driven from the same secondary shaft that drives the governor. This pump is fitted with a double strainer, supplies oil for lubricating purposes, and is located in an oil reservoir which is below the floor level.

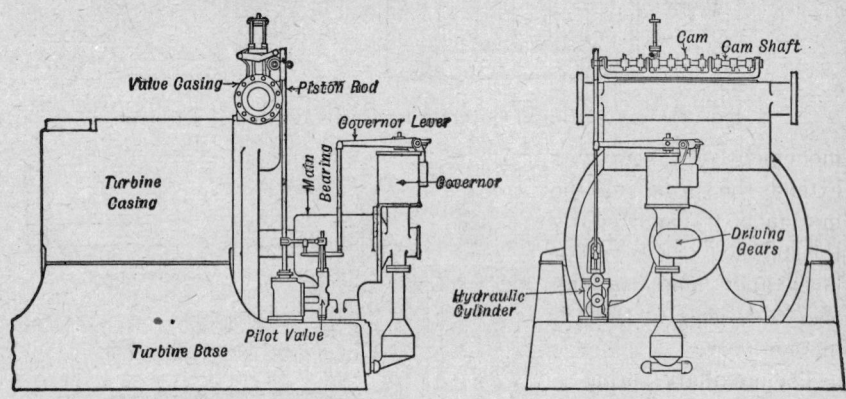

FIG. 404. — Arrangement of Hydraulic Valve Gear on Horizontal Curtis Turbines.

456. Curtis Hydraulic-operated Valve Mechanism. — This valve gear operates the steam valves by the movement of a piston in a cylinder, to which oil under pressure is admitted by a **pilot valve** controlled by the governor. The piston rod has a **rack** at its upper end which engages with a pinion on the end of a cam shaft. A series of cams and levers are arranged to raise the steam valves in succession as the cam shaft is turned.

The arrangement of the parts of this gear is shown in Fig. 404. As previously mentioned, a centrifugal governor is used. A gear oil pump attached to the lower end of the governor spindle, furnishes oil under a pressure of 75 pounds per square inch, for the oil cylinder of the valve gear, and at 15 to 25 pounds per square inch, with the aid of a reducing valve, for the lubricating system. The **governor lever** is attached to one end of the **floating lever,** which is pivoted on a pin carried by the pilot valve stem, Fig. 405. The opposite end of the floating lever is connected by links to the piston rod of the oil cylinder. When the speed changes, the governor lever moves the pistons of the pilot valve away from their normal position, which is over the ports of the oil cylinder and oil is admitted under pressure to move the piston. This moves the rack and pinion and turns the **cam bar or shaft,** which operates the levers that raise or lower the steam valves. At the same time, the movement of the piston rod returns the pilot valve to its central position, ready for the next change in speed.

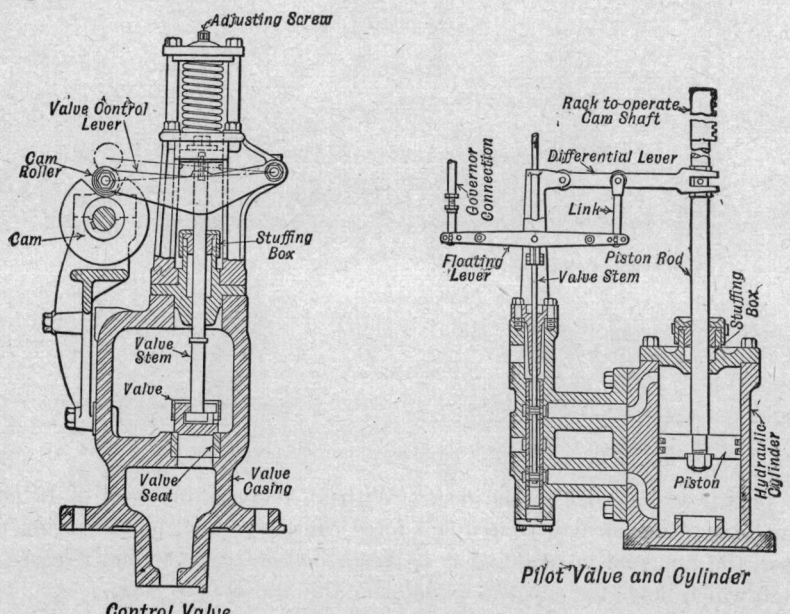

Fig. 405. — Hydraulic-operated Control Valve with operating Cylinder and Pilot Valve.

If the speed of the turbine shaft drops, the governor beam raises the end of the floating lever, and the pilot valve stem admits oil above the piston. This lowers the piston, piston rod and rack, turns the cam bar and raises one or more steam valves. Lowering the oil piston also lowers the pilot valve to its stationary position.

457. Poppet Type Valve Gear and Secondary Oil Relay. — This valve gear consists of the parts shown in Fig. 406. Poppet valves are attached to a **lifting bar** so arranged that it opens or closes the valves in sequence. As the primary relay is moved by the governor to correspond with the variations in speed, it operates a secondary relay through a system of connecting levers, and admits or discharges oil to or from the chamber beneath the operating piston. As the piston moves, it raises or lowers the lifting bar to cut in or out sufficient nozzles to carry the load. A compression spring located at the upper end of the lifting rods attached to the lifting beam tends to keep the **lifting rods** in their lowest position.

458. Curtis Emergency Governors. — On small Curtis turbines, two spiral clock springs, in tension, are carried by spring posts projecting from the face of a disk, Fig. 407, on the shaft. The free ends of the springs rest

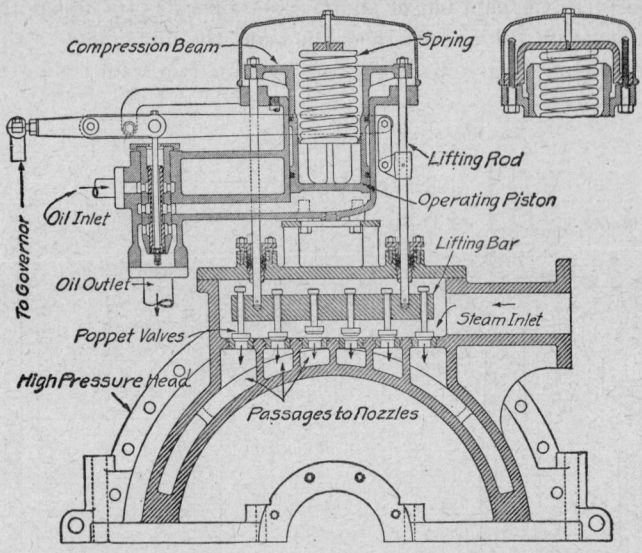

Fig. 406. — Poppet Valve Gear and Secondary Oil Relay.

on pins near the edge of the disk. With an increase in speed of 10 per cent above the normal, centrifugal force causes the end of the spring to leave the pin, and in so doing it strikes a bell-crank lever and releases a latch which holds the throttle valve open and allows it to close.

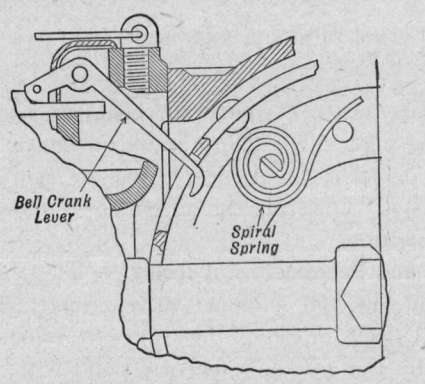

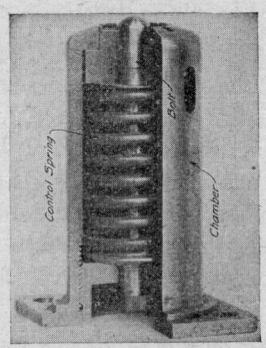

Fig. 407. — Curtis Spring Type Emergency Stop.

Fig. 408. — Bolt Type Emergency Stop.

An emergency governor of the bolt type, Fig. 408, consists of a chamber in which is located a bolt and controlling spring, the whole being mounted in the extension of the main shaft, and is entirely separate from the main governor and valve system. The bolt is normally held concentric with the main shaft by the spring, and as the speed increases above normal, centrifugal force overcomes the force of the spring and the bolt strikes a trip, releasing the latch which holds the throttle valve open, and thus closing the valve.

459. Throttle and Emergency Valve. — A Schutte and Koerting balanced throttle and trip valve, used on the larger Curtis turbines, is shown in Fig. 409, with the more important parts named. The **valve spool** and balancing piston form a single casting and provide a seat for the by-pass valve attached to the valve stem. A stuffing box prevents leakage around the valve stem. Below the stuffing box, the valve stem is threaded through a sliding nut which slides through the bracket on the valve yoke. A lever pinned to this sliding nut normally holds the valve open. With the valve closed, turning the handwheel, " to open," immediately opens the auxiliary valve, and steam above the balancing piston is discharged, thus relieving the pressure above the valve, which is then easily opened. The emergency governor releases the latch at speeds above normal, and the valve is automatically closed by the unbalanced pressure below the balancing piston.

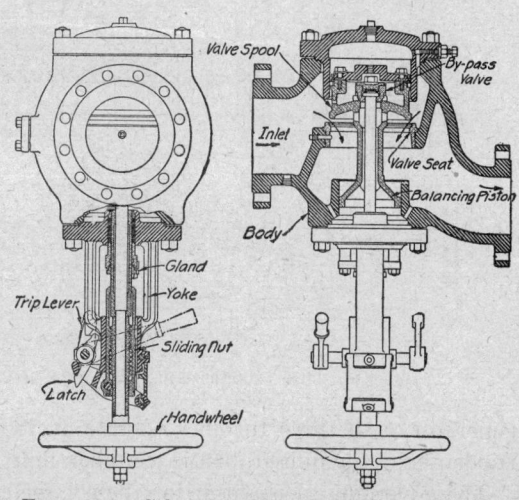

Fig. 409. — Schutte and Koerting Throttle Valve.

460. Large-capacity and Multi-cylinder Curtis Turbines. — The general construction of a 65,000-kilowatt, 1800-r.p.m., 60-cycle,[1] 17-stage, single-cylinder turbine installed at the Edgar Station, in Boston, is shown in Fig. 410, and normally operates at 350 pounds pressure and 720 deg. fahr. temperature, with a steam rate of 9.3 pounds of steam per kilowatt-hour when operating at capacity. This turbine is arranged for extraction of a portion of the steam for feedwater heating at four points. An external view of a 94,000-kilowatt, tandem-compound unit installed in the Long Beach Station is shown in Fig. 514, page 609, and a similar view of the 208,000-kilowatt, triple-cross-compound unit built for the State Line Generating Company is shown in Fig. 411. The *center or high-pressure* turbine is a 76,000-kilowatt turbine which exhausts through steam reheaters

[1] An alternating current is said to complete a cycle when it rises to a maximum in one direction, falls to zero, and then rises to a maximum in the opposite direction. A 60-cycle current would reverse its direction of flow sixty times a second.

into two low-pressure turbines, each of which is of double-flow construction and drives a 62,000-kilowatt main generator and a 4000-kilowatt service

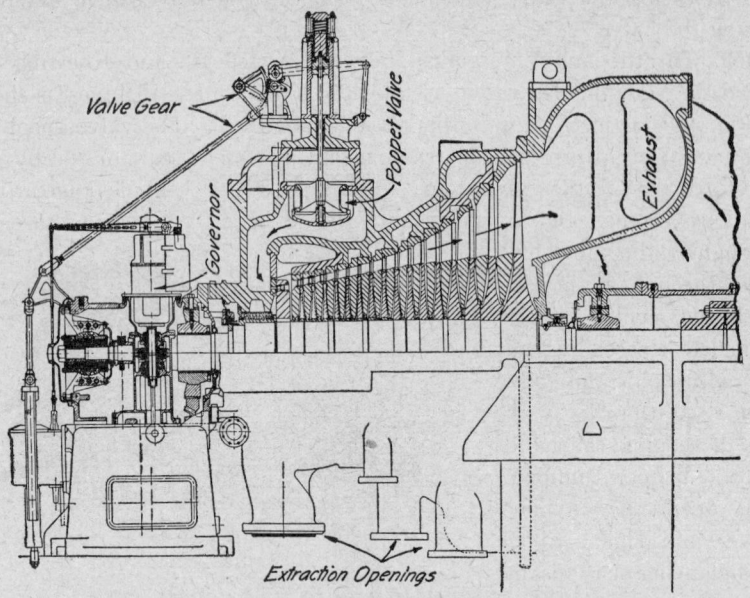

Fig. 410. — 65,000-kw., 17-stage Curtis Impulse Turbine.

generator. All three turbines operate at 1800 r.p.m., and eight vertical condensers are required for the complete unit.

The governing mechanism for these large-capacity machines resembles

Fig. 411. — 208,000-kw., Triple Cross-compound Curtis Turbine.

that installed on the smaller-sized machines, with special provisions to control the speed and the load of the set, and in addition to protect it from all dangers of overspeeding which might arise in the parallel operations of the units. In the case of the triple-compound turbines, arrangements are usually made to bring any one of the three units up to speed, synchronize and load it, when the other two are carrying the load.

461. High-pressure Curtis Turbine. — The general construction of the 1200-pound-pressure turbines is similar to the low-pressure turbines, with the exception that the casings are made of steel and of much heavier construction than the turbines for low pressures. A 1200-pound turbine installed in the Edgar Station of the Boston Edison Co., Fig. 412, has a capacity of 12,500 kilowatts. Whereas the older turbines of this type were superimposed upon the lower-pressure turbines of the plant, the newer units expand the steam from 1200 pounds down to back pressure in a

FIG. 412. — 12,500-kw. High-pressure Curtis Turbine.

single machine. Several recent installations are known as **vertical-compound turbines**, Fig. 413, and are constructed with the high-pressure units mounted vertically above the low-pressure units. One type has the high-pressure element, consisting of turbine and generator operating at 3600 r.p.m., and comprising approximately 25 per cent of the total capacity of the unit, mounted on top of the generator of the low-pressure element which operates at 1800 r.p.m. By this arrangement space is conserved, some piping and one foundation saved, and one set of air coolers can be used for both generators.

A somewhat different design for a 110,000-kilowatt unit, operating at 1200 pounds pressure, has the high-pressure turbine mounted on top of the low-pressure turbine, both elements operating at 1800 r.p.m. The generators are duplicates, with the generator driven by the high-pressure element mounted directly on top of the generator driven by the low-pressure element. Steam enters the upper or high-pressure element of the turbine at 1200 pounds pressure and 725 deg. fahr. total temperature, and exhausts into two adjacent steam reheaters in which its temperature is raised to 550

Fig. 413. — 50,000-kw. Vertical or Steeple-compound High-pressure Turbine.

deg. fahr. Then the reheated steam enters the double-flow, low-pressure element of the turbine and exhausts at one inch back pressure.

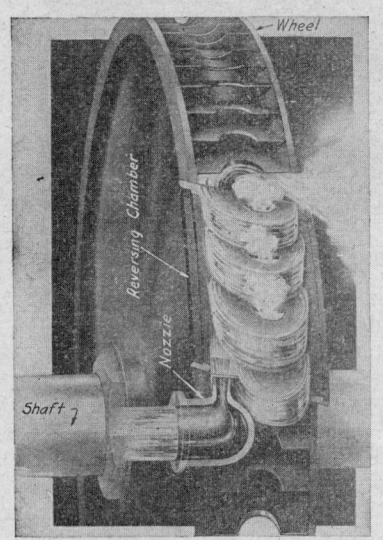

Fig. 414. — Action of Steam in Nozzle and Buckets — Terry Turbine.

462. Terry Turbine, Non-condensing. — This turbine is of the tangential-flow impulse type, using multi-velocity staging and is well adapted for direct drive for blowers, fans, generators, and pumps. The moving buckets are semi-circular and milled in the solid rim of the wheel. The steam expanded in the nozzle enters the buckets at one side, and after leaving the buckets is re-directed on the wheel by a similar set of stationary reversing buckets. It is given several reversals before being discharged, as is shown diagrammatically in Fig. 414. By this method of using the steam, the speed of rotation is reduced. The reversing buckets are arranged in several groups, at intervals around the wheel, and are bolted to the casing as illustrated in Fig. 415, where the top half is thrown back for purposes of inspection. The nozzles are separate from the reversing bucket and have hand-controlled valves to permit using only the number required by the load and thus maintaining high efficiency. The speed is controlled by a centrifugal governor connected to a double-beat poppet valve.

The Terry condensing turbine uses the Curtis type of element in the high-pressure stages and the Rateau type of element in the lower-pressure stages. The governor used on the smaller sizes of this turbine is a centrifugal one, direct-connected to the governor valve, and on the larger sizes an oil relay system is used to operate the governor valve. Carbon

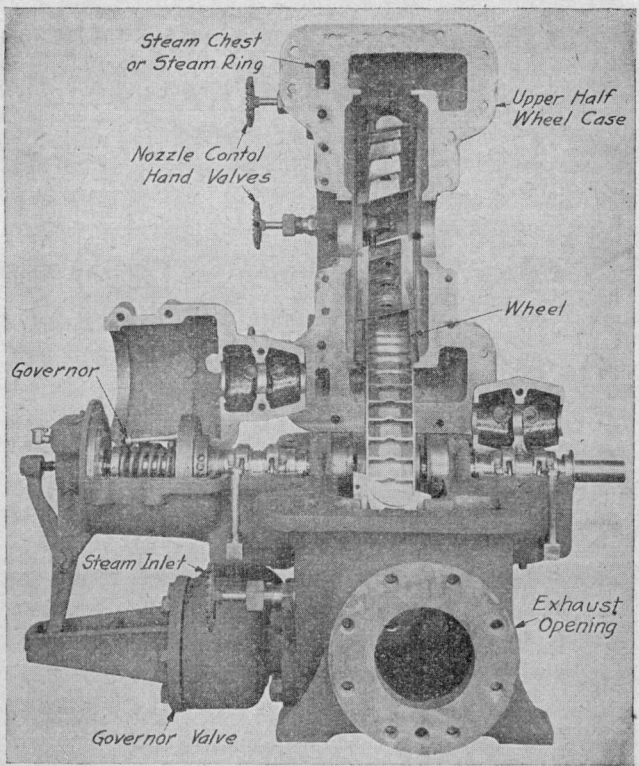

FIG. 415. — Terry Non-condensing Turbine.

ring packing is used to prevent leakage of steam in the smaller sizes, and labyrinth packing in the larger sizes.

A view of the **Sturtevant pelton multi-velocity type turbine**, Fig. 416, shows the arrangement of the nozzle and reversing buckets and the method of attaching them to the casing. The nozzle is made as a part of the reversing bucket casting, Fig. 417. The other parts of this turbine are similar to corresponding parts of the Terry turbine.

A Sturtevant turbine arranged for automatic control is shown in Fig. 418. It is used for driving auxiliaries such as fans and pumps, and is similar in general construction to the turbine shown in Fig. 416, except that it has a relatively large number of independent nozzles, and each

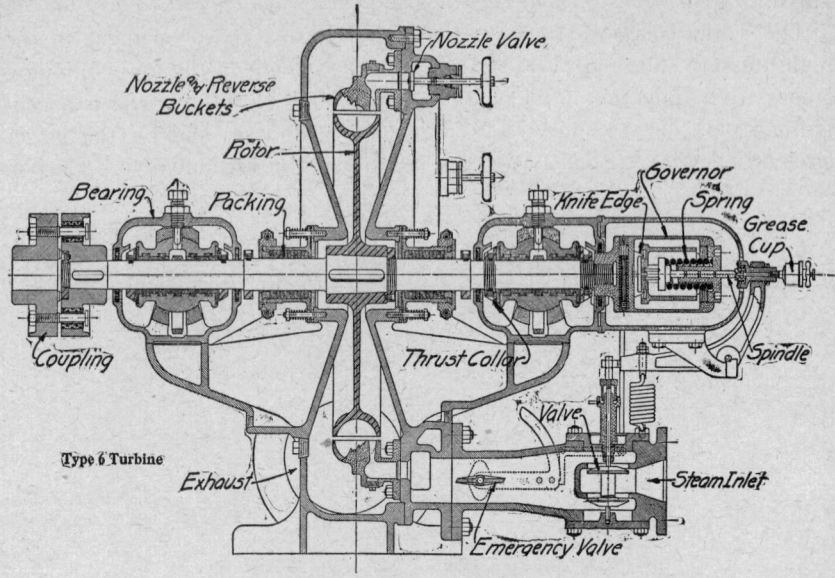

Fig. 416. — Sturtevant Impulse Turbine.

nozzle is controlled by a single control valve which automatically opens and closes to meet changing load requirements. The nozzle valves are operated from a cam shaft which is actuated by a hydraulic relay cylinder.

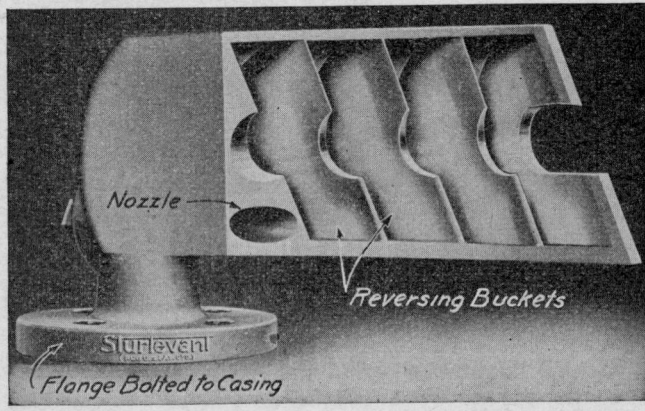

Fig. 417. — Sturtevant Turbine Nozzle and Reversing Buckets.

By this method of control it is claimed that the efficiency of the auxiliary drive is increased since immediate and accurate control of the nozzle is obtained without manual labor.

ELEMENTARY THEORY, IMPULSE TURBINE NOZZLE

Fig. 418. — Sturtevant Turbine arranged for Automatic Control.

463. Westinghouse Impulse Turbine. — The construction, Fig. 419, used for the nozzles, buckets, and reversing buckets enables units to be made as small as one kilowatt. The non-condensing unit has a single row of moving buckets with a single set of reversing buckets arranged at the side of the wheel. By this method, the velocity of the steam is reduced in steps as in the velocity-stage Curtis turbine.

464. Elementary Theory, Impulse Turbine Nozzle. — The action of steam in a nozzle may be briefly described as follows: steam supplied to the **bowl**, or entrance, of the nozzle, Fig. 420, passes through the **throat**, or smallest section, and expands through the conical divergent part of the nozzle, attaining a high velocity at the instant of discharge against the buckets. *The pressure of the steam at the throat drops to about* 0.58 *of the*

Fig. 419. — Westinghouse Impulse Turbine — Casing Removed.

pressure in the bowl, and then decreases to the pressure of discharge at the **mouth**. *The volume of the steam and its velocity increase from the throat to the mouth of the nozzle.*

The velocity at which steam is discharged from a nozzle can be calculated, since all the energy in a pound of steam in the bowl is present in the same pound passing out of the nozzle, neglecting the slight amount of heat lost by radiation during its rapid passage through the nozzle.

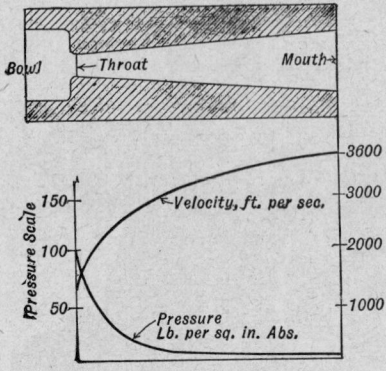

FIG. 420. — Curves showing Pressure and Velocity Changes in an Expanding Nozzle.

The total energy per pound of steam is the sum of its *kinetic energy*, $\dfrac{AV^2}{2g}$, its *internal energy*, $h - APu$, and its *pressure energy*, APu, in which $V =$ velocity in feet per second, $A =$ reciprocal of the heat equivalent of work $= 1/778$, $h =$ total heat per pound of steam, and $APu =$ the external work performed in maintaining constant pressure, as explained in Art. 82, page 98. Denoting the bowl and mouth condition respectively by subscripts 1 and 2, there results, for saturated steam,

$$\frac{AV_1^2}{2g} + (h_1 - AP_1u_1x_1) + AP_1u_1x_1 = \frac{AV_2^2}{2g} + (h_2 - AP_2u_2x_2) + AP_2u_2x_2$$

Since V_1 is very small compared to V_2, and the external work items cancel, the equation becomes

$$\frac{AV^2}{2g} = h_1 - h_2,$$

or

$$V_2 = \sqrt{2 \times 778 \times 32.2\,(h_1 - h_2)} = 223.8\sqrt{h_1 - h_2} \quad . \quad . \quad (108)$$

in which h_1 and h_2 equal the total heat per pound of steam at the bowl and mouth pressures respectively, taking into account the quality of the steam at those points. Since the passage of steam is almost instantaneous, the change in pressure and volume is assumed to be adiabatic, neglecting radiation. The quality after expansion can be calculated with this assumption, or the value h_2 can be found by using the total-heat-entropy diagram.

The actual conditions, however, do not give an adiabatic change on account of the friction of the steam against the surface of the nozzle. The energy lost in friction is given back to the jet of steam in the form of heat, so that the steam at the mouth of the nozzle has more heat per pound than it would have if the expansion were adiabatic. If the effect of friction in decreasing the velocity of the jet is y part of the theoretical heat energy drop, then the velocity, considering friction, becomes

$$V_2 = 223.8 \sqrt{(h_1 - h_2)(1 - y)} \quad \ldots \ldots \quad (109)$$

The value of y varies under different conditions. With the form of nozzle shown it is about 10 per cent for steam discharged to atmospheric pressure.

Example 56. — The steam pressure in the bowl of a nozzle is 100 lb. per sq. in. abs., and the vacuum at the exit is 28 in. of mercury, barometer 29.92. Neglecting friction, find the theoretical velocity of the jet of steam leaving the nozzle, if the initial condition of steam is dry.

Solution. — Using Equation (108) and *steam table*, page 94.

Velocity in feet per second = $223.8 \sqrt{h_1 - h_2}$
= $223.8 \sqrt{1186.6 - 899.7} = 3780$

h at 100 lb. per sq. in. abs. = 1186.6

$h_2 = x_2 L_2 + h_{f2}$ at 1.92 in. mercury absolute = $0.803 \times 1036.6 + 67.33 = 899.7$

From Art. 104 $\quad s_{f_1} + \dfrac{x_1 L_1}{T_1} = s_{f_2} + \dfrac{x_2 L_2}{T_2}$, or $x_2 = \left[s_{f_1} - s_{f_2} + \dfrac{x L_1}{T_1} \right] \div \dfrac{L_2}{T_2}$

Referring to *steam table*, page 94, for entropy values

$$x_2 \text{ at 1.92 in. mercury abs.} = \frac{0.4742 - 0.1283 + 1.1280}{1.8447} = 0.80$$

The weight of steam passing through the nozzle, under ordinary operating conditions, depends upon the area of the throat and the corresponding velocity of the steam at the throat. The area of the nozzle at any point depends upon the weight (W), density (d), and velocity of the steam flowing, and since $Q = \dfrac{W}{d}$, the area equals

$$A = Q \div v \quad \ldots \ldots \ldots \ldots \quad (110)$$

in which Q = volume in cubic feet per second, A = area in square feet, and v = velocity in feet per second.

The **impulse force** of the steam can be found from the equation $F = Ma = \dfrac{Wv}{gt}$ in which t = time in seconds, g = acceleration of gravity, feet per second per second and other symbols as above.

Example 57. — Find the impulsive force of 4 nozzles each discharging 1 lb. of steam at a velocity of 3600 ft. per sec.

Solution. — Using the above equation,

$$F = \frac{Wv}{gt} = \frac{4 \times 1 \times 3600}{32.2 \times 1} = \frac{14,400}{32.2} = 447.2 \text{ lb.}$$

465. Elementary Theory, Impulse Turbine Buckets. — The kinetic energy of the steam jets leaving the nozzles is changed into energy of rotation in the buckets by the direct action or impact of the steam, and by the reaction of the steam leaving the buckets after being deflected in its course.

Consider a turbine wheel to be running with a velocity at the center of the buckets of u feet per second. Steam enters the buckets with velocity V, Fig. 421, at the absolute entrance or nozzle angle α (about 20 degrees). Relative to the bucket, steam enters at angle β and with velocity W. It is deflected by the bucket and leaves at an angle β_1 and with a velocity W_1 relative to the bucket. While in the buckets it has the velocity of the buckets, u. Its direction and velocity of exit with regard to a stationary point, that is, its absolute exit angle and velocity, are found by combining the velocities W_1 and u, giving V_1 as the exit velocity. Work has been done by the change in velocity of the steam. On striking the buckets the steam gives to the buckets its component of velocity in direction of rotation, bd. On leaving the bucket the steam emerges with an absolute velocity V_1 found as previously explained, of which gf is the component in direction opposite to the direction of the bucket. Since the reaction due to exit is equal to the action but in opposite direction, gf is the useful change in velocity. The sum of bd and gf is the complete useful change in velocity of the steam, or acceleration, imparted to the bucket wheel. Since the force, F, equals the mass, M, times the acceleration, a:

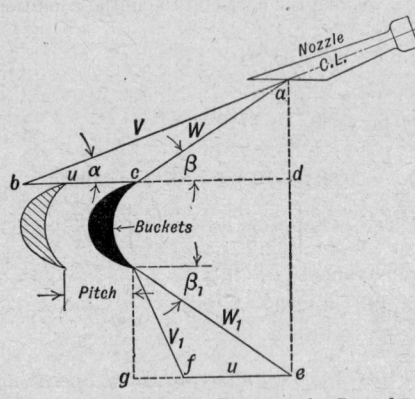

FIG. 421. — Velocity Diagram for Impulse Turbine Buckets.

$$F = Ma = \frac{1}{g}(bd + gf) \text{ per pound of steam} \quad \ldots \quad (111)$$

and since the force was exerted through a distance, s, equal to the velocity (u) of the bucket, or (bc), in feet per second

$$F \times s = \frac{(bd + gf)}{g} bc \text{ foot pounds per second} \quad \ldots \quad (112)$$

To find the work done, V and u are laid out to scale at the angle α and, after the construction is completed, the values of bd and gf are measured to the same scale.

The significance of the terms **absolute** and **relative**, as applied to the steam entering and leaving the buckets of a turbine, may be made clearer by considering the bucket to be a boat moving in a stream at u feet per second. A person on the bank would throw a ball to a person on the boat by aiming ahead of the boat. The actual or absolute direction of motion of the ball would make the angle α with the center of the stream. The person on the boat would receive the ball at the angle β. If the ball were allowed to rebound to the opposite side it would leave at the angle β_1 and with a velocity W_1, relative to the boat; but since the person on the boat is moving at a velocity u the ball would have an actual or absolute angle of leaving greater than β, and an actual velocity V_1 at the opposite shore.

For best turbine economy the nozzle and exit bucket angles should be small with the bucket proportions determined by the volume, velocity and direction of steam flow. For high pressures and velocities the buckets are short, and for low pressures and moderate velocities they are long. The maximum capacity of condensing turbines is usually determined by the effective areas of the low-pressure buckets. Considerations of centrifugal force, with resultant allowable stresses, limit the diameters of the bucket-carrying elements, while considerations of steam flow and bucket fastening limit the height of a bucket which may be applied to a given diameter.

466. Reaction Turbines. — One of the best-known types of reaction turbines is the Parsons steam turbine, introduced in the United States in 1895 and manufactured, by the Westinghouse Electric and Mfg. Co., as the Westinghouse-Parsons turbine. This type of turbine is also manufactured by the Allis Chalmers Co., which has the manufacturing rights for the American Brown-Boveri Co. Reaction turbines, depending upon the capacity of the machine at a given speed, are built: **single-flow,** with the steam passing through the whole turbine in one axial direction; straight **double-flow,** with the steam entering at the center of the turbine and passing to exhaust in two directions; or **semi-double-flow,** with the high-pressure cylinder being single-flow, and the low-pressure cylinders double-flow. The general arrangement of this turbine and its later forms are shown in the diagrams in Figs. 422 to 424.

1. *Single-flow turbine.* — The general construction may be likened to an expanding nozzle made up of a number of transverse sections, alternate sections being carried by a shaft through the center, and the other stationary sections supported by the outside or frame of the expanding nozzle. Each stationary section consists of a row of radial buckets inserted into a dovetailed, circumferential groove in the frame or cylinder, or into blade rings carried by the cylinder. The rotating rows of buckets are carried by comparatively thin cylindrical drums having similar dovetailed grooves in their circumferences, in which the blading is securely fastened, Fig. 422. Steam, after passing the emergency and governor valve, enters an annular chamber in the turbine casing and passes through the stationary and moving buckets, which are made of increasing length to allow the necessary area for the steam to pass, until it reaches the exhaust connection. **Dummy pistons,** connected with **equilibrium pipes** to the exhaust spaces of the second and third drums, are provided to equalize the unbalanced end-thrust along the shaft, caused by the passage of the steam. The dummy pistons are usually grooved to join with corresponding projections

on the casing. This forms a labyrinth packing and minimizes the leakage of steam around the balance piston. The running clearances for both the dummy pistons and the bucket tips must be small. The use of high steam

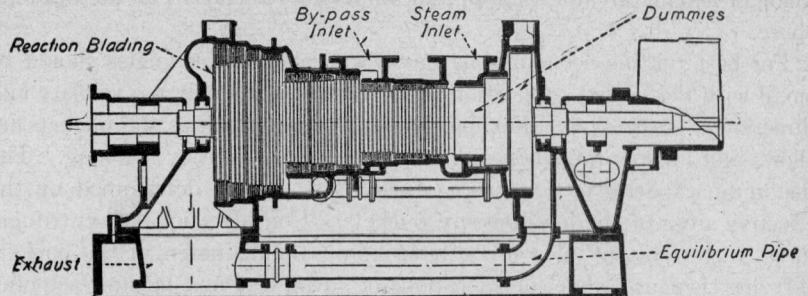

Fig. 422. — Westinghouse Single-flow Reaction Turbine.

pressure has resulted in the substitution of an impulse velocity stage at the high pressure end of the single-flow turbine.

2. *Single-flow, impulse, and reaction turbine.* — An impulse element in this turbine, Fig. 423, replaces the high-pressure drum in the single-flow reaction turbine, generally shortens the machine and makes it possible to

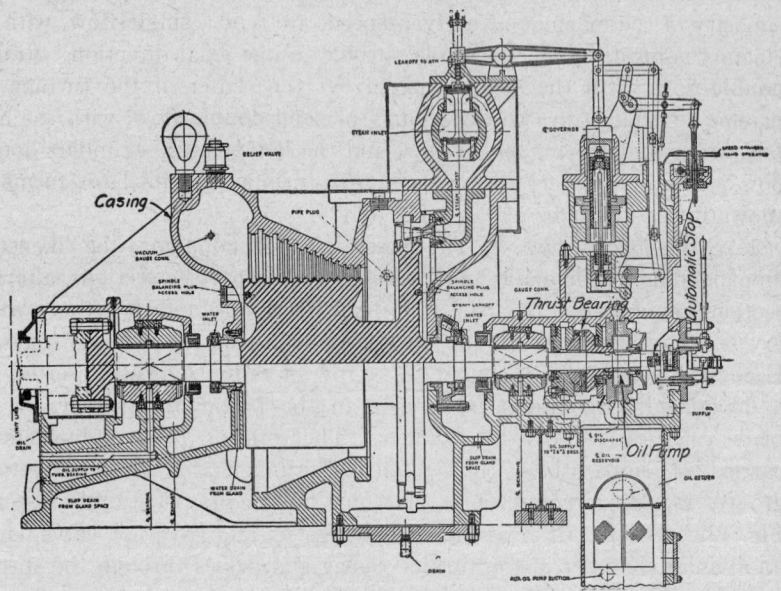

Fig. 423. — Westinghouse Single-flow Impulse and Reaction Turbine.

handle the high-pressure steam more efficiently. The dummy piston is usually omitted, and the axial thrust balanced by a thrust bearing, although in some machines of this type a single dummy piston is used to reduce the load on the thrust bearing.

3. *Double-flow, reaction turbine.* — This construction is ordinarily used as the low-pressure element in the multi-cylinder machine, the steam entering from the high-pressure cylinder at the center, as shown in Fig. 440, page 522, and then divides and passes toward the two ends of the machine through the reaction blading. This is, in effect, two single-flow reaction turbines, no dummy pistons being required to balance the axial thrust. A unit of about double the power is obtained at the same speed. To obtain the increased capacity in a single-flow machine, under the same conditions, the length of the buckets would have to be increased, thus increasing the mean diameter of the bucket ring and the stresses caused by

FIG. 424. — Lower Half of Casing with Rotor, Westinghouse Reaction Turbine.

centrifugal force. The machine may be regarded as analogous to the combination of two multiple-expansion reciprocating steam engines working on the same shaft.

4. *Semi-double flow, impulse, and reaction turbine.* — In this type of turbine, steam first enters a velocity stage impulse element, passes to a small drum, and divides, going in opposite directions through the buckets of the large drums. A dummy ring is used to balance the end-thrust produced by the intermediate single-flow section. This design may be considered analogous to a four-cylinder triple-expansion reciprocating engine, having high pressure, an intermediate pressure and two low-pressure cylinders.

The rotors usually consist of hollow-steel drums, finished inside and outside and fastened to flanged or enlarged ends on the shaft, or may be completely machined from a one-piece solid-steel forging. The drums of larger diameter may consist of separate steel rings balanced and pressed on the central drums. The end-thrust of a single-flow turbine was balanced by steam pressure against dummy pistons in the early type of turbine. The use of the Kingsbury thrust bearing has made it possible to do without the two larger dummy pistons shown in Fig. 422. In turbines combining impulse, or single-flow sections, with double-flow sections, dummy pistons or rings are used for balancing. A rotor for the combination impulse and reaction single-flow turbine is shown in Fig. 424, located in the turbine cylinder or casing, which is conically bored to provide a smooth passage of continually increasing area for the flow of steam. It is made either of cast-iron or steel, depending upon the steam operating conditions, and is divided horizontally at the center line of the shaft permitting ready access to all internal parts.

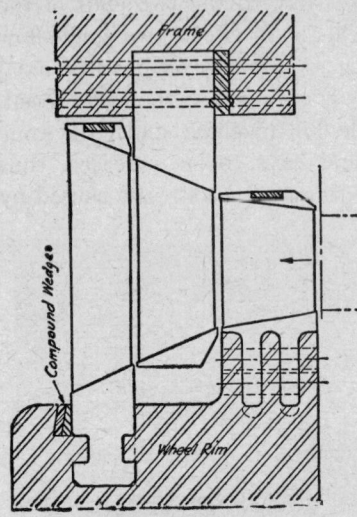

FIG. 425. — Method of installing Impulse Buckets.

The method of securing the impulse buckets in the cylinder and rotor is shown in Fig. 425. The purpose of the compound wedges is to allow the buckets to be properly and securely fixed in position on the rotor without the necessity of extreme accuracy in machine work required for a piece filling the slot. Recent types of impulse buckets are made as shown in Fig. 426, and are fastened to the rotor by suitable pins. The reaction buckets, Fig. 427, are made of phosphor bronze, heat treated, and formed with a shoulder on one side at the root end. At the bottom of the dovetail grooves, on the rotor and cylinder, a rectangular slot is milled

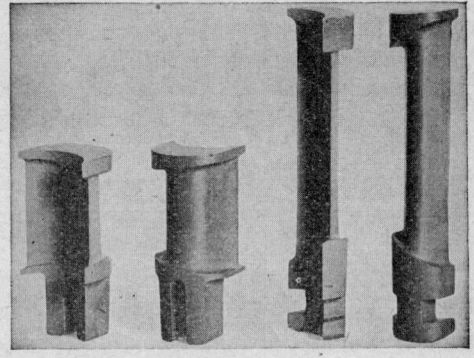

FIG. 426. — Impulse Buckets — Westinghouse Turbine.

to receive the shoulder on the buckets. Steel distance pieces with beveled edges are fitted into the dovetailed grooves above this projection on the buckets, thus holding them securely. The buckets are made thicker at the root, and tapered to the standard section to increase strength and distribute vibration. With the larger sizes of buckets, steel wedges are calked into the space between the buckets and packing pieces and the side of the dovetailed groove.

To support the outer ends of the buckets and to maintain a proper distance between them, a **lashing wire** of " comma " section is passed through holes of similar section punched in the buckets. After the buckets are straightened and spaced, the tail of the comma is bent over. The shape of the reaction bucket is different from that of the impulse bucket, because it is shaped to allow the steam to expand.

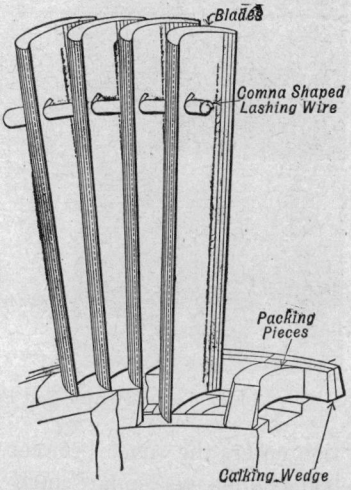

Fig. 427. — Method of Assembling Reaction Buckets.

The bearings, Fig. 428, consist of a cast-iron shell lined with high-grade babbitt, anchored in place by dove-tail grooves running across the inside of the shell. Each bearing is fitted with four machined pads bolted to the shell at the top, bottom, and sides which rest in the spherical bore of the pedestal and make the bearing self-aligning. Liners located underneath the pads provide easy and accurate means of adjusting the radial position of the revolving element. These bearings are supplied with oil under pressure from the oil purifier.

At the ends where the shaft passes through the casing, **water-seal packing**, Fig. 429, prevents the escape of steam from the high-pressure end or entrance of steam into the low-pressure end. Packing by this method is accomplished by means of a disk or impeller, revolving with the shaft in

Fig. 428. — Bearing for Westinghouse Reaction Turbine.

a small recess in the turbine casing to which water is supplied. Centrifugal force carries the water to the outer edge of the disk, and the water effectually prevents the passage of air or steam.

For marine service, at speeds below that required to produce a centrifugal force sufficient to maintain the water seal, the governor is arranged to operate a relay, turning water off and steam into the packing chamber; and a labyrinth packing, located inside the water-seal packing, is added.

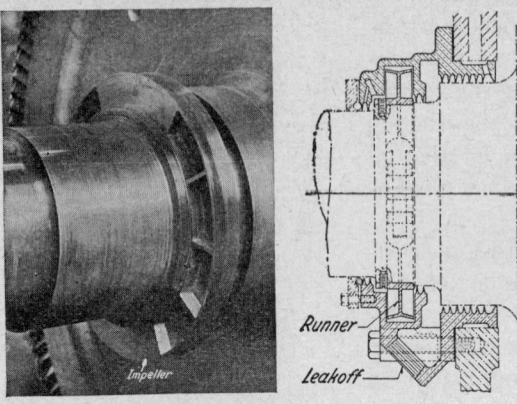

In the modern turbine the end-thrust caused by the difference in pressure at the entrance and exit of the several stages is taken by a thrust bearing, preferably of the **Kingsbury type.** In this bearing, Fig. 430, several **pivoted segments,** or **shoes,** receive the pressure of the **thrust collar** through oil films of wedge section, the thicker part of the film being where the oil drawn in by rotation enters the space between segment and collar. Pressures as large as 350 pounds per square inch for moderate speeds, and 500 pounds per square inch for high speeds, are successfully carried, as compared with 50 pounds per square inch which is a high value for the ordinary thrust bearing with parallel contact surfaces. The axial position of the rotor is fixed accurately by using a Kingsbury thrust bearing.

Fig. 429. — Water-seal Packing.

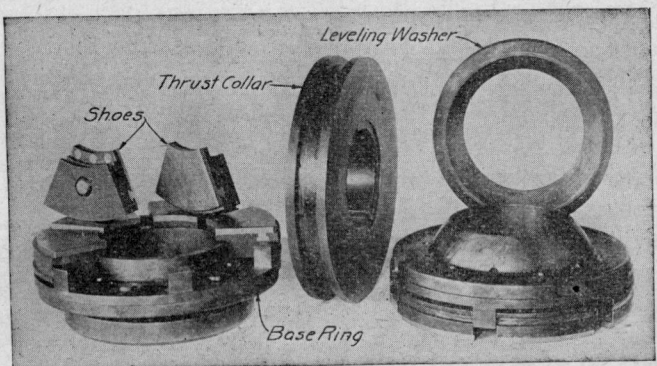

Fig. 430. — Kingsbury Thrust Bearing.

467. Governors and Valve Gears for Reaction Turbines. — The governing mechanism consists of either a **mechanical** or a **hydraulic governor** operating in conjunction with a **hydraulic relay** for operating the steam valves. This arrangement has replaced the *direct lever connection and steam-operated relay* formerly used. Overspeeding is prevented by an auto-stop mechanism located at the end of the turbine shaft which operates to close the throttle valve in case the turbine overspeeds (see Fig. 423).

Centrifugal flyball governor. — This mechanical type governor, Fig. 431, is used on some of the large Westinghouse turbines. It is driven by a straight-faced worm gear, keyed to the end of the turbine spindle, and meshes with a worm wheel keyed to the governor spindle. With the governor at rest, the compression of the main governor spring holds the governor weights at their inmost position and the spring sleeve at its lowest point. This, in turn, is connected to the clutch sleeve by two sleeve bolts (not shown in figure) and holds the **clutch lever** at the lowest point of its travel. The clutch lever moves the governor linkage which moves the steam chest oil relay for opening and closing the valves. With an increase in speed, the centrifugal force of the governor weights overcomes the compression in the main governor spring and causes the weights to move outward. This raises the spring sleeve, the clutch sleeve and the clutch lever, thus moving the governor linkage and operating the oil relay plunger to close the valves. With increase in load, the speed drops off, the governor weights move in, and through the governor linkage moves the relay plunger to open the valves. The relay plunger is ordinarily given an oscillating movement in order to make it more sensitive. A synchronizing device attached to the governor permits variation of the compression of the main spring to raise or lower the operating speed of the turbine a small amount, ordinarily from 8 to 9 per cent.

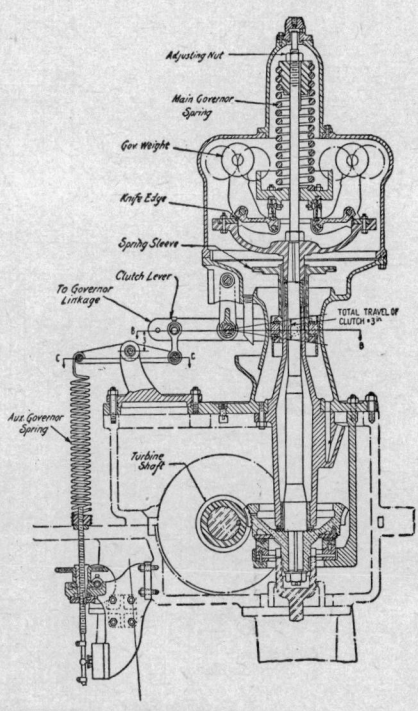

Fig. 431. — Centrifugal Flyball Governor.

468. Impeller Type Oil Governor. — This type of governor eliminates the many wearing parts of the flyball governor and consequently has greater reliability. It consists of the parts shown in the longitudinal section, Fig. 432, with the governor at the mid-point of its travel. *The operation of the governor depends upon the variation in oil pressure produced by changes in speed of the impeller of the centrifugal pump carried on the turbine spindle between the thrust bearing collar and the automatic stop body.* With an increase in load the speed decreases and the oil pressure decreases as the square of the change in speed. This opens the inlet valves to admit more steam to the turbine. The converse is true in the case of the decrease in the load.

Oil is supplied to the inlet side of the impeller by an oil ejector, and is discharged from the impeller to act directly on the governor. Sufficient oil is withdrawn from the impeller discharge to operate the ejector, prime the impeller and furnish oil to the bearings. Any air entering the system is permitted to escape through several small holes drilled in the thrust retaining ring.

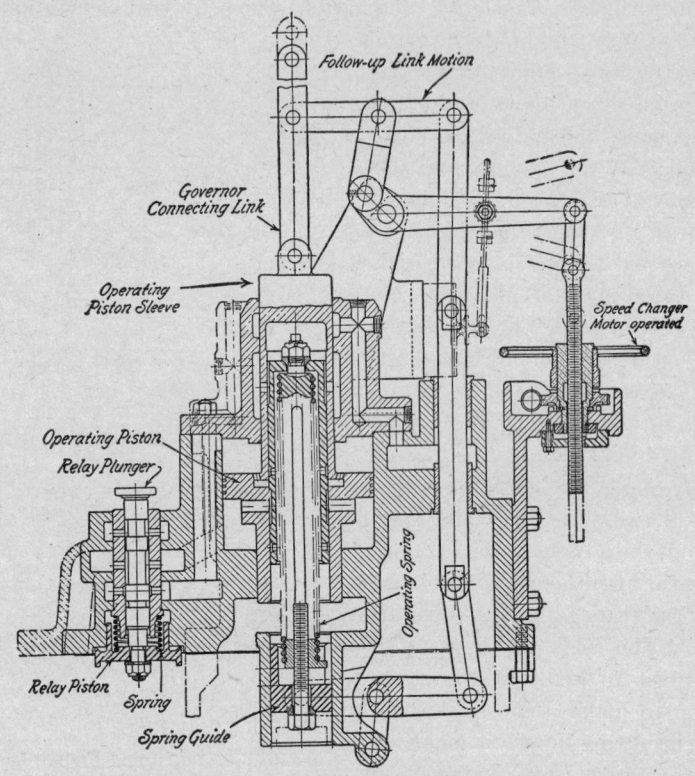

FIG. 432. — Impeller Type Oil Governor.

The lower half of the governor housing is filled with oil at the impeller discharge pressure which at normal speed is approximately 65 to 70 pounds per square inch. This pressure pushes upward on the **operating piston sleeve,** and also on the **annular ring** at the bottom of the operating piston. The pressure acting on the operating piston is continually trying to close the inlet valves, while the upward force on the sleeve is balanced by a tension spring connecting the piston sleeve to the operating piston through a system of levers, as shown, so arranged that the oil pressure acting upward on the piston sleeve exerts a force which gives a further closing to the valves. *The piston sleeve controls the admission or discharge of oil to or* *from the upper side of the operating piston only.* With an increase in load and decrease in speed the pressure drops slightly causing the sleeve to drop a small amount and open the ports thus increasing the oil pressure above the operating piston. This results in a downward movement of the piston which opens the steam inlet valves to carry the increased load. In the case of a decrease in load, the increased speed and oil pressure moves the sleeve upwards and reduces the pressure on top of the operating piston, allowing the valves to close as much as required.

The oil pressure on the lower side of the operating piston is that of the bearing supply and impeller inlet. This space is connected to the bearing supply through cored passages in the governor housing.

The speed variation is dependent upon the scale, and the change in extension of the governor spring caused by the governor movement, and may be increased or decreased between narrow limits by adjusting the spring nuts to increase or decrease the number of acting coils. However, any adjustment of speed variation required will normally be made by changing the ratios of levers. Normally the ratio of these levers is such as to cause the extension of the spring to increase 0.3 inch for a 1 inch travel of the governor. By decreasing the amount of this spring extension the speed variation is decreased and conversely increased by increasing the change in spring extension for a certain movement of the governor. This can be done either with a hand or electric-speed changer acting through the screw shown.

469. Steam Chest for Medium-sized Reaction Turbines. — The high-pressure steam chest and its valve gear are shown in Fig. 433. It consists of a steel body into which is shrunken and dowled a steel sleeve containing the high-pressure inlet valves. The **secondary and auxiliary nozzle chambers** are suspended in the turbine cylinder cover from the lower face of the steam chest. The **primary nozzle chamber** is located in the turbine cylinder base and covers nearly 50 per cent of the periphery of the impulse wheel. The three nozzle chambers combined give about 80 per cent steam admission to the impulse wheel.

The valve bushings contain ports which admit steam to the nozzles when the valves are lifted by the governor. The inlet valves are balanced, steam being admitted from the center of the sleeve with the upper and lower-valve discs being approximately the same diameter. The upper disc forms a piston valve, and the lower disc forms a poppet valve with a 45-degree seat on the lower bushings and a 90-degree edge on the valve disc. The valve stem has no packing other than that obtained from grooves cut in the stems. Steam enters the steam chest through the inlet pipe on the left-hand side and surrounds both primary and secondary valves at approximately throttle pressure. As the primary valve is lifted by the governor, steam passes from the center of sleeve, out around the two-valve discs, through the ports in the valve bushings, and into the connection on the back of the steam chest where it is led to the nozzle chamber in the cylinder base through the primary inlet pipe. The path of the steam passing through the secondary and primary supplementary valves is shown by the arrows.

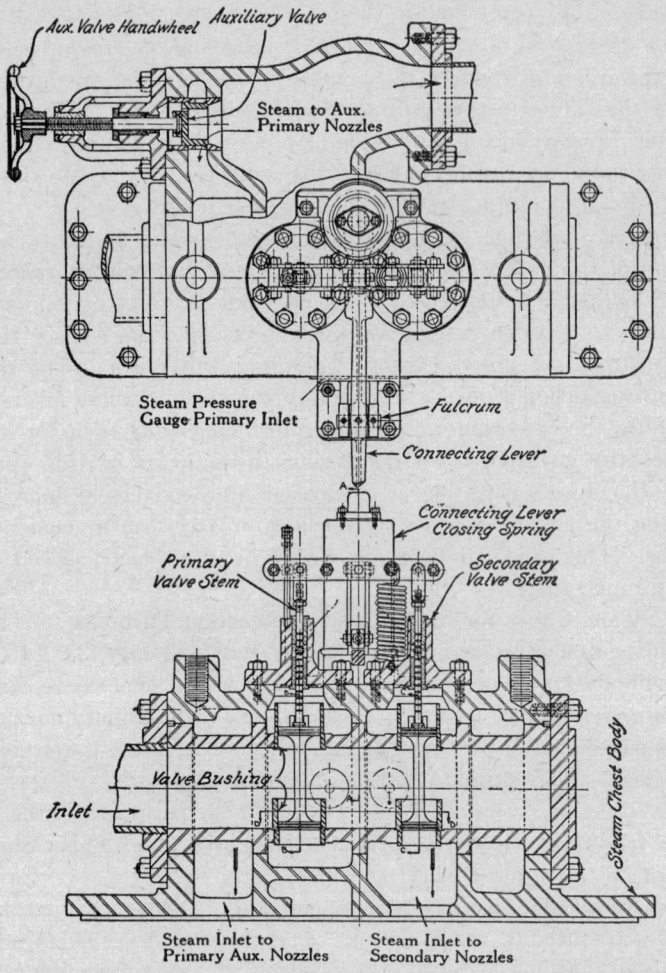

FIG. 433. — Steam Chest for Westinghouse Medium-sized Reaction Turbine.

The valves are connected to the governor operating piston by a system of levers. The lever attached to the governor is fulcrumed so that a downward movement of the governor operating piston, corresponding to a decrease in speed, will open the valves. A spring attached to the valve operating lever is under tension and holds the secondary valve closed until the primary valve has been fully opened. Any further movement of the governor will open the secondary valve. There is one hand-operated auxiliary valve furnished with this steam chest which works in series with the primary inlet valve. The steam must first pass through the primary valve before entering the auxiliary valve, and when the governor acts to close the primary valve, the auxiliary valve is automatically cut out of service. This valve is considered as an overload valve and would not be necessary when carrying the full rated load of the turbine. A spring under compression bears against an extension on the connecting lever and aids to overcome the governor mechanism and helps to close the valves in case they tend to stick.

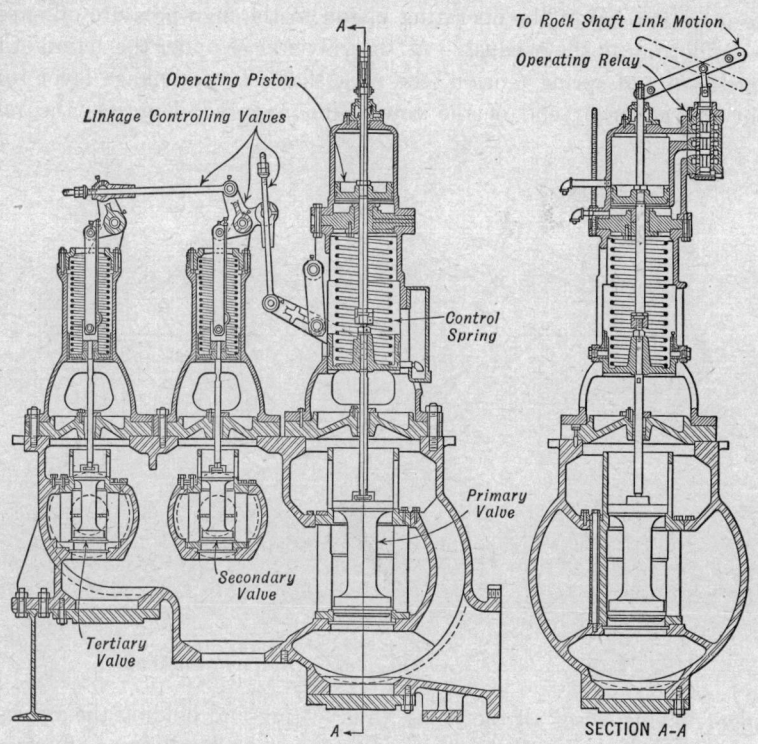

Fig. 434. — Oil-operated Relay Mechanism.

470. Hydraulic or Oil-operated Relay. — For the larger turbines, an oil relay mechanism, Fig. 434, is used to control the steam admission valves. It consists of a steam chest in which are located the primary, secondary, and tertiary valves, the two latter being in series with the primary valve but in parallel in respect to one another. All the steam entering the turbine passes through the primary valve, with only sufficient passing through the other two valves to maintain the speed. When the turbine is completely shut down all the valves are held tightly on their seats by the valve springs, with the governor holding the relay plunger in a position such that the ports connecting the bottom of the operating piston with the high-pressure oil pipe are opened. When the auxiliary oil pump is started, preparatory to starting up the turbine, oil flows to the under side of the operating piston, and when the pressure becomes high enough to overcome the valve spring, the primary valve is opened. After the primary valve has risen about 5 inches, it is wide open, and any further movement of the operating piston opens the secondary valve, and when the secondary bell crank lever has been turned $2\frac{3}{16}$ inches it will open the tertiary valve. With an increase in speed the governor, through the connecting governor linkage (not shown), raises the relay plunger uncovering the proper ports to connect the top of the operating piston to the high-pressure oil supply and the bottom to the exhaust. As the valves close under the action of the oil pressure and spring tension, the rock shaft elastic linkage lever turns about the governor end of the connecting rod and depresses the relay

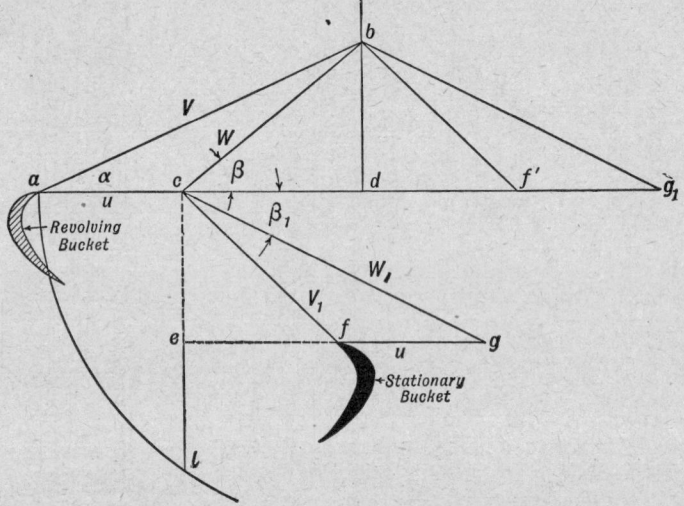

Fig. 435. — Velocity Diagram for Reaction Buckets.

plunger. This closes all the ports, thus setting and holding the valves at any desired opening. With a decrease in load, the speed drops off, causing the governor balls to drop inward, lowering the governor clutch, and producing a reversal of the operations just described.

471. Action of Steam in the Buckets. — In the reaction turbine, steam expands continuously in the stationary vanes and moving buckets. Its velocity and volume increase, requiring increased area between the buckets. As it is inconvenient to make a large number of different diameter drums and lengths of buckets, a small number of drums are used, with buckets of one length on a drum.

Steam is directed against the moving buckets by the stationary guide vanes or buckets, with velocity V, at an angle α, as represented in Fig. 435. If the bucket velocity is u feet per second, the relative entrance velocity is W. The steam expands between the moving buckets, increasing its velocity and leaving with relative velocity W_1, and exit angle β_1, approximately equal to its entrance velocity V and entrance angle α. Compounding the relative velocity W_1 with the velocity u, which it has while on the buckets, the absolute exit velocity is V_1. The steam entering the next set of stationary vanes with this velocity is further expanded and is redirected against the next row of moving buckets with a slightly increased velocity.

The work done per pound of steam per stage is found by laying out the velocities and angles as in Fig. 435, and measuring the components of the absolute velocities V and V_1 in direction of rotation, that is, $(ad + ef)$; and the work per pound of steam, $F \times s$ is as follows:

$$F \times s = \frac{1}{g}(ad + ef)ac \quad \ldots \ldots \ldots \quad (113)$$

A simple geometric construction in this case shows that $(cl)^2$ is equal to $(ad + ef)\,ac$, since $ad + ef = cg_1$ by construction.

FIG. 436. — 40,000-kw. Single-cylinder Reaction Turbine.

472. Large Capacity Westinghouse Single-cylinder Reaction Turbine. — An external view of a 40,000-kilowatt single-cylinder unit installed at the Colfax Station of the Duquesne Light Co. of Pittsburgh is shown in Fig. 436. It operates at 260 pounds pressure with 214 deg. fahr. superheat and a back pressure of 1 inch of mercury absolute, and is arranged for bleeding at four points for feedwater heating. A sectional view of the general design is shown in Fig. 437. The cylinder consists of a steel high-pressure portion divided into upper and lower halves, and a cast-iron low-pressure portion similarly divided, and also divided vertically in the transverse plane through the center of the exhaust opening. The blading is of the combined impulse and reaction type, with the low-pressure blading of the *multi-exhaust type*, with the last row of guide vanes cast in sections which are held in place in the cylinder by means of an interlocking joint and capscrew. Provision for differential expansion is made in the case of the nozzle chambers, and the design is such that the steam inlet pipes are bolted to flanges on the chambers rather than directly to the cylinder. The turbine parts are arranged to permit free expansion with reference to the supporting members, and without imposing undue stress upon the parts and without disturbing the alignment of turbine and generator. In other respects the construction of this turbine is similar to the turbine described in Art. 466, page 507.

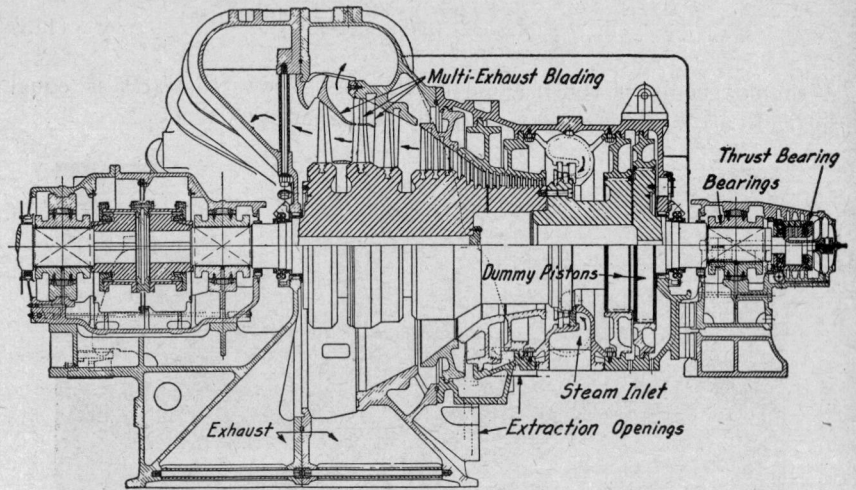

Fig. 437. — Single-cylinder Reaction Turbine.

473. High-pressure Westinghouse Reaction Turbine. — A section through a 10,000-kilowatt high-pressure combined impulse and reaction turbine is shown in Fig. 438. The turbine cylinder is machined from steel forgings as is also the steam chest which sits alongside the unit. The throttle valve and single governing valve are located in this chest which is connected by four flexible pipes to the turbine cylinder. The cylinder is supported from the exhaust casting, which is suspended by steam-heated steel links from stationary brackets attached to the foundation. This allows free cylinder expansion and at the same time maintains a fixed center line regardless of what pipe strain might be imposed upon the exhaust casting by the exhaust pipe.

The impulse element of the turbine consists of two rows of revolving blading, and one row of stationary blading made of stainless steel. This is followed by 14 rows of reaction blading, made of pure nickel. All blades or buckets are shrouded and of the end-tightening type. A stainless-steel built-up nozzle block expands the 1200 pounds steam down to 845 pounds before it enters the first impulse blades.

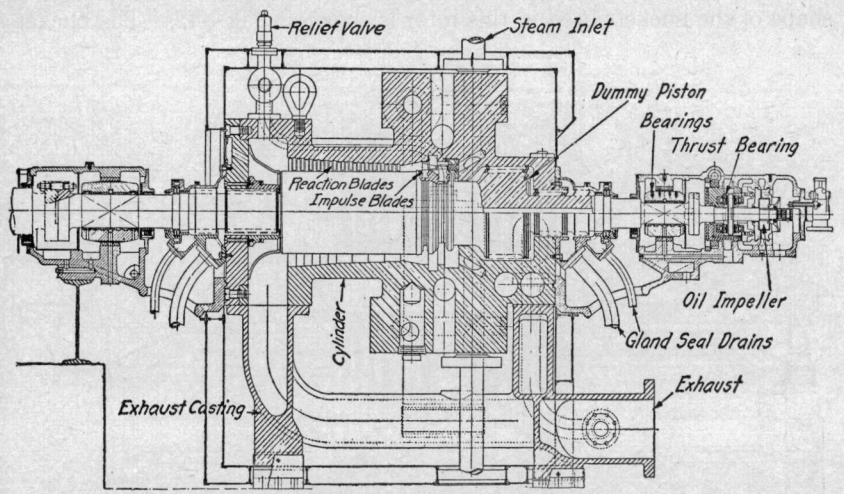

FIG. 438. — Westinghouse High-pressure Reaction Turbine.

474. Multi-cylinder Reaction Turbines. — Multi-cylinder turbines are constructed with the cylinders arranged either *tandem* or *cross-compound*. A cross-compound unit of 160,000-kilowatt capacity, using complete reaction blading, and having a high-pressure cylinder with 22 stages, is shown in Fig. 439. The rotor in this cylinder operates at 1800 r.p.m. when using steam at a pressure of 270 pounds per square inch. The low-pressure

FIG. 439. — 160,000-kw. Cross-compound Brown Boveri Reaction Turbine.

cylinder, Fig. 440, is arranged double flow, with 8 stages on each side of the entering steam.

A double-flow rotor for the low-pressure reaction cylinder of a 110,000, kilowatt cross-compound unit at the Brooklyn Edison Company's plant, in which both rotors operate at 1800 r.p.m., is shown in Fig. 441. The shape of the buckets used in this rotor is shown in Fig. 442. The buckets

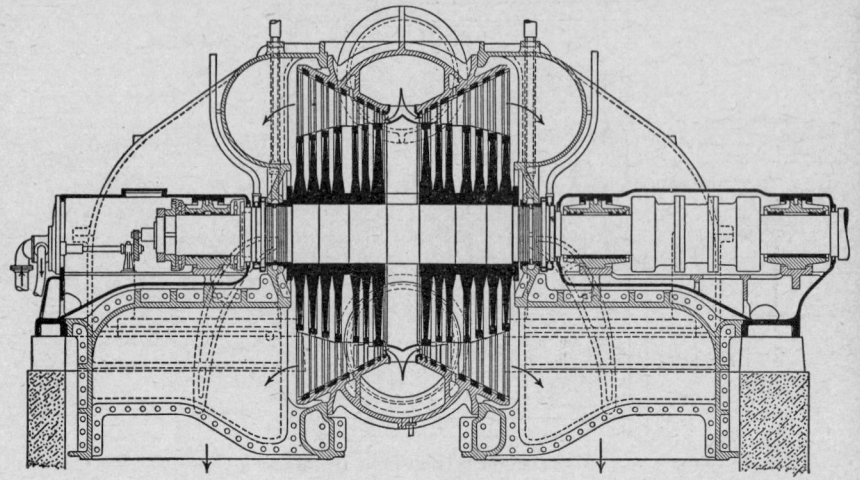

Fig. 440. — Low-pressure Cylinder 160,000-kw. Unit. — Hell Gate Station.

are spoken of as warped buckets which are tapered to reduce the stresses caused by centrifugal force, and are twisted in order that the steam may be received tangentially at all points throughout the buckets length; and also

Fig. 441. — Rotor for Double-flow Westinghouse Turbine.

that the passages between the buckets may be of such shape as will give the necessary nozzle effect in guiding the steam through the buckets for best efficiency. A three-cylinder cross-compound unit of 60,000-kilowatt capacity, installed at the Colfax Station of the Duquesne Light Co. is shown in Fig. 443.

LOW-PRESSURE, MIXED-PRESSURE, BLEEDER TURBINES

A section through a three-cylinder tandem-compound 30,000-kilowatt Brown Boveri turbine, operating at 3000 r.p.m., is shown in Fig. 444 and is of interest because the cylinders are so arranged that the steam flow takes place in such a manner that balancing or dummy pistons are unnecessary. This machine operates at 500 pounds per square inch, with 800 deg. fahr. temperature. Steam is admitted to the high-pressure cylinder at two points and extracted from the intermediate cylinder at two points, steam flowing to the right through the high-pressure cylinder, to the left through the intermediate, and double flow through the low-pressure cylinder.

475. Low-pressure, Mixed-pressure, and Bleeder Turbines. — The general construction of *low-pressure, mixed-pressure, bleeder, and high-back-pressure turbines* is the same as that of the standard forms, except for the special valves and connections required.

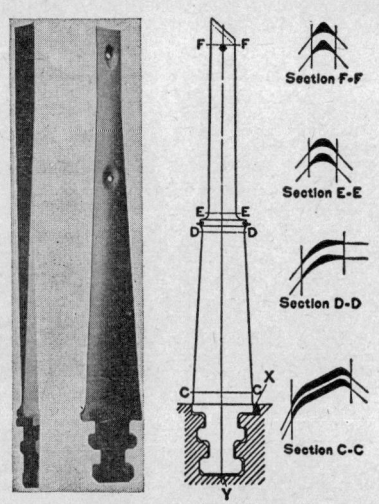

Fig. 442. — Warped Buckets.

Low-pressure turbines use exhaust steam from engines, pumps, or other steam-driven apparatus. The exhaust steam is generally supplied at

Fig. 443. — 60,000-kw. Three-cylinder Cross-compound Reaction Turbine.

slightly more than atmospheric pressure and expanded to as low a pressure as the condensing apparatus and conditions will permit.

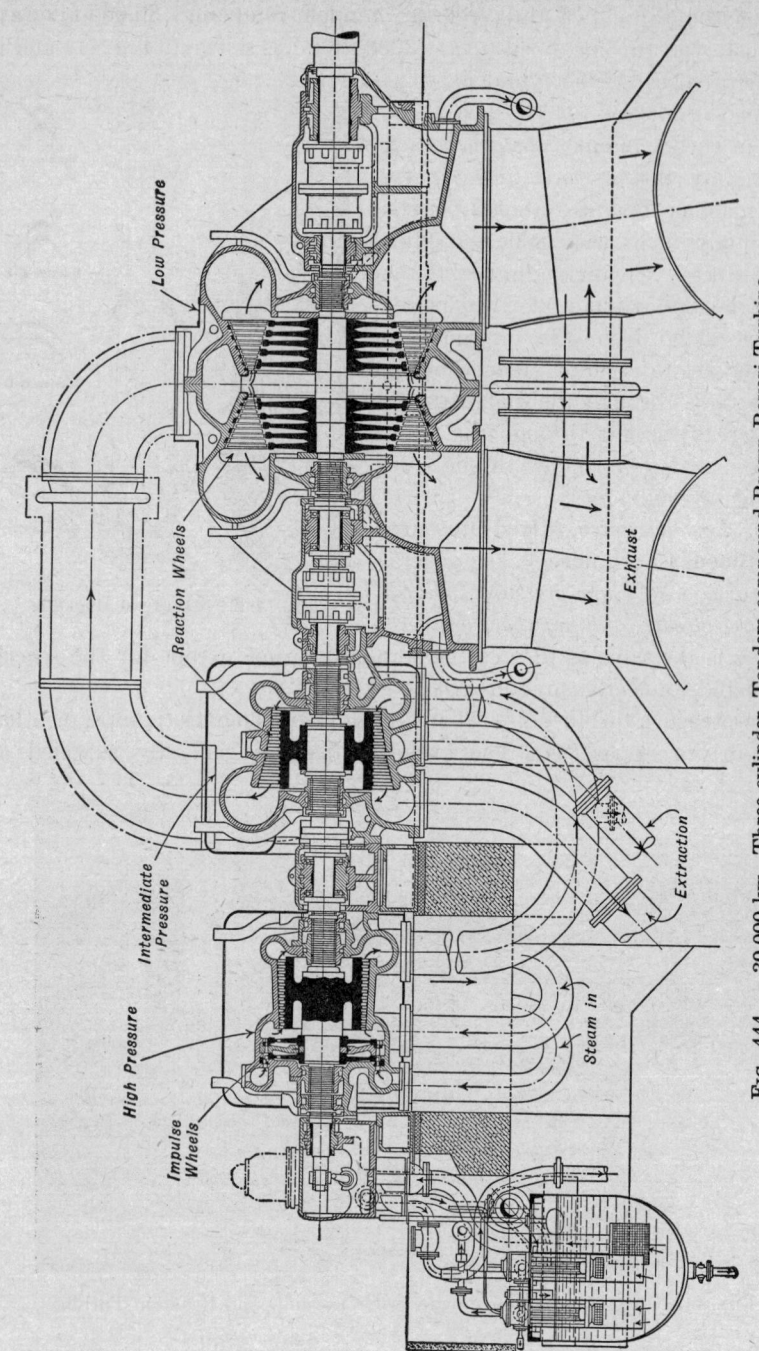

Fig. 444.—30,000-kw., Three-cylinder, Tandem-compound Brown Boveri Turbine.

LOW-PRESSURE, MIXED-PRESSURE, BLEEDER TURBINES

If the steam is taken from non-condensing engines, the additional power developed may be 80 to 100 per cent of that developed by the engines. If the engines have been running condensing, the use of low-pressure turbines may increase the power from 25 to 40 per cent, since the steam is expanded to a lower pressure, before leaving the turbine buckets, than is practical in the reciprocating engine, on account of the large size of the low-pressure cylinder that would be required. Also, the condensation of entering steam caused by contact with cooler surfaces, as in the engine cylinder, is decreased, since the flow of steam is continuous in the turbine.

The **mixed-pressure turbine** is a combination of a high- and low-pressure turbine which normally operates as a low-pressure turbine, using steam admitted to the low-pressure stages under governor control. When suffi-

FIG. 445. — Automatic Extraction Type F, G.E. Turbine — External View.

cient low-pressure steam is not available to develop the electrical output demanded of the unit, additional steam, sufficient to carry the load, is supplied to the high-pressure end and is expanded through the high-pressure and low-pressure elements in the same manner as it would be in the turbine of a high-pressure condensing type. It may carry all or any part of its rated load on high-pressure steam.

A **bleeder,** or **extraction, turbine** is a combination of a condensing and non-condensing turbine, and is used in industrial and central station plants where there is either a demand for medium or low-pressure steam for process work, or for heating buildings or feedwater. A portion of the steam is bled or extracted from an intermediate stage of the turbine, the unbled steam passing on through the turbine to the exhaust. The amount and pressure of the steam withdrawn is controlled by an automatic valve of the **ring, gridiron, Corliss** or **poppet type.**

An external and sectional view of an extraction turbine of the impulse type is shown in Figs. 445 and 446. The general construction resembles a straight condensing turbine, with the addition of a device for the automatic control of the extracted steam. The valve, Fig. 447, is of annular gridiron form and bears against a diaphragm arranged with ports corresponding to the ports in the valve. As the ports in the diaphragm are opened more steam passes through to the later stages of the turbine and the tendency is to decrease the pressure in the extraction stage. *The regulating mechanism, consisting of a* **bellows diaphragm, pilot valve, operating cylinder and extraction valve,** *automatically adjusts the port opening to maintain the required pressure of extracted steam when the demand for such*

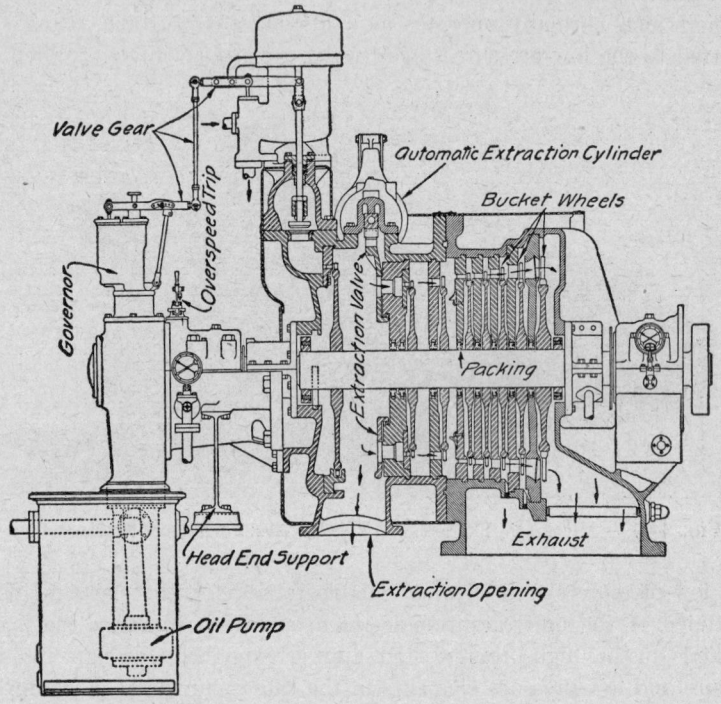

FIG. 446. — 8-stage Condensing Extraction Impulse Turbine — Sectional View.

steam varies. With an increase in pressure in the extraction line the bellows diaphragm is compressed and the levers to the pilot valve are moved to the right, thus admitting steam to the cylinder, moving the piston rod and rotating the extraction valve to increase the opening of the ports. This allows more steam to pass through the stages of the turbine beyond the extraction stage and reduces the flow of steam to the extraction line, restoring the normal extraction pressure. The pilot valve is returned to its normal position after each readjustment by means of the restoring levers operated by the main control piston rod. The bellows is set for the desired extraction pressure by adjusting the spring tension to balance the thrust in the bellows.

LUBRICATION OF STEAM TURBINES

When large turbines are bled for heating the boiler feedwater, it becomes necessary to provide some means of preventing the water, in case of a tube failure in the heater, from getting into the turbine. It is also necessary to guard against re-evaporation of the condensed steam in the long extraction pipe and heater itself, as this steam could, in case the turbine lost its load and the throttle valve were tripped shut by the emergency governor, further increase the overspeed to the danger point. This protection is ordinarily afforded by a properly designed bleeder line valve which will close when there is an accumulation of water in the heater shell or when the main throttle heater valve is tripped shut. This valve may be operated by oil or by electricity.

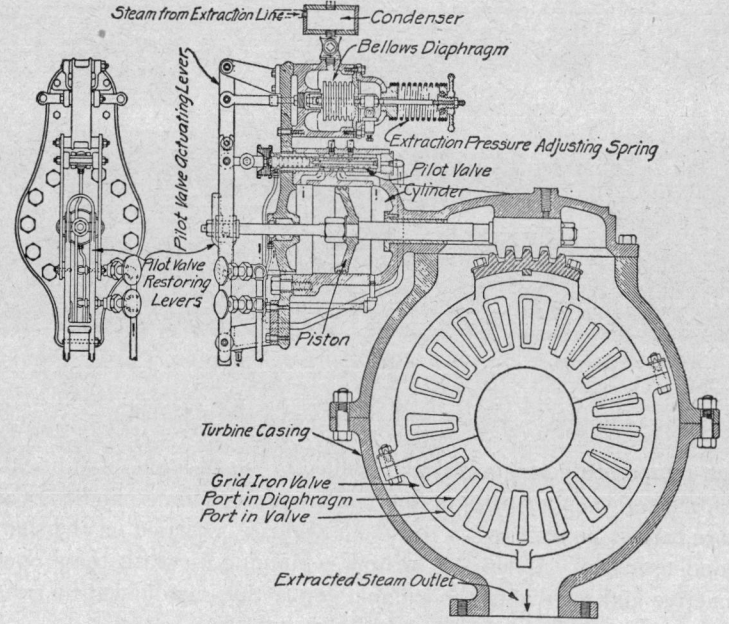

FIG. 447. — Assembly of Turbine Extraction Device.

476. Lubrication of Steam Turbines. — Because of the high journal speed used in the bearings of steam turbines, and of the temperature conducted to the bearings from the steam end of the turbine, it is highly important for continuous operation that the best methods of lubrication be used, and that the efficiency of the oil be maintained by removing impurities from the oil, by using either an oil filter or centrifugal type of purifier. The bearings of large turbines and gears are usually supplied with oil by either using *gravity* or *forced feed*, with the smaller turbines using ring oil bearings.

In the **forced-feed system,** Fig. 448, a rotary oil pump, driven by an extension of the governor spindle, forces oil from an oil reservoir at the base of the turbine, through a water-cooled coil, directly to the pipes distributing oil to the bearings, or to an elevated tank. An unloading valve returns oil to the suction tank if the delivery pressure is too high. Oil leaving the bearings passes through a filter to the suction tank, although a **centrifugal purifier** may be used for this purpose.

477. Marine Steam Turbines. — The steam turbine is extensively used on shipboard, principally because of: the smaller space required as compared with reciprocating engines, the decreased expense for attendance and supplies, the decreased weight for the same power, and the greater cruising radius for the same weight of fuel. Its superior economy is accounted for, as in all steam turbines, by its ability to take advantage

FIG. 448. — Forced Feed Oiling System for a Turbine.

of high-temperature steam without difficulty on account of lubrication, and of greater vacuum without the limitation of volume handled. Since turbines cannot be reversed, a reversing element is carried on the shaft at the condenser end. When the turbine is running forward, these buckets are inactive and revolve in the exhaust connection surrounded by exhaust steam.

A 3000 horsepower **De Laval impulse marine turbine**, Fig. 449, is a multi-pressure stage turbine having eight wheels and eight sets of nozzles for the **ahead turbine** and two wheels and two sets of nozzles for the **astern turbine**. These wheels are inclosed by a single casing split horizontally to allow easy access. The exhaust connection is to the upper half of the casing. The forward turbine has two velocity stages in the first pressure stage, and in each stage of the astern turbine. High- and low-pressure turbines with cross-compound cylinders are generally used.

A high-pressure **Westinghouse reaction marine turbine**, Fig. 450, shows the arrangement of the impulse element and the reaction element. The general construction is similar to the land reaction turbine except that a backing-turbine impulse wheel, having two velocity stages, is included in the casing. Balanced pistons are used to take the thrust resulting from the steam pressure, and a Kingsbury thrust bearing is used to take the thrust in either direction.

The economical speed of the steam turbine is too high for direct connection to the propeller shaft of vessels, where the speed is less than 17

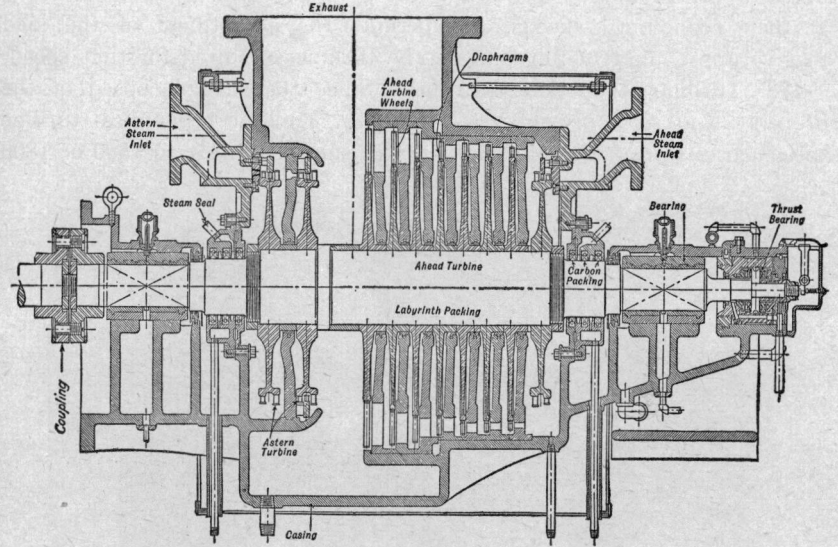

Fig. 449. — De Laval Marine Impulse Turbine.

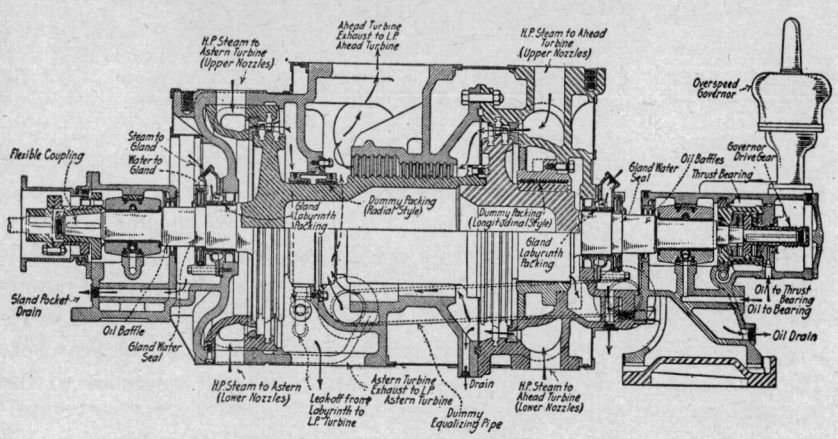

Fig. 450. — Westinghouse High-pressure Marine Reaction Turbine.

knots[1] per hour. Since a large proportion of merchant vessels, as well as battleships at cruising speed, run at a lower speed, the successful application of steam turbines to vessels requires some suitable means of speed reduc-

[1] A knot is equal to a speed of 6080 ft. per hour.

tion. As mentioned in Chapter I, the two methods used are (1) electrical transmission and (2) gearing.

The electrical transmission is used on some battleships and employs **turbo-generator units** to furnish current for **electric motors** directly connected to propeller shafts. The power is generated by several turbine units, an arrangement which allows the propellers and turbines to be run at their economical speeds, and permits the adjustment of the load to give approximately full load on the turbines required for that speed.

478. Turbine Reduction Gearing. — Since the suitable speed of the propeller shaft is approximately 100 r.p.m., and the economical turbine speed varies from 5000 to 6000 r.p.m. for small turbines, to 1500 or 1800

FIG. 451. — Westinghouse Two-pinion Single-reduction Gear.

r.p.m. for large turbines, double-reduction or single-reduction gears are required, the former being used for speed reductions greater than 20 to 1.

A two-pinion single-reduction gear, Fig. 451, consists of pinions driven by the turbine shafts meshing with a gear on the propeller shaft. Helical, or twisted tooth, gears are used to insure continuous transmission of power. They are arranged in pairs, with the teeth inclined at equal angles to remove axial thrust and run in oil supplied under pressure to the gear box. The location of a typical double-reduction gear, with reference to the turbine, is shown in Fig. 452.

The principal difficulty in the use of gearing, as applied to marine service, in order to prevent breaking the gears by concentrated load, lies in maintaining alignment under severe working conditions. In the gear, Fig. 451, the pinion is supported by bearings that are held in alignment by a heavy frame which is supported in the middle of its length by an I-beam placed at right angles to the axis of the pinion. The flexing of the web of the I-beam allows the whole frame to adjust itself in a plane through the axis of the pinion. The rocking of this frame allows the pinion to adjust its alignment to the changed helical angle caused by the torque, and also allows the pinion to follow the gear wheel, thus automatically maintaining alignment and distribution of tooth pressure over the entire width of face.

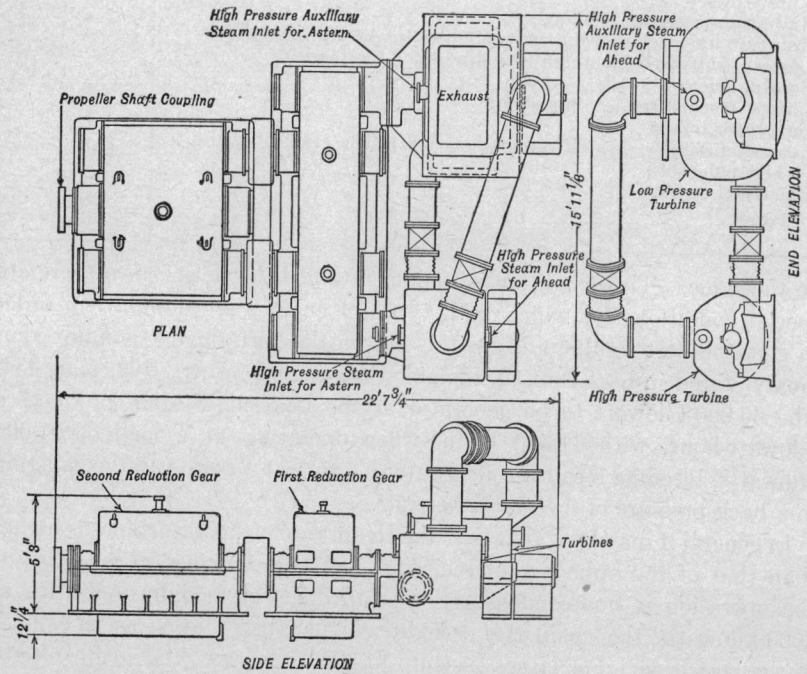

Fig. 452. — De Laval High- and Low-pressure Marine Turbine with Double Reduction Gear; 8000 hp.; 90 R.P.M.

479. Losses in Steam Turbines. — The losses in steam turbines are of two kinds, namely, (1) *heat or thermodynamic losses* and (2) *mechanical losses*. The **thermodynamic losses** are the result of radiation, friction of the steam in the nozzles and buckets, loss by leakage past the buckets, stray motion and energy left in the steam, and loss to exhaust. The **mechanical losses** vary as some function of the size of the unit, and are caused by friction in bearings, gland and pump resistance, and by rotation of the wheels in vapor-filled compartments. The distribution of the losses in a 200-kilowatt turbine-generator is given in Table 33, from " Steam Turbines," by Moyer.

480. Economy of Steam Turbines. — The performance of steam turbines is stated as for steam engines, and is explained in Art. 400, page 433.

When using the steam rate to compare the performance of turbines and reciprocating engines, the conditions of operation with regard to initial steam pressure, quality or superheat of steam, and final pressure should

TABLE 33. — LOSSES IN A 200-KW. De LAVAL TURBINE GENERATOR

Nature of Loss	Amount of Loss Per Cent
Nozzle losses..	12
Radiation losses and leakage...............................	1
Rotation losses caused by turbine wheel revolving in steam......	4
Losses resulting from steam traveling over the blades...........	9
Bearing friction losses.....................................	1
Losses in speed-reduction gearing...........................	2
Generator losses...	4
Losses caused by residual kinetic energy in the steam passing to the condenser..	8
Electrical output..	59
Total..	100

be the same. This condition is seldom attained, however, since there are special conditions for which both types of engines are designed in order to give the best results, and for this reason the performance is more commonly stated in heat rate, as explained for engines in Art. 402, page 434. The 60,000-kilowatt turbo generator at the Lakeside Station produces a kilowatt-hour, with 11,510 B.t.u. when operating at a load of 61,600 kilowatts, bleeding steam for feedwater heating at 4 points and exhausting to a back pressure of 0.75 inch mercury.

In general it may be stated that the steam rate of small turbines is higher than that of the same size of reciprocating steam engines, particularly if the operation is non-condensing. For turbines of medium sizes, up to 3000 kilowatts, the steam rate is slightly higher than that of piston engines of the best type. For larger sizes up to 6000 kilowatts, the steam rate of turbines and engines is about the same, the difference being in favor of the turbine at 7500 kilowatts.

A statement of the performance of steam turbines at their rated capacity is given in Table 34, taken from "Steam Power Plant Engineering," by GEBHARDT.

481. Influence of Vacuum, Superheat, and Initial Pressure upon Steam Rate. — The effect of increased vacuum is to decrease the steam rate by approximately 5 per cent between 27 and 28 inches, and nearly 10 per cent between 28 and 29 inches vacuum. In modern stations, a vacuum of $29\frac{1}{2}$ inches is quite common. A comparison of the steam rates for a turbine operating condensing and non-condensing, with various outputs, is given in Table 35, taken from "Steam Turbines," by MOYER.

Experience indicates that a high degree of superheat increases the work done per pound of steam, and results in other savings because of the increased volume and decreased frictional losses. The steam rate of a turbine is decreased approximately 1 per cent for each 10 degrees of superheat up to 100 deg. fahr. of superheat. The saving is slightly less for higher

INFLUENCE OF VACUUM, SUPERHEAT, INITIAL PRESSURE

TABLE 34. — PERFORMANCE OF MODERN STEAM TURBINES AT RATED CAPACITY

Make of Turbine	Rated Capacity	Operating Conditions					
		R.p.m.	Initial Pressure Lb. Abs.	Back Pressure In. Mercury	Superheat °F.	Lb. of Steam per kw.-hr.	Rankine Cycle Ratio
Westinghouse	500 kw.	3600	165	2.0	0	19.8	53.0
	1,000 kw.	3600	165	2.0	0	18.1	58.0
	5,000 kw.	1800	165	2.0	0	16.0	65.6
	15,000 kw.	1800	215	1.5	125	12.6	71.0
	30,000 kw.	1200	235	1.0	200	10.65	75.0
	45,000 kw.	1200	215	1.0	200	10.65	76.0
Curtis	500 kw.	3600	215	2.0	0	18.5	54.0
	1,000 kw.	3600	215	2.0	0	17.5	57.1
	5,000 kw.	1800	215	2.0	125	14.3	64.6
	15,000 kw.	1800	215	2.0	125	12.5	74.0
	30,000 kw.	1200	215	2.0	125	12.2	75.8
	45,000 kw.	1200	215	2.0	125	11.9	77.6
Kerr	25 hp.	3600	165	atmos.	0	43.0	31.0
	50 hp.	3600	165	atmos.	0	38.0	34.2
	100 hp.	3600	165	atmos.	0	32.0	40.6
	500 hp.	3600	165	atmos.	0	27.0	48.1
	1,000 hp.	3600	165	atmos.	0	24.75	57.5
	1,500 hp.	3600	165	atmos.	0	24.0	59.1

NOTE. — Rankine cycle ratio based on electrical horsepower for Westinghouse and Curtis, and on developed horsepower for Kerr.

TABLE 35. — COMPARISON OF STEAM RATES FOR 500-KW. TURBINE OPERATING CONDENSING AND NON-CONDENSING AT 3600 R.P.M.

Brake Horsepower	Steam Pressure Lb. per sq. in., Gage	Superheat deg. fahr.	Vacuum, inches	Pounds per b.hp.-hr.
383.5	152.6	0.2	28.2	14.15
755.6	149.2	1.2	27.8	13.28
1121.9	148.8	5.1	26.5	14.32
385.6	148.2	2.7	0.8	24.94
766.8	147.3	2.6	0.8	22.10
1144.4	126.1	11.4	0.8	24.36

degrees of superheat. With high pressures, reheating is necessary to prevent extensive moisture in the lower stages of the turbine because such moisture increases the windage and friction losses, decreases the economy about 2 per cent for each 1 per cent of moisture, and in addition has a corrosive and erosive action on the buckets of the turbine.

An increase in pressure decreases the steam rate, the decrease being 1 per cent for 10 pounds change in pressure. The present tendency is toward higher steam pressures, and the upper limit seems to be set by the mechanical difficulties caused by high temperature.

482. Testing of Steam Turbines.

— In general, the method used to test steam turbines is the same as that used for steam engines, with the modifications which the difference in method of operation requires. When guarantee tests of turbines are made at other conditions of pressure, quality, superheat, and vacuum than those stated, it is customary to correct the test to the guarantee conditions, provided the variation is not too great, by using curves furnished by the manufacturers, which show the correction to be made in each factor. Consult "Steam Turbines," by Moyer, for a discussion of the method to be used in making these corrections. For detailed information on testing of steam turbines consult A. S. M. E. Power Test Code.

REFERENCES

Mechanical Engineers Handbook, Marks.
Steam Power Plant Engineering, Gebhardt.
Applied Thermodynamics, Ennis.
Steam Turbines, Roe.
Steam Turbines, Moyer.
Elements of Heat Power Engineering, Hirshfeld and Barnard.
Publications of the following companies: General Electric Co.; De Laval Steam Turbine Co.; Westinghouse Electric and Mfg. Co.; Terry Turbine Co.

REVIEW QUESTIONS AND PROBLEMS

1. In what fundamental respects does a steam turbine differ from a reciprocating steam engine?
2. State the essential difference between the impulse and reaction types of turbines.
3. What is meant by a stage in a turbine, and what two methods of staging are used? What is the prime purpose of staging?
4. Describe the construction of the following turbines: (a) Simple type of De Laval, (b) Curtis horizontal, (c) Westinghouse reaction, (d) Terry.
5. What methods are used to govern steam turbines? Describe the method of governing (a) De Laval, (b) Curtis, (c) Westinghouse, turbines.
6. What types of gland packing are used, and what is the purpose of such packing?
7. Describe the forced-feed oiling system, as applied to a turbine.
8. The steam pressure on a turbine nozzle is 150 lb. per sq. in. abs., and the pressure at discharge 14.7 lb. per sq. in. abs. Find the theoretical velocity of the issuing jet, assuming 10 per cent friction loss and dry steam.
9. What would be the impulse force produced by 3.5 lb. of steam discharged per sec. from the nozzle in Problem 8?
10. Explain the meaning of (a) low pressure, (b) bleeder, and (c) mixed pressure as applied to turbines.
11. What is a double-flow turbine, and why are turbines made with this construction?
12. Describe the construction of a marine impulse turbine.
13. Name the losses occurring in a steam turbine.
14. What effect have the following upon the water rate of a turbine: (a) increase in steam pressure, (b) decrease in superheat, (c) increase in vacuum?
15. The performance of a certain line of turbines, in 1899, was as follows: load, 300 kw.; steam pressure, 125 lb. gage; superheat, 0 deg. fahr.; vacuum, 27 in. mercury; steam per kw. per hr., 22 lb. In 1920 this type of turbine gave: load, 26,505 kw.;

steam pressure, 233.1 lb. gage; superheat, 124.3 deg. fahr.; vacuum, 28.85 in. mercury; steam per kw. per hr., 11.27 lb. The barometer in both cases was 30.05 in. mercury. Find the efficiency, or Rankine Cycle, ratio for the two turbines.

16. Check the Rankine Cycle ratios for any two turbines given in Table 34, page 533.

17. Given the following data: steam pressure, 260 lb. per sq. in. abs.; exhaust pressure, 15 lb. per sq. in. abs.; steam temperature, 460 deg. fahr.; weight of steam discharged, 1800 lb. per hr. Assuming a nozzle without friction find: (a) the theoretical velocity of the steam in the throat and at the exit of the nozzle, (b) the diameter at the throat of the nozzle.

18. What is the impulsive force produced by five nozzles each discharging 0.96 lb. of steam at a velocity of 3500 ft. per sec.?

19. 1800 lb. of steam per hr. is discharged through a frictionless nozzle. Initial steam pressure, 250 lb. per sq. in. abs.; temperature of steam, 450 deg. fahr.; back pressure, 20 lb. per sq. in. abs. Find: (a) the theoretical velocity of the steam at the throat and exit of the nozzle; (b) diameter of the throat and exit of the nozzle; (c) the length of the nozzle from throat to exit assuming that the length in inches equals $\sqrt{15a_0}$, where a_0 = area of throat in sq. in.

20. A turbo-generator has an output of 3000 kw. and a steam rate of 15 lb. per kw-hr. Initial pressure, 250 lb. per sq. in. abs.; superheat, 100 deg. fahr. Find the throat diameter if 10 nozzles are used and the drop in pressure for the first stage is 150 lb.

21. An 18,500-kw. turbo-generator has a steam rate of 9.5 lb. per kw-hr. Steam pressure, 350 lb. per sq. in. abs.; superheat, 250 deg.; exhaust pressure, 2 in. of mercury abs.; generator efficiency, 96 per cent. Find: (a) total steam used per hour; (b) brake horsepower of the turbine; (c) total heat per pound of steam at throttle and exhaust; (d) adiabatic heat drop per pound of steam; (e) heat consumption, B.t.u. per kilowatt-hour; (f) steam rate of the ideal turbine, pounds per kilowatt-hour; (g) efficiency ratio.

CHAPTER XXIII

STEAM- AND POWER-DRIVEN PUMPS

483. Foreword. — Pumps are used in power plants for the following purposes: to pump water into boilers; to circulate cooling water in, and to remove condensate, air, and vapors from condensers; to pump lubricating oil; and to supply water for fire protection.

Formerly, the piston pump, driven by a small uneconomical steam engine, was commonly used and still is in small power plants. It is reliable, gives satisfactory service, and the exhaust steam is available for heating the boiler feedwater, which overcomes the loss resulting from the poor economy of the driving engine.

The perfecting of the steam turbine and the centrifugal air compressor also brought about the perfecting of the centrifugal pump. This type of pump is now used exclusively in *central stations* to feed boilers, to circulate cooling water in, and to remove condensate from condensers. It has few operating parts, requires small space, can be operated at high speeds, is economical in use of steam, requires a minimum of attention, and can be operated under practically any head and capacity.

Air pumps are described in Chapter XXIV.

484. Classification of Pumps. — Pumps used for water, in modern power plant practice, may be classified according to the principle of operation as: (1) *piston*, (2) *centrifugal*, (3) *rotary*, (4) *jet*, and (5) *direct pressure*.

485. Piston and Plunger Pump. — In this type of pump, motion, and pressure are imparted to the fluid by direct contact. The action is positive, and a definite amount of fluid is pumped per stroke for a given pressure and velocity. Piston pumps may be subdivided as follows:

1. Direct-acting, steam-driven { Simplex / Duplex
2. Flywheel
3. Power-driven, electrically, or by belt { Simplex / Duplex / Triplex

486. Direct-acting Steam-driven Piston Pump. — *A steam pump that has no flywheel is known as a direct-acting pump.* Such pumps have a small number of working parts, require little attention, and are generally reliable. For general power plant service, they are used extensively. They may have a single steam cylinder and a single water cylinder, in which case they are **simplex;** or they may have two steam and two water cylinders, in which case they are **duplex.** The **size** of this type of pump is stated by giving first the diameter of the steam cylinder, then the diameter of the water cylinder, and finally the length of the stroke, all in inches; thus, 8 in. × 6 in. × 10 in.

487. Simplex Steam-driven Pump. — The cylinders of this pump are generally arranged with the pistons at opposite ends of the piston rod, and a connecting casting between. Reversal of the stroke is obtained by using a **steam-thrown steam valve,** of which there are many types, and which takes the place of a flywheel. The principle of operation is the same in all types, and will be illustrated by the description of two typical pumps.

488. Knowles Steam-driven Simplex Pump. — This pump is shown in Fig. 453, with the essential parts named. The steam cylinder and valve chest of the pump are practically the same as those described under the steam engine, with the addition of a special control of the slide valve.

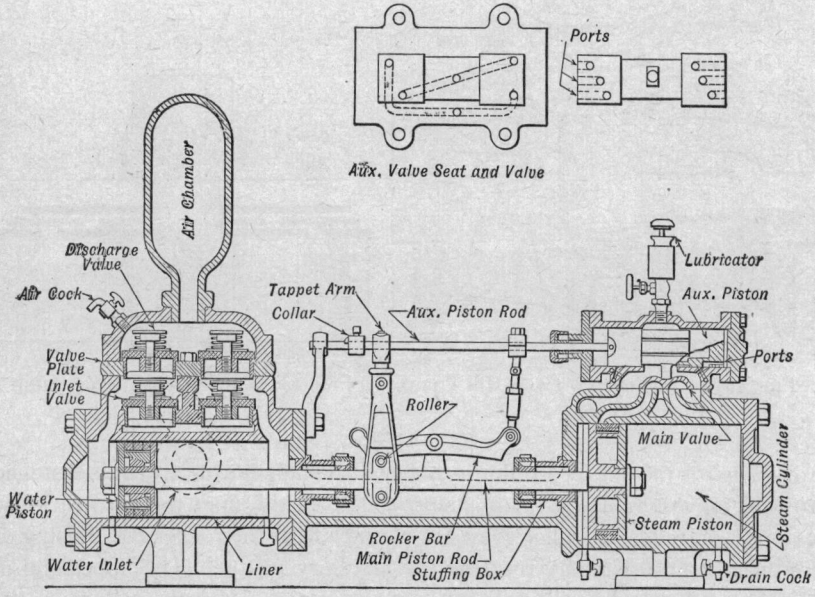

Fig. 453. — Knowles Simplex Direct-acting Steam-driven Pump.

The operation of the steam end of the pump is as follows: The **auxiliary piston** is attached to and drives the **main steam valve** back and forth, upon its flat seat; steam is thus admitted to the cylinder, back of the main steam piston, and the pump operated. The main steam valve is a plain slide valve having the form of the letter B. It does not have any lap, and hence steam at full pressure is used for the entire length of the stroke.

Movement of the auxiliary piston is accomplished by giving it a slight rotary motion at the proper time. This rotation, relative to the steam chest, brings the small steam ports, located on the under side of the auxiliary piston, to coincide with the proper openings cut in the walls of the steam chest. When the ports coincide, steam from the steam chest is admitted to that end, back of the auxiliary piston, and this piston, together with the main steam valve, is moved toward the opposite end. This movement of the main valve connects one side of the main piston to admission and the other to exhaust. To prevent the auxiliary piston from striking the steam chest head, a port is so located that it is uncovered by the auxiliary piston after it has traveled a certain distance, and steam is admitted to cushion the piston.

Rotation of the auxiliary piston is produced by a rocker bar, connected to the side of the frame and attached to the auxiliary piston rod. The

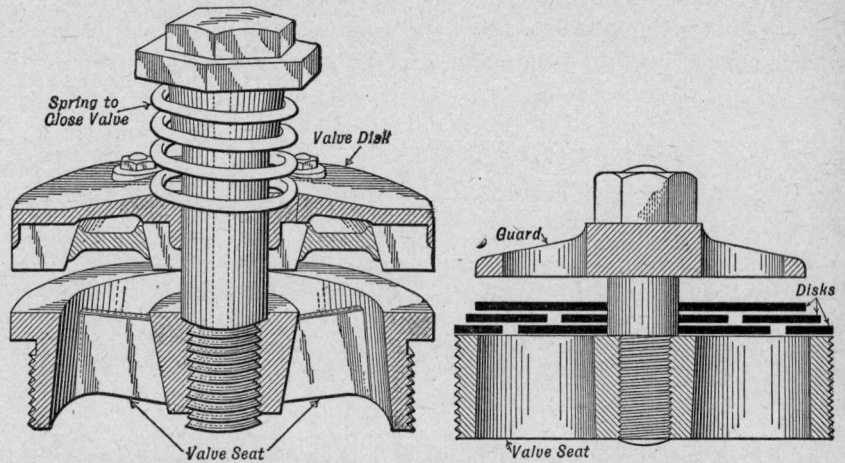

Fig. 454. — Sectional View of Hill Pump Valve.

Fig. 455. — Multi-disk Type Pump Valve.

rocker bar is moved by a **roller** located on the **tappet arm,** which is fastened to and moves with the main piston rod. As the main piston nears the end of its stroke, the roller comes in contact with the rocker bar, raising or lowering it and partially rotating the auxiliary piston. The upper end of the tappet bar ordinarily does not come in contact with the **collars** on the auxiliary valve rod. If the steam pressure for some reason should fail to move the auxiliary piston, the tappet arm strikes the collars and moves the auxiliary piston mechanically.

At the position of the steam end shown in Fig. 453, steam is about to be admitted to the right of the auxiliary piston. When this occurs the main valve will be moved to the left; steam will be admitted to the left-hand side of the main piston, its motion will be reversed, and the space to the right of the main piston will be connected to exhaust.

The operation of the water end is as follows: at the position of the water end shown, water has just ceased to flow through the water passage at the head end and upward past the discharge valves located in a removable **valve plate** above the inlet valves. During the discharge from the head end of the pump, the crank-end discharge valves are held closed by the pressure of the water in the discharge chamber and the inlet valves are opened, by the unbalanced pressure produced by the movement of the piston to the left, and water enters to fill the space behind the moving piston. When the direction of movement of the piston is reversed, by the action of the valves in the steam cylinder, the crank-end inlet valves are closed by the pressure of the discharge water, which flows through the crank-end discharge valves into the outlet pipe, shown dotted in the figure. During discharge at the crank end, the head-end discharge valves are closed and the inlet valves are forced open by the pressure on the water in the suction chamber, and water again enters to fill the space behind the piston. The discharge pressure for a pump of this type depends upon the relative areas of the steam and water pistons.

An **air chamber** is attached to the discharge, to maintain a uniform pressure in the discharge line and to produce a nearly uniform rate of flow by storing water when there is an excess and delivering water when there is a deficiency.

The **water** or **pump valve** used on this type of pump, and shown in Fig. 454, consists of a rubber-composition disk supported by a brass casting and held closed by a coil spring. The **removable seat** is made of brass to prevent corrosion and rusting. Several small valves are used instead of one large one, because they give more satisfactory service. The discharge valves are located in a removable **valve plate**.

A recent type of pump valve is shown in Fig. 455. It consists of a number of superimposed metallic disks, each being free to move on a central stud. The bottom disks contain a number of holes, so placed that when all the disks are closed no passageway is formed. The disks rise from their seats upon each other, independently or together, to permit water and air to pass.

489. Cameron Steam-driven Simplex Pump. — The water end of the Cameron pump, Fig. 456, has the same general construction as the Knowles pump, but the operation of the steam end is different, in that the main piston is reversed by using two plain **tappet valves** which control exhaust passages to the auxiliary piston. All the operating parts are enclosed.

The main slide valve is moved by the auxiliary piston, which is hollow at each end and filled with steam from the main steam chest. Steam passes through a small hole in each end of the auxiliary piston, and fills the space between the ends of the piston and the heads of the steam chest in which it moves. The pressure acting on each end of the auxiliary piston is thus equal, and the piston is ordinarily balanced and motionless.

When the main piston has traveled until it strikes the reversing tappet, at either end, steam is discharged through a small exhaust port, E, from that end of the piston valve, to the main exhaust passage. This causes an unbalanced pressure on the piston valve, which moves, carrying with it the main valve and thus reversing the piston of the pump. When the piston valve passes the port E, it encloses steam and forms a cushion. As soon as the main piston moves out of contact with the reversing tappets, they are closed by steam pressure conveyed directly from the steam chest through ports K shown by broken lines.

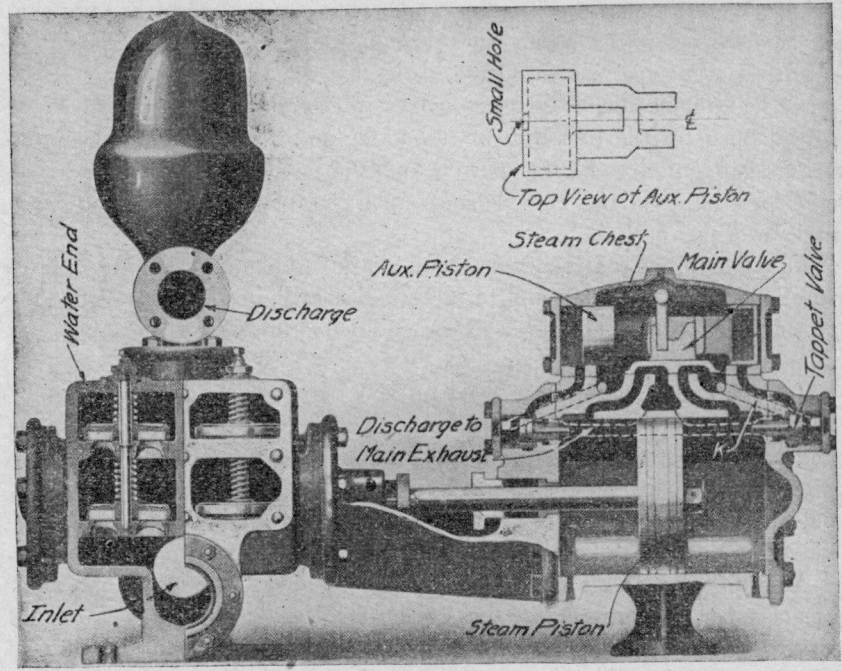

FIG. 456. — Cameron Steam-driven Simplex Pump.

490. Direct-acting Duplex Steam-driven Pump. — A sectional view of a Deane Brothers duplex direct-acting steam pump is shown in Fig. 457. This pump has two steam cylinders and two water cylinders set side by side, and is, in effect, made up of two pump cylinders in one casting.

The steam valve of each side is operated by a connection to the piston rod of the opposite side, thus making the operation of the steam valves positive. To a person standing at the steam end, the pump at the right is known as the **right-hand side** and the other as the **left-hand side**.

As one piston moves to the end of its stroke and is gradually brought to rest, it moves the slide valve of the opposite steam cylinder, admitting steam back of the piston, which is at rest, and causing it to move forward to the opposite end of its stroke.

The valves are plain slide valves, without lap or lead. Lost motion is allowed between the valve and its operating mechanism, which causes the

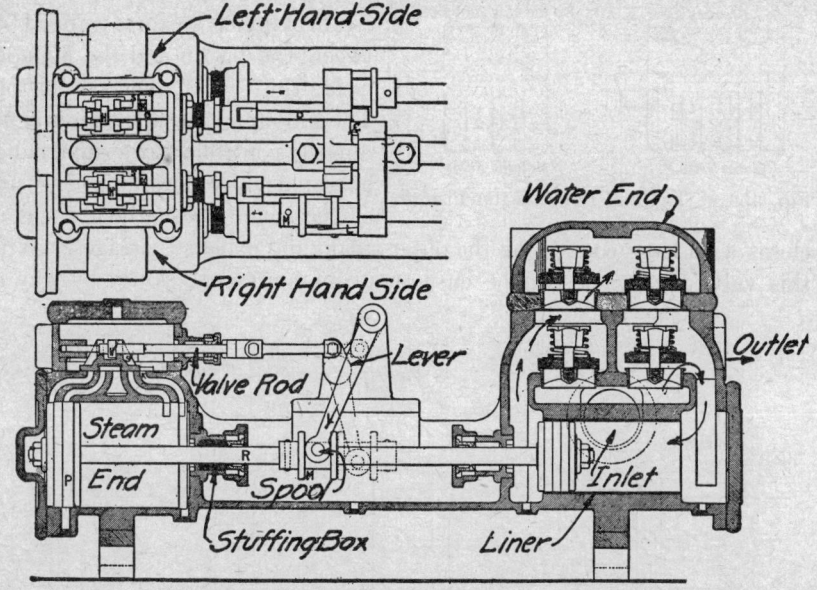

Fig. 457. — Duplex Steam-driven Pump.

piston to pause at the completion of the stroke. The pistons are in motion only about five-eighths of the time. The length of stroke is varied by

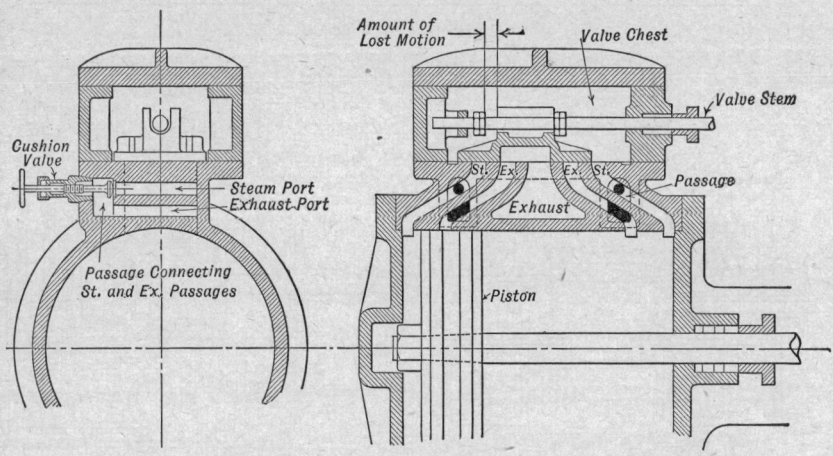

Fig. 458. — Steam Cylinder of Duplex Pump Showing Cushion Valves.

changing the lost motion at the valve, and is never constant, but changes with the discharge pressure.

To prevent the piston striking the cylinder head at each end of the stroke, the steam cylinder is made with five ports, Fig. 458, two outside end ports for admission of steam and three inner ones for the exhaust. When the piston closes the exhaust port in the steam cylinder, steam is trapped between the piston and the cylinder head, thus cushioning the piston. The amount of this cushion is ordinarily controlled by a small hand-operated valve, located at the side of the steam chest which opens or closes a passage connecting the outer steam and exhaust passages. With this valve nearly closed, the cushion steam can escape slowly by way of

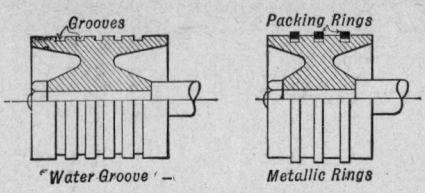

FIG. 459. — Types of Pump Water Pistons.

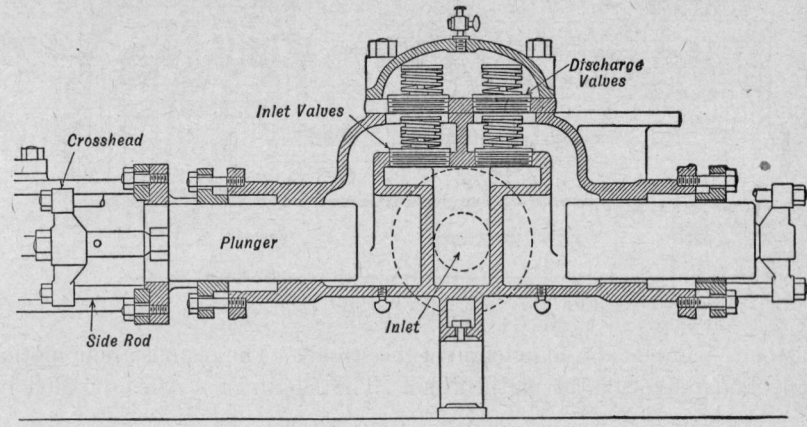

(a) End Outside Packed.

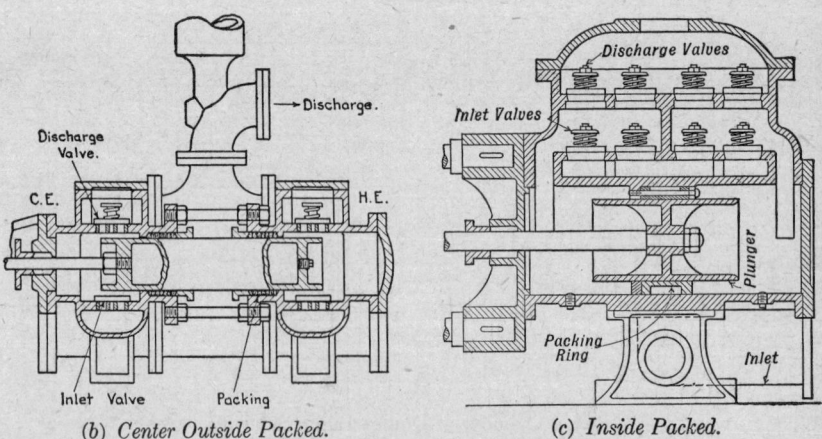

(b) Center Outside Packed.   (c) Inside Packed.

FIG. 460. — Types of Plunger Pumps.

the steam passage through the cushion valve, to the exhaust passage. This should be the condition when running rapidly and at light loads. With the cushion valves wide open there will be only a slight cushioning; and the length of the stroke will be a maximum.

491. Types of Pump Water Ends. — Most of the simplex pumps of the less efficient types have pistons, which are constructed as shown in Fig. 453, and have a space into which durable water packing, made of **layer canvas,** is placed. The cylinder is generally fitted with a **brass liner** to prevent rust,

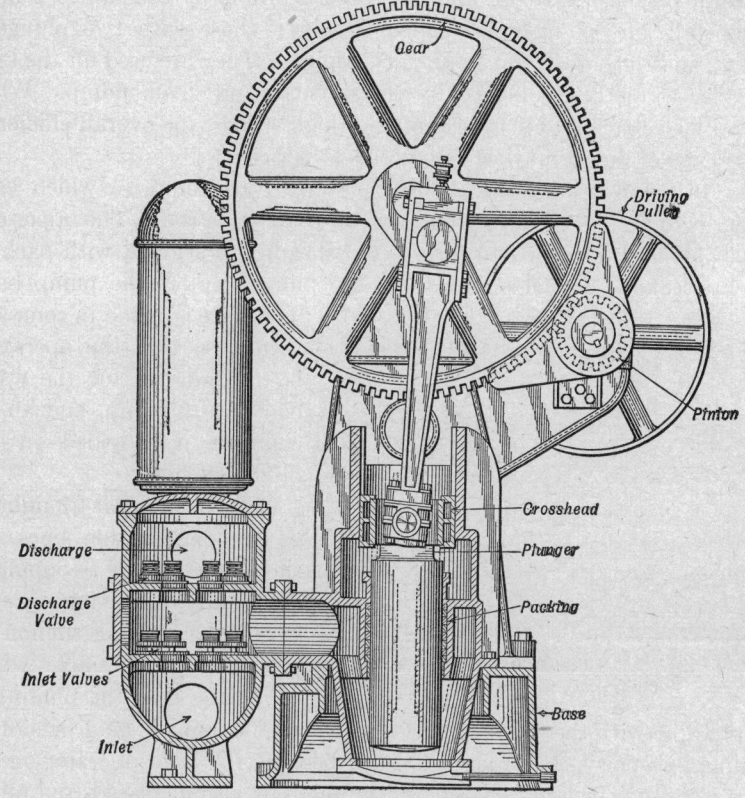

FIG. 461. — Goulds Single Acting Triplex Plunger Pump.

and the consequent rapid wear of the packing; and when thus made, a wide piston with grooves, Fig. 459, made on its face is often used. The fit between the piston and liner and the effect of the grooves are relied upon to prevent leakage. A plunger, Fig. 460c, is often substituted for the piston, and leakage is prevented by a packing held in place by a gland through which the plunger slides. In the above types the packing is inside the cylinder, where leakage cannot be easily detected, and to renew the packing the cylinder head must be removed.

To make leakage past the plunger visible, outside packed plunger pumps are used. The leakage is then easily prevented by adjusting the gland against the packing. In Fig. 460b, only one plunger is used with the packing at the inner end of the cylinder; while in Fig. 460a two plungers, connected by a yoke so that they move together, are used, and the packing is placed at each outside end of the cylinder. In this latter pump a partition separates the compartments in which the plungers work.

492. Power-driven Pumps. — Piston pumps driven by belting, chains or gearing are classed as power-driven pumps. The source of the power may be steam, gas, or electricity, and the speed is ordinarily constant. Pumps having only one plunger are called **simplex;** those with two plungers, **duplex;** and those with three plungers, **triplex.** They are used for the kind of service formerly performed by the direct-acting steam pump. When driven by electric motors and running at high speeds, the **overall efficiency** of this type of pump may be as high as 83 per cent.

Fig. 461 shows a typical triplex power-driven pump, in which each plunger is driven by means of a crank and connecting rod. The upper end of each plunger is shaped to form a crosshead, and a gland with packing prevents leakage at the point where the plunger enters the pump base. The base is made in compartments, so that the operation is the same as for the water end of any pump, and an air chamber is provided on the discharge side.

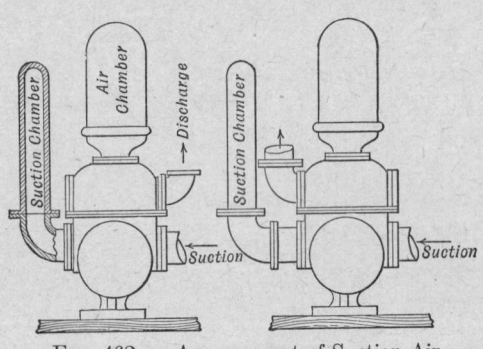

Fig. 462. — Arrangement of Suction Air Chambers for Pumps.

493. Suction Air Chambers. — On long suction lines, the shock caused by stopping a large body of moving water is taken up by using **suction air chambers,** Fig. 462, which should be near the pump and in direct line with the current. By this means the pump is made smoother running, less noisy, and more efficient, because the head of water on the pump is fully utilized. The life of the pump is also increased and repair bills are saved.

494. Pump Governor. — Steam-driven pumps are generally equipped with a pump governor which maintains a nearly constant pressure in the discharge pipe of the pump, irrespective of the quantity of water flowing. It does this by controlling the speed of the pump. A typical governor, Fig. 463, consists of a pressure-reducing valve located in the steam supply-pipe of the pump and moved by slight variations in the discharge water pressure which acts upon the top of a diaphragm attached to the double balanced valve in the steam line. A spring attached to the valve stem normally holds the valve open, and a small regulating hand wheel can be adjusted to change the amount of discharge pressure. A change in this pressure, acting on the diaphragm at the upper end of the valve stem,

CENTRIFUGAL PUMPS 545

causes a difference in pressure between that caused by the water and that resulting from the spring. The balanced valve then adjusts itself to this differential pressure and regulates the discharge pressure by changing the amount of steam admitted to the pump. This changes the speed and gives the desired discharge pressure.

Power-driven pumps generally run continuously, at a constant speed, and deliver sufficient water to supply the maximum demand. To allow for variation in the demand, a by-pass connection, containing a gate valve, a relief valve, and a check valve, is provided. The amount of feed may be regulated by hand, or by means of a pressure regulator similar to the pump governor. The surplus water returns to the source of supply.

495. Centrifugal Pumps. — This type of pump, Fig. 464, is particularly adapted for low heads and large volumes and by using stages for high pressure. The average mechanical efficiency of centrifugal pumps varies from 70 to 82 per cent, according to the load. The cost is low, and this offsets the higher mechanical efficiency of high-grade pumping engines. They are generally direct-driven by a steam turbine or either a variable or constant speed electric motor. When the impeller revolves in a right-handed direction, viewed from the driving side, the pump is known as a **right-hand centrifugal pump**. When revolving in the opposite direction, the pump is **left-hand**.

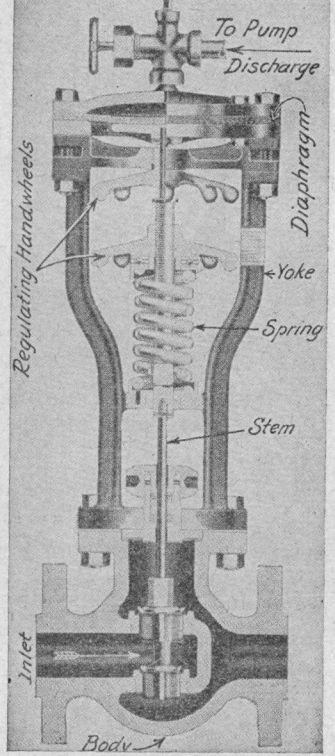

Fig. 463. — Swartwout Pump Governor.

A centrifugal pump has the following essential parts:

1. A **rotary impeller** to increase the velocity and pressure of the water, which enters at the center of the impeller and is discharged at its circumference with increased velocity and slightly increased pressure.

2. A cast-iron **casing**, which may be **split** or **solid**, having inlet and outlet passages. The casing guides the water from the circumference of the impeller to the discharge outlet, and is often shaped to convert the velocity, or kinetic energy, of the water to pressure, or potential energy.

3. A **shaft**, which supports the impeller; and either ball or babbitt bearings in which the shaft revolves.

The impeller may be of the open or the closed type, Fig. 465. The **open**

impeller consists of a circular central disk partition to which are attached blades having an **involute form**. This type of impeller does not fit closely in the casing within which it revolves, and consequently the guidance of the

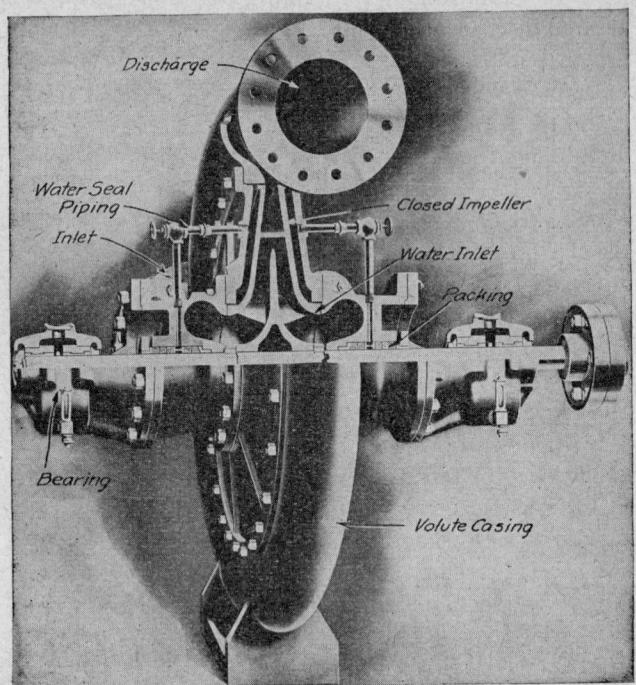

Fig. 464. — Wheeler Closed Impeller Centrifugal Pump.

water and the flow lines are poor. In addition, a large amount of water slips between the revolving impeller blades and the casing wall, thereby reducing the efficiency.

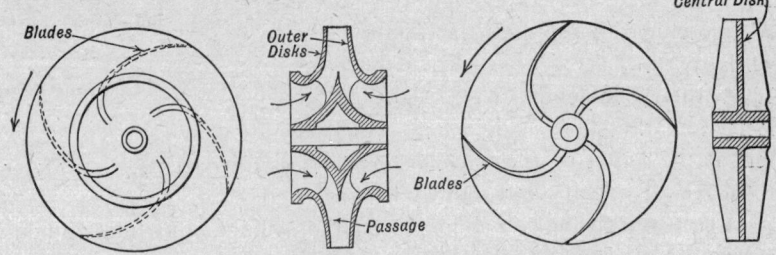

Fig. 465. — Types of Impellers for Centrifugal Pumps.

The **closed impeller** has two outer disks separated by blades, similar in form to the open-type blades, which form passages for the water. The water enters the passages at the center, and is guided to the outlet. The side walls of the impeller prevent water from slipping past, and wearing rings of either the lantern or labyrinth type prevent leakage past the impeller, thereby improving the efficiency.

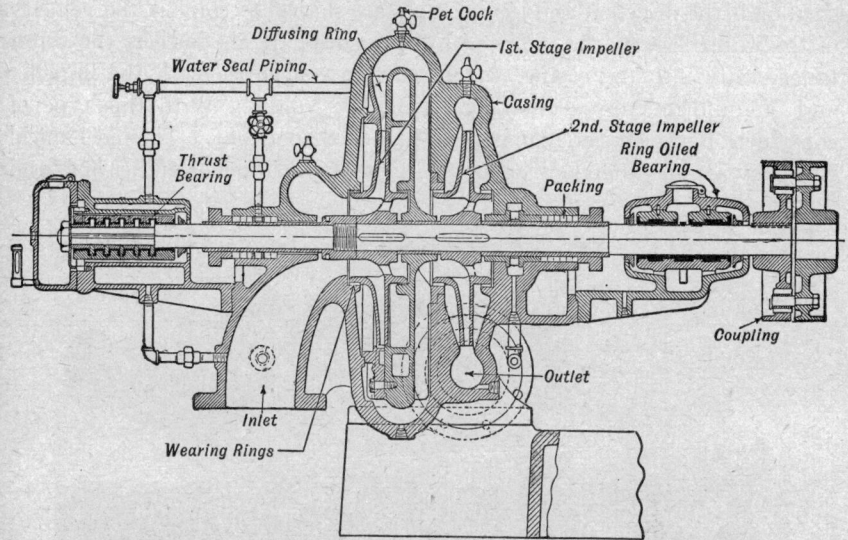

Fig. 466. — Cameron Turbine-type Centrifugal Pump — 2 stage.

With either of these types of impellers, the water may enter from one or both sides. The **double-suction** type is preferable, as it minimizes the end-thrust on the shaft, produced by the unbalanced water pressure, which exists when water enters from one side only. The **single-suction** impeller must be balanced, to overcome the end-thrust.

Centrifugal pumps are generally classified, according to the construction of the casing, as: (1) *volute*, (2) *turbine*.

The **volute pump** does not use a diffusing, or guide, vane surrounding the impeller, but has a **spiral**, or **volute, casing** to guide the water, as it leaves the impeller on its way to the discharge pipe. The shape of the volute is such that it converts the velocity of the water leaving the impeller into pressure. In this type of pump, the water leaving the impeller is thrown across the stream of water in the casing, and the flow is disturbed.

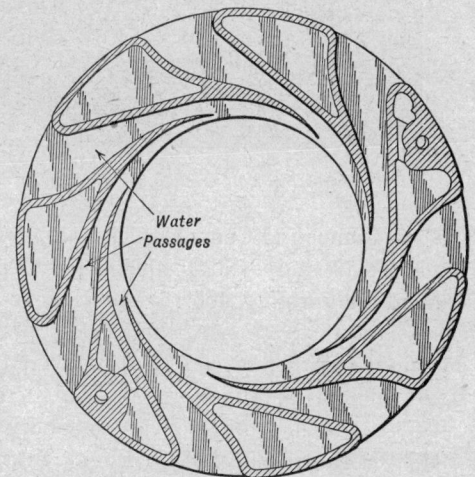

Fig. 467. — Diffusing Ring for Turbine-type Centrifugal Pump.

The **turbine pump**, Fig. 466, uses a circular diffusing ring, Fig. 467, containing ribs which form passages for the discharge of water. This ring surrounds the impeller, and the passages are shaped to convert the velocity of the water leaving the impeller into pressure. It also directs the water tangentially into the casing, which may be concentric with the impeller and of uniform cross section, or it may be volute. With this type of construction, the efficiency may be as high as 82 per cent. The mechanical efficiency attained depends upon the shape of the impeller and casing, and upon the number of stages used.

Fig. 468. — Cameron 6-stage High-pressure Feedwater Pump — External View.

496. Compound Centrifugal Pump. — When it is desired to increase the discharge pressure above 100 pounds per square inch, a compound centrifugal pump, Fig. 466, is used. Several impellers are mounted on a shaft, and revolve in separate compartments, or stages, of the main casing. The discharge from each stage is delivered to the suction of the next higher stage, and the final discharge pressure depends upon the number of stages. Closed single-inlet impellers are generally used.

497. High-pressure, Multi-stage Centrifugal Pump. — An external view and a sectional view of the Cameron six-stage high-pressure pump is shown in Figs. 468 and 469 and is equipped with a balancing drum and a Kingsbury thrust bearing. The general construction resembles the centrifugal pumps previously described. The single-inlet impellers discharging through a diffusion vane so arranged that the 180-degree turn before entering the inlet of the next stage is accomplished without serious friction loss. This process is repeated until the last stage is reached, at which the diffusion vane is omitted and a volute chamber is substituted in order to simplify the construction of the pump. Pumps of this type are used up to six stages in a single casing, and when a greater number of stages are necessary to generate the desired discharge head, the stages are generally divided into two multi-stage pumps so arranged that half the pressure is generated

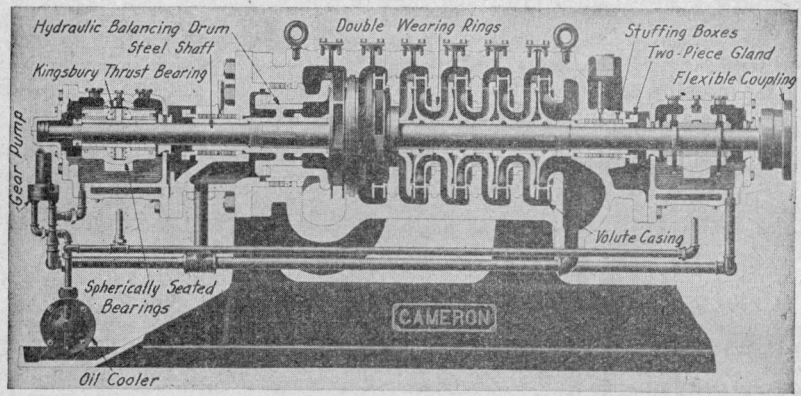

Fig. 469. — Cameron 6-stage High-pressure Pump — Sectional View.

by the first group of stages in the first pump, and the remaining pressure in the second group by the second pump. Leakage out of or into the pump casing is prevented by a water-sealed connection made at the middle of the stuffing box.

498. Rotary Pump. — This pump, Fig. 470, is used for the same purposes as the centrifugal pump; the small sizes are also used to pump oil for lubricating purposes. It can be operated at a low speed of rotation, the pressure attained is moderate, the volume discharged is large, and the space required is small. At high speeds, however, this type of pump is noisy. It consists of an accurately machined casing, having an inlet and outlet opening. Within the casing are two rotors attached to separate shafts, which are connected by gearing and to one of which the power is applied. Each rotor, or impeller, is accurately machined to run with a small clearance within the casing, and to mesh with the other rotor. Water enters

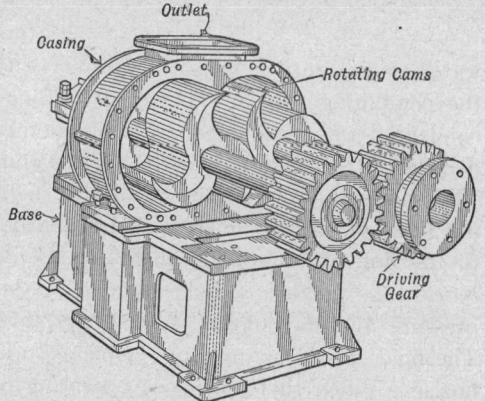

Fig. 470. — Goulds Rotary Pump, Cover removed.

at the bottom, is enclosed within the pocket formed between the rotors, and discharged at the top, the volume delivered varying with the speed at which the pump is operated.

499. Jet Pump. — This type of pump is used mainly for feeding water into a boiler, where its high thermal efficiency justifies its use, because nearly all the heat used to operate the pump is returned to the boiler as warm feedwater. Multi-jet pumps, for removing air from steam condensers, are described in Art. 530, page 575. For general pumping service the mechanical efficiency of a jet pump is low.

The **injector** is a typical jet pump. It is efficient, convenient, cheap, compact, has no moving parts, delivers warm water into a boiler without preheating, and has no exhaust to be disposed of. It is universally used on locomotive boilers, and in connection with feed pumps as a reserve on large land boilers. It will not handle hot water. The essential parts of the injector are illustrated in Fig. 471. Steam from the boiler enters at the top and flows to the atmosphere through the **steam tube, combining tube** and **overflow.** *The steam tube is so shaped that it converts the pressure energy of the steam into velocity energy.* As a result of the high velocity attained in the steam tube, the air is partially exhausted from the inlet pipe, thereby causing the

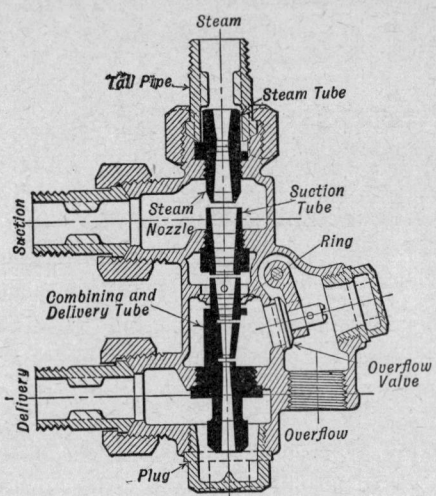

FIG. 471. — Interior View Metropolitan Automatic Injector.

water to rise until it comes in contact with the steam at the entrance to the combining tube. The steam, coming in contact with the water, is condensed and imparts considerable momentum to the water. The condensing of the steam reduces its volume and serves to maintain the vacuum. The shape of the combining tube is such that the velocity of the moving mass of water is increased sufficiently to carry it across the opening to the **delivery tube,** in which the *velocity of the water is partially converted into pressure.* The water lifts the check valve, because of this pressure and momentum, and enters the boiler against boiler pressure. The final discharge pressure depends upon the shape of the discharge nozzle. When the injector is operating properly, water will not show at the overflow.

Injectors are either **hand-starting** or **automatic,** and either **single-** or **double-tube.** The automatic injector will resume its flow, after an interruption, without any attention from the operator.

500. Automatic Injector. — A single-tube automatic injector is shown in Fig. 471 with the operating parts so marked as to require no description. The operation is similar to the general description in the previous article, with the exception that the combining tube is surrounded by a loose ring, which normally remains closed when in operation. Should the discharge be interrupted, this ring opens and passes water to the overflow until the vacuum is again established, then it closes and the injector again delivers water. It is started by opening the steam valve in the steam pipe connecting the injector to the boiler.

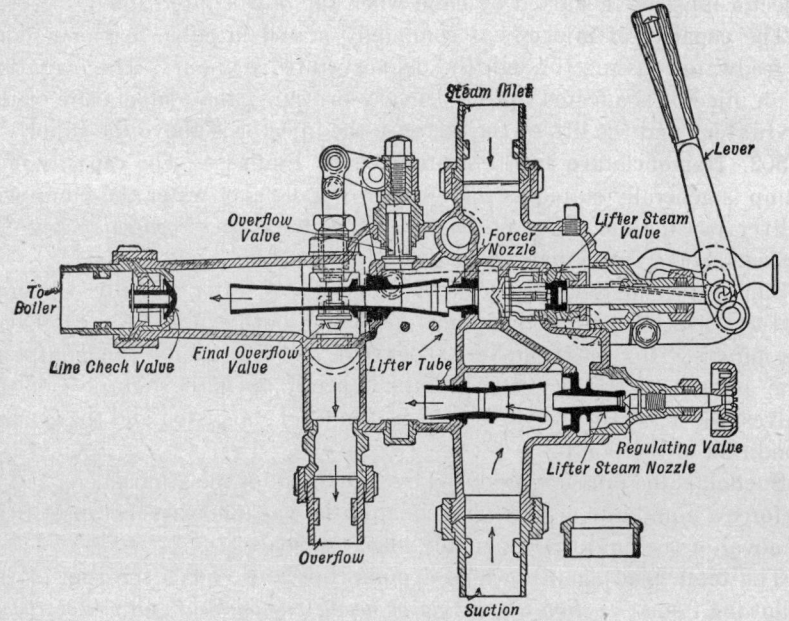

Fig. 472. — Hancock Double-tube Injector.

501. Double-tube Injector. — The Hancock double-tube injector, Fig. 472, illustrates this type, which consists of a double set of nozzles, called the **lifting set** and the **forcing set**. The lifting set is the lower set and consists of a steam nozzle and a combining tube. The forcing set of tubes consists of a steam nozzle, combining, and delivery tube. The valve which controls the steam flow to the forcing set is attached to a lever, to which the overflow valve is also attached. When starting, the lever is only partly open and this holds the overflow open. When started, and with the lever wide open, the overflow is closed. This type of overflow is called a **closed overflow.**

To operate, the *hand-operated regulating steam valve, which controls the amount of steam used, is partly opened.* This admits steam to the lifting steam nozzle, then through the lifting combining tube into the overflow chamber, to the atmosphere. The rapid passage of steam creates a vacuum in the suction chamber, whereupon water flows in, mingles with and condenses the steam. When water appears at the overflow, the lever is pulled back to the stop and the forcing nozzle put into operation. The water discharged from the lifting combining tube is delivered to the forcing combining tube, where the forcing jet meets and mingles with it and delivers the water into the boiler, as in the injector previously described. *The use of a double-tube injector increases the delivery pressure and permits varying the capacity by regulating the steam inlet valve to the lifting steam nozzle.* This injector must be re-started by hand when the flow is interrupted.

The **capacity of injectors** is commonly stated in cubic feet or gallons of feedwater passing through the delivery tube per hour. The operation of an injector is affected by the delivery pressure; the temperature of the feedwater; and the **lift,** or the distance the injector is above the supply.

502. Nomenclature and Performance of Pumps. — The **capacity** of a pump is generally stated as the number of gallons of water the pump will deliver per minute. *For piston pumps, manufacturers generally state the capacity as the piston displacement per minute.*

Slip is the difference between the quantity of water actually delivered and the displacement or theoretical discharge, both expressed in cubic feet per minute. It varies from 5 to 20 per cent, and is greater at small outputs. The amount of slip in piston pumps depends upon the tightness of the valves, the action of the valve springs, the fit of the piston and the general condition of the pump.

Suction is the pressure produced by the weight of the atmosphere, acting to force a liquid into a space wherein a partial vacuum exists because of the removal of the fluid that originally filled the space.

The **total head** against which a pump operates equals the sum of the following items: *suction lift, discharge head, friction head* and *velocity head.* The **suction lift** is the vertical distance from the level of the water being pumped to the center line of the pump. In case the water level is above the center line the distance is subtracted from the other three items. The **discharge head** is the vertical distance between the center line of the pump and the level to which the water is elevated. The **friction head** is caused by piping, elbows, and valves, and is ordinarily determined from Charts and Tables found in standard textbooks on Hydraulics. The **velocity head** in feet per second is calculated from the fundamental relation, $h = \dfrac{v^2}{2g}$, as explained on page 78. This head is small and is usually neglected except where the total head is low, the suction lift high, or in accurate test work.

The **horsepower** of a pump is generally called **water horsepower,** and equals the weight of water in pounds actually delivered per minute multiplied by the distance through which the water is lifted divided by 33,000, or

NOMENCLATURE AND PERFORMANCE OF PUMPS

$$Water\ horsepower = \frac{Wh}{33,000} \quad \ldots \ldots \ldots \quad (114)$$

in which W = pounds of water delivered per minute.

h = head in feet = pressure in pounds per square inch $\times$ 144 $\div$ density of water pumped.

Example 58. — During the test of a 432,000-gallon Fairbanks Morse compound duplex pump, 133,250 lb. of water were delivered per hr. against a pressure of 121.8 lb. per sq. in. gage; temperature of water, 70 deg. fahr. Find: (a) the water horsepower, considering the discharge pressure as the only resistance to be overcome; (b) the head caused by the change in velocity when the diameter of the discharge pipe is 4 in. and the suction pipe 6 in.

Solution. — (a) Using Equation (114),

$$Water\ horsepower = \frac{Wh}{33,000} = \frac{2221 \times 282}{33,000} = 18.9$$

$$W = \frac{133,250}{60} = 2221\ \text{lb. per min.}$$

$$h = \frac{121.8 \times 144}{\text{density at 70 deg. fahr.}} = \frac{121.8 \times 144}{62.3} = 121.8 \times 2.30 = 282\ \text{ft.}$$

(b) Velocity head in discharge = $\frac{v_1^2}{2g}$; velocity head in inlet = $\frac{v_2^2}{2g}$

$$\text{Head caused by velocity change} = \frac{v_1^2 - v_2^2}{2g} = \frac{(6.72)^2 - (2.96)^2}{2 \times 32.2} = 0.565\ \text{ft.}$$

$$v_1 = \frac{\text{cu. ft. of water per sec.}}{\text{area of 4 in. discharge pipe}} = \frac{0.5950}{0.0884} = 6.72\ \text{ft. per sec.}$$

$$v_2 = \frac{\text{cu. ft. of water per sec.}}{\text{area of 6 in. inlet pipe}} = \frac{0.5950}{0.2006} = 2.96\ \text{ft. per sec.}$$

The **indicated water horsepower** is found from the data obtained from indicator diagrams taken from the water end of a piston type pump and worked up in the manner explained in Art. 391, page 424. Care should be taken to use an indicator having a piston properly fitted for use with water, otherwise incorrect results will be obtained on account of leakage past the piston. The shape of the indicator diagrams taken from a steam driven pump is nearly rectangular for both the steam and water ends, and the inlet and discharge lines are wavy because of the imperfect action of the valves.

Thermal efficiency of a steam-driven pump is generally expressed as the number of foot-pounds of work done by the pump per 1000 pounds of steam or per 1,000,000 B.t.u. consumed by the driving engine, and is called **duty**.

Example 59. — The following data were taken during the test of a 30-million-gallon steam-turbine-driven reduction-gear centrifugal pump: pressures, barometer 29.22 in. mercury, steam 160.1 lb. per sq. in. gage; total head on pump, 140.94 ft. of water; quality of steam, 96.88 per cent; temperature of water pumped, 34 deg. fahr.; water pumped per 24 hr., 21.3 million gallons; vacuum on turbine, 14.16 lb. per sq. in.; temperature corresponding to vacuum, 51.7 deg. fahr.; weight of condensate per hr., 9389 lb. Find the duty of the pump under conditions of test.

Solution. — According to definition

$$\text{Duty} = \frac{\text{Foot-pounds of work} \times 1{,}000{,}000 \text{ B.t.u.}}{\text{Weight of steam} \times \text{B.t.u. per pound of steam above exhaust pressure}}$$

$$= \frac{1{,}042{,}687{,}750 \times 1{,}000{,}000}{9389 \times 1148.7} = 96{,}800{,}000 \text{ ft-lb. per million B.t.u.}$$

Ft-lb. of work per hour $= \dfrac{21{,}300{,}000 \times 8.33}{24} \times 140.94 = 1{,}042{,}687{,}750$.

Weight of steam per hour $= 9389$ lb.

B.t.u. per pound of steam $= xL + h_{f_1} - h_{f_2} = 0.968 \times 852.5 + 343.3 - 19.78 = 1148.7$

L corresponding to 174.5 lb. per sq. in. abs. $= 852.5$ B.t.u.
h_{f_1} corresponding to 174.5 lb. per sq. in. abs. $= 343.3$ B.t.u.
h_{f_2} corresponding to 0.187 lb. per sq. in. abs. $= 19.78$ B.t.u.

Characteristic curves such as those shown in Fig. 473 are supplied by manufacturers of centrifugal pumps. *These curves show the manner in which the total head, brake horsepower and efficiency change with variation in*

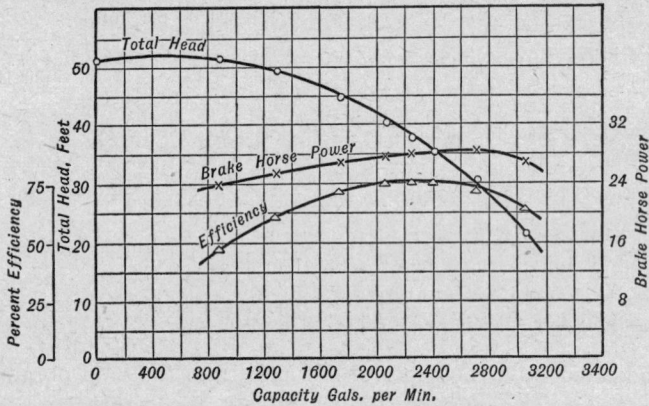

FIG. 473. — Characteristics for a 10-inch Wheeler Centrifugal Pump — 900 R.P.M.

capacity, when the pump is operated at constant speed. The **capacity rating** *of the pump is generally made for the point at which the efficiency is a maximum, hence for best results the pump should be operated at this capacity.*

503. Installation of Pumps. — The foundations of pumps must be rigid, with the suction pipe short, and as direct as possible. The suction lift should never be over 15 feet, and when pumping warm water the pump should be located below the supply, as otherwise the formation of vapor at low pressure destroys the vacuum and prevents the pump from lifting the water. The suction piping should be tight, to prevent destruction of the vacuum by air leakage. The piping for a centrifugal pump should be self supporting, and provision should be made for priming the pump when starting. The pump may be primed in the following ways:

1. Locating pump below water level of supply.
2. Using of a foot valve, Art. 504, with connection to discharge pipe, if the discharge pipe is left full of water.

3. Attaching a vacuum pipe or injector to the opening at top of casing, to remove air from casing, so that water will rise into pump through suction pipe.
4. Filling pipe casing with water from an external source. In this case, a foot valve must be used at foot of suction pipe.

A check valve should be placed on the discharge line of the centrifugal pump to prevent breakage from water hammer and a gate valve to control the capacity.

504. Foot Valve and Strainer. — The foot valve, Fig. 474, is used to hold the water in the suction or discharge pipe of a pump. When used in the suction pipe, it permits starting the pump without priming. It consists of a cast-iron body having a number of lift, or hinged, rubber-faced valves, opening against a spring. The upper end of the body is attached to the suction pipe, and when the pump is in operation the valves remain open; but when it is idle the weight of water in the pipe closes the valves and keeps the pipe full of water.

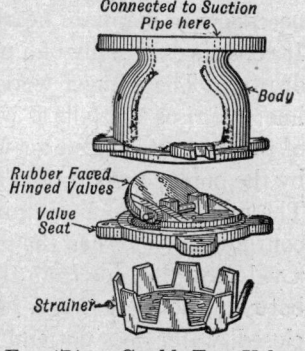

Fig. 474. — Goulds Foot Valve and Strainer.

The **strainer** consists of a frame supporting a removable wire-mesh strainer. It is attached to the suction pipe to keep out all foreign material which would clog the pump. When a foot valve is used, the strainer is attached to the lower part of the foot valve and is often an integral part of it.

505. Method of Testing a Pump. — The method of testing a steam pump is essentially the same as for a steam engine, with the addition of some suitable means of measuring the water delivered. For details regarding testing of pumps consult A. S. M. E. POWER TEST CODE.

It is customary to measure the quantity of water flowing in closed conduits, by using any of the following methods: pitot tube, venturi meter, water meter, salt-velocity and pressure-time or Gibson. For open flumes some form of weir is used. These methods with the exception of the salt-velocity and Gibson methods have been explained elsewhere in the text.

The **salt-velocity method** is based on the fact that salt in solution increases the electrical conductivity of water. Brine is introduced at any point in a conduit and its passage past one or more pairs of electrodes is timed. *The discharge in cubic feet per second equals the volume in cubic feet between the electrodes divided by the time in seconds.* The **Gibson method** is based on the principle that impulse is equal to momentum. The pressure in a pipe is changed by closing a valve and the changes in pressure are photographically recorded to produce a " pressure-time " diagram — the area of which is a measure of the flow in the conduit at the instant prior to the interruption of the flow.

506. Valve Setting of Steam-driven Pumps. — To set the slide valve or valves of a steam-driven pump, the piston is usually placed midway between its extreme positions, and the valves placed centrally over both steam ports. When the lost motion allowed between the lug of the valve and the valve rod connection, which moves it, is made adjustable, the amount of the lost motion is made equal at each side of the lug by moving the nuts on the valve rod. The length of the stroke is controlled on most slide valve pumps by the amount of this lost motion. The exact amount of lost motion to give the desired stroke must be found by trial.

507. Accumulator. — In the discharge pipe lines of pumps, where it is necessary to maintain nearly a constant pressure, an accumulator is used. It consists essentially of a plunger of small area loaded with a large amount of iron. The plunger works in a cylinder connected directly into the pipe line. It rises and falls as water is forced in or passes out with slight change of pressure, storing water when the amount used is less than that supplied by the pump, and giving it out when the demand is greater than the supply. The pressure is thus maintained constant and the pump is allowed to run continuously. When the plunger rises to the extreme of its travel, it closes a valve on the pipe from the pump, and at the same time opens a by-pass between the suction and discharge pipes of the pump. While this valve is closed, water does not enter the plunger cylinder.

REFERENCES

Cyclopedia of Engineering, AMERICAN TECHNICAL SOCIETY.
Steam Power Plant Engineering, GEBHARDT.
Mechanical Equipment of Buildings, Vol. II, HARDING and WILLARD.
Centrifugal Pumps, DAUGHERTY.
Practice and Theory of the Injector, KNEASS.
Test Code for Reciprocating Steam-Driven Pumps, A. S. M. E.

REVIEW QUESTIONS AND PROBLEMS

1. Classify steam-driven piston pumps.
2. Describe the method of reversing the stroke on the Cameron pump.
3. How does a turbine centrifugal pump differ from a volute pump?
4. What is meant by a power-driven pump?
5. Explain the operation of a pump governor, and state why one is used.
6. Describe the construction and explain the operation of a single-tube injector.
7. Define: slip, water horsepower, pressure head, suction.
8. A 15.8-million-gallon vertical, triple-expansion plunger type pumping engine delivered 16.0 million gallons of water at 67 deg. fahr. through a total head of 298.7 ft. per 24 hr. Average steam pressure, 161.1 lb. gage; average quality of steam, 98.5; barometric pressure, 30.13 in. mercury; vacuum in exhaust pipe, 28.5 in. mercury; temperature corresponding to vacuum in exhaust pipe, 119 deg. fahr.; weight of steam condensed per hr., 9620 lb.; average i.hp., 871.5.
Calculate the duty in foot-pounds per million B.t.u., and per 1000 lb. steam.

9. Calculate the water horsepower and the mechanical efficiency of the pump in Problem 8.

10. A 1000 gal. per min. centrifugal pump works against a 100 ft. total head. The brake horsepower is 32.3. Find: (*a*) the water horsepower; (*b*) the mechanical efficiency.

11. A pump delivers 1600 gal. per min. against a discharge pressure of 45 lb. per sq. in. Suction lift 2 ft. Efficiency of pump 65 per cent. Find the amount of horsepower required to drive the pump.

12. A pump delivers 160 gal. per min. at 60 deg. fahr. against a discharge pressure of 45 lb. per sq. in. Center of discharge gage 1.5 ft. below the pump center line. Diameter of discharge pipe 6 in., of suction, 8 in. Vacuum on suction 5 in. mercury with gage attached 2 ft. below pump center line. Find the total head.

CHAPTER XXIV

CONDENSERS AND CONDENSER AUXILIARIES

508. Foreword. — *The primary object in operating engines and turbines "condensing" is to obtain a greater amount of useful work from a given weight of steam than could be obtained without the condensers, however, in some cases recovery of the condensate for boiler feedwater is of equal importance.*

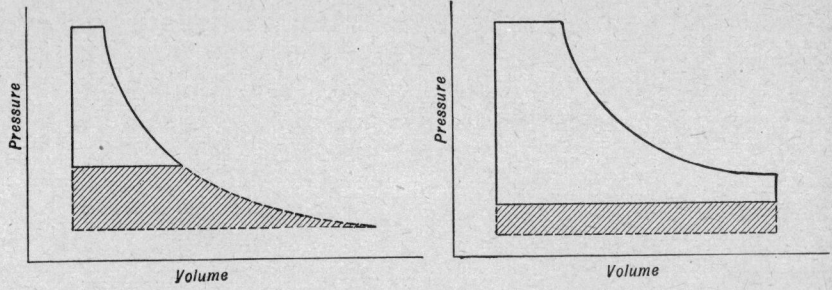

Fig. 475. — Diagrams showing Gain resulting from Condensing Operation.

The gain in power resulting from the use of a condenser is illustrated by Fig. 475 in which the full lines represent the theoretical diagrams for non-condensing operation. If the expansion could be continued or the back-pressure line lowered, as shown by the dotted lines, the area of the diagrams would be increased by the cross-hatched area. In steam engine practice the gain in power amounts to about 25 per cent for a vacuum of 26 inches of mercury, which is nearly the maximum for steam engine practice, because of the excessive size of cylinders required for the low pressure. The high cost, internal friction, and condensation losses of lower vacuums with reciprocating engines offset the gain in energy. An increase in power means an increase in economy, as shown by the curves in Fig. 476 for an engine operating condensing

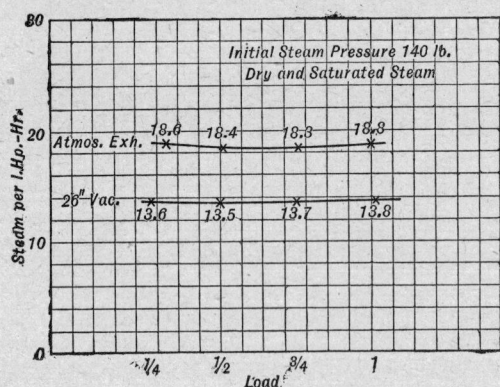

Fig. 476. — Steam Rate Curves for Condensing and Non-condensing operation of a 21″ × 22″ Universal Unaflow Engine.

and non-condensing, with the same initial steam pressure in each case. In marine practice the condenser is essential, to maintain the purity of the water in the boilers.

The advent of the steam turbine, operating either with high-pressure steam or with low-pressure steam from a reciprocating engine, has created a demand for a much higher vacuum than was required by the reciprocating steam engine. This has revolutionized condenser practice, and vacua corresponding to 28 or 29 inches of mercury are now common. The percentage of fuel saved for each of the last few inches in vacuum varies from

FIG. 477. — Turbine Room — Richmond Station.

5 to 8 per cent, depending upon the type of turbine. In general, a point is reached between 28 and $28\frac{1}{2}$ inches vacuum at which the gain in economy is offset by the increased cost of condensing apparatus and cooling water.

An application of a 70,000 square foot surface condenser to a modern high-capacity 50,000-kilowatt turbine installation, at the Richmond Station, Philadelphia, is shown in Fig. 477. This illustration clearly shows the motor-driven circulating pumps and the motor-operated valves, and also gives an excellent idea of the interior appearance of a modern turbine room.

509. Classification of Condensers. — Condensers may be roughly classified as (1) jet condensers, and (2) surface condensers. In the jet condenser the steam and water mingle, the steam being condensed by direct contact with water. In the surface condenser, the steam and cooling water do not come in direct contact with each other, the heat of the steam being extracted by conduction.

510. Jet Condenser. — This class of condensers may be subdivided into:

(1) Low-level condensers { Standard
                                 High-vacuum

(2) Barometric condensers
(3) Ejector condensers

Jet condensers are constructed on the parallel- or counter-current principle. In a **parallel-current condenser,** the steam, cooling water, and non-condensable vapors flow in the same direction, downward. Steam enters at the top or side of the condenser, and the condensing water immediately

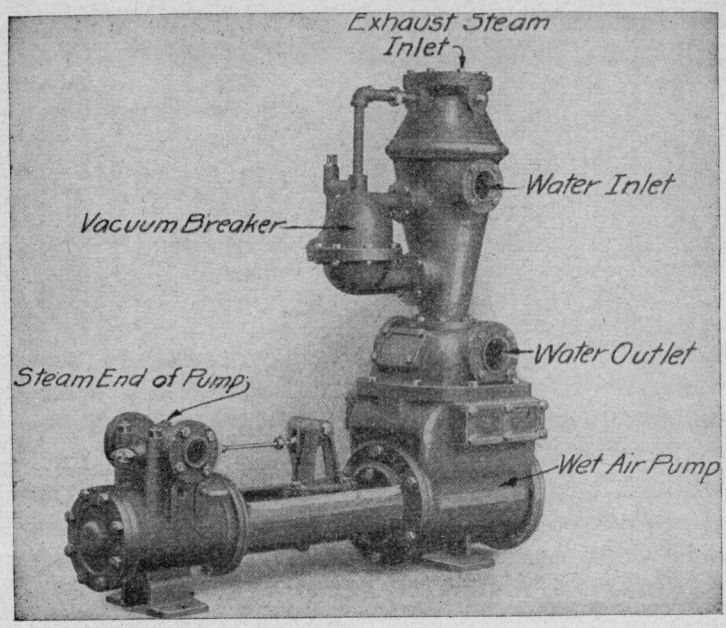

FIG. 478. — Standard Low-level Jet Condenser — Outside View.

below. In a **counter-current condenser,** the steam enters the lower part of the condenser, and rises through the falling condensing water, which enters at the side near the top of the condenser. Air and other non-condensable vapors are withdrawn near the top of the counter-current condenser, after they have been cooled by passing through the cooling water, and at the bottom of the parallel-current condenser, from a location directly over the surface of the water.

511. Standard Low-level Jet Condenser. — In this condenser, Figs. 478 and 479, the cooling water and a portion of the air are removed by some form of pump known as a wet-vacuum pump. The condensing water enters at the top of the pear-shaped condensing chamber and flows through a spray head which breaks the stream of water into a fine spray. Exhaust

BAROMETRIC CONDENSER

steam enters from the right and, mingling with the condensing water, gives up its latent heat of evaporation to the water. The condensed steam, water, and air are removed from the bottom by the **wet-vacuum pump,** i.e. a pump which removes both condensate and air. The condensing water rises into the chamber, as a result of the vacuum maintained by the wet-vacuum pump, when the suction head is below 18 feet. In this type of condenser a vacuum-breaking device is used; it consists of a float located in a chamber attached near the top of the condensing chamber, and below the exhaust connection. The float opens a valve to the atmosphere, when the water enters faster than the pump removes it, and thus breaks the

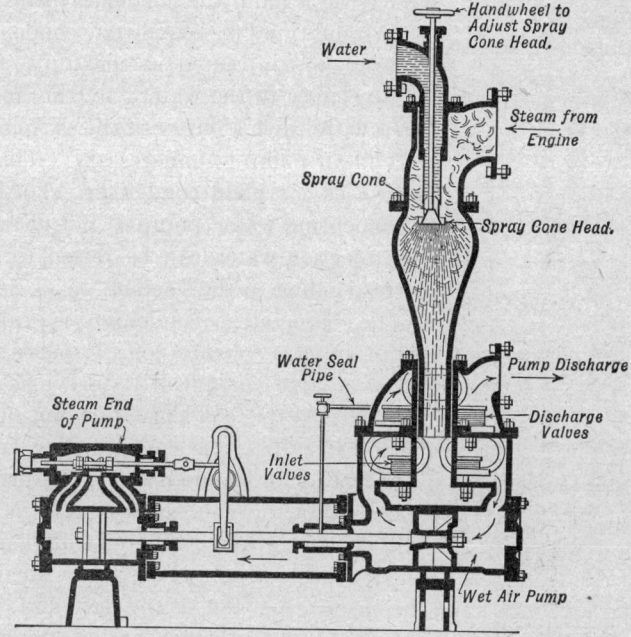

FIG. 479. — Standard Low-level Jet Condenser — Sectional View.

vacuum which prevents water from backing up into the exhaust pipe of the engine. This condenser has a low first cost, and is used with reciprocating steam engines where the vacuum required is not above 24 to 26 inches of mercury.

512. Barometric Condenser. — The barometric jet condenser differs from the standard jet condenser in that the condensing chamber is elevated about 35 feet above the discharge; the water is then discharged through the **tail pipe** without the aid of a water removal pump. Water stands in the tail pipe to a height corresponding to the vacuum; this height would be about 30 feet for normal vacua.

A section through an Ingersoll-Rand counter-current barometric condenser is shown in Fig. 480 with all parts clearly named. Steam enters into the center of the condensing chamber through a steam inlet nozzle and curved internal hood, and is condensed as it rises through a falling sheet of condensing water. The cooling water is pumped into a pool in the condensing chamber at a point above the steam entrance, and in its passage downward to the tail pipe is broken into fine streams by suitably arranged baffles. The air, released from the steam when the latter is condensed at the bottom of the condensing chamber, passes upward through the descending sheets of cold water, and after being thoroughly cooled and de-vaporized, is removed by the vacuum pump at a point above the level of the condenser pool.

In some types of barometric condensers, the tail pipe is contracted at the top to impart a high velocity to the water; by this means the air is withdrawn along with the escaping water, and an air pump is unnecessary. This type is known as a **syphon condenser**. For low lifts, a condensing water pump is not necessary, as the injection water can be raised to a height corresponding to the vacuum.

The barometric condenser is capable of maintaining a vacuum of 27 inches, but requires a large amount of water which must be handled by the condensing water pump. A typical installation is shown in Fig. 481.

513. Ejector Condenser. — The principle of operation of this condenser is that the momentum of flowing water ejects the discharge without the aid of a pump. Exhaust steam enters the ejector, Fig. 482, at the right and surrounds a central tube, in which are located numerous passageways. The cooling water enters the nozzle at the top and, because of the shape of the nozzle, attains considerable velocity. In flowing past the openings this water produces a vacuum, which causes the steam to flow through the openings and mix with the condensing water. The cooling water enters continuously at the top, because of the vacuum, and sufficient velocity is given the jet to discharge the combined mass of condensed steam, cooling water and air, against the pressure of the atmosphere.

There is about 3 feet of pipe above the nozzle and at least 2 feet below the discharge, which should be water-sealed. The vacua attained are from 20 to 25 inches for the single jet types, and up to the highest vacua for the multi-jet type.

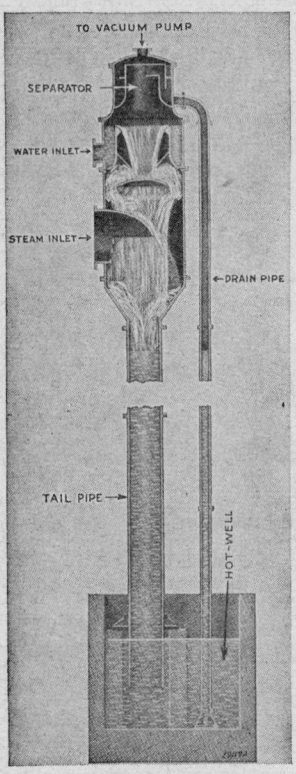

Fig. 480. — Barometric Condenser.

EJECTOR CONDENSER

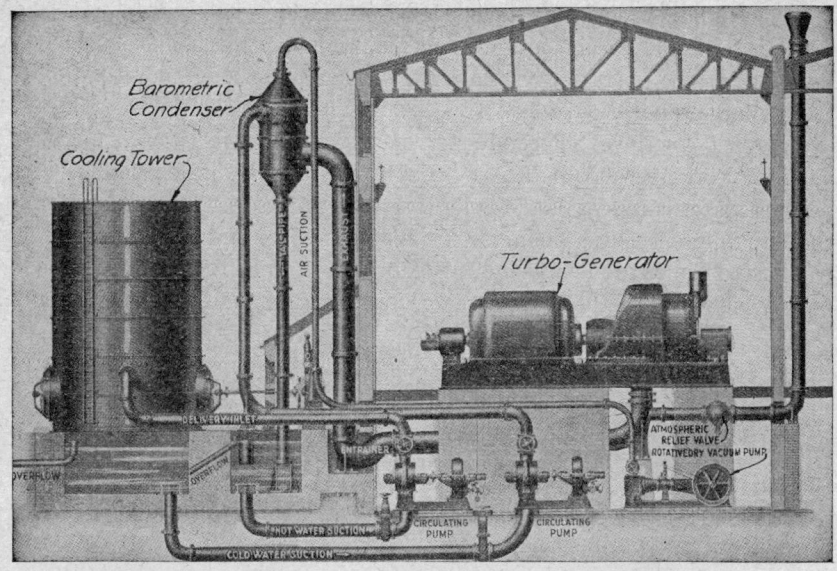

Fig. 481. — Worthington Barometric Condenser Installation:

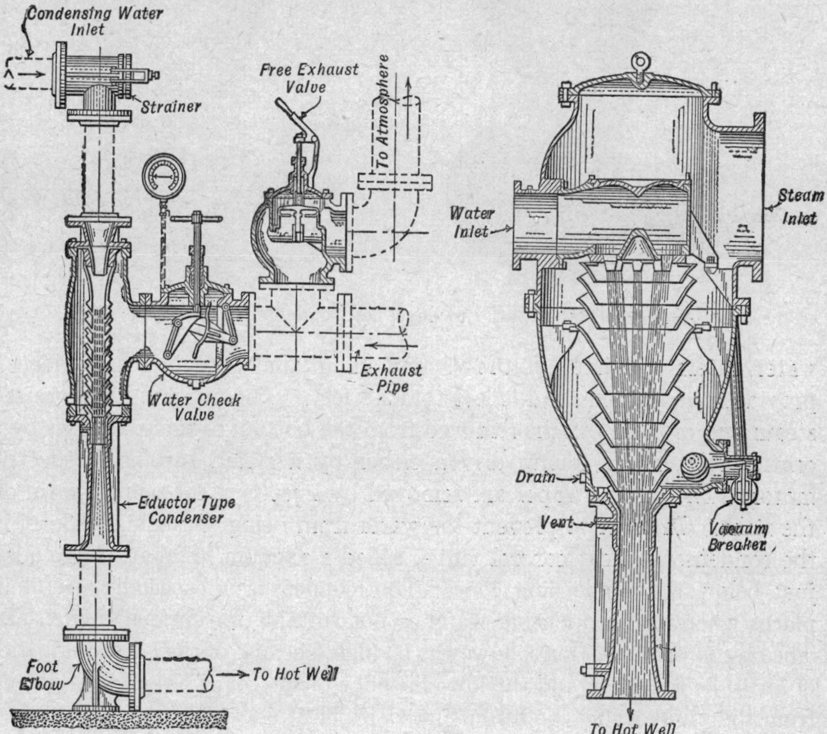

(a) Standard Type.   (b) Multi-jet Type:

Fig. 482. — Ejector Condensers.

514. High-vacuum, Low-level Jet Condenser. — The jet condensers previously described are not suitable for producing the high vacua necessary for turbine operation, because the air pumps used are not adapted to handle large quantities of air.

A typical **low-level jet condenser** of a type extensively used for high-vacuum work, because of its low first cost, is shown in Fig. 483. Condensing water enters at the left side near the top and is distributed around the circumference by a chamber, from which **spiral nozzles** discharge the water into the **condensing cone.** Steam enters from the top and is thoroughly mixed with the condensing water from the nozzles. The cooling

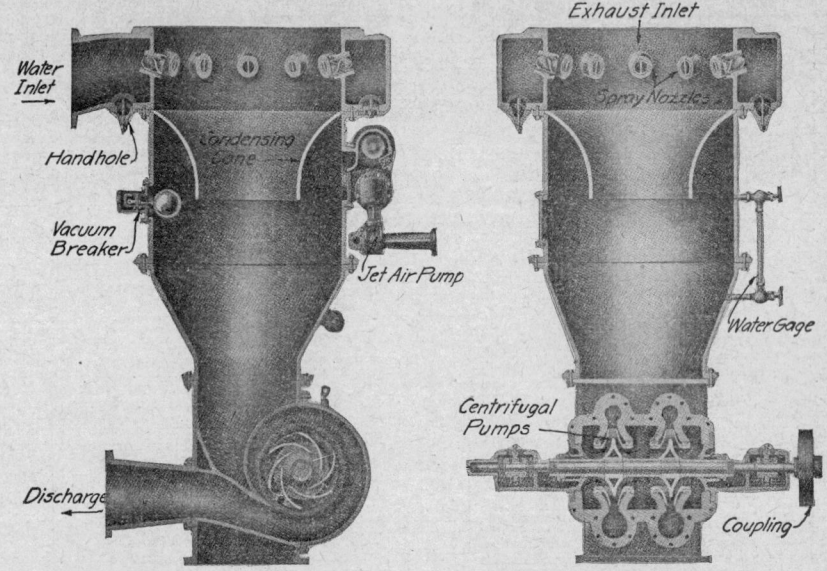

FIG. 483. — Elliott Erhart Low-level Jet Condenser — Sectional Views.

water enters by virtue of the vacuum maintained within the condenser, provided the suction head is less than 18 feet. The mixture of condensed steam and cooling water is removed from the bottom of the condenser by a centrifugal removal pump driven either by a steam turbine or electric motor. The air and vapor are removed by a jet type air pump shown on the right. In order to prevent the water from rising to too high a level in the condenser, a float control valve, called a **vacuum breaker,** is provided just below the condensing cone. This condenser is especially useful in places where the condensing water is not suitable for surface condensers. The cost of operation may, however, be high, because of the large quantities of air to be removed, and the low efficiency of the removal pump, resulting from the unfavorable conditions under which it operates.

Centrifugal pumps, when used for circulating water for condensers, are designed for large capacity against comparatively low heads, the capacity being determined by the size and design of the condenser, the steam rate and the temperature of the circulating water. The normal head is from 15 to 40 feet, and consists of a **static head** which is fairly constant, and a **friction head** resulting from the fluid friction in the condenser and piping. The latter friction depends upon the length and number of passes and size of tubes. For hot-well service this type of pump must deliver its water against a head corresponding to the vacuum in the condenser, plus the friction head and the static head. For this service, reliability of operation and simplicity are of prime importance. Pumps used are generally of the single-stage, double-suction, type.

515. Surface Condensers. — There are two general types of surface condensers, namely, (1) those in which the cooling water is inside the tubes and the exhaust steam on the outside, known as the **standard type,** and (2) those in which the steam is inside and the water outside the tubes, known as the **water-works type.** The first type is most common and generally operates on the **counter-current principle;** that is, the water and steam flow in opposite directions. Either type may operate on the **wet system** in which both the air and condensate are withdrawn by a single pump, or on the **dry system** in which two pumps are used — one to handle the air and one to handle the water. The dry system is used with large condensers where operating economy is of prime importance.

516. Wheeler Standard Surface Condenser. — This condenser, Fig. 484, which is adapted for steam engine service, consists of a cast-iron shell closed at each end by a head arranged to form a **water box** or chamber from or to which the condensing water enters or leaves the tubes. Between each head and the shell is located a **tube sheet,** into which tubes are fastened. Packing is placed in the tube sheet surrounding each tube, and a **ferrule** is screwed tightly in place against the packing. The tube may be either packed and ferruled at one end only and expanded into the tube sheet at the other end, or, as in recent designs, the tube may be expanded and rolled in at both ends. In either case the joint should be tight, against leakage of either air or water. The tubes are made of copper or brass, and are supported by a plate at the middle of their length. Water enters the water box at the bottom right, passes to the left in the lower bank of tubes, and returns to the top right in the upper bank of tubes, thus passing the length of the condenser twice. Steam enters at the top and strikes against a **distributing baffle** which prevents excessive short-circuiting of the steam. The condensed steam and air are withdrawn at the lower left-hand side by a **wet-vacuum pump** of either the reciprocating or centrifugal type.

517. High-vacuum Surface Condenser. — The surface condenser previously described is suitable for a vacuum of about 26 inches. To obtain a vacuum of about 28 or 29 inches the following conditions must be fulfilled: (1) the air entering the condenser must be rapidly removed, because when allowed to accumulate it acts as a blanket around the tubes and interferes

566   CONDENSERS AND CONDENSER AUXILIARIES

with the heat transmission; (2) the steam distribution must be made effective, by arranging the tubes to form **steam lanes** which conduct the steam

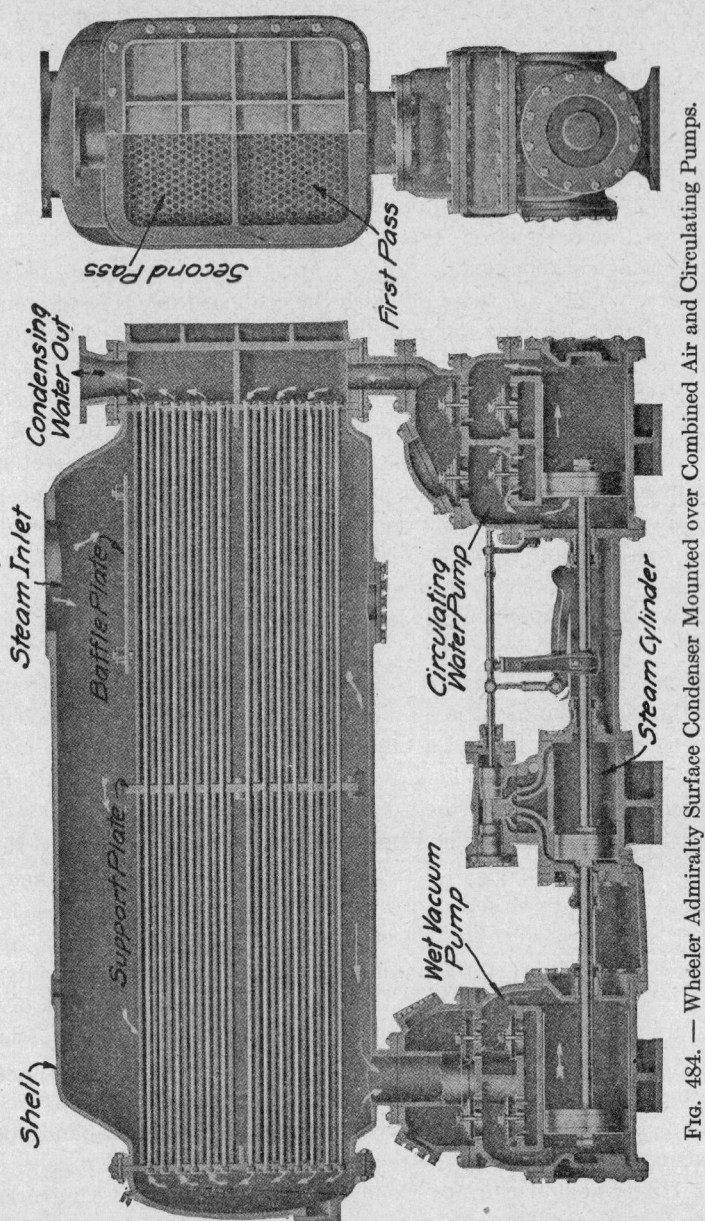

Fig. 484. — Wheeler Admiralty Surface Condenser Mounted over Combined Air and Circulating Pumps.

to short banks of tubes so arranged that all the tubes are interposed in the direction of steam or vapor flow between the exhaust steam inlet and the air outlet, with a wider spacing at the steam inlet and a narrow spacing at the air outlet, to correspond to the reduction in the vapor volume, and the formation of dead air spots thus prevented; (3) the air must be removed by a separate air pump, with an air connection located at a point where the air is the coolest and has the greatest density and hence the smallest volume; (4) baffles must be inserted between the banks of tubes to drain off the condensate, prevent flooding of the tubes, and protect the air outlets; (5) condensing water should pass through the condenser with the least possible friction, but at a velocity consistent with high efficiency; (6) thorough longitudinal distribution of the exhaust steam throughout the entire length of the condenser, for which purpose some condensers are

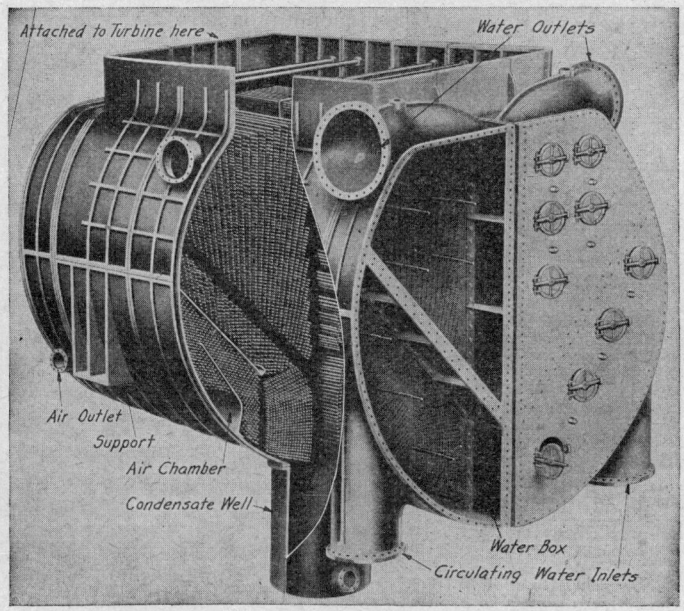

FIG. 485. — Wheeler Surface Condenser — Cut-away Section.

constructed with a large clear space, called the **dome** of the condenser, between the exhaust steam inlet and the condenser tubes; and in other condensers are divided into sections by the tube support with separate air removal connections for each section.

Condensers of this type are made in sizes as large as 90,000 square feet of condensing surface, using either a horizontal or vertical shell, and may be arranged as a **single-pass condenser** or as a **two-pass condenser**. In the single-pass construction the condensing water after entering the water box enters all the tubes from the water entry end, and after passing through the tubes is discharged at the other end; or the water box may be divided into sections with the water entering one of the sections, passing to the water box at the other end through a portion of the tubes, and returning to the entry water box through the remaining tubes. Where conditions call for a relatively large quantity of condensing water, as compared with the amount of condensing service, the tendency is to use a single-pass condenser, but where a small quantity of water is employed, the tendency is toward a multiple-pass condenser. For average conditions, the two-pass condenser is most commonly used.

A cut-away view of the *two-pass, radial-flow, double-bank* condenser is shown in Fig. 485. The first or colder pass consisting of the two lower sections or tubes, and the tubes of the second or warmer pass being those of the two upper sections. The division of the tubes into two distinct banks, with a wide central lane between them as shown, provides a condensate temperature very nearly equal to the vacuum temperature. The condensed

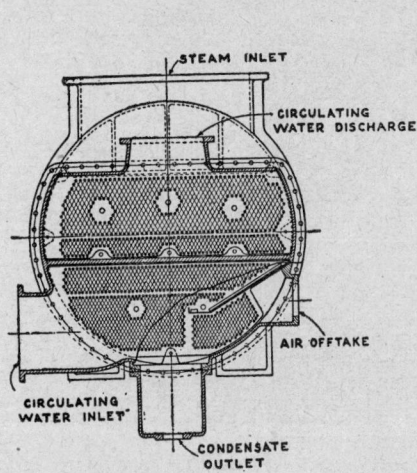

FIG. 486. — Down Flow Surface Condenser.

FIG. 487. — Divided-waterbox Surface Condenser.

steam collects in the hot- or condensate-well, at the bottom of the condenser, arranged to maintain the water at a sufficiently high temperature to thoroughly deaërate the condensed steam.

Condensers of this type may also be arranged with a downward steam flow, Fig. 486; and some condensers are arranged with divided water boxes at each end, Fig. 487. With the latter arrangement, it is possible to clean one-half of the condenser and still keep the other half in operation, which is an advantage in some installations.

518. Conditions Necessary for Heat Transfer. — The transfer of heat from steam to the cooling water in a surface condenser depends, (1) upon the condition of the tube surface, (2) upon the proportion of air in the steam and the water, (3) upon the velocity and temperature of the cooling water, and (4) to some extent upon the material of which the tubes are made. For ordinary types of surface condensers, from 250 to 300 B.t.u. per hour are transmitted per square foot of tube surface for each deg. fahr. rise in temperature of the circulating water. In the high vacuum condensers, this value may amount to 600 or 800 B.t.u. per square foot per deg. fahr. per hour. Each square foot of cooling surface will condense about 10 pounds of steam per hour, for ordinary condensers at 24 to 26 inches vacuum. For small turbine installations, $2\frac{1}{2}$ to 4 square feet of cooling surface per kilowatt are provided, and for large plants with high efficiency condensers, 0.95 to $2\frac{1}{2}$ square feet per kilowatt rated capacity.

519. Comparison of Types. — Jet condensers have low first cost, occupy small space, frequently require more pump capacity, and have more air to remove, thus requiring larger air pumps. Surface condensers occupy large space, give high efficiency, and provide pure feedwater where the condensing water is impure.

520. Elementary Theory of Condensers. — The vacuum within a condenser is customarily referred to a 30-inch barometer. The height of the standard barometer, at sea level and latitude 45 degrees is 29.92 inches when the temperature of the mercury is 32 deg. fahr. For a 30-inch barometer the temperature of the mercury is increased to 58.4[1] deg. fahr. To correct a mercury column for change in temperature, the following equation may be used:

$$h = h_1[1 - 0.000101(t_1 - t)] \quad \ldots \ldots \quad (115)$$

in which h = height of mercury column corrected to temperature t.
h_1 = observed height of mercury column, inches.
t_1 = observed temperature of mercury column, deg. fahr.
t = temperature to which column is to be referred, deg. fahr.

Example 60. — The height of mercury in a manometer used to measure vacuum is 28.52 in., at a temperature of 80 deg. fahr., barometer reading 29.85 in. at a temperature of 42 deg. fahr. Determine the vacuum, referred to a 30-in. barometer.
Solution. — Vacuum corrected, $h = 28.52[1 - 0.000101(80 - 58.4)]$
= 28.46 in. mercury.
Barometer corrected, $h = 29.85[1 - 0.000101(42 - 58.4)]$
= 29.89 in. mercury.

Absolute pressure in inches mercury = 29.89 − 28.46 = 1.44 in. mercury.
Vacuum referred to a 30-in. barometer = 30.00 − 1.44 = 28.56.

521. Condenser Pressure. — *The pressure existing within a condenser is made up of the pressure of the air plus that of the water vapor.* This is in accordance with **Dalton's law,** which states that in *a mixture of gas and vapor enclosed in a given space, the total pressure is equal to the sum of the partial pressures which each gas would exert if occupying the space alone.* As the vacuum increases, the proportion by weight of air to vapor increases for

---

[1] This value is sometimes taken as 58.1 deg. fahr. See also discussion in A. S. M. E. Power Test Code, Instruments and Apparatus, Part 2.

a given air pressure, and there is a corresponding increase in the amount of air that must be removed to maintain the vacuum.

Example 61. — The absolute pressure in a condenser is 2.03 in. of mercury, and the temperature of the air-vapor mixture is 100.6 deg. fahr. Find the pressure produced by the air within the condenser.

Solution. — From the Steam Table, the pressure of the vapor at 100.6 deg. fahr. is 1.97 in. mercury, and therefore, by Dalton's law,

$$P_c = P_a + P_v, \quad \text{or} \quad P_a = P_c - P_v$$

in which P_c = condenser pressure, in. mercury = 2.03
P_a = air pressure, in. mercury.
P_v = vapor pressure = 1.97 in. mercury.
$P_a = P_c - P_v = 2.03 - 1.97 = 0.06$ in. mercury.

522. Weight of Cooling Water Required by a Condenser. — The quantity of cooling water per pound of steam condensed depends upon the degree of vacuum obtained and the inlet and outlet temperatures of the cooling water. The weight of cooling water per pound of steam varies from 25 to 30 pounds for vacua of 25 to 26 inches of mercury, and increases to above 50 pounds for high vacua.

Neglecting the effect of radiation and leakage, the heat absorbed by the cooling water will equal that given up by the steam in the condenser. *For a surface condenser*, the above relation may be written.

$$W = \frac{xL + h_f - h_{fc}}{h_{f_2} - h_{f_1}} = \frac{xL + h_f - (t_c - 32)^1}{t_2 - t_1} \quad \ldots \quad (116)$$

in which W = weight of cooling water per hour per pound of steam condensed, pound.
x = quality of exhaust steam; assumed as 90 per cent or figured from Equation (101), page 440.
L = latent heat at the absolute pressure in the condenser, B.t.u.
h_f = heat of liquid at the absolute pressure in the condenser, B.t.u.
h_{fc} = heat of liquid at the temperature of the condensate (t_c), B.t.u.
t_c = from 5 to 20 deg. fahr. below the temperature of the exhaust, deg. fahr.
h_{f_2} = heat of liquid at the temperature (t_2) of the outlet condensing water, B.t.u.
t_2 = from 10 to 25 deg. fahr. below the temperature corresponding to the absolute pressure in the condenser. For an average value 15 deg. fahr. may be used.
h_{f_1} = heat of liquid at inlet temperature (t_1).
t_1 = temperature of inlet condensing water.

[1] The second form of this equation holds for condensers, provided the influence of the entrained air on the heat content of the exhaust steam is neglected, and the mean specific heat of the water at condenser temperatures is taken as unity. W is usually increased from 5 to 15 per cent, to allow for cooling of the air-vapor mixture and inefficient absorption of heat.

Example 62. — The following data were taken from the test of a condenser attached to a 30,000-kw. turbine: Duration of test, 3 hr.; steam condensed per hr., 317,000 lb.; steam pressure, 237 lb. sq. in. abs.; superheat, 155 deg. fahr.; vacuum, 28.17 in. mercury; quality at exhaust, 0.90; initial temperature of cooling water, 76.5 deg. fahr. Find weight of cooling water per hour. Temperature of condensate, 91.3 deg. fahr.; barometer, 29.93 in. mercury.

Solution. — Using the second form of Equation (116) with proper substitutions,

$$W = \frac{xL + h_f - (t_c - 32)}{t_2 - t_1} = \frac{0.90 \times 1038.8 + 64.93 - (91.3 - 32)}{86.9 - 76.5}$$

$$= \frac{940.55}{10.4} = 90.5 \text{ lb. water per pound of steam.}$$

L at 1.76 in. mercury, abs. = 1038.8 B.t.u.; h_f at 1.76 in. mercury, abs. = 64.93 B.t.u.; t_c = 91.3 deg. fahr.; t_2 = 96.9 − 10 = 86.9 deg. fahr.; t_1 = 76.5 deg. fahr.
Total weight of cooling water per hour = 90.5 × 317,000 = 28,688,500 lb.

For a jet condenser, the final temperature of the cooling water equals that of the condensate, and t_c varies from 5 to 20 deg. fahr. below the temperature corresponding to the vacuum, because of the pressure exerted by air and similar gases. Making these changes in Equation (116), there results:

$$W = \frac{xL + h_f - h_{fc}}{h_{f_2} - h_{f_1}} = \frac{xL + h_f - (t_c - 32)}{t_c - t_1} \quad \ldots \quad (117)$$

Example 63. — Using the data of Example 62, find the weight of cooling water required if a jet condenser were used.

Solution. — Using Equation (117), second form, and making proper substitutions,

$$W = \frac{xL + h - (t_c - 32)}{t_c - t_1} = \frac{0.90 \times 1038.8 + 64.93 - (91.3 - 32)}{91.3 - 76.5}$$

$$= \frac{940.55}{14.8} = 63.5 \text{ lb. water per pound of steam.}$$

Total weight of cooling water per hour = 63.5 × 317,000 = 20,129,500 lb.

523. Condenser Auxiliaries and Accessories. — The principal auxiliaries used with condensers are water-circulating pumps, dry- and wet-air or vacuum pumps, relief valves, expansion joints, gate valves, and water-cooling apparatus.

524. Circulating Pumps. — The types of circulating pumps used with condensers are direct-acting and centrifugal. The latter are now being used almost entirely because of their simplicity.

525. Air Pumps. — The function of the air pump is to maintain the vacuum formed by the condenser. Non-condensable gases enter with the steam and leak in through the various parts; in the case of a jet condenser they also enter in solution with the cooling water. An air pump should be capable of removing a large volume of air under all conditions of operation, for highest condenser efficiency. Four different types of air pumps are commonly used; namely, *reciprocating, rotary displacement, centrifugal entrainment, and steam-ejector vacuum pumps.*

526. Wheeler-Edwards Air Pump. — This pump, which is shown in section in Fig. 488, is a wet-air pump consisting of a plunger driven by a steam piston located directly above the pump. The pump has no inlet valves, as the inlet is through ports uncovered by the piston when on its down stroke. Water of condensation flows continuously, by gravity, from the condenser into the base of the pump. Upon descent of the conical plunger, this water is projected at a high velocity through the ports into the working barrel. As soon as the ports are uncovered by the piston, air is also drawn into the barrel, along with the water, because of the vacuum formed

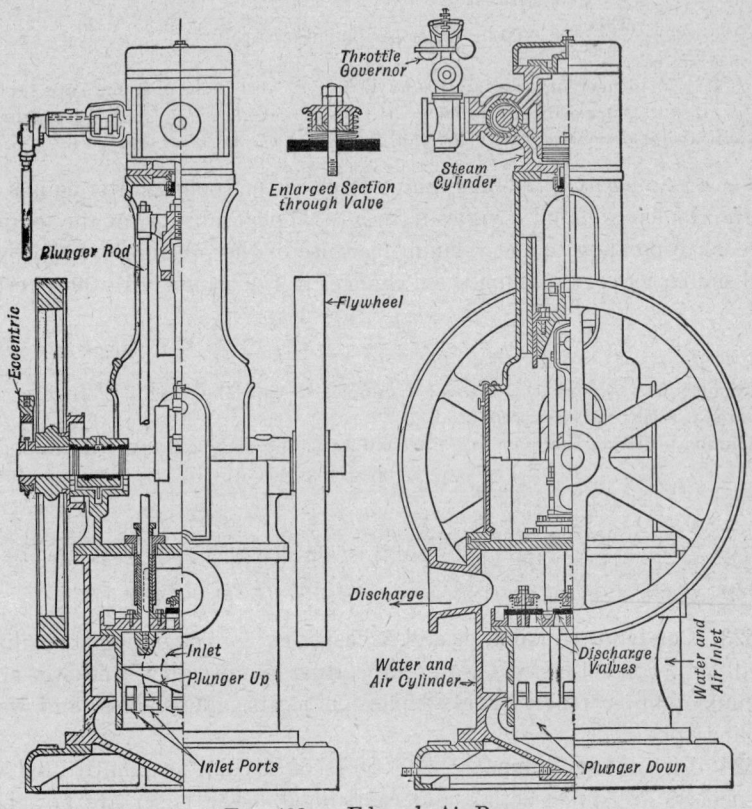

FIG. 488. — Edwards Air Pump.

by the descending plunger. The plunger, upon rising, discharges the mixture through the valves at the top of the barrel, the air passing out first.

This pump is capable of producing a 28-inch vacuum when used in connection with a **dry-tube surface condenser**. It is especially adaptable to marine work.

527. Mullan Displacement Vacuum Pump. — The Mullan pump, Fig. 489, is double-acting, operating upon the wet-vacuum principle and without suction valves. The piston, at the end of its suction stroke, uncovers a series of ports distributed around the middle of the cylinder, through which the vapors are drawn into the cylinder. The discharge valves are of the **Gutermuth type,** Fig. 490, consisting of a strip of phosphor bronze coiled at one end and attached to the valve stem. They are mounted on a plate, or **deck,** attached to each end of the pump barrel. The condensate enters the cylinder at the top or side, and is forced through the discharge valves, together with the air, on the discharge stroke of the piston.

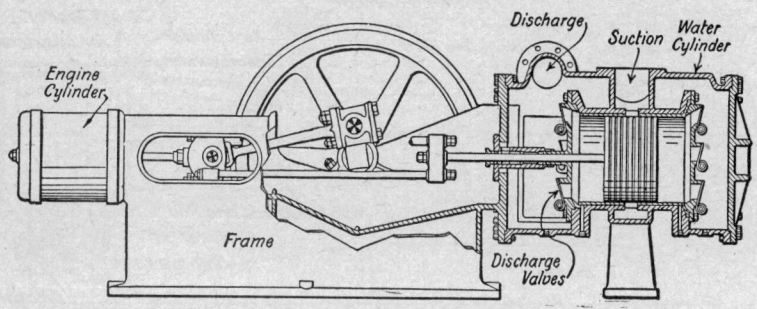

Fig. 489. — Mullan Vacuum Pump.

The water cylinder is lined with hard brass. The piston is of iron, brass covered, cast very light and with ample surface to ensure minimum pressure on the liner. The actual bearing surface of the piston on the liner consists of a number of end-grain **lignum vitae plugs,** which are turned off slightly larger in diameter than the piston cover. This combination of lignum vitae and brass, with water lubrication, is virtually wear proof and has an important bearing on the durability of the Mullan pump. A series of **water grooves** turned in the piston surface make it vacuum tight without the use of piston rings. The piston rod of Tobin bronze is secured in the piston by a taper end, lock nut, and pin. It works through a long double stuffing box containing a water seal in addition to the packing, eliminating all possibility of air leakage at that point.

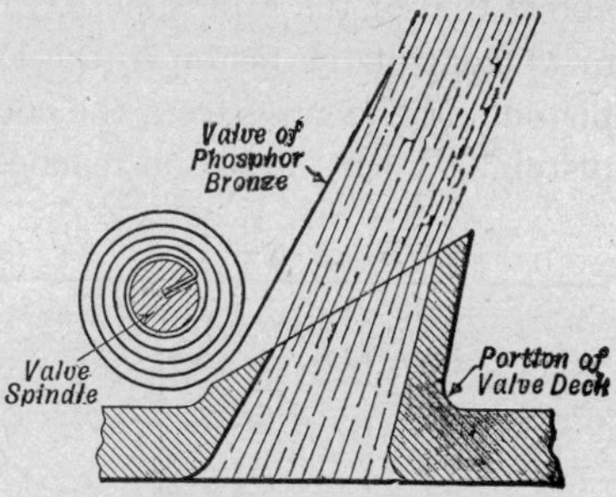

The valve gear for the steam end comprises an eccentric, working a " D " slide valve for the smaller pumps, or a balanced piston valve with self-adjusting snap rings for the larger pumps.

Fig. 490. — Gutermuth Flexible Metallic Valve.

A throttling governor with a safety stop ensures against overspeeding.

This pump is suitable to maintain a vacuum within a small fraction of an inch of the theoretically perfect vacuum.

528. Rotative Dry-vacuum Pump. — This type of vacuum pump resembles an air compressor in mechanical construction and operation. It is capable of producing a high vacuum, operates with high efficiency, and is used where a high degree of vacuum is essential as with condensers attached to turbines.

The Wheeler steam-driven pump shown in Fig. 491 has the air and steam cylinders arranged in tandem with the steam cylinder nearer the crank.

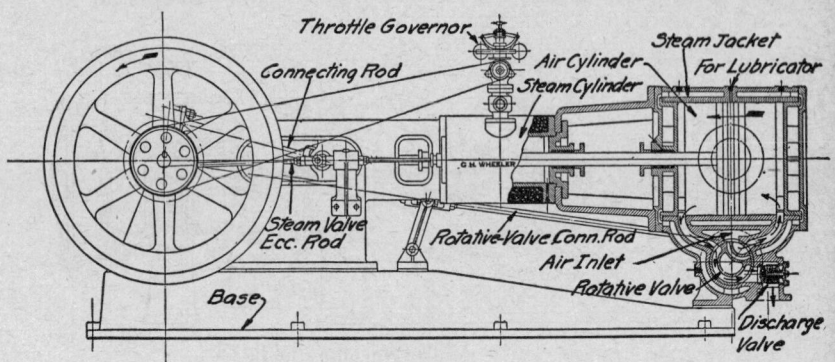

Fig. 491. — Wheeler Rotative Steam-driven Dry-vacuum Pump.

The steam cylinder has a piston valve operated by an eccentric connected to the main shaft. The air cylinder is completely water-jacketed and the semi-rotative inlet valve is mechanically driven from the crank shaft, by means of an eccentric operating a crank attached to the valve rod; while the discharge valves are of the poppet spring-loaded type.

For the position of the air piston and valve shown, the piston is moving to the left compressing the gases and discharging them, at a pressure slightly above that of the atmosphere, through the poppet discharge valves. The movement of the piston to the left lowers the pressure on the right of the piston and the gases from the condenser flow in to fill the space behind the piston. When the piston reaches the end of the stroke, the rotative valve is moved to close the suction port and open an auxiliary passage connecting each end of the cylinder. This equalizes the pressure and, by eliminating the effect of the clearance space, prevents the loss in efficiency which would result from the re-expansion of the gases in the clearance space. The cycle of events for a typical rotative dry-vacuum pump is shown in Fig. 492.

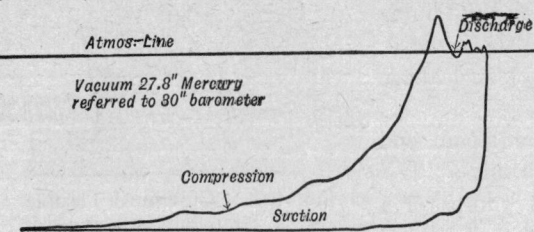

Fig. 492. — Diagram from Dry-vacuum Pump.

Valves of the multi-disk type are used on many rotative dry-vacuum pumps and are claimed to operate at high efficiency.

STEAM JET OR EJECTOR AIR PUMP

529. Centrifugal Entrainment Pump. — One type of this pump, used on the Westinghouse-Leblanc low-level jet condenser is shown in Fig. 493, in which the rotors of the air and condensate pumps are mounted on the same shaft. The air pump consists of a stationary nozzle, rotating vanes located in a casing, and an inlet and discharge pipe. The operation is as follows: water for operating the pump enters the inner chamber, surrounds the shaft, and flows out through the nozzle. The rotating vanes on the impeller rotate clockwise and cut off layers of water, which are projected continuously into

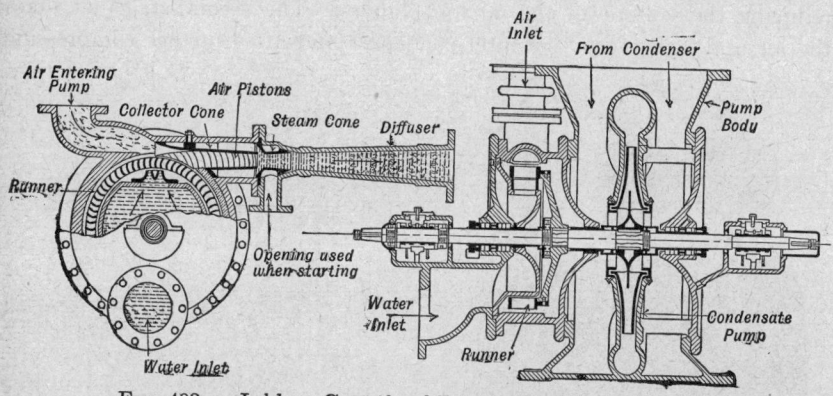

Fig. 493. — Leblanc Centrifugal Entrainment Vacuum Pump.

the cone, at high velocity. Air is caught, or **entrained,** between successive layers of water and is forced at approximately atmospheric pressure into the discharge cone. The high velocity of the water pistons is converted into pressure by means of the diffusing cone, which permits discharging against pressure. A steam ejector is used to put the condenser into service quickly.

530. Steam Jet or Ejector Air Pump. — This pump uses a jet or jets of high-velocity steam to accomplish the removal of air or other gases by entraining and compressing the gas. It is the most efficient type of high-vacuum pump and is in almost universal use for condenser work. It operates on the dry-air principle, and is a substitute for air pumps of the reciprocating, rotative, or hydraulic entrainment type. It has no moving parts; its weight and its space requirements are small; its efficiency is high and does not change after long periods of operation; it does not require lubrication or attention during operation; and it is simple and rugged in construction.

Steam jet air pumps are classified according to the number of stages used, as **single-stage** or **multi-stage.** The single-stage pump, in which the gas is compressed in one stage, is used for vacua under 26 inches, and is applied to condensers for reciprocating engines and evaporator condensers. The two-stage pump, Fig. 494, is suitable for vacua between 26 and $29\frac{1}{2}$ inches. This air pump consists of two multi-nozzle steam ejectors arranged to work in series with an inter-condenser and an after-condenser incorporated in a shell upon which the steam jets, together with the necessary valve and inner connections, are mounted.

In operation live steam is delivered to each stage of the ejector. In the first stage the steam passes through the expansion nozzles and across the suction opening, entraining and compressing the air and vapors and delivering them to the inter-condenser through which condensate or boiler feedwater flows, thus condensing the steam used in the first stage and reducing the volume of the air and vapors. The second-stage jet takes the air and gases from the inter-condenser and after further compressing

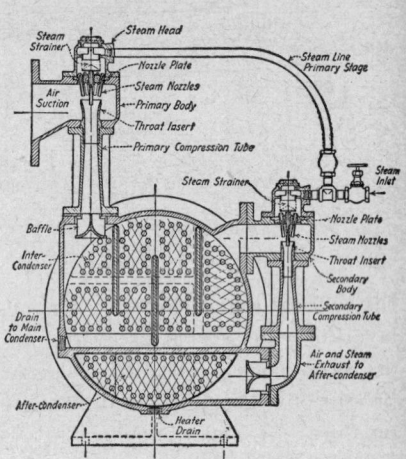

FIG. 494. — Wheeler Steam Jet Air Pump with Inter-and-after Surface Condensers.

them delivers them to the after-condenser in which the heat of the second stage jet is recovered. The quantity of air which can be entrained by the steam jets depends upon the surface of contact between the jet and the air.

The steam nozzles and throat insert are designed to give the highest overall efficiency, and are made from monel metal in order to withstand all conditions of superheat.

The steam consumption of this pump is low, and the thermal efficiency close to 95 per cent.

A typical installation of an ejector pump is shown in Fig. 483, page 564.

531. Exhaust Connections. — The exhaust connections to a condenser should contain, (1) an atmospheric exhaust relief valve, to give free exhaust to the atmosphere should the vacuum fail; (2) an expansion joint, to make a flexible link between the prime mover and condenser; and (3) a gate valve to isolate the prime mover when necessary.

Atmospheric Relief Valve. — A typical valve of this type is shown in Fig. 495. The relief valve is placed in a branch taken from the main exhaust line, leading to the atmosphere. As long as the vacuum is maintained in the condenser, the relief valve remains tightly closed and is water-sealed to prevent leakage. As soon as the vacuum is lost, the valve promptly opens and the engine exhausts to the atmosphere until the vacuum is restored, when the valve closes, automatically. A dash pot is provided to permit the valve to close quietly, and a lip is placed around the valve seat for water-sealing.

Two methods are ordinarily used to compensate for the expansion of the condenser. **One** *is to bolt the condenser to the exhaust flange of the turbine and to support its weight upon springs.* The **other** *is to support the extension of the turbine on separate foundations and to connect the two units with an* **expansion joint.** This joint uses as a flexible medium a rubber diaphragm, separated by metal flanges of a simple design, or may be made of a corrugated copper joint.

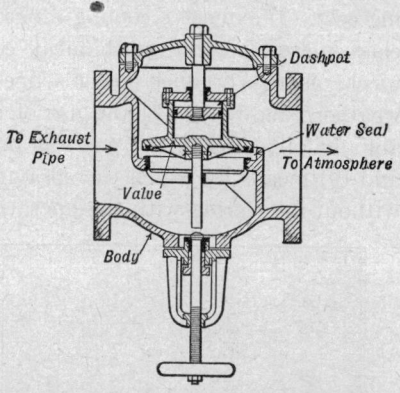

Fig. 495. — Atmospheric Relief Valve.

532. Cooling Ponds and Towers. — Where a power plant or manufacturing plant is not located adjacent to, or within a reasonable distance of, a river, lake, or other source of natural water supply, the cold water required for condensers and for many other industrial purposes must be obtained by means of a **water re-cooling system.** It also frequently occurs that such natural water as is available is not suitable for the required purpose, because of the presence of free acid or sewage contamination, resulting in the rapid deterioration of the metal parts of the condensers or other apparatus through which the water flows. Also, the available water may contain considerable quantities of foreign matter which render it unsuitable for condensing or cooling purposes. In either one of these cases, some water re-cooling system must be employed. *All such systems depend upon the exposure of the warmed water to the evaporation and consequent cooling effect of the atmosphere, using either cooling ponds or cooling towers.*

Of the two methods, the **cooling pond** is much the cheaper, provided an adjacent and suitable ground or roof area, of sufficient extent, is available. The cooling results obtained from a cooling pond, however, are not equal to those obtained from a tower, either of the forced or natural draft type.

The simplest method of cooling water is to discharge it through a single pipe line to a pond of sufficient area, so that the water will be cooled to the temperature desired, by contact with the atmosphere on the surface of the water. Such a pond, however, must be of large dimensions, and, to reduce the area required, it is necessary to add some device by which additional contact of the water with the atmosphere is obtained. This device consists of a spraying system, of which there are a number of types in use at the present time.

The most desirable features of any spraying system, as given by the C. H. Wheeler Mfg. Co., are as follows: (1) low initial cost, (2) low operating cost, (3) extensive cooling range, (4) low maintenance charge, (5) first-class materials and mechanical construction, (6) adjustability of the mechanism to produce either a fine or a coarse spray, as required by the weather conditions, (7) the use of a small pond area and of as small an amount of piping in the pond as possible, (8) the elimination of driftage, and (9) means by which the spraying mechanism may be quickly cleaned without interfering with the operation of the cooling system.

FIG. 496. — Spray Cooling Pond — In Operation and Idle.

533. Spray Cooling System. — A cooling pond using sprays is shown in operation and, when not in operation, in Fig. 496. The latter shows the distributing pipe system and the arrangement of the spraying apparatus over the pond. This type of cooling pond may be constructed of concrete, or it may be of natural soil having sod banks. Such ponds need not be deep, except when depth is required for storage purposes, a depth of 3 feet or even less ordinarily being sufficient.

The spray is produced by some type of nozzle, which aims to break the water up into a spray suitable for the weather conditions under which it is working. In all types, the shape of the nozzle is such that it gives to the water a whirling motion, thus producing more effective atomizing of the water.

Spray nozzles are constructed with (a) rigid heads, and (b) adjustable heads. A few typical nozzles of each class are shown in Fig. 497. The construction of each type is evident from the illustrations.

An adjustable spray head, which has given satisfactory service, was designed by Professor Carl C. Thomas and described in TRANS. AM. SOC. M. E., 1917, page 625. This head consists of a cast-iron supporting base containing the water-entry opening and carrying a bronze tube wound spirally. This tube is held between the base and a cap, which fits the top, by a central bronze stem passing down through a close-fitting bushing in the base. This stem is movable and is operated by a bell crank, having an extended vertical arm which gives accurate control of the position of the stem. By moving the stem slightly, the opening of the spiral, through which the water is discharged, is varied. The result of the motion, which is made by a system of levers located at some convenient point on the bank

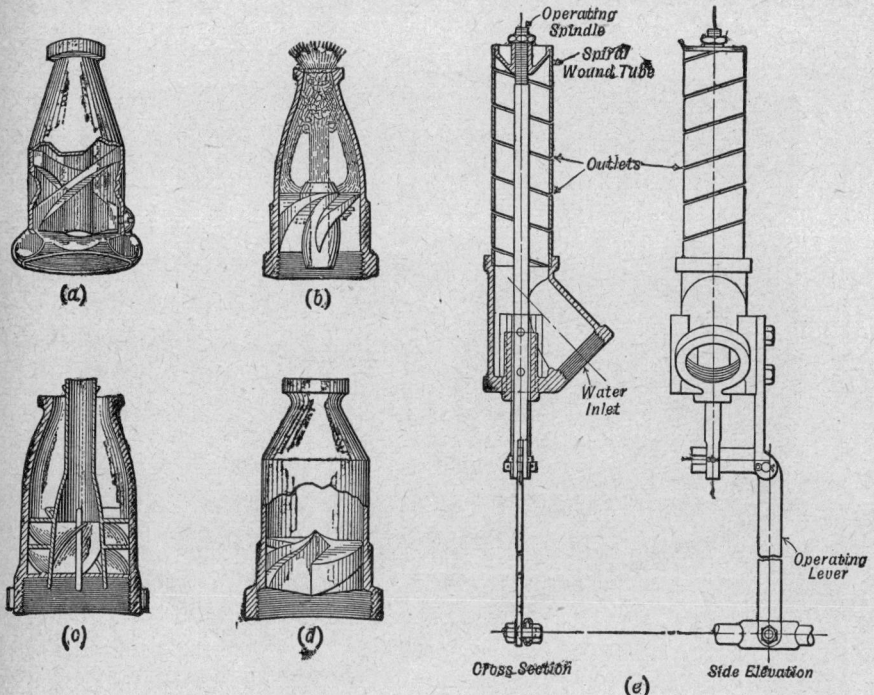

Fig. 497. — Rigid Head and Adjustable Head Spray Nozzles.

of the pond, is that the fineness of the film of water is varied. Water is discharged from the head in a continuous stream, during the time of operation, with an upward inclination. The water film spreads, and becomes thinner on account of its increasing diameter. Finally, a point is reached where the *surface tension* of the water is overcome, and the sheet of water breaks into a fine spray, a mist, or an infinite number of small drops, depending upon the adjustment of the size of the spiral opening.

The nozzles and spray heads are attached to a system of piping, into which the cooling water is pumped at a pressure varying from 3 to 8 pounds, depending upon the type of nozzle and the fineness of the spray desired.

The cooling effect produced varies with the weather conditions. In general, it varies from 20 to 40 deg. fahr. The final temperature is often within 3 or 4 degrees of the temperature of the air, and sometimes, under suitable conditions, may be below the temperature of the air.

534. Water-cooling Towers. — In the case of cooling towers, the water is generally delivered to the top of the tower, and is then permitted to fall to a tank below the tower. The falling water is broken into a spray, by some means that does not require pressure.

Water cooling towers may be classified as follows: (1) *natural draft — open or atmospheric type*, (2) *natural draft — closed or flue type*, (3) *forced draft*, and (4) *combined forced and natural draft*.

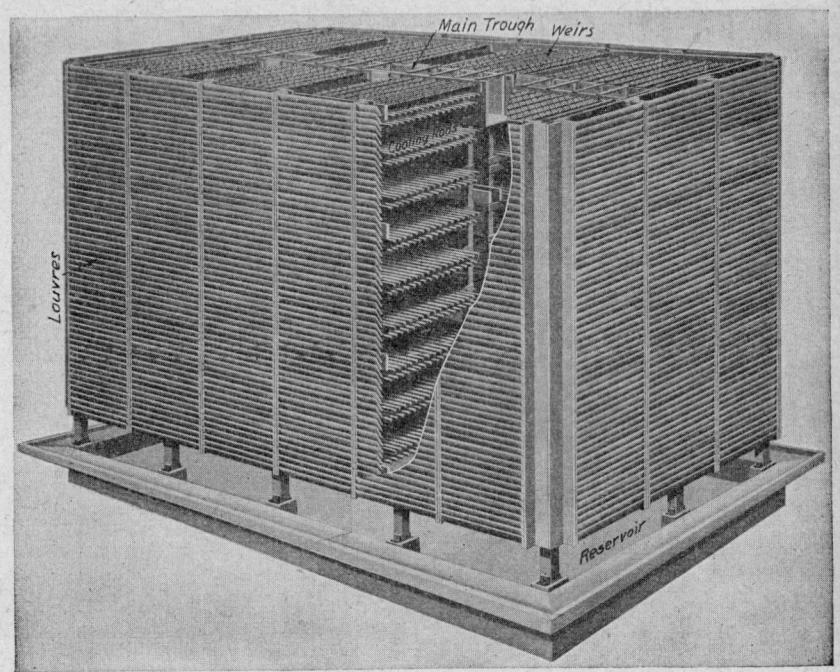

Fig. 498. — A Wheeler Natural-draft Cooling Tower — Cut-away View.

535. Natural-draft Cooling Towers. — The **open or atmospheric** type of tower, Fig. 498, consists of a strong wooden frame supporting the remaining parts of the tower. The walls of the tower are constructed of **louvres**, or slats, with openings between them and are made of long-leaf yellow pine, treated to prolong its life. At the top of the tower is an open main trough extending the entire length of the tower, and to it are connected a number of subsidiary troughs in which are cut small triangular weirs from which the water flows. From this distributing system the water falls to the cooling surface which consists of triangular cross-section strips of cypress or redwood, arranged in horizontal rows, the rows being arranged in tiers of two rows each. As the water strikes the strips it is broken up into a fine spray. Evaporation and heat-conduction areas are provided between the tiers with the strips in each tier so placed that no water can pass from one evaporation and heat-conduction area to another without striking a cooling strip. This retards the downward travel of the water which is broken again and again to produce a spray effect and thus give a maximum contact between the air and the water. The cooling strips are supported on horizontals, and are carefully fitted into the notches therein.

In operation the water to be cooled is pumped to the system of troughs at the top of the tower, and passing through the weirs, is cooled as it falls by gravity from the distributing system at the top of the tower over the entire cooling surface to the reservoir or tank over which the tower is erected.

The natural-draft tower is sometimes constructed with closed sides and the addition of a considerable height above the portion of the tower containing the cooling surfaces, and when thus constructed is called a **closed or chimney cooling tower.**

536. Forced-draft Cooling Tower. — The Wheeler forced-draft tower, Fig. 499, is similar to the open natural-draft type so far as the interior construction is concerned, but the sides of the tower are closed and form an air and water tight structure, except for the fan openings for the inlet of the fresh air at the base

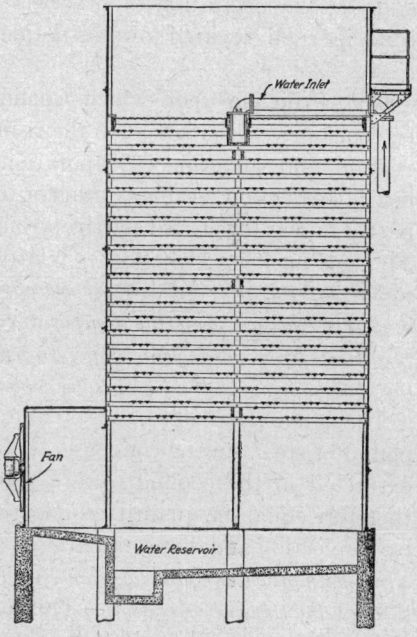

FIG. 499. — Forced-draft Cooling Tower.

of the tower and the outlet for the exit of the heated and vapor saturated air at the top of the tower. The necessary air for cooling is delivered under pressure to the interior of, and is forced through, the tower by fans or propellers especially designed for the purpose of producing the required air flow through the tower. The fans are located in a fan hood or hoods projecting from the main portion of the tower at its base, in order not to decrease the space provided for the air and water flow.

537. Comparisons of Cooling Towers. — The forced-draft tower requires small space and is adapted for use where: (1) the space available for a tower is limited; (2) the atmospheric conditions are unfavorable; and (3) natural draft is not obtainable on account of the location of the tower. The cost of this type of tower is in excess of the open, natural-draft type.

The open, natural-draft tower requires greater space than the forced-draft tower, and is adapted to about 75 per cent of all installations.

538. Location of Cooling Tower. — Cooling towers are located at the ground level, on a roof, or on some other elevated structure, wherever the space is available. The ground-level installation has the following advantages: (1) simplicity of foundation and reservoir construction; (2) shorter pipe lines, resulting in lower first and operating costs; and (3) localization of possible spray during high winds.

With cooling towers of the natural-draft type, an elevated location is often preferred, for the sake of (1) unimpeded circulation of air currents, and (2) the utilization of otherwise unoccupied space. With a surface condenser, an elevated tower is not desirable, because of increased pumping costs.

539. Principle upon which Cooling Towers Operate. — The water to be cooled gives up its heat to the rising column of air, by evaporation, convection, and radiation. Evaporation absorbs from 75 to 85 per cent of the heat; convection, or direct transfer to the air, comes next; while radiation, partly in the tower and partly through the piping, makes up the balance. The cooling from radiation is relatively small. *The amount of vapor absorbed by the air, because of evaporation, depends upon the moisture already in the air and the temperature of the air. The amount of cooling resulting from convection depends upon the temperature difference between the cooling air and the water to be cooled.*

The latent heat absorbed by the cooling water, while condensing one pound of steam in the condenser, equals the quantity of heat that must be extracted in the cooling tower. The quantity of water evaporated will therefore equal the quantity condensed, less the percentage of heat removed by convection and direct radiation; that is, *the cooling tower has to evaporate a quantity of water equaling from 75 to 85 per cent of the weight of steam passing through the turbine or engine.* This loss must be replaced by a fresh supply. Expressed as percentage of the total quantity of cooling water supplied, it equals nearly 4 per cent. The water to be cooled may be lowered in temperature 40 to 50 deg. fahr.

Capacity. — The capacity of nozzles and cooling towers is stated in gallons of water cooled per minute.

REFERENCES

Spray Cooling for Industrial Purposes, VARNALL WARING Co.
Cooling Water with Sprays, E. B. BADGER and SONS Co.
The Spraco System, SPRAY ENGINEERING Co.
Condensers and Auxiliaries, C. H. WHEELER MFG. Co.
Steam Power Plant Engineering, GEBHARDT.
Mechanical Equipment of Buildings, Vol. II, HARDING and WILLARD.
Test Code for Condensing Apparatus, A. S. M. E.
Reports on Condensing Equipment, NATIONAL ELECTRIC LIGHT ASSOC.

REVIEW QUESTIONS AND PROBLEMS

1. Name the principal reason for operating engines with a condenser.
2. Classify condensers, and explain the essential differences in the fundamental types.
3. Describe the operation of (a) a jet condenser, (b) a surface condenser.
4. What type of condenser should be used to produce a high vacuum? Describe a typical condenser of this type.
5. Describe the Mullan displacement vacuum pump. Why is such a pump used?
6. Describe the operation of the ejector vacuum pump.
7. What two methods are used to support condensers?
8. What are steam lanes in a surface condenser and why are they used?
9. Why is it necessary to cool water used for power plant or industrial work? Name two methods used.
10. Describe the operation of a forced-draft cooling tower.
11. A 26,000 sq. ft. surface condenser condenses 215,000 lb. of steam per hr. at a corrected vacuum of 28.58 in. Condensate temperature 90 deg. fahr., temperature of cooling water in 70.5 deg. fahr., out 83.5 deg. fahr. Find the gallons of circulating water per minute.

CHAPTER XXV

MODERN POWER PLANTS AND FACTORS RELATED TO THEIR DESIGN AND OPERATION

540. Foreword. — The advances made in both central station and industrial power plants have been so rapid that apparatus and operating conditions which are now modern will probably not be considered so in a few years — general tendencies, however, may be discussed. The amount of heat required to produce a kilowatt hour, using the **straight steam cycle,** has been lowered from about 22,500 B.t.u. in 1913 to less than 13,000 at the present time. These results have been brought about by using: (1) higher steam pressures and temperatures, (2) improvements in methods of burning coal with more efficient stokers and powdered fuel-burning equipment, and (3) large capacity steam generating and turbine equipment. As a result of these developments the limit of the straight steam cycle is being approached and it may be that future development will take place along the lines of: (1) the **mercury-vapor steam cycle,** page 621; (2) the operation at higher steam temperatures; and (3) the cooling of turbo-generators with hydrogen in order to increase their capacity.

541. Boilers and Boiler Auxiliaries. — Refinements in equipment and operation, increased size of boiler units, more intelligent supervision, more general use of instruments, and improved design to bring about a better coördination of the individual pieces of the boiler unit has steadily advanced the standards of steam-generating plant performance. In 1910 the

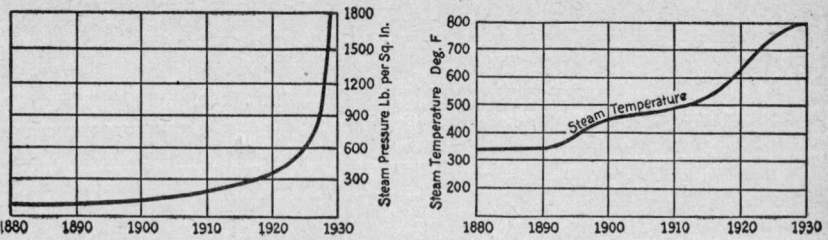

Fig. 500. — Trend of Steam Pressures and Temperatures.

maximum efficiency obtainable from boilers was about 75 per cent, whereas today boilers are operating at efficiencies as high as 93 per cent. Future improvement in boiler operation will undoubtedly be made by using higher steam pressures and temperatures, and still larger capacity units. The trend of steam pressures and temperatures is shown by the curve in Fig. 500. For present large station practice 400 to 600 pounds pressure is the established standard, while a number of stations are operating at from 800 to 1500 pounds. Steam temperatures range from 600 to 750 deg. fahr., with experimental work being done at the Detroit Edison Company on the use of a steam temperature of 1000 deg. fahr.

The boilers which are used for high-pressure units have developed along the lines of the low-pressure units, and are mostly of the bent-tube or cross-drum type, with the drums set high to give large furnace volume. In some cases the boiler outlets are 65 feet above the floor. Some engineers

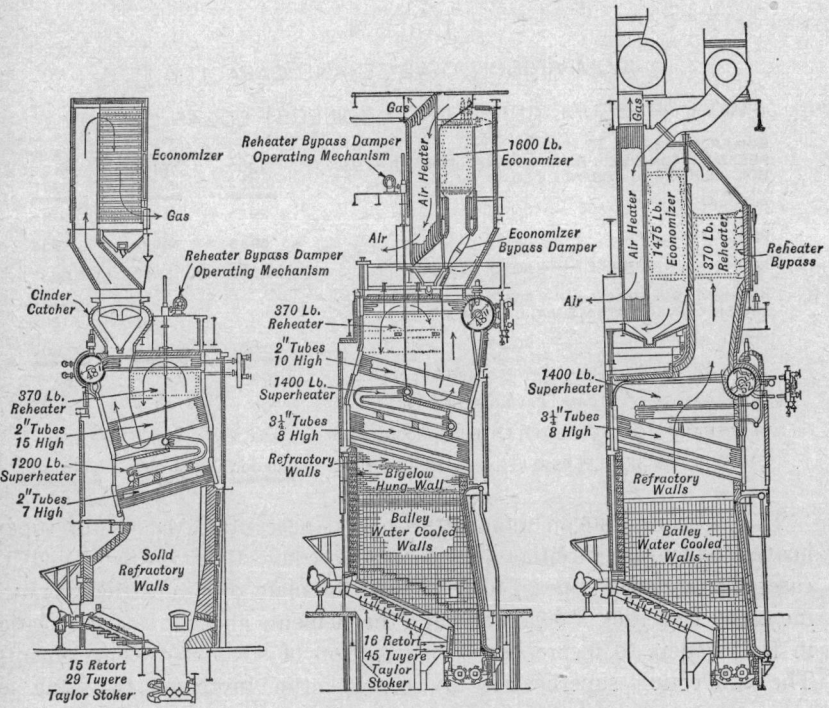

FIG. 501. — High-pressure Boiler Units, Edgar Station.

feel that the present boiler is only a transition between the older type boiler and a boiler along the lines of the drumless high-pressure boiler being developed in Europe. Increase in pressures has made it necessary to use other heating surface than the boiler proper, which has resulted in what is now known as a **steam-generating unit,** several of which are described in Art. 549, page 617, and which is a combination of stoker or pulverized fuel system, water-cooled walls, air preheater, economizer, superheater and in some cases a reheater. As the pressure has increased, there has been a redistribution of heating surface between the various units forming the steam generator, the surface of the boiler proper decreasing as the amount of surface in the other portions of the generating unit increase until in some cases it amounts to only 12 per cent of the total generating surface.

This tendency is clearly shown by the various high-pressure boiler units installed in the Edgar Station of the Edison Electric Illuminating Co. of Boston as shown in Fig. 501. The distribution of the heating surface between the various pieces of equipment, that go to make up these high-pressure units, is shown in Table 36. The capacity of generating units has increased until units are available which are capable of operating a turbine of over 80,000-kilowatt capacity, and have a reliability nearly equal that of turbo-generators.

TABLE 36
COMPARISON OF SIZE AND CAPACITY
OF
HIGH PRESSURE CROSS DRUM BOILERS – EDGAR STATION

BOILER NUMBER	3	5 & 6	7 & 8
PRESSURE FOR WHICH BOILERS WERE DESIGNED – LBS.	1200	1400	1400
BOILER HEATING SURFACE – SQ. FT.	15,732	15,093	6971
SUPERHEATER SURFACE —— SQ. FT.	2923	3483	5096
REHEATER SURFACE —— SQ. FT.	5938	8879	6936
ECONOMIZER SURFACE —— SQ. FT.	11,091	5596	9634
AIR HEATER SURFACE —— SQ. FT.	0	33,032	29,665
WATER COOLED FURNACE WALL SURFACE – SQ. FT.	0	1235	1206
TOTAL SURFACE —— SQ. FT.	35,684	67,318	59,508
TEMPERATURE OF STEAM LEAVING SUPERHEATER	700° F.	733° F.	750° F.
TEMPERATURE OF STEAM LEAVING REHEATER	700° F.	750° F.	750° F.
TEMPERATURE OF FEED WATER ENTERING ECONOMIZER	250° F.	420° F.	420° F.
CAPACITY IN POUNDS PER HOUR	143,000	250,000	300,000

The maximum steam temperatures used necessitate the use of superheaters which are of either the radiant or convective type, and in both cases the tendency appears to be toward the plain tube construction, with increased velocities of steam through the elements and decreased velocities in the headers to insure better distribution of steam between elements. The convection superheaters are being better protected by using an intervening bank of boiler tubes.

The use of high-steam pressures has increased the importance of the economizer which is now being used in nearly all installations; and there is also an increasing use of the air preheater with a consequent lowering of the exit gas temperatures, in some cases, to below 250 deg. fahr.

Large capacity boilers and the use of both economizers and air preheaters have required increases in fan capacity and pressure. The largest induced-draft fan reported in operation is at the Gorgas Plant of the Alabama Power Company. Each 30,315 square foot boiler is served by one fan of 358,000 c.f.m. capacity at 18 inches static pressure. The fans are each driven by two motors, one of 900 horsepower and the other of 1800 horsepower, and are equipped with vane control.

Practically all stations using a pressure over 450 pounds and a temperature of 750 degrees use a single stage of steam reheating. In a few stations a combination of a steam reheater with a gas reheater is being used, the reheat temperature being maintained nearly constant by thermostatic control.

In the majority of recent boiler installations, the furnace walls are cooled using either air or water, which has resulted in a decided decrease in maintenance costs of refractories, and at the same time has increased the steaming capacity. In some water-cooled furnaces as much as 40 to 65 per cent of the total heat in the steam is being absorbed by the water-walls.

There appears to be a distinct trend toward the use of the rotary car dumper for unloading railroad coal cars, the hydraulic flume for handling ashes, and the use of the drag scraper for storage of coal.

The trend in fuel firing appears to be towards that of pulverized fuel, with the bin system predominating in central stations, and the unit system in the industrial field. Three methods of firing pulverized fuel in general use are: vertical firing, horizontal firing, and corner or tangential firing. Horizontal firing is gaining in favor since this type is more simple and somewhat cheaper than vertical firing, and the burners can be made in capacities as high as 8000 pounds of coal per hour. These burners are generally placed in the front wall with all the secondary air, as well as the primary air, being supplied through the burner under a pressure from 2 to 3 inches of water, and in order to produce intensive mixing, two sets of horizontal burners are sometimes used, one in each of two opposing walls. In installations of the vertical method of firing, the secondary air is supplied through air ports in the front wall of the furnace under a pressure of 1 to 2 inches of water which makes it possible to admit the air into the furnace at high velocity. For pulverizing, roller and ball mills are used more extensively than the high-speed impact mills, since in general they pulverize coal to a higher degree of fineness at greater capacity, and have low power consumption and maintenance. Some of these mills have capacities as high as 40 tons per hour.

Rapid progress has been made in the development of stokers of the under-feed and forced-draft chain-grate type. Large stokers have advanced principally in size, while the smaller units have been made more convenient. Stokers of the under-feed type are available which are capable of producing 500,000 pounds of steam per hour. The under-feed stokers at the new Delray Plant of the Detroit Edison Company are 15 retorts wide and 57 tuyeres long, are capable of burning 17 tons of coal an hour, and are equipped with automatic air-control dampers to regulate the air supply to the stoker sections. The operation of these large units is being improved by the reduction of the wind box air pressures, better methods of conditioning to provide a light and spongy fuel bed, and in the case of forced-draft chain-grate stokers the injection of air over the fire.

The use of air preheaters as a final heat-absorbing medium is widespread, and consequently stokers have had to be adapted for use of preheated air.

This has been done by using improved materials, allowing for structural expansion, and increasing the size of the air openings in the various types of grates. The maximum air temperature allowable for use with stokers depends to some extent upon the fuel used. Some coals have a tendency to fuse at comparatively low temperatures, obstructing the passage of air through the grates. In some cases an air temperature of 300 deg. fahr. has been found to be the maximum allowable on this score. Many of the stoker-fired furnaces are equipped with water-cooled block-covered surfaces along the rear and side walls of the furnace.

The rate of heat release in recent furnaces varies from 18,000 to 30,000 B.t.u. per cubic foot of furnace volume, while slag-tap furnaces are in operation with a heat release rate of 40,000 B.t.u. As yet there appears to be no completely satisfactory method for the removal of ash carried up the stack, although the electrostatic precipitator is proving effective in the removal of fly ash from stacks, and the use of cloth bags and mechanical shaking in giving satisfactory service in the removal of coal dust discharged by cyclone separators.

As a result of increased steam pressures, more and more attention is being given to the **Station Heat Balance.** The auxiliaries are chosen to fit into the heat balance and the feedwater is being heated by the regenerative cycle. Four-stage heating is quite common, and one station is using five stages. In some cases a feedwater temperature of 430 deg. fahr. is being used.

Since pure feedwater is so important in operating at high pressures, increasing attention is being given to feedwater treatment. Changes in combinations of chemical and zeolite softening, addition of coagulants, continuous blow-down systems, and steam purifiers have changed the methods of feedwater treatment. The use of sodium aluminate and phosphates to control corrosion, and inhibit embrittlement, has increased.

Instruments are being increasingly used to indicate to the fireman conditions involved in the efficient burning of fuel. An instrument and control board in a recent plant has the following instruments mounted upon it: (1) pressure gage for boiler drum and feed line, (2) Bailey multiple-draft gages for compartment pressures, auxiliary air pressure and drafts at outlet of furnace, boiler, economizer, and preheater, (3) Bailey air-flow steam-flow recording meter, (4) stoker-speed controller, (5) push-button controls for forced draft, induced draft, and auxiliary fans, and (6) position indicator for forced-draft blast damper.

542. Turbines and Turbine Auxiliaries. — The growth of power systems has made the use of large turbines economical. The largest single generating unit is a 150,000-kilowatt unit operating at 1800 r.p.m., and the largest tandem-compound unit with a single generator has a capacity of 160,000 kilowatts, while the largest three-cylinder, single-impulse unit in operation is a 208,000-kilowatt machine operating at 1800 r.p.m. In

TURBINES AND TURBINE AUXILIARIES

nearly all cases these large turbine units are tailor made, being built for special conditions to fit in with the power systems. The steam rates of these large turbines are in some cases as low as 9.5 pounds of steam per kilowatt hour.

In nearly all the high-pressure plants the high-pressure turbines are superimposed on the standard equipment, and are termed reducing valve turbines which are run at constant load, and consequently the speed governor does not present any particular difficulties. In some installations, however, these turbines are arranged to care for a variable load by using several control valves.

Multi-stage small turbines are being used increasingly in a range from 50 to 300 horsepower, with a resulting increase in economy, and in addition are being operated with automatic control devices which include automatic nozzle control as well as remote control of the speed. In a typical small turbine installation the load, back pressure, and extraction pressure are all under automatic or remote control, and in addition safety devices have been included which make it possible to operate without attendance. Several small turbine units have been constructed using an arc-welded casing of plate steel.

Fig. 502. — Payne Dean Turbine Signal.

The generator windings in the large turbo-electric generators are being cooled with surface-type air coolers in enclosed ducts, the air being used over and over again in cooling the windings. This eliminates dust from the windings, controls the moisture content, reduces the fire hazard, and saves heat. Either raw water or turbine condensate may be used as the cooling medium. In some recent installations hydrogen gas is being used in place of air for cooling. It is claimed that the output of the generator unit can be materially increased by this method.

A signal system, Fig. 502, for communication between turbine room and switchboard is commonly used. One such system has a pedestal usually mounted in the switchboard room and consists of a cast-iron pedestal completely finished on which is mounted a combination " Send " and " Repeat " instrument. Mounted on the turbine gage board is an instrument similar to the one shown, but without the pedestal, and provided with an electric contactor device for sounding an audible signal, preferably an electrically operated air whistle, for each change of position, it being impossible to move the indicator hand from one point to another without notifying the turbine room operators.

Surface condensers having 90,000 square feet or more, of condensing surface in a single shell are now in operation. In some cases the condensing surface used to serve a large main unit is made in several vertical shells of about 40,000 square feet each. Where the condensing water is taken from rivers in which the water level fluctuates widely, the condensing equipment is located in waterproof concrete wells, to obviate the pumping of large quantities of condensing water through high heads. In some cases these wells are 75 feet deep and 50 feet in diameter. The pipe connecting the condenser and turbine, in this case, is large and long, and a special type of expansion joint is used at the turbine end of the exhaust pipe. Jet condensers capable of handling 13,000,000 gallons of water per hour are giving satisfaction.

Turbine- or motor-driven centrifugal pumps are used in sizes ranging from twenty up to several hundred horsepower per unit for boiler-feed pump, circulating water pump, and in some cases for fire service.

The auxiliaries, such as boiler feed, circulating and condensate, and air pumps, and also the forced- and induced-draft fans, stokers, clinker grinders, surface pumps, air compressors, coal handling equipment, pulverized fuel feeders and blowers, in nearly all central stations are electrically driven, although in some cases a combination of motor and turbines are used. In the majority of stations the current for the motors is taken from the main bus-bar through transformers. Some stations, however, obtain current from a **house turbine,** a **shaft generator,** or a transformer connected to the generator leads. When the auxiliaries are motor-driven at constant speed, alternating-current motors are used, with a voltage of 2300 for motors above 75 horsepower and 440 for smaller motors. With auxiliaries requiring variable speed, direct-current motors are used, at a voltage of 250.

Air compressors are used to supply air for coal transport, blowing boiler tubes, cleaning and miscellaneous work about the plant.

To illustrate the tendencies in various types of *Central Station Steam Power Plants,* the following will be discussed:

1. *Edgar Station,* Boston Edison using *underfeed stokers.*
2. *Trenton Channel,* Detroit Edison using *pulverized fuel.*
3. *Long Beach No.* 3, Southern California Edison using *oil.*

543. Edgar Station. — This power station of the Edison Electric Illuminating Co. of Boston, Fig. 503, will have a capacity of 300,000 kilowatts, and was the first station to use a pressure of 1400 pounds per square inch. The steam generated in the high-pressure boilers passes through the high-pressure turbines, which serve as reducing valves and exhaust at 375 pounds pressure to the main steam header from which the normal pressure turbo-generator units take their supply of steam.

The power station buildings, Fig. 504, are of selected red brick, with the interior of the turbine room, Fig. 506, finished with a white and gray enamel brick, laid up in a simple paneled design without ledges and dirt-collecting corners. The floors are of red tile with black borders. In the boiler room, Fig. 505, a glazed buff brick has been used for the interior walls to give ease of cleaning and improve the lighting. All exposed steel and iron work is covered with a light shade of paint.

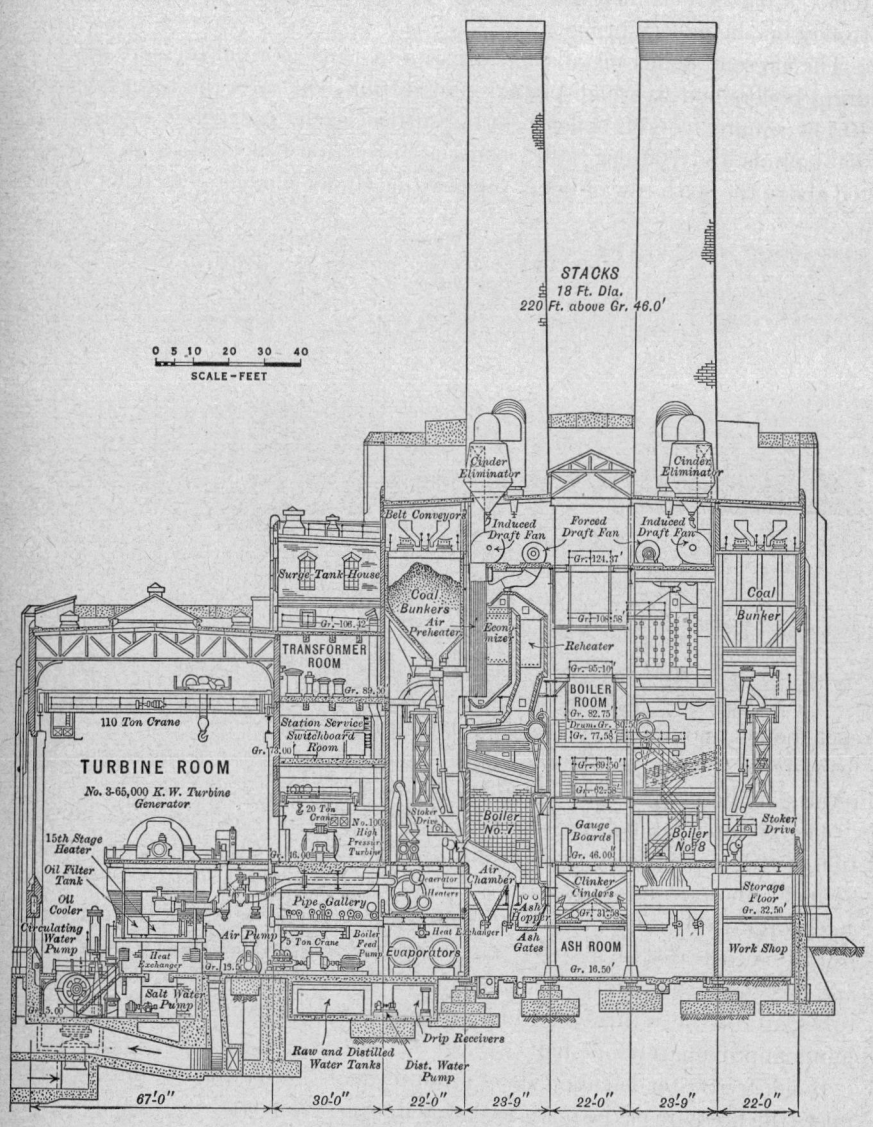

Fig. 503. — Edgar Station — Sectional View.

Coal is received in colliers and is unloaded by two electrically operated hoisting towers, each of which has an unloading capacity of 400 tons per hour. The towers have shuttle discharge conveyors which normally deliver the coal to conveyor belts on the elevated trestle 30 feet high, and is carried either directly to the power station, passing through a Bradford breaker on the way, or to a 250-foot gantry bridge for delivery to any part of the storage pile which at present provides a storage capacity of 100,000 tons. Coal is reclaimed from storage by means of a 6-ton man-operated trolley on the gantry bridge.

The present boiler installation consists of three normal-pressure cross-drum boilers and five high-pressure cross-drum boilers. The former have 19,743 square feet of boiler heating surface each, and deliver steam at 350 pounds and 700 deg. fahr., using a convection type superheater located above the sixth row of boiler tubes with a cinder catcher located between

FIG. 504. — Edgar Station, Boston Edison — External View.

each boiler and its economizer. The furnace arrangement provides for firing from the low end of the boilers, using under-feed stokers having 16 retorts, and equipped with clinker grinders.

The 1400-pound high-pressure boilers, Fig. 501, page 585, are a modification of the conventional cross-drum type, the heating surface consisting of $3\frac{1}{4}$-inch tubes arranged single-pass. The drums are made from a seamless steel forging 39 feet long and 52 inches in diameter, with a wall 4 inches thick. A reheater is located above the boiler opposite the economizer, and designed for an outlet temperature of 750 deg. fahr. The economizer tubes are of the return-bend-hairpin type, which reduces the number of joints approximately one-half.

Boiler-feedwater is supplied by a motor-driven centrifugal pump which raises the pressure of the main feed system from 425 pounds to 1300 pounds for high-pressure service. This pump is automatically controlled and is driven by a slip-ring motor operating at 3600 r.p.m. For preliminary operation and as stand-by equipment, a steam-driven centrifugal boiler-feed pump is mounted near the boiler so as to provide a source of boiler feed under the direct control of the fireman.

The forced-draft fan equipment consists of one fan and motor for each boiler. This equipment is in a chamber partitioned off from the ash basement. Air for combustion is primarily supplied through louvres in the outside wall. Warm air from the upper part of the boiler room can also be discharged into this chamber and drawn in by the suction of the forced-draft fans. The discharge ducts are arranged so that any fan can be used to supply its individual boiler, any other boiler or group of boilers. Nor-

FIG. 505. — Boiler Room, Firing Aisle — Edgar Station.

mally, it is expected to operate a fan on its own boiler. The motors are of the alternating-current brush-shifting variable-speed type.

Each boiler has an induced-draft fan beside the economizer with slip-ring motor on one end of its shaft and a smaller similar type motor on the other end. The controls are arranged so that the smaller motor carries the load to the limit of its speed range at which point the larger motor automatically cuts in and carries it from there on.

The boilers are arranged with their back or uptake ends facing a central operating aisle and their fronts or stokers facing outside firing aisles, Fig. 505. The firing aisles are partitioned off from the remainder of the boiler room and are used for coal feed only. Coal is normally fed to the stoker hoppers by means of stationary chutes supplied from bunkers over the firing aisles. A 6-ton traveling weigh larry is installed in each firing aisle. The central aisle contains the boiler control board on which are the necessary instruments to guide the operator in controlling the output and efficiency of each boiler.

The main generating equipment consists of two 32,000-kilowatt turbines, each driving a 30,000-kilowatt, 14,000-volt, 60-cycle main generator with a 2000-kilowatt, 2300-volt, 60-cycle auxiliary generator direct connected to the shaft of the main generator, and a 65,000-kilowatt, 17-stage turbine, Fig. 410, page 498, driving a 60,000-kilowatt generator direct connected to an auxiliary generator of 5000-kilowatt capacity. The turbines are supplied

FIG. 506. — Turbine Room, Edgar Station.

with steam at a nominal pressure of 350 pounds and a total temperature of 700 deg. fahr. There are also a 3000-kilowatt, 10,000-kilowatt, and two 12,500-kilowatt high-back-pressure turbines, which take steam at 1200 pounds pressure and 700 deg. fahr., and exhaust through a reheater into the 350-pound steam main supplying the main generator unit. These high-pressure machines are built along standard lines, except that the parts are strengthened to withstand the combined high pressure and temperature to which they are subjected. They are of the horizontal type, straight impulse, multi-stage type, using four bearings and operating at 3600 r.p.m.

The main units are installed on a platform with their axes parallel to the length of the turbine room, Fig. 506, thus giving light and ease of access to the auxiliaries below and also approximately equivalent growth between the turbine and boiler room as the plant increases in size. The steam ends of each two units are adjacent, with one condensing water intake and tunnel serving two units.

The generators are kept cool by air circulating in a closed system of ventilation, which reduces the fire risk in the generators and also gives an opportunity of checking an incipient fire by "flooding" the ventilation chamber with the inert gas CO_2. The heat is normally removed by condensate flowing through surface coolers.

The hot-well pumps are in duplicate on the condenser floor. Water from the hot wells is pumped to the generator air coolers and the turbine oil coolers. After flowing through this apparatus, the water passes through a small condenser installed for condensing gland leak-off steam from the high-pressure turbo-generator whenever that unit is running. The water flows next through the low-pressure 15th-stage extraction heater, then through the gland steam condenser and to the deaërators. The deaërators are drained by the deaërator pumps, which discharge through the 12th-stage high-pressure extraction heater and from there to the surge tank. This tank is entirely closed to eliminate re-absorption of air and is intended to operate on hot water. At normal operating loads the water will give off vapor and form its own steam blanket. At very light loads there would be a tendency to form a slight vacuum in the tank. In this case the steam blanket is formed by steam extracted from the gland leak-off. Relief valves have been provided and the tank covered to prevent heat loss. From the surge tank the water flows to the suction header of the boiler-feed pumps. From these pumps it is discharged through the high-heat-level condenser of the high-pressure-three-effect evaporators, to the economizers and into the boilers.

In the power station itself the water system for power generation is laid out for two sources of supply; a distilled water circuit forming the boiler-feed lines with make-up and draw-off connections, and the power service system which supplies raw fresh water for various power users and for boiler feed during emergencies.

All the normally operating station auxiliaries are electrically driven, using alternating-current motors throughout. Motors of 25 horsepower and less are, in general, supplied at 550 volts; those of greater capacity at 2300 volts.

Normal and high-pressure steam piping is welded throughout, flanged work having Sargol welded joints and the small piping having the threads sealed with a welded bead. The flanged joints in the 350-pound boiler-feed discharge are of the Van Stone type with composition asbestos gaskets.

Important control and sectionalizing valves in main and auxiliary normal and high-pressure steam lines are motor operated, with their opening and closing stations located near the valves. Remote closing stations

TABLE 37. — PRINCIPAL EQUIPMENT OF UNIT NO. 1002 EDGAR STATION, BOSTON EDISON COMPANY

No.	Equipment	Kind	Size	Use	Operating Conditions	Maker
2	Boilers	Cross-drum	15,093 sq. ft. H.S.	Steam generator	1300 lb. press. 733° F.	Babcock and Wilcox Co.
2	Superheaters	Interdeck, 3 pass	3483 sq. ft. H.S.	Superheating steam	155° superheat, 250,000 lb. per hr.	Babcock and Wilcox Co.
2	Reheaters	Tubular, 2 section	8879 sq. ft. H.S.	Reheating steam	750° F. @ 375 lb. per sq. in.	Babcock and Wilcox Co.
2	Economizers	Horizontal steel tube	5596 sq. ft. H.S.	Heating feedwater		Babcock and Wilcox Co.
2	Air preheaters	Tubular	33,032 sq. ft. H.S.	Heating air for combustion		Amer. Eng. Co.
2	Stokers	Taylor — Underfeed	16 retorts, 585 sq. ft.	Fire coal	26,900 lb. per hr. forced draft	B. F. Sturtevant
4	Fans	Hor. — double inlet	47,500 c.f.m. @ 90° F.	Induced draft	Driven by slip-ring motor	
4	Fans	Hor. — double inlet	66,000 c.f.m. @ 330° F.	Forced draft	11.2 in. water pressure	
4	Feed pumps	Centrifugal	1620 g.p.m.	Primary boiler feed	Motor drive, 625 hp., 1800 r.p.m.	Worthington Pump & Machinery Corp.
2	Feed pumps	Centrifugal	2133 g.p.m.	Secondary boiler feed	Motor drive, 2030 hp., 1800 r.p.m.	Worthington Pump & Machinery Corp.
1	Feed pumps	Centrifugal	1910 g.p.m.	Boiler feed	Turbine drive, 3670 r.p.m., 1600 lb.	Ingersoll Rand Co.
1	Turbine	Impulse — single cylin.	2450 hp.	Feedpump drive	350 lb. press. @ 725° F.	General Electric Co.
3	Heaters, extraction	Hor. str. tube	2245 sq. ft.	Heating feedwater	535,000 lb. per hr., 90° F. to 290° F.	Griscom Russell Co.
1	Turbo-generator	16-stage, single-cylinder, impulse	10,000 kw.	Electric generator	1400 lb. press., 3600 r.p.m.	General Electric Co.
1	Air cooler	Tubular — horizontal	26,000 c.f.m.	Air for 10,000-kw. gen.	Air cooled from 143.5° F to 97° F.	General Electric Co.
1	Turbine	17-stage, single cylinder	65,000 kw.	Generator drive	350 lb., 725° F., 1 in. back pressure	General Electric Co.
1	Generator	60 cycle	75,000 kva.	Electric generator	0.80 power factor, 3 phase, 14,000 volts	General Electric Co.
1	Generator	60 cycle	6250 kva.	Auxiliary generator	3 phase, 2300 volts	
2	Fans	Multi-vane	110,000 c.f.m.	Gen. air cooler	200 hp., 22,000-volt drive	B. F. Sturtevant
1	Air cooler	Tubular	110,000 c.f.m.	Air cooler for 66,000-kw.	Air cooled from 142° F. to 99.5° F.	General Electric Co.
1	Deaërator	Open — contact	635,000 lb. per hr.	Air removal	29.9 lb. working pressure	Cochrane Corp.
1	Vent condenser	Horizontal tubular	370 sq. ft. H.S.	Condensing vapor	0 to 80 lb. abs. working pressure	Cochrane Corp.
1	Condenser	Surface — single pass	51,200 sq. ft.	Main condenser		
2	Pumps	54-in. volute	105,000 g.p.m.	Circulating water	195 r.p.m., 275 hp. motor drive	Worthington Pump & Machinery Corp.
2	Pumps	8-in. × 3-in. stage	1350 g.p.m.	Hot well	247 ft. head at 870 r.p.m.	Griscom Russell Co.
2	Evaporators	Reilly — coil	28,000 lb. per hr.	Feedwater purifier	29.3 lb. press., outlet 201° F.	Griscom Russell Co.
1	Evaporator condenser	Tubular — horizontal	570,000 lb. per hr.	Serves evaporators		

Miscellaneous equipment: Link Belt water screens; Mason reducing valves; Ruggles-Klingeman pressure regulators; Northern Equipment Co.'s feedwater regulators; Foxboro and Leeds-Northrup recording thermometers; Bailey combustion control.

controlling all the valves are mounted on an emergency operating board in the outside firing aisle where the boiler-room partition isolates it from the steam piping. Additional remote closing stations controlling the valves located in the turbine room have been provided near the throttle valves of the first pair of main units.

The size and type of equipment installed in this plant are tabulated in Table 37.

544. Trenton Channel Power Station. — This station of the Detroit Edison Company, Fig. 507, is located on the Trenton Channel of the Detroit River, and is of particular interest because the plant was designed for consistency of size, arrangement and type of equipment throughout. The installed capacity is 300,000 kilowatts. The generating units are supplied with steam at 375 pounds per square inch gage pressure and 700 deg. fahr. from 14 Stirling Type W boilers each having 30,307 square feet of heating surface and fired with pulverized fuel.

The generating units, the boilers and auxiliary plant are all of the same size and type, which decreases the operating difficulties. *The general intention has been not to run after the very latest practice and highest thermal economy, but to make a uniformly good job, a station low in total cost, simple and easy to work, and one which would easily become so well known to its staff that any irregular working might be quickly revealed and as promptly rectified.* This policy reduces the number of spare parts carried, with a consequent reduction of cost in this item and in maintenance. One of the controlling factors fixed upon when the plant was planned was to adopt the largest single cylinder turbines then considered practicable, coupled with the largest alternators of reasonable weight and speed.

Coal is received by rail, and after being weighed on track scales is either sent to storage or dumped into track hoppers. The coal going into storage is handled by steam locomotive cranes with grab buckets. Storage is provided for 140,000 tons.

Coal entering into the plant is dumped into concrete hoppers from two tracks, with a special car unloading machine installed over one of the tracks for breaking up frozen coal. From the bottom of the track hoppers, coal is carried upward on apron conveyors and delivered to the feeding hopper of the Bradford breakers which break the coal to pass through a $1\frac{1}{4}$-inch ring. The broken coal is fed to four-bucket conveyors by four apron conveyors. It is then elevated and delivered into a 3100-ton crushed-coal bunker at the top of the coal preparation house. The coal next passes by gravity downward through the driers which, instead of using flue gas from the boilers to dry the coal, use air drawn through the lane of coal and heated by hollow steam heated grids. From the bottom of the dryer the coal passes to 28 roller type pulverizers. Carried upward by the stream of air passing through the mill, the pulverized coal is separated from the air in a cyclone separator, and passes to a screw conveyor which

598   MODERN POWER PLANTS

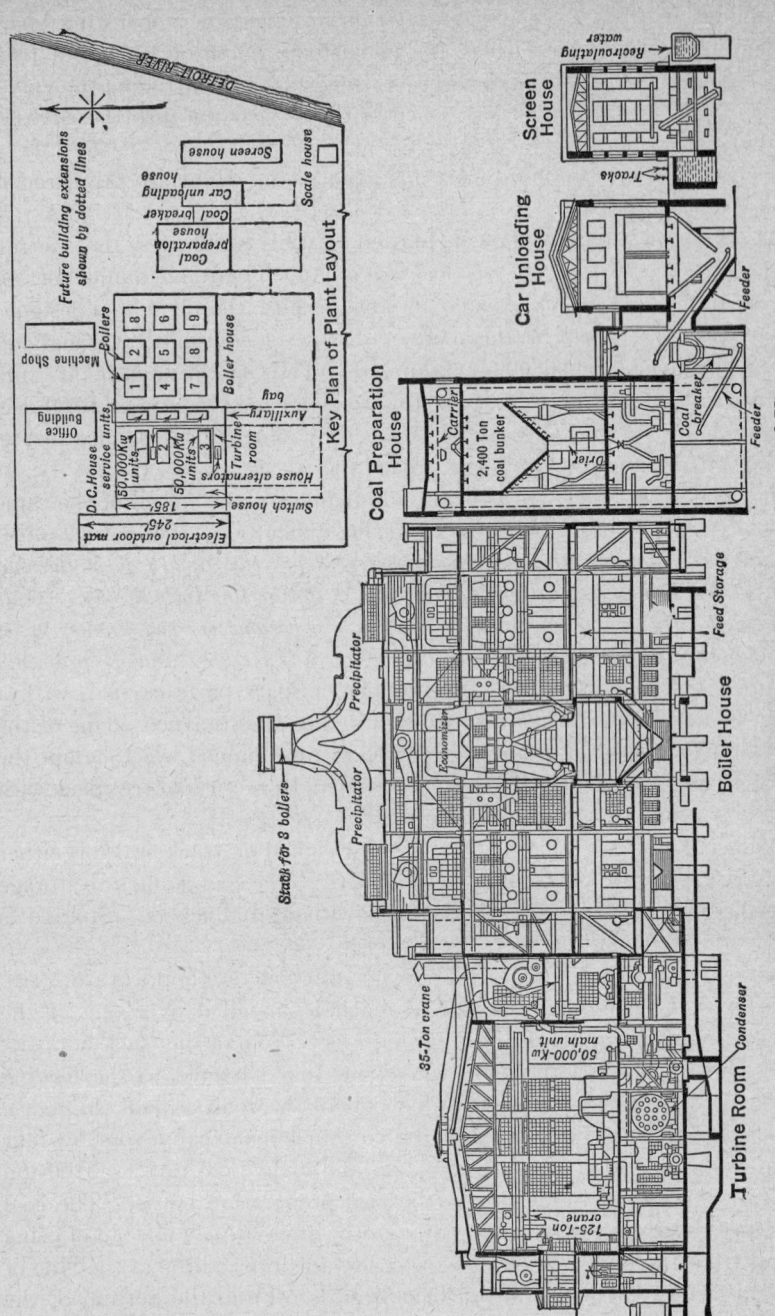

Fig. 507. — Trenton Channel Power Plant — Sectional View.

carries it into the feeding hoppers of a pulverized fuel-transport system for distribution to the individual boiler bins.

In the process of coal preparation and transportation dust from the pulverizing mill circuit vents, Bradford breaker, and cyclone separators is recovered in sets of bag filters which effectively remove the dust from the discharged air, while the grit carried by the flue gases is removed by Cottrell precipitators.

The boiler house (*see Frontispiece*) is a six-story brick building with the boilers arranged in three rows. On the ground floor are situated the ash hoppers and sluicing troughs; on the first floor are the main piping arrange-

FIG. 508. — Boiler Room, Firing Aisle — Trenton Channel Power Plant.

ments and the upper part of the water screen for the boilers. The main boiler floor and firing control room, Fig. 508, is 41 feet above ground level. Above this at a height of 65 feet is a floor, Fig. 509, on which are the feeders for the burners. At 78 feet above the ground floor is another gallery, and at 95 feet a floor level with the fuel bunkers and economizers. The floor at 114 feet carries the draft fans, and on the roof at 130 feet are dust precipitators and the stack bases.

The boilers are fitted with two 3-pass superheaters providing 6070 square feet of heating surface and two steel-tube economizers with 18,984 square feet of heating surface and a pair of hearth screens having 1220 square feet of heating surface. Water screens are used primarily to prevent slagging at the furnace bottom, but at the same time, by the absorption of radiant heat, they reduce the average temperature in the combustion chamber, increase the life of the refractories, and make it possible to operate with less excess air.

The furnace is rectangular in plan, and measures 23 feet by 34 feet inside, its volume being 25,140 cubic feet, and when delivering 360,000 pounds of steam per hour, coal is burned at a rate of 1.43 pounds per cubic foot of furnace volume per hour.

Each boiler is provided with two steel plate pulverized coal bunkers of 75 tons capacity each, vented by pipes carried above the roof. At the bottom of each bunker is a hopper fitted with a gate valve and two Lopulco

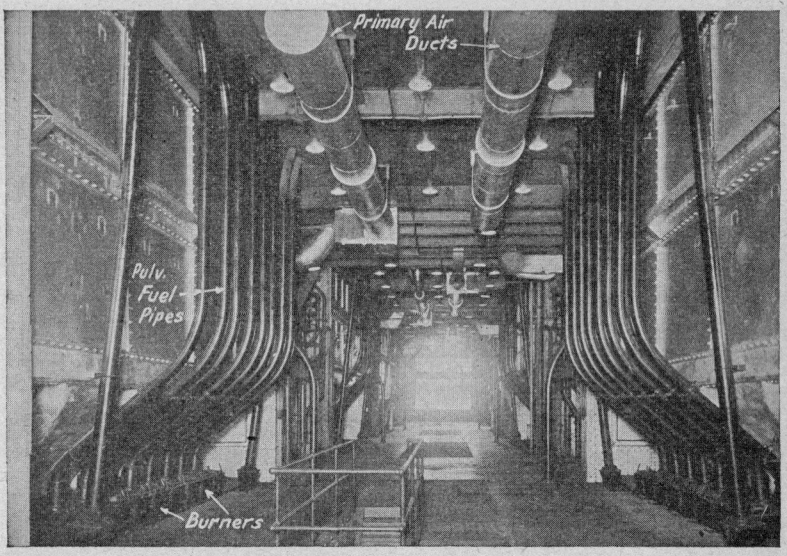

FIG. 509. — Pulverized Coal Feeders — Trenton Channel Power Plant.

feeders. Under each main drum of the main boiler is a Liptak arch, through which 8 fish-tail pulverized coal burners shoot their flame vertically downward across the face of an air cooled wall in which are scattered many openings through which the secondary air is admitted, after being warmed by passing between the inner refractory lining and the outer casing. The primary air for each set of 8 burners is supplied by a motor-driven blower and amounts to from 12 to 28 per cent of the total air required for combustion. This air is warmed by being drawn through some of the ducts through the upper part of the setting. There are eight secondary air openings per burner, having an area of about 4 square feet. The amount of air drawn into the furnace is controlled by the speed of the induced-draft fan, and by the position of the dampers in the air entrance ports. The induced-draft fans, which are located in the upper floor of the boiler house and above the economizers, discharge the flue gases into breechings so arranged that one stack serves three boilers. The stacks are steel, lined to the top with common brick, which are grouted to the steel with 3 inches of grout. Each stack is 16 and 21 feet in diameter inside at throat and top, and 104 feet high above the roof girders on which it rests.

The main steam piping is seamless-steel tubing, with all joints of the Van Stone type, and all fittings and valves of cast steel. At strategic points in the main steam piping system motor operated valves are located. These are operated by push buttons located in a room between the plant and the office building. The valves may be closed individually from this

Fig. 510. — Turbine Room, Trenton Channel Power Plant.

remote point, but cannot be opened from there. Expansion is provided for by using bends and loops.

The combustion instruments and controls are mounted on a board, Fig. 508, for each boiler unit. The instruments comprise pressure gages for feedwater, main boiler saturated steam, tube screen saturated steam, and superheated steam, gages indicating draft at five points within the setting, ammeters indicating the current drawn by blower and feeder motors, and air-flow steam-flow meters. On this board are mounted controls for the motors driving the coal feeders, the primary air blowers and the induced-draft fans. A reversing switch controls a small motor that actuates the dampers at the economizer outlet.

The main turbines, Fig. 510, are 21-stage impulse machines, running at a speed of 1200 r.p.m., with provision for extracting steam from the 14th, 17th, and 19th stages, and are arranged with their axes crosswise of the turbine room and grouped in three pairs, with large hatches between the pairs for light and for crane service to the condenser room. On the main floor of the turbine room in line with the main units are three 2000-kilowatt turbo-alternators which supply power for operation of coal preparation machinery and all building lighting.

In a bay between the main turbine room and the boiler room are located four turbine-driven direct-current generating sets, having a normal output of 4000 kilowatts. These turbines are of the two-barrel type, two turbine rotors driving separate pinions on opposite sides of the main gear. The high-pressure turbine runs at 4000, the low-pressure at 3000, and the generator at 360 r.p.m. These house turbines are operated straight condensing, with provisions for steam extraction for building, heating and coal drying.

The main condensers are bolted directly to the turbine exhaust flanges, with car-spring supports to carry a large share of the condenser weight. They are the single-pass type, each having 50,000 square feet of heating surface. The auxiliary condensers are standard design, two-pass condensers.

The circulating pumps for the two main condensers are 48-inch pumps, rated at 60,000 gallons per minute, driven by 325 horsepower direct-current motors, with a range of speed from 175 to 255 r.p.m. The auxiliary condensers are served by smaller centrifugal pumps, one for each condenser — those serving the house turbo-alternators being driven by alternating-current motors, and those serving the direct-current machines by direct-current motors.

The main condensers and the condensers serving the house service direct-current units have reciprocating vacuum pumps driven by direct-connected direct-current motors. The main pumps are of the vertical type while the auxiliaries are horizontal. All are of the single-cylinder, single-acting two-stage type, with feather valves. For the house alternating-current units the condensers are served by water-jet vacuum pumps.

Condenser cooled water is taken through the screen house located near the river bank, and housing 15 trash racks and traveling screens. The river water passing through the trash racks and screens enters the forebay which extends the full length of the screen house. The intake canal runs from the end of the forebay as a submerged conduit, around the end of the coal preparation house, and under the boiler room, to the turbine room, where it meets at right angles two intake tunnels which extend lengthwise of the turbine room, one beneath the main units, and one beneath the auxiliary direct-current units. Two pumping units are installed for each main condenser, each unit consisting of a hot-well pump and a boiler-feed pump, coupled together to form what is called a combination pump. For each unit there is one combination pump driven by a direct-current motor, with a standby set driven by a steam turbine. For each combination set the hot-well pump draws water from the condenser hot well and discharges

TRENTON CHANNEL POWER STATION

it through the stage heaters to the suction of the boiler-feed pump which delivers to the main and auxiliary feed headers which are of the straight tube type, with the tubes expanded into tube sheets at the bottom, one tube sheet being mounted on a floating box to care for expansion. Together with the evaporator and its condenser each set of heaters is located just below the turbine-room floor, as close to the turbine as possible, and thus making the connecting piping as short as possible.

The boiler-feed piping is made up of 8-inch extra-heavy pipe, with Van Stone joints throughout. From each pair of boiler-feed pumps there is a complete duplicate system of piping to and including duplicate nozzles entering each boiler drum.

In order to leave an effective margin for the operation of the economizers, the feedwater heating is confined to the lower temperatures. With condensate at about 75 deg. fahr., the temperature of the feed is raised in the first heater, fed from the 19th stage, to about 124 deg. fahr. In the next heater, fed from the 17th stage, the feed temperature is raised to 181.5 deg. fahr. The feed next passes through a small condenser, in which it is used to condense steam vapor required for make-up purposes, and in this way its temperature is raised to about 201 deg. fahr. From this condenser the feed passes through a third and last heater, fed with steam bled from the 14th stage of the main turbine. On leaving this heater, its temperature being about 245 deg. fahr., it passes to the motor-driven boiler-feed pump.

Make-up water is condensate from the evaporator which uses steam bled from the 14th stage of the main unit. The vapor from the evaporator is condensed in a special evaporator condenser, the combination being designed to supply $1\frac{1}{4}$ per cent make-up continuously. The condensed bleeder steam from all three feedwater heaters, and from the evaporator, as well as the condensed make-up supply from the evaporator condenser, are fed into the main condenser. It is found that the closed system with condensed make-up passed through the main condenser allows of quite satisfactory deaëration. Trenton Channel is one of a few large plants using steel-tube economizers and not having deaërators in service. The system described is so effective as practically to eliminate corrosion of these tubes, but careful attention is required to this matter and observations are constantly made on the percentage of dissolved oxygen present. The oxygen is not allowed to rise above 0.035 centimeters per liter, and checks are constantly made to ensure this.

The steam bled for the feed heaters amounts to about 5 per cent of the steam supplied to the turbine at each of the three points at which it is taken off.

In order to allow for irregularity of load on the main units, and consequently variation of feed supply to the boilers, each system is provided with a storage surge tank. In times of decreasing load, water will be

TABLE 38. — PRINCIPAL EQUIPMENT OF TRENTON CHANNEL PLANT, DETROIT EDISON COMPANY

No.	Equipment	Kind	Size	Use	Operating Conditions	Maker
12	Boilers	Sterling — Class W	29,087 sq. ft. H.S.	Steam generator	390 lb. press., 700° F.	Babcock and Wilcox Co.
12	Superheaters	Convection, U-tube	6070 sq. ft. H.S.	Superheating steam	250 deg. superheat	Babcock and Wilcox Co.
24	Economizers	Duratex	9425 sq. ft. H.S.	Heating feedwater		Babcock and Wilcox Co.
88	Feeders	Lopulco duplex		Feed powdered coal	8 per boiler	Combustion Eng. Corp.
128	Burners	Lopulco fishtail		Burn powdered coal	16 per boiler	Combustion Eng. Corp.
54	Burners	Lopulco forced draft		Burn powdered coal	16 per boiler	Combustion Eng. Corp.
6	Chimneys	Steel, Venturi shaped	190 ft. high, 16 in. dia. throughout	Serve boilers	1 for 3 boilers	Combustion Eng. Corp.
22	Blowers		6500 c.f.m. at 15 in. water	Primary air	2 per boiler, 30 hp., 1200 r.p.m.	Buffalo Forge Co.
	Dryers	Steam heated		Dry raw coal		
28	Pulverizers	Raymond ball	6 ton per hr.	Pulverizes coal	100 hp., 450 r.p.m.	Combustion Eng. Corp.
28	Exhauster	Screw	6000 c.f.m.	For pulverizer	Motor drive, 60 hp., 1200 r.p.m.	Combustion Eng. Corp.
8	Conveyors	Fuller-Kinyon	100 ton per hr.	Feed pulverized coal		Fuller-Lehigh Co.
4	Precipitators	Cottrell	50 ton per hr.	Convey pulverized coal Remove dust from flue gases	2 per boiler	The Research Corp.
2	Coal breakers	Bradford	315 ton per hr.	Break raw coal	Motor drive, 150 hp.	Pennsylvania Crusher Co.
	Ash handling	Hydraulic sluicing	2000 g.p.m.		60 hp., 3-single stage pumps	Baker Dunbar Co.
22	Feed pump	6 stage, centrifugal	1300 g.p.m.	Feedwater and hotwell	1300 ft. head, 1 steam, and 1 motor	De Laval and Worthington
6	Turbines	21 stage, Curtis	50,000 kw.	Generator drive	385 lb., 245° superheat, 1200 r.p.m.	General Electric Co.
6	Generators		62,500 kva.	Electric generator	3 phase, 60 cycle, 12,200 volts	General Electric Co.
6	Condensers	Surface, single pass	50,000 sq. ft.	Serve turbine		Worthington Pump & Mach. Corp.
12	Pumps	Centrifugal	60,000 g.p.m.	Circulating water	325 hp., 175-255 r.p.m.	Worthington Pump & Mach. Corp.
11	Evaporators	Reilly, coil tube	6625 lb. per hr.	Purify feedwater		Griscom Russell Co.
	Pumps	Rotative dry vacuum	30-in X 18-in. vertical			Laidlaw-Dunn Gordon
4	Turbo-generators	Multi-stage	4000 kw., 11 stages high, 6 stage low-press.	Auxiliary power, D.C.	Steam extracted between high and low-pressure turbines	De Laval Steam Turbine Co.

Miscellaneous equipment: Bury air compressor; Chain Belt Co.'s traveling screens; Nash vacuum pumps; Hopkinson steam traps; Dean motor operated gage valves; Davis atmospheric relief valves; Maxwell, Manning and Moore safety valves; Bailey air and steam flow meters; Foxboro gages and thermometers; Copes feedwater regulators.

accumulated in this tank, while with increasing load, this storage is drawn upon to supplement the normal flow. These operations are controlled automatically by float valves fitted near the main condenser.

The type and size of the equipment installed is tabulated in Table 38.

545. Long Beach Power Station, No. 3. — A sectional view of this plant of the Southern California Edison Co. is shown in Fig. 511. It is located on an island on the outskirts of Long Beach, has an unlimited supply of condensing water available from the Pacific Ocean, and will have an ultimate capacity of 1,000,000 horsepower, with a present installed capacity of 125,000 horsepower, generated in a tandem-compound, single-shaft, turbo-generator unit which has an overall length of 103 feet, and at the time of installation was the largest single-shaft unit ever installed.

In the design of the buildings the possibility of earthquake stresses has been anticipated, and all the structures, including the furnaces, the superimposed concrete stacks, the turbine and condenser foundations, and other important parts, have been designed to resist the effects of seismic disturbances. The buildings, Fig. 512, have been designed to produce a pleasing appearance, but no ornamentation is provided that is not justified by utility. The floors of the turbine rooms, condenser pits, the oil pump room, and the firing floor of the boiler house all are surfaced with quarry tile.

Each of the three Babcock & Wilcox cross-drum boilers, Fig. 513, is designed for operation at 460 pounds steam pressure with a maximum evaporating capacity of 450,000 pounds of steam per hour. The furnaces are completely constructed of Bailey refractory-faced blocks mounted on water-cooled tubes. The boilers are equipped with interdeck superheaters of the three-pass convection type, which deliver steam at a temperature of 750 deg. fahr. at 350 per cent of boiler rating. The tubes in the last pass of the superheater are of chromium alloy steel to withstand the high temperatures encountered. Steam strainers are installed in the superheated steam leads at all boilers to preclude the possibility of foreign material lodging in and rupturing valves or equipment.

Natural gas and oil are used for fuel, piped directly to the plant from nearby oil fields. These fuels are burned in 20 insulated-front wide-range Peabody combined oil and gas burners per boiler. These burners project through a hollow front wall which distributes the incoming air. Oil is delivered to the burners at 200 pounds pressure, and 250 deg. fahr. Provision has been made to change from the present fuels to pulverized coal whenever such change may be economically justified, and the present furnaces and station have been so designed that the change can be easily made.

Forced-draft fans placed above the boilers deliver air for combustion through air preheaters which raise the temperature of this air to 450 deg. fahr. Induced-draft fans also placed above the boilers remove the gases

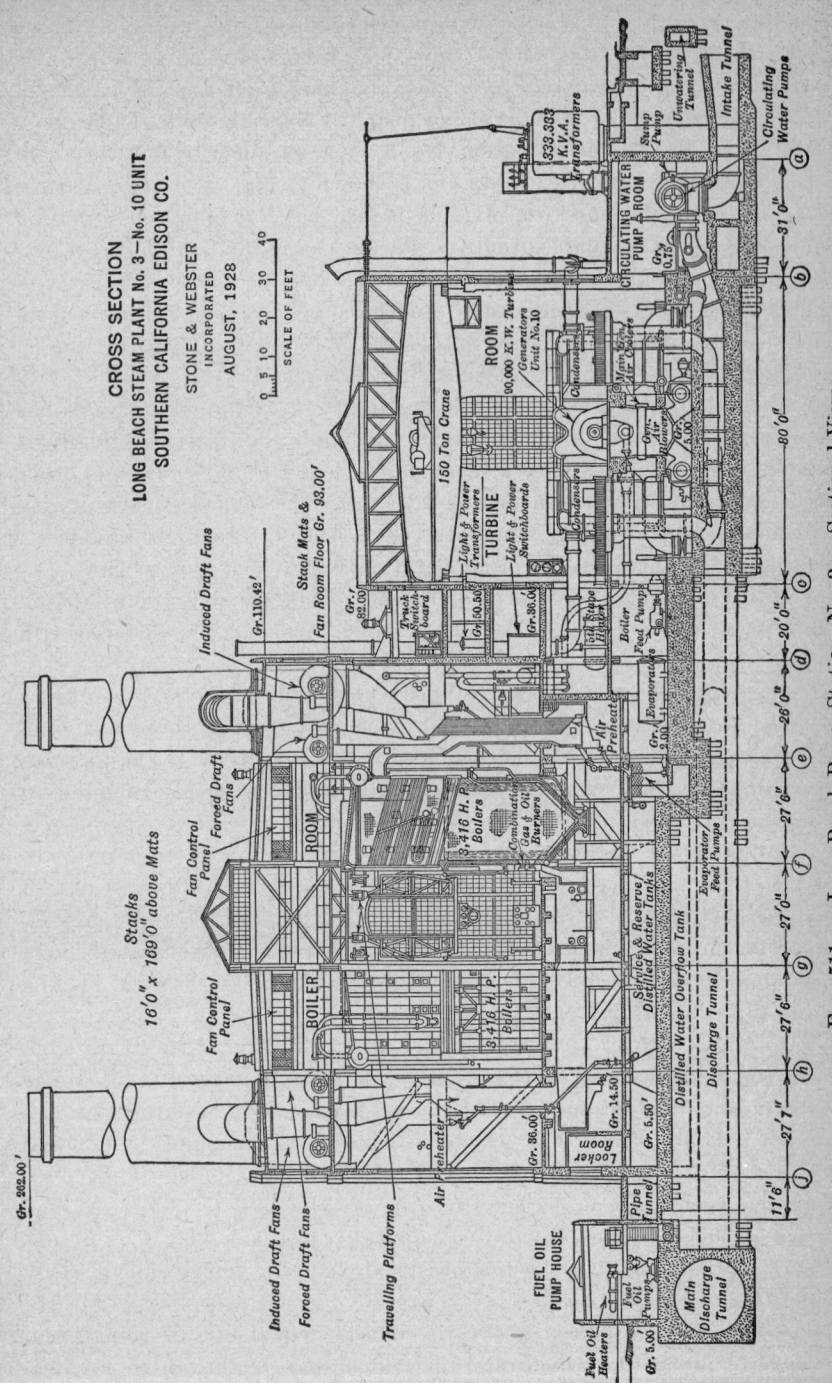

Fig. 511.— Long Beach Power Station, No. 3 — Sectional View.

LONG BEACH POWER STATION, NO. 3

from the boiler chambers and discharge them into the two superimposed reinforced concrete stacks.

A Bailey automatic combustion control system is used to provide a constant balance between fuel and air as the steam demand changes with load. The flow of fuel oil is controlled by by-pass valves and the flow of gas by throttle valves, all electrically operated. To allow economic operation and to assure accurate control over a wide range of steam demand, three motor-driven boiler-feed pumps have been installed. A new type of compensator in the form of an automatically controlled valve mounted ahead of the float valve in the boiler drum has been installed in one boiler in addition to the usual feedwater regulators. With this equipment the excess pressure of water over steam may be varied to suit the demand on the particular boiler as indicated by the pressure drop through the preheater. To assure a constant water supply to all boilers a common type of regulator changes the pump speed to maintain a constant excess pressure in the main feedwater header.

A new control house has been provided for the

FIG. 512. — Long Beach Power Station, No. 3.

centralized control of the Long Beach No. 3 steam plant. This building is entirely separated from other buildings in the plant group and is about 250 feet from the generator room. On the main control board in this building are all of the related relays and instruments. Communication to the turbine engineer and to other important points is secured through an automatic electric dial-signal system similar to those commonly employed on large ships. This is supplemented by an ordinary dial-type automatic intercommunicating telephone system which completely covers all important points in the entire plant layout. Further to assist in coördinating operation, especially at times of trouble, large load and frequency indicators have been mounted in the turbine and boiler rooms as the best method of conveying the nature of system trouble to all persons concerned.

Gage boards are installed at the turbine, in the boiler room, and adjacent to the related principal auxiliary equipment. On these boards graphic recording instruments are used extensively to indicate the trend of operations as well as to indicate instantaneous values. To assure accurate coördination of these records the charts are all driven by means of synchronous motors. Each gage board is equipped with an annunciator system to assist in the prompt and definite location of trouble.

Fig. 513. — Boiler Room — Long Beach Power Station, No. 3.

Particular attention has been given to the design of the steam piping system to insure reliability, ease of operation and low maintenance charges. The main steam header is a one-piece forging formed from a single steel billet and all of the main steam line piping is seamless steel tubing. All joints in the high-pressure piping are of the Van Stone lap with Sargol seal, and the high-pressure steam valves are of cast steel with Monel metal disc facing and stainless steel seats and stems.

The General Electric Curtis turbine, Fig. 514, is of tandem-compound design, comprising an 18-stage high-pressure rotor, and a 3-stage double-flow low-pressure rotor. It is rated at 125,000 horsepower at 1500 r.p.m., and is directly connected to a 90,000-kilowatt generator, and a 4000-kilowatt auxiliary generator. At full load, the steam consumption of the turbine is 1,037,000 pounds per hour.

Steam at throttle pressure of 400 pounds gage and 725 deg. fahr. is delivered to the turbine throttle through two 18-inch pipes, and after passing through the high-pressure cylinder is delivered to the center of the low-pressure cylinder through an overhead connection to the exhaust side of the high-pressure turbine.

FIG. 514. — Turbine Room, Long Beach Power Station, No. 3.

The unusual physical appearance of the turbine is due to the use of four vertical condensers which extend from the condenser floor, 15 feet below the turbine floor, to a point higher than the top of the turbine casing. This arrangement permitted important construction and operating economies.

Steam is extracted from the turbine at the 5th, 10th, 14th, and 18th stages with corresponding pressures of 229, 100.2, 36, and 7.5 pounds absolute. The condensate from the high-pressure heaters is trapped and sent to the flash tank and the condensate from the low-pressure heaters is discharged into the feedwater lines after the respective heaters. Low-pressure heaters are of the vertical type and the high-pressure heaters are horizontal. Feedwater is discharged into the boilers directly from the high-pressure heaters.

TABLE 39.—PRINCIPAL EQUIPMENT OF LONG BEACH STATION NO. 3 CALIFORNIA EDISON COMPANY

No.	Equipment	Kind	Size	Use	Operating Conditions	Maker
3	Boilers	Cross-drum, 51 tubes wide	31,416 bl. hp.	Steam generator	450 lb. press., 750° F. temp.	Babcock and Wilcox Co.
3	Air heaters	Tubular	51,232 sq. ft.	Preheat air		Babcock and Wilcox Co.
6	Fans		53,000 c.f.m.	Forced draft		Green Fuel Econ. Co.
3	Fans		180,000 c.f.m.	Induced draft		Green Fuel Econ. Co.
60	Burners	Combination gas and oil	1750 lb. oil per hr. 28,000 cu.ft. gas per hr.	Serve boiler	20 burners set in rows of 10 each	Peabody Eng. Co.
1	Stack	Reinforced concrete	16 ft. dia., 226 ft. high	Draft		Heine Chimney Co.
1	Turbine	Tandem-compound impulse. 18 stages H.P. 3 stages L.P.	94,000 kw.	Drives main generator	1500 r.p.m., 415 lb. press., 0.75 in. exhaust press.; temp. 750° F., 4 extraction points	General Electric Co.
4	Condensers	Vertical, single pass	20,000 sq. ft.	Serves turbine	16 feet head	Ingersoll Rand Co.
2	Pumps	Horizontal, volute	82,500 g.p.m.	Circulating water		Ingersoll Rand Co.
3	Pumps	Horizontal, 2 stage	1200 g.p.m.	Condensate		
1	Pump	Jet	3-2 in. primary 4-2 in. secondary	Air removal	4 sets primary jets, 1 set secondary	Ingersoll Rand Co.
1	Generator	Direct connected	90,000 kw.	Main generator	16,500 volts, 3-phase, 50 cycles, 0.90 P.F.	General Electric Co.
1	Generator	Direct connected		Auxiliary generator	2300 volts, 3-phase, 50 cycles, 0.80P.F.	General Electric Co.
1	Turbo-generator	Horizontal, single-stage, impulse	3200 kw.	Carries auxiliary load with frequency below 49½ cycles	Non-condensing, 3000 r.p.m., 400 lb. 700° F., exhaust pressure, 17.7 lb. abs., 2300 volts, 50 cycles	General Electric Co.
1	Evaporator and condenser	Horizontal, high pressure, submerged	16,000 lb. per hr.	Make-up feedwater supply		Griscom Russell Co.
4	Heaters	2 high pressure, horizontal 2 low pressure, vertical	4040 and 3910 sq. ft. 2781 and 3160 sq. ft.	Feedwater	Pressures 229, 100, 36 and 7.6 lb. abs.	Griscom Russell Co.
4	Pumps	Centrifugal — 5 stage	2000 g.p.m.	Boiler feed	Turbine driven, 415 hp.	Ingersoll Rand Co.
4	Pumps	Centrifugal — 6 stage	1000 g.p.m.		Motor driven	Terry Steam Turbine Co.
1	Pump		2000 g.p.m.	Fire		
1	Water softener	Zeolite	60,000 gal.	Heater drips		Permutit Co.
6	Pumps	Centrifugal				Ingersoll Rand Co.
1	Pump	Centrifugal	10,000 g.p.m.	Condenser back wash		Ingersoll Rand Co.

Miscellaneous equipment: Bailey automatic control; Permutit continuous blow-down equipment; Ruggles-Klingeman motor control; Diamond soot blowers — 10 per boiler; Smoot make-up control; Braun fuel oil heaters; De Laval oil purifiers.

Make-up water to the evaporators, which amounts to about 0.5 per cent of the water used, is taken from wells and is treated in Zeolite water softeners. A continuous blow-down system to keep the concentration in the evaporator shell always below the point where priming takes place is used. The feedwater is handled by four feed pumps, two motor-driven 2000-gallon pumps, one turbine-driven 1000-gallon pump, and one motor-driven 1000-gallon pump, all operating under a discharge pressure of 550 pounds per square inch.

Each of the four vertical Ingersoll-Rand condensers contains 20,000 square feet of cooling surface and is of the single-flow type with the cooling water passing upward. The condenser tubes are protected from corrosion by a large number of cast-iron electrodes suspended over the tube-ends in the circulating water and maintained at a positive electrical potential with respect to the tubes by means of a 10-volt motor-generator set which delivers about 40 amperes to each condenser. Cooling water for the condensers is taken from the Long Beach ship channel at the rear of the plant by two circulating-water pumps, with a total capacity of 160,000 g.p.m., and is supplied to the condensers through pipes embedded in the concrete foundation and so arranged that the condensers may be backwashed to clean out any accumulation of growth and trash, and is discharged directly into the ocean through a 1300-foot circular concrete discharge tunnel 22 feet in diameter. Settling basins and trash racks to protect against sand and floating débris are provided at the channel end of two 1350-foot rectangular tunnels which convey the cooling water from the channel to the screen house and thence 60 feet further to the circulating-pump intake. Traveling screens collect the sea growth which develops in the intake tunnels, as well as other material which may come through, and prevents the fouling of pumps or condensers.

All the auxiliary equipment at Long Beach No. 3 steam plant is electrically driven because it has been found that this system adds to the efficiency and simplifies the operation of the plant. With the use of extraction steam for heating the feedwater, the auxiliary equipment has become of vital importance as well as more complicated. Station power circuits serving the many motors concerned consequently have become nearly as important as the main outgoing power circuits, and a flexible and reliable station power system has become an absolute necessity.

All of the pumps throughout the station are installed in duplicate motor-driven pairs to insure against service interruption from operating troubles in any one unit and to allow inspection and maintenance of one unit of each pair without interfering with plant operation. Electrical circuits supplying the station auxiliary equipment have been carefully laid out so that the two motors in each duplicate installation are supplied from separate circuits, and in addition several sources of power are provided for each of the auxiliary buses.

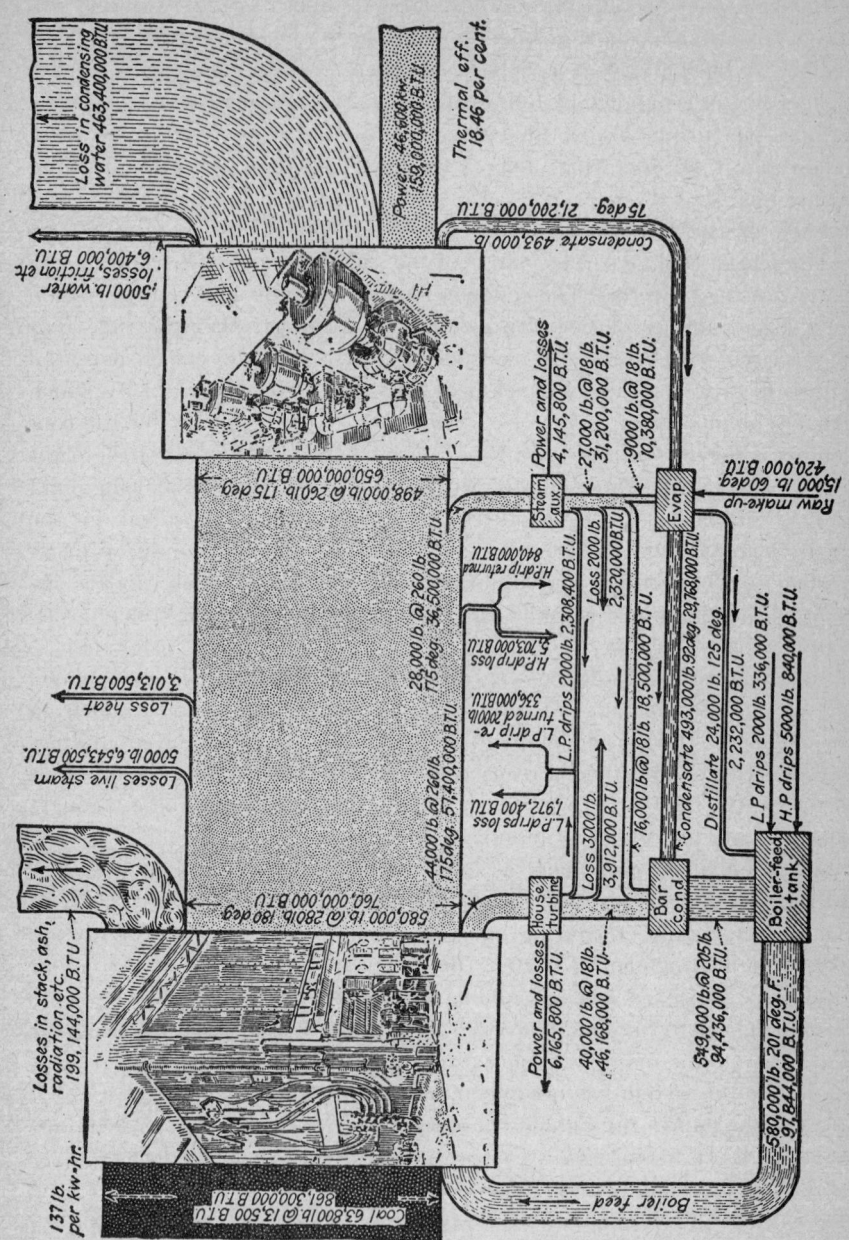

Fig. 515.—Energy Distribution in a Modern Central Station.

POWER PLANT BUILDINGS AND ARRANGEMENT OF EQUIPMENT 613

To expedite maintenance and to permit rapid changes in case of trouble, the 2300-volt station service switchboard is built up of truck-type oil circuit breaker panels, and is used for the control of the auxiliary generators and for all motors of 100 horsepower or more. A separate 440-volt bus serves the smaller units through a switchboard whereon carbon-circuit breakers have been installed to eliminate the complication and possible hazards of oil circuit breakers. All motors are controlled from push-button stations at the units and all of the motors are of the type that can be started at full line voltage. Constant-speed induction motors are used on all of the auxiliaries except the boiler-feed pumps and the draft fans where accurate and automatic speed control is desired. Slip-ring motors are used on the feed pumps and brush-shifting motors are used on the fans, the latter to provide a wider speed range.

No special test equipment has been included in the layout of the Long Beach No. 3 steam plant, but particular attention has been given to arrangement of the miscellaneous gage and small power-control boards located at many convenient points throughout the plant. There is a liberal use of graphic instruments on the various gage boards throughout the plant.

The type and size of equipment installed is tabulated in Table 39.

546. Energy Distribution in a Modern Central Station Plant. — An energy distribution chart, Fig. 515, of the Colfax Steam Power Plant, shows where the greatest losses occur when developing power with steam. A careful study of this chart shows that in the modern central station the efficiency of the plant as a whole is high.

547. Power Plant Buildings and Arrangement of Equipment. — The buildings used to house power plant equipment should be fire-proof, and should have a type of exterior and interior architecture adapted to the building and the purpose for which the building is to be used. Recently there has been a decided effort on the part of a few utility companies to build power plants in such a manner as to produce architectural beauty, both inside and out. This beauty is obtained by simplicity and proportion, and for that reason is not costly. This also applies to the landscaping and general physical appearance of the generating, transmission and distribution stations. There is a decided psychological effect of this beauty upon the personnel who are unconsciously affected by it, and put greater effort and pride in their work and it also aids in maintaining and cementing public relations and goodwill. The Richmond Station of the Philadelphia Electric Co., Fig. 516, which is situated on the Delaware River, is a good illustration of this trend. The building is architecturally patterned after the Graeco-Roman classical buildings as shown in the English Renaissance after the Napoleonic wars. In order to set off the huge mass of building to the best advantage, it is built so that it appears to stand on a terrace well above the river. At night it is flood-lighted, Fig. 517, which brings out the fine classic lines of the building and reveals details not seen in the daytime. Inside, the lofty turbine hall, Fig. 477, page 559, with its high vaulted roof, has a cathedral-like beauty; and the lighting and painting add to the beauty of the room without detracting in any way from the use to which any turbine hall is put.

Fig. 516. — Richmond Station, Philadelphia Electric Company.

By reference to the power plants which were described previously in this chapter, it will be seen that the machinery is symmetrically arranged which results in better appearance and simplifies piping connections. It also promotes convenience and accessibility. There should always be sufficient clearance around machines so that repairs can be easily made, with a crane provided to handle the equipment when making repairs.

Fig. 517. — Richmond Station at Night.

In some recent designs efforts have been made to reduce the vibration noise set up by the rotating machinery, from being transmitted through the building frame, by building the machine and building foundations as independent structures, the only contact between the two being through a mastic expansion joint. The operating room has been made of hollow construction, the center being filled with Sil-o-cel, and all windows are double. The walls are covered with acoustical felt, the floors with one-quarter inch linoleum, and the control panels constructed of asbestos ebony. This arrangement reduced the noise disturbances to a minimum.

In addition, ideal air and lighting conditions are maintained by indirect lighting, and by a forced ventilating and air filtering system.

The main boiler and generating equipment installed in the various types of power plants, is generally arranged **back to back, end to end,** or **multiple-story.**

In the *back to back arrangement* the boilers are so placed that their axes are parallel and in line with the axes of the engines or turbines, as shown in Figs. 1, 2, and 3 of Chapter I. In this case additional engine and boiler equipment may be added without the necessity of a change of the steam main. In some cases in using this arrangement the engine room floor is at a higher elevation than the boiler room floor.

In the *end to end arrangement* the axes of the boilers and engines are parallel but not in line. Such a plant arrangement is only used when the land available demands this type of plant because such a plant can only be enlarged by adding engines at one end and boilers at the other, and connecting these by a larger steam pipe.

The *multiple-story arrangement* is used where land has a high value, such as in large cities. By this arrangement the generating machinery is placed on the ground floor and the boilers are arranged on one or more floors above, with the coal bunkers above the upper boiler room floor. In some plants the boiler house is made multiple-story, with the one-story engine or turbine room alongside, and the machines arranged lengthwise of the building.

Power houses for manufacturing plants generally have a separate building with plenty of light and ventilation, and are usually located as near as possible to the center of distribution, and where possible near a railroad track or wharf for ease in handling coal and ashes. The power plants in office buildings are usually placed in the basement or sub-basement, and in many cases the space available for machinery is limited so that it is not possible to give as much room as is really advisable for the various apparatus. The ventilation of this type of a power plant is sometimes quite difficult.

548. Industrial and Miscellaneous Types of Power Plants. — Industrial power plant practice has, in the main, followed the example of central stations in the matter of higher boiler pressures, together with superheated steam, larger boilers, improved methods of firing, using modern type stokers, or unit pulverizers, modern furnace construction, higher boiler ratings, air preheaters, combustion control equipment, feedwater treatment, and operating methods; and in addition, more attention has been given to balancing power and process loads by using high back-pressure or extraction turbines. For example, the Anheuser Busch plant in St. Louis is equipped with 465 pound boilers supplying steam to non-condensing turbines which will generate all the power required while bleeding steam at 150 pounds to process through an accumulator and exhausting at 5 pounds pressure. A number of plants have gone to still higher pressures, and for some installations, pressures not far from the critical are being considered. One industrial plant is using 1800 pounds pressure which operates two vertical 6000-horsepower triple-expansion engines using steam at 1455 pounds pressure and 797 deg. fahr. These engines consist of two single-acting high-pressure cylinders, two single-acting intermediate cylinders and one double-acting low-pressure cylinder, with the high and intermediate cylinders arranged tandem compound. In some instances where large amounts of process steam are used, central stations are being built to supply both power and steam in large quantities; and in some cases the interchange of power between large industrial and central stations is being practiced. For instance, the Deepwater Station supplies the entire exhaust output of one of its high-pressure units to an evaporator which supplies process steam to the neighboring DuPont industrial plant.

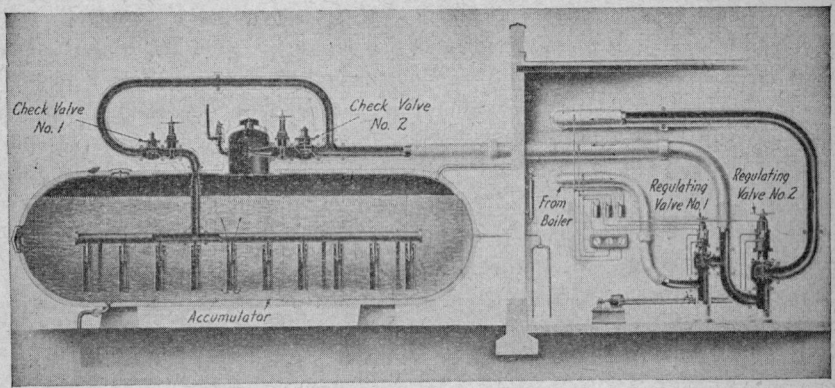

FIG. 518. — Ruths Steam Accumulator.

Since the year 1918 office and hotel buildings have generally depended upon outside sources for power, but recently there has been a re-examination of the economies obtainable through generating power where large amounts of steam are needed for heating and cooking, and as a result many public buildings are installing power plants. The power plant of the Hotel New Yorker, described in Volume 70, page 830, *Power*, is an example of present practice.

Since most industrial power plants operate at loads which fluctuate between rather wide limits, and since it is desirable that the boiler operate at as nearly constant rating as possible, increased attention has been devoted to the question of steam storage in order to smooth out the load curve and permit boiler operation at a constant load. The **Ruths Steam Accumulator** shown in Fig. 518 is often used for this purpose. It consists of a large storage tank called an "Accumulator," which is partly filled with water, suitable steam piping connecting the boiler, accumulator and load with the necessary control valves for automatic operation. *The accumulator works upon the principle that a portion of the water, confined under pressure and at saturation temperature in the accumulator, will be evaporated into steam when the pressure in the accumulator is reduced.* The pressure fluctuations which may vary between wide limits are confined to the accumulator, only, by using regulating valves to maintain constant pressures in the steam boiler and in the mains carrying steam for power and process work. These valves also provide for automatic charging and discharging of the accumulator, so that steam is stored during periods when demand for energy is lower than the supply, and discharged again at peak load periods. The steam line is connected directly to the boiler, and all steam in excess of that used by the high-pressure consumers is passed through the **automatic regulator valve** of the steam switchboard to the accumulator line. While valve No. 1 thus maintains a constant pressure in the high-pressure line, the regulator valve No. 2 works as the reducing valve maintaining a constant pressure in the low-pressure line supplying steam to low-pressure consumers. Should more steam be passed through valve No. 1 than valve No. 2, the excess will be passed through the accumulator pipe check-valve, distributing pipe and steam nozzles into the water space of the accumulator where it will be stored by heating the water content of the accumulator. If, on the other hand, the steam demand in the low-pressure line exceeds the supply of steam through valve No. 1, pressure in the accumulator line will drop, the check-valve No. 1 will close and the check-valve No. 2 will open, and as much of the boiling water in the accumulator will be flashed into steam as will supply the deficiency in the low-pressure line. Fluctuations in the supply of steam from the boilers, as well as fluctuations in the consumption of high- and low-pressure steam will thus automatically be transmitted to the accumulator. The accumulator is provided with circulating pipes around the steam nozzles and safety-steam nozzle, water level gauge, and valves for the adjusting of the water level in the accumulator. In some instances the accumulator has been used in connection with large central stations to take care of the peak load.

549. Large Steam Generating Units. — The steam generating unit, Fig. 519a, at the East River Station of the New York Edison Company has a maximum capacity of 800,000 pounds of steam per hour, and normally operates at 425 pounds gage with Elesco superheaters delivering steam at a total temperature of approximately 725 deg. fahr. The boiler is of the double-ended, bent-tube type, with four upper water and steam drums and four lower water drums for the complete unit. Steam flows from the two 54-inch diameter upper rear drums into two 48-inch diameter dry drums, and then passes through a header to the superheater. The boiler, together with the integral economizer, water-cooled furnace walls, and water-cooled slag screen, comprise a total area of 67,470 square feet of heat-absorbing surface; and is equipped with two plate-type air preheaters, each of which

618 MODERN POWER PLANTS

has a total area of 41,360 square feet. The gas temperature leaving the air preheater ranges from 365 deg. fahr. at 500,000 pounds per hour, up to approximately 400 deg. fahr. at 800,000 pounds per hour. The unit is fired by twenty vertical Lopulco burners, ten burners on each side of the furnace which has a furnace volume of approximately 39,000 cubic feet.

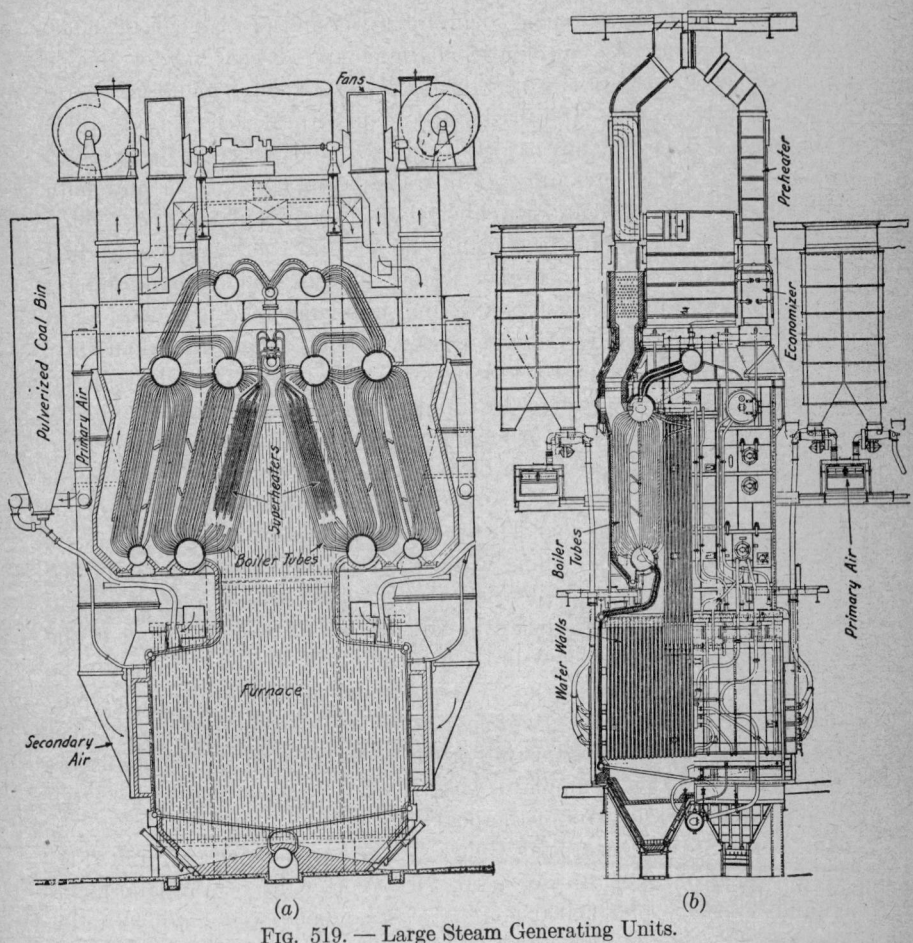

Fig. 519. — Large Steam Generating Units.

The boiler unit installed in the Kip's Bay plant of the New York Steam Corporation, and shown in Fig. 519b, consists of a boiler, economizer, and air preheater, having a total surface of 71,682 square feet, made up as follows: boiler, 10,670 feet; air heater, 46,800 feet; economizer, 10,080 feet, and water screen, 3632 feet. The steam is generated at 275 pounds and discharged through reducing valves to the distribution system at 150 pounds. Guaranteed efficiency of each unit is 88 per cent at 325,000 pounds of steam power.

LARGE STEAM GENERATING UNITS 619

The boiler is a six-drum Ladd boiler, including the water screen drum, with two main steaming sections placed vertically, and with a dry top drum connected by bent tubes to the two upper drums but not located in the path of the furnace gases. The lower drum forms the mud drum and is connected through the floor screen to the vertical front water walls of the furnace. The furnace is entirely water-cooled, with bare fin tubes which are extended up the two sides of the boiler to the center line of the upper steam and water drums. The steam generated in the floor screen, the front wall tubes and arch tubes, is carried to the upper steam and water drums of the boiler.

The furnace volume is approximately 19,000 cubic feet which will give a heat release of approximately 22,000 B.t.u. per cubic foot of volume. The combustion chamber, boiler, economizers, air heater, and induced-draft fans are directly in line vertically with

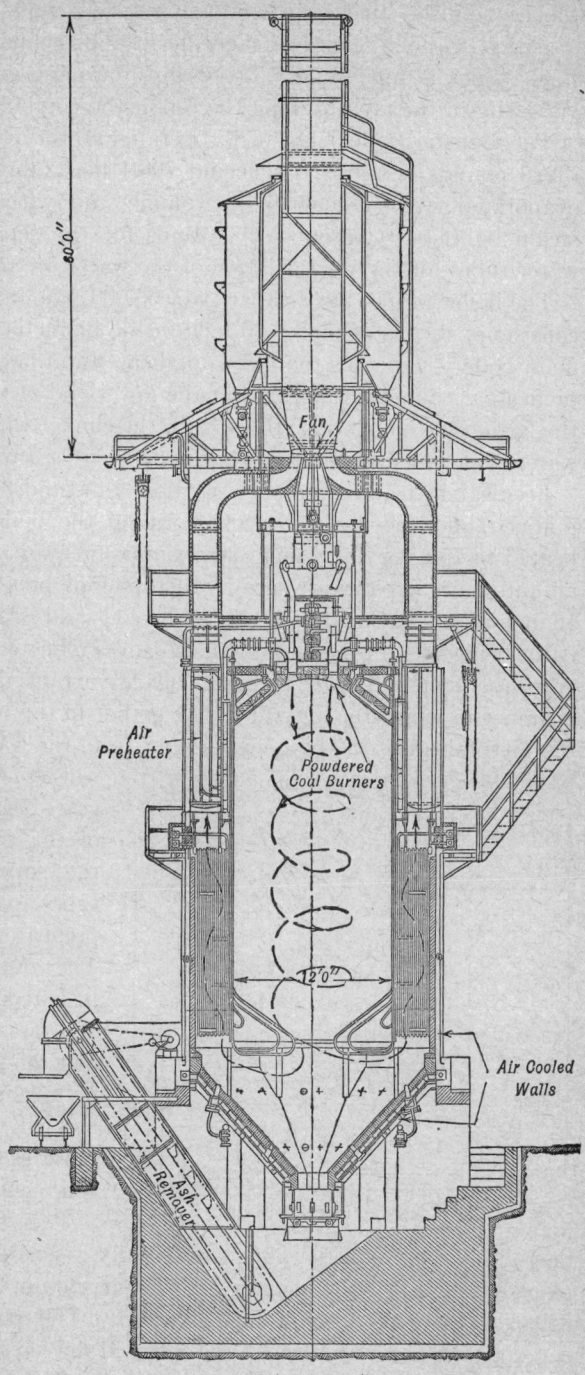

FIG. 520. — Benson Boiler.

the economizer, air heater, and fans in two units per boiler. The fans discharge through cinder catchers into the breeching leading to the stack. Each boiler is fired by 12 forced-draft burners, six on each side placed vertically in the arch forming the top of the combustion chambers.

The **Benson boiler**, Fig. 520, operates at the critical steam condition (3226 pounds per square inch and 706.1 deg. fahr.). The water is thus evaporated without change in volume and without the formation of steam bubbles. Consequently, drums for the separation of steam, and water tubes for the re-circulation of the water are not needed.

The boiler proper has a capacity of 60,000 pounds of steam per hour and consists of 8 coils of pipe 0.80 inch inside diameter. Four coils form the main boiler; two coils the first superheater and two coils form the second preheater. Some of the tubes at the lower end of the coil are bent out in the form of an ell in order to break up the clinker which then drops into the water pocket and is removed continuously by a screw conveyor.

Feedwater at approximately 212 deg. fahr. and 3500 pounds pressure is pumped into the first four coils forming the boiler proper. Here it is heated to 752 deg. fahr. and 3200 pounds pressure, and expanded through an automatic expansion valve to 1500 pounds pressure and 626 deg. fahr. It then passes back through two of the coils and is heated to 840 deg. fahr. After being used in the turbine the steam is reheated by returning it to the two superheater coils. The eight coils are arranged to form a combustion chamber in which the only fire brick is that in the top arch and in front of the air preheaters. The boiler is fired from the top, and the combustion products are swirled by air currents from the secondary air inlets. The gases pass down around the front of the tubes, then up through the superheater back of the tubes and finally through the air preheaters to the stack above. The outside of the boiler is jacketed by air which enters the air passages around the lower part of the boiler, then passes up around the outside, down through the air heaters, and then to the secondary and primary air combustion nozzles.

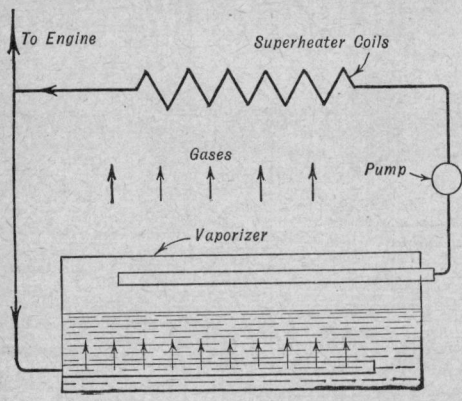

FIG. 521. — Loeffler Boiler.

The **Loeffler boiler**, shown diagrammatically in Fig. 521, is essentially a separately fired superheater, placed in series with a steam circulating pump, and an unfired drum, called a **vaporizer**, partly full of water. The pump forces bubbles of superheated steam up through the water in the vaporizer, where the superheat is absorbed in the generation of additional steam. This automatically increases the pressure until steam equal in amount to that generated is drawn off from the circuit for outside use. The economizer in the setting, with the superheater, preheats the feedwater going to the drum. This boiler is adapted for pressures in excess of 700 pounds and at 1850 pounds pressure the pump absorbs only about two per cent of the total steam generated in the boiler. It has been built to deliver 310,000 pounds of steam per hour at 200 pounds pressure and 930 deg. fahr.

550. Emmet Mercury-vapor Process. — In this process mercury is vaporized in many short dead-ended tubes, which project in the combustion space of the furnace and at temperatures that can be much higher than those which are practicable with steam. These tubes discharge into their respective drums, which in turn discharge the mercury vapor into a vapor dome. It is then carried through a mercury turbine, doing useful work, and is discharged into a surface condenser, where it gives up its latent heat to make steam at the pressure required for power use. The condensed mercury flows back to the boiler by gravity.

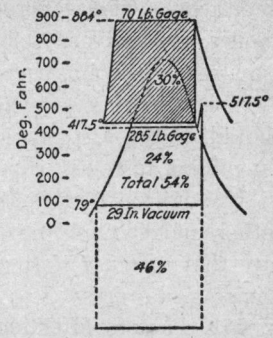

Fig. 522. — Mercury-steam Cycle.

Fig. 523. — Emmet Mercury-vapor Boiler Installation.

This combined process affords a means of increasing the temperature ranges with steam under conditions which afford large gains in efficiency. The theoretical possibilities of the mercury-steam cycle without reheating is indicated in Fig. 522, using mercury vaporized at 70 pounds gage, steam made at 285 pounds gage, superheated 100 deg. fahr., and expanded according to the Rankine cycle. The efficiency can be further increased by heating the feedwater, using extracted steam.

The present equipment, Fig. 523, installed at the Hartford Electric Light Company's plant operates at 70 pounds gage pressure on the mercury boiler, and delivers 10,000 kilowatts from the mercury turbine, and makes steam at 280 pounds pressure, with a temperature of 735 deg. fahr.

The furnace gases pass in order through the mercury boiler, mercury liquid heater, a steam superheater and a feedwater heater. The unit, as at present installed, can produce a kilowatt-hour on about 9140 B.t.u.

In addition to the use of mercury for this type of process, diphenyl ether and sulphur have been suggested. Present indications, however, are that mercury is the only liquid which is well adapted for the steam generation process.

551. Factors Affecting the Cost of Power. — The real function of the power station engineer is to deliver power on the station bus bars at the lowest possible cost per kilowatt-hour. This total cost is made up of **operating expenses** such as: *fuel, removal of ash and refuse, water, labor, oil, waste and other supplies, repairs and maintenance,* and **fixed charges** *consisting of interest on investment, depreciation, taxes, insurance, and management.* Each of these items, with the possible exception of depreciation, are fairly well understood. It is obvious with the interest and taxes paid in connection with an investment in power stations, and the money that must be set aside each year to provide for the cost of making renewals, to replace equipment wearing out, are as tangible elements entering into the costs of the kilowatt-hour as the money that must be paid for the operating expenses. **Depreciation,** or *loss in value, occurs because of the wear and tear, exposure to the elements, obsolescence and functional decay.* The amount charged off each year for depreciation depends upon the probable life of the equipment and the interest rate, and is ordinarily calculated by the methods explained in *Engineering Economics* by Professor J. C. L. Fish, and other similar books.

In the past ten years there has been a marked reduction in fuel costs per kilowatt-hour resulting from improvements in power plant design, and the decline in the price of coal. On the other hand, the fixed charges have rapidly mounted during the same period. The indications appear to be that in striving for lower fuel costs by using more efficient stations, the fixed charges per kilowatt-hour will rise still higher and the increase in fixed charges per kilowatt-hour will more than offset the decrease in the fuel cost per kilowatt-hour. It apparently would be more profitable and certainly more effective in the reduction of cost of power if concentration were made on the question of using the knowledge and equipment which are available. The problems which appear to offer a possible reduction in the total cost of power and steam are: choosing equipment that will be available for operation during the largest possible percentage of the hours in the year; reducing the amount invested in plant, since fixed charges in many cases exceed the fuel cost by at least 50 per cent; so locating the power station as to reduce the cost of construction, the cost of fuel, the cost of operating labor and the investment in the transmission system; and simplifying operation by having uniformity of design throughout a given station, and the concentration for the responsibility of operation in the hands of fewer men.

552. Elements in the Design of a Power Plant. — When considering the design of a power plant, it is the **function of the engineer** *to estimate the future needs of the plant; to study load curves and their tendencies; to study present practice as regards plant design, kinds, types, and size of equipment; to study available locations in the case of new plants, and finally to estimate the probable cost.*

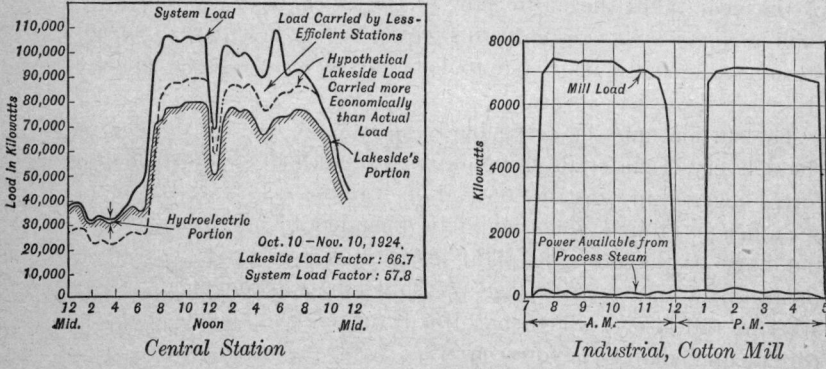

Fig. 524. — Load Curves.

The determination of the probable **load curve** is perhaps the most important factor in designing a power plant, since the size of the plant and the machines to be used are in a large measure determined by the kind of a load curve required by the plant. Load curves for a central station and for an industrial plant using process steam are shown in Fig. 524. These curves show the demand on the station for each hour in the day, and from them the **load factor;** *that is, the ratio of the average load to the maximum load during a certain period of time,* may be found. Special study should be given to the maximum daily load and minimum daily load, temporary peak loads, the probable average yearly load, and the possible future increase in load. **The station load factor** and the **yearly load factor** have considerable bearing on the cost of operation, and may be closely approximated from the daily load curves, since the area under any one of these curves represents the energy output for one day. The load factors vary from 25 to 50 per cent of the possible output of the station, and in exceptional cases may be as high as 90 per cent.

The general plan should provide for construction in units, small relative to the ultimate station capacity, with a minimum of idle investment. The

three objectives of *good operation* — *reliability, high efficiency and low first cost* — are interrelated in the sense that the application of a high standard of reliability cannot fail to raise the first cost, and in a lesser degree the operating cost, above the lowest possible value. *The most suitable plan, of course, all of the other requirements having been met, is that in which the sum of the investment and operating charges is a minimum.*

553. Power Plant Location. — The location of a power plant *is influenced* by the amount of condensing water available, railroad and barge facilities for coal deliveries, land values, land available for storage of coal, and the center of load distribution. Modern high-vacuum plants require large amounts of cooling water for condensing and is the most important limitation as regards locating plants at the mouth of the mine. The Springdale Plant of the West Penn Power Company, near Pittsburgh, at certain times of the year takes the entire flow of the Allegheny River for condensing cooling water. As a result of this requirement, the choice of sites for large stations is usually restricted to locations near an ocean, lake, or large river.

The station must be accessible by all the modes of traffic necessary for the delivery of materials and equipment, and for the transportation of the forces engaged in construction and operation.

Sufficient ground area should be included to accommodate all of the buildings in the ultimate plant as initially projected, with coal-storage facilities, and, if high-voltage transmission is planned, an outdoor substation, and also to allow space for future developments, many of which cannot immediately be foreseen.

Neighboring properties should introduce no conditions dangerous or injurious to equipment or men — for instance, fire hazard, emission of corrosive fumes or interference with natural lighting and ventilation.

Geographically, the station should stand in a well-balanced relation to the distribution system, both as it exists and as it is forecast to grow, so that excessive investment and operating charges shall not be incurred in the transmission of the delivered energy. Its position should also be favorable for interconnection with other stations, whether of the same system or of outside companies.

The present tendency is to concentrate power generation in large central stations, because of: (1) higher economy in large generating units; (2) saving in cost of supervision; (3) ease of transmitting power electrically from plant to point of use. These, together with the factors mentioned above, have led to the consideration of what is known as the " **Super Power System.**" This is a plan providing for the power needs for the district between Boston, Mass. and Washington, D. C. and from the Atlantic Ocean to, roughly, 150 miles inland by a system of large plants with inter-connected lines. Stations are to range from 60,000- to 300,000-kilowatt capacity with no generating unit smaller than 30,000 kilowatts.

MISCELLANEOUS TABLES

REFERENCES

Vol. 53, pages 532, 554, 806, and 841; Vol. 54, page 622; Vol. 55, pages 2, 604, 678, 716, 760, and 846; POWER.

Vol. 46, pages 397 and 447; POWER PLANT ENGINEERING.

Reports of Prime Movers Committee, NATIONAL ELECTRIC LIGHT ASSOCIATION.

REVIEW QUESTIONS

1. What is the present tendency in choosing modern power plant equipment?
2. Sketch the arrangement of the equipment in the Edgar Station.
3. Describe the method of handling coal at the Trenton Channel Plant of the Detroit Edison Co.
4. Draw a chart showing the distribution of the energy in a modern central station using coal to generate steam.
5. Describe the operation of the mercury-vapor steam cycle.
6. What changes are necessary in a coal-fired plant, to accommodate it to a pulverized-coal-burning plant?
7. Name a common method of supplying the circulating water to the condenser.
8. What is the purpose of a house turbine?
9. Describe the arrangement of the equipment in some power plant which you have visited.
10. Describe the construction of the Kip's Bay boiler.
11. Name some of the factors which determine the location of a power plant.
12. What items are included in the cost of power?
13. What is meant by the term **load factor** and what bearing does it have on the design of a power station?
14. Describe (a) the Loeffler boiler, (b) the Benson Boiler.

MISCELLANEOUS TABLES

TABLE 40. — AREA OF CIRCLES

Diam.	Area	Diam.	Area	Diam.	Area	Diam.	Area	Diam.	Area	Diam.	Area
$\tfrac{1}{8}$	0.0123	$4\tfrac{1}{2}$	15.904	$16\tfrac{1}{2}$	213.82	32	804.24	56	2463.0	80	5026.5
$\tfrac{1}{4}$	0.0491	5	19.635	17	226.98	33	855.30	57	2551.7	81	5153.0
$\tfrac{3}{8}$	0.1104	$5\tfrac{1}{2}$	23.758	$17\tfrac{1}{2}$	240.52	34	907.92	58	2642.0	82	5281.0
$\tfrac{1}{2}$	0.1963	6	28.274	18	254.46	35	962.11	59	2733.9	83	5410.6
$\tfrac{5}{8}$	0.3067	$6\tfrac{1}{2}$	33.183	$18\tfrac{1}{2}$	268.80	36	1017.8	60	2827.4	84	5541.7
$\tfrac{3}{4}$	0.4417	7	38.484	19	283.52	37	1075.2	61	2922.4	85	5674.5
$\tfrac{7}{8}$	0.6013	$7\tfrac{1}{2}$	44.178	$19\tfrac{1}{2}$	298.64	38	1134.1	62	3019.0	86	5808.8
1	0.7854	8	50.265	20	314.16	39	1194.5	63	3117.2	87	5944.6
$1\tfrac{1}{8}$	0.9940	$8\tfrac{1}{2}$	56.745	$20\tfrac{1}{2}$	330.06	40	1256.6	64	3216.9	88	6082.1
$1\tfrac{1}{4}$	1.227	9	63.617	21	346.36	41	1320.2	65	3318.3	89	6221.1
$1\tfrac{3}{8}$	1.484	$9\tfrac{1}{2}$	70.882	$21\tfrac{1}{2}$	363.05	42	1385.4	66	3421.2	90	6361.7
$1\tfrac{1}{2}$	1.767	10	78.54	22	380.13	43	1452.2	67	3525.6	91	6503.8
$1\tfrac{5}{8}$	2.073	$10\tfrac{1}{2}$	86.59	$22\tfrac{1}{2}$	397.60	44	1520.5	68	3631.6	92	6647.6
$1\tfrac{3}{4}$	2.405	11	95.03	23	415.47	45	1590.4	69	3739.2	93	6792.9
$1\tfrac{7}{8}$	2.761	$11\tfrac{1}{2}$	103.86	$23\tfrac{1}{2}$	433.73	46	1661.9	70	3848.4	94	6939.7
2	3.141	12	113.09	24	452.39	47	1734.9	71	3959.2	95	7088.2
$2\tfrac{1}{4}$	3.976	$12\tfrac{1}{2}$	122.71	$24\tfrac{1}{2}$	471.43	48	1809.5	72	4071.5	96	7238.2
$2\tfrac{1}{2}$	4.908	13	132.73	25	490.87	49	1885.7	73	4185.3	97	7389.8
$2\tfrac{3}{4}$	5.939	$13\tfrac{1}{2}$	143.13	26	530.93	50	1963.5	74	4300.8	98	7542.9
3	7.068	14	153.93	27	572.55	51	2042.8	75	4417.8	99	7697.7
$3\tfrac{1}{4}$	8.295	$14\tfrac{1}{2}$	165.13	28	615.75	52	2123.7	76	4536.4	100	7854.0
$3\tfrac{1}{2}$	9.621	15	176.71	29	660.52	53	2206.1	77	4656.0	101	8011.8
$3\tfrac{3}{4}$	11.044	$15\tfrac{1}{2}$	188.69	30	706.86	54	2290.2	78	4778.3	102	8171.3
4	12.566	16	201.06	31	754.76	55	2375.8	79	4901.6	103	8332.3

TABLE 41. — COMMON LOGARITHMS (Log_{10})

Nat. Nos.	0	1	2	3	4	5	6	7	8	9	Proportional Parts							
											1	2	3	4	5	6	7	8	9
10	0000	0043	0086	0128	0170	0212	0253	0294	0334	0374	4	8	12	17	21	25	29	33	37
11	0414	0453	0492	0531	0569	0607	0645	0682	0719	0755	4	8	11	15	19	23	26	30	34
12	0792	0828	0864	0899	0934	0969	1004	1038	1072	1106	3	7	10	14	17	21	24	28	31
13	1139	1173	1206	1239	1271	1303	1335	1367	1399	1430	3	6	10	13	16	19	23	26	29
14	1461	1492	1523	1553	1584	1614	1644	1673	1703	1732	3	6	9	12	15	18	21	24	27
15	1761	1790	1818	1847	1875	1903	1931	1959	1987	2014	3	6	8	11	14	17	20	22	25
16	2041	2068	2095	2122	2148	2175	2201	2227	2253	2279	3	5	8	11	13	16	18	21	24
17	2304	2330	2355	2380	2405	2430	2455	2480	2504	2529	2	5	7	10	12	15	17	20	22
18	2553	2577	2601	2625	2648	2672	2695	2718	2742	2765	2	5	7	9	12	14	16	19	21
19	2788	2810	2833	2856	2878	2900	2923	2945	2967	2989	2	4	7	9	11	13	16	18	20
20	3010	3032	3054	3075	3096	3118	3139	3160	3181	3201	2	4	6	8	11	13	15	17	19
21	3222	3243	3263	3284	3304	3324	3345	3365	3385	3404	2	4	6	8	10	12	14	16	18
22	3424	3444	3464	3483	3502	3522	3541	3560	3579	3598	2	4	6	8	10	12	14	15	17
23	3617	3636	3655	3674	3692	3711	3729	3747	3766	3784	2	4	6	7	9	11	13	15	17
24	3802	3820	3838	3856	3874	3892	3909	3927	3945	3962	2	4	5	7	9	11	12	14	16
25	3979	3997	4014	4031	4048	4065	4082	4099	4116	4133	2	3	5	7	9	10	12	14	15
26	4150	4166	4183	4200	4216	4232	4249	4265	4281	4298	2	3	5	7	8	10	11	13	15
27	4314	4330	4346	4362	4378	4393	4409	4425	4440	4456	2	3	5	6	8	9	11	13	14
28	4472	4487	4502	4518	4533	4548	4564	4579	4594	4609	2	3	5	6	8	9	11	12	14
29	4624	4639	4654	4669	4683	4698	4713	4728	4742	4757	1	3	4	6	7	9	10	12	13
30	4771	4786	4800	4814	4829	4843	4857	4871	4886	4900	1	3	4	6	7	9	10	11	13
31	4914	4928	4942	4955	4969	4983	4997	5011	5024	5038	1	3	4	6	7	8	10	11	12
32	5051	5065	5079	5092	5105	5119	5132	5145	5159	5172	1	3	4	5	7	8	9	11	12
33	5185	5198	5211	5224	5237	5250	5263	5276	5289	5302	1	3	4	5	6	8	9	10	12
34	5315	5328	5340	5353	5366	5378	5391	5403	5416	5428	1	3	4	5	6	8	9	10	11
35	5441	5453	5465	5478	5490	5502	5514	5527	5539	5551	1	2	4	5	6	7	9	10	11
36	5563	5575	5587	5599	5611	5623	5635	5647	5658	5670	1	2	4	5	6	7	8	10	11
37	5682	5694	5705	5717	5729	5740	5752	5763	5775	5786	1	2	3	5	6	7	8	9	10
38	5798	5809	5821	5832	5843	5855	5866	5877	5888	5899	1	2	3	5	6	7	8	9	10
39	5911	5922	5933	5944	5955	5966	5977	5988	5999	6010	1	2	3	4	5	7	8	9	10
40	6021	6031	6042	6053	6064	6075	6085	6096	6107	6117	1	2	3	4	5	6	8	9	10
41	6128	6138	6149	6160	6170	6180	6191	6201	6212	6222	1	2	3	4	5	6	7	8	9
42	6232	6243	6253	6263	6274	6284	6294	6304	6314	6325	1	2	3	4	5	6	7	8	9
43	6335	6345	6355	6365	6375	6385	6395	6405	6415	6425	1	2	3	4	5	6	7	8	9
44	6435	6444	6454	6464	6474	6484	6493	6503	6513	6522	1	2	3	4	5	6	7	8	9
45	6532	6542	6551	6561	6571	6580	6590	6599	6609	6618	1	2	3	4	5	6	7	8	9
46	6628	6637	6646	6656	6665	6675	6684	6693	6702	6712	1	2	3	4	5	6	7	7	8
47	6721	6730	6739	6749	6758	6767	6776	6785	6794	6803	1	2	3	4	5	5	6	7	8
48	6812	6821	6830	6839	6848	6857	6866	6875	6884	6893	1	2	3	4	4	5	6	7	8
49	6902	6911	6920	6928	6937	6946	6955	6964	6972	6981	1	2	3	4	4	5	6	7	8
50	6990	6998	7007	7016	7024	7033	7042	7050	7059	7067	1	2	3	3	4	5	6	7	8
51	7076	7084	7093	7101	7110	7118	7126	7135	7143	7152	1	2	3	3	4	5	6	7	8
52	7160	7168	7177	7185	7193	7202	7210	7218	7226	7235	1	2	2	3	4	5	6	7	7
53	7243	7251	7259	7267	7275	7284	7292	7300	7308	7316	1	2	2	3	4	5	6	6	7
54	7324	7332	7340	7348	7356	7364	7372	7380	7388	7396	1	2	2	3	4	5	6	6	7

$$e = 2.71828.$$

TABLE 41. — (Continued.) COMMON LOGARITHMS (Log_{10})

Nat. Nos.	0	1	2	3	4	5	6	7	8	9	Proportional Parts							
											1	2	3	4	5	6	7	8	9
55	7404	7412	7419	7427	7435	7443	7451	7459	7466	7474	1	2	2	3	4	5	5	6	7
56	7482	7490	7497	7505	7513	7520	7528	7536	7543	7551	1	2	2	3	4	5	6	6	7
57	7559	7566	7574	7582	7589	7597	7604	7612	7619	7627	1	2	2	3	4	5	5	6	7
58	7634	7642	7649	7657	7664	7672	7679	7686	7694	7701	1	1	2	3	4	4	5	6	7
59	7709	7716	7723	7731	7738	7745	7752	7760	7767	7774	1	1	2	3	4	4	5	6	7
60	7782	7789	7796	7803	7810	7818	7825	7832	7839	7846	1	1	2	3	4	4	5	6	6
61	7853	7860	7868	7875	7882	7889	7896	7903	7910	7917	1	1	2	3	4	4	5	6	6
62	7924	7931	7938	7945	7952	7959	7966	7973	7980	7987	1	1	2	3	3	4	5	6	6
63	7993	8000	8007	8014	8021	8028	8035	8041	8048	8055	1	1	2	3	3	4	5	5	6
64	8062	8069	8075	8082	8089	8096	8102	8109	8116	8122	1	1	2	3	3	4	5	5	6
65	8129	8136	8142	8149	8156	8162	8169	8176	8182	8189	1	1	2	3	3	4	5	5	6
66	8195	8202	8209	8215	8222	8228	8235	8241	8248	8254	1	1	2	3	3	4	5	5	6
67	8261	8267	8274	8280	8287	8293	8299	8306	8312	8319	1	1	2	3	3	4	5	5	6
68	8325	8331	8338	8344	8351	8357	8363	8370	8376	8382	1	1	2	3	3	4	4	5	6
69	8388	8395	8401	8407	8414	8420	8426	8432	8439	8445	1	1	2	2	3	4	4	5	6
70	8451	8457	8463	8470	8476	8482	8488	8494	8500	8506	1	1	2	2	3	4	4	5	6
71	8513	8519	8525	8531	8537	8543	8549	8555	8561	8567	1	1	2	2	3	4	4	5	5
72	8573	8579	8585	8591	8597	8603	8609	8615	8621	8627	1	1	2	2	3	4	4	5	5
73	8633	8639	8645	8651	8657	8663	8669	8675	8681	8686	1	1	2	2	3	4	4	5	5
74	8692	8698	8704	8710	8716	8722	8727	8733	8739	8745	1	1	2	2	3	4	4	5	5
75	8751	8756	8762	8768	8774	8779	8785	8791	8797	8802	1	1	2	2	3	3	4	5	5
76	8808	8814	8820	8825	8831	8837	8842	8848	8854	8859	1	1	2	2	3	3	4	5	5
77	8865	8871	8876	8882	8887	8893	8899	8904	8910	8915	1	1	2	2	3	3	4	4	5
78	8921	8927	8932	8938	8943	8949	8954	8960	8965	8971	1	1	2	2	3	3	4	4	5
79	8976	8982	8987	8993	8998	9004	9009	9015	9020	9025	1	1	2	2	3	3	4	4	5
80	9031	9036	9042	9047	9053	9058	9063	9069	9074	9079	1	1	2	2	3	3	4	4	5
81	9085	9090	9096	9101	9106	9112	9117	9122	9128	9133	1	1	2	2	3	3	4	4	5
82	9138	9143	9149	9154	9159	9165	9170	9175	9180	9186	1	1	2	2	3	3	4	4	5
83	9191	9196	9201	9206	9212	9217	9222	9227	9232	9238	1	1	2	2	3	3	4	4	5
84	9243	9248	9253	9258	9263	9269	9274	9279	9284	9289	1	1	2	2	3	3	4	4	5
85	9294	9299	9304	9309	9315	9320	9325	9330	9335	9340	1	1	2	2	3	3	4	4	5
86	9345	9350	9355	9360	9365	9370	9375	9380	9385	9390	1	1	2	2	3	3	4	4	5
87	9395	9400	9405	9410	9415	9420	9425	9430	9435	9440	0	1	1	2	2	3	3	4	4
88	9445	9450	9455	9460	9465	9469	9474	9479	9484	9489	0	1	1	2	2	3	3	4	4
89	9494	9499	9504	9509	9513	9518	9523	9528	9533	9538	0	1	1	2	2	3	3	4	4
90	9542	9547	9552	9557	9562	9566	9571	9576	9581	9586	0	1	1	2	2	3	3	4	4
91	9590	9595	9600	9605	9609	9614	9619	9624	9628	9633	0	1	1	2	2	3	3	4	4
92	9638	9643	9647	9652	9657	9661	9666	9671	9675	9680	0	1	1	2	2	3	3	4	4
93	9685	9689	9694	9699	9703	9708	9713	9717	9722	9727	0	1	1	2	2	3	3	4	4
94	9731	9736	9741	9745	9750	9754	9759	9763	9768	9773	0	1	1	2	2	3	3	4	4
95	9777	9782	9786	9791	9795	9800	9805	9809	9814	9818	0	1	1	2	2	3	3	4	4
96	9823	9827	9832	9836	9841	9845	9850	9854	9859	9863	0	1	1	2	2	3	3	4	4
97	9868	9872	9877	9881	9886	9890	9894	9899	9903	9908	0	1	1	2	2	3	3	4	4
98	9912	9917	9921	9926	9930	9934	9939	9943	9948	9952	0	1	1	2	2	3	3	4	4
99	9956	9961	9965	9969	9974	9978	9983	9987	9991	9996	0	1	1	2	2	3	3	3	4

Naperian $\log_e = 2.302 \log_{10}$.

TABLE 42. — DECIMAL EQUIVALENTS OF FRACTIONS OF ONE INCH

$\frac{1}{64}$	.015625	$\frac{17}{64}$	.285625	$\frac{33}{64}$	.515625	$\frac{49}{64}$	.765625
$\frac{1}{32}$	.03125	$\frac{9}{32}$	.28125	$\frac{17}{32}$	.53125	$\frac{25}{32}$	.78125
$\frac{3}{64}$	.046875	$\frac{19}{64}$	.296875	$\frac{35}{64}$	.546875	$\frac{51}{64}$	.796875
$\frac{1}{16}$	.0625	$\frac{5}{16}$	.3125	$\frac{9}{16}$	.5625	$\frac{13}{16}$	.8125
$\frac{5}{64}$	.078125	$\frac{21}{64}$	.328125	$\frac{37}{64}$	.578125	$\frac{53}{64}$	.828125
$\frac{3}{32}$	.09375	$\frac{11}{32}$	.34375	$\frac{19}{32}$	.59375	$\frac{27}{32}$	.84375
$\frac{7}{64}$	.109375	$\frac{23}{64}$	.359375	$\frac{39}{64}$	.609375	$\frac{55}{64}$	.859375
$\frac{1}{8}$	.125	$\frac{3}{8}$	.375	$\frac{5}{8}$	.625	$\frac{7}{8}$	.875
$\frac{9}{64}$	.140625	$\frac{25}{64}$	.390625	$\frac{41}{64}$	.640625	$\frac{57}{64}$	.890625
$\frac{5}{32}$	.15625	$\frac{13}{32}$	.40625	$\frac{21}{32}$	.65625	$\frac{29}{32}$	.90625
$\frac{11}{64}$	.171875	$\frac{27}{64}$	.421875	$\frac{43}{64}$	.671875	$\frac{59}{64}$	.921875
$\frac{3}{16}$	.1875	$\frac{7}{16}$	.4375	$\frac{11}{16}$	.6875	$\frac{15}{16}$	.9375
$\frac{13}{64}$	.203125	$\frac{29}{64}$	.453125	$\frac{45}{64}$	.703125	$\frac{61}{64}$	.953125
$\frac{7}{32}$	.21875	$\frac{15}{32}$	.46875	$\frac{23}{32}$	.71875	$\frac{31}{32}$	.96875
$\frac{15}{64}$	.234375	$\frac{31}{64}$	.484375	$\frac{47}{64}$	.734375	$\frac{63}{64}$	.984375
$\frac{1}{4}$	.25	$\frac{1}{2}$	.50	$\frac{3}{4}$	.75	1	1.

INDEX

Numbers refer to Pages.

A

Absolute, pressure, 78.
  temperature, 71.
  vacuum, 78.
  velocity, 506.
  zero, 77.
Acceleration, 65.
Accumulator, 556.
Adiabatic changes, 110, 436.
  expansion, 85.
Admission in steam engines, 358.
Ahead and astern turbine, 528.
Air, chamber on pumps, 539, 544.
  compressor, vertical, 13.
    use in power station, 590.
  cooled furnace arch, 197.
  density of, 81.
  heat lost by excess, 179.
  horsepower, 315.
  infiltration of, 176.
  motors, 204.
  preheaters, 8, 45, 263, 275.
    types, 276.
  primary, 165.
  properties of, 81.
  pumps (*see vacuum pumps*)
  required for combustion, 162–165.
    actual, 165.
    theoretical, 162.
  spaces in boiler settings, 26.
  spaces in grate bars, 182.
  specific heat of, 82, 84.
  steam and water cooled walls, 216.
  washer, 555.
  weight and volume, 164.
Alarm, high and low water, 54.
Alberger barometric condenser, 526.
Allen trick valve, 390.
American Standard pipe fittings, 252.
A.S.M.E., boiler code, 17.
  boiler rating, 220.
  test code, 227.
Ampere, 67.

Analyses, coal, 153:
  feedwater, 286.
  flue gas, 166.
  oil, 153.
Anderson steam trap, 247.
Anemometer, 230.
Anthracite, coal, 139.
  sizes, 140.
Angle and T-bracket, 47.
Angle of lap and lead, 368.
Angle of advance, 368.
Angle valve, 257.
Angularity of connecting rod, 367.
Apron coal feeder, 328.
Arches for boiler furnace, 204.
Arch plate, 195.
Area, unit of, 64.
Ash, bunker or bin, 7, 336.
  effect of, in coal, 143.
  handling methods, 332.
    drag chain, 335.
    hydraulic, 335.
    skip hoist, 334.
    steam jet, 334.
    V-bucket, 322.
  hoe, 184.
  pit, 9, 203.
  samples for boiler test, 232.
Atoms, 63.
Atomizer for cylinder oil, 456.
Atmospheric, relief valve, 7, 462, 577.
  heater, 264.
  line on indicator card, 361.
  pressure, 77.
Automatic, combustion control, 318.
  controlled valves, 259.
  flue gas recorders, 169.
  high-speed engine, 375.
  injector, 517.
  stop valve, 258.
Available, draft, 305.
  hydrogen, 137, 164.
Avogadro's Law, 84, 162.

B

Babbitt metal, 349.
Babcock's pipe equation, 242.
Babcock and Wilcox, superheater, 32.
  cross-drum boiler, 34.
  longitudinal drum boiler 33.
  marine, 52.
    water circulation in, 53.
Back connection for a boiler, 28.
Back pressure, 436.
Back-pressure valve, 462.
Badenhousen boiler, 41.
Baffle, box, 435.
  cross, 42.
  horizontal, 52.
  inclined, 185.
  plates, 36, 53.
  ring, 486.
  tile, Heine boiler, 38, 39.
    Stirling boiler, 42.
  vertical, 32.
  wall, Wickes boiler, 43.
Bagasse as a fuel, 113.
Bailey, combustion control, 319, **607.**
  burner, 213.
  feeder, 212.
Balanced draft, 306, 309.
Ball, pipe joint, 253.
Barometer, 77.
  aneroid, 77.
  mercurial, 77.
  standard type, **77.**
Barometric condenser, 561.
Beading tool, 22.
Bearer bars, 27, 29, 182, 183, 192, **197.**
Bearings, engine, 353.
  lubrication of, 452–459.
  pedestal leg for, 382.
  turbine, Curtis, 491.
Belpaire locomotive firebox, 46.
Benson boiler, 620.
Bigelow-Hornsby boiler, 45.
Bilgram valve diagram, 372.
Bin system, pulverized fuel, 144.
Bituminous coal, 141.
  coking and non-coking, 143.
  size of, 141.
Blast area, 315.
Blast-furnace gas, 153.
Bleeder or extraction turbine, 523.
Blow down, 56, 290.

Blow down, continuous, 291.
Blowers, Coppus, 308.
  soot, 278.
Blow-off, cock, 59.
  connections to boiler, 58–59.
  flange, 24, 58.
  pipe, 24.
  surface, 59.
  tank, 59.
  valve, 58.
Boiler, boilers, 17–52.
  accessories, 5, 53.
  and furnace heat balance, 176.
  A.S.M.E. code, 17.
  auxiliaries, 584.
  Babcock and Wilcox, 33.
  bent-tube, 41.
  brackets, 24.
  capacity, 220.
  classification, 18.
  combined, or overall, efficiency, **225.**
  compounds, 285.
  course, 18, 21.
  cross-drum, 34, 45.
  data chart, 233.
  dry back, 51.
  drum, 18.
  double-end, 51.
  economy, 224.
  efficiency, 225, 226.
    effect of capacity on, 227.
    method of computing, 225.
  feed pipe, 24.
  feedwater connections to, 58.
  fire-tube, 3, 20–31.
  fittings, 53.
  flange, 24.
  front, 29, 36.
  hangers, 26.
  heating, 17.
  heat balance for, **179.**
  heat losses, 176–179.
  horsepower, 220
  internally fired, 13, 49.
  large capacity units, 44, **617.**
  locomotive, 45.
  marine, 48–53.
  material, 17.
  New York, 22.
  nomenclature, 18.
  nozzle, 23.
  output, 220.

INDEX

Boiler, boilers, performance, 225.
  power, 17, 18.
  rating, 220-222.
  saddle, 51.
  scale, 280.
    effect of, 281.
  settings (*see settings*)
  set in battery, 27.
  shell, 18.
  stays (*see stays and braces*)
  suspension, 3-point, 26.
  testing (*see testing, boiler*)
  trends, 584.
  tubes, 22, 46.
    size of, 223.
  vertical, 29-32, 42.
  water-tube, 7, 32-45.
    classification, 32.
Boiling point, 91.
Booster engine, 13.
Bourdon pressure gage, 60.
Boyle's Law, 82.
Bradford coal breaker, 327, 592.
Brake, horsepower, 427.
  tare, 428.
Brakes, friction, 427, 470.
Breeching, 3, 316.
Brick chimneys, 299.
Bridge wall, 27, 36.
Briggs pipe thread, 248.
British thermal unit, 73.
  mean value of, 73.
Brown steam-flow meter, 467.
Bucket, conveyor, 322-323.
  trap, 246.
Buckets, steam turbine, 488-489.
  impulse, 480, 489, 500.
  reaction, 511, 522.
  theory, 505, 518.
  warped, 510.
Buckeye, engine valve, 393.
  governor, 380.
Buck bar, 29.
Bullhead tee, 252.
Bunsen burner, 155.
Burners, fuel oil, 207-212.
  gas, 206.
  oil and gas, combined, 210.
  powdered coal, 212.
Butt strap, 34, 38.
  weld, 254.

C

Calking, 21.
  tool, 21.
Calibration, gages, 75.
  thermometers, 70.
Calorimeter, steam, 102, 107.
  errors and limits, 105, 107.
  sampling nozzle, 103, 107.
  separating, 105.
  throttling, 102.
  types of, 102.
Calumet burner, 214.
Cameron simplex pump, 539.
  multi-stage, high-pressure pump, 548.
Capacity, boiler, 220.
  fan, 315.
  injector, 552.
  nominal, rated for boiler, 222.
  oil burners, 210.
  pump, 552.
Carbon dioxide, 160.
  monoxide, 160, 171.
Car dumper, 321.
Caustic embrittlement, 282.
Centennial boiler rating, 230.
Centigrade degrees, 69.
Central Station (*see power plants*)
  types, 2.
Centrifugal force, 64.
Centrifugal governor, 377.
  Armstrong, 377.
  Curtis, 492.
  Westinghouse, 513.
Centrifugal, oil purifier, 528.
  crank-pin oiler, 454.
  separator, 461.
Centrifugal pumps, 545-549.
  compound, 548.
  installation of, 554.
  performance of, 552.
  turbine, 548.
  volute, 547.
Charles' Law for gases, 82.
Check valve, 58, 256.
Chicago wing wall setting, 186.
Chain grate stoker, 3, 188.
  forced draft type, 191.
Chemicals for water softening, 284.
Chemistry of water softening, 288.
  reactions for combustion, 162.
Chimney, or stack, 299-306.
  cap, 300.

Chimney, comparison of types, 301.
  concrete, 301.
  costs, 303.
  determination of size, 303.
  empirical formulae for, 305.
  foundation, 300.
  lining, 301.
  Prat or Evasé, 307.
  radial brick, 299.
  stability of, 300.
  steel, 300.
    guyed and self supporting, 300–301.
  Thermix, 308.
Cinder catcher and trap, 333.
Circulation oiling system, 455.
Clean-out, door, 28, 36, 39.
  plugs, 46.
Clearance, cut-off valve, 393.
  method of finding, 471.
  volumetric of engine, 341.
Clear well, 287.
Clinker, 143.
  grinder, 197.
  methods of preventing, 143.
Closed feedwater heater, 266.
Coal, 132–144.
  air required to burn, 165, 172–174.
  analysis, 113–115.
    method of reporting, 134.
  anthracite, 139.
  ash in, 133, 143.
  bins or bunkers, 321.
  bituminous, 141.
  briquetted, 144.
  calorimeter, 135.
  cannel, 141.
  carbon content of, 139.
  chute, 7.
  classification of, 138.
  clinkering and non-clinkering, 143.
  coking and non-coking, 141.
  combustible constituents of, 160.
  compared with oil, 153.
  composition of, 133.
  conveyors (*see conveyors, coal*)
  crusher, types, 327
  effect of, moisture in, 143.
    wetting, 177.
  fields, 142.
  fired by hand, 184.
  firing methods, 184.
  fixed carbon in, 133.

Coal, formation of, 132.
  handling from ground storage, 328.
  high and low ash, 142.
  incombustible constituents of, 160.
  moisture in, 133.
  physical and burning characteristics, 139–141.
  pulverized, 144–150.
  purchase of, 142.
  rank of, 139.
  ratio of carbon to hydrogen in, 139.
  sampling during boiler test, 232.
  semi-bituminous, 141.
  storage and weathering, 142.
  sulphur in, 134.
  valves, 328.
  volatile matter in, 133.
Cochrane, feedwater heater, 264.
  hot process softener, 289.
  V-notch weir, 228.
Cock, 58.
Coefficient of expansion, 250.
Coil tube heater, 260.
Coke, 151–153.
Coking arch, 42, 186, 190, 204.
  plate, 195.
Colfax Station, 522, 613.
Colloidal state, 283.
Combined law of gases, 83.
Combustion, 158–166.
  actual air required for, 165.
  chemistry of, 160–164.
  control for fan, 316.
  indicator, 172.
  of pulverized coal, 160.
  progress in, 158.
  rates, 191, 195, 198, 202.
  requirements for perfect, 158.
  smokeless, 186.
  space, or chamber, 19, 49.
  stages for solid fuel, 159.
  theoretical air required for, 162.
Common grate bar, 183.
Compound engines, 443–451.
  economy of, 450.
  governing of, 450.
Compound steam gage, 463.
Compression in steam engines, 360.
Condensate, 2.
  well, 568.
Condenser, condensers, 558–571.
  air pumps for, (*see vacuum pumps*)

INDEX

Condenser, condensers, auxiliaries, 571.
  barometric, 561.
  circulating pump for, (*see pumps*)
  classification, 559.
  comparison of types, 569.
  cooling water for, 570.
  counter current, 560.
  ejector, 562.
  elementary theory of, 569–571.
  gain by using, 558, 559.
  heat transfer in, 568.
  jet, 560–564.
    high vacuum, 564.
    standard low level, 560.
  parallel current, 560.
  pressure in, 569.
  single pass, 10, 567.
  surface, 10, 565–568.
    down flow, 568.
    high vacuum, 565.
    method of support, 577.
    radial flow, 568.
    steam lanes in, 566.
    two-pass, 567.
    wet and dry tube, 565, 572.
  syphon, 562.
  vent, 265.
Conduction of heat, 74.
Concrete chimney, 301.
Connelly boiler, 41.
Conoidal fan, 311.
Constant pressure, specific heat, 82.
  expansion, 85.
Constant quality lines, 111.
  total heat lines, 111.
  volume specific heat, 82.
Convection of heat, 75.
Conveyors, coal, 322–326.
  belt, 325.
  buckets for, 324.
  classification, 322.
  flight, 7, 322.
  pivoted bucket, 7, 324.
  portable, 330.
  scraper, power drag, 329.
  screw, 559.
  skirt for, 325.
  telpherage, 328.
  V-bucket, 322.
Cooling tower, 580–582.
  capacity of, 582.
  chimney or closed, 581.

Cooling tower, comparison of types, 581.
  forced draft, 581.
  location of, 582.
  natural draft, 580.
  principle of operation, 582.
Cooling water by spraying, 578.
Copes feedwater regulator, 245.
Coping, 299.
Coppus blower, 308.
Corliss engines, 398–410.
  cylinder, 344.
Corliss valve, 399.
  laps and leads for, 408.
  setting, 407.
Corliss valve gears, 399–402, 404–407
  four-valve, 406.
  releasing, 400.
  non-releasing, 406.
Corrosion in boilers, 281.
  methods of prevention, 282–283.
Cost of chimneys, 303.
Cottrell precipitator, 334.
Counter weight, 247.
Covering strip, 21.
Critical pressure and temperature, 92.
Croll-Reynolds heater, 274.
Cross-box, 33.
  girder, B. and W. boiler, 36.
Cross-compound engine, 445.
Crown bar, 47,
Crown sheet, 29, 46, 49.
Crusher plate, 326.
Curtis, steam turbine, 487–500.
  bleeder, 525.
  emergency governor, 496.
  governor, 492.
  horizontal, 487.
  hydraulic valve gear for, 494.
  large capacity, 497.
  multi-cylinder, 497.
  poppet valve gear, 495.
  primary relay, 493.
  secondary relay, 495.
  throttle valve for, 497.
Cushion steam, 430.
Cut-off, 359.
  governing, 377.
  nominal, 473.
Custodias radial brick chimney, 300.
Cycle, definition of, 497.
Cycles, steam, 360.
Cyclone separator, 9, 145.

Cylinder, condensation, 438.
   feed, 430.
   for steam engine, 343–344.
   head, 345.
   oil, 457.
   ratio for compound engine, 447.

D

Dahl mechanical oil burner, 208.
Dalton's Law, 534.
Damper, 3, 29.
   frame, 29.
   regulator, 317.
Dashpot, 363.
   for Corliss valve gear, 402.
Deactivator, 282.
Dead center, for engine, 341.
   method of finding, 394.
Dead plate, 183, 193.
Deaërator, 282.
   heater, 10, 265, 282.
Deane Bros. duplex pump, 540.
Deconcentrator, 290.
Deflecting plate, 37, 44.
Degradation of heat, 283.
Degree, 69.
   of superheat, 101.
De Laval, steam turbine, 480–486.
   buckets, 484.
   diaphragm, 484.
   governor, 482.
   marine, 526.
   nozzles, 484.
   pressure stage, 484.
   range of horsepower, 484.
   speed, 480.
   velocity stage, 483.
   reduction gearing, 482, 531.
Delta, Greek letter, 364.
Density (*see material concerned*)
   of common substances, 64.
   specific, 81.
Depreciation, 622.
Desuperheater, 124.
   types, 126.
Diagram factor, 433.
   for compound engine, 450.
Diaphragms, 488.
Diffusing ring, 548.
Discharger, 123.
Discharge, tunnel, 10.

Disengaging surface, boiler, 9.
Disk water meter, 374.
Displacement, piston, 341.
   of slide valve, 363.
   relative, of valve and piston, 366.
Dissociate, 177.
Dome of condenser, 567.
Double-flow steam turbine, 509.
   -ported valves, 371.
Downcomer tubes, 44.
Downtake tubes, 53.
Downdraft furnace, 186.
Draft, available, 305.
   balanced, 309.
   chimney, 299–305.
   definition of, 297.
   fans, 310.
   forced, 9, 306.
   gages, 76, 297.
   induced, 9, 306.
   kinds of, 299.
   loss in boilers, 298.
      chimneys, 305.
   mechanical, 306.
   method of expressing, 297.
   natural, 299.
   Prat system, 307.
   regulator, 310.
   required for various coals, 298.
   with closed ashpit and boiler room, 309.
Drips, high pressure, 246.
Driving wheels, locomotive, 15.
Dry-air pumps, 571–576.
Dry-pipe in boiler, 35, 42.
Dry-back Scotch boiler, 51.
Dulong's equation, 137.
Dummy piston on turbine, 507.
Duplex, steam pump, 540.
   stoker, 203.
Dusting door, boiler setting, 36, 53.
Dutch oven furnace, 185, 195.
Duty of a pump, 553.
Dynamic pressure fan, 312, 314.
Dynamometer, 470.
Dyson express marine boiler, 49.

E

East River, steam unit, 617.
Eccentric, or sheave, 357.
   rod, crossed and open, 383.
   shifting, 377.

Eccentric, or sheave, strap, 358.
Eccentricity, 310.
Economy, boiler, 224.
  engine, 434.
  turbine, 531.
Economizer, 9, 263, 269.
  Foster, steel tube, 272.
  Green, 269.
  integral type, 271.
  pressure drop through, 298.
  unit type, 272.
Edwards air pump, 571.
Electrical units, 67.
Electrons, 63.
Elesco, superheater, 122, 123.
Efficiency (*see name of apparatus in question*)
Ejector condenser, 562.
Electric motors, 530.
Electrolytic action, 281.
Elementary steam power plants, 1–16.
  central station, 9.
  comparison of condensing and non-condensing, 7.
  condensing, 7.
  locomotive, 11.
  marine, 15.
  non-condensing, 3.
Elements of P. P. design, 623.
Elevator for coal, 322.
Elliott-Erhart jet condenser, 564.
Emmet mercury-vapor process, 621.
Energy, definition of, 65.
  distribution in a central station, 613.
  electrical, 67.
  internal, or intrinsic, 68, 85–87.
  kinetic, 65.
  mechanical, 65.
  potential, 65.
Engine, engines, steam, 337–476.
  accessories, 459–464.
  automatic high speed, 375.
  bearings, 353.
  bore and counterbore, 343.
  classification, 337.
    of compound, 443.
  clearance, 341.
  compound, multi-expansion, 443–451.
  constant, 425.
  condensing, economy of, 450.
  connecting rod, 350.
  Corliss, 344, 398.

Engine, engines, crank, 352.
  cross-compound, 445.
  crosshead, 349.
  cylinder, 343.
    head, 345.
  double and single acting, 361.
  economy of compound, 450.
  effect of leakage and clearance on, 438.
  effect of incomplete expansion, friction and moisture on, 439.
  efficiency, 435.
    ratio, 435.
  flywheel, 7, 354.
  fore and aft, 444.
  foundations, 355.
  friction, 452.
  frames, 342.
  function of parts, 339.
  governor, 7, 361, 377, 403, 411, 450.
  heat balance, 473.
  heat rate of, 434.
  heat loss in, 437.
  indicator, 417.
  junk ring for, 348.
  leads, piping, 240.
  Lentz, poppet valve, 410.
  lubrication of, 452–459.
  marine, 16, 387.
  mechanical efficiency of, 429.
  method of stating size, simple, 340.
    compound, 443.
  method of improving economy in, 440.
  multi-valve, 398–416.
  nomenclature, 340.
  operation "*D*" slide valve, 358.
  packing, 345.
  parts, 338.
  performance of, 433.
  piston, 347.
  piston, displacement for, 341.
    ring, 347.
    rod, 349.
  Rankine efficiency, 435.
  rating, 433.
  reciprocating, 337.
  speed, 341.
  stuffing box, 345.
  thermal efficiency, 435.
  tandem-compound, 444.
  testing, 465.
  unaflow, 7, 412–415.
  valves, balanced, 375.

Engine, valves, Corliss, 399.
    double-ported, 376, 390.
    "D" slide, 357–374.
    vertical, small, 390.
    water, or steam, rate for, 434.
    Woolf compound, 444.
Entropy, 107.
Enthalpy, 98.
Equivalent evaporation, 224.
Equilibrium pipe, turbine, 507.
Erie City boiler, 41.
Evaporator, 291.
    arrangement of, 293.
    condenser, 10, 292.
    single and multiple effect, 293.
    systems, 292.
    types, 292.
Evaporation, factor of, 225.
    latent heat of, 91, 98.
Evasé stack, 308.
Excess air, 165.
Exhaust, clearance, 364.
    head, 5, 462.
    piping in power plants, 242.
Exhaust steam, 2.
    heat loss in, 439.
Exhauster, 9.
Expansion, in engine, 359, 361.
    joints, 249.
    of piping, 250.
    polytropic, 85.
Expansion curve for engine, actual, 432.
    theoretical, 432.
Explosion diaphragm, 459.
Extra-strong pipe, 248.
Extraction heater, 274.
Extraction valves, turbine, 525.

F

Factor of evaporation, 225.
Fahrenheit degrees, 69.
Fantail burner, 24.
Fan, blast area of, 315.
    characteristics, 315.
    conoidal, 311.
    draft, 310.
    forced draft, 9.
    mechanical efficiency of, 315.
    multi-vane, 312.
    parts, 311.
    performance, 315.
    radial flow, 311.

Fan, single and double inlet, 311.
    static efficiency of, 315.
    steel plate, 311.
    theory, 313.
Feedpipe for boilers, 24.
    arrangements, 243.
Feedwater, boiler, 280.
    analyses, 285.
    boiler compounds for, 285.
    distillation of, 291.
    effect of heat upon, 286.
    hardness, how measured, 90.
        temporary and permanent, 90.
    heaters, 263–269.
        classification of, 263.
        closed, vertical, 7, 266.
        coil tube, 267.
        comparison of open and closed, 268.
        extraction, 274.
        Hoppes horizontal, 265.
        induced, 264.
        jet type, 265.
        live steam, 268.
        open, vertical, 264.
        primary, 263.
        secondary, 264.
        through, 264.
    impurities in, 90, 280.
    log for test, 231.
    piping, 35, 243.
    regulators, 244.
        Bailey, 244.
        Copes, 245.
    saving by heating, 268.
    softeners and purifiers, 284–293.
    treatment, 280.
    weighing on test, 228.
Filters, oil, 461.
Firebox, 19.
    for locomotives, 46.
    on vertical boilers, 29.
Firebrick, 27.
Firedoor arches, 28.
Firedoor liner, 29.
Fire, thickness of, 191, 195, 198, 199.
Fire-tube cleaner, 295.
Firing methods, hand, 184.
Fittings, screwed pipe, 252.
Fixed carbon in coal, 160.
Fixed charges, 622.
Flanged pipe fittings, 243.
Flanges, boiler, 24.

Flanges, method of facing, 254.
  pipe, 253.
Flap coal valve, 328.
Flight coal conveyor, 323.
Float trap, 246.
  valve, 10.
Flue gas, 165-172.
  analysis of, 166.
  apparatus to find, 167.
    Orsat, 167.
  constituents of, 166.
  deductions from analysis, 170.
  recorders, 169.
    Brown, 170.
    Hays, 169.
  sampling, 166.
  tube, 167.
  weight per lb. coal, 174.
Flues, locomotive boiler, 47.
Flush front boiler setting, 25.
Flywheel (see engine)
Foaming, 289.
Foot valve for pump, 555.
Foot-pound-second system, 63.
Force, 64.
  centrifugal, 64.
Forced draft, 9, 16, 306, 307.
  Howden system, 309.
Forced feed lubrication, 492.
Foster, automatic non-return valve, 258.
  attached superheater, 121.
Foundation, engine, 355.
  chimney, 299.
Four-valve engines, 406.
Fraction cold, 216.
Free burning coal, 140.
Free hydrogen, 137.
Friction, head, 552.
  horsepower, 428.
  in engines, 452.
  loss in flues, 297.
Fuel, fuels, 132-143.
  bed zones, 160, 203.
  calorimeter, 135, 155.
    gas, 155.
    Mahler bomb, 135.
  character for boiler test, 228.
  classification of, 138.
  definition of, 135.
  economizer (see economizer)
  gaseous, 153.
  heat value (see heat)

Fuel, fuels, high and low heat value, 136.
  liquid, 152.
Fuel oil, 152.
  advantages of, 153.
  analyses, 153.
  burners, 210.
  methods of burning, 207.
Furnace, air cooled, 204.
  corrugated, 49.
  definition of, 19.
  design to prevent smoke, 185.
  externally fired, 27, 49.
  extended type, 203.
  flush-front, 25.
  for gas fuel, 207.
  for oil fuel, 211.
  for pulverized fuel, 215.
  hand fired, 182.
  internally fired, 30.
  pressure regulator for, 310.
  refractories, for, 216.
  steam cooled, 204.
  stoker fired, 192.
  tools, 184.
  volume, 204, 215.
  water cooled, 204.
Fusible plug, 29, 48, 55.
  location of, 55.
Fusibility of ash, 143.

G

Gage, Bourdon pressure, 60.
  calibration of, 75.
  cocks, 19, 54.
    on vertical boiler, 31.
  compound, 463.
  dial for, 61.
  diaphragm, 61.
  double spring, 61.
  draft, 77.
  glass, 19, 32.
  recording, 62.
  standard test, 75.
  steam, 5, 60.
  testers, 75.
  vacuum, 463.
Galvanometer, 71.
Garter spring, 492.
Gas analysis, 166.
  deductions from, 170-172.
Gas, blast furnace, 153.

Gas, by-product coke oven, 153.
  calorimeter, 155.
  casing head, 153.
  expansions, 85.
  fuels, 153–155.
  furnaces for, 207.
  illuminating, 154.
  laws, 82–83.
  method of burning, 206.
  natural, 153.
  producer, 154.
  scrubbers, 334.
Gate valves, 256.
  rising and non-rising stem, 256.
Gaskets, boiler, 24.
  engine, 339.
  pipes, 255.
Gearing, helical, 482.
  reduction, 482, 530.
  spur, 326.
Generator, electric, 5, 7.
  turbine driven, 13.
  turbo-, 16, 53.
Gibson method of measuring water, 555.
Globe valves, 58, 257.
Goulds triplex pump, 544.
Governor, automatic, 7, 377.
  Corliss engine, 403.
  "D" slide valve, 361.
  hunting in, 363.
  inertia, 380.
  Lentz engine, 411.
  pump, 544.
  sensitiveness, 363.
  shaft, 377.
  stability, 362.
  throttling, 361.
  turbine, 482, 486, 492, 513.
Graphite in boilers, 285.
Grate-bar, 183.
  Herringbone, 183.
  Tupper, 183.
  rocking, 192.
Grates, 19, 182–183.
  circular, 182.
  definition, 19.
  rocking, 182.
  shaking, 183.
  stationary, 3, 182.
  traveling, 188.
Grate surface, 19, 223.
  method of expressing, 19.

Grease cups, 455.
Green, chain grate, 188.
  fuel economizer, 269.
Grid-iron valve, 391.
Guard plate, 194.
Gudgeon pin, 351.
Guide vanes, 486.
Gutermuth air-pump valve, 573.
Guyed steel stack, 301.

H

Hammel oil burner, 208.
Hammer coal pulverizing mill, 147.
Hancock injector, 551.
Handhole, opening, 22, 29, 34, 46.
  cap, 38.
Hand-fired furnace, 185.
Hand-operated stokers, 185.
Hand revolution counter, 468.
Hand stoking, 182.
  tools, 184.
Hardness of water, 90.
Hardinge ball mill, 150.
Harmonic motion, 366.
Harrington stoker, 191.
Hays boiler efficiency meter, 172.
Head, exhaust, 5, 462.
  on pump, 552.
  on weir, 466.
  pressure, 78.
  velocity, 78.
Headers, boiler, 33, 34, 52.
  box type, 37.
  side, 52.
  size of pipe, 241.
Heat, 67.
  content of steam, 98.
  definition of, 67.
  external latent, 98.
  internal latent, 98.
  kinds of, 67.
  latent, 67.
  liberated per lb. carbon, 163.
  loss in chimney gases, 176.
  loss in exhaust steam, 439.
  methods of expressing heat value, 137.
  of evaporation, 98.
  of liquid, 89, 91, 97.
  per lb. of dry steam, 98.
  per lb. of superheated steam, 101.

INDEX

639

Heat, per lb. of wet steam, 100.
  quantity of, 73.
  rate of engines, 402.
  release, 204, 216.
  sensible, 67.
  specific, 74.
  total, 68, 98, 100.
  transmission, 74.
  value of solid fuel, 135, 137, 138.
    of fuel oil, 153.
    of gas, 154.
Heat balance, boiler, 176.
  engine, 473.
Hearth screens, 599.
Heat cycle, 10, 128.
Heater, feedwater (*see feedwater heater*)
Heating, boiler, 17.
  feedwater, 286.
Heating surface, definition of, 20.
  for Manning boiler, 31.
  method of computing, 222.
  for H.R.T. boiler, 222.
  for Heine boiler, 223.
  steam generator, 223.
  water and superheating, 20.
Height of chimneys, 303–305.
Heine, boiler, 37, 41.
Helical, gear, 482.
  spring, 481.
Herringbone grate bar, 183.
High and low water alarm, 54.
  heat values of coals, 137.
High-pressure drips, 246.
High-pressure, turbine, 499, 520.
  pump, 548.
High-speed engine, 341, 375.
High-vacuum condensers, 564, 565.
Hook gage, 466.
Horizontal return tubular boiler, 3, 20.
Hoppers, ash, 337.
  coal weighing, 328.
  track, 7.
Hot well, 7.
Horsepower, definition of, 67.
  boiler, 220–221.
  hour, 67.
  of engine, 424–429.
  of pumps, 552.
  rated for compound engine, 449.
House turbine, 590.
Hunting in a governor, 363.
Hydrogen, available, 137, 164.

Hydrostatic lubricator, 457.
Hyperbolic curve, 431.

I

Ignition, spontaneous, 142.
  temperatures, 159.
Impeller, for centrifugal pump, 10, 545.
  open and closed, 545–546.
  single and double suction, 546.
  type oil governor, 513.
Impulse force, 505.
Impulse turbines, 478–507.
  and reaction, 508.
  pressure stage, 479, 484.
  velocity stage, 479, 483.
Incomplete combustion, 163, 178.
Indicator, for steam engine, 417–421.
  classification of, 417.
  cocks, 421.
  Crosby, continuous, 421.
    inside and outside, 418–420.
  springs, 419.
Indicator diagram, 361.
  accuracy of, 423.
  Corliss, 404.
  for dry-vacuum pump, 574.
  method of taking, 423.
    for shaft governor, 379.
  steam estimated from, 394.
  theoretical, 432.
Indicated horsepower, 424.
  for compound engine, 448.
Induced draft, 16, 306.
Industrial power plants, 615.
Inertia, 63.
  governor for engine, 377.
Ingersoll-Rand counter-current condenser, 561.
Initial condensation, 412.
  pressure, 532.
Injector, 13, 550–552.
  automatic, 550.
  capacity of, 552.
  double-tube, 551.
Inlet tunnel, 10.
Inside packed plunger pump, 543.
Interpolation from steam table, 99.
Internally-fired boiler, 30, 45.
Isolated power plant, 3, 7.
Isothermal expansion, 85.
  heat change, 82.

J

Jacket, steam, 344.
Jacobs-Shupert firebox, 46.
Jahn's turbine governor, 486.
Jet condensers, 560–564.
Joints, boiler, 21.
    lap and butt, 21.
    longitudinal, 21.
    ring, 21.
Joints, pipe, 252–255.
    phonographic, 255.
    serrated, 255.
Jones underfeed stoker, 198.
Joule, definition of, 66.
Joule's equivalent, 74.
Journal box, 354.
Junker gas calorimeter, 155.

K

Keenan's steam table, 94, 118.
Kent's equation, 306.
Kerosene in boilers, 285.
Kidwell boiler, 41.
Kieley, reducing valve, 259.
Kilgour boiler setting, 186.
Kilowatt, 67.
    horsepower equivalent, 74.
    hour, 67.
Kips Bay boiler, 618.
Kindling temperatures, 159.
Kingsbury thrust bearing, 512.
Knowles simplex pump, 537.
Koerting ejector condenser, 562.
    oil furnace, 211.
Kohlenscheidungs Gesellschaft carbonizing process, 152.

L

Ladd boiler, 619.
Lakeside Station, 532.
Lap, and lead, lever, 384.
    effect of, 365.
    on steam engines, 363.
    steam and exhaust, 363.
Lay shaft, 411.
Lashing wire for turbine blades, 511.
Latent heat, 67.
    of evaporation, 91, 98.
    internal and external, 98.

Lazy bar, 184.
Lead for engines, 364.
Leads, boiler and engine, 241.
Leblanc air pump, 575.
Length, unit of, 64.
Leveling bottle, 168.
Lignite, 141.
Lignum vitae, 513.
Lime, 287.
Load factor, 623.
Loaded governor, 404.
Location of power plants, 624.
Locomotive, 11.
    boiler, 45.
    classification, 15.
    crane, 329.
    crank, 353.
    engine, 381–382.
    frame, 381.
    running gear, 382.
    stoker, 205.
    superheater, 123.
    valve gear, 382.
Loeffler boiler, 620.
Logarithms (*Table*), 626.
Long Beach station, 605.
    heat balance, 275.
Loop header pipe system, 239.
Loss of pressure in steam pipe, 242.
Losses, engine, 437.
    turbine, 531.
Louvres, 580.
Low-load factor, 185
Low-pressure turbine, 523.
Low-speed engine, 341.
Low-temperature carbonization, 151.
    kind of processes, 151.
    retorts, 151.
Lubricants, 452.
Lubricator, cylinder, 457–458.
    hydrostatic, 457.
    mechanically operated, 458.
Ljungström, air preheater, 277.

M

Magnesia, 14.
Mahler-bomb calorimeter, 135.
Make-up water, 610.
Manhole, 22, 34.
    ring, 24.
Manometer, 75.

INDEX

Manning fire-tube boiler, 29.
Manufacturers rating of a boiler, 222.
Marine, boilers, 48–53.
    dry back, 51.
    engines, 387.
        crank shaft for, 353.
        turning gear for, 387.
    turbines, 529.
Matter, 63.
    law of the conservation of, 63.
    measurement of, 63.
    units of, 63.
Mass, 63, 65.
Maximum port-opening, 364.
Mean effective pressure, 424.
    theoretical and probable, 432–433.
Measuring burette, 170.
Mechanical, boiler tube cleaner, 293.
    draft, 306.
    efficiency of engine, 429.
    fan, 315.
    losses, in turbine, 531.
    oil burner, 207.
    reduction gearing, 16, 530.
    stokers, 187–204.
    stoking, 182.
Mercury-vapor cycle, 584, 621.
Metallic pyrometer, 71.
Metropolitan injector, 550.
Meyer valve for steam engine, 392.
Mid-position of valve, 363.
Mixed pressure turbine, 525.
Moisture, in air, 178.
    in coal, 143.
    in steam, 100, 102.
Mol, 84, 162.
Molecular form, 162.
Molecule, 63.
Mollier diagram, 112.
Motion, uniform and non-uniform, 64.
Motor, electric, 16.
    brush-shifting, 593.
    slip-ring, 593.
Motor-driven auxiliaries, 590.
Mud drum, 29, 35.
    ring, 46.
Mullan vacuum pump, 487, 572.
Multi-cylinder turbines, 497, 521.
Multi-jet condenser, 562.
Multi-stage centrifugal pump, 547, 548.
Murphy side-overfeed stoker, 195.
Myriawatt, 220.

N

Napier's equation, 106.
Napier's law, 106.
Narrow locomotive firebox, 46.
Natural gas, 153.
Navy standard boiler compound, 285.
Non-pressure oiling system, 456.
Non-return traps, 247.
    valve, 258.
Nordberg unaflow engine, 415.
Nozzle loss in steam turbine, 504.
Nozzle, turbine, 481, 503.
    bowl, mouth and throat of, 503–504.
    plate, 490.
    theory of impulse, 503.

O

Oil, burning installation, 211, 605.
    characteristics for a lubricant, 452.
    filter, 461.
    furnace, 32, 210.
    in suspension and emulsion, 459.
    kind of burner, 207.
    kinds of, 152.
    method of burning, 207.
    pump, 456.
    separator, 5, 461.
    wiper, 455.
Oilers, types, 453–454.
Oil burners, 207–211.
    capacity of, 210.
    mechanical, 209.
    steam spray, 207.
Ogee ring, 29.
Open feedwater heater, 264.
Operating expenses, 622.
Optical pyrometer, 71.
Orsat gas apparatus, 167–169.
Outside packed plunger pump, 543.
Overhung-front boiler setting, 29.
Overtravel of a valve, 364.
Oxygen, in air, 81.
    effect of, in coal, 144.
    in flue gas, 172.

P

Packing, carbon, 484, 492.
    fibrous for engine, 345.
    labyrinth, 492.

INDEX

Packing, metallic, 346.
   water seal, 511.
Pantograph, 419.
Parallel-current condenser, 560.
Parsons turbine, 507–517.
Passes in a boiler setting, 36.
Peabody calorimeter, 102.
Peabody, oil burner, 208.
   combined oil and gas burner, 209.
   Fisher oil burner, 210.
Peat, 141.
Perfect gases, laws of, 83.
Permanent hardness, 90.
p**H**, 282.
Phonographic joint, 254.
Physical units, 63–78.
Piezometer, 314.
Pillow block, 354.
Pinch cocks, 169.
Pipe, bends, 249.
   columns, 260.
   covering, 260.
   creased, 250.
   fittings, screwed, 252.
      American Standard, 252.
   hangers, rolls and supports, 260.
   joints, 253.
   material, 247.
   names, or grades, 248.
   threading, 248.
   unions, 252.
Piping, for steam, 3, 247.
   allowable velocity, 241.
   blow-off, 5, 58.
   commercial classification, 248.
   exhaust, 5, 242.
   expansion in, 249.
      joints, 249.
   flanges, 243.
   feedwater, 58, 243.
   high-pressure drip, 246.
   high-pressure system, 238.
   joints and fittings, 253.
   loop header system, 239.
   points to be considered in, 238.
   society, 254.
   single header system, 238.
   spider system, 239.
   size of, 243.
   table, 248.
   unit system, 242–243.
Piston, for steam engine, 347.

Piston, for pump, 543.
Pitting, 281.
Pivoted bucket conveyor, 324.
Planimeter, 425.
Plugs for pipe, 252.
Plumb bob, 408.
Polonceau valve, 393.
Polytropic path, 68.
   expansions, 85, 88.
Pond, cooling, 577.
Pony truck, 15.
Pop safety valve, 56, 57.
Port opening, 364.
   maximum, 364.
Potential efficiency, 435.
Pound-mol, 162.
Power, 67.
   boilers, 17.
   driven pumps, 544.
   station heat balance, 272, 588.
Power plant, steam, 2.
   accessories, 5.
   arrangement of equipment in, 612.
   buildings, 613.
   central station, 9.
   classification, 1, 2.
   Colfax, 522, 613.
   Edgar, 590.
   essential equipment, 3.
   elements in design of, 623.
   energy distribution in, 613.
   factors which determine type of,
      affecting cost of, 622.
   Hartford Electric Co., 621.
   isolated, 7.
   Lakeside, 532.
   locomotive, 11.
   Long Beach No. 3, 605.
   marine, 15.
   non-condensing, 3.
   oil burning plant, 605.
   pulverized fuel, 597.
   Richmond Station, 559, 613.
   trends, 584.
   Trenton Channel, 597.
   types of steam, 2.
   soundproofing of, 614.
   water, gas, oil, and steam, 1.
Poppet valve engines, 410–412.
Prat or Evasé system of induced draft, 308.
Pressure, 67.
   conversion of, 78.

Pressure, measurement, 75.
  oiling system, 456.
  plate, 375.
  total or dynamic, 314.
  volume-diagram, 85.
Primary feedwater heater, 263.
Prime mover, 3.
Priming, in boilers, 20, 289, 291.
  of centrifugal pump, 554.
Prony brake, 427.
Propeller, shaft, 16.
  ship, 16.
Properties, of air, 81.
  of steam, 81.
  of water, 89.
Prosser tube expander, 22.
Protons, 63.
Proximate analysis of a fuel, 133.
Pulverized coal, 144.
  mills, 8, 147.
    types of, 147.
  pumps, 330.
  systems, 144.
Pulverized fuel burner, 212.
  types, 212.
Purchasing coal, 142.
Purifiers for feedwater, 263, 269.
Pump, pumps, 536–556.
  air chamber for, 539.
  boiler feed, 5, 10.
  capacity of, 552.
  centrifugal, 10, 545.
    characteristics, 554.
    combination, 602.
    compound, 548.
    diffusing ring for, 548.
    efficiency of, 545.
    performance of, 552.
    removal, 564.
    right and left hand, 545.
    turbine, 548.
    volute, 547.
    wearing rings for, 546.
  classification of, 536.
  direct acting, duplex, 540.
    reciprocating, 536–544.
    simplex, 537.
    size, how stated, 537.
  displacement of, 552.
  duty, 553.
  governor, 544.
  head on, 552.

Pump, high-pressure, multi-stage, 548.
  installation of, 554.
  jet (*see injector*)
  left-hand side of duplex, 540.
  lift, 552.
  nomenclature, 552.
  piston and plunger, 536.
  power driven, 544.
    efficiency of, 544.
    simplex, duplex and triplex, 544.
    rated capacity, 554.
  right-hand side of duplex, 540.
  rotary, 549.
  slip of a, 552.
  suction, 552.
  testing, 555.
  vacuum, 7.
  vacuum chamber for, 544.
  valves and plate for, 539.
  water ends for, 543.
  wet and dry air, 7, 571.
Pyrometers, 71–73.
  accuracy of, 73.

Q

Quality of steam, 100.
  determination of, 102–107.
  method of computing, 104, 107.

R

Radial, brick chimney, 299.
  boiler stay, 47.
  flow, fan, 312.
Radian, 64.
Radiation, of heat, 74.
  losses in boilers, 179.
  pyrometer, 71.
Radiator, 5.
Rake, coal, 184.
Randolph drier, 144.
Rankine cycle, 435.
  efficiency, 432.
  ratio, 435.
Rateau type turbine, 484.
Rating, boilers, 220.
  engines, 433.
  traps, 247.
Ratio, connecting rod to crank, 367.
  heating surface to grate surface for boilers, 224.

Ratio, of expansion, 432.
    for compound engines, 447.
Reaction turbines, 507–522.
    multi-cylinder, 521.
    double-flow, 507.
    semi-flow, 507.
    single-flow, 507.
Réaumur, thermometer, 69.
Rear wall brace, 29.
Receivers for compound engines, 444.
Reciprocating engines, 337–416.
Rectangular hyperbola, 432.
Reduction gearing, 16, 530.
    location of, 530.
Reduction zone, 160.
Reducing motion for indicators, 422.
    pantograph, 422.
    reducing wheel, 423.
Reducing valve, 258.
Regulation of engine speed, 7.
    (*see governor*)
Refractories, 216.
Regulator, feedwater, 244.
    damper, 317.
    fineness, 148.
Reheaters, 8, 125.
    types, 127.
Reilly feedwater heater, 267.
Reinforced-concrete chimney, 301.
Relation between efficiency and capacity of a boiler, 227.
Release in a steam engine, 359.
Relative position of crank and eccentric, 359.
Relative velocity, 506.
Relief valve, 459.
Retorts for stokers, 198, 200.
Return-tubular boiler, 20.
    heating surface for, 222.
Return plate for boiler setting, 29.
Return-trap, 247.
Returns tank, 5.
Reverse lever, locomotive, 366.
Richmond Station, turbine room, 559, 613.
Riding cut-off valve, 380, 391.
Riley underfeed stoker, 199.
Ring oiled bearing, 455, 484, 485.
Ringelmann smoke chart, 218.
Riser tubes for Wickes boiler, 44.
Rising and non-rising spindle gate valve, 256.
Rites inertia governor, 380.

Rivets, ring seam, 21.
    longitudinal seam, 21.
Riveted joints, lap and butt, 21.
Roller to support pipe, 261.
Rocker arms, 363, 366.
    bar, 193.
    shaft, 197.
Roof sheet of boiler, 46.
Rotary car dumper, 587.
Rotary pump, 549.
Ruth's steam accumulator, 616.

S

Safety cams, 325.
    stop, 404.
Safety valve, 55–58.
    adjusting ring for, 56.
    direct spring loaded, 56.
    size and capacity of, 57.
    twin, 57.
Salt-velocity method, 555.
Sampling tube, or nozzle, flue gas, 167.
    steam, 103, 107.
Sand dome for locomotive, 47.
Sargol joint, 253.
Saturated steam, 91.
Saturation curve, 99.
Scale, boiler, 280.
    effect of, 281.
    loss caused by, 281.
    of indicator spring, 419.
    prevention, 284.
    removal, 293.
Scales, automatic coal, 7.
    platform, 228.
Scaife-feedwater purifier, 287.
Scarfed boiler plate, 21.
Scotch marine fire-tube boiler, 49.
Scraper conveyor, 322.
Screenhouse, 10, 602.
Secondary feedwater heater, 264.
Section, boiler, 33.
Seger cone pyrometer, 71.
Self-supporting steel stack, 299.
Semi-anthracite coal, 140.
Semi-double flow turbine, 509.
Separately-fired superheater, 117.
Separating calorimeter, 105.
Separators, steam, 5, 459.
    oil, 5, 461.
Serrated pipe joint, 254.

Setting, fixtures, 18, 26, 28.
  casing, 28, 44.
  foundation, 26.
  fronts, 26, 42.
  Kilgour, 186.
  standard, 26, 36, 38, 41.
  wing-wall, 186.
Shaft generator, 590.
Shaft governor, 337–382.
  Armstrong, 377.
  Buckeye, 380.
  centrifugal, 377–380.
  Rites-inertia, 380.
Shaking grates, 183.
Shell — one course, 31, 34.
Shifting eccentric, 378.
Shims, 351.
Shrouding strips, 489.
Sight feed oiler, 454.
Side rods, 383.
Silocel for boiler setting, 26.
Simplex, coal valve, 328.
  steam pump, 537.
Single-flow reaction turbine, 507, 508.
Skinner unaflow engine, 413.
  auxiliary exhaust valve for, 415.
  limit to compression pressure in, 413, 415.
Slag screen, 204.
Slice bar, 184.
Slide valve, engines, 337.
  cycle of events for, 361.
  "D" slide valve, 357.
  setting, 393–396, 431.
  terms, 363.
Sling strap for boiler, 35.
Smoke, and its cause, 181.
  box, 13, 31, 47.
  box front, 47.
  connection to boiler, 3, 21.
  density of, 218.
    measuring, 218.
  prevention, 181.
    in hand-fired furnace, 185.
    in mechanically fired furnace, 187.
  unit, 219.
Soda ash, 287.
Soot blower, 39.
  hand operated, 278.
  mechanically operated, 278.
Soot removal, 278.
Specific heat, 74.

Specific heat, constant pressure, 74.
  constant volume, 74.
  mean, 74.
  of air, 82.
  of mixture of gases, 82.
  of superheated steam, 102.
  of water, 90.
Specific gravity, 89.
  coal, 139–141.
  of oil, 153.
Specific volume, saturated steam, 99.
  superheated steam, 101.
  wet saturated steam, 100.
Speed, 64.
  measuring of, 468.
  reduction, 530.
Spider system of piping, 239.
Splash, guard for engine, 459.
  lubrication, 455.
Spray, burners for oil, 207.
  nozzles for cooling pond, 578.
  pond, 578.
Stack (*see chimney*)
  locomotive, 13.
Static pressure for fan, 313.
Station heat balance, 272, 588.
Stay, stays, and braces for boilers (*boiler chapter*)
  bolt, 29, 38, 48.
    hollow, 38.
  diagonal, 23.
  girder, 49.
  gusset, 40, 48.
  longitudinal, 23.
  radial, 47.
  sling, 47.
  throat, 37.
  through, 23, 49.
Stay sheet, 46.
Stay tube, 49.
Steam, accumulator, 616.
  calorimeter, 102, 105.
  circulating tubes, 41.
  computed from indicator diagram, 429.
  critical, temperature of, 92.
    pressure of, 92.
  cycles, 9, 128.
    mercury-vapor, 584, 621.
    straight, 584.
  density of dry, 99.
  dome, 13, 47.
  dry and saturated, 92.

Steam, engine (*see engine, steam*)
   entropy of, 107.
   exhaust, 2.
   flow meter, 467.
   generating unit, 585.
   header, 3, 241.
      size of, 242.
   indicator diagrams, 361.
   jackets, 344.
   jet air pump, 575.
   jets in furnace, 186.
   lance, 53, 278.
   lead, 3.
   live, 5.
   loop, 246.
   measuring on test, 467.
   nozzle, 3, 26, 34, 51, 53.
   passages, 339.
   pipe in locomotive, 47.
   primed, 20.
   properties of, 93, 117.
   purifiers, 123.
   rate, 434.
   separator, 459.
      receiver and baffle, 460.
   space of boiler, 19.
   superheated, 9, 92, 101.
   total heat per lb. of, 100, 101.
   tables, 94, 118.
   trap (*see traps*)
   turbine (*see turbines*)
   volume of dry, 99.
   wet, 100.
Steel for boilers, firebox, and open-hearth, 17.
Steel chimneys, 300.
   guyed and self supported, 300, 301.
Stephenson link motion, 383–384.
Stirling boiler, 41.
   method of support, 41.
   water circulation in, 42.
Stop watch, 468.
Stokers, 7, 187–204.
   advantages of, 187.
   classification of, 187.
   chain grate, 188–191.
   comparison of, 205.
   disadvantages of, 187.
   hand operated, 185.
   hydraulic operated, 203.
   overfeed, 192–198.
   screw type, 13.

Stokers, underfeed, 198–204.
Storage tank, 10.
Strainer for pump, 555.
Stroke counter, 468.
Sturtevant impulse turbine, 501.
   preheater, 276.
Sub-bituminous coal, 141.
Suction lift of a pump, 552.
Sulphur, in coal, 134.
   effect on coal, 143.
Superheat, 101.
   advantages of, 116.
   economy of, 116, 440.
Superheaters, 9, 116–123.
   advantages, 120.
   attached, 121–123.
   convection, 121.
   interdeck, 120.
   radiant, 120, 123.
   saving by using, 116.
   separately fired, 117.
Superpower, 2, 624.
Surge tank, 10, 595.
Swartwout pump governor, 544.
Sweet balanced valve, 376.

T

Tables (*see list in front part of text*):
Tachometers, 469.
   centrifugal, 469.
   chronometric, 470.
   electric, 469.
Tail pipe, 561.
Tandem-compound engines, 444.
Taylor underfeed stoker, 45, 203.
Telpherage coal handling, 328.
Telescopic oiler, 455.
Temperature, 69, 97.
   absolute, 71.
   of evaporation, 97.
Temperature-entropy diagram, 109.
   constant quality lines on, 111.
   evaporation line on, 110.
   liquid line on, 109.
   saturation line on, 110.
   steam region on, 110.
   superheat line on, 110.
   total heat lines on, 111.
Temporary hardness in feedwater, 286.
Tender, locomotive, 13.
Terry turbine, 500.

Testing of boilers, 227–235.
  apparatus for, 228.
  records of, 231.
  report of, 233.
  smoke observations during, 233.
Testing of engines, 465–476.
  apparatus for, 465.
  conduct and preparation for, 471–472.
  results of, 472.
Testing of pumps, 555.
Thermo-couple, 71, 72.
Thermo-electric pyrometer, 71.
Thermometers, 69.
  calibration of, 70.
  types of, 70.
Thermostatic control, 8.
Thermal efficiency, 1.
  of engines and turbines, 435.
    jet pumps, 576.
    power plants, 1, 9, 11, 13.
    pumps, 553.
Thermodynamics, first law, 74.
Thermodynamic losses, turbine, 531.
Threading dies, 248.
Thornycroft boiler, 49.
Three point suspension of boiler, 26.
Throat sheet, 46.
Throat stay, 37.
Throttle governor, 361.
Throttling calorimeter, 102–104.
Throttle valve, 7, 459.
Thrust bearing, 16.
  marine engine, 387.
  turbine, 491, 512.
Time, unit of, 63.
Toggle motion, 407.
Tools for hand-fired furnaces, 184.
Total-heat entropy diagram, 111.
Track scales, 597.
Trailing truck, 14.
Tram, 394.
Traps, steam, 3, 246.
  bucket, 246.
  float, 246.
  tilting, 247.
  types of, 246.
  return, 247.
Traveling coal hopper, 327.
Triple swing pipe connections, 204.
Triple expansion engines, 387.
Triplex pump, 544.
Try, or gage, cocks, 54.

Tube, boiler, 3.
  cleaners, 293–296.
  dimension of, 223.
  expanders, 22.
  ferrules, 49.
  return, 52.
  sheet, 22.
Turbine centrifugal pump, 548.
Turbine, steam, 3, 9.
  advantages of, 477.
  arc welded casing for, 589.
  auxiliaries for, 589.
  bleeder, 525.
  buckets, 477, 480, 488, 510, 523.
    reversing, 502.
    theory of impulse, 505.
    theory of reaction, 518.
  classification, 477.
  double-flow, 509.
  economy of, 531.
  general types, 478.
  governor for, 492, 512.
  high-pressure, 9, 499, 520, 589.
  high back pressure, 523.
  impulse, 480–505.
  influence of vacuum, superheat and pressure upon, 532.
  losses, 531.
  low-pressure, 10, 523.
  lubrication of, 527.
  marine, 528.
  mixed pressure, 523.
  multi-cylinder, 497.
  multi-pressure, 479.
  multi-velocity, 479.
  nozzle, terms, 503.
    plate, 490.
    theory of impulse, 503.
  reaction, 507–522.
  reduction gearing, 530.
  signal system, 589.
  single-flow reaction, 507.
  single-flow impulse and reaction, 508.
  superimposed, 499, 589.
  tailor made, 589.
  testing, 534.
  trend, 588.
  throttle and emergency valve for, 497.
  types, 478.
  vertical compound, 499.
Turbine valve gears, Curtis, 492–497.
  Westinghouse reaction, 512–518.

Turbo generator, 16, 530.
  cooling of, 589.
Turn buckle, 51.
Turner baffle wall, 36.
Tuyere boxes, 198.

U

U-loop, 38.
Ultimate analysis of a fuel, 134.
Unaflow engine, 412–415.
Underfeed stoker, 198–204.
Unions, screwed, 252.
Unconsumed carbon in ash, 178.
Universal gas constant, 84.
Unit, of evaporation, 222.
  of smoke, 218.
  system of piping, 208.
    pulverizing coal, 146.
Uptake tubes, marine boiler, 53.

V

Vacuum, 78.
  breaker, 561, 564.
  chamber for pumps, 544.
  corrections to standard conditions, 569.
  effect of air upon, 569.
  influence on economy of turbine, 532.
Vacuum, or air, pump, 571–576.
  centrifugal entrainment, 575.
  ejector, 575.
  Mullan displacement, 572.
  rotative dry-vacuum, 573.
  steam jet, 576.
  types, 571.
  Wheeler-Edwards, 571.
Valve, valves, automatically controlled, 359.
  for vacuum pumps, 573.
  for piping, 255–260.
    atmospheric relief, 6, 459, 462.
    back pressure, 5, 462.
    blow-off, 58.
    check, 57, 243, 555.
    gate, 256.
    globe, 257.
    non-return, 258.
    reducing, 5, 258.
    relief, 7, 462.
    stop, 58, 256–257.
  for pumps, 539.

Valve, for pumps, disc, 539.
  foot, 555.
  steam thrown, 537.
  for steam engines, Allen trick, 390.
    balanced, 361.
    Corliss, 399.
    double-ported, 390.
    grid-iron, 391.
    in mid-gear, 384.
    in full gear, 384.
    multi-ported, 390.
    poppet, 410.
  for turbine, 481, 492.
  plate, 539.
  safety, 5, 55.
  seat, 344.
  snifting, 402.
  thermostatic, 5.
  types of, 256.
Valve central, 363.
Valve diagram, 368–374.
  Bilgram, 372–373.
  ellipse, 369.
  Zeuner, 369–372.
    for Armstrong governor, 379.
    for Meyer valve, 392.
Valve setting, 393, 407.
  Corliss, 407.
  on pump, 556.
  slide valve, 393–396.
  special types, 396.
Valve terms, 363, 369.
Valve travel, 363.
Valve gear, 357.
  De Laval impulse turbine, 482.
  Corliss, 399–405.
  Curtis, impulse turbine, 492–497.
  engine, 357, 376, 387, 399, 400, 405, 406.
  locomotive, 382.
  Westinghouse reaction turbine, 512–518.
Vanstone flange joint, 253.
Vaporizer, 620.
V-bucket conveyor, 322.
V-notch weir, 465.
  head on, 466.
Velocity, 64.
  absolute and relative, 506.
  angular, 64.
  head, 552.
  pressure, 313.
  tangential, 64.
Vent condenser, 265, 283.

INDEX

Venturi water meter, 228.
Vernier on planimeter, 426.
Vertical tubular boiler, 29–32.
Viscosity, 452.
Volatile matter in coal, 160.
Volt, 66, 67.
Volume, unit of, 64.
Volute centrifugal pump, 513.

W

Wainwright closed feedwater heater, 266.
Wall bracket for pipe, 260.
Walschaert radial valve gear, 384–387.
Warped turbine bucket, 522.
Waste heat drier, Randolph, 144.
Water, analysis, 285.
  back, 190.
  box, condenser, 565.
  brake, 470.
  circulation in boiler, 35.
  column, 5, 19, 24, 30, 38, 53.
  cooled walls, furnace, 8, 45, 192, 217.
  equivalent of calorimeter, 136.
  gage, 5, 54.
  glass, 19, 32.
  horsepower, 552.
  leg of boiler, 29, 38, 46.
  level, 19.
  measurement of, 555.
  meter, 228, 466.
  power plant, 1.
  properties of, 89.
  rate, or steam, 434.
  screens, 9, 216.
  siphon, 61.
  softening apparatus, 284–290.
  space of boiler, 19.
  specific heat of, 90.
  storage tank, 13.
  weight per cubic foot, 90.
  tube cleaners, 295.
Watt, 67.
  pendulum governor, 403.
Wearing rings, 546.
Weathering of coal, 125.
Weight, unit of, 64.
Welded joints and flanges, 254.
We-Fu-Go, purifying system, 287.
Westinghouse-Parsons turbine, 503, 507.

Westinghouse-Parsons turbine, bearings for, 511.
  high-pressure, 520.
  large capacity, 519.
  marine, 428.
  valve gears for, 512–518.
    direct connected, 512.
    hydraulic, 517.
Westinghouse-Leblanc air pump, 575.
  -Roney stoker, 192.
Wet-air or vacuum, pump, 7, 560.
Weighing, hopper, 327.
  water on test, 228.
Weight of air per lb. of coal, 172.
Weir, V-notch, 466.
Wheeler, cooling tower, 580.
  Edwards rotative pump, 571.
  steam jet air pump, 576.
    thermal efficiency of, 576.
  surface condenser, 484.
Whyte classification of locomotives, 15.
Wickes water tube boiler, 42.
  method of support for, 42.
Wide firebox for locomotives, 46.
Willans line, 434.
Woolf engine, 444.
Work, 66.
  at constant pressure, 85.
Work diagrams, 66.
  adiabatic, 87.
  constant pressure, 85.
  isothermal, 86.
Wing-wall boiler setting, 186.
Wire drawing, 439.
Woolson arch, 27.
Wootten firebox, locomotive, 46.
Wrist plate, 400.
  central position of, 408.

Y

Yarrow boiler, 49.

Z

Zeolite water softening process, 288.
Zero hardness of water, 286.
Zeuner valve diagram, 369–372.
  applied to Armstrong governor, 379.
  applied to Meyer valve, 392.
Zinc used in boilers, 282.